Handbook of Green Information and Communication Systems

This book is dedicated to our families.

Handbook of Green Information and Communication Systems

Editors

Mohammad S. Obaidat, Alagan Anpalagan,

and Isaac Woungang

AMSTERDAM • BOSTON • HEIDELBERG • LONDON
NEW YORK • OXFORD • PARIS • SAN DIEGO
SAN FRANCISCO • SINGAPORE • SYDNEY • TOKYO

Academic Press is an Imprint of Elsevier

Academic Press is an imprint of Elsevier
225 Wyman Street, Waltham, 02451, USA
The Boulevard, Langford lane, Kidlington, Oxford OX5 1GB, UK

First edition 2013

Library of Congress Cataloging-in-Publication Data
A catalog record for this book is available from the Library of Congress

British Library Cataloguing in Publication Data
A catalogue record for this book is available from the British Library

ISBN: 978-0-1241-5844-3

Printed and bound in the United States of America
13 14 10 9 8 7 6 5 4 3 2 1

CONTENTS

Preface

Overview and Goals

As the world's economic activities are expanding, energy comes to the forefront in order to have sustainable growth. Nowadays, with escalating energy costs and the greater awareness of the wireless communications and networking industry and global warming impact on the environment, there is a clear demand for developing more energy efficient 'green communications' as well as improving the power efficiency in information, communication and network systems. The 'green communications' paradigm advocates the idea that the entire communications networks – i.e. from the physical layer to medium access, and through end-global networking systems - should be integrated while accounting for the interdependence among networking layers and the relationships between all aspects of the wireless communications life cycle. The ultimate goal is to significantly reduce energy consumption. This vision has become a reality thanks to the ubiquity of wireless communications and systems combined with the advent of improved processer capabilities and wireless network technologies and services. A variety of civilian applications worldwide such as greenhouse gases reduction, low-carbon economy development, energy consumption of mobile networks, to name a few, have already demonstrated the feasibility and versatility of the 'green information and communications' paradigm.

The book provides a comprehensive guide to selected topics, both ongoing and emerging, in green communications, computing, information, and networking systems, using a treatment approach suitable for pedagogical purposes. Consisting of contributions from worldwide well-known and high profile researchers in their respective specialties, selected topics that are covered in this book are related to the aforementioned areas, include smart grid technologies and communications, spectrum management, cognitive and autonomous radio systems, computing and communication architectures, data centers, distributed networking, cloud computing, next generation wireless communication systems, 4G access networking, optical core networks, cooperation transmission, security and privacy, among others.

The book has been prepared keeping in mind that it needs to prove itself to be a valuable resource, dealing with both the important core and the specialized issues in green communications, computing, information and networking. We hope that it will be a valuable resource and reference for instructors, researchers, students, engineers, scientists, managers, and industry practitioners in these fascinating fields. All chapters are integrated in a manner that renders the book as a supplementary reference volume and textbook for use in both undergraduate and graduate courses on green communications, computing and networking. Each chapter is of an expository, but also of a scholarly or survey style, on a particular topic within the scope of green information and communication systems.

Organization and Features

The book is organized into 27 chapters, each chapter is written by topical area experts. These chapters are grouped into four parts.

PART I is devoted to topics related to *green communications*, and is composed of seven chapters: Chapters 1–7.

Chapter 1 advocates the idea that introducing cognitive principles in co-tier network design may contribute to the successful deployment of femtocell networks. An overview of recent research works on cognitive two-tier networks by both industrial and academic research communities is presented, with emphasis on spectrum awareness, victim detection, resource allocation, and spectrum sharing techniques. From an energy efficiency perspective, the two-tier networks are examined and approaches to achieve green communication through femtocell deployment are reviewed in-depth.

Chapter 2 focuses on the need for smart energy management in the residential sector for sustainable energy and monetary savings. A comprehensive overview of the state-of-the-art in-home area communications and networking technologies for energy management is provided. Few challenges related to the design of future energy management systems, such as the need for interoperability and network security are also discussed.

Chapter 3 focuses on the need for designing novel and efficient embedded processors for the smart grid. Different applications that are expected to execute on different distributed computing nodes of the smart grid, and their real-time performance and power demands, are discussed. A thorough review of existing and emerging embedded platforms for the smart grid, along with available simulation and architecture platforms that can be used to design future embedded processors for smart grids, is presented.

Chapter 4 argues that wireless sensor networks (WSNs) are anticipated to play a critical role from different endpoints in the smart grid to support applications from transformer and substation monitoring to home and building energy management. The IEEE 802.15 standard wireless personal area network (WPAN) is introduced. Network measurements as well as challenges encountered with IEEE 802.15.4/ZigBee when designing WSNs in the smart grid are discussed.

Chapter 5 gives an insight view of the Smart Grid Communications Network (SGCN) by presenting its layered architecture, the typical types of traffic that it may carry and associated quality of service requirements. Candidate wireless communications technologies that can be employed for its implementation are also presented. Networking issues that the neighbor area network segment of SGCN needs to tackle are highlighted. Finally, a number of standards for smart grid inter-operability are reviewed in-depth.

Chapter 6 focuses on the intercell interference (ICI) issue in cellular communication systems. State-of-the-art techniques in controlling ICI under different multicell network settings are surveyed. For the conventional homogeneous network, various coordinated multi-point transmission and reception (CoMP) schemes to efficiently coordinate or take advantage of the ICI are presented. For heterogenous networks, the implementation of femtocell networks and their interference management mechanisms are discussed in-depth, as well as current advances in ICI coordination to improve the energy efficiency of cellular networks while maintaining a good tradeoff with the spectral efficiency goal.

Chapter 7 focuses on green radio communications for delay tolerant applications. The energy optimization issue for cross-layer packet adaptations in wireless networks is discussed in-depth. A system where time-varying packets are coming from higher layer applications for which instantaneous state may be unknown at the transmitter is also studied, and two packet schedulers over wireless channels, which may possess memory, are discussed. It is shown through analytical and simulation models that taking a suitable action based on cross-layer interaction and information can result in significant energy saving form the scheduler side, thus making the transmitter green.

PART II focuses on topics related to *green computing*. It is composed of five chapters: Chapters 8–12.

Chapter 8 discusses the main concepts of green computing and communication architectures from various viewpoints. These include green computing and governmental effort, energy-based monitoring architecture for green computing, green communication protocols and models, and design of green computing models.

Chapter 9 focuses on issues involved in designing 'green' processing platforms for biomedical systems. The power and real-time performance requirements of embedded platforms in biomedical computation are discussed. Existing embedded processors are surveyed and a method for identifying their computational characteristics using the pair-wire correlation as example application is proposed, both for implantable devices and external portable computing platforms. Finally, available simulation and design space exploration platforms for different processor micro-architectures that can be used to design the green biomedical computation platform are discussed.

Chapter 10 discusses the current challenges and research trends for datacenter infrastructures. A new energy-oriented model based on server consolidation is presented and discussed.

Chapter 11 focuses on cloud computing, by investigating current solutions towards energy-efficient cloud computing. These solutions are grouped into three categories, namely: (1) Energy-efficient data processing and storage in the high performance data centers with focus on thermal/cooling-aware workload placement techniques and energy-saving storage technologies, (2) Optimal data center placement by considering renewable resources, and (3) Energy-aware anycast and many-cast algorithms running at the Internet backbone. A detailed comparison of the existing solutions to the problem of thermal monitoring of data centers is also provided, as well as a discussion on open issues and research challenges in energy efficient cloud computing. The studied topics are presented at a level of detail that is not found elsewhere in the literature.

Chapter 12 focuses on the issue of quantify of the power consumption of data centers, by providing an in-depth review of energy-efficiency metrics that have been proposed so far for data centers. A discussion on the techniques to improve energy efficiency of data centers is also provided.

Part III focuses on *green networking* and related issues. It is composed of ten chapters: Chapters 13–22.

Chapter 13 discusses the issue of power conservation in wireless sensor networks and underwater sensor networks. A survey of some of the energy efficient routing protocols that have been proposed for terrestrial and underwater sensor networks is provided, with focus on the network layer and its protocols.

Chapter 14 discusses the challenges of balancing energy efficiency, connectivity and performance in wireless networks. Techniques to improve wireless energy efficiency including power control, topology control, relay nodes, handover techniques, caching, and other energy aware communications protocols, are overviewed and contrasted. A qualitative comparison is also provided of how effective each approach is, applications that a particular approach may be suited for, and how they may be integrated into next generation wireless networks.

Chapter 15 presents a review the state-of-the-art research on multiple input multiple output orthogonal frequency division multiplexing (MIMO-OFDM) systems with focus on physical layer optimization of energy efficiency. In addition, a discussion on future research trends and directions on this topic is provided.

Chapter 16 investigates the problem of resource allocation and deployment of green base stations (i.e. BSs powered by green energy) in sustainable wireless networks. The problem is formulated as a minimal BS placement problem and a reduced time complexity heuristic algorithm based on Voronoi diagram is proposed, which approaches the optimal solution under a variety of network settings.

Chapter 17 focuses on the problem of energy consumption in access networks. The energy consumptions of both broadband wireless and wireline access networks are compared, and various techniques that have been proposed so far to reduce energy consumptions of broadband access networks are discussed. Existing techniques on greening cellular networks, and the opportunities and challenges on greening broadband wireless networks through cooperative networking, are also discussed in-depth. The wireline access solutions and their related energy consumption are highlighted. Finally, existing schemes that have been proposed for reducing the energy consumption in passive optical access networks are discussed.

Chapter 18 also investigates the problem of energy consumption in access networks. Emergent solutions to curb the growing energy demands of access networks are detailed. The approaches are wide-ranging in scope, detailing promising techniques across the full gamut of challenges, ranging from network-wide issues, such as deployment, architecture and topology, to link level techniques. This chapter also offers a system-wide perspective, highlighting the fundamental principles of energy efficient communications and contextualizing each approach.

Chapter 19 promotes the idea that the design and development of new network equipment and solutions must fundamentally be dominated by energy efficiency objective, in addition to performance-related ones. Inspired by the introduction of energy models that characterize the energy consumption of various networking devices under different traffic loads for both optical and electronic network layers, several ideas related to new emerging paradigms, energy-efficient architectures and energy-aware algorithms and protocols for next generation networks are discussed.

Chapter 20 investigates the issue of energy consumption in peer-to-peer networks. Existing peer-to-peer networks and overlay protocols and approaches in the context of their energy requirements are reviewed in some level of details. Recent research works on peer-to-peer systems that have energy efficiency as their main design goal, are explored, and a taxonomy of such systems is proposed, with the goal of establishing a base of reference.

Chapter 21 deals with the problem of power management in 4th generation (4G) broadband wireless access networks by proposing an overview of power management mechanisms that are incorporated in the 4G standards. The latest findings on the issues associated with the original power management mechanisms in 4G are also discussed. Novel mechanisms that are currently explored for potential inclusion into next generation 4G standards in order to efficiently accommodate both current and future diverse data applications, are described.

Chapter 22 deals with energy efficient network designs and energy-aware network, with focus on how modular structure of switches, mixed line rates, dynamic networks, and optimum repeater spacing can reduce the running power consumption of network equipment. A discussion is given on how to design energy efficient optical core networks while reducing their energy consumption.

Part IV deals with **green innovative** applications in the areas of communications, computing and networking. It is composed of 5 chapters: Chapters 23–27.

Chapter 23 focuses on security issues related to modern Internet applications. The requirements and features of social applications are analyzed, under the goal of optimizing the energy consumption of the most critical security mechanisms. In this perspective, possible enhancements in the application design phase of the aforementioned mechanisms are discussed, and a first attempt to model the energy consumption of a security mechanism in terms of security and network costs, is provided.

Chapter 24 provides an overview of representative approaches to achieve energy-efficient routing in wireless ad hoc and sensor networks is provided, with focus on Ant colony optimization (ACO)-based approaches. An ACO-based energy efficient protocol for mobile ad hoc networks is also introduced and validated through simulation analysis.

Chapter 25 focuses on recent advances in the power line communication technologies including the emerging IEEE P1901 standard, in addition to wireless communication standards such as IEEE 802.11, IEEE 802.15.4, IEEE 802.16 and emerging wireless standards such as IEEE 802.22 and LTE-A. The communication-based applications in various parts of the smart grid including residential demand response and electric vehicle charging infrastructure, are reviewed in-depth. The challenges related to smart grid communication protocols with focus on security aspects are discussed.

Chapter 26 presents the offered capabilities of smart grids such as automated metering; as well as a description of its market and social aspects. In addition, the current state of standardization activities conducted by industry, academia and government are discussed.

Chapter 27 investigates the problem of radio resource management in heterogeneous wireless networks (HetNets) from an energy efficiency perspective. The problem is tackled by using the dynamic coverage management (DCM) approach. Insights on how the mathematical framework of queueing theory can be utilized to model this problem and identify the green opportunities in HetNets are presented.

Below are some of the important features of this book, which, we believe, would make it a valuable resource for our readers:

- This book is designed, in structure and content, with the intention of making the book useful at all learning levels.
- Most of the chapters of the book are authored by prominent academicians, researchers and practitioners, with solid experience and exposure to research on green communications, computing and networking areas. These contributors have been working with in these areas for many years and have a thorough understanding of the concepts and practical applications.
- The authors of this book are distributed in a large number of countries and most of them are affiliated with institutions of worldwide reputation. This gives this book an international flavor.
- The authors of each chapter have attempted to provide a comprehensive bibliography section, which should greatly help the readers interested further to dig into the aforementioned topics.
- Throughout the chapters of this book, most of the core research topics of green communications, computing and networking have been covered. This makes the book particularly useful for industry practitioners working directly with the practical aspects behind enabling the technologies in the field.

We have attempted to make the different chapters of the book look as coherent and synchronized as possible. However, it cannot be denied that due to the fact chapters were written by different authors, it was not always possible to achieve this task. We believe that this is a limitation of most edited books of this sort.

Target Audience

The book is written primarily for the student community. This includes students of both senior undergraduate and graduate levels—as well as students having an intermediate level of knowledge of the topics, and those having extensive knowledge about many of the topics. To keep to this goal, we have attempted to design the overall structure and content of the book in such a manner that makes it useful at all learning levels. The secondary audience for this book is the research community, in academia or in the industry. Finally, we have also taken into consideration the needs of those readers, typically from the industries, who wish to gain an insight into the practical significance of the topics, in order to discover how the spectrum of knowledge and the ideas are relevant for real-life applications of the 'green communications' paradigm and technologies.

Acknowledgments

We are extremely thankful to the 63 authors of the 27 chapters of this book, who have worked very hard to bring forward this unique resource for helping students, researchers, and community practitioners. We feel it is appropriate to mention that as the individual chapters of this book are written by different authors, the responsibility of the contents of each of the chapters lies with the respective authors.

We would like to thank Mr. Tim Pitts, the Elsevier Senior Commissioning Editor, who worked with us on the project from the beginning, for many of his suggestions. We also thank Elsevier Publishing and Marketing staff members, in particular, Ms. Charlie Kent, Editorial Project Manager (Elsevier), who tirelessly worked with us and helped us in the publication process. Finally, we would also like to thank our respective families, for their continuous support and encouragement during the course of this project.

Mohammad S. Obaidat
Alagan Anpalagan
Isaac Woungang

About the Editors

Prof. Mohammad S. Obaidat (Fellow of IEEE and Fellow of SCS) is an internationally well-known academic/researcher/ scientist. He received his Ph.D. and M. S. degrees in Computer Engineering with a minor in Computer Science from The Ohio State University, Columbus, Ohio, USA. Dr. Obaidat is currently a full Professor of Computer Science at Monmouth University, NJ, USA. Among his previous positions are Chair of the Department of Computer Science and Director of the Graduate Program at Monmouth University and a faculty member at the City University of New York. He has received extensive research funding and has published over Ten (10) books and over Five hundred (500) refereed technical articles in scholarly international journals and proceedings of international conferences, and currently working on three more books. Professor Obaidat has served as a consultant for several corporations and organizations worldwide. Mohammad is the Editor-in-Chief of the Wiley International Journal of Communication Systems, the FTRA Journal of Convergence and the KSIP Journal of Information Processing. He served as an Editor of IEEE Wireless Communications from 2007–2010. Between 1991–2006, he served as a Technical Editor and an Area Editor of Simulation: Transactions of the Society for Modeling and Simulations (SCS) International, TSCS. He also served on the Editorial Advisory Board of Simulation. He is now an editor of the Wiley Security and Communication Networks Journal, Journal of Networks, International Journal of Information Technology, Communications and Convergence, IJITCC, Inderscience. He served on the International Advisory Board of the International Journal of Wireless Networks and Broadband Technologies, IGI-global. Prod. Obaidat is an associate editor/ editorial board member of seven other refereed scholarly journals including two IEEE Transactions, Elsevier Computer Communications Journal, Kluwer Journal of Supercomputing, SCS Journal of Defense Modeling and Simulation, Elsevier Journal of Computers and EE, International Journal of Communication Networks and Distributed Systems, The Academy Journal of Communications, International Journal of BioSciences and Technology and International Journal of Information Technology. He has guest edited numerous special issues of scholarly journals such as IEEE Transactions on Systems, Man and Cybernetics, SMC, IEEE Wireless Communications, IEEE Systems Journal, SIMULATION: Transactions of SCS, Elsevier Computer Communications Journal, Journal of C & EE, Wiley Security and Communication Networks, Journal of Networks, and International Journal of Communication Systems, among others. Obaidat has served as the steering committee chair, advisory Committee Chair and program chair of numerous international conferences. He is the founder of two well known international conferences: The International Conference on Computer, Information and Telecommunication

Systems (CITS) and the International Symposium on Performance Evaluation of Computer and Telecommunication Systems (SPECTS). He is also the co-founder of the International Conference on Data Communication Networking (DCNET).

Between 1994–1997, Obaidat has served as distinguished speaker/visitor of IEEE Computer Society. Since 1995 he has been serving as an ACM distinguished Lecturer. He is also an SCS distinguished Lecturer. Between 1996–1999, Dr. Obaidat served as an IEEE/ACM program evaluator of the Computing Sciences Accreditation Board/Commission, CSAB/CSAC. Obaidat is the founder and first Chairman of SCS Technical Chapter (Committee) on PECTS (Performance Evaluation of Computer and Telecommunication Systems). He has served as the Scientific Advisor for the World Bank/UN Digital Inclusion Workshop- The Role of Information and Communication Technology in Development. Between 1995–2002, he has served as a member of the board of directors of the Society for Computer Simulation International.

Between 2002–2004, he has served as Vice President of Conferences of the Society for Modeling and Simulation International SCS. Between 2004–2006, Prof. Obaidat has served as Vice President of Membership of the Society for Modeling and Simulation International SCS. Between 2006–2009, he has served as the Senior Vice President of SCS. Between 2009–2011, he served as the President of SCS. One of his recent co-authored papers has received the best paper award in the IEEE AICCSA 2009 international conference. He also received the best paper award for one of his papers accepted in IEEE GLOBCOM 2009 conference. Prof. Obaidat has been awarded a Nokia Research Fellowship and the distinguished Fulbright Scholar Award. Dr. Obaidat received very recently the Society for Modeling and Simulation Intentional (SCS) prestigious McLeod Founder's Award in recognition of his outstanding technical and professional contributions to modeling and simulation. He received in Dec 2010, the IEEE ComSoc- GLOBECOM 2010 Outstanding Leadership Award for his outstanding leadership of Communication Software Services and Multimedia Applications Symposium, CSSMA 2010. He received very recently the Society for Modeling and Simulation International's (SCS) prestigious Presidential Service Award for his outstanding unique, long-term technical contributions and services to the profession and society.

He has been invited to lecture and give keynote speeches worldwide. His research interests are: wireless communications and networks, telecommunications and Networking systems, security of network, information and computer systems, security of e-based systems, performance evaluation of computer systems, algorithms and networks, high performance and parallel computing/computers, applied neural networks and pattern recognition, adaptive learning and speech processing. During the 2004/2005, he was on sabbatical leave as Fulbright Distinguished Professor and Advisor to the President of Philadelphia University in

Jordan, Dr. Adnan Badran. The latter became the Prime Minister of Jordan in April 2005 and served earlier as Deputy Director General of UNESCO. Prof. Obaidat is a Fellow of the Society for Modeling and Simulation International SCS, and a Fellow of the Institute of Electrical and Electronics Engineers (IEEE). For more info; see: http://bluehawk.monmouth.edu/mobaidat/.

Prof. Alagan Anpalagan received the B.A.Sc. (H), M.A.Sc. and Ph.D. degrees in Electrical Engineering from the University of Toronto, Canada. He joined the Department of Electrical and Computer Engineering at Ryerson University in 2001 and was promoted to Full Professor in 2010. He served the department as Graduate Program Director (2004–09) and the Interim Electrical Engineering Program Director (2009–10). During his sabbatical (2010–11), he was a Visiting Professor at Asian Institute of Technology and Visiting Researcher at Kyoto University. Dr. Anpalagan's industrial experience includes working at Bell Mobility on 1xRTT system deployment studies (2001), at Nortel Networks on SECORE R&D projects (1997) and at IBM Canada as IIP Intern (1994).

Dr. Anpalagan directs a research group working on radio resource management (RRM) and radio access & networking (RAN) areas within the WINCORE Lab. His current research interests include cognitive radio resource allocation and management, wireless cross layer design and optimization, collaborative communication, green communications technologies and QoE-aware femtocells. Dr. Anpalagan serves as Associate Editor for the IEEE Communications Surveys & Tutorials (2012), IEEE Communications Letters (2010) and Springer Wireless Personal Communications (2009), and past Editor for EURASIP Journal of Wireless Communications and Networking (2004–2009). He also served as EURASIP Guest Editor for two special issues in Radio Resource Management in 3G+ Systems (2006) and Fairness in Radio Resource Management for Wireless Networks (2008). Dr. Anpalagan served as TPC Co-Chair of: IEEE WPMC'12 Wireless Networks, IEEE PIMRC'11 Cognitive Radio and Spectrum Management, IEEE IWCMC'11 Workshop on Cooperative and Cognitive Networks, IEEE CCECE'04/08 and WirelessCom'05 Symposium on Radio Resource Management.

Dr. Anpalagan served as IEEE Toronto Section Chair (2006–07), ComSoc Toronto Chapter Chair (2004–05), Chair of IEEE Canada Professional Activities Committee (2009–11). He is the recipient of the Dean's Teaching Award (2011), Faculty Scholastic, Research and Creativity Award (2010), Faculty Service Award (2010) at Ryerson University. Dr. Anpalagan also completed a course on Project Management for Scientist and Engineers at the University of Oxford CPD Center. He is a registered Professional Engineer in the province of Ontario, Canada.

Dr. Isaac Woungang received his M.S. & Ph. D degrees, all in Mathematics, from the Université de la Méditerranée-Aix Marseille II, France, and Université du Sud, Toulon & Var, France, in 1990 and 1994 respectively. In 1999, he received a M.S degree from the INRS-Materials and Telecommunications, University of Quebec, Montreal, Canada. From 1999 to 2002, he worked as a software engineer at Nortel Networks. Since 2002, he has been with Ryerson University, where he is now an Associate Professor of Computer Science and Coordinator of the Distributed Applications and Broadband (DABNEL) Lab at Ryerson University. During his sabbatical (2008–09), he was a Visiting Professor at National Ilan University, Taiwan. His current research interests include network resource allocation and management, green communications technologies, network security, computer communication networks, and mobile communication systems. Dr. Woungang has published six (6) book chapters and over seventy (70) refereed technical articles in scholarly international journals and proceedings of international conferences, and currently working on two more books.

Dr. Woungang serves as Editor in Chief of the International Journal of Communication Networks and Distributed Systems (IJCNDS), Inderscience, U.K, and the International Journal of Information and Coding Theory (IJICoT), Inderscience, U.K, as Associate Editor of the International Journal of Communication Systems (IJCS), Wiley, the Computers & Electrical Engineering (C&EE), An International Journal, Elsevier, as Associate Editor of the International Journal of Wireless Networks & Broadcasting Technologies (IGI Global), as Associate Editor of the Journal of Convergence (FTRA, http://ftrai.org/joc/index.htm), as Associate Editor of the Human-centric computing and Information Sciences, Springer. He also served as Guest Editor for several special issues with several journals such as IET Information Security, International Journal of Internet Protocol Technology (Inderscience); Computer and Electrical Engineering (Elsevier), Mathematical and Computer Modelling (Elsevier), Computer Communications (Elsevier); The Journal of Supercomputing (Springer), Telecommunication Systems (Springer), International Journal of Communication Systems (Elsevier).

Dr. Woungang edited six (6) books in the areas of wireless ad hoc networks, wireless sensor networks, wireless mesh networks, pervasive computing, and coding theory, published by reputed publishers such as Springer, Elsevier, Wiley, and World Scientific. He is currently working on two more books. Also, he has been serving as Symposium Co-Chair of the 29th IEEE Symposium on Reliable Distributed Systems (SRDS 2010), Organizing Chair of the 2nd, 3rd, and 4th International workshop on Dependable Network Computing and Mobile Systems (DNCMS), Track Chair, and Program Chair of several International conferences; and in the Program Committees of over a dozen International conferences. Since January 2012, Dr. Woungang serves as Chair of Computer Chapter, IEEE Toronto Section.

Contributors

Hamid Aghvami
Centre for Telecommunications Research, King's College London, Strand, London, WC2R 2LS, UK; e-mail: hamid.aghvami@kcl.ac.uk

Alagan Anpalagan
Department of Electrical and Computer Engineering, Ryerson University, 350 Victoria Street, Toronto, ON, M5B 2K3, Canada; e-mail: alagan@ryerson.ca

Nirwan Ansari
Advanced Networking Laboratory, Department of Electrical and Computer Engineering, New Jersey Institute of Technology, Newark, NJ, 07102, USA; e-mail: nirwan.ansari@njit.edu

Hanna Bogucka
Poznán University of Technology, ul. Polanka 3, 60-965 Poznán, Poland; e-mail: Bogucka@et.put.poznan.pl

Lin X. Cai
School of Engineering and Applied Science, Princeton University, 2530 Jessica Ln. #410, Schaumburg, Illinois 60173, USA; e-mail: lincai@princeton.edu

Davide Careglio
Edifici D6 Despatx 103 C. Jordi Girona, 1-3 08034 Barcelona, Spain; e-mail: careglio@ac.upc.edu

Glaucio H. S. Carvalho
Faculty of Computation, Federal University of Pará, Belém, Pará, Brazil; e-mail: ghsc@ufpa.br

Luca Caviglione
CNR - ISSIA Via de Marini 6, Genova I – 16149, Italia; e-mail: luca.caviglione@ge.issia.cnr.it

Valentina Cecchi
Department of Electrical and Computer Engineering, University of North Carolina at Charlotte, 9201 University City Blvd., Charlotte, NC 28223, USA; e-mail: vvcecchi@uncc.edu

Periklis Chatzimisios
CSSN Research Lab, Department of Informatics, Alexander TEI of Thessaloniki, Greece; e-mail: pchatzimisios@ieee.org

Antonio De Domenico
CEA, LETI, Minatec Campus, 17, rue des Martyrs 38054 Grenoble, France; e-mail: antonio.de-domenico@cea.fr

Sanjay K. Dhurandher
Division of InformationTechnology, Netaji Subas Institute of Technology, University of Delhi Azad Hing Fauj Marg Sector – 3 Dwarka (Pappankalan) New Delhi - 110 078, India; e-mail: dhurandher@rediffmail.com

Maria-Gabriella Di Benedetto
Department of Information Engineering, Electronics, and Telecommunications (DIET), University of Rome, La Sapienza 00184 Rome, Italy; e-mail: gaby@acts.ing.uniroma1.it

Beniamino Di Martino
Seconda Università di Napoli, Dipartimento di Ingegneria dell'Informazione, Via Roma, 29 - I-81031 Aversa (CE), Italy; e-mail: beniamino.dimartino@unina.it

Jason B. Ernst
School of Computer Science, University of Guelph, 50 Stone Rd. East, Guelph, Ontario N1G 2W1, Canada; e-mail: jernst@uoguelph.ca

Melike Erol-Kantarci
School of Electrical Engineering and Computer Science, University of Ottawa, 800 King Edward Avenue Ottawa, Ontario, K1N 6N5, Canada; e-mail: melike.erolkantarci@uottawa.ca

Ugo Fiore
Università degli studi di Napoli-Federico II, Complesso universitario di Monte Sant'Angelo, Via Cinthia 80126, Napoli; e-mail: ufiore@unina.it

Tarik Guelzim
Dept. of Computer Science and Software Engineering, Monmouth University, West Long Branch, NJ 07764, USA; e-mail: tarik.guelzim@gmail.com

Maruti Gupta
Intel Corporation 2111 NE 25th Ave, JF3-206 Hillsboro, OR 97124, USA; e-mail: Maruti.gupta@intel.com

Megha Gupta
Division of Computer Engineering, Netaji Subhas Institute of Technology, University of Delhi, Azad Hing Fauj Marg Sector – 3 Dwarka (Pappankalan) New Delhi, - 110 078, India; e-mail: meghabis@gmail.com

Tao Han
Advanced Networking Laboratory, Department of Electrical and Computer Engineering, New Jersey Institute of Technology, Newark, NJ, 07102, USA; e-mail: th36@njit.edu

Shibo He
Department of Electrical and Computer Engineering, University of Waterloo, 350 Columbia Street West, Unit 173 Waterloo, Ontario, N2L 6P8, Canada; e-mail: s28he@uwaterloo.ca

Béat Hirsbrunner
Intelligence Research Group, Department of Informatics, University of Fribourg, Boulevard de Pérolles 90, CH-1700, Fribourg, Switzerland; e-mail: hirsbrunner@unifr.ch

Quang-Dung Ho
Department of Electrical and Computer Engineering, McGill University, 3480 University Street, Montreal, Quebec, H3A 2A7, Canada; e-mail: quang.ho@mcgill.ca

Oliver Holland
Centre for Telecommunications Research, King's College London, Strand, London, WC2R 2LS, UK; e-mail: oliver.holland@kcl.ac.uk

Bharat S. Joshi
Department of Electrical and Computer Engineering University of North Carolina at Charlotte 9531, University Terrace Drive, Apt A Charlotte, NC 28262, USA; e-mail: bsjoshi@uncc.edu

Aravind Kailas
Department of Electrical and Computer Engineering, University of North Carolina at Charlotte, 9201 University City Bvd., Charlotte, NC 28223, USA; e-mail: Aravind.Kailas@uncc.edu

Burak Kantarci
School Electrical Engineering and Computer Science, University of Ottawa, Ottawa, Ontario, K1N 6N5, Canada; e-mail: kantarci@site.uottawa.ca

Ashok Karmokar
Department of Electrical and Computer Engineering, Ryerson University, 350 Victoria Street Toronto, ON, M5B 2K3, Canada; e-mail: ashok@ryerson.ca

Ali T. Koc
Intel Corporation 2111 NE 25th Ave, JF3-206 Hillsboro, OR 97124, USA; e-mail: ali.t.koc@intel.com

Vinay Vijendra Kumar Lakshmi
Department of Electrical and Computer Engineering, University of North Carolina at Charlotte 9531, University Terrace Drive, Apt A Charlotte, NC 28262, USA

Tho Le-Ngoc
Department of Electrical and Computer Engineering, McGill University, 3480 University Street Montreal, Quebec, H3A 2A7, Canada; e-mail: tho.le-ngoc@mcgill.ca

Chun-Hao Lo (Thomas Lo)
Advanced Networking Laboratory, Department of Electrical and Computer Engineering New Jersey Institute of Technology, Newark, NJ, 07102, USA; e-mail: cl96@njit.edu

Apostolos Malatras
Intelligence Research Group, Department of Informatics, University of Fribourg, Boulevard de Pérolles 90, CH-1700, Fribourg, Switzerland; e-mail: malatras@unifr.ch

Alessio Merlo
DIST – University of Genova AILab, Via F. Causa 13, Genova I – 16145, Italia; e-mail: alessio.merlo@dist.unige.it

Mohammad Ali Mohseni
Department of Electrical Engineering, Sahand University of Technology Sahand New Town, Tabriz, Iran; e-mail: mohammadali.mohseni@gmail.com

Hussein T. Mouftah
School Electrical Engineering and Computer Science, University of Ottawa, Ottawa, Ontario, K1N 6N5, Canada; e-mail: mouftah@site.uottawa.ca

Arindam Mukherjee
Department of Electrical and Computer Engineering, University of North Carolina at Charlotte, 9201 University City Blvd, Charlotte, NC 28223, USA; e-mail: amukherj@uncc.edu

Duy Trong Ngo
Department of Electrical and Computer Engineering, McGill University, Room 633, McConnell Engineering Building, 3480 University Street, Montreal, QC, H3A 2A7, Canada; e-mail: duy.ngo@mail.mcgill.ca

Duy Huu Ngoc Nguyen
Department of Electrical and Computer Engineering, McGill University, Room 633, McConnell Engineering Building, 3480 University Street, Montreal, QC, H3A 2A7, Canada; e-mail: huu.n.nguyen@mail.mcgill.ca

Mohammad S. Obaidat
Dept. of Computer Science and Software Engineering, Monmouth University, West Long Branch, NJ 07764, USA; e-mail: msobaidat@gmail.com, m_obaidat@yahoo.com, Obaidat@monmouth.edu

Francesco Palmieri
Seconda Università di Napoli, Dipartimento di Ingegneria dell'Informazione, Via Roma, 29 - I-81031 Aversa (CE), Italy; e-mail: Francesco.palmieri@unina.it

Ashish Panday
Department of Electrical and Computer Engineering, University of North Carolina at Charlotte 9531, University Terrace Drive, Apt A Charlotte, NC 28262, USA; e-mail: apanday@uncc.edu

Fei Peng
Intelligence Research Group, Department of Informatics, University of Fribourg, Boulevard de Pérolles 90, CH-1700, Fribourg, Switzerland; e-mail: fei.peng@unifr.ch

Diogo Quintas
Centre for Telecommunications Research, King's College London, Strand, London, WC2R 2LS, UK Phone: 44 (0) 20 7848-2898; e-mail: diogo.quintas@kcl.ac.uk

Zimran Rafique
Department of Electrical & Electronic Engineering, School of Engineering, Auckland University of Technology, Private Bag 92006, Auckland 1142, New Zealand; e-mail: zrafique@aut.ac.nz

Akbar Ghaffarpour Rahbar
Department of Electrical Engineering, Sahand University of Technology, Sahand, New Town, Tabriz, Iran; e-mail: akbar_rahbar92@yahoo.com

Sergio Ricciardi
Universitat Politècnica de Catalunya, Departament d'Arquitectura de Computadors, Edifici C6 DESPATX 217 C. Jordi Girona, 1-3 08034 Barcelone, Spain; e-mail: sergio.ricciardi@ac.upc.edu

Germán Santos-Boada
Universitat Politècnica de Catalunya, Departament d'Arquitectura de Computadors, Edifici C6 DESPATX 217 C. Jordi Girona, 1-3 08034 Barcelone, Spain; e-mail: german@ac.upc.edu

Boon-Chong Seet
Department of Electrical & Electronic Engineering, School of Engineering, Auckland University of Technology, Private Bag 92006, Auckland 1142 , New Zealand

Xuemin (Sherman) Shen
Department of Electrical and Computer Engineering, University of Waterloo, 200 University Avenue, West Waterloo, Ontario, N2L 3G1; e-mail: xshen@bbcr.uwaterloo.ca

Josep Solé-Pareta
Universitat Politècnica de Catalunya, Departament d'Arquitectura de Computadors, Edifici D6 DESPATX 110 C. Jordi Girona, 1-3 08034 Barcelona, Spain; e-mail: pareta@ac.upc.edu

Georgios Stavrou
CSSN Research Lab, Department of Informatics, Alexander TEI of Thessaloniki Greece; e-mail: gitsirop@mail.ntua.gr

Dimitrios G. Stratogiannis
School of Electrical and Computer Engineering, National Technical University of Athens, Athens, Greece; e-mail: dstratog@mail.ntua.gr

Emilio Calvanese Strinati
CEA, LETI, Minatec Campus, 17, rue des Martyrs 38054 Grenoble, France; e-mail: emilio.calvanese-strinati@cea.fr

Rohith Tenneti
Department of Electrical and Computer Engineering, University of North Carolina at Charlotte, 9531, University Terrace Drive, Apt A Charlotte, NC 28262, USA; e-mail: rtenneti@uncc.edu

Jordi Torres-Viñals
Universitat Politècnica de Catalunya, Departament d'Arquitectura de Computadors, Edifici C6 DESPATX 217 C. Jordi Girona, 1-3 08034 Barcelona, Spain; e-mail: torres@ac.upc.edu

Georgios I. Tsiropoulos
School of Electrical and Computer Engineering, National Technical University of Athens, Athens, Greece; e-mail: gitsirop@mail.ntua.gr

Rath Vannithamby
Intel Corporation, 2111 NE 25th Ave, JF3-206 Hillsboro, OR 97124, USA; e-mail: rath.vannithamby@intel.com

Isaac Woungang
Department of Computer Science, Ryerson University, 350 Victoria Street Toronto, Ontario, M5B 2K3, Canada; e-mail: iwoungan@scs.ryerson.ca

Jingjing Zhang
Advanced Networking Laboratory, Department of Electrical and Computer Engineering, New Jersey Institute of Technology, Newark, NJ, 07102, USA; e-mail: th36@njit.edu

Yan Zhang
Advanced Networking Laboratory, Department of Electrical and Computer Engineering, New Jersey Institute of Technology, Newark, NJ, 07102, USA; e-mail: yz45@njit.edu

Zhongming Zheng
Department of Electrical and Computer Engineering, University of Waterloo, 350 Columbia Street West, Unit 339 Waterloo, Ontario, N2L 6P8, Canada; e-mail: z25zheng@bbcr.uwaterloo.ca

CHAPTER

1 Cognitive Strategies for Green Two-Tier Cellular Networks: A Critical Overview

Antonio De Domenico[a], Emilio Calvanese Strinati[a], Maria-Gabriella Di Benedetto[b]

[a]CEA, LETI, Minatec Campus, 17, rue des Martyrs, 38054 Grenoble, France,
[b]Department of Information Engineering, Electronics, and Telecommunications (DIET), University of Rome, La Sapienza 00184 Rome, Italy

1.1 INTRODUCTION

The success of mobile cellular networks has resulted in wide proliferation and demand for ubiquitous heterogeneous broadband mobile wireless services. While recent studies confirm that more than 60% of mobile traffic is generated indoors [1], customers still access mobile networks by connecting to Macro Base Stations (M-BSs). Nowadays, indoor and cell edge users usually experience very poor performance (see Figure 1.1); therefore, the growth in traffic rate demand and services requires cellular operators to further ameliorate and extend their infrastructure.

Femtocell Access Points (FAPs) are low power access points that offer radio coverage through a given wireless technology (such as LTE and WiMAX) while a broadband wired link connects them to the backhaul network of a cellular operator (see Figure 1.2) [2]. Such a technical solution presents several benefits both to operators and end consumers. Originally envisioned as a means to provide better voice coverage in the home, FAPs represent an efficient way of offloading data traffic from the macrocell network.

In a cellular network, traffic is carried from an end-user device to the cell site and then to the core network using the backhaul of the Mobile Service Provider (MSP). With network offload, cellular traffic from the UE is redirected to a local AP; then, it is carried over a fixed broadband connection, either to the MSP's core network or to another Internet destination. This reduces the traffic carried over the MSP's radio and backhaul networks, thereby increasing available capacity and limiting OPerational EXpense (OPEX) at the mobile operators. Juniper Research forecasts that by 2015, 63% of mobile data traffic will be offloaded to fixed networks through femtocells and WiFi APs [3]. On the other hand, femto users (F-UEs) may obtain larger coverage, better support for high data rate services, and prolonged battery life for their devices. The advantages mainly come from the reduced distance between an end-user terminal and the AP, the mitigation of interference due to propagation and penetration losses, and the limited number of users served by a FAP.

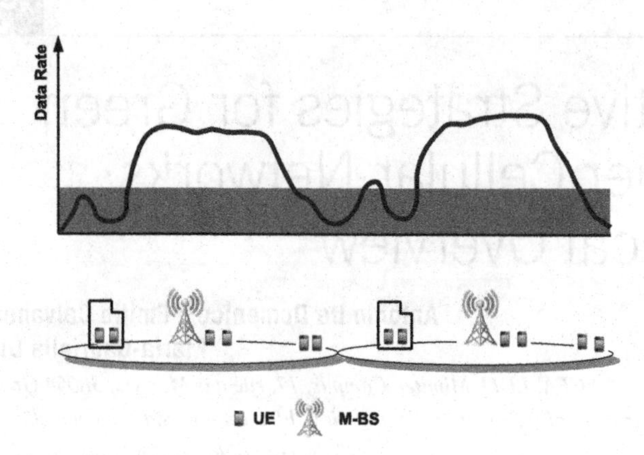

Figure 1.1 User experience challenges for a uniform broadband wireless service.

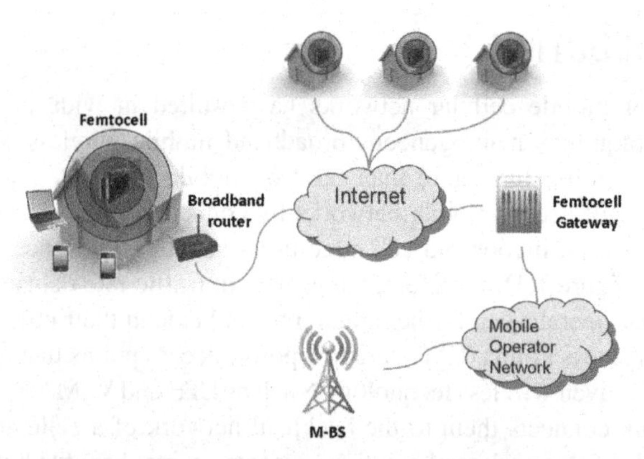

Figure 1.2 A generic femtocell architecture.

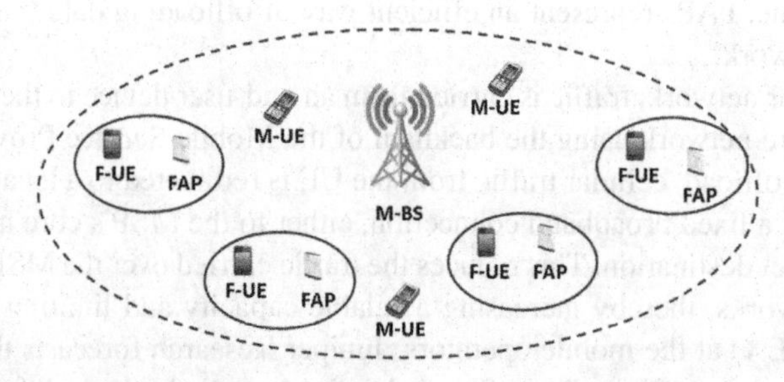

Figure 1.3 Two-tier cellular network.

In this novel network architecture, macrocells and femtocells may share the same spectrum in a given geographical area as a two-tier network (see Figure 1.3). Thus, *cross-tier interference* may drastically corrupt the reliability of communications.

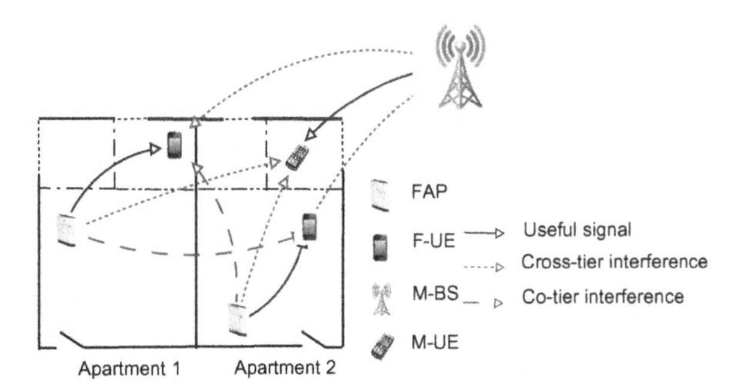

Figure 1.4 Downlink interference scenarios in two-tier cellular networks.

Similarly, neighbor femtocells belonging to same operators may also interfere with each other thus generating *co-tier interference*. These interference scenarios are further illustrated in Figure 1.4.

The impact of interference is highly related to the femtocell access control mechanism. Three different approaches have been investigated in the past: *closed access*, *open access*, and *hybrid access* [4]. In closed access, only a restricted set of users is allowed to connect to the femtocell; in open access femtocells, a subscriber is always allowed to connect to the closer FAP; in the hybrid access approach, femtocells allow access to all users but a certain group of subscribers maintain higher access priority. In closed access femtocells, the issue of interference can become an important bottleneck with respect to QoS and performance of communications. On the contrary, open access limits the interference problem while security issues and high signaling due to handover procedures can reduce the overall performance. Furthermore, due to resource sharing, open access limits benefits for the femtocell owner. Henceforth, according to a recent market research [5], the closed access scheme is the favourite access method of potential customers. Thus, further solutions that limit both *cross-tier* and *co-tier interference* have to be investigated.

In recent years, Cognitive Radio (CR) has been proposed as a powerful instrument to improve the Spectral Efficiency (SE) and permit the coexistence of heterogeneous wireless networks in the same region and spectrum [6]. A CR terminal can monitor the wireless environment and inform the resource allocation controller about local and temporal spectrum availability and quality. Thus, agile AP can dynamically select available channels and adapt transmission parameters to avoid harmful interference between contending users. In CR taxonomy, nodes that are licensed to operate in a certain spectrum band are usually named as primary users, while opportunistic users are often referred to as secondary users. Associating secondary users to F-UEs and primary users to M-UEs is straightforward, although recent work [7] investigates a scenario in which opportunistic M-UEs coexist with licensee F-UEs.

This chapter aims to describe the benefits of cognitive principles towards the successful deployment of femtocells. The main functionalities of a CR (i.e., spectrum sensing, dynamic resource allocation, spectrum sharing, and spectrum mobility) represent the natural answer to issues that rely on the ad hoc nature of FAPs. Spectrum sensing enables cognitive devices to be aware of spectrum usage at nearby macro and femtocells, and to maintain a dynamic picture of available resource. In the investigated scenarios, the heterogeneous nature of the interference results in channel availability time–space dependency, which introduces new challenges with respect to the classic cellular network. Dynamic resource allocation schemes take advantage of spectrum awareness to satisfy QoS constraints while limiting the generated interference and power consumption. Spectrum access strategies deal with contention between heterogeneous users in order to avoid harmful interference. A cognitive FAP can exploit the characteristic of an M-UE's transmissions to detect the presence of nearby victims. Spectrum mobility allows a FAP to vacate its channel when an M-UE is detected, and to access an idle band where it can reestablish the communication link.

It is important to note that all these functionalities should be distributed at both the FAP and its serving UEs. Accordingly, the appropriate reconfiguration process should be cooperatively implemented in order to achieve a robust decision. An example of such cooperation is given in [8], where FAPs and associated UEs cooperate to gather reliable information on the radio environment. Furthermore, the femtocell architecture can reliably assist the configuration of cognitive networks. For instance, the cellular network backhaul could be used to deploy infrastructure-based agile solutions such as the Cognitive Pilot Channel (CPC) [9] or a geolocation database [10].

Few frameworks that assess the major benefits and research challenges in cognitive femtocell networks have been presented in literature. In [11], the authors focus on the characteristics of an infrastructure-based architecture that enables dynamic access in the next generation broadband wireless system. In [8, 12], the authors discuss the drawbacks of traditional interference management schemes in heterogeneous cellular networks; therefore, they show the advantages of implementing cognitive interference mitigation schemes in such environment.

In this chapter, we asses the fundamental goals of cognitive two-tier networks and review existing works in the field. We focus on both industrial and academic contributions.

In particular, Section 1.2 critically presents existing algorithms dealing with spectrum usage awareness and victim detection. Section 1.3 analyses Radio Resource Management (RRM) strategies proposed for cognitive two-tier networks. Section 1.4 discusses spectrum sharing solutions and Section 1.5 analyses green communication issues in two-tier cellular networks and gives an overview of the strategies proposed to enhance the system Energy Efficiency (EE). Finally in Section 1.6, we conclude the chapter by discussing some important open issues.

1.2 SPECTRUM AWARENESS AND VICTIM DETECTION

In this section we overview CR-based functionalities that enable FAPs to be aware of spectrum usage at nearby (macro and femto) cells and detect the presence of UEs that are served by neighboring APs. Accordingly, cognitive APs can dynamically select available channels and adapt transmission parameters to avoid harmful interference towards contending cells.

A major misconception in CR literature is that detecting the signal of the legacy transmitter is equivalent to discovering transmission opportunities [13]. On the contrary, even when such a signal can be perfectly detected, this discovery is affected by three main problems: the *hidden transmitter*, the *exposed transmitter*, and the *hidden receiver*. These are well-known issues and have been investigated in depth in ad hoc wireless networks literature [14]. Nevertheless, although the former problems have been solved, there are still not feasible solutions for the hidden receiver. Due to wall attenuation, the occurrence of these problems is even higher in femtocell deployment scenarios. Furthermore, it is expected that the M-BS likely allocates all frequency resources when there is a high number of end-users to serve. Hence, a sensing analysis based on the classic *energy detection*[15] operations may detect few spectrum opportunities. However, since many M-UEs can be located far away from the FAP, their channels can be effectively reused in the femtocell [16]. More sophisticated and effective detection schemes, based on the knowledge of legacy users' signal characteristics, have been presented in CR literature (cf. [17,18]). Furthermore, infrastructure-based solutions such as the Cognitive Pilot Channel (CPC) [9] and the geolocation database [10] have been proposed to assist cognitive networks across different Radio Access Technologies (RATs) and available spectrum resources.

1.2.1 Sensing and LTE User Detection

Sensing performance is limited by hardware and physical constraints. For instance, only cognitive devices equipped with two transceivers can transmit and sense simultaneously (such as in [19]). Moreover, in order to limit sensing overhead, only a partial state of the network is usually monitored. There is a fundamental trade-off between the undesired overhead and *spectrum holes* detection effectiveness: the more bands that are sensed, the higher the number and quality of the available resource. In order to improve the sensing process reliability in two-tier deployment scenarios, several approaches have been investigated in literature.

Lien et al. propose a mechanism that optimizes both sensing period and frequency resource allocation to statistically guarantee users' QoS [20]. Barbarossa et al. introduce a strategy that jointly optimizes the energy detection thresholds and the power allocation under a constraint on the maximum generated interference [21]. Sahin et al., on the contrary, investigate an agile strategy that jointly exploits spectrum sensing and scheduling information obtained by the M-BS [16]. Uplink sensing is used to individuate frequency resources that have been allocated to nearby M-UEs. Then uplink/downlink scheduling information is exploited to identify these M-UEs and their downlink frequency resources, respectively. Hence, this algorithm permits a reliable detection of the available spectrum opportunities at FAPs. However, it presents some drawbacks: first, the proposed coordination between M-BS and FAPs and limited availability of backhaul bandwidth result in scalability and security issues; second, the technique presented above may be ineffective in practice.

This is due to dense femtocells deployment with a consequent large population of interferers expected to be coordinated by the cellular operator. Due to such unsolved problems, direct coordination among femto and macrocells was not implemented in 3GPP Release 10 [22].

Lotze et al. consider a scenario in which LTE-like femtocells are deployed in the GSM spectrum in order to avoid cross-tier interference [23] towards LTE M-UEs. In this scenario, neighboring femtocells have to dynamically adapt their spectrum usage to prevent harmful co-tier interference. The authors propose a spectrum sensing technique that permits a FAP to detect the presence of neighboring interferers without decoding their signals. The proposed approach combines the classic energy detection operations with a feature detection approach that enables descrimination between LTE transmissions (that have to be avoided) and other signals (that have no priorities). This technique allows the detection of weak signals at a complexity cost slightly higher than the classic energy detector. The proposed detection algorithm has been successfully implemented in the frame of the Iris platform [24].

Several studies related to *victim* aware interference management have been recently proposed in the 3GPP frame. DoCoMo investigates a method for determining whether there are M-UEs in close proximity of a FAP [25,?]. In this scheme M-UEs detect the cell IDentification (ID) of interferers by listening to the Broadcast CHannel (BCH) of neighboring FAPs. Then, based on the received Reference Signal Received Power (RSRP), each M-UE identifies the most harmful FAPs and feedbacks this information to its serving M-BS. PicoChip and Kyocera propose a method where FAPs determine the presence of an M-UE by detecting its uplink reference signal [27]. Furthermore, the authors propose to exploit the properties of the uplink reference signal in order to discriminate between a single dominant transmission of a nearby victim and the aggregated power due to further away users. Note that a victim M-UE is often easy to detect because it likely transmits with relatively high power due to the experienced attenuation (composed mainly of path loss and wall losses). However, the above solutions are ineffective in the presence of an idle mode M-UE. An idle M-UE neither transmits nor is able to report the presence of neighboring femtocells, thus, protecting M-BS downlink control channels is necessary. For instance, Qualcomm proposes to introduce orthogonality between FAPs and M-BS control channels [28]. Further potential solutions are presented in [29].

1.2.2 Architectures towards Geographical-Based Interference Mapping

SpectrumHarvest is a novel architecture that manages spectrum access in cognitive femtocell networks [30]. This architecture is composed of four components: a Multi-Operator Spectrum Server (MOSS), a Femto Coordination/controller Server (FCS), a cognitive FAP, and associated end-user terminals. The MOSS combines information on spectrum availability at different cellular operators with local measurements performed by different femtocells. After processing these inputs, the MOSS indicates to each FCS local available resources. The FCS has the role of providing the spectrum usage information to each of its served FAPs, along with additional information such as power level of neighboring femtocells and the location of M-BSs/M-UEs. Each FAP is characterized by a Spectrum Usage Decision Unit (SUDU) that exploits information received by both the FCS and MOSS to allocate spectrum for each transmission. Furthermore, the SUDU performs local spectrum measurements to detect the presence/absence of neighboring mobile terminals. Finally, end-user devices support a new air interface operating in noncontinuous channels across a multi-operator and multi-services broadband.

Kawade and Nekovee propose to use TV White Spaces (TVWS) in order to support home networking services [31]. The TVWS spectrum comprises large portions of the UHF/VHF spectrum that became available on a geographical basis for dynamic access as a result of the switchover from analog to digital TV.

The proposed investigation is carried out using a database approach: a centrally managed database containing the information of available TV channels is made available to femtocells. Based on the geolocation data and QoS requirements, FAPs query the central database for channel availability through the fixed-line connection. The database returns information about various operating parameters such as number of channels, centre frequencies and associated power levels for use in specific location. Simulation results show how TVWS can be an effective solution for low/medium rate services, while it should be used as a complementary interface for highly loaded traffic scenarios. Furthermore, the study underlines that, due to the lower propagation losses, operating in TVWS bands may result in a significant energy saving.

Similarly, Haines discusses advantages and drawbacks of implementing the CPC in the femtocell deployment scenarios [32]. The CPC is a dedicated logic channel, and is used to disseminate radio context information permitting terminals configuration and interference mitigation. Moreover, the distributed deployment of the CPC (DCPC) is under investigation [33] to improve the coexistence of heterogeneous systems in a shared band. In this scenario, several networks (such as WiFi and Bluetooth), which exploit different technologies and bands, may coexist in the femtocell coverage area. In each household a Smart Femto Cell Controller (SFC-C) is proposed as an interface with a centralized database managed by the cellular operator. The SFC-C is also able to collect and process the sensing outputs of different cognitive devices deployed in the area. Then, it sends interference management information to neighbouring terminals in order to coordinate the access to common spectral resources and avoid mutual interference. Furthermore, a peer to peer link is established to permit neighboring SFC-Cs, which coordinate different heterogeneous networks, to exchange local spectrum measurement and interference policies.

1.3 DYNAMIC RADIO RESOURCE MANAGEMENT

Static RRM algorithms allocate different parts of the available spectral resource between macrocell and femtocell tiers [34]. The goal of these techniques is to avoid in-band concurrent transmissions using full time/frequency orthogonalization of transmissions. However, these approaches are far from the SE targets of operators. The system performance can be enhanced by exploiting more flexible approaches.

Frequency Reuse (FR) schemes have been proposed to geographically reallocate to femtocells part of the spectral resources used by the macrocell (see, for instance [35]). However, due to the high number of expected interferers in dense femtocell deployments, FR schemes can be ineffective. A CR network, on the contrary, can exploit awareness on the spectrum usage and dynamic RRM algorithms to break the capacity limit of classic macrocell networks.

Li et al. propose a dynamic channel reuse scheduler that limits both *macro-to-femto interference* and *femto-to-femto interference* in the downlink scenario [36]. In this scheme, each femtocell exploits sensing outputs to construct a two-dimensional interference signature matrix. This model describes the local network environment in the time/frequency domains and attempts to avoid high peaks of interference. When there is a user to serve, the scheduler assigns the channel and power according to QoS and power constraints. First, it picks up the best available channel with respect to the user interference perception; then, the optimal transmission power is allocated in order to limit femtocell power consumption. Such a cognitive approach is further extended to the uplink scenario [37], where the presence of M-UEs in the household of the femtocell may cause a high level of macro-to-femto interference. This is a well-known issue, and it has been referred as the kitchen table problem [38]. In the above discussed scheme, the absence of coordination between neighboring femtocells results in high scalability and low complexity. However, nearby cognitive FAPs, which experience the same level of interference, may simultaneously access the same channels, thus interfering with each other. Furthermore, FAPs located at the macrocell edge may not be able to detect M-BS transmissions and subsequently cause strong cross-tier interference towards nearby M-UEs.

Wayan Mustika et al. propose a game-theoretic approach to deal with resource allocation in self-organizing closed access femtocells [39]. In the considered uplink interference scenario FAPs are affected by both co-tier and cross-tier interference. The resource allocation problem is modelled as a noncooperative potential game where F-UEs are represented by the set of players, the selected Resource Blocks (RBs) are the associated strategy, and the utility function takes into account both the perceived and generated interference. Hence, each F-UE iteratively acquires information about its environment and distributively selects the most appropriate set of RBs that improve its utility function. The authors demonstrate that the proposed approach converges to a Nash Equilibrium (NE). In such a state, no player gains any advantages from deviating from the selected strategy [40]. However, according to the simulation results the number of iterations necessary to meet the NE is quite high. Therefore, such a state may not be reached during the wireless channel coherence time and poor performance may be experienced during the *long* iterative process. Finally, the proposed approach requires knowledge of the link gains between F-UEs and nearby FAPs and M-BS, which is a hard task in the considered scenario.

Xiang et al. investigate a scenario in which cognitive femtocells access spectrum bands that are licensed to different legacy systems (such as macrocell and TVWS) to increase the aggregate network capacity [19]. Authors propose a joint channel allocation and power control scheme that aims to maximize the downlink femtocell capacity. The optimal allocation is found formulating a mixed integer nonlinear problem, which is decomposed [41] and distributively solved at each femtocell. F-UEs are allowed to use only the channels that are temporally not used by the primary system. Each FAP is equipped with two transceivers, with one dedicated to data transmission and the other used to continuously monitor the radio environment (i.e., sensing and transmission can be performed in parallel). Note that this approach ameliorates the protection of the legacy system, however, it also increases the cost of the devices and the complexity of the proposed approach. Whenever a FAP detects concurrent transmissions of the legacy system, it stops its activity and updates channel assignments accordingly. In order to increase spectral reuse and capacity, the proposed approach permits neighboring femtocells to access the same channels as long as generated interference is not harmful. Furthermore, neighboring FAPs share information about the presence of legacy transmissions in order to cooperatively improve the M-UEs' detection. However, this signaling exchange is realized on the cellular backhaul, thus the delay introduced by the Internet limits the benefits of the cooperative detection. Finally, as opposed to the classic cellular technologies, in the proposed scheme, each UEs can use only one channel, and the FAP is constrained to exploit a number of channels that is equal to the number of its served UEs.

As already mentioned, open access femtocells may limit harmful interference in two-tier networks. In fact, open access FAPs do not restrict access, and mobile users are allowed to connect to the closer femtocell in the vicinity. Hence, open access femtocell deployment results in higher macrocell offload and enhanced network capacity [42]. As drawbacks, network signalling and frequency of handover increase. Furthermore, security issues emerge in this type of access. Torregoza et al. consider a scenario in which both the M-BS and open access FAPs offer WiMAX services in their coverage areas [43]. Each femtocell has some private customers although public M-UEs, in order to improve their performance, can access both macro- and femtocells. In the presented model, a backhaul connection is introduced to permit communications between the M-BS and FAPs. A femtocell, which serves public M-UEs, receives a certain amount of the backhaul capacity as compensation for the poorer performance perceived by its private clients. Thus, a joint power control, base station assignment, and channel allocation scheme is proposed. This scheme improves the aggregate throughput while minimizing the need for femtocell compensation. In order to find the Pareto optimal solution for both downlink and uplink scenarios, two multi-objective problems are formulated and solved through a sum-weighted approach.

Jin and Li analyse further potential benefits of applying CR techniques in WiMAX-based two-tier networks [7]. In the investigated scenario, F-UEs connected to WiMAX femtocells via a dedicated channel experience guaranteed QoS. Oppositely, M-UEs are allowed to connect only to the M-BS with best effort services. However, this deployment scenario permits a high number of spatial reuse opportunities. Femtocells are characterized by small coverage areas, few UEs per cell, and low transmission power; hence, outdoor M-UEs can opportunistically reuse channels utilized by faraway femtocells. Therefore, in order to exploit the spatial diversity experienced by cognitive M-UEs, a two-hop cooperative transmission scheme is proposed (see Figure 1.5). In classical CR networks, sender and receiver periodically exchange their sets of available channels; then, in order to find a common available resource through which to communicate, a channel filtering procedure is realized [6]. In the considered scenario, when there are no channels free from interference, the presence of a relay introduces further possibilities to establish a reliable communication between the M-BS and indoor M-UEs. For instance, in Figure 1.5, the M-BS is allowed to communicate with M-UE 1 and M-UE 4 through relays M-UE 2 and M-UE 3, respectively. Furthermore, cooperative communication may permit reduction of the M-BS transmission power, which results in lower interference perceived at F-UEs and energy saving. Finally, the authors propose an RRM protocol that maximizes the aggregated throughput and includes routing strategies, power assignment, and channel allocation. Using stochastic optimization, nonlinear integer programming problems are formulated. At the best of our knowledge, the proposed cognitive framework is the only protocol that exploits the location dependency of spectrum opportunities to enhance the two-tier network performance. However, the high level of signaling and sensing overhead may result in low scalability.

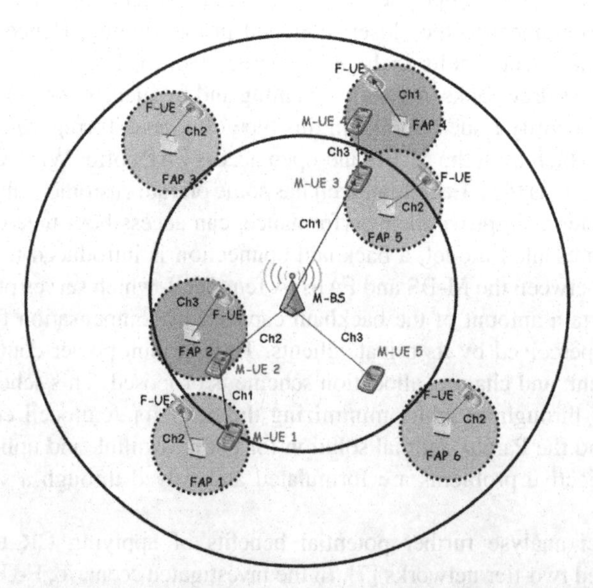

Figure 1.5 Two-hop cooperative transmission scheme for two-tier cellular networks [7]: the M-BS is allowed to communicate with M-UE 1 and M-UE 4 through relays M-UE 2 and M-UE 3, respectively. Furthermore, due to the cooperative communication scheme, neighboring FAPs are not interfered with by macrocell transmissions.

1.4 DYNAMIC SPECTRUM SHARING

Spectrum sharing functionalities aim to improve the coexistence of heterogeneous users accessing the radio resource. Three different cognitive transmission access paradigms are currently investigated in literature [44]: *underlay*, *overlay*, and *interweave*. In underlay transmissions, cognitive users are allowed to operate in the band of the primary system while generated interference stays below a given threshold (see Figure 1.6).

In overlay transmissions, cognitive devices exploit some specific information to either cancel or mitigate perceived/generated interference on concurrent transmissions (see Figure 1.7).

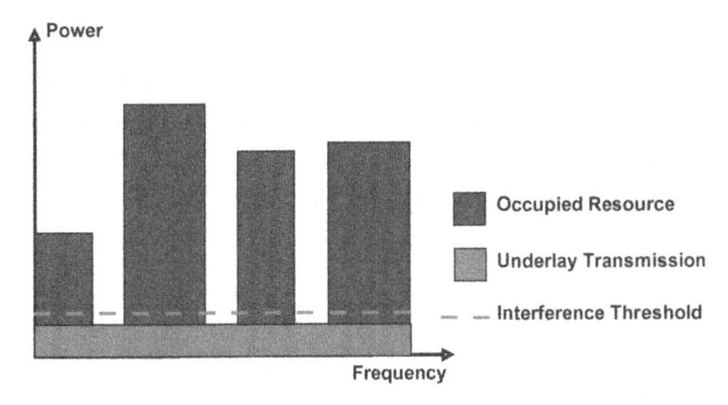

Figure 1.6 Underlay transmission scheme [44].

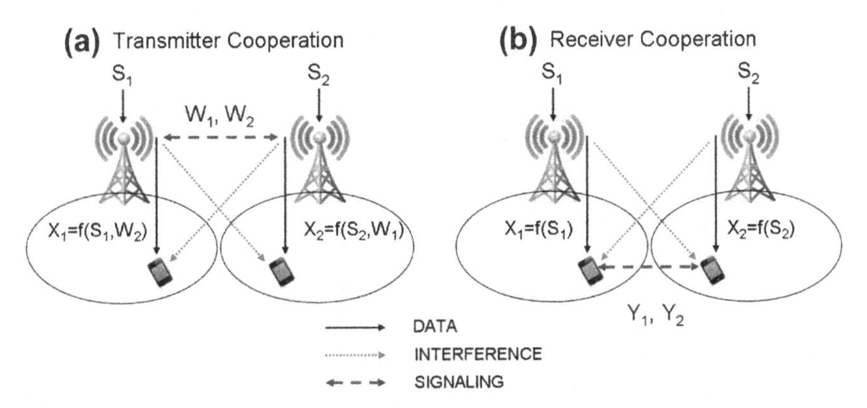

Figure 1.7 Overlay transmission scheme in transmitter/receiver cooperation scenarios [44](a) Overlay transmission scheme in transmitter cooperation scenario: the two BSs exchange information in order to acquire a priori knowledge about the concurrent transmissions. Such information is then exploited to either cancel or mitigate mutual interference [44]. (b) Overlay transmission scheme in receiver cooperation scenario: the two UEs jointly process received signals to correctly decode desired information.

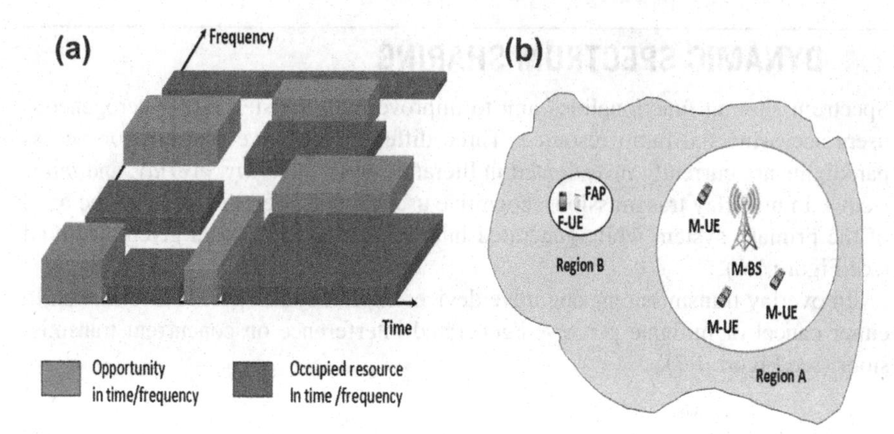

Figure 1.8 (a) Interweave transmission approach in time/frequency domain: FAPs are not allowed to transmit while M-BS is transmitting. (b) Interweave transmission approach in space domain: some area around the M-BS (i.e., Region A) cannot be reused by FAP transmissions; however Region B can be used without interfering M-UEs.

In interweave transmissions, opportunistic users transmit only in spectrum holes; if during in-band sensing a cognitive user detects a primary one, it vacates its channel to avoid harmful interference (see Figure 1.8).

In order to improve the coexistence between macrocells and femtocells, these schemes have been implemented also in cognitive-based two-tier networks.

Cheng et al. investigate the theoretical downlink capacity of a two-tier network, for each of the above-mentioned approaches [45]. The authors conclude that underlay and overlay schemes can result in better spectral reuse than the interweave scheme. However, former mechanisms require a better awareness of the network state and higher level information (such as the position of neighbouring interfered users, generated interference, scheduling information, and channel gains).

1.4.1 Underlay Spectrum Access

Galindo-Serrano et al. propose an algorithm that controls the aggregated *femto-to-macro interference* [46] by using a distributed Q-learning based mechanism [47]. Femtocells find the optimal policy by exploiting sensing outputs and periodic reports transmitted by the M-BS. However, the successful implementation of the proposed mechanism in a broadband cellular network can be constrained by the length of the learning process (see also the discussion related to the work of Mustika et al. in Section 1.3). Hence, the authors propose a novel cognitive concept, named *docition*, which permits FAPs to exchange their knowledge and experience. Femtocells are able to identify the most appropriate *teachers*. In fact, it is fundamental that cognitive terminals learn from *experts* that are in the same radio environment. This cooperative process increases convergence speed and accuracy. Depending on the degree of cooperation, two docitive schemes are investigated: in *Startup Docition*, cognitive femtocells share their policies only when a new node joins the network; in *IQ-Driven Docition*, cognitive femtocells periodically share their knowledge. The docitive approach may significantly reduce the sensing period resulting in energy saving and higher throughput. However, in the proposed scheme, additional overhead and complexity are introduced by coordination between femtocells and M-BSs, which periodically feedback the interference perceived at their associated M-UEs.

Chandrasekhar et al. consider a different approach to limit the femto-to-macro interference [48]. The authors first investigate the consequences of the near-far effect, which may result in excessive cross-tier interference in two-tier cellular networks (see Figure 1.9). Simulation results show that, due to mutual interference, achieving high SINR in one tier constricts the achievable SINR in the other tier. In order to deal with this issue, a closed loop power control is implemented at femtocells. The proposed algorithm iteratively reduces the F-UE uplink power until the SINR target at M-BSs is met. The authors propose that each F-UE distributively adapts its power in order to maximize a utility function, which results in lower complexity and overhead. This function is made up of two components: a reward and a penalty. The reward describes the gain achieved by the F-UE as a function of the difference between the experienced link quality and the minimum SINR target. The penalty represents the cost that femto transmission implies for the macrocell network as a function of the interference perceived at the M-BS. The proposed power adaptation mechanism does not require explicit cooperation between macrocell and femtocell networks. Nevertheless, in order to evaluate the generated interference at the M-BS, each F-UE needs to estimate the channel gain characterizing its link to the M-BS. However, the proposed power control scheme may not be sufficient to guarantee reliability of macrocell transmissions. In such cases, cross-tier coordination is introduced, and based on periodic M-BS reports, stronger femto interferers iteratively decrease their SINR target until the generated interference is adequately lowered.

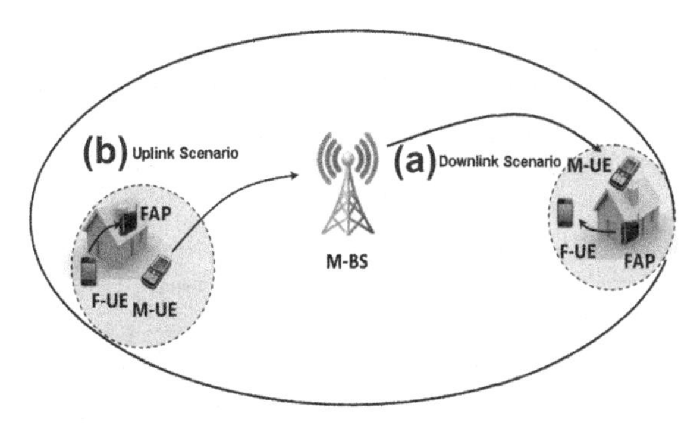

Figure 1.9 Near-far scenarios in two-tier cellular networks: (a) Downlink Scenario: FAP transmits to its F-UE and creates a dead zone for the cell edge M-UE. (b) Uplink Scenario: cell edge M-UE transmits with high power to its serving M-BS and creates a dead zone for the nearby FAP.

1.4.2 Interweave Spectrum Access

In order to allow distributed intercell spectrum access, da Costa et al. propose a Game-based Resource Allocation in Cognitive Environment (GRACE) [49]. In the proposed approach spectrum access is autonomously managed at each femtocell. Hence intercell coordination is avoided, leading to better scalability and SE. The aggregate femtocell strategy is modelled as a game $\Gamma = \langle \Im, (\Sigma_i)_{i \in \Im}, (\Pi_i)_{i \in \Im} \rangle$, where $\Im$ is the set of femtocell players, Σ_i is the access strategy of the ith player, and Π_i is the utility function of the ith player. The utility function indicates the preferences of each player with respect to the possible access strategies. In GRACE, this function is constructed to jointly optimize capacity, load balance, and femto-to-femto interference. Simulation results show that GRACE achieves higher throughput than classic FR schemes especially at cell edge users. GRACE avoids inter-cell coordination, leading to high scalability and SE; however, this approach decreases the speed of the learning process and reduces the accuracy of the algorithm.

Garcia et al. propose a Self-Organizing Coalitions for Conflict Evaluation and Resolution (SOCCER) mechanism, which fairly distributes available resources between cognitive femtocells [50]. This approach is based on both graph and coalitional game theories and atempts to avoid harmful femto-to-femto interference. Soccer is composed of two main phases (see Figure 1.10): in the first phase, FAPs, which join the network or seek for more band, detect the presence of possible conflicts.

In particular, based on the RSRP measured at F-UEs, each FAP estimates which neighboring FAPs are currently strong interferers. In SOCCER, an interferer is strong whenever, due to the mutual interference, and an orthogonal access results to be more effective than a reused one's scheme. In the second phase, coalitions are formed and resources are distributed accordingly. The authors show through simulation that, even in high density deployment scenarios, the degree of interference (i.e., the number of neighbor interferers) of a FAP is rarely higher than three. Thus, in order to limit the algorithm complexity and the network overhead, SOCCER considers the case in which a new entrant BS sends a Coalition Formation Request (CFR) to at most two coalition candidates. After receiving the CFR message, coalition candidates fairly reorganize the available spectrum to avoid mutual interference. It is important to note that interference towards the legacy system is not considered in either the work of da Costa or Garcia, which may result in poor performance at M-UEs.

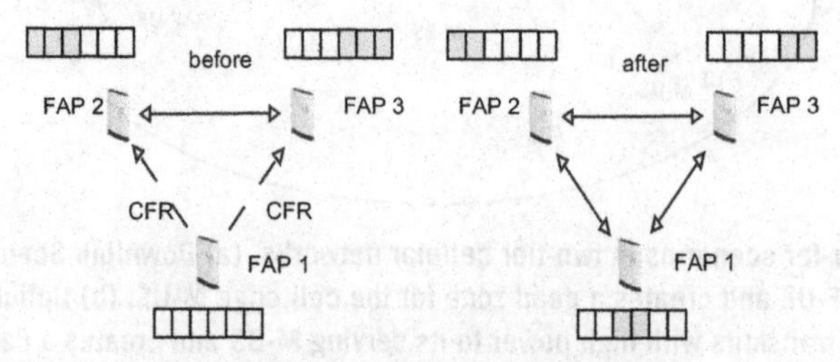

Figure 1.10 Dynamic spectrum sharing according to SOCCER [50].

On the contrary, Pantisano et al. propose a cooperative spectrum sharing approach, which may limit the effect of both cross-tier and co-tier interference [51]. In the proposed scheme, FAPs dynamically access the bandwidth used by the macrocell uplink transmissions, in order to avoid interference towards nearby M-UEs (see Figure 1.11). Furthermore, neighboring FAPs are able to form coalitions in which available frequency resources are cooperatively shared and femto-to-femto interference is limited. In fact, although isolated FAPs select their channels according to the perceived level of interference, FAPs in the same coalition use a Time Division Multiple Access (TDMA) scheme, which avoids having neighbouring FAPs simultaneously transmit on the same channel. However, co-tier interference is not fully eliminated because different coalitions generate mutual interference with one other. The proposed cooperation algorithm is divided into three main phases: neighbor discovery, recursive coalition formation, and coalition-level scheduling. During the first phase, each FAP uses a neighbor discovery technique to identify potential coalition partners. The second step is iteratively performed: First, each FAP finds coalitions characterized by an acceptable cost. This cost is related to power consumption due to the signaling exchange among the coalition members and depends on spatial distribution of the coalition members. Then, each FAP joins the coalition that ensures the maximum payoff. The coalition payoff is related to the achieved throughput and corresponding power consumption. Finally, the coalition formation ends when a stable partition is reached (i.e., FAPs have no incentive to leave the partition they blong to). In the third phase, FAPs within the same coalition first exchange information about their scheduling preferences on a defined in-band common control channel; then, a graph-coloring based resource allocation algorithm [52] is used in each coalition. The proposed approach may enable reliable co-channel deployment of femtocells and macrocells in the same geographic region. However, perfect sensing is supposed at FAPs, although missing detection of the M-UE transmission can result in harmful macro-to-femto interference.

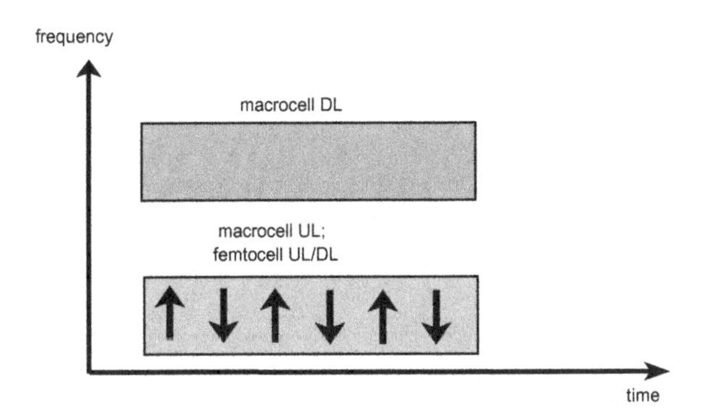

Figure 1.11 The TDD transmission scheme implemented in [51].

1.4.3 Overlay Spectrum Access

In suburban or low density urban deployment scenarios a femtocell generally operates in a very high SINR regime. However, we can identify two scenarios in which macro-to-femto interference may affect transmission reliability. In the first case, harmful downlink interference is experienced at F-UEs when the femtocell is very close to an M-BS. In the second case, an indoor M-UE that is far from its serving BS adapts it transmission power using fractional power control [53]. This mechanism may generate harmful interference at neighbouring FAPs (see Figure 1.9). However, in both cases, the harmful transmission is generally characterized by both high power and low rate. Thus, with high probability, the receiver (either a FAP or a F-UE) can process and cancel the perceived interference and subsequently correctly decode the useful message (a more detailed discussion on Interference Cancellation (IC) theory can be found in [54]). The general assumption of overlay transmission schemes is that the cognitive network possesses the necessary information (such as the channel gain related to the interferer transmission) to either cancel or mitigate interference.

The amount of signaling and coordination classically required in overlay access schemes may result in excessive complexity and overhead. Therefore, Rangan proposes an overlay approach that limits cross-tier cooperation [55]. The proposed method assumes Fractional Frequency Reuse (FFR) at macrocells. The overall band is divided in four contiguous bands (f_0, f_1, f_2, and f_3) as shown in Figure 1.12.

Each cell site is three way sectorized, and each of this sectors is referred to as a cell. Each cell is further divided in two regions, an inner region and an outer region. Subband f_0 is reused in all cells and it is allotted to M-UEs that are closer to the M-BS. However, remaining bands are used in only one cell per site and are allocated to cell edge M-UEs. Although this scheme reduces the system SE, it limits both the macro-to-macro and the femto-to-macro interference. The subband partitioning permits femtocells, regardless of their positions, to access a part of the spectrum where the generated cross-tier interference is minimal. In order to effectively reduce femto-to-macro interference, joint femtocell channel selection and power allocation is based on a *load spillage* power control method.This method avoids the operaion

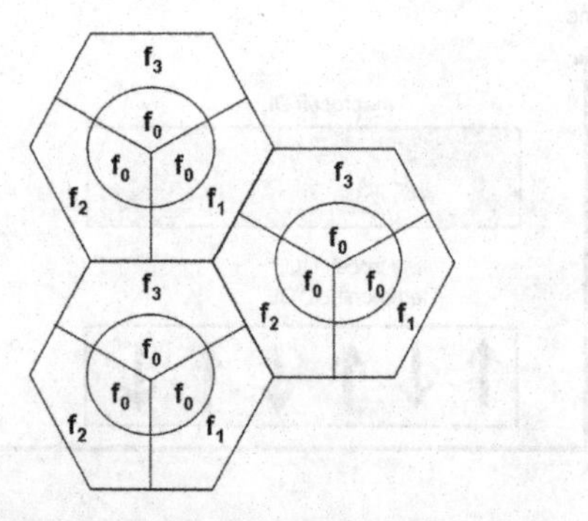

Figure 1.12 A FFR scheme for two-tier cellular networks [55].

of femtocells in bands either where the operating macro receiver is close or high load traffic is transmitted. However, signaling between macro and femto networks is required to allow each femtocell to be aware of the macro load factors. Furthermore, to successfully implement the load spillage power control, femto transmitters need to estimate channel gains that characterize the femto-to-macro links. Nevertheless, this scheme does not reduce the macro-to-femto interference experienced at F-UEs close to the M-BS. Thus, the author proposes implementation of the above-mentioned IC technique to jointly decode and cancel undesired signals. However, femto-to-femto interference is not mitigated by the proposed approach; on the contrary, due to the geographic based frequency partition, this interference can strongly affect femtocell performance especially in a dense deployment scenario.

Zubin et al. propose a dynamic resource partitioning scheme between femto and macrocells that does not rely on IC [26]. In this scheme, M-UEs identify the cell IDs of interferers by listening to the BCH of neighbor FAPs. Then, based on the corresponding RSRP, each M-UE identifies the most interfering femtocells and feedbacks this information to its serving M-BS. Hence, the M-BS indicates to interfering FAPs the channels that they should refrain from allocating in order to avoid *cross-tier interference*. This coordination message can be disseminated via X2 and S1 interfaces [56]. However, whenever the macrocell scheduling pattern changes, the M-BS should transmit new information to each interferer that is located in its region. The length of the scheduling period in modern cellular systems is related to the wireless channel coherence time. Hence, in medium/high density deployment scenarios, this scheme may result in excessive overhead.

Kaimaletu et al. extend the idea previously presented to the femto-to-femto interference scenario [57]. In this scheme, both M-BSs and FAPs schedule frequency resources by considering the potential interference generated towards each others' UEs. Each UE classifies interfering cells according to the strength of its RSRP. Therefore, it feedbacks this information to its serving cell, which reports to neighboring cells the number of victim UEs they create and the total number of UEs it serves. According to this information, macro/femtocells cooperatively block a subset of their frequency channels so that the victim UEs are protected. At the end of this iterative process, each cell uses a Proportional Fair (PF) scheduler [58] to serve its UE with a minimum amount of perceived/generated interference. In order to realize the proposed scheme, the authors assume perfect synchronization between all cells in the system both in time and frequency. In the proposed scheme, signaling exchange is not performed in small time scale; thus, it results in lower overhead with respect to Zubin's proposition [26]. However, in a dense deployment scenario, sources of interference can frequently change. This may increase the need for coordination, increasing overhead and reducing system scalability.

Pantisano et al. investigate a cooperative framework for uplink transmissions, where F-UEs act as a relay for neighboring M-UEs [59]. In this framework, each M-UE can autonomously decide to lease part of its allotted bandwidth to a cooperative F-UE. The latter split these channels into two parts: the first part of the band is used to forward the M-UE's message to its serving FAP; the second part of the

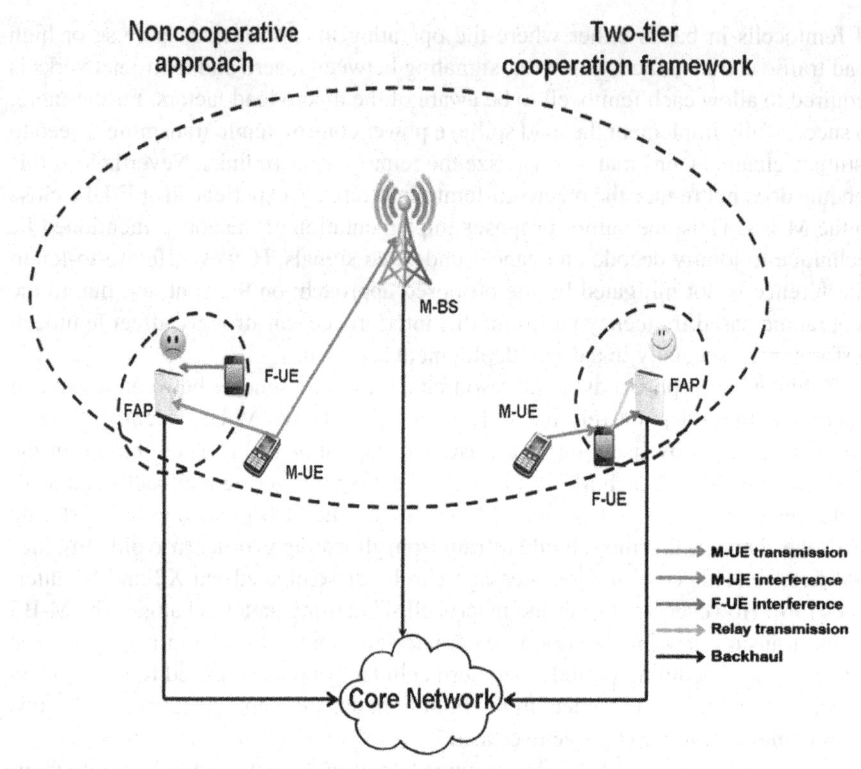

Figure 1.13 Two-tier network cooperation framework [59].

band represents a reward for the relaying F-UE, which can transmit its own traffic avoiding interference from the cooperative M-UE. Coordination can be beneficial for both M-UEs and FAPs, which may avoid excessive retransmissions (i.e., latency) and reduce the perceived cross-tier interference, respectively (see Figure 1.13). In fact, cell edge and indoor M-UEs likely experience poor performance due to penetration and propagation losses; hence, they are expected to transmit with relative high power to avoid excessive outage events. Therefore, M-UEs' transmissions result in strong interference at neighboring FAPs (see Figure 1.9). In order to ameliorate perceived performance, F-UEs can decide to cooperate with a group of M-UEs by forming a coalition, where transmissions are managed at the F-UE and separated in time in a TDMA fashion. Each member of the coalition receives a payoff that is measured as the ratio between the experienced capacity and related latency. Note that such a payoff depends on the amount of power/band that the F-UE uses to relay M-UEs messages and the remaining part that is used to transmit its own packets. To the best of our knowledge, this proposal is the first to consider device-to-device communication to enable cooperative transmission in a two-tier cellular network. However, two main challenges arise in such a work: first, this kind of cooperation results in security problems due to the exchange of data among coalition partners; second, relaying through the FAP introduces additional latency, due to transmission over the IP-based backhaul, that can results in poor performance at the M-UEs.

Cheng et al. propose a mixed transmission strategy that enhances the system SE by exploiting both interweave and overlay paradigms [60]. In fact, in such a strategy, F-UEs access idle RBs (as in the interweave approach) and further exploit transmission opportunities that arise during M-BS retransmissions. In particular, each FAP overhears communications originated at the M-BS so that during the retransmissions, F-UEs can transmit their data and FAPs are able to eliminate or mitigate the perceived interference. In order to be aware of retransmission events, it perfect synchronization is assumed between femtocells and the macrocell; hence, FAPs are able to detect the Automatic Repeat reQuest (ARQ) feedback sent by neighboring M-UEs. Moreover, the authors suppose that each F-UE knows the channel statistics of the links between M-BS and M-UE, between M-BS and itself, and between M-UE and itself. The process of interference mitigation is divided into two stages as illustrated in Figure 1.14. In stage S0, a FAP overhears the data transmitted by the M-BS while its F-UE is idle. When the M-UE does not correctly decode the received message, it sends a Negative ACKnowledgement (NACK) to the M-BS requiring a retransmission. This NACK is received also at the FAP, which schedules its F-UE in the next slot. Thus, in stage S1 the M-BS and the F-UE transmit simultaneously. Furthermore, the FAP exploits the message received at stage S0 to improve its decoding capability. Whenever it is able to correctly decode the M-BS message, the FAP completely eliminates the perceived interference; otherwise, it optimally combines data received in S0 and S1 to maximize the experienced SINR. Femtocells decide to access retransmission slots according to a given probability p. The optimal value of p is numerically computed to maximize the femtocell SE. Furthermore, in order to limit the femto-to-macro interference during retransmissions, the transmission power at the F-UE is constrained by a maximum value. This power is computed such that the outage probability constraint at the M-UE is satisfied. The proposed strategy strongly ameliorates the SE achieved with the classic interweave approach, however its efficacy is related to the precision of synchrony between the macrocell and femtocells and also to the channel statistics awareness at FAPs. These are complex tasks that require high overhead and may need signaling exchange between the M-BS and FAPs.

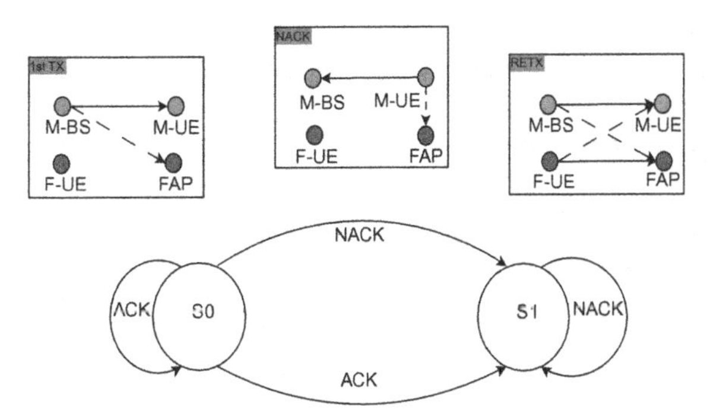

Figure 1.14 The overlay transmission scheme proposed in [60].

1.5 GREEN COGNITIVE FEMTOCELL NETWORKS

Femtocell networks have been proposed as an efficient and cost-effective approach to enhance cellular network capacity and coverage. Recent economic investigations claim that femtocell deployment might reduce both the OPEX and CAPital EXpenditure (CAPEX) for cellular operators [61]. A recent study [62] shows that expenses scale from $60,000/year/macrocell to $200/year/femtocell. However, according to the ABI Research [4], by the end of 2012 more than 36 million femtocells are expected to be sold worldwide with 150 million customers. Cellular network energy consumption might be drastically increased by the dense and unplanned deployment of additional BSs. The growth of energy consumption will cause an increase in global CO_2 emissions and impose more and more challenging operational costs.

In order to investigate the relationship between the BS load and its power consumption, the EARTH Energy Efficiency Evaluation Framework (E^3F) maps the radiated RF power to the power supply of a BS site [63]. Furthermore, the impact of the different components of the BS transceivers on the aggregate power consumption is analysed. Such a study in based on the analysis of the power consumption of various LTE BS types as of 2010. The effect of the various components of the BS transceivers is considered: antenna interface, power amplifier, the small-signal RF transceiver, baseband interface, DC-DC power supply, cooling, and AC-DC supply. Therefore, E^3F proposes a linear power consumption model that approximates the dependency of the BS power consumption to the cell load:

$$P^* = \begin{cases} P_0 + \Delta_p P^{RF}, & 0 < P^{RF} \leqslant P_{max}; \\ P_{sleep}, & P_{out} = 0. \end{cases} \tag{1.1}$$

where P^* is the BS input power required to generate the irradiated P^{RF} power, and Δ_p is the slope of the load dependent power consumption. Moreover, P_{max}, P_0, and P_{sleep} indicate the RF output power at maximum load, minimum load, and in sleep mode, respectively.

Table 1.1 shows the classical values of P_{max}, P_0, and Δ_p for M-BSs and FAPs. Note that the value of P_{sleep} depends on the hardware components that are deactivated during BS sleep intervals. However, more deactivated hardware components result in a slower reactivation process.

Figure 1.15 shows the operating power consumption of M-BSs (left) and FAPs (right) with respect to the traffic load.

Such representations evidence that:

Table 1.1 BS Power Model Parameters

BS type	P_{max} [W]	P_0 [W]	Δ_p
M-BS	40	712	14.5
FAP	0.01	10.1	15

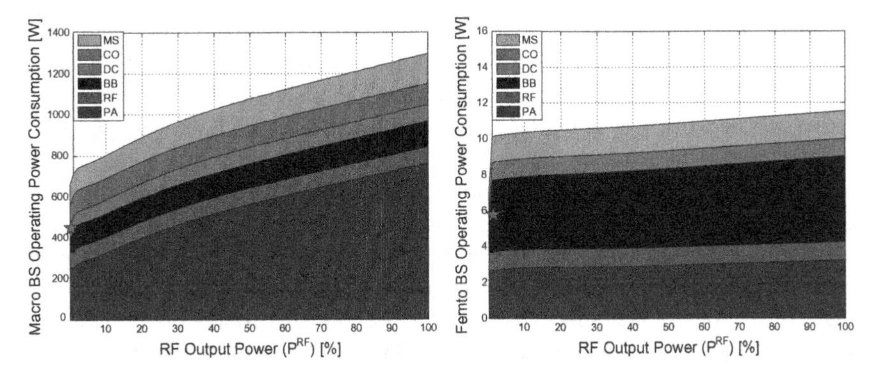

Figure 1.15 M-BS (left) and FAP (right) system power consumption dependency on relative output power [63]. Legend: PA=power amplifier, RF=small signal RF transceiver, BB=base-band processor, DC= DC–DC converters, CO=Cooling (only applicable to the M-BS), PS=AC/DC power supply; the red star mark indicates the BS power consumption in sleep mode.

- M-BS power consumption is strongly related to the load, thus, macro offloading via femtocell deployment can greatly enhance the overall cellular network EE;
- FAP power consumption does not vary much with the load, thus, the EE of femtocells is reduced in lightly loaded scenarios;
- Retransmissions have a higher impact on macrocell performance and slightly affect femto EE; therefore, retransmissions towards M-UEs should be performed by neighbouring small cells;
- Low cost power amplifiers designed to scale their power consumption with the load could improve the energy performance of femtocell networks;
- Dynamic cell switch-off techniques can adapt femtocell activity to the load in order to operate only in high EE state.

It is important to note that femtocells normally work in low load scenarios. Due to the limited number of UEs that can be simultaneously served by a FAP and the short distance between the AP and the user terminal, spectrum/power resources are often underutilized at FAPs. Furthermore, the femtocell density in urban scenarios is expected to be very high. A high number of low energy efficient FAPs can have a detrimental effect on the aggregate cellular network performance.

Although the E^3F model allows us to understand the trade-offs related to femtocell deployment, the relationships between EE, service constraints, and deployment efficiency are not straightforward and reducing the overall energy consumption while adapting the target of SE to the actual load of the system and QoS emerges as a new challenge in wireless cellular networks.

Furthermore, most of the literature aims to underline how much energy gain is achievable by deploying femtocells in the macrocell region (see for instance [64]), and few practical algorithms have been proposed to enhance the two-tier network EE.

A general classification of such energy-aware mechanisms can be realized according to the temporal scale in which they operate (cf. Figure 1.16). Deployment of additional femtocells, which offload the neighbouring macrocell and improve the network EE, is realized in a long-time scale (such as weeks). Due to the static characteristic of the indoor femtocell deployment scenarios, mechanisms that depend on the cell load, such as dynamic cell switch-off [65] and cell zooming [66] operate in mid-time scale (such as hours). Finally, energy-aware resource allocation schemes are implemented in short-time scale (e.g., the system scheduling period).

Ghost Femtocells is a short-time-scale algorithm that trades off transmission energy for frequency resources [67]. This RRM strategy profits from the inherent characteristics of co-channel femtocell deployment: due to the limited number of UEs that can be simultaneously served by a FAP (typically this number is less than four) and the short distance between the AP and the user terminal, spectrum resources are often underutilized at femtocells. The proposed algorithm is mainly composed by two steps: first, the femtocell scheduler attempts to serve as many UEs as possible according to their QoS constraints; second, further available resources are smartly exploited to spread the original message and lower the associated Modulation and Coding Scheme (MCS). Decreasing the MCS permits the subsequent strong limitation of the transmission power and enhancement of the transmission robustness. Simulation results show that the proposed approach limits both cross-tier and co-tier interference and increases the femtocell EE especially in low load traffic scenarios. However, this algorithm optimizes only the femtocell RF output power, which slightly impacts on the overall system power consumption.

López-Pérez et al. propose a similar approach, where self-organizing femtocells independently assign MCSs, RBs, and transmission power levels to UEs, while minimizing the cell RF output power and meeting QoS constraints [68]. In such a scenario, FAPs are aware of the spectrum usage at nearby femtocells

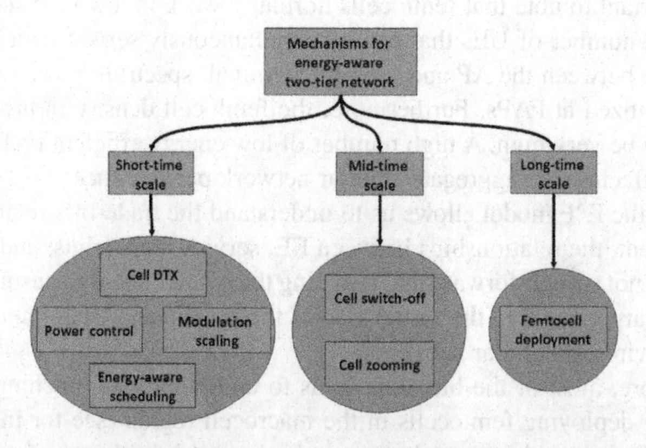

Figure 1.16 Time-scale chart of energy-aware mechanisms for two-tier cellular networks.

and tend to allocate less power to those UEs that are located in the proximity of the serving FAP or have low data-rate requirements. Subsequently, nearby FAPs allocate UEs with bad channel conditions or high data-rate constraints on those RBs characterized by low interference. Thus, in the proposed algorithm, neighboring femtocells dynamically control inter-cell interference without any coordination or static frequency planning. However, the authors do not investigate the impact of the proposed scheme in terms of femto-to-macro interference. M-UEs are affected by the aggregate interference produced by nearby FAPs. As showed in Figure 1.17, the discussed approach may create spikes of interference, which can corrupt macrocell transmissions, especially in high density femtocells scenarios.

Cheng et al. propose a more effective power optimization strategy [69]. The authors investigate a scenario in which femtocells and macrocells are deployed on orthogonal bands in order to avoid cross-tier interference. Furthermore, the spectral reuse between femtocells is limited to control the co-tier interference. System EE is measured through the *green factor*, which is defined as

$$\text{Green factor} = \frac{W(rT_{\text{m}} + (1-r)N_f r_f T_f)}{P^T_{\text{system}}}, \tag{1.2}$$

where W is the cellular bandwidth, r is the ratio between the number of channels dedicated at the macrocell and the total available channels, T_{m} is the macrocell downlink throughput, N_f is the number of deployed FAPs, r_f is the spectral reuse factor at femtocells, T_f is the femtocell downlink throughput, and P^T_{system} is the aggregate system power consumption of both macro and femtocells. Therefore, the proposed strategy aims to efficiently share the spectrum between M-BSs and FAPs in order to maximize the green factor while guaranteeing a certain SE at both M-UEs and F-UEs. However, although the proposed approach results in limited interference and

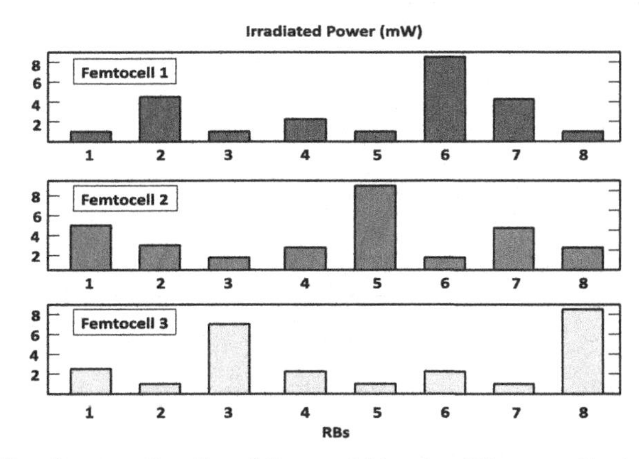

Figure 1.17 RB and power allocation of three neighbouring FAPs according to the scheme proposed in paper [68].

low complexity it may not be suitable in realistic scenarios, where SE constraints of different UEs can strongly vary. Furthermore, due to the orthogonal bandwidth deployment, it can result in low SE performance for both M-UEs and F-UEs, especially in deployment scenarios characterized by a high density of femtocells.

Higher EE can be achieved by dynamically switching off those FAPs that are not serving active users. Idle FAPs disable pilot transmissions and associated radio processing that represent the strongest contribution to the femtocell system power consumption. Dynamic femtocell switch-on/off is capturing the attention of both operators and researchers because it can introduce an important energy gain without seriously affecting UE performance. In fact, in this scenario, cellular coverage is guarantee by the active macrocell, while femtocells dynamically create high capacity zones adapting their activity status to the UE deployment.

Ashraf et al. propose to equip FAPs with an energy detector that permits *sniffing* the presence of nearby M-UEs [65]. As previously discussed, an indoor UE, which is served by the M-BS, likely transmits with high power; hence it is easy to detect. The detection threshold is computed at a femtocell by estimating the path loss to the M-BS, such that UEs located at the femtocell edge can be correctly detected (see Figure 1.18). Henceforth, when an UE is detected, the FAP switches to active mode and if the UE has the right to access the femtocell, the handover process is initiated. Otherwise, the FAP reverts to the switch-off mode. Simulation results show the proposed approach can lead to high energy gain. However, in high density scenarios the aggregate energy received from different sources of interference can cause false alarm events that affect detection reliability.

In order to avoid additional hardware at FAPs, the authors extend the energy detector based proposal by introducing two novel algorithms that control femtocell activity [70]. In the first scheme, the femtocell status is managed by the core network that is in charge to transmit, via the backhaul, a specific message that controls FAP activation/deactivation. The core network exploits the knowledge about the UE position to find FAPs to which the UE is able to connect. Moreover, this solution has the

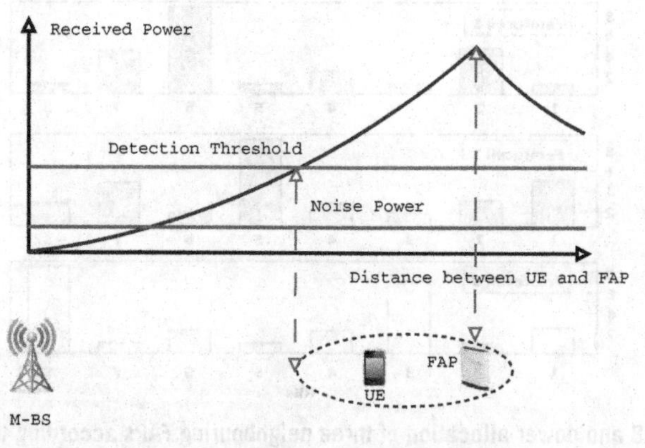

Figure 1.18 UE detection scheme in cell switch-off strategy [65].

advantage of distinguishing between registered UEs that can be served by closed access FAPs and unregistered UEs that can be served only by open/hybrid access femtocells. Furthermore, this approach also implements a centralized decision that considers global knowledge on the status of the netwok. In the second scheme, the femtocell activity is controlled directly by the UE. Two approaches are feasible: In the first case, a UE served by the M-BS periodically broadcasts a wake-up message to find idle FAPs in its range. In the second case, a reactive scheme is followed: a UE sends the wake-up message either when it experiences poor performance from the M-BS or when it requires higher data rate. In both the schemes FAPs are required to be able to detect the wake-up message during sleep mode. This message could include identification information such that closed access femtocells wake up only for registered UEs. The UE-controlled scheme suffers mainly from two drawbacks: First it increases UE battery consumption, especially in the proactive version. Furthermore, it requires the specification of a robust physical/logical wireless control channel where the UE can send the wake-up message.

An alternative solution is proposed by Telefonica [71] where the UE detection is based on the usage of a short range radio (SRR) interface, such as Bluetooth Low Power. In order to reduce power consumption, the SRR interface is maintained in standby as much as possible, and it is activated only when the UE is located nearby its serving FAP. In fact, UEs store a database, named the femto-overlapping macrocells list, that includes the IDs of M-BSs located in the serving FAP neighborhood. A UE camping in any of these M-BSs actives its SRR interface (shifting from standby mode to initiating mode) and starts searching for advertising packets broadcast by its FAP. Subsequently, if the UE is allowed to access the FAP, the two SRRs change in connect status, and after the connection is created, the FAP switches on its RF apparatus. This method is reliable because it is based on a point-to-point connection between the FAP and the UE, however, the main drawback is that currently there are no practical solutions to switch the FAPs from standby mode to advertising mode. Hence, FAPs should always keep their SRRs activated, decreasing the system EE and increasing interference in the already overcrowded ISM bands.

Dynamic cell switch-off mechanisms are inherent to FAPs that are not serving active UEs; however, cell zooming [66] and cell DTX [72, 73] have been recently proposed to enhance EE of lightly loaded systems. Cell zooming adaptively adjusts the cell size according to traffic load, user requirements, and channel conditions. Therefore, FAPs under light load self-deactivate to reduce network energy consumption; subsequently UEs located in coverage areas of such idle FAPs have to connect to the nearby M-BS. Alternatively, when open access femtocells are deployed, active FAPs may dynamically increase their irradiated power to guarantee service in the regions of neighboring idle FAPs. However, cell zooming can still create holes in the network coverage and strongly affect the system performance.

Cell DTX is implemented on a faster time scale and allows the FAP to immediately switch off cell specific signaling during subframes where there are no user data transmissions. Such a fast adaptation mechanism may allow for great energy savings especially in low traffic scenarios. However, depending on signals that are not transmitted, connectivity issues arise as UEs may not be able to detect a cell in DTX mode. Therefore, this mechanism may result in excessive handover latency and packet loss.

1.6 CONCLUSIONS AND FUTURE LINES OF RESEARCH

A comprehensive overview on cognitive strategies for two-tier cellular networks was presented. First, we critically discussed spectrum awareness, victim detection, resource allocation, and spectrum sharing techniques, which enable the effective coexistence of macrocells and femtocells in the same radio environment. Then, we examined the two-tier network in the EE perspective and we reviewed approaches proposed to achieve green communication through femtocell deployment.

Table 1.2 synthesizes the main features of the analyzed RRM schemes. The first column indicates the bibliography reference of such schemes and the second column describes the investigated interference scenario. The third and the fourth columns indicate the cellular technology and the transmission scenario (either uplink or downlink) in which the algorithm is implemented. The fifth and the sixth columns describe the FAP access type and the kind of cooperation that is exploited by the FAPs. A FAP can cooperate with the underlying M-BS as in [16], with neighboring FAPs as in [19], or both types of coordination can be implemented as in [46]. Obviously, cooperation has a direct impact on both the overhead and the system complexity, which are indicated in columns seven and eight, respectively. Note that in order to give a qualitative description of the algorithm complexity, we have taken into account also the number of transceivers required at the cognitive device, the type of information (location of the M-UE, scheduling, channel gain, etc.) needed, and the architecture (i.e., centralized or distributed) required.

Including cognitive principles in two-tier networks is producing great expectation. However, the design of efficient and robust cognitive-aware strategies for two-tier networks is still an open research field. With this chapter we aim to underline some of the major issues in the domain:

- In broadband mobile wireless scenarios, channel availability and quality change with space and time. When a licensed user is detected, to realize seamless transmission, a cognitive device vacates its channel and reconstructs a transmission link on a different channel. The procedure that permits this transition from a channel to another with minimum performance degradation is called *spectrum mobility*. While this functionality is fundamental in CR, to the best of our knowledge, it has not been investigated yet in cognitive femtocell scenarios. Specific solutions that reduce delay and loss during spectrum mobility are necessary. Furthermore, these algorithms should be aware of the running applications and adapt to QoS constraints. For instance, FTP traffic requires tight constraints on packet error rate: a retransmission protocol should be implemented to refrain from outage. Voice communication

Table 1.2 Characteristics of Analysed CR-Based RRM Schemes

Ref.	Interference	Technology	Scenario	Access	Cooperation	Overhead	Complexity
[7]	Cross-tier	WiMax	DL	Closed	No	High	High
[16]	Cross-tier	LTE	UL/DL	Closed	Macro	High	High
[19]	Co-tier	–	DL	Closed	Femto	Medium	High
[26]	Cross-tier	LTE	DL	Closed	Macro	High	High
[36, 37]	Cross-tier/co-tier	LTE	UL/DL	Closed	No	Low	Low
[43]	Cross-tier/co-tier	WiMax	UL/DL	Open	Macro	Average	High
[39]	Cross-tier/co-tier	LTE	UL	Closed	No	Low	High
[46]	Cross-tier	–	DL	Closed	Macro/femto	High	High
[48]	Cross-tier	–	UL	Closed	Macro	Medium	High
[49]	Co-tier	–	UL/DL	Closed	No	Low	Medium
[50]	Co-tier	LTE	UL/DL	Closed	Femto	Medium	Medium
[51]	Cross-tier/co-tier	–	UL/DL	Closed	Femto	High	High
[55]	Cross-tier	LTE	UL/DL	Closed	Macro	Medium	High
[60]	Cross-tier	LTE	UL	Closed	Macro	Medium	High
[67]	Cross-tier/co-tier	LTE	DL	Closed	Femto	Medium	Medium
[68]	Co-tier	LTE	DL	Closed	No	Low	Low
[69]	Cross-tier/co-tier	LTE	DL	Closed	No	Low	Low
[59]	Cross-tier	–	UL	Closed	Macro	Medium	High

allows, however, a limited delay for the channel mobility to avoid call interruption.

- Backhaul connection between femtocells and the cellular network is a potential means for coordination between macro/femto cells. A cognitive control channel could be established over this fixed-line connection either to implement cooperative sensing techniques or coordinate the spectrum access. Moreover, cellular operators might broadcast information about the state of coexisting networks. Knowledge about users' traffic, location, and QoS constraints could greatly enhance the performance of cognitive two-tier networks. Nevertheless, increasing the amount of signalling augments the network overhead. Hence, to limit overhead, it is fundamental to exchange the most effective information, and also the frequency of these reports has to be limited. Furthermore, backhaul reliability and security should be considered.

- Cooperative sensing may greatly enhance the effectiveness of M-UEs' detection wireless fading channels [74]. Collaborative detection is, however, affected by spatially correlated shadowing. For a given SNR, a larger number of correlated sensing nodes is needed to achieve the same detection probability of a few independent users. Future solutions should investigate the effectiveness of cooperative sensing in two-tier network deployment. Furthermore, correlation between different detectors has to be taken into account to develop more efficient sensing schemes.
- Classically, researchers have tried to develop spectrally efficient systems to enable heterogeneous networks to coexist within the same spectrum. Nevertheless, recent studies showed how spectrum scarcity is almost all due to static resource allocation strategies and that CR can notably improve the spectrum usage. However, femtocell deployment requires a new paradigm for two main reasons. First, F-UEs can benefit from a high quality downlink signal enabled by short range communications characterizing femtocell deployments. Second, only a few users locally compete for a large amount of the frequency resources in a femtocell. Therefore, a femtocell may benefit from a huge amount of available spectral/power resources. In this context, novel *green* cognitive approaches should be investigated in order to save power consumption, reduce interference, and improve the battery life of customer's devices.
- Cell discontinuous transmission (DTX) [72] is emerging as a means to greatly enhance the system EE especially in light load scenarios. Although DTX can be hard to implement at macrocells due to the need to continuously transmit the control channels, it can be more easily implemented at femtocells where the cellular coverage is always guaranteed by the nearby M-BS. However the implementation of this technique would completely change the characteristics of the perceived interference. Classical schemes to measure the quality of the wireless links may result in unreliable estimation. Cognitive techniques can be implemented to improve the awareness about interference behavior and permit reliable transmissions.
- Most of the CR literature for two-tier networks is based on the association secondary users to F-UEs and primary users to M-UEs. To the best of our knowledge only the authors of [7] have investigated a different approach. In our view, both macro and femto users should dynamically access the cellular spectrum; this deployment could result in more flexible and efficient systems.
- A multi-operator spectrum sharing approach might permit to the realization of novel and effective RRM techniques [30]. This strategy can potentially solve problems related to both co-tier and cross-tier interference and increase macrocell offloading. However, new business and pricing models are required

to implement scenarios in which competitors cooperate towards a more efficient usage of the aggregate resources.

- Amongst the discussed studies only the authors of [43] have considered the open access femtocell case. However, this is a very promising scenario in terms of both SE (due to the less perceived/generated interference) and EE (due to the increased macrocell offloading). Cognitive algorithms can represent a powerful instrument to deal with the problems inherent to this scenario such as the frequency of handover, femto-to-femto interference, etc.

ACKNOWLEDGMENT

This work has been partially supported by the European Commission in the framework of the BeFEMTO project (ICT-FP7–248523).

REFERENCES

[1] Mobile Broadband Access at Home, Informa Telecoms & Media, August 2008.
[2] V. Chandrasekhar, J. Andrews, A. Gatherer, Femtocell networks: a survey, IEEE Communications Magazine 46 (9) (2008) 59–67.
[3] Alcatel-Lucent, Metro cells: a cost-effective option for meeting growing capacity demands, 2011. <www.alcatel-lucent.com>.
[4] G. De La Roche, A. Valcarce, D. López-Pérez, J. Zhang, Access control mechanisms for femtocells, IEEE Communications Magazine 48 (1) (2010) 33–39.
[5] S. Carlaw, IPR and the Potential Effect on Femtocell Markets, FemtoCells Europe, ABIresearch, 2008.
[6] A. De Domenico, E. Calvanese Strinati, M.G. Di Benedetto, A survey on MAC strategies for cognitive radio networks, IEEE Communication Surveys and in press. Available at IEEExplore.
[7] J. Jin, B. Li, Cooperative resource management in cognitive WiMAX with femto cells, in: Proceedings of the 29th Conference on Information Communications (INFOCOM 2010), San Diego, CA, USA, March 2010, pp. 1–9.
[8] A. Attar, V. Krishnamurthy, O.N. Gharehshiran, Interference management using cognitive base-stations for UMTS LTE, IEEE Communications Magazine 49 (8) (2011) 152–159.
[9] J. Perez-Romero, O. Salient, R. Agusti, L. Giupponi, A novel on-demand cognitive pilot channel enabling dynamic spectrum allocation, in: Second IEEE International Symposium on New Frontiers in Dynamic Spectrum Access Networks (DySPAN 2007), Dublin, Ireland, April 2007, pp. 46–54.
[10] C. Stevenson, G. Chouinard, Z. Lei, W. Hu, S. Shellhammer, W. Caldwell, IEEE 802.22: the first cognitive radio wireless regional area network standard, IEEE Communications Magazine 47 (1) (2009) 130–138.
[11] G. Gur, S. Bayhan, F. Alagoz, Cognitive femtocell networks: an overlay architecture for localized dynamic spectrum access dynamic spectrum management, IEEE Wireless Communications 17 (4) (2010) 62–70.

[12] S.M. Cheng, S.Y. Lien, F.S. Chu, K.C. Chen, On exploiting cognitive radio to mitigate interference in macro/femto heterogeneous networks, IEEE Wireless Communications 18 (3) (2011) 40–47.

[13] W. Ren, Q. Zhao, A. Swami, Power control in cognitive radio networks: how to cross a multi-lane highway, IEEE Journal on Selected Areas in Communications 27 (7) (2009) 1283–1296.

[14] A.C.V. Gummalla, J.O. Limb, Wireless medium access control protocols, IEEE Communications Surveys and Tutorials 3 (2) (2009) 2–15.

[15] H. Urkowitz, Energy detection of unknown deterministic signals, Proceedings of the IEEE 55 (1967), 523–531.

[16] M.E. Sahin, I. Guvenc, M.R. Jeong, H. Arslan, Handling CCI and ICI in OFDMA femtocell networks through frequency scheduling, IEEE Transactions on Consumer Electronics 55 (4) (2009) 1936–1944.

[17] T. Yücek, H. Arslan, A survey of spectrum sensing algorithms for cognitive radio applications, IEEE Communication Surveys and Tutorials 11 (1) (2009) 116–130.

[18] W.A. Gardner, Signal interception: a unifying theoretical framework for feature detection, IEEE Transactions on Communications 36 (1988) 897–906.

[19] J. Xiang, Y. Zhang, T. Skeie, L. Xie, Downlink spectrum sharing for cognitive radio femtocell networks, IEEE Systems Journal 4 (4) (2010) 524–534.

[20] S.Y. Lien, C.C. Tseng, K.C. Chen, C.W. Su, Cognitive radio resource management for QoS guarantees in autonomous femtocell networks, in: IEEE International Conference on Communications (ICC 2010), Cape Town, South Africa, May 2010, pp. 1–6.

[21] S. Barbarossa, S. Sardellitti, A. Carfagna, P. Vecchiarelli, Decentralized interference management in femtocells: a game-theoretic approach, in: Proceedings of the Fifth International Cognitive Radio Oriented Wireless Networks Communications (CROWNCOM 2010), Cannes, France, June 2010, pp. 1–5.

[22] 3GPP TSG-RAN1#62, R1-105082, Way forward on eICIC for non-CA based HetNets, August 2010.

[23] J. Lotze, S.A. Fahmy, B. Özgül, J. Noguera, L.E. Doyle, Spectrum sensing on LTE femtocells for GSM spectrum re-farming using Xilinx FPGAs, in: Software-Defined Radio Forum Technical Conference (SDR Forum), USA, December 2009.

[24] P.D. Sutton, J. Lotze, H. Lahlou, S.A. Fahmy, K.E. Nolan, B. Ozgul, T.W. Rondeau, J. Noguera, L.E. Doyle, Iris: an architecture for cognitive radio networking testbeds, IEEE Communications Magazine 48 (9) (2010) 114–122.

[25] 3GPP TSG-Ran4#52 and NTT DOCOMO, R4-093244, Downlink Interference Coordination Between eNodeB and Home eNodeB, 24–28 August 2009.

[26] B. Zubin, S. Andreas, A. Gunther, H. Harald, Dynamic resource partitioning for downlink femto-to-macro-cell interference avoidance, EURASIP Journal on Wireless Communications and Networking 2010 (2010).

[27] 3GPP TSG-Ran4 Ad hoc #2010-01, picoChip Design, and Kyocera, R4-100193 Victim UE Aware Downlink Interference Management, January 2010.

[28] 3GPP TSG-Ran Wg4 #51 and Qualcomm Europe, R4-091908, Partial Bandwidth Control Channel Performance.

[29] 3GPP TSG-Ran WG4 Meeting #52bis and picoChip Designs, R4-093668, Victim UE Aware Downlink Interference Management.

[30] M.M. Buddhikot, I. Kennedy, F. Mullany, H. Viswanathan, Ultra-broadband femtocells via opportunistic reuse of multi-operator and multi-service spectrum, Bell Labs Technical Journal 13 (4) (2009) 129–143.

[31] S. Kawade, M. Nekovee, Can cognitive radio access to TV white spaces support future home networks? in: IEEE Symposium on New Frontiers in Dynamic Spectrum (DySPAN 2010), Singapore, April 2010, pp. 1–8.
[32] R.J. Haines, Cognitive pilot channels for femto-cell deployment, in: Seventh International Symposium on Wireless Communication Systems (ISWCS 2010), NewYork, UK, September 2010, pp. 631–635.
[33] M. Mueck, C. Rom, Wen Xu, A. Polydoros, N. Dimitriou, A.S. Diaz, R. Piesiewicz, H. Bogucka, S. Zeisberg, H. Jaekel, T. Renk, F. Jondral, P. Jung, Smart femto-cell controller based distributed cognitive pilot channel, in: Fourth International Conference on Cognitive Radio Oriented Wireless Networks and Communications (CROWNCOM'09), Hannover, Germany, June 2009, pp. 1–5.
[34] V. Chandrasekhar, J. Andrews, Spectrum allocation in tiered cellular networks, IEEE Transactions on Communications 57 (10) (2009) 3059–3068.
[35] I. Guvenc, M.R. Jeong, F. Watanabe, H. Inamura, A hybrid frequency assignment for femtocells and coverage area analysis for co-channel operation, IEEE Communications Letters 12 (12) (2008) 880–882.
[36] Y.Y. Li, M. Macuha, E.S. Sousa, T. Sato, M. Nanri, Cognitive interference management in 3G femtocells, in: IEEE International Symposium on Personal, Indoor and Mobile Radio Communications PIMRC'09, Tokyo, Japan, September 2009, pp. 1118–1122.
[37] Y.Y. Li, E.S. Sousa, Cognitive uplink interference management in 4G cellular femtocells, in: IEEE 21st International Symposium on Personal Indoor and Mobile Radio Communications (PIMRC 2010), Instanbul, Turkey, September 2010, pp. 1567–1571.
[38] Z. Shi, M.C. Reed, M. Zhao, On uplink interference scenarios in two-tier macro and femto co-existing UMTS networks, EURASIP Journal on Wireless Communications and Networking 2010 (2010).
[39] I. Wayan Mustika, Koji Yamamoto, Hidekazu Murata, Susumu Yoshida, Potential game approach for self-organized interference management in closed access femtocell networks, in: IEEE 73rd Vehicular Technology Conference (VTC Spring 2011), May 2011, pp. 1–5.
[40] K. Akkarajitsakul, E. Hossain, D. Niyato, D. Kim, Game theoretic approaches for multiple access in wireless networks: a survey, IEEE Communications Surveys Tutorials, (99) (2011), 1–24.
[41] D.P. Palomar, M. Chiang, A tutorial on decomposition methods for network utility maximization, IEEE Journal on Selected Areas in Communications 24 (8) (2006) 1439–1451.
[42] D. Calin, H. Claussen, H. Uzunalioglu, On femto deployment architectures and macrocell offloading benefits in joint macro-femto deployments, IEEE Communications Magazine 48 (1) (2010) 26–32.
[43] J.P.M. Torregoza, R. Enkhbat, W.J. Hwang, Joint power control, base station assignment, and channel assignment in cognitive femtocell networks, EURASIP Journal on Wireless Communications and Networking, 2010 (2010).
[44] A. Goldsmith, S.A. Jafar, I. Maric, S. Srinivasa, Breaking spectrum gridlock with cognitive radios: an information theoretic perspective, Proceedings of the IEEE 97 (5) (2009) 894–914.
[45] S.M. Cheng, W.C. Ao, K.C. Chen, Downlink capacity of two-tier cognitive femto networks, in: IEEE 21st International Symposium on Personal Indoor and Mobile Radio Communications (PIMRC 2010), Instanbul, Turkey, September 2010, pp. 1303–1308.

[46] A. Galindo-Serrano, L. Giupponi, M. Dohler, Cognition and docition in OFDMA-based femtocell networks, in: IEEE Global Telecommunications Conference (GLOBECOM 2010), Miami, FL, USA, December 2010, pp. 1–6.

[47] C.J.C.H. Watkins, P. Dayan, Q-learning, Machine Learning, 8 (3) (1992), 279–292.

[48] V. Chandrasekhar, J.G. Andrews, T. Muharemovic, Z. Shen, A. Gatherer, Power control in two-tier femtocell networks, IEEE Transactions on Wireless Communications 8 (8) (2009) 4316–4328.

[49] G.W.O. da Costa, A.F. Cattoni, I.Z. Kovacs, P.E. Mogensen, A scalable spectrum-sharing mechanism for local area network deployment, IEEE Transactions on Vehicular Technology 59 (4) (2010) 1630–1645.

[50] L.G.U. Garcia, G.W.O. Costa, A.F. Cattoni, K.I. Pedersen, P.E. Mogensen, Self-organizing coalitions for conflict evaluation and resolution in femtocells, in: IEEE Global Telecommunications Conference (GLOBECOM 2010), Miami, FL, USA, December 2010, pp. 1–6.

[51] F. Pantisano, M. Bennis, W. Saad, R. Verdone, M. Latva-aho, Coalition formation games for femtocell interference management: a recursive core approach, in: IEEE Wireless Communications and Networking Conference (WCNC 2011), March 2011, pp. 1161–1166.

[52] F. Pantisano, K. Ghaboosi, M. Bennis, M. Latva-Aho, Interference avoidance via resource scheduling in TDD underlay femtocells, in: IEEE 21st International Symposium on Personal, Indoor and Mobile Radio Communications Workshops (PIMRC Workshops 2010), September 2010, pp. 175–179.

[53] C.U. Castellanos, D.L. Villa, C. Rosa, K.I. Pedersen, F.D. Calabrese, P.-H. Michaelsen, J. Michel, Performance of uplink fractional power control in UTRAN LTE, in: IEEE Vehicular Technology Conference (VTC Spring 2008), Calgary, Canada, May 2008, pp. 2517–2521.

[54] R.H. Etkin, D.N.C. Tse, Hua Wang, Gaussian interference channel capacity to within one bit, IEEE Transactions on Information Theory 54 (12) (2008) 5534–5562.

[55] S. Rangan, Femto-macro cellular interference control with subband scheduling and interference cancelation, in: IEEE GLOBECOM Workshops 2010, Miami, FL, 2010, pp. 695–700.

[56] I. Widjaja, H. La Roche, Sizing x2 bandwidth for inter-connected ENBS, in: IEEE 70th Vehicular Technology Conference Fall (VTC 2009-Fall), September 2009, pp. 1–5.

[57] S. Kaimaletu, R. Krishnan, S. Kalyani, N. Akhtar, B. Ramamurthi, Cognitive interference management in heterogeneous femto-macro cell networks, in: IEEE International Conference on Communications (ICC 2011), June 2011, pp. 1–6.

[58] K. Norlund, T. Ottosson, A. Brunstrom, Fairness measures for best effort traffic in wireless networks, in: 15th IEEE International Symposium on Personal, Indoor and Mobile Radio Communications (PIMRC 2004), September 2004, vol. 4, pp. 2953–2957.

[59] F. Pantisano, M. Bennis, W. Saad, M. Debbah, Spectrum leasing as an incentive towards uplink macrocell and femtocell cooperation, IEEE JSAC Special Issue on Femtocell Networks, April 2012.

[60] S.M. Cheng, W.C. Ao, K.C. Chen, Efficiency of a cognitive radio link with opportunistic interference mitigation, IEEE Transactions on Wireless Communications 10 (6) (2011) 1715–1720.

[61] Airvana Inc, How femtocells change the economics of mobile service delivery, <http://www.airvana.com/>.

[62] M. Heath et al., Picocells and femtocells: will indoor base stations transform the telecoms industry? 2007. <http://research.analysys.com>.

[63] G. Auer, V. Giannini, C. Desset, I. Godor, P. Skillermark, M. Olsson, M.A. Imran, D. Sabella, M.J. Gonzalez, O. Blume, et al., How much energy is needed to run a wireless network? IEEE Wireless Communications 18 (5) (2011) 40–49.

[64] F. Cao, Z. Fan, The tradeoff between energy efficiency and system performance of femtocell deployment, in: IEEE seventh International Symposium on Wireless Communication Systems (ISWCS 2010), pp. 315–319.

[65] I. Ashraf, L.T.W. Ho, H. Claussen, Improving energy efficiency of femtocell base stations via user activity detection, in: IEEE Wireless Communications and Networking Conference, WCNC 2010, 2010, pp. 1–5.

[66] Z. Niu, Y. Wu, J. Gong, Z. Yang, Cell zooming for cost-efficient green cellular networks, IEEE Communications Magazine 48 (11) (2010) 74–79.

[67] E. Calvanese Strinati, A. De Domenico, A. Duda, Ghost femtocells: a novel radio resource management scheme for OFDMA based networks, in: IEEE Wireless Communications and Networking Conference (WCNC 2011), Cancun, Mexico, 2011.

[68] D. López-Pérez, X. Chu, A.V. Vasilakos, H. Claussen, Minimising cell transmit power: towards self-organized resource allocation in OFDMA femtocells, in: ACM SIGCOMM 2011, Toronto, Canada, August 2011.

[69] W. Cheng, H. Zhang, L. Zhao, Y. Li, Energy efficient spectrum allocation for green radio in two-tier cellular networks, in: IEEE Global Telecommunications Conference (GLOBECOM 2010), Miami, FL, USA, December 2010, pp. 1–5.

[70] I. Ashraf, F. Boccardi, L. Ho, Sleep mode techniques for small cell deployments, IEEE Communications Magazine 49 (8) (2011) 72–79.

[71] R3-110030, Dynamic H(e)NB Switching by Means of a Low Power Radio Interface for Energy Savings and Interference Reduction, 3GPP TSG RAN WG3 Meeting, Dublin, Ireland, January 2011.

[72] P. Frenger, P. Moberg, J. Malmodin, Y. Jading, I. Godor, Reducing energy consumption in LTE with cell DTX, in: IEEE 73rd Vehicular Technology Conference (VTC Spring 2011), May 2011, pp. 1–5.

[73] A. De Domenico, R. Gupta, E. Calvanese Strinati, Dynamic traffic management for green open access femtocell networks, in: IEEE 75th Vehicular Technology Conference Fall (VTC 2012-Spring), Yokohama, Japan, May 2012.

[74] A. Ghasemi, E.S. Sousa, Collaborative spectrum sensing for opportunistic access in fading environments, in: Proceedings of the Symposium on Dynamic Spectrum Access Networks (DySPAN'05), Baltimore, MD, USA, November 2005, pp. 131–136.

CHAPTER

2 A Survey of Contemporary Technologies for Smart Home Energy Management

Aravind Kailas, Valentina Cecchi, Arindam Mukherjee

Department of Electrical and Computer Engineering, University of North Carolina at Charlotte, 9201 University City Blvd., Charlotte, NC 28223-0001, USA

2.1 INTRODUCTION

Residential energy consumption and the amount of pollution emitted from electric generators creates side effects that are not beneficial to public health and well-being, including increased pollution in the air and water (CO_2 and other greenhouse gases, mercury, and other trace elements and particulate matter), and the depletion of finite resources [1]. "Green Smart Home Technologies" are aimed at reducing the footprint of greenhouse gases by efficient energy management in residential buildings. Studies have shown that the display of real-time information on consumption can result in reductions of up to 30% by enabling end users to consume responsibly and manage effectively [2]. In recent times, more so than ever, the consumer has become more "green" conscious and therefore is looking for real-time visibility of energy consumption [3]. Further, the market for residential energy management is poised to grow dramatically due to increased consumer demand and new government and industry initiatives [4]. Smart homes have been studied since 1990s, and their primary focus has been resident comfort [5]. They employ energy efficiency by occupancy check or adaptability to outside conditions. However, they are not automatically a component of the smart grid. Their integration to the smart grid is an active topic [6–8]. With this in mind, this article motivates future research in the field of home area networking by revisiting the concepts of smart grids and smart homes, and summarizing the state-of-the-art in home energy management (HEM) communications and control technologies.

2.1.1 Bringing Smart Grids to Green Smart Homes

The smart grid is accelerating the transformation of energy-value change, and will enable electricity distribution systems to manage alternative energy sources (e.g., solar and wind), improve reliability, facilitate faster response rates to outages, and manage peak-load demands. Building a smart digital meter, the advanced metering infrastructure (AMI), is a first step, and would enable processing and reporting usage data to providers and households via two-way communication with the utility offices [9–11]. In recent years, there have been a lot of initiatives on the part of the government, utilities companies, and technology groups (e.g., standards committees, industries, alliances, etc.) for realizing smart grids for green smart homes [12]. Government initiatives include mandating upgrades to the grid, and adding intelligence to meters that measure water, gas, and heat. The market for smart home products, such as lighting and heating, ventilating, and air conditioning (HVAC) controls, in-home utility monitors, and home security systems, is also on the rise, driven in part by the desire to conserve energy and by the expansion of home automation services and standards-based wireless technologies. Further, energy directives and smart grid initiatives have attracted hundreds of companies with energy management systems (EMSs) including General Electric, Cisco, Google, and Microsoft. Efforts are underway to design new standards, protocols, and optimization methods that efficiently utilize supply resources (i.e., conventional generation, renewable resources, and storage systems) to minimize costs in real time. In other words, smart grid technologies so far focus on integrating the renewable energy resources to the grid to reduce the cost of power generation and integrating these resources requires storage systems. Smart grids can be potent tools in helping consumers reduce their energy costs, but consumers have several concerns that could inhibit rapid adoption. In order to maximize smart grids, utilities and suppliers of energy management solutions must first educate consumers about the benefits of these advanced systems and then package these solutions so that capabilities and advantages are obvious to consumers and easily integrated into their lifestyles.

2.1.2 Home Energy Management and Home Area Networks

The term home area networks (HANs) has been used loosely to describe all the intelligence and activity that occurs in HEM systems, and this section describes the concepts of HEM systems and HANs. Stated simply, HANs are extensions of the smart grid and communications frameworks, much like the familiar local area networks (LANs), but within a home [10]. Instead of a network of servers, printers, copiers, and computers, the HAN connects devices that are capable of sending and receiving signals from a meter and/or HEM system (HEMS) applications. Wired or wireless, there are trade-offs that involve power consumption, signaling distance, sensitivity to interference, and security. The main point here is that HANs are *not* energy management applications, but enable energy management applications to monitor and control the devices on the home network.

With limited data input and display capabilities, in-home displays (IHDs) function as a visual indicator of the electricity rates at any point in time. Moreover, IHDs are one-way communication devices, meaning the user can only monitor, but not take real-time actions and provide feedback to the HAN like the HEM systems. So, HANs and IHDs still need an energy management application, a HEM solution [13,14], in order to gain the most benefit from these smart grid components. A web-based portal for a HEM system is the best interface to the utility billing and demand response (DR) programs, because it enables the easiest execution

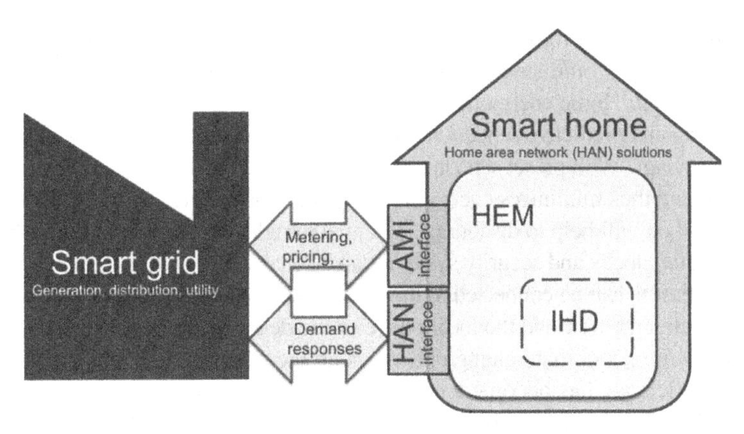

Figure 2.1 Realizing smart grids in smart homes.

and control of intelligent appliances that can be "enrolled" into such programs. A HEM solution would enable the user to recall the optimized presets for sustainable energy saving, get suggestions on energy efficiency improvements, and see how one's energy management compares to others in one's peer group or neighborhood [10]. A basic representation of a smart grid-smart home interface that uses a variety of different networking topologies across the different domains and subdomains is illustrated in Figure 2.1, and the focus of this paper is HEM systems and the HAN technologies.

2.1.3 Benefits of HEM

1. *Minimize energy wastage:* Home automation and real-time energy monitoring makes energy savings feasible. For example, lighting control is not about reducing light, but facilitating the correct light when and where required, while reducing wastage. Energy savings can also be realized according to occupancy, light level, time of day, temperature, and demand levels. For example, blinds and shutters could open and close automatically, based on the time of day and amount of light to optimize the mix of natural light and artificial light; or according to the temperature difference between indoor and outdoor to optimize heating, ventilating, and air conditioning (HVAC) power consumption. Home automation also minimizes energy wastage without affecting usage with occupancy detection, and planning appliance control and various settings depending on demand response levels. Energy savings come off the highest rate, and HEM systems help in monitoring the usage to track the highest electricity rate one is paying. For example, consider a 100 W light bulb (in the state of California, USA), a common electrical device in homes. Leaving the light on all month would cost $7.92 if the overall energy usage is very low, but

increases to $25.20 for substantially higher overall usage [2]. So, keeping that 100W light off would save the household $25.20 per month [2].

2. *Peace-of-mind:* Home energy management is important because it provisions time scheduling and predictive scheduling that ensures peace of mind while yielding energy savings. With preset scheduling, the user does not need to turn them on all the time and thus minimizes energy consumption. Further, having the lights and TV turned on will help to discourage potential intruders while you are away from home. Safety locks and security systems can be enabled as well; lighting and sound/motion sensors can be connected to the HEM system that track activity 24 hours a day, and alert the user and the local police or fire department if and when needed.
3. *Eco-friendly:* As climate change becomes an increasingly real concern, energy efficiency has become a top priority in homes and businesses alike. When describing a green home, energy efficiency refers to every aspect of energy consumption, from the source of electricity to the style of lightbulbs. Reducing energy consumption requires long-term behavioral change, the first step being an investigation of the current carbon footprint of residential and office buildings. HEM systems aid in this change by helping the user monitor the usage (e.g., heat, light, and power in homes), and by offering suggestions on how to cut down CO_2 emissions, a primary cause of global climate change. Continuing with the same example from [2], and using the "Terra Pass Carbon Footprint Calculator," a reduction in the CO_2 emissions (from the household) by a factor of more than 60% would be possible.
4. *Well-being of residents:* With the average family spending a huge amount annually on gas and electricity supplies alone, it certainly makes sense to do everything possible to reduce household utility bills. Good energy management within the home brings about this reduction, thereby increasing available capital. HEM systems also increase transparency and improve billing service. Such systems make life easy by providing the user with control and management, which will help manage one's time better and thus help to reduce stress. Reducing the energy consumption in a household by about 23% cut the monthly bills by over a third [2].
5. *Public good:* In terms of public good, four things can occur simultaneously when homes are energy efficient: (i) finite energy supplies are not depleted as quickly, (ii) emissions are reduced (including all the corresponding benefits associated with reduced emissions), (iii) consumers save money, and (iv) consumers increase net disposable income. With low-to-moderate income residents, saving money on utilities and spending those savings elsewhere can significantly affect quality of life. An additional public benefit can result from energy efficient housing. When government agencies serve as the housing provider for low-income residents, energy efficiency can contribute to taxpayer savings. Money can be saved when the government does not have to finance wasteful energy practices with public housing. An example of a governmental agency collaboration designed to reduce energy use in public housing is the partnership between local housing agencies (LHAs) [agencies who receive

program funding from the Department of Housing and Urban Development (HUD)] and the DOE Rebuild America program.

In summary, HEM systems are a step more advanced than previous energy-saving appliances that provides even more eco-friendly performance through the use of sensor technology. HEM systems allow energy monitoring, and automation of appliances and control system settings to respond to demand response levels. Thus, planning personal energy consumption plans in advance is encouraged to leverage rebates/incentives for green homes, and to benchmark oneself on a community level. To the best knowledge of the authors, this is the first comprehensive tutorial on the state-of-the-art in home area communications and networking technologies for energy and power management. This article also presents a classification of the many affordable smart energy products offered by different companies that are available in the market.

2.2 BACKGROUND ON DEMAND RESPONSE (DR) AND DEMAND-SIDE MANAGEMENT (DSM) PROGRAMS

In support of "smart grid" initiatives, several emerging technologies and techniques have been presented in the past decade. These techniques include, among others, advanced metering infrastructure (AMI) and two-way communication, integration of home area network (HAN) and home automation, and a push to invest in renewable micro-generation. The traditionally inelastic demand curve has begun to change, as these technologies enable consumers (industrial to residential) to respond to energy market behavior, reducing their consumption at peak prices, supplying reserves on an as-needed basis, and reducing demand on the electric grid [15]. Therefore, the development of smart grid-related activities has resulted in an increased interest in demand response and demand-side management programs.

Demand response programs are used to manage and alter electricity consumption based on the supply, e.g., during a reliability event (emergency DR), or based on market price (economic DR) (e.g., [16,17]). These programs can involve curtailing electric load, as well as utilizing local micro-generation (customer owned distributed generation). DR programs can be incentive-based programs (IBP), classical and market-based programs, or price-based programs (PBP) [17].

Demand-side management (DSM) refers to planning, implementation, and evaluation techniques, including policies and measures, which are designed to either encourage or mandate customers to modify their electricity consumption, in terms of timing patterns of energy usage as well as level of demand. The main objective is to reduce energy consumption and peak electricity demand.

DR and DSM initiatives can benefit customers and utilities, as well as society as a whole. From the customer perspective, these programs can help reduce the electric

Table 2.1 DSM Benefits

Customer benefits	Utility benefits	Societal benefits
Satisfy demand for electricity	Lower cost of service	Conserve resources
Reduce costs	Improve efficiency and flexibility	Reduce environmental impact
Improve service	Reduce capital needs	Protect environment
Improve lifestyle and productivity	Improve customer service	Maximize customer welfare

bill and is possibly incentivized by the utility (e.g., through tax credits). From a utility perspective, in addition to reducing supply costs (generation, transmission, and distribution), benefits also include deferral of capital expenditure on increasing system capacity, improved system operating efficiency and reliability, and better/more data to be used for planning and load forecasting. Society as a whole benefits also through the reduction of greenhouse gas emissions, due to the decrease (or non-increase) in energy consumption and peak demand and the avoided expansion of grid generation capacity. Major benefits of DSM are summarized in Table 2.1.

Several objectives are included in DSM initiatives, mainly focused on load management and energy efficiency [19] (refer to Figure 2.2). Under the load management objectives, we have peak clipping, valley filling, and load shifting. Energy efficiency, or conservation, involves a reduction in overall electricity usage. Electrification and flexible load shape, also shown in Figure 2.2, involve, respectively, programs for customer retention and development of new markets, and programs that utilities set up to modify consumption on an as-needed basis (i.e., customers in these programs will be treated as curtailable loads). DSM concepts have been studied since the 1980s and

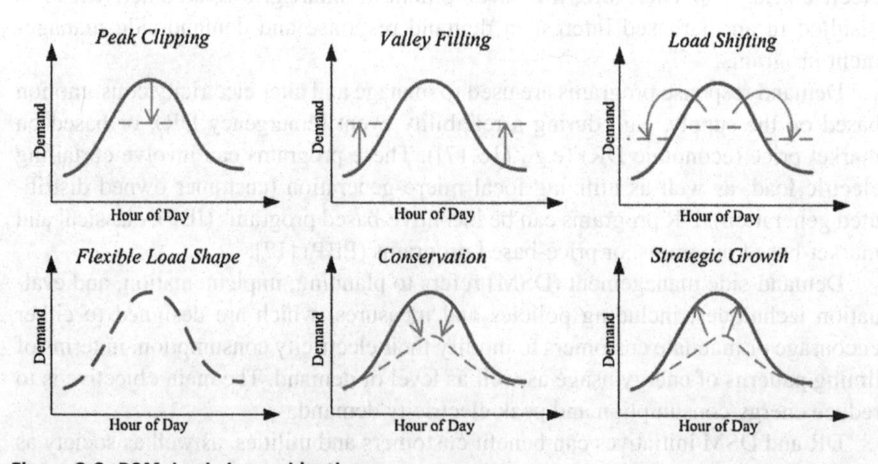

Figure 2.2 DSM–load shape objectives.

early 1990s; reports and surveys on the subject were published by the Electric Power Research Institute (EPRI) and the North American Electric Reliability Corporation (NERC) [20–22], among others.

In the past decade, the focus on smart grid applications and progress in communication protocols and technologies has improved the communication ability between electricity suppliers and end-use consumers, which would allow active deployment of DR at all times (demand dispatch [23]), not just event-based DR. Customers are then able to monitor and control their load in real time, and to possibly trade in the energy market. This requires the use of a sophisticated energy management system (EMS) to control equipment and appliances [24].

In [25], an optimized operational scheme for household appliances is introduced through the use of a demand-side management (DSM)-based simulation tool. The tool uses a particle swarm optimization algorithm to minimize the cost to customers and determine a source management technique. In a 1989 paper [26], the authors describe a system used to control electricity usage in homes or small businesses, by shifting some of the load from the peak to the valley and using a real-time variable pricing scheme. The proposed system uses a telephone to power line carrier (PLC) interface, a meter that measures energy with variable pricing, and a controller that adjusts energy utilization based on price. In [27], the authors developed, using mixed integer linear programming, a home energy management system (HEMS), which provides optimum scheduling for operation of electric appliances and controls the amount of power provided back to the grid from the excess local photovoltaic generation. In [28], the authors present a common service architecture developed to allow end-user interaction with other consumers and suppliers in an integrated energy management system. The architecture would facilitate users with renewable micro-generation to integrate with the electric grid through the use of a central coordinator inside their home gateway. In [29], a multi-scale optimization technique for demand-side management is presented. A home automation system is proposed that dynamically takes into account user comfort level as well as limits on power consumption. In [30], a novel strategy for control of appliances is proposed and utilizes a home automation and communication network. The goal of the proposed technique is to provide continued service, at possibly reduced power consumption levels, during power shortages. In [31], communication methodologies among control devices in home automation systems are demonstrated. Specifically, communication over a power line is presented to enable control of appliances in building/home energy management systems. In [32], a home automation system that controls household energy consumption is proposed. The system takes into account predicted/anticipated events, and uses a tabu search to maximize user comfort and minimize consumption costs. In [33], control mechanisms to optimize electricity consumption within a home and across multiple homes in a neighborhood are presented and evaluated. Energy management controllers (EMC) are assumed to control appliance operation based on energy prices and consumers' preset preferences. The authors first show that a simple optimization model used for determining appliance time of operation purely based on energy price may actually result in higher peak demand. An EMC optimization model, based on dynamic programming, which also accounts for electricity capacity constraints, is then presented. A distributed scheduling mechanism is also proposed to reduce peak demand within a neighborhood.

To summarize, DSM refers to planning, implementation, and evaluation techniques, including policies and measures, which are designed to either encourage or mandate customers to modify their electricity consumption, in terms of timing patterns of energy usage as well as level of demand. The main objective is to reduce energy consumption and peak electricity demand. Potential research in this area should focus on identifying optimized system-level hardware-software co-designed solutions to implement the DSM functionalities in the most energy efficient manner to respond to the dynamically changing operating environment of the HAN under real-time constraints.

2.3 HAN COMMUNICATIONS AND NETWORK TECHNOLOGIES

The energy management system is at the heart of green buildings, and enables home energy control and monitoring, providing benefits to both consumers and utilities. The HEM system intelligently monitors and adjusts energy usage by interfacing with smart meters, intelligent devices, appliances, and smart plugs, thereby providing effective energy and peak-load management. The platform for this communication is the HAN, and this section reviews the communications and network technologies for the HAN

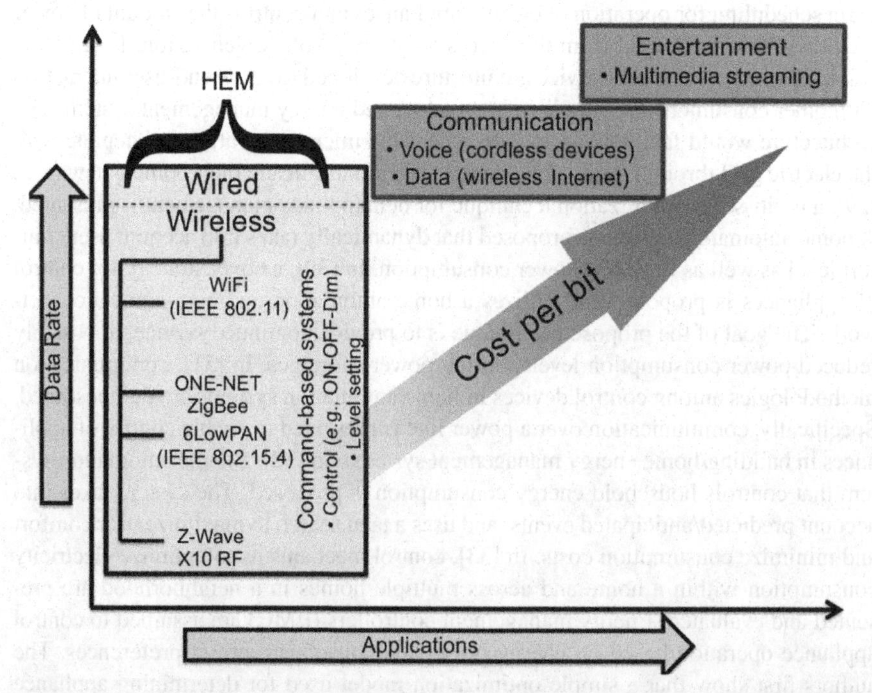

Figure 2.3 The cost and use of wired and wireless technologies for different home applications.

that can connect the HEM system to endpoints and smart meters [34]. The cost associated with HEM applications, is significantly lower compared to other home applications because of the differing functionalities (Figure 2.3). For example, HANs comprise command-based systems that require very short acquisition time for sending data to multiple destinations, and this cuts down the data rate and the bandwidth requirements compared to link-based systems (e.g., communication and entertainment systems) that need a reliable point-to-point communication link for longer periods of time.

Internet Protocol (*IP*) is a protocol used for communicating data within a packet-switched internetwork, and it is responsible for delivering data from source to destination based on an IP address. Being the foundation on which the Internet is built and communication is achieved, IP is a single layer within a multi-layer suite known as the TCP/IP stack. Due to this abstraction, IP can be used across a number of different heterogeneous network technologies. Due to the ease of interoperability, ubiquitous nature, widespread adoption, and work being performed to create a lightweight interface, IP is being seen as essential to the success of HAN and smart grid development. As the significance of devices communicating within the HAN increases, so does the requirement for usable IP addresses. Very broadly, the different technologies (comprising specifications for the physical and network layers) can be classified based on the *transmission medium* into wired and wireless, as shown in Figure 2.4.

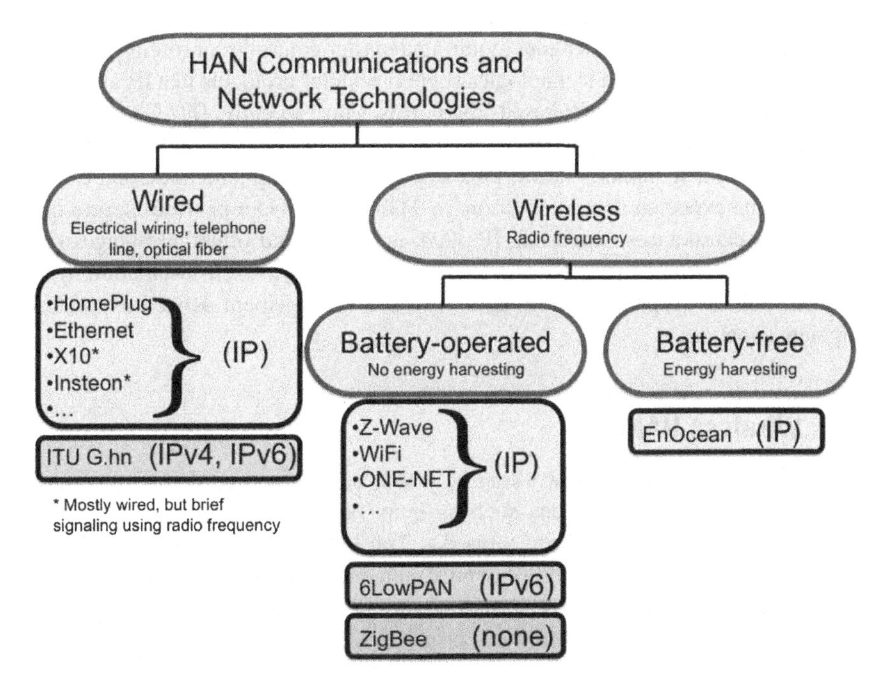

Figure 2.4 Communications and networking possibilities for a home area network.

2.3.1 Wired HANs

First, we discuss the technologies in which the transmission mediums are electronic wiring, telephone lines, coaxial cables, unshielded twisted pairs, and/or optical fibers. *HomePlug*, a power line communications technology that uses the existing home electricity wiring to communicate, is widely adopted for high-speed wired communication applications (e.g., high-quality, multi-stream entertainment networking) with a mature set of standards. *Ethernet* is a very common technology and supports a range of data rates using either unshielded twisted pairs (10 Mbps–1 Gbps), or optical fibers (as high as 10 Gbps). It utilizes a common interface found in numerous pieces of household equipment, including laptops, servers, printers, audio-video (AV) equipment and game consoles. Ethernet may not be appropriate for connecting all devices in the HAN (especially appliances) due to the high cost and power requirements plus the need for separate cabling back to a central point.

X10 is a technology (and an international and open industry standard) that uses power line wiring for signaling and control of home devices, where the signals involve brief radio frequency (RF) bursts representing digital information. However, it suffers from some issues such as incompatibility with installed wiring and appliances, interference, slow speeds, and lack of encryption. *Insteon* addresses these limitations while preserving backward compatibility with X10, and enables the networking of simple devices such as light switches using the power line [and/or RF]. All Insteon devices are peers, meaning each device can transmit, receive, and repeat any message of the Insteon protocol, without requiring a master controller or routing software. All the previously described technologies support popular protocols like IP, and hence can easily be integrated with IP-based smart grids. More recently, *ITU G.hn* has been developed by the International Telecommunication Union (ITU) and promoted by HomeGrid Forum. It supports networking over power lines, phone lines, and coaxial cables, and the expected data rates are up to 1 Gbps. ITU G.hn provides secure connections between devices supporting IP, IPv4, and IPv6, and offers advantages such as the ability to connect to any room regardless of wiring type, self-installation by the consumer, built-in diagnostic information, and self-management as well as multiple equipment suppliers.

2.3.2 Wireless HANs

Next, we discuss wireless networking of low-cost, low-power (battery-operated) control networks for applications such as home automation, security and monitoring, device control, and sensor networks. The low-cost *ZigBee*-based solutions allow wide deployment in wireless control and monitoring applications; the low power usage allows longer life with smaller batteries (up to 10 years) and the mesh networking provides high reliability and broader range.

Z-Wave, a proprietary wireless communications technology designed specifically to remote control applications in residential and light commercial environments

[35] is popular for the following reasons. Unlike WiFi and other IEEE 802.11-based wireless LAN systems that are designed primarily for high-bandwidth data flow, the Z-Wave radio frequency (RF) system operates in the sub-gigahertz frequency range (≈900 MHz), and is optimized for low-overhead commands such as ON-OFF-DIM (as in a light switch or an appliance), raise-lower (as in a volume control), and cool-warm-temp (as in a HVAC) with the ability to include device metadata in the communications. As a result of its low-power consumption and low cost of manufacture, Z-Wave is easily embedded in consumer electronics products, including battery-operated devices such as remote controls, smoke alarms, and security sensors. More importantly, Z-Wave devices can also be monitored and controlled from outside of the home by way of a gateway that combines Z-Wave with broadband Internet access.

WiFi is a popular IP-based wireless technology used in home networks, mobile phones, video games, and other electronic devices. Support is widespread with nearly every modern personal computer, laptop, game console, and peripheral device providing a means to wirelessly access the network via WiFi. Another IP-based wireless technology is the *ONE-NET*, also an open-source standard that is not tied to any proprietary hardware or software; it can be deployed using a variety of low-cost off-the-shelf radio transceivers and microcontrollers from various manufacturers.

6LoWPAN [also a standard from the Internet Engineering Task Force (IETF)] optimizes *IPv6*, the next generation IP communication protocol for internetworks and the Internet [36], for use with low-power communications technologies such as IEEE 802.15.4-based radios [37], enabling transfer of small packet sizes using low bandwidth. It is primarily aimed at evolving the current IPv4 protocol, which is predicted to be exhausted of address space in 2011. Operation of the 6LoWPAN involves compressing 60 bytes of headers down to just 7 bytes. The target for IP networking for low-power radio communication are the applications that need wireless Internet connectivity at lower data rates for devices with very limited form factor. 6LoWPAN allows communication with devices across the Internet without having to go through ZigBee-to-IP translation.

Table 2.2 Summary of Communications and Networking Technologies for Home Area Networks

Connectivity	Technology	Max Speed per Channel	Range	Adoption Rate
Wired	HomePlug (IEEE P1901)	14–200 Mbps	300 m	Medium
	Ethernet (IEEE 802.3)	10–1000 Mbps	100 m	Extremely high
	X10 (X10 standard)	50–60 kbps	300 m	Medium

Continued

Table 2.2 Summary of Communications and Networking Technologies for Home Area Networks (*continued*)

Connectivity	Technology	Max Speed per Channel	Range	Adoption Rate
	Insteon (X10 standard)	1.2 kbps	3000 m	Medium
	ITU G.hn (G.hn)	Up to 1 Gbps	–	Not widely
Wireless	Z-Wave (Zensys, IEEE 802.15.4)	40 kbps	30 m	Widely
	WiFi (IEEE 802.11, IEEE 802.15.4)	11–300 Mbps	100 m	Extremely high
	ONE-NET (open-source)	38.4–230 kbps	500 m (outdoors) 60–70 m (indoors)	Not widely
	6LowPAN (IEEE 802.15.4)	250 kbps (2.4 GHz) 40 kbps (915 MHz) 20 kbps (868 MHz)	10–75 m	Medium
	ZigBee (IEEE 802.15.4)	250 kbps (2.4 GHz) 40 kbps (915 MHz)	10–75 m	Widely
	EnOcean (EnOcean standard)	120 kbps	30 m	Not widely

Finally, *EnOcean* technology efficiently exploits applied slight mechanical excitation and other potentials from the ambiance (motion, pressure, light, and temperature) using the principles of energy harvesting for networking self-powered wireless sensors, actuators, and transmitters. In order to transform such energy fluctuations into usable electrical energy, electromagnetic, piezo-generators, solar cells, thermocouples, and other energy converters are used. The transmission range is around 30 m inside the building, and this technology allows for wireless gateway connectivity with common automation systems. The wired and wireless communications described thus far have been summarized in Table 2.2.

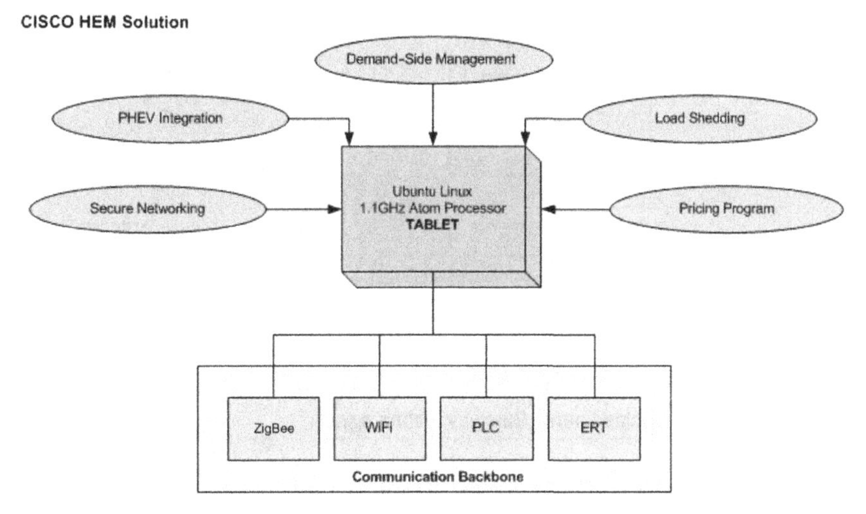

Figure 2.5 Cisco's HEC architecture.

2.4 HEM HARDWARE

The imminent penetration of HEM systems in green homes has created a new market segment for embedded hardware providers. In June 2010, Cisco Systems unveiled its home energy controller (HEC), which is part of a much larger smart grid infrastructure that spans solutions for utilities, substation networks, smart meter networks, and the home network. The HEC has a 7-in. user interface tablet that runs Ubuntu Linux, powered by a 1.1 GHz Intel Atom processor. Supplementing the HEC on the utility side is Cisco's Home Energy Management Solution, which gives utility companies the right tools to enhance customer satisfaction and effectively implement demand management, load shedding, and pricing programs for residential deployments. Figure 2.5 shows Cisco's HEC architecture.

Using the HEC, consumers can take advantage of special energy pricing programs, demand response can be managed, and electric vehicle integration becomes a reality. The HEC provides (1) user-engaging and easy-to-use energy management applications to monitor and budget energy use and control thermostats and appliances, (2) a utility with the ability to provision and manage a home area network (HAN) that monitors and controls energy loads, and (3) highly secure end-to-end data communications across wired and wireless media and networking protocols. The HEC is a networking device that coordinates with the networks in the home and the associated security protocols, such as ZigBee (communication with smart appliances), WiFi (communication with the home network), and PLC and ERT (communications with utilities). To monitor and control energy loads such as heating, ventilating, and air conditioning (HVAC) systems, pool pump, water heaters, TVs, computers, and other devices, consumers will need to wirelessly connect the appropriate compatible, tested

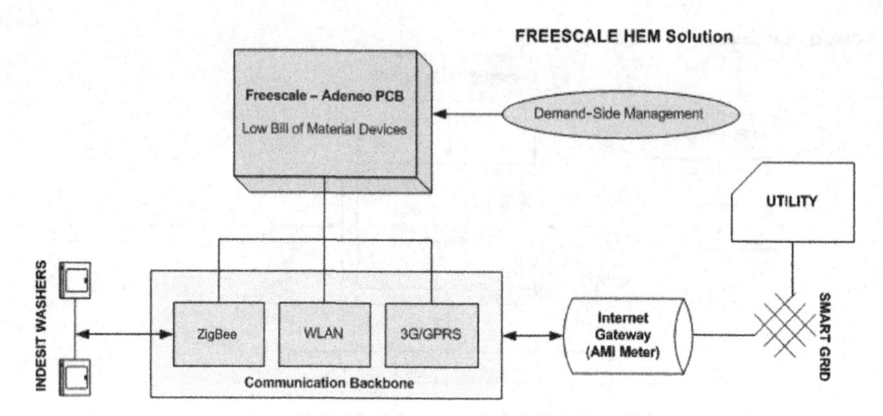

Figure 2.6 Freescale's Home Energy Gateway technology.

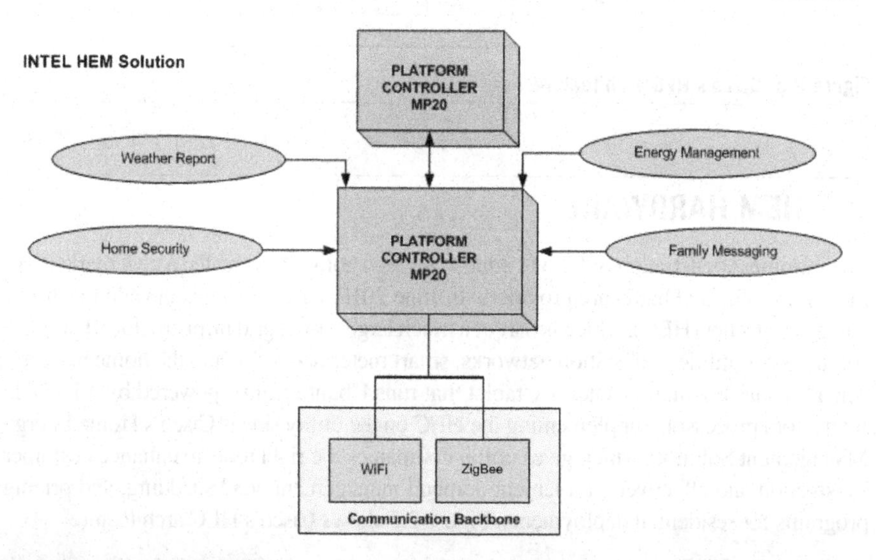

Figure 2.7 Intel's HEM solution.

peripherals to the HEC. Cisco is currently in trials with utilities for the home energy controller.

To scale and support devices implemented in residential deployments, Cisco's Energy Management Software is deployed in utility facilities, and its hosted services help utilities provide personalization and data to increase customer satisfaction for energy programs. These services include

- Provisioning and management capabilities
- Unique, customized look and feel for devices
- Mass firmware updates to thousands of devices
- Integration with utility back-end applications and third-party software

During the last quarter of 2010, both Freescale Semiconductor and Intel Corp. have announced reference designs targeting the HEM market (see Figures 2.6 and 2.7). Freescale demonstrated its Home Energy Gateway (HEG) reference platform in September 2010 in Europe. The Freescale Home Energy Gateway reference platform is based on the i.MX ARM9 SoC that is both flexible and scalable and based on ZigBee Smart Energy 1.0 mesh architecture for bidirectional control. The HEG's controller integration allows for a low bill-of-materials cost. Freescale's HEG includes a central hub that links smart meters, smart appliances, and smart devices in the home area network (HAN) and collects and reports power usage data. The Freescale HEG allows every point of the smart home to be connected and controlled from a central point, enabling power efficiency and energy optimization. The HEG links to a WAN for remote control and monitoring by the utility and communications service provider.

Functions of the Home Energy Gateway include

- Collecting real-time energy consumption from smart meter and power consumption data from various in-house objects.
- Controlling activation/deactivation of home appliances.
- Generating a dashboard to provide feedback about power usage.
- Providing control menus to control appliances.
- Providing a ubiquitous link to the broadband Internet.

Freescale's reference platform is available now through its systems integrator partner Adeneo Embedded, which will provide hardware manufacturing and board support package (BSP) customization and support. The HEG uses a four-layer PCB and boasts a low-cost bill of materials.

In Europe, Freescale announced this summer a smart grid demonstration project with the Indesit Company, an Italian maker of smart appliances. Indesit's Smart Washer was equipped with a Freescale ZigBee node that enables it to adjust its cycle starting time according to energy cost and availability of green power. The washer retrieves this information from the local utility via a ZigBee-enabled Internet connection to the smart grid.

Close on the heels of Freescale, Intel announced its Home Energy Management reference design earlier in October 2010. Intel's HEM reference design is based on the Atom processor Z6XX series and Intel's Platform Controller Hub MP20. The reference design is manufacturing-ready and supports both WiFi and ZigBee. The processor integrates a DDR2 memory controller that can accommodate up to 2 Gbytes of memory.

Intel is marketing the reference design as providing more than just energy management, with the ability to add new applications as they are available. Embedded apps on the dashboard currently include a family message board, weather reports, and home security.

The existing commercial platforms outlined above are the first generation platforms for HEM. As standardization of control and communication protocols is better adumbrated, and the penetration of HEM use among consumer households

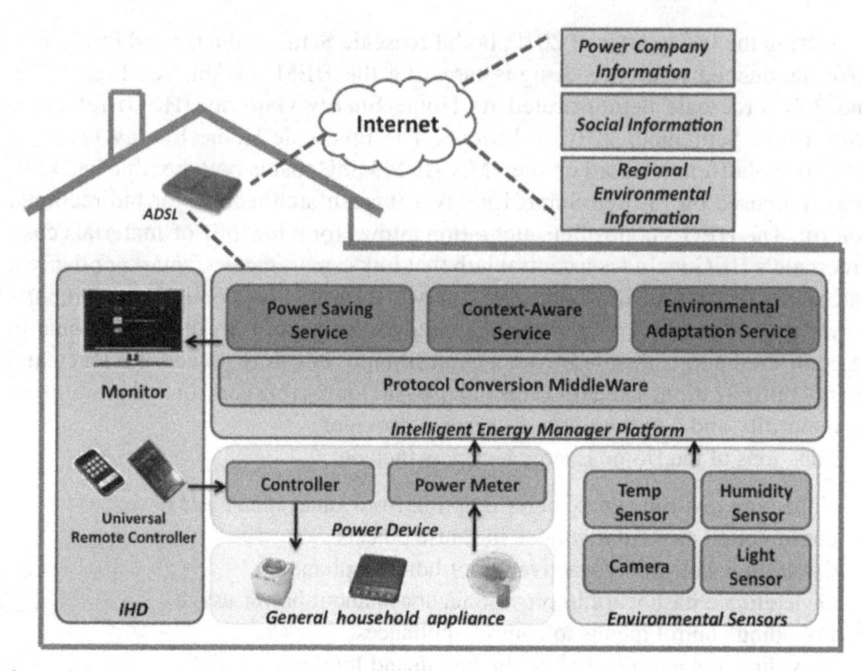

Figure 2.8 Future HEM system architecture.

increases exponentially in the near future, research into designing the optimally efficient and scalable hardware platforms for the next generation HEM hardware will be paramount. We believe that the next generation HEM devices will also provide various value-added services to the consumers, such as bill payment and security monitoring for example, besides the expected DSM. Furthermore, these HEM devices will be truly embedded in the HANs, and as is the case with such platforms, the applications and the operating system (OS) which will run on these platforms should be co-developed and co-optimized with the emerging HEM device architectures.

2.5 SYSTEM ARCHITECTURE AND CHALLENGES IN DESIGNING FUTURE HEMS

In this section, an architecture for a futuristic HEM system is introduced and the challenges and solutions facing the design and deployment of this system are presented. Figure 2.8 shows an architecture for a future HEM system. Going forward, it is envisioned that a HEM system will be based on an open, nonproprietary, and standards-based platform. This will facilitate the ability to control and network intelligent appliances manufactured by different vendors. The main HEM system can be classified into three subsystems, namely, the sensor and control devices, the monitoring and control system, and the intelligent energy management platform. What follows is a detailed description of these subsystems. It is remarked that while some of these capabilities are available now from a number of HEMS providers, others are future possibilities and it will be quite some time before they hit the market.

2.5.1 Sensor and Control Devices

This subsystem concerns the basic devices in a HEM system. It is envisioned that future smart home architectures will comprise self-powered (energy scavenging) devices that will facilitate generation of power and energy storage management and diagnostics at a microscale. Other than the power detector it also needs to include the environmental sensor. In addition to detecting power efficiency, the goal is to detect environmental parameters, such as temperature, humidity, and whether there are people around, and to allow this information to be sent through the HAN and be utilized by an intelligent management platform. The controller is used to receive the remote controller commands to control home appliances. The main challenges facing the deployment of a HEM system are summarized below:

1. *Accuracy:* For a HEM system, the power detection device should not only give an approximation of the current value, but also accurately measure the current value in the device to enable the intelligent management platform to effectively perform its appliance recognition function, and to determine whether the appliance is operating efficiently using the appliance power source data.
2. *Compatibility:* Networking normal home appliances entails integrating the infrared transfer method to the HAN. For example, one can deploy the bridge device discussed in [38] that encodes the received HAN signal into an infrared signal making it compatible with most home appliances. This encoding enables a bidirectional control link between the sensor (on the appliance) and the control device.
3. *Low power cost:* Detecting the power consumption in a house and the surroundings along with the cost of power consumption requires many strategically deployed detection devices. However, it is essential that these detection devices have low power consumption and cost and good power management standards, thereby avoiding excessive sensing devices that would escalate the cost of power consumption in the HEM system.

2.5.2 Intelligent Power Management Platform (IPMP)

An intelligent power management platform (IPMP) is at the heart of a HEM system. This is because it exploits the received sensor data and external Internet data (power company information, regional environment information, social information, to

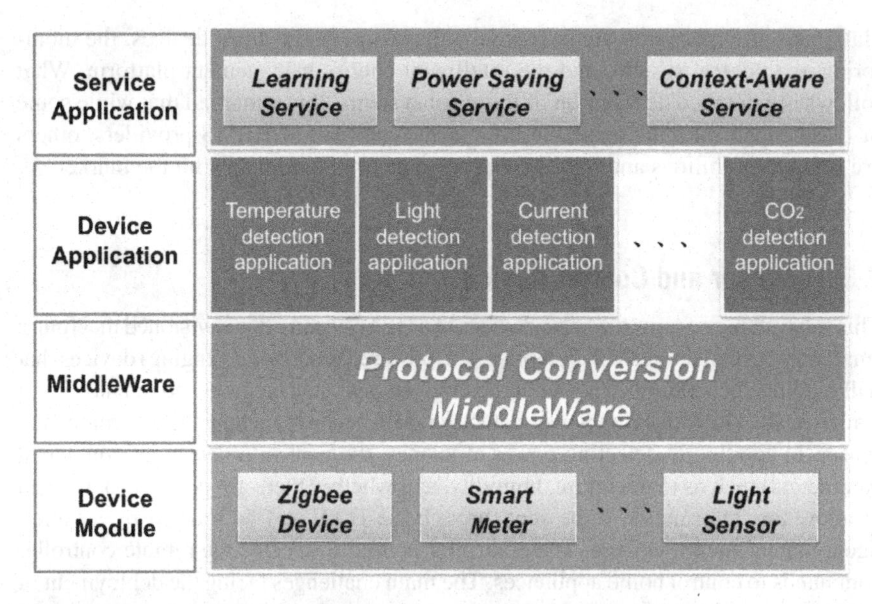

Figure 2.9 Software stack of intelligent power management platform (IPMP).

name a few), and transfers the data to the IHD display for the user. Alternatively, the IPMP automates home control after processing the sensor data in accordance with the "recent" historic sensor data or external information. The IPMP provides middle-ware conversion software and allows upper level device and service applications to communicate with each other, thereby facilitating the transfer of data and control signals to lower level devices. The three key services offered by an IPMP are as follows:

1. *Power management service:* In addition to recording the power usage of each device/appliance, the power management service includes transmitting the power consumption information to the IHD display, and providing appliance recognition and self-managing functions. Through the power sensing device, the power usage of every appliance during the different states of operation is recorded, thereby generating a personalized power consumption profile for each appliance. This power profile can then be used to track and predict the OFF states of an appliance, and this enables a reduction in the power consumed in the whole system by cutting off the power to the appliance during its OFF state. Furthermore, using the power profile of an appliance, one can perform fault analysis to detect a broken or malfunctioning appliance and report it to the user via the IHD (see Figure 2.9).
2. *Context-aware service:* Context-awareness enables procuring regional environmental information (such as position, climate, and humidity) through the sensor network. This service facilitates recording sensor readings at any time to determine the users' habits, and through further processing and analysis,

to automatically control the system under different situations, or using states of the user to prevent wasted power.

3. *Social network service:* A good intelligent managing platform should also be equipped with a social networking function that uses the Internet to send the power consumption profiles in a home, and to receive information for the accompanying social networks, including power company data, power costs, and power consumption of each appliance in neighboring homes. Using this information, not only can the user become aware of the power usage in one's home but also the usage within neighborhood. This social network service data can be used to achieve a more detailed power management function. However, there exist security concerns to keep the user information private, and so designing secure and reliable communication links for metering, pricing, control, and billing purposes are areas for future research.

2.5.3 Monitor and Control System

The main function of this subsystem is to provide a visual interface (such as displaying on the IHD) for the useful information (e.g., power consumption, costs, etc.) for the user to facilitate timely action and control of the HEM system. The design challenge then is to devise a user-friendly and simple integrated control interface for the numerous networked appliances at home. Even though one would envision that universal control panels (i.e., centralized) could offer a good choice for integrating controls, there are still two key challenges:

1. *Integration:* Designing an integrated platform that will make the appliances from different vendors operating under different standards interoperable is an open research issue. Using universal controllers entails significant dependence on learning or letting the user record different sets of control signals from different manufacturers to suit each function, thereby limiting convenience and making the deployment of new devices (i.e., scalability of the HAN) harder and expensive.
2. *User friendliness:* Trying to incorporate a number of appliance controls and functions on a single control panel may result in a panel with numerous control buttons. This might not be the optimal design even for normal users, and more so for senior citizens or children. The simplicity and intuitiveness of the user interface will be of paramount importance to the success of smart grids and HEMs in homes. Further, the ease of deployment and upgrading when necessary will preserve the customer base for smart home technologies.

2.6 CONCLUSIONS

On a concluding note, the need for smart energy management in the residential sector for sustainable energy efficiency and monetary savings was revisited in this article. As the smart grid extends out to homes and businesses, wireless sensors and mobile control devices become important elements in monitoring and managing energy use. There are several challenges of which smart energy system designers need to be aware. One challenge is the fragmentation of the HAN market. There are several wireless standards that are currently used in HANs including WiFi, ZigBee, Z-Wave, and Bluetooth, however, despite the emergence of many wireless standards for HANs, there is no clear winner at this point, and so it is up to the system designers to select a wireless technology that best fits their application while addressing the potential problem of interoperability with other HAN devices. A comprehensive summary of the state-of-the-art in home area communications and networking technologies for energy management was provided in the paper, followed by a review of the affordable smart energy products offered by different companies. The paper also shed light on the challenges facing the design of future energy management systems, such as the need for interoperability and network security. Our discussions will hopefully inspire future efforts to develop standardized and more user-friendly smart energy monitoring systems that are suitable for wide-scale deployment in homes.

REFERENCES

[1] G.T. Gardner, P.C. Stern, The short list: the most effective actions US households can take to curb climate change, Environment Magazine 50 (5) (2008) 12–21.

[2] (Online). <http://wiki.micasaverde.com/index.php/Energy_Savings>.

[3] M. Fitzgerald, Finding and Fixing a Home's Power Hogs. New York Times, 2008, July 27. <http://www.nytimes.com/2008/07/27/technology/27proto.html>.

[4] C.W. Gellings, The Smart Grid: Enabling Energy Efficiency and Demand Response, CRC Press, (2009).

[5] D.J. Cook, S.K. Das, How smart are our environments? An updated look at the state of the art, Pervasive and Mobile Computing 3 (2) (2007) 53–73.

[6] K. Kok, S. Karnouskos, D. Nestle, A. Dimeas, A. Weidlich, C. Warmer, P. Strauss, B. Buchholz, S. Drenkard, N. Hatziargyriou, V. Lioliou, Smart houses for a smart grid, in: 20th International Conference on Electricity Distribution (CIRED), June, Czech Republic, Prague, 2009.

[7] J.R. Roncero, Integration is key to Smart Grid management, in: IET-CIRED Seminar on SmartGrids for Distribution Frankfurt, Germany, June 2008, pp. 1–4.

[8] M.Erol-Kantarci, H.T. Mouftah, Wireless sensor networks for domestic energy management in smart grids, in: 25th Biennial Symposium on Communications, Kingston, ON, Canada, May 2010.

[9] D.G. Hart, Using AMI to realize the Smart Grid, in: Proceedings of IEEE Power and Energy Society General Meeting – Conversion and Delivery of Electrical Energy in the 21st Century, July 2008, pp. 1–2.

[10] C. Hertzog, Smart Grid Dictionary, GreenSpring Marketing LLC, 2009.

[11] H. Farhangi, The path of the smart grid, IEEE Power and Energy Magazine 8 (1) (2010) 18–28.

[12] K. Wacks, The Gridwise Vision for a Smart Grid, Parks Associates' CONNECTIONS, June 2009.

[13] M. Inoue, T. Higuma, Y. Ito, N. Kushiro, H. Kubota, Network architecture for home energy management, IEEE Transactions on Consumer Electronics 49 (3) (2003) 606–613.

[14] N. Kushiro, S. Suzuki, M. Nakata, H. Takahara, M. Inoue, Integrated residential gateway controller for home energy management system, IEEE Transactions on Consumer Electronics 49 (3) (2003) 629–636.

[15] R. Masiello, Demand response: the other side of the curve, IEEE Power and Energy Magazine 8 (3) (2010) 18 (guest editorial).

[16] US Department of Energy, Benefits of demand response in electricity markets and recommendations for achieving them, Report to the United States Congress, February 2006. <http://eetd.lbl.gov>.

[17] M.H. Albadi, E.F. El-Saadany, Demand response in electricity markets: an overview, IEEE Power Engineering Society General Meeting, 2007, pp. 1–5.

[18] IIEC, Demand side management best practices guidebook for Pacific island power utilities, Prepared for South Pacific Applied Geoscience Commission and United Nations Department of Economic and Social Affairs by International Institute for Energy Conservation (IIEC), July 2006.

[19] Charles River Associates, Primer on demand-side management with an emphasis on price-responsive programs, Prepared for The World Bank by Charles River Associates, February 2005.

[20] EPRI, Survey of utility demand-side management programs, EPRI TR-102193, 1993.

[21] NERC, Electricity Supply and Demand, for 1993–2002, North American Electric Reliability Council, Princeton, NJ, 1993.

[22] Barakat & Chamberlin, Inc., Principles and Practice of Demand-Side Management, EPRI TR-102556, Palo Alto, California, August 1993.

[23] A. Brooks, E. Lu, D. Reicher, C. Spirakis, B. Weihl, Demand dispatch, IEEE Power and Energy Magazine 8 (3) (2010) 20–29.

[24] A. Vojdani, Smart integration, IEEE Power and Energy Magazine 6 (6) (2008) 71–79.

[25] N. Gudi, Lingfeng Wang, V. Devabhaktuni, S.S.S.R. Depuru, Demand response simulation implementing heuristic optimization for home energy management, in: North American Power Symposium (NAPS), 2010, pp. 1–6.

[26] T.C. Matty, Advanced energy management for home use, IEEE Transactions on Consumer Electronics 35 (3) (1989) 584–588.

[27] T. Ikegami, Y. Iwafune, K. Ogimoto, Optimum operation scheduling model of domestic electric appliances for balancing power supply and demand, in: International Conference on Power System Technology (POWERCON), 2010, pp. 1–8.

[28] T. Verschueren, W. Haerick, K. Mets, C. Develder, F. De Turck, T. Pollet, Architectures for smart end-user services in the power grid, in: IEEE/IFIP Network Operations and Management Symposium Workshops, 2010, pp. 316–322.

[29] Duy Long Ha, F.F. de Lamotte, Quoc Hung Huynh, Real-time dynamic multilevel optimization for demand-side load management, in: IEEE International Conference on Industrial Engineering and Engineering Management, 2007, pp. 945–949.

[30] K.P. Wacks, Utility load management using home automation, IEEE Transactions on Consumer Electronics 37 (2) (1991) 168–174.

[31] M.H. Shwehdi, A.Z. Khan, A power line data communication interface using spread spectrum technology in home automation, IEEE Transactions on Power Delivery 11 (3) (1996) 1232–1237.

[32] Long Duy Ha, S. Ploix, E. Zamai, M. Jacomino, Tabu search for the optimization of household energy consumption, in: IEEE International Conference on Information Reuse and Integration, 2006, pp. 86–92.
[33] S. Kishore, L.V. Snyder, Control mechanisms for residential electricity demand in smartgrids, in: Proceedings of 2010 First IEEE International Conference on Smart Grid Communication (SmartGridComm), 2010, pp. 443–448.
[34] B. Ablondi, Residential Energy Management, Parks Associates' CONNECTIONS, June (2009).
[35] Z-Wave Alliance (Online). <http://www.z-wavealliance.org/modules/AllianceStart/>.
[36] U. Saif, D. Gordon, D. Greaves, Internet access to a home area network, IEEE Internet Computing 5 (1) (2001) 54–63.
[37] E. Callaway et al., Home networking with IEEE 802.15.4: a developing standard for low-rate wireless personal area networks, IEEE Communications Magazine 40 (8) (2002) 70–77.
[38] W.K. Park, I. Han, K.R. Park, ZigBee-based dynamic control scheme for multiple legacy IR controllable digital consumer devices, IEEE Transactions on Electric Appliances 53 (1) (2007) 172–177.

CHAPTER

Embedded Computing in the Emerging Smart Grid

3

Arindam Mukherjee, Valentina Cecchi, Rohith Tenneti, Aravind Kailas

Department of Electrical and Computer Engineering, University of North Carolina at Charlotte, 9201 University City Blvd., Charlotte, NC 28223-0001, USA

3.1 INTRODUCTION

The emerging smart grid requires a high-speed control and information technology (IT) backbone, to enable the widespread penetration of new technologies that the electrical grid needs to support. These new technologies include cutting-edge advancements in metering, transmission, distribution, and electricity storage, all of which will provide observability and controllability to consumers and utilities for better planning of infrastructural investment and use, more efficient and reliable operation of the smart grids, smarter provisioning of electricity and integration of renewable sources, informed market trading of power by the utilities, and ultimately smarter consumption of power at the user end to reduce our CO_2 footprint. The smart grid is expected to reduce CO_2 emission by 25% in the near future [1–3].

We envision that embedded computing platforms will have to be seamlessly integrated within the smart grid to implement the required high speed, high data volume, and real-time constrained information processing and control operations. With data from advanced metering infrastructures (AMI) and dispatchable loads expected to grow exponentially in the near future, moving around the sensor and actuator data in the smart grid under real-time conditions for acceptable responsiveness and reliable operation will be a major challenge. We believe that the paradigm of taking computation to data, as opposed to bringing data to computation, will gain traction and revolutionize computation in the smart grid by moving away from centralized supervisory control and data acquisition (SCADA) center servers to distributed embedded computing platforms throughout the grid.

The aim of this chapter is to provide the basis for exploring designs of embedded processing platforms in the emerging smart grid. Different applications and algorithms running on the smart grid will be discussed, including power analysis and control algorithms, security algorithms, and signal processing algorithms at the interface of sensors and actuators within the grid, to name a few. The different methods to characterize these applications with respect to their performance, bandwidth, and time constraint requirements are presented. These algorithms and their characterized data will then be used as benchmarks to evaluate the performance of existing commercial embedded processors as smart grid computing platforms, and also to motivate the case for designing new embedded processors for smart grids. A novel methodology to explore microarchitectural solutions for future embedded processors in smart grids, using existing processor architecture simulation platforms and optimal design space exploration algorithms, is outlined in the chapter.

- Automatic generation control (AGC)—controls the operation of generation units
- Economic dispatch (ED) and optimal power flow (OPF)—optimal electrical system operation
- State estimator—estimate of the current state in the power grid
- Supervisory control and data acquisition (SCADA) and remote terminal units (RTU)—communication between these two for actuator control according to sensor data.

Figure 3.1 shows some of the interconnected components of the smart grid that are potential locations for the insertion of embedded computing platforms. An embedded processor in the RTU will need to ensure secure communication and control under real-time constraints which will be system specific. This processor will also need to have signal processing capabilities and a limited point-to-point communication chipset. The SCADA centers require powerful embedded computation servers to operate on all the data coming from the RTUs and AGCs with round trip time constraints of the order of 5 ms [4, 5]. These servers execute a variety of applications varying from database management to grid control. At other grid points like the AGC, computation of power flow (PF) and optimal power flow (OPF) is required, while distributed state estimation requires those computing platforms to have substantial computation and networking capabilities.

This chapter is organized as follows. In Section 3.2 we explain the major applications in the smart grid in more detail. In Section 3.3, state-of-the-art commercial processors have been discussed in the context of their effectiveness in executing smart grid algorithms and provide a qualitative analysis of how the performance can be improved. In Section 3.4 we present our methodology of designing an embedded processor for smart grid applications and our current work in this area. Open research questions are discussed in Section 3.5. We hope that the ideas presented in this chapter will spawn a new paradigm of research in the field of computation in smart grids.

3.2 OVERVIEW OF SMART GRID FEATURES AND APPLICATIONS

This section introduces aspects and functionalities of the smart grid that drive the need for a specific set of characteristics of the embedded computing infrastructure. Increased sensing and automation capabilities in the electric power grid in conjunction with communication systems and enabling technologies affect the computing environment and require a redesign of the embedded processors used for these applications.

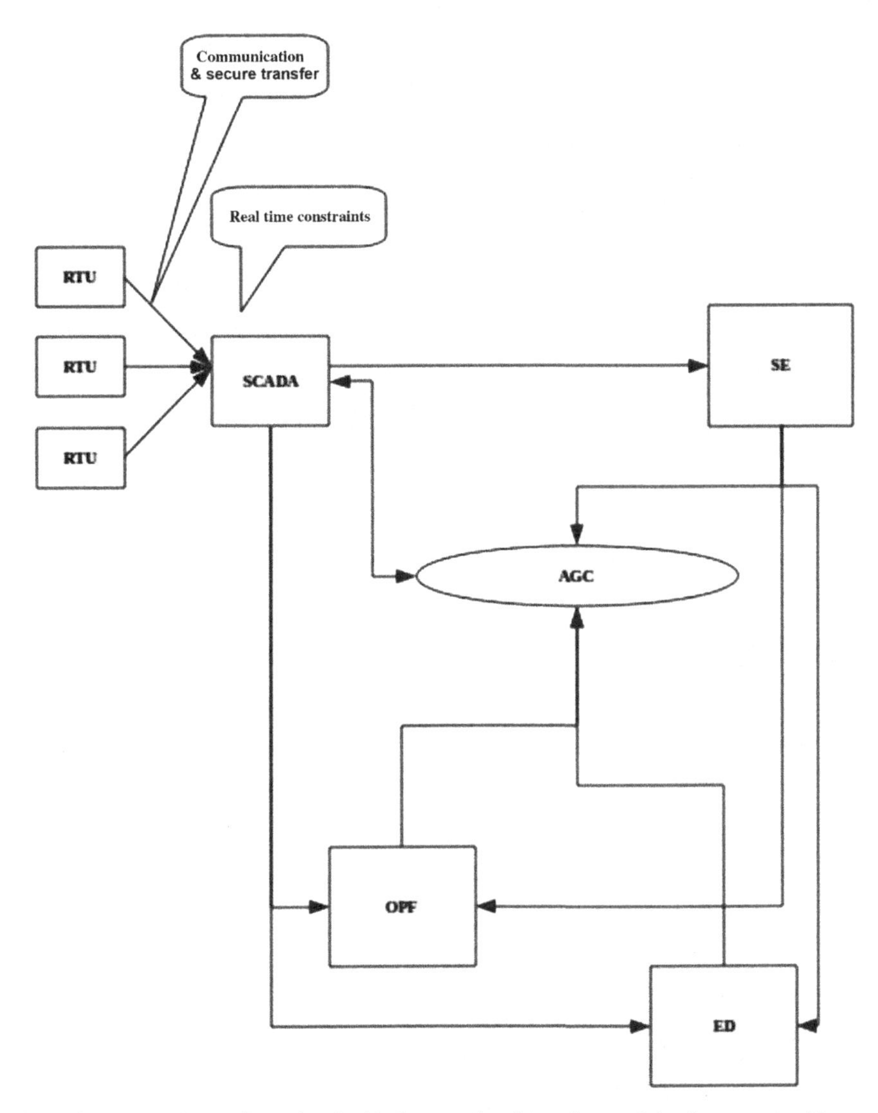

Figure 3.1 Dependency flow of embedded computing-intensive modules in a smart grid.

Emerging aspects of the smart grid that will affect the computing infrastructure include distribution automation (DA), advanced metering, demand-side management (DSM) and demand response (DR) programs, and wide-area measurement and control systems. Moreover, the ability of the grid to self-heal following a fault or disturbance, and the need for increased reliability and power quality as well as for asset management through monitoring, will play a role in the determination and design of an optimized computing environment for smart grids. Therefore this section forms the basis for choosing the applications that we think are appropriate to test on the embedded processors for their optimal performance in terms of speed of execution and power consumption.

Before discussing the applications and their computational properties, a conceptual view of the entire grid is illustrated in Figure 3.2 to point out a few of the many possible locations where embedded processor solutions can solve the computation

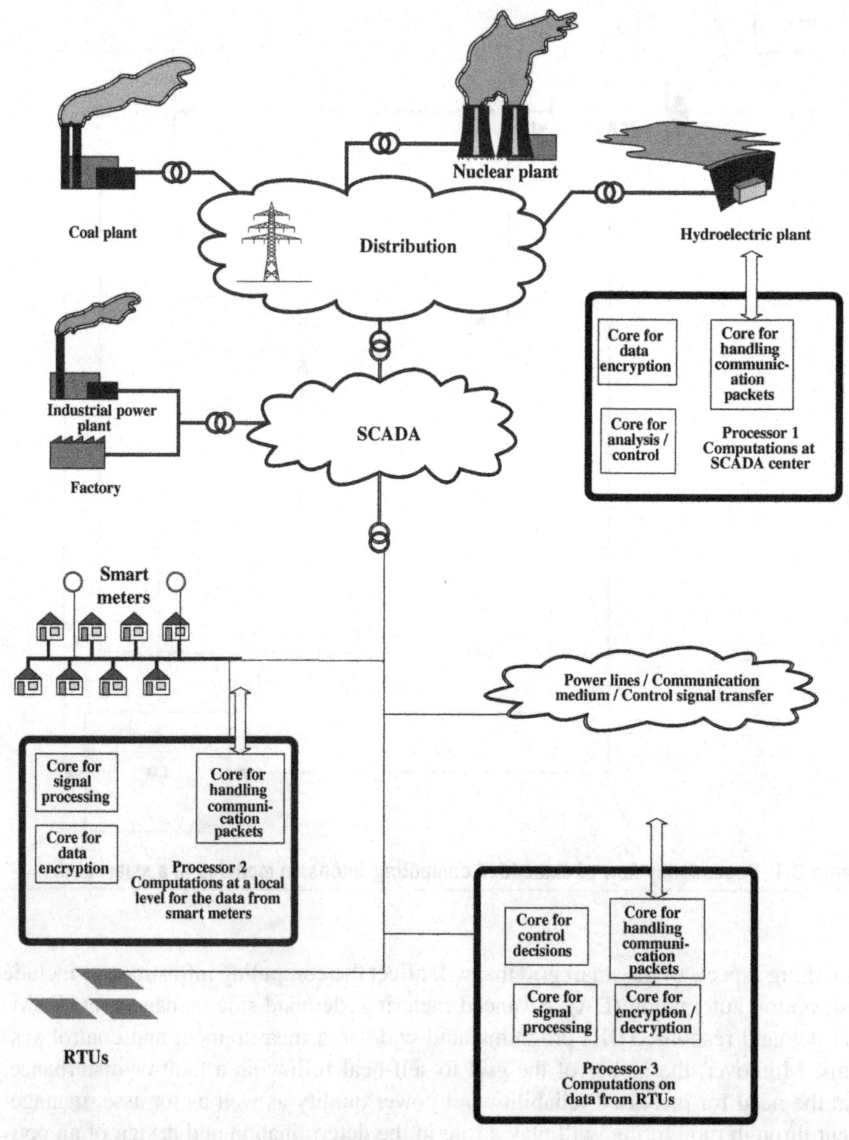

Figure 3.2 The power grid and embedded processors.

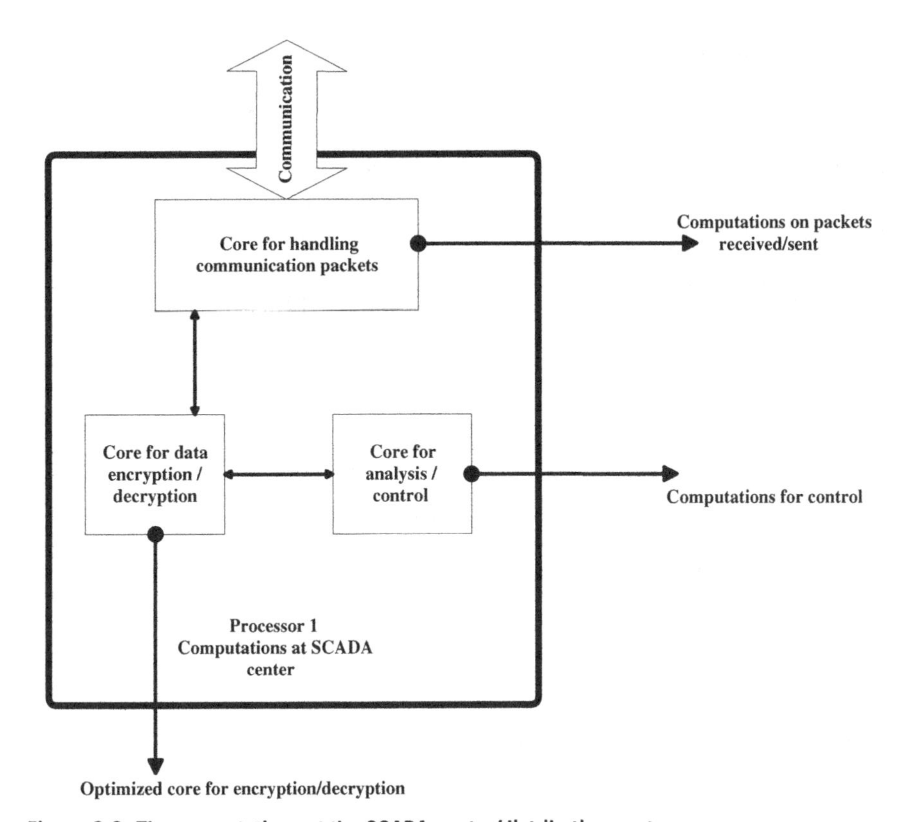

Figure 3.3 The computations at the SCADA center/distribution center.

problems. The applications and features in this section are discussed with the computational problems associated with them in mind.

Three main parts are identified in the above figure depicting the power grid, where embedded processors are required to handle the computations in control, analysis, and communication needs for better management of the power resources. The computation needs are different at different levels, and hence the architectures of the processors proposed will be different.

- *Analysis/control core*: The SCADA center makes a decision based on the signals received from the RTUs and initiates control signals back to the RTUs. This computation resource can also be used to calculate the power flow values, thereby enabling load control at the distribution center.
- *Encryption/decryption core:* In order to communicate the data in a secure way the data is encrypted/decrypted in an efficient manner based on the method adopted.
- *Communication core*: The communication packet processing has to be done in an efficient manner in order to reduce the delay in response, and follow the real-time deadlines for control. See Figure 3.4.

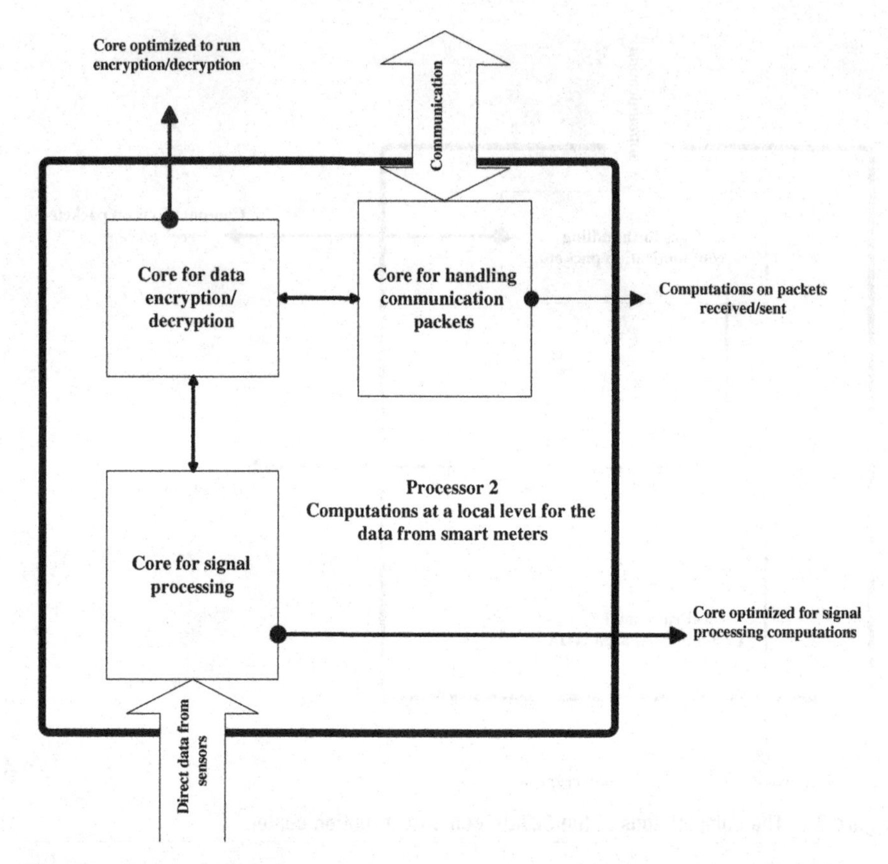

Figure 3.4 The computations on data coming in from the smart meters in a home network.

The smart meters send power consumption data from various devices like heaters, electric vehicles, etc. The processor reads in the signals and sends it to the server at the next higher level or if required handles the switching on/off of the household devices depending on the functionality adopted.

- *Signal processing core:* The incoming data is digital/analog and the information requires fast processing to initiate prompt responses for optimally utilizing the power.
- *Encryption/decryption core:* Encryption/decryption is required in order to keep the data safe while transferring personal information of a particular house.
- *Communication core:* The communication core enables fast unwrapping/ packaging and processing of data. See Figure 3.5.
- *Control core:* RTUs receive data from the field sensors, and send them to the SCADA center to analyze and make a decision. The burden on the main

SCADA center can be reduced by incorporating control capability to the embedded processors at the RTU level.

- *Signal processing core:* The signals (analog/digital) received from the sensors have to be processed and transferred to the control core in order to make a decision about the control.
- *Communication core:* If the processor at the RTU requires sending the information acquired from sensors to the SCADA, it follows methods similar to earlier processors via the encryption and communication cores.

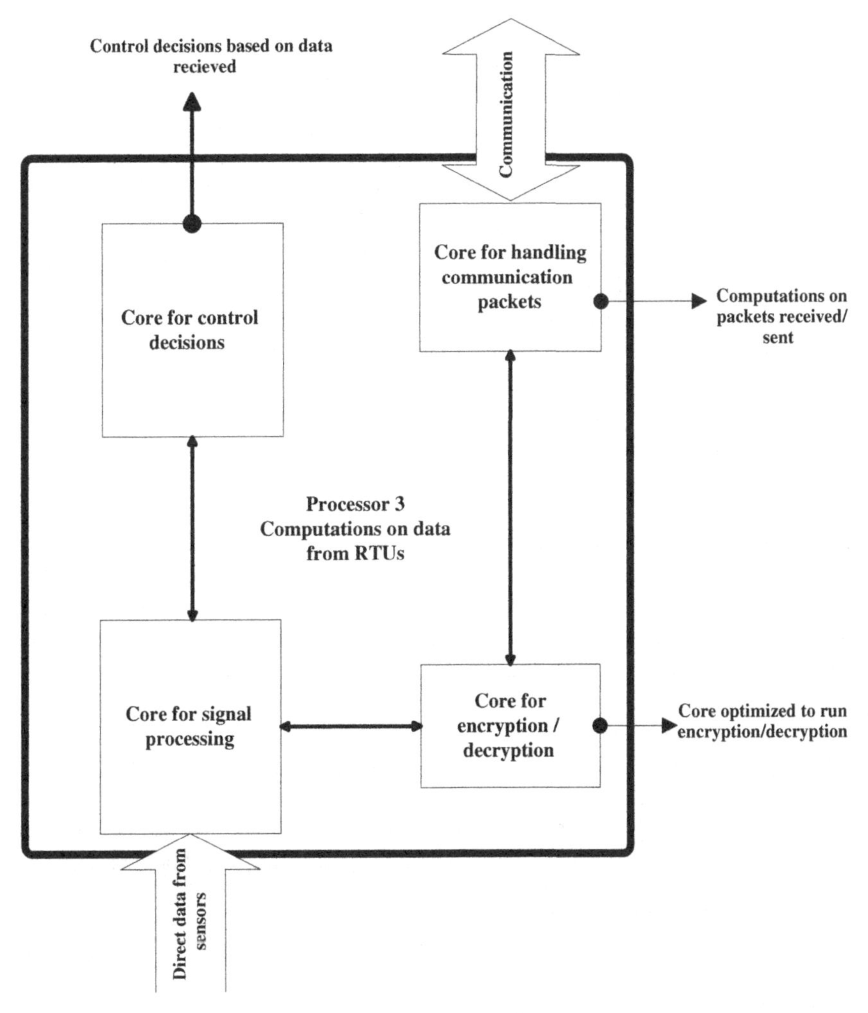

Figure 3.5 Computation for control based on data received from the RTUs.

3.2.1 Analysis and Control in the Smart Grid

The smart grid is a stride forward in making the power grid autonomous. In order to achieve this, there is interaction between the power producer and the consumer in the form of sensor data, control signals, device functionality status, resource availability, and economic status, etc. Based on the data being collected from characteristically different components, the computation performed will be different.

As discussed earlier the smart meters collect data and send the information to the appliance controllers. After the data collection phase, a decision is made about the functionality of the appliance. The factors such as maximum power load acceptable by a house (or locality), distribution of power to various devices in the house (or locality), dynamic price of the power, past power usage, etc. affect the decision to switch on/off the appliance. These factors are continuously changing and there are numerical methods to solve such problems.

The SCADA center has the responsibility of analyzing the data coming in from the RTUs or smart meters and then sending a control action to be taken on the actuators or the appliances being monitored. Heuristic techniques are applicable in cases where the past data from the meters and RTUs is collected and analyzed. This is where numerical methods in computational intelligence (CI) are required. CI deals with situations where the environment is complex, changing, and uncertain and the adaptive mechanisms include artificial and bio-inspired intelligence paradigms that exhibit an ability to learn or adapt to new situations, to generalize, abstract, discover, and associate. These methods are ideal in the situation described earlier. CI includes fuzzy logic systems, neural networks, and evolutionary computation. CI also deals with systems that are hybrids of the before-mentioned areas, e.g., swarm intelligence and artificial immune systems, which can be seen to be built around evolutionary computation; Dempster-Shafer theory, chaos theory, and multivalued logic, which can be seen as part of fuzzy logic systems, etc. [36]. These actions take place at the economic dispatch unit present in the distribution center. The economic dispatch units calculate and provide power to the load in a manner that minimizes the cost of production. For each dispatchable generator, the program calculates the optimum base load and regulating participation factors for use by the closed loop control algorithms. Each generator may be assigned up to three sets of economic parameters, consisting of fuel rate curves, rate limits, and high and low economic limits. The dispatcher selects the appropriate economic parameter set at will.

PF studies provide complete information for optimal operation, which includes complete voltage angle and magnitude information for each bus in a power system for specified load and generator real power and voltage conditions. There are many methods by which a PF problem can be solved [1]. A few are listed below:

- Linear programming method
- Newton–Raphson method
- Quadratic programming method
- Nonlinear programming method
- Interior point method

- Artificial intelligence methods
  - Fuzzy logic method
  - Genetic algorithm method
  - Evolutionary programming
  - Ant colony optimization
  - Particle swarm optimization
  - Artificial neural networks method

Each bus in the distribution system has four parameters: the voltage magnitude, phase angle, real power, and reactive power. The buses in the system are classified into three bus types: slack, load, and regulated.

Slack bus system: This is the reference bus for the magnitude and phase angle. This is selected arbitrarily; this bus makes up for the difference between scheduled loads and generated power that are caused by the losses in the network.

Load bus system: A carrier bus that has no generator attached to it. Active and reactive powers are known in this case and the magnitude and the phase angle are unknown.

Regulated bus system: Real power and voltage magnitude are specified at this bus and a limit on reactive power is imposed; phase angle is an unknown.

As an example, algorithm to solve a power flow problem using the Newton–Raphson method is described below; this involves iterative matrix inversion computations.

ALGORITHM 3.1

Power flow

```
Axx² + Bxy² =a
Dxx² + Exy² =b
C = [a b]
1.Initial values to voltage angles and magnitudes, generally the
  angles are set to zero and the magnitudes to 1.0 p.u.
2.Calculate the power balance matrix with the current voltage angle
  and magnitude. (F)
3.The Jacobian of the power balance matrix is calculated at the
  current voltage phase and magnitude values. (J)
4.The change in voltage angle and magnitude is calculated by J⁻¹ × DC
  where, DC = C - F
  The voltage magnitude and angle are updated and steps 2 to 4 are
  repeated until the change in step 4 is small enough [35]
```

According to [6], in solving the power flow using the Newton–Raphson method the computation effort in calculating the elements of the Jacobian varies as $O(n)$, the required factorization varies as $O(n^{1.4})$, and that of the necessary repeat solution varies as $O(n^{1.2})$. As there are lengthy trigonometric calculations involved in calculating the elements of the matrix $O(n)$ is also a notable factor when compared to $O(n^{1.4})$ and $O(n^{1.2})$. The storage requirements are bound to grow at $O(n^{1.2})$.

Power flow calculations involve compute-intensive matrix inversion and when dealing with large data sets, it also requires faster data access. The computations could be on integers or floating point values, and thus the need arises for separate integer and floating point units in the architecture for faster execution. The advanced methods of matrix inversion methods involve parallel execution of the code. The code would also have lesser branch instructions eliminating the need for a separate branch prediction unit and hence reducing the power and area requirements. The results of power flow calculations are used for resource allocation, which makes a case for real-time deadlines and hence lower latency for the executing instructions.

Another area where the smart grid evolves in having better control over the grid is the self-healing property. Self-healing is where the grid detects power supply errors in real time and then reroutes or limits the power supply. For the grid to have such an ability we need a wide-area monitoring network of sensors linked together in a number of ways: through time signals from the global positioning service, commands through secure Internet, and sensor-to-sensor communication via dedicated optics [7].

The information from the sensory level would continually be passed on to the control level, which might consist of another network of smart devices, in this case semiautonomous control mechanisms coordinated with a myriad of grid components, such as transformers, generators, parts of generators (responsible for such things as boiler temperature, steam pressure, etc.) switches, and circuit breakers.

To summarize, the design of the self-healing grid must have the following abilities:

- Identify and evaluate the impact of impending failures in the power or communication systems.
- Perform a system-wide vulnerability assessment incorporating the power system, protection system, and the communication system.
- Enable the power system to take self-healing actions through reconfiguration.
- Perform power system stabilization on a wide-area basis.
- Monitor and control the power grid with a multiagent system designed to reduce power system vulnerability.

We now discuss one of the strategies to support a self-healing grid and implement the above-mentioned properties [8], using a distributed autonomous real-time (DART) system.

The following are main features of a DART system:

- Computing and communication systems: This involves a large number of distributed embedded processors communicating with each other with a dedicated network to exchange data and enable control.

- Autonomous systems: These are required to work intelligently and in real-time without human intervention. Some of the important tasks would include gathering necessary information, analyzing the performance of the relevant power system subject, alerting agents at higher levels based on the results of analysis, etc.
- Integrated messaging and data environment: Data is made available to all the functional units in real time from multiple sources in distributed coherent databases.
- Temporal coordination of tasks: Based on the control response times, computational burden, operating needs, physical phenomena, etc. the time duration of the tasks range from 10 ms–1 hr.

3.2.2 Sensing and Measurement Infrastructure

As discussed in Section 3.2.1, a device status signal to the controller will trigger the actuator into operation. The sensor data could be digital/analog depending on the device and the purpose of the device. The device could be a smart meter reporting the total power consumed by a house or it could also be an instrument reporting the amount of energy available at the power source. Based on the data and source the subsequent action taken is different.

Advanced metering infrastructure is being deployed in conjunction with the smart meters. An important characteristic of this advanced metering system is its capability for two-way communication between the meters and the utility, enabling remote meter readings through direct communication into the customers' homes, remote connect and disconnect of service, and outage reporting. The power usage data of the devices in the area being monitored is collected and based on usage pattern an intelligent algorithm is run on the embedded processors. This helps in creating the cost function and this information is sent to the device power controller, which will effectively manage the power consuming devices.

Before the control action is executed, the signals received from the sensors have to be processed. The processing required could vary from case to case. Here, an algorithm to solve the discrete Fourier transform (DFT) is provided, which decomposes series of values to discrete components of frequencies. An efficient method to implement this is the fast Fourier transform (FFT); the pseudocode is provided below [37].

ALGORITHM 3.2
Fast Fourier Transform

```
FFT (A, n, w)
A is a Vector of length n
n is a power of 2
w is a primitive nth root of unity
Vector A represents a polynomial: A(z) = A[1] + A[2]*z + . . +
   A[n]*z^(n-1)
```

```
The value returned is a Vector of the values: [a(1),a(w),..,a(w^(n-1))]-
  Computed via a recursive FFT algorithm
The pseudocode is presented below
If n = 1 then
Return A
Else
Aodd ← Vector (A[1], A[3],...,A[n-1])
Aeven ← Vector (A[2], A[4],..,A[n])
Veven ← FFT(Aeven, n/2, w^2)
Vodd ← FFT(Aodd, n/2, w^2)
V ←Vector(n) #Define a vector of length n
For i from 1 to n/2 do
V[i] ←Veven[i] + w^(i-1) * Vodd[i]
V[n/2 + i] ← Veven[i] - w^(i-1)*Vodd[i]
End do
Return V
End if
```

FFT takes O(nlogn) operations to complete the computation. For the computational efficiency, in the for loop we build up the powers of w using just one multiplication each pass through the loop. Similarly for the recursive FFT calls, w^2 should be computed only once.

By using methods such as loop unrolling the FFT algorithm can be parallelized and an architecture supporting the parallel implementation such as a single instruction multiple data (SIMD) system can be used to improve the performance in terms of speed of execution. When the data being operated on is high, there develops a need for a processor with good multi-level cache system.

Apart from FFT, there could be algorithms such as finite impulse response (FIR), infinite impulse response (IIR), etc. that could possibly run on the embedded processor. Depending on the place where the embedded processor is installed in the grid, the algorithm for the signal processing could differ and hence the processor with the best microarchitecture to implement the corresponding algorithm has to be designed.

3.2.3 Communication and Security

The smart grid as discussed earlier requires communication of information for almost all possible actions to control, deliver power, or get feedback from devices. Many network technologies can be used to implement this, but none of them suits all the applications and there is always a best fit of a technology or a subset of technologies that may be chosen for a group of power system applications, either operating in the same domain or having similar communication requirements. Before a communication technology is chosen for a particular power system application, a thorough analysis is required to match the application requirements with the technology properties. The available network technologies include the following categories:

Power line communication: The power lines are mainly used for electrical power transmissions, but they can also be utilized for data transmissions [9–15]. The power line communication systems operate by sending modulated carrier signals on the power transmission wires. Typically data signals cannot propagate through transformers and hence the power line communication is limited within each line segment between transformers. Data rates on power lines vary from a few hundred of bits per second to millions of bits per second, in a reverse proportional relation to the power line distance. Hence, power line communication is mainly used for in-door environments [16] to provide an alternative broadband networking infrastructure [5, 17] without installing dedicated network wires.

Wireline network: Dedicated wireline cables can be used to construct data communication networks that are separate from the electrical power lines. These dedicated networks require extra investment on the cable deployment, but they can offer higher communication capacities and shorter communication delay. Depending on the transmission medium used, the wireline networks include Synchronous Optical Networking (SONET)/Synchronous Digital Hierarchy (SDH) [18, 19], Ethernet [20], digital subscriber line (DSL), and coaxial cable access network. SONET/SDH networks transmit high-speed data packets through optical fibers with supported data rate between 155 Mbps and 160 Gbps. Ethernet is popularly used in our homes and at workplaces, providing a data rate between 10 Mbps and 10 Gbps. DSL and coaxial cable can be used for Internet access. The currently available technology facilitates data rates upto 10 Mbps.

Wireless network: Advancement in wireless networking technology has enabled us to connect devices in a wireless way, eliminating the installation of wirelines. In general, wireless signals are significantly subject to transmission attenuation and environmental interference. As the result, wireless networks usually provide short distance connections with comparatively low data rates. The 802.11 networks [21] are the most popularly used local area wireless networks, which can communicate at a maximum data rate up to 150 Mbps and a maximum distance up to 250 m. In a smaller personal networking area around 10 m in distance, the 802.15 networks [22] provide wireless data exchange connections at rates ranging from 20 kbps to 55 Mbps. For broadband wireless Internet access, 802.16 networks [23] can support data transmissions up to 100 Mbps in a range of 50 km.

The communication infrastructure in the smart grid undertakes important information exchange responsibilities, which are the foundations for the function-diversified and location-distributed electric power devices to work synergistically. Any malfunction in communication performance not only limits the smart grid from achieving its full energy efficiency and service quality, but also could potentially cause damage to the grid system.

One potential area of concern is security. In future power systems, an electricity distribution network will spread over a considerably large area, e.g., tens or hundreds of miles in dimension. Hence, physical and cyber security from intruders is of the utmost importance. Moreover, if a wireless communication medium (like WiFi or ZigBee) is used as part of the communication network, security concerns are increased because of the shared and accessible nature of the medium. Hence, to provide security protection for the power systems, we need to identify various communication use cases (e.g., demand-side management, advanced meter reading, communication between intelligent energy management (IEM) and intelligent fault management (IFM) devices, and local area communication by IEM devices) and find appropriate security solutions for each use case, for example, authorized access to the real-time data and control functions, and use of encryption algorithms for wide-area communications to prevent spoofing [4].

We provide the pseudocode for one such cipher encryption method called Blowfish, which could possibly run on the embedded processors installed at various locations in the smart grid. It is shown to be efficient whenever there is a continuous data stream or packet stream. It is also shown to be efficient when implemented on a microprocessor [38, 46].

ALGORITHM 3.3
Blowfish

```
Cipher Working:
Consists of two parts—key expansion and data encryption
Key expansion converts a key of at most 448 bits to several sub-key
   arrays totaling 4168 bytes.
Data encryption is 16 rounds. Each round consists of a key dependent
   permutation and a key-data dependent substitution. All operations
   are XORs and addition on 32-bit words (efficient on Intel and
   Motorola architectures).
Sub-key:
P-Array: 18 32-bit sub-keys
S-boxes: 4 boxes with 256 entries of 32bit numbers.
Encryption:
Input is a 64-bit data element (x).
Divide x into two 32-bit halves: xL, xR
For i = 1 to 16:
xL = xL XOR Pi
xR = F(xL) XOR xR
Swap xL and xR
Next i
Swap xL and xR (Undo the last swap.)
xR = xR XOR P17
```

```
xL = xL XOR P18
Recombine xL and xR
Function F: Divide xL into four eight-bit quarters: a, b, c, and d
F(xL) = ((S1,a + S2,b mod 232) XOR S3,c) + S4,d mod 232
Decryption:
Same process with P-Array is used in reverse order.
```

For fast implementation the loop can be unrolled and the sub-keys stored in the highest level cache. Also, since the number of rounds does not change with the number of blocks the computation complexity for Blowfish is $O(n)$.

Major modifications have been proposed by various authors, which help Blowfish retain the avalanche effect desired in an encryption algorithm but make it possible to execute compute intensive steps in parallel [45].

A GPU kind of architecture has been proposed to implement Blowfish in an optimized and efficient way in [46]

3.3 STATE-OF-THE-ART PROCESSORS

This section aims to describe the reverse engineered microarchitecture of processors that the manufacturers claim to be useful for smart grid devices and then evaluate the performance of running smart grid algorithms on them.

3.3.1 Intel Atom

Intel launched an ultra-low power processor named Intel Atom, which targets smart grid devices. The aim in designing the atom was to sufficiently reduce power consumption, provide complete Internet experience, and not deviate from x86 compatibility. Atom has been implemented in home automation devices, mobile Internet devices, nettops, netbooks, etc.

Atom is based on an entirely new microarchitecture, having the same instruction-set of CPUs based on the Core microarchitecture, like Core 2 Duo. The main difference of the microarchitecture used on Atom is that it processes instructions in order, which is the way CPUs up to the first Pentium used to work. CPUs starting with the Pentium Pro and Pentium II use an out-of-order engine. This change was made in order to save energy, since the components in charge of issuing and controlling microinstructions execution could be removed. Atom can decode two instructions per clock cycle.

Atom has a 16-stage pipeline, which is a little bit longer than current two-core CPUs. There are several reasons this was done. First, this allows for better power efficiency. More stages implies more units, that can be spread across the chip thereby facilitating better heat spreading, instead of having fewer units, which would concentrate heat on a single point. With more units the probability of having some of them idle is higher compared to a CPU with fewer units, meaning that they can be turned

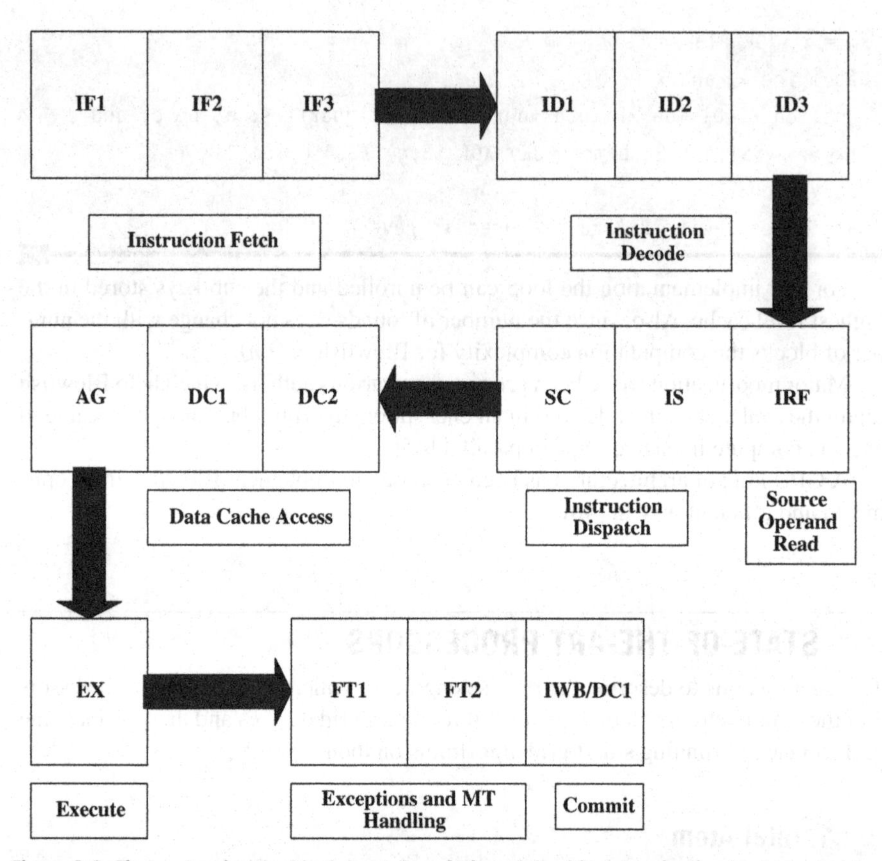

Figure 3.6 Figure showing the block level description of the 16-stage pipeline implemented in Intel Atom. IF(instruction fetch), ID(instruction decode), AG(address generate), DC(data cache), SC(switch context), IS (instruction schedule), IRF(instruction register file), EX(execute), FT (fault tolerant), IWB/DC1(write back).

off for power savings. Another advantage in a longer pipeline is that the microarchitecture can achieve higher clock rates. That is the reason each unit will have fewer transistors, making it easier to pump clock rate [39]. See Figure 3.6.

IF (Instruction fetch), ID (Instruction decode), AG (Address generate), DC (Data cache), SC (Switch context), IS (Instruction schedule), IRF (Instruction register file), EX (Execute), FT (Fault tolerant), IWB/DC1 (Write back)

The pipeline of the Intel Atom is explained in some detail below [39]:

- *Instruction fetch:* The instruction fetch rate is approximately 8 bytes per clock cycle on average when running a single thread. The fetch rate can get as high as 10.5 bytes per clock cycle in rare cases, such as when all instructions are 8-bytes long and aligned by 8, but in most situations the fetch rate is slightly less than 8 bytes per clock when running a single thread. The fetch rate is lower when running two threads in the same core. The instruction fetch rate is likely to be a bottleneck when the average instruction length is more than 4 bytes. The instruction fetcher can catch up and fill the instruction queue in case execution is stalled for some other reason.

- *Instruction decoding:* The two instruction decoders are identical. Instructions with up to three prefixes can be decoded in a single clock cycle. There are severe delays for instructions with more than three prefixes (which would almost never occur unless prefixes are used for padding). Most instructions generate only a single mop. There is no penalty for length-changing prefixes. Decoded instructions go into a queue with 16 entries per thread. The two 16-entry queues can be combined to a single 32 entries queue if one thread is disabled.
- *Execution units:* There are two clusters of execution units: an integer cluster that handles all instructions on general purpose registers and a floating point and SIMD cluster that handles all instructions on floating point registers and SIMD vector registers. A memory access cluster is connected to the integer unit cluster. Moving data between the clusters is slow. The micro operations from the decoders can be dispatched to two execution ports, which are called Port 0 and Port 1. Each execution port has access to part of the integer cluster and part of the floating point/SIMD cluster. The two parts of the integer cluster are called ALU0 and ALU1 and the two parts of the floating point/SIMD cluster are called FP0 and FP1; the numbers refer to the corresponding ports. The two execution ports can thus handle two parallel streams of mops, with the work divided as follows:
  Instructions that can be handled by both Port 0 and Port 1:
  - Register-to-register moves
  - Integer addition in general purpose or SIMD registers
  - Boolean operations in general purpose or SIMD registers

  Instructions that can be handled only by Port 0:
  - Memory read or write
  - Integer shift, shuffle, pack in general purpose or SIMD registers
  - Multiply
  - Divide
  - Various complex instructions

  Instructions that can be handled only by Port 1:
  - Floating point addition
  - Jumps and branches
  - LEA instruction

The four units ALU0, ALU1, FP0, and FP1 probably have one integer ALU each, though it cannot be ruled out that there are only two integer ALUs, which are shared between ALU0 and FP0 and between ALU1 and FP1, respectively. There is one multiply unit that is shared between ALU0 and FP0, and one division unit shared between ALU1 and FP1. The SIMD integer adders and shift units have full 128-bit widths and one clock latency. The floating point adder has full 128-bit capability for single precision vectors, but only 64-bit capability for double precision. The multiplier and the divider are 64 bits wide. The floating point adder has a latency of five clock cycles and is fully pipelined to give a throughput of one single precision vector addition per clock cycle. The multiplier is partially pipelined with a latency of four clocks and a throughput of one single precision multiplication per clock cycle. Double precision and integer multiplications have longer latencies and a lower throughput. The time from when one multiplication starts till the next multiplication can start varies from one clock cycle in the most favorable cases, to two clock cycles or more in less favorable cases. Double precision vector multiplication and some integer multiplications cannot overlap in time. Division is slow and not pipelined. A single precision scalar floating point division takes 30 clock cycles. Double precision takes 60 clock cycles. A 64-bit integer division takes 207 clock cycles.

- *Instruction pairing:* The maximum throughput of two instructions per clock cycle can only be obtained when instructions are ordered so that they can execute two at a time. Two instructions can execute simultaneously when the following rules are obeyed:
  - The core runs only one thread. The other thread, if any, must be idle or stalled by a cache miss, etc.
  - The two instructions must be consecutive with no other instructions between.
  - The two instructions do not have to be contiguous. A predicted taken branch instruction can pair with the first instruction at the branch target.
  - The second instruction does not read a register that the first instruction writes to. This rule has one exception: a branch instruction that reads the flags can pair with a preceding instruction that modifies the flags.
  - The two instructions do not write to the same register, except for the flags register. As an example, INC EAX/MOV EAX,0 cannot pair because both modify EAX. INC EAX/INC EBX pair OK even though both modify the flags.
  - The two instructions do not use the same execution port. The first instruction goes to Port 0 and the second instruction to Port 1; or the first instruction goes to Port 1 and the second instruction to Port 0.
  - An instruction that uses resources from both ports/pipelines cannot pair with any other instruction. For example, a floating point add instruction with a memory operand uses FP1 under Port 1 for floating point addition and the memory unit under Port 0 for the memory operand. It follows from these rules that it is not possible to do a memory read and a memory write at the same time because both use the memory unit under Port 0. But it is possible to do a floating point addition (without memory operand) and a floating point multiplication simultaneously because they use FP1 and FP0 respectively.
- *Instruction latencies:* Simple integer instructions have a latency of one clock cycle. Multiplications, divisions, and floating point instructions have longer latencies. Unlike most other processors, no delays were found in the Atom processor when mixing instructions with different latencies in the same pipeline. The LEA instruction uses the address generation unit (AGU) rather

than the ALU. This causes a latency of four clock cycles when dependent on a pointer register or index register because of the distance between the AGU and the ALU. It is therefore faster to use addition and shift instructions than to use the LEA instruction in most cases. Instructions that move data between a SIMD vector register and a general purpose register or flag have a latency of four to five clock cycles because the integer execution cluster and the floating point/SIMD cluster have separate register files. There is no penalty for using XMM move, shuffle, and Boolean instructions for other types of data than they are intended for. For example, you may use PSHUFD for floating point data or MOVAPS for integer data.

- *Memory access:* Each core has three caches:
  - Level 1 instruction cache: 32 kB, 8-way, set associative, 64 B line size
  - Level 1 data cache: 24 kB, 6-way, set associative, 64 B line size
  - Level 2 cache: 512 kB or 1 MB, 8-way, set associative, 64 B line size

Each cache is shared between two threads, but not between cores. All caches have a hardware prefetcher. An instruction with a memory operand takes no more time to execute than a similar instruction with a register operand, provided that the memory operand is cached. It is not totally "free" to use memory operands, though, for two reasons. First, the memory operand uses the memory unit under Port 0 so that the instruction cannot pair with another instruction that would require Port 0. And second, the memory operand may make the instruction code longer, especially if it has a full 4 byte address. This can be a bottleneck when instruction fetching is limited to 8 bytes per clock cycle. Cache access is fast for instructions running in the integer execution cluster, but slower for instructions running in the floating point/SIMD cluster because the memory unit is connected to the integer cluster only. Instructions using floating point or XMM registers typically take four to five clock cycles to read or write memory while integer instructions have only one clock cycle of effective cache latency thanks to store forwarding, as described below. The latency for a memory read that depends on a recently changed pointer register is three clock cycles. Store forwarding is very efficient. A memory operand that is written in one clock cycle can be read back in the next clock cycle. Unlike most other processors, the Atom can do store forwarding even if the read operand is larger than the preceding write operand or differently aligned. The only situation we have found where store forwarding fails is when a cache line boundary is crossed. Misaligned memory accesses are very costly when a cache line boundary is crossed. A misaligned memory read or write that crosses a 64-byte boundary takes 16 clock cycles. The performance monitor counters indicate that the misaligned memory access involves four accesses to the level 1 cache, where two accesses would suffice. There is no cost to misaligned memory accesses when no 64-byte boundary is crossed.

- *Branches and loops:* The throughput for jumps and taken branches is one jump per two clock cycles. Not-taken branches go at one per clock cycle. The minimum execution time for a loop is therefore two clock cycles if the loop contains no 16-byte boundary, and three to four clock cycles if a 16 byte boundary is crossed inside the loop. Branch prediction uses an 8 bit global history register. This gives reasonably good predictions, but the branch target buffer (BTB) has only 128 entries. In some of our tests, there were more BTB misses than hits. A branch misprediction costs up to 13 clock cycles, sometimes a little less. If a branch is correctly predicted taken, but it fails to predict a target because the BTB entry has been evicted, then the penalty is approximately seven clock cycles. This happens very often because the pattern history table has 4096 entries while the BTB has only 128.

- *Multithreading:* Each processor core can run two threads. The two treads are competing for the same resources so that both threads run slower than they would when running alone. The caches, decoders, ports, and execution units are shared between the two threads of the core, while the prefetch buffers, instruction queues, and register files are separate. The maximum throughput for the whole core is still two instructions per clock cycle, which gives one instruction per clock cycle in each thread on average. If both threads need the same resource, for example memory access, then each thread will get the contested resource half of the time. In other words, you will have one memory access every two clock cycles in each thread if there are no cache misses. It was found that the instruction fetch rate for each thread when running two threads is more than half the fetch rate for a single thread but less than the full single-thread rate. The instruction fetch rate per thread when running two threads is between 4 and 8 bytes per clock cycle, but never more than 8. The numbers depend heavily on instruction lengths and alignment. These findings indicate that there may be two instruction fetchers that are capable, to a limited extent, of serving the same thread when the other thread is idle. The branch target buffer (BTB) and the pattern history table are shared between the two threads. If the two threads are running the same code (with different data) then we might expect the two threads to share identical entries in these two tables. However, this doesn't happen. The two tables are apparently indexed by some simple hash function of the branch address and the thread number, so that identical entries in the two threads don't use the same table index. Tests indicate that two threads running the same code have slightly more branch mispredictions and significantly more BTB misses than a single thread running alone. In the worst case of a code with many branches, each thread may run at less than half the speed of a single thread running alone. These resource conflicts apply only to the case where two threads are running in the same processor core, of course. Some versions of the Atom processor have two cores capable of running two threads each, giving a maximum of four threads running simultaneously. If each core runs only one thread then there are no resource conflicts.

3.3.1.1 Bottlenecks in Atom

Some of the execution units in the Atom processor are quite powerful. It can handle two full 128-bit integer vector ALU instructions per clock cycle, though this capacity would rarely be fully utilized because of bottlenecks elsewhere in the system. The floating point addition unit is also reasonably good, while multiplication and division are slower. The execution is likely to be limited by other factors than the execution units in most cases, which would affect the efficiency of running our smart grid algorithms.

Some of the more plausible follow:

- In-order execution. The processor can do nothing while it is waiting for a cache miss or a long-latency instruction, unless another thread can use its resources in the meantime. If the algorithm exhibits high instruction level parallelism an in order core would be highly inefficient and way slower.
- The instruction fetch rate is less than 8 bytes per clock cycle in most cases. This is insufficient if the average instruction length is more than 4 bytes.
- Memory access is limited to one read or one write per clock cycle. It cannot read and write simultaneously.
- Memory access has long latencies for floating point and SIMD instructions. If algorithms like Blowfish encryption were implemented in SIMD fashion, which has faster execution time, this processor would produce poor results.
- The branch target buffer is rather small.
- x87 style floating-point code executes much slower than SSE style code.
- The maximum throughput of two instructions per clock cycle can only be achieved if the code is optimized specifically for the Atom and instructions are ordered in a way that allows pairing.
- The throughput is halved if two threads are running simultaneously in the same core.
- The resources most likely to be bottlenecks, such as cache, memory port, and branch target buffer, are shared between two threads.

The conclusion is that the Atom may be insufficient for highly CPU-intensive and memory-intensive applications, such as running the Blowfish and FFT algorithms simultaneously.

3.3.2 ARM

3.3.2.1 ARM926EJ-S

ARM designed processors for devices that handle complex smart grid and real-time problems such as power monitoring/analytics and support smart power algorithms such as total harmonic distortion, RMS (voltage/current), and transient detection.

ARM9EJ-S has a five-stage simple pipeline; 1. Instruction Fetch, 2. Instruction Decode and Register Read, 3. Execute Shift and ALU or Address Calculate, 4. or Multiply, 5. Memory Access, 6. and Multiply and Register Write. By overlapping the various stages of execution, ARM9EJ-S maximizes the clock rate achievable to execute each instruction. It delivers a throughput approaching one instruction per cycle. It also has 32-bit data buses between the processor core and instruction/data caches, which is helpful for data parallel movement. The caches are virtually addressed, with each line containing eight words. Cache lines are allocated on a read–miss basis. Victim replacement can be configured to either round-robin or pseudorandom. A cachable memory region can be defined as either being write-back or write-through. This is controlled by the memory region attributes contained in the page table entry for the region.

Due to the smaller pipeline most of the instructions execute in a cycle, with a few more needed for instructions like branch, multiply, etc. ARM9EJ-S, like a few other ARM processors, has Jazelle DBX, which allows it to execute Java byte codes directly. This helps in executing Java applications on handheld devices at a faster rate [40].

3.3.2.2 ARM Cortex A-8

ARM Cortex is one of the most sophisticated low-power, high-performance architecture designs. It has dual-issue, in-order, and statically scheduled ARM integer pipeline. Pruning the logic circuits during static scheduling also allows for reduced power during execution. It has twin ALUs that are symmetric and both can handle most arithmetic instructions. ALU pipe 0 always carries the older one in the pair of issued instructions. The Cortex-A8 processor also has multiplier and load-store pipelines, but these do not carry additional instructions to the two ALU pipelines. These can be thought of as "dependent" pipelines. Their use requires simultaneous use of one of the ALU pipelines. The multiplier pipeline can only be coupled with instructions that are in ALUO pipeline, whereas the load-store pipeline can be coupled with instructions in either ALU. See Figure 3.7.

The branch prediction unit in the 13-stage pipeline helps in the operation at higher frequencies. To minimize the branch penalties typically associated with a deeper pipeline, the Cortex-A8 processor implements a two-level global history branch predictor. It consists of two different structures: the BTB and the global history buffer (GHB), which are accessed in parallel with instruction fetches. The BTB indicates whether or not the current fetch address will return a branch instruction and its branch target address and contains 512 entries. On a hit in the BTB a branch is predicted and the GHB is accessed. The GHB consists of 4096 2-bit saturating counters that encode the strength and direction information of branches. The GHB is indexed by a 10-bit history of the direction of the last 10 branches encountered and

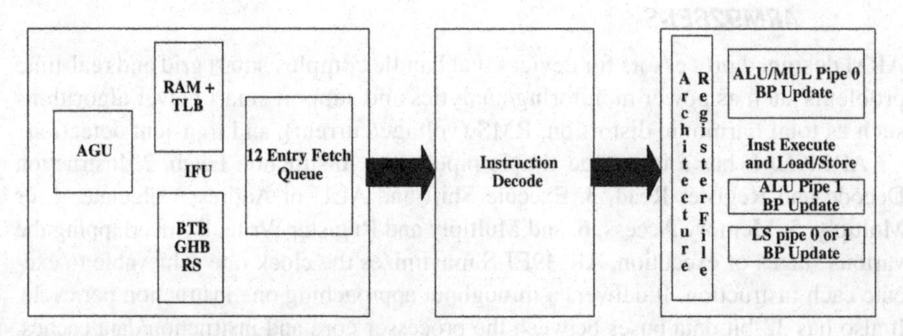

Figure 3.7 ARM Cortex-A8 pipeline stages.

4 bits of the PC. In addition to the dynamic branch predictor, a return stack is used to predict subroutine return addresses. The return stack has eight 32-bit entries that store the link register value in register 14 (r14) and the ARM or Thumb state of the calling function. When a return-type instruction is predicted taken, the return stack provides the last pushed address and state.

The Cortex-A8 processor has a single-cycle load-use penalty for fast access to the Level-1 caches. The data and instruction caches are configurable to 16k or 32k. Each is 4-way set associative and uses a Hash Virtual Address Buffer (HVAB) way prediction scheme to improve timing and reduce power consumption. The caches are physically addressed (virtual index, physical tag) and have hardware support for avoiding aliased entries. Parity is supported with one parity bit per byte. The replacement policy for the data cache is write-back with no write allocates. Also included is a store buffer for data merging before writing to main memory. The HVAB is a novel approach to reducing the power required for accessing the caches. It uses a prediction scheme to determine which way of the RAM to enable before an access.

The Cortex-A8 processor includes an integrated Level-2 cache. This gives the Level-2 cache a dedicated low-latency, high-bandwidth interface to the Level-l cache. This minimizes the latency of Level-1 cache line fills and does not conflict with traffic on the main system bus. It can be configured in sizes from 64k to 2M. The Level-2 cache is physically addressed and 8-way set associative. It is a unified data and instruction cache, and supports optional ECC and parity. Write-back, write-through, and write-allocate policies are followed according to page table settings. A pseudorandom allocation policy is used. The contents of the Level-1 data cache are exclusive with the Level-2 cache, whereas the contents of the Level-1 instruction cache are a subset of the Level-2 cache. The tag and data RAMs of the Level-2 cache are accessed serially for power savings.

The Cortex-A8 processor's NEON media processing engine pipeline starts at the end of the main integer pipeline. As a result, all exceptions and branch mispredictions are resolved before instructions reach it. More importantly, there is a zero load-use penalty for data in the Level-1 cache. The ARM integer unit generates the addresses for NEON loads and stores as they pass through the pipeline, thus allowing data to be fetched from the Level-1 cache before it is required by a NEON data processing operation. Deep instruction and load-data buffering between the NEON engine, the ARM integer unit, and the memory system allow the latency of Level-2 accesses to be hidden for streamed data. A store buffer prevents NEON stores from blocking the pipeline and detects address collisions with the ARM integer unit accesses and NEON loads.

The NEON unit is decoupled from the main ARM integer pipeline by the NEON instruction queue (NIQ). The ARM Instruction Execute Unit can issue up to two valid instructions to the NEON unit each clock cycle. NEON has 128-bit wide load and store paths to the Level-1 and Level-2 cache, and supports streaming from both. The NEON media engine has its own 10-stage pipeline that begins at the end ARM integer pipeline. Since all mispredicts and exceptions have been resolved in the ARM integer unit, once an instruction has been issued to the NEON media engine it must be completed as it cannot generate exceptions. NEON has three

SIMD integer pipelines, a load-store/permute pipeline, two SIMD single-precision floating-point pipelines, and a non-pipelined Vector Floating-Point unit (VFPLite). NEON instructions are issued and retired in-order. A data processing instruction is either a NEON integer instruction or a NEON floating-point instruction. The Cortex-A8 NEON unit does not parallel issue two data-processing instructions to avoid the area overhead with duplicating the data-processing functional blocks, and to avoid timing critical paths and complexity overhead associated with the muxing of the read and write register ports. The NEON integer datapath consists of three pipelines: an integer multiply/accumulate pipeline (MAC), an integer shift pipeline, and an integer ALU pipeline. A load-store/permute pipeline is responsible for all NEON load/stores, data transfers to/from the integer unit, and data permute operations such as interleave and de-interleave. The NEON floating-point (NFP) datapath has two main pipelines: a multiply pipeline and an add pipeline. The separate VFPLite unit is a non-pipelined implementation of the ARM VFPv3 floating point specification targeted for medium performance IEEE 754 compliant floating point support. VFPLite is used to provide backwards compatibility with existing ARM floating point code and to provide IEEE 754 compliant single and double precision arithmetic [41].

Most likely bottlenecks in ARM processors 2.2 are as follows:

- In-order execution and multicore unavailability: both the processors have in-order pipelines to decrease the power and area consumption but in the case of exploiting the instruction level parallelism available this is a deterring factor. Availability of multiple cores greatly increases the speed of execution but also increases the power requirement. In cases such as parallel implementation of matrix inversion, multicore processors will be beneficial for performance.
- Branch prediction when not utilized due to lesser percentage of branch instructions is a waste in terms of area and power consumed.
- Deep pipelines also reduce the latency we are looking for in real-time applications.

3.4 PROCESSOR DESIGN METHODOLOGY

From the observations made in the previous section, we can infer that there is a need for a possibly heterogeneous multicore embedded computing platform for executing smart grid applications. The steps involved in a typical processor design are outlined in Figure 3.8 below.

- The application for which we are designing the process is first identified.
- To identify the benchmark codes to run on the embedded processor, we examine the signals and the kind of data that the embedded processor will possibly handle.

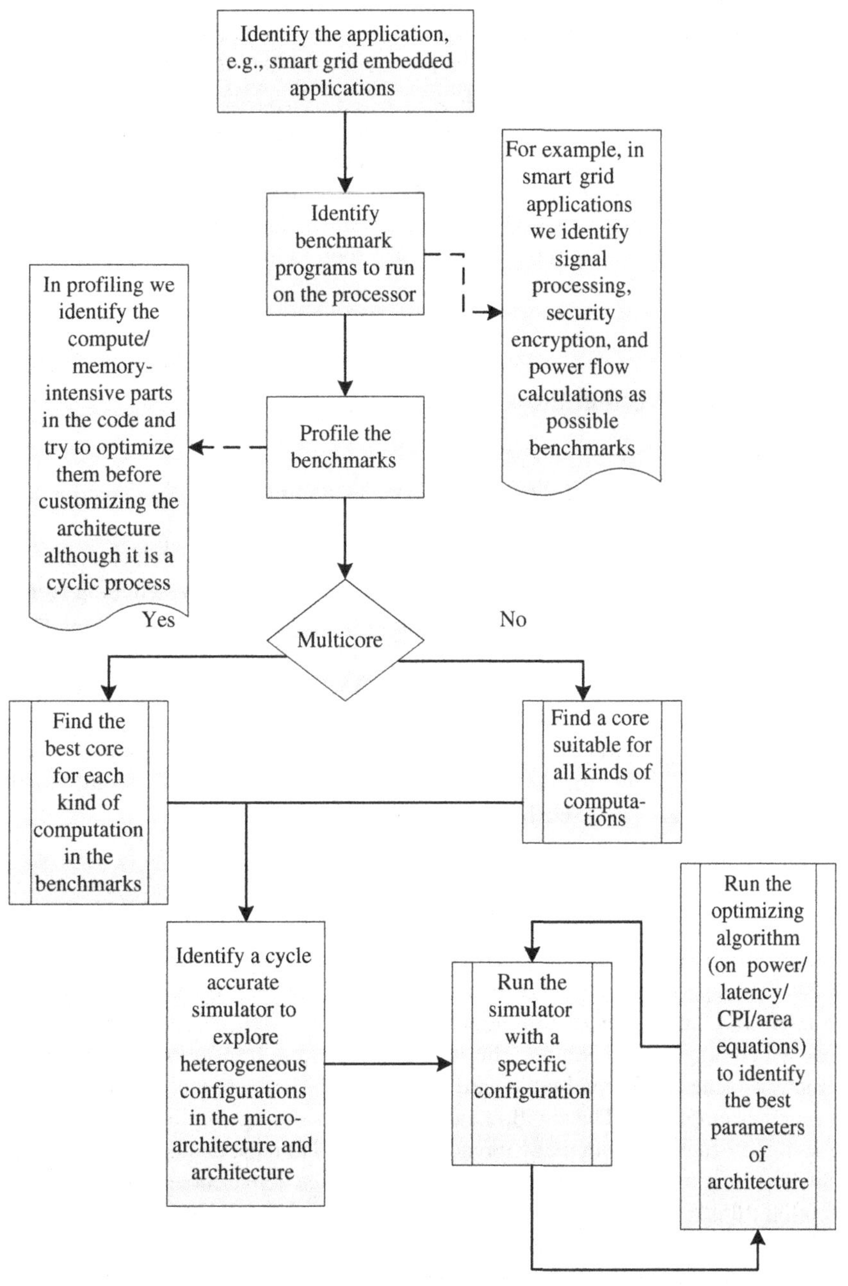

Figure 3.8 Design space exploration flow.

- The computations in the benchmark code are characterized and the complexities are calculated, to identify the type of execution units required in the processor.
- A cycle accurate simulator having the capability to test and run the benchmarks to provide performance statistics like power, cache hits/misses, CPI, area, latency, throughput, etc. is selected.
- Depending on the optimizing algorithm chosen (genetic algorithm, linear regression, etc.) the statistics from the simulator are provided to the optimizer for the corresponding design parameters. The optimizing algorithm finds the best fit in the relation between the design parameters and performance results.
- The simulator and the optimizer are run in a loop to find the best possible configuration.

3.4.1 Processor Simulation Frameworks

3.4.1.1 *CASPER*

CASPER—A SPARCV9-based Cycle-accurate chip multithreaded Architecture Simulator for Performance, Energy, and aRea analysis.

CASPER uses deep chip vision technology to estimate the area, delay, power dissipation, and energy consumption of processor architecture. It uses switching factors of microarchitectural components that are known from cycle-accurate simulation and precharacterized HDL libraries of the microarchitectural components, which are scalable over the range of different architectural parameters. CASPER models the open-source Sun Ultra Sparc T1 architecture model in C++. The modular coding style in CASPER gives the flexibility to modify parts of code and customize its execution. Before we go into the details of the architecture we present the configurable parts of the simulator [42]. See Table 3.1.

3.4.1.1.1 Microarchitecture

The pipeline has six main stages: Instruction Fetch (F-stage), Thread Schedule stage (S-stage), Branch and Decode stage (D-stage), Execution stage (E-stage), Memory Access stage (M-stage), and finally Write-Back stage (W-stage). The instruction fetch unit has the instruction address translation buffer, instruction cache, and thread scheduling state machine. The instruction address translation buffer and the instruction cache are shared by the hardware threads. There are mainly two scheduling schemes employed. The small latency thread scheduling employed in CASPER allows the instructions from ready threads to be issued into the D-stage at every clock cycle. The long latency scheduling scheme allows one active thread to continue its execution till it is complete or interrupted by higher priority threads.

The Decode stage implements full SPARCV9 instruction set decoder; it also supports the special set of hyper-privileged registers and instructions used by Hypervisor—the virtualization layer implemented in UltraSPARC processors.

Table 3.1 Configurable Design Parameters in CASPER [44]

Parameter	Range	Parameter Description
Cores	1:NC	Number of cores on chip
Strands	1:NS	Hardware threads per core
FPU	1 or 0	FPU can be shared between the cores or threads
I$_C/D$_C	4:64 (KB)	Size of L1 I-D cache
I$_B/D$_B	4:64	Size of L1 I-D cache block
I$_A/D$_A	2:8	Associativity of L1 I-D cache
I$/D$ Hit Latency	2:4 clock cycles	Measured in Cacti for 45nm technology
IFQ	1NS:8NS	Size of Instruction Fetch Queue
MIL	1NS:8NS	Size of Missed Instruction List
BBUFF	4NS:16NS entries	Size of Branch Address Buffer
LMQ	1NS:8NS	Size of Load Miss Queue
DFQ	1NS:8NS	Size of Data Fill Queue
SB	1NS:16NS	Size of Store Buffer (store-ordering)
L2$_C	256KB:16MB	Size of L2 cache
L2$_B	8:24	Size of L2 cache block
L2$_A	4:16	Associativity of L2 cache
L2$_NB	4:16	Number of L2 cache banks
L2_FB	8:16	Size of L2 cache Fill Buffer

The Execution unit has the RISC 64-bit ALU, an integer multiplier and divider. The Load Store Unit (LSU) has two modules, M-stage and W-stage. It also includes the data translation look aside buffer (D-TLB) and data cache (D$). D-TLB size and hit latency, along with D$ size, associativity, block size, and hit latency, are also configurable in CASPER. The Missed Instruction list controls the I$ miss path, while that of the D$ is controlled through Load Miss Queue (LMQ), which maintains cache misses separately for each thread and are similar in organization to that of UltraSPARC T1.

In CASPER we can measure the CPI (per thread and per core). The reported statistic events are pipeline stalls, wait time of threads due to Missed Instruction List and Load Miss Queue, wait time for threads due to Store Buffer being full, I-Cache and D-Cache misses, etc. Casper uses the Cacti tool to report cache hit latencies and access times [42].

3.4.1.2 *MPTLSim/PTLSim*

PTLsim [43] like CASPER is a cycle-accurate simulator; it also has the capability of full system simulation and targets the out-of-order x86-64 microprocessor at a configurable level of detail ranging from RTL-level models of all key pipeline structures, caches, and devices up to full-speed native execution on the host CPU.

PTLsim implements the main superscalar out-of-order, an SMT (simultaneous multithreaded) version of that core, and an in-order sequential core used for rapid testing and microcode debugging. The source code for PTLsim is developed in C++, hence any CPU model can be added by plugging in the class and recompiling PTLsim. The default model provided is an inspiration mainly of Intel Pentium 4, AMD K8, and Intel Core 2. It also incorporates some ideas from IBM Power4/Power5 and Alpha EV8.

The simulator directly fetches pre-decoded micro-operations from the basic block cache although it can simulate the cache access as if x86 instructions are being decoded as they are being fetched. Functional units, the mapping of functional units to clusters, issue ports and issue queues, physical register files, and μop (micro-operation) latencies are all configurable. The commit unit supports x86 specific semantics, including atomic commits and exception handling and recovery. The cache sizes (I-cache and D-cache), associativity, latency, bandwidth, etc. can be configured; the hierarchy consists of private L1 D-cache, L1 I-cache, unified L2 cache, and unified L3 cache along with data translation lookaside buffer (DTLB) and instruction translation lookaside buffer (ITLB). PTLsim also has configurable branch prediction; it supports hybrid g-share based predictor, bimodal predictor, saturating counters, etc. In the Simultaneous Multithreading mode it can support up to 16 threads per core with separate per-thread fetch queue, re-order buffer, load queue, store queue, and other structures, but shared caches, issue queues, and functional units.

For the full system implementation PTLsim makes use of Xen hypervisor which is an open-source x86 virtual machine monitor. Each virtual machine is called a "domain," where domain 0 is privileged, runs Linux, and accesses all hardware devices using unmodified drivers; it can also create and directly manipulate other domains. Guest domains typically use Xen-specific virtual device drivers, and make hyper-calls into the hypervisor to request certain services that cannot be easily or quickly virtualized. Each guest can have up to 32 VCPUs (virtual CPUs). Xen has essentially zero overhead due to its unique and well-planned design. All major Linux distributions now fully support Xen, along with Solaris and FreeBSD.

MPTLSim is the multicore version of PTLSim. The OOO core in PTLsim was modified to implement an Intel P4 style hyper-threaded data path and some variants. Each VCPU maintains a context structure that holds all information about the VCPU/thread context. This includes the values of the x86 architectural registers, x86 machine state registers (MSRs), page tables (as PTLsim implements its own memory management mechanism), and various PTLsim internal state variables. MPTLSim supports up to 16 VCPUs; the challenges that they faced were in designing the memory hierarchy, which supports up to two levels of cache (shared and distributed), and implements the MESI cache coherency protocol and various interconnection networks.

3.4.1.3 MV5

CASPER simulates in-order cores of the Ultra Sparc T1 model and PTLsim models out-of-order with x86-64 instructions as the base. However, future processors that require heterogeneity in terms of fundamental design principles cannot be modeled on these simulators. MV5 tries to solve this problem.

Highly parallel codes involving simple computations when run on a number of simple in-order cores can give a good result; codes that are sequential when run on a single out-of-order core can provide high performance. Embedded processors for the smart grid might require a processor that can handle parts of the code that are sequential and parts that are parallel; hence exploring a processor design that can handle both sequential and parallel parts of the code will be helpful, and hence we provide details of the MV5 simulator.

MV5 offers all the customizations available in PTLsim and CASPER except full system simulation. Apart from the usual configuration abilities, it provides array style SIMD (single instruction, multiple data) cores, coherent caches, and on-chip networks. MV5 uses its own runtime threading library to manage SIMD threads. An advantage provided by MV5 as discussed earlier is that the in-order and out-of-order cores can be configured on the same chip.

For accurate power modeling MV5 uses Cacti for calculating the dynamic energy of reads, writes, and leakage power for the caches. Energy consumption of the cores is calculated using watch. The pipeline energy is divided into seven parts including fetch and decode, integer ALUs, floating point ALUs, register files, result bus, clock, and leakage. Dynamic energy is accumulated each time a unit is accessed [44]. See Figure 3.9.

The figure shows a heterogeneous configuration having private Level-1 and Level-2 caches and a shared level-3 cache. The different microarchitectures of the out-of-order and SIMD cores add to the heterogeneous nature of the processor.

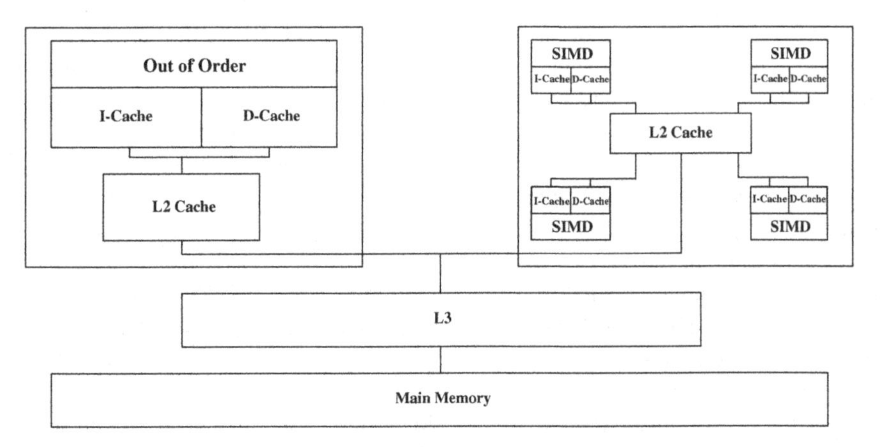

Figure 3.9 Possible heterogeneous configuration of the processor.

3.4.2 Microarchitectural Design Space Exploration

Once the simulators are in place, Figure 3.8 shows a systematic approach to explore the design space. To find the best configuration, all the possible configurations of the design parameters can be explored. Each design parameter will have a range of values. For instance, if we have ten design parameters and each parameter can have four values, we will still have to explore 1048576 (4^{10}) combinations!

Design space exploration in general consists of a multiobjective optimization problem. Although several heuristic techniques have been proposed so far to address this problem, they are all characterized by low efficiency to identify the Pareto front of feasible solutions. Among those heuristics, evolutionary or sensitivity based algorithms represent the most notable, state-of-the-art techniques.

Several methods have been recently proposed in literature to reduce the design space exploration complexity by using conventional statistic techniques and advanced exploration algorithms. Among the most recent heuristics to support power/performance architecture exploration, we can find [24, 25]. In [25], the authors compare Pareto simulated annealing, Pareto reactive taboo search and random search exploration to identify energy-performance trade-offs for a parametric superscalar architecture executing a set of multimedia kernels. In [24] a combined genetic-fuzzy system approach is proposed. The technique is applied to a highly parameterized system on chip platform based on a very large instruction word (VLIW) processor in order to optimize both power dissipation and execution time. The technique is based on a strength Pareto evolutionary algorithm coupled with fuzzy system rules in order to speed up the evaluation of the system configurations. State-of-the-art system performance optimization has been presented in [26–29]. A common trend among those methods is the combined usage of response surface modeling and design of experiments techniques. In [27], a radial basis function has been used to estimate the performance of a superscalar architecture; the approach is critically combined with an initial training sample set that is representative of the whole design space, in order to obtain good estimation accuracy. The authors propose to use a variant of the Latin hypercube method in order to derive an optimal, initial set. In [28, 29] linear regression has been used for performance prediction and assessment. The authors analyze the main effects and the interaction effects among the processor architectural parameters. In both cases, random sampling has been used to derive an initial set of points to train the linear model. A different approach is proposed in [26], where the authors tackle performance prediction by using an artificial neural network paradigm to estimate the system performance of a chip multiprocessor. In [30, 31], a Plackett–Burman design of experiments is applied to the system architecture to identify the key input parameters. The exploration is then reduced by exploiting this information. The approach shown in [32] is directed towards the optimization of an FPGA, soft-core-based design, while in [31] the target problem is more oriented to a single, superscalar processor.

According to the analysis done earlier in this section, we predict that different applications require different microarchitectures. To achieve the best configuration, we are using the exploration technique in the genetically programmed response surface tool [33]. The response surface expresses the performance metric that we choose in terms of equations using genetic programming. The equations are built using the design parameters and results of test runs. This tool avoids the cumbersome task of conducting several time-consuming simulations to arrive at the equations. The design points to be given to the tool are distributed uniformly within the design variable domain using the Latin hypercube sampling method described in [34]. According to [34], this is The design of experiments (DoE) for N variables and P experiments which is independent of the application under consideration, so once the design is formulated for P points and N design variables, it is stored in a matrix and need not be formulated again.

In our research as we need heterogeneity in core architectures, we bind the benchmarks to specific cores that we think will give the best performance according to the qualitative analysis done earlier.

3.5 OPEN RESEARCH QUESTIONS

In this chapter, we have identified computation and data bound applications typical to the emerging smart grid, motivated the need for designing new embedded processors for the smart grid, and described existing simulation platforms and efficient optimization tools to enable the exploration of novel microarchitectures. For any processor architecture exploration effort, the first step involves developing a benchmark suite for the target application area. Our research shows a dearth of such benchmarks for smart grids, and part of our current effort is developing one for the community. The benchmark suite will aim to capture the computations that would run on the embedded processor if it were installed in the smart grid; this is expected to bridge the gap between computer and electrical engineers and enable embedded processor architecture research in the context of smart grids.

Furthermore, we believe that identifying the optimal operating system (OS), virtualization schemes, and scheduling algorithms for smart grid applications are other critical research directions. The computer science community should actively work in this area to identify optimizations that can be done at the kernel level, together with possible compiler optimizations and auto-tuning, to realize the optimal operation of the smart grid under real-time operating constraints. We hope the ideas presented in this chapter have paved the way for new research directions and have helped out current research in perspective.

REFERENCES

[1] 20% Wind Energy by 2030 Increasing Wind Energy's Contribution to US Electricity Supply. 20percentwind.org. October 28, 2011. <http://www.20percentwind.org/20p.aspx?page=Overview>.

[2] What is the Smart Grid?. Smartgrid.gov. October 28, 2011. <http://www.smartgrid.gov/the_smart_grid#smart_grid>.

[3] Smart Grid Initiative. October 28, 2011, Greensmartgridinitiative.org. <http://www.greensmartgridinitiative.org/did_you_know/index.php>.
[4] IEEE, IEEE Standard Communication Delivery Time Performance Requirements for Electric Power Substation Automation, IEEE Std 1646–2004, 2005, pp. 0_1–24, doi: 10.1109/IEEESTD.2005.95748.
[5] IEC, IEC 61850-5 Communication networks and systems in substations—Part 5: Communication requirements for functions and device models.
[6] F.L. Alvarado, Computational complexity in power systems, IEEE Transactions on Power Apparatus and Systems 95 (4) (1976) 1028–1037.. doi:10.1109/T-PAS.1976.32193.
[7] S.M. Amin, For the good of the grid, IEEE Power and Energy Magazine 6 (6) (2008) 48–59.
[8] K. Moslehi, A.B.R. Kumar, D. Shurtleff, M. Laufenberg, A. Bose, P. Hirsch, Framework for a self-healing power grid, in: Power Engineering Society General Meeting, 2005. vol. 3, IEEE, 12–16 June 2005, p. 3027.
[9] N. Ginot, M.A. Mannah, C. Batard, M. Machmoum, Application of power line communication for data transmission over PWM network, IEEE Transactions on Smart Grid 1 (2) (2010) 178–185.
[10] C. Konate, A. Kosonen, J. Ahola, M. Machmoum, J.-F. Diouris, Power line communication in motor cables of inverter-fed electric drives, IEEE Transactions on Power Delivery 25 (1) (2010) 125–131.
[11] F. Gianaroli, A. Barbieri, F. Pancaldi, A. Mazzanti, G.M. Vitetta, A novel approach to power-line channel modeling, IEEE Transactions on Power Delivery 25 (1) (2010) 132–140.
[12] A. Kosonen, J. Ahola, Communication concept for sensors at an inverter-fed electric motor utilizing power-line communication and energy harvesting, IEEE Transactions on Power Delivery 25 (4) (2010) 2406–2413.
[13] Z. Marijic, Z. Ilic, A. Bazant, Fixed-data-rate power minimization algorithm for OFDM-based power-line communication networks, IEEE Transactions on Power Delivery 25 (1) (2010) 141–149.
[14] N. Andreadou, F.-N. Pavlidou, Modeling the noise on the OFDM power-line communications system, IEEE Transactions on Power Delivery 25 (1) (2010) 150–157.
[15] J. Zhang, J. Meng, Robust narrowband interference rejection for power-line communication systems using IS-OFDM, IEEE Transactions on Power Delivery 25 (2) (2010) 680–692.
[16] J. Anatory, N. Theethayi, R. Thottappillil, Channel characterization for indoor power-line networks, IEEE Transactions on Power Delivery 24 (4) (2009) 1883–1888.
[17] J. Anatory, N. Theethayi, R. Thottappillil, Performance of underground cables that use OFDM systems for broadband power-line communications, IEEE Transactions on Power Delivery 24 (4) (2009) 1889–1897.
[18] American National Standards Institute, Synchronous Optical Network (SONET)—sub-STS-1 interface rates and formats specification.
[19] American National Standards Institute, Transmission and Multiplexing (TM)—Digital Radio Relay Systems (DRRS); Synchronous Digital Hierarchy (SDH); System performance monitoring parameters of SDH DRRS.
[20] 802.3av-2009 IEEE Standard for Information Technology—Part 3: (CSMA/CD) Access Method and Physical Layer Specifications Amendment 1: Physical Layer Specifications and Management Parameters for 10 Gb/s Passive Optical Networks, IEEE, October 2009.

[21] IEEE 802.11 Working Group, IEEE 802.11-2007: Wireless LAN Medium Access Control (MAC) and Physical Layer (PHY) Specifications, June 2007.

[22] IEEE P802.15.6/D01, Wireless Medium Access Control (MAC) and Physical Layer (PHY) Specifications for Wireless Personal Area Networks (WPANs) Used in or around a Body, May 2010.

[23] D. Johnston, J. Walker, Overview of IEEE 802.16 security, IEEE Security and Privacy 2 (3) (2004) 40–48.

[24] G. Ascia, V. Catania, A.G. Di Nuovo, M. Palesi, D. Patti, Efficient design space exploration for application specific systems-on-a-chip, Journal of Systems Architecture 53 (10) (2007) 733–750.

[25] G. Palermo, C. Silvano, V. Zaccaria, Multiobjective design space exploration of embedded system, Journal of Embedded Computing 1 (3) (2006) 305–316.

[26] E. İpek, S.A. McKee, R. Caruana, B.R. de Supinski, M. Schulz, Efficiently exploring architectural design spaces via predictive modeling, in: Proceedings of the 12th International Conference on Architectural Support for Programming Languages and Operating Systems, Vol. 40, No. 5, 2006, pp. 195–206.

[27] P.J. Joseph, Kapil Vaswani, Matthew J. Thazhuthaveetil, A predictive performance model for superscalar processors, MICRO 39: Proceedings of the 39th Annual IEEE/ACM International Symposium on Microarchitecture, IEEE Computer Society, Washington, DC, USA, 2006, pp. 161–170.

[28] P.J. Joseph, Kapil Vaswani, M.J. Thazhuthaveeti, Construction and use of linear regression models for processor performance analysis, in: The 12th International Symposium on High-Performance Computer Architecture, 2006, pp. 99–108.

[29] B.C. Lee, D.M. Brooks, Accurate and efficient regression modeling for microarchitectural performance and power prediction, in: Proceedings of the 12th International Conference on Architectural Support for Programming Languages and Operating Systems, vol. 40, No. 5, 2006, pp. 185–194.

[30] D. Sheldon, F. Vahid, S. Lonardi, Soft-core processor customization using the design of experiments paradigm, in: DATE '07: Proceedings of the Conference on Design, Automation and Test in Europe, 2007, pp. 821–826.

[31] J.J. Yi, D.J. Lilja, D.M. Hawkins, A statistically rigorous approach for improving simulation methodology, in: The Ninth International Symposium on High-Performance Computer Architecture, 2003, HPCA-9 2003, Proceedings, 2003, pp. 281–291.

[32] David Sheldon, Frank Vahid, Stefano Lonardi, Soft-core processor customization using the design of experiments paradigm, in: DATE '07: Proceedings of the Conference on Design, Automation and Test in Europe, 2007 pp. 821–826.

[33] H. Cook, K. Skadron, Predictive design space exploration using genetically programmed response surfaces, Design Automation Conference, 2008, DAC 2008, 45th ACM/IEEE, 8–13 June 2008, pp. 960–965.

[34] Steven J.E. Wilton, Norman P Jouppi, CACTI: an enhanced cache access and cycle time model, IEEE Journal of Solid-State Circuits 31 (1996) 677–688.

[35] C.R. Fuerte-Esquivel, E. Acha, H. Ambriz-Perez, A comprehensive Newton–Raphson UPFC model for the quadratic power flow solution of practical power networks, IEEE Transactions on Power Systems 15 (1) (2000) 102–109.. doi:10.1109/59.852107.

[36] G.K. Venayagamoorthy, Potentials and promises of computational intelligence for smart grids, in: Proceedings of IEEE PES GM, No. 1443, 2009, Calgary, Canada.

[37] S. Bouguezel, M.O. Ahmad, M.N.S. Swamy, A general class of split-radix FFT algorithms for the computation of the DFT of length-2m, IEEE Transactions on Signal Processing 55 (8) (2007) 4127–4138.. doi:10.1109/TSP.2007.896110.

[38] B. Schneier, Fast software encryption, Cambridge Security Workshop Proceedings (December 1993), Springer-Verlag., pp. 191–204.

[39] B. Beavers, The story behind the Intel Atom processor success, IEEE Design and Test of Computers 26 (2) (2009) 8–13.. doi:10.1109/MDT.2009.44.

[40] ARM9 Processor Family, ARM.com, October 28, 2011. <http://www.arm.com/products/processors/classic/arm9/>

[41] ARM A-8 Processor, ARM.com, October 28, 2011. <http://www.arm.com/products/processors/cortex-a/cortex-a8.php>.

[42] K. Datta, A. Mukherjee, G. Cao, R. Tenneti, V. Vijendra Kumar Lakshmi, A. Ravindran, B.S. Joshi, CASPER: embedding power estimation and hardware-controlled power management in a cycle-accurate microarchitecture simulation platform for many-core multi-threading heterogeneous processors, Journal of Low Power Electron 2 (2012) 30–68.

[43] M.T. Yourst, PTLsim: a cycle accurate full system x86–64 microarchitectural simulator, in: IEEE International Symposium on Performance Analysis of Systems & Software, 2007, ISPASS 2007, 25–27 April 2007, pp. 23–34, http://dx.doi.org/10.1109/ISPASS.2007.363733.

[44] Jiayuan Meng, K. Skadron, A reconfigurable simulator for large-scale heterogeneous multicore architectures, in: IEEE International Symposium on Performance Analysis of Systems and Software (ISPASS), 10–12 April 2011, pp. 119–120, http://dx.doi.org/10.1109/ISPASS.2011.5762722.

[45] G.N. Krishnamurthy, V. Ramaswamy, G.H. Leela, M.E. Ashalatha, Performance enhancement of Blowfish and CAST-128 algorithms and security analysis of improved Blowfish algorithm using Avalanche effect, International Journal of Computer Science and Network Security 8 (3) 244–250.

[46] Zhu Wang, Josh Graham, Noura Ajam, Hai Jiang, Design and optimization of hybrid MD5-blowfish encryption on GPUs, in: Proceedings of 2011 International Conference on Parallel and Distributed Processing Techniques and Applications (PDPTA), July 18–21, 2011, Las Vegas, Nevada, USA.

CHAPTER

IEEE 802.15.4 Based Wireless Sensor Network Design for Smart Grid Communications

4

Chun-Hao Lo, Nirwan Ansari

Advanced Networking Laboratory, Department of Electrical and Computer Engineering, New Jersey Institute of Technology, Newark, NJ 07102, USA

4.1 INTRODUCTION

The national electric power system is anticipated to evolve into a smart grid to provide green energy with the capabilities of intelligent fault detection and energy efficiency in the next 20 years or so. The existing power grid lacks standardized communications protocols and cooperative communications despite the fact that the system is 99.97% reliable [1]. It has encountered power congestion in the network because of rising power demand and power generation along with equipment failures attributed to hardware/software malfunctions and natural disasters. Responsiveness to fault detection as well as time for system restoration is exigent to grid reliability and stability in such a critical infrastructure. Integrating state-of-the-art information and communications technologies (ICT) and smart grid technologies into the legacy system will provide greater transparency with data collection and aggregation throughout the entire network. Nevertheless, the implementation of the new cyberphysical system involves foreseeable challenges such as interconnection of heterogeneous networks, interoperability of various technologies, and vulnerability to cybersecurity threats. The heterogeneity ranges from wide area network (WAN) and neighborhood/field area network (NAN and FAN), to home area network (HAN), as well as from traditional power plants and renewable energy sources (RESs), to (hybrid) electric vehicles [2]. Communications technologies predominantly tailored in the power system can be power line communications (PLC), wired or wireline communications (coaxial cables and optical fibers), and wireless communications (cellular, WiFi, and WiMAX). In this chapter, we focus on wireless sensor networks (WSNs) deployed in different endpoints throughout the power grid system. In Section 4.2, we first present a number of smart grid applications entailed in the power grid system, and particularly discuss the shortcomings of PLC in comparison with the IEEE 802.15.4 standard implemented in WSN. In Section 4.3, we introduce a list of Task Groups corresponding to the IEEE 802.15 standards, as well as the proposal by the IEEE 802.15.4g Task Group for the smart utility network design. In Section 4.4, we discuss the essential network measurements and associated design challenges in IEEE 802.15.4 communications protocols. In Section 4.5, we summarize the focal points and draw a conclusion.

4.2 THE ROLE OF WSN IN SMART GRID COMMUNICATIONS

Smart grid applications essentially entail intelligent sensing and monitoring, equipment fault diagnosis, and meter reading through power generation, transmission and distribution (T&D), and end-use sectors. The conventional supervisory control and data acquisition and energy management systems (SCADA/EMS), the sophisticated phasor management units and phasor data concentrators (PMU/PDC), the advanced metering infrastructure (AMI), and a wide range of remote terminal units (RTUs) are typical components at the T&D level in the electric power system. These critical applications are supervised by utility operators and supported in sensor networks and WSNs, which can be extended further to wireless multimedia sensor networks (WMSNs) [3] and wireless multimedia sensor and actor networks (WMSANs) [4]. WSNs consist of multimedia sensors and scalar sensors with wireless capability that retrieve video and audio streams, still images, and scalar data from the physical environments. Processing nodes (e.g., actors, actuators) with rich resources in WSNs may act as clusters and process data collected from the resource-constrained sensors and send the processed data to the associated wireless gateways and sink nodes, before entering the WAN. Types of sensors deployed in different scenarios and locations such as chemical, electrical, environmental, and pressure sensors in the power system have been vastly used to monitor the status of electric components. For example, sensors installed in harsh environments such as extremely high temperature fuel cells and internal combustion engines in power plants [5], sensors mounted on power lines to measure current and voltage conditions [6], and sensors deployed in houses to control households energy consumption [7]. Figure 4.1 illustrates the five domains of

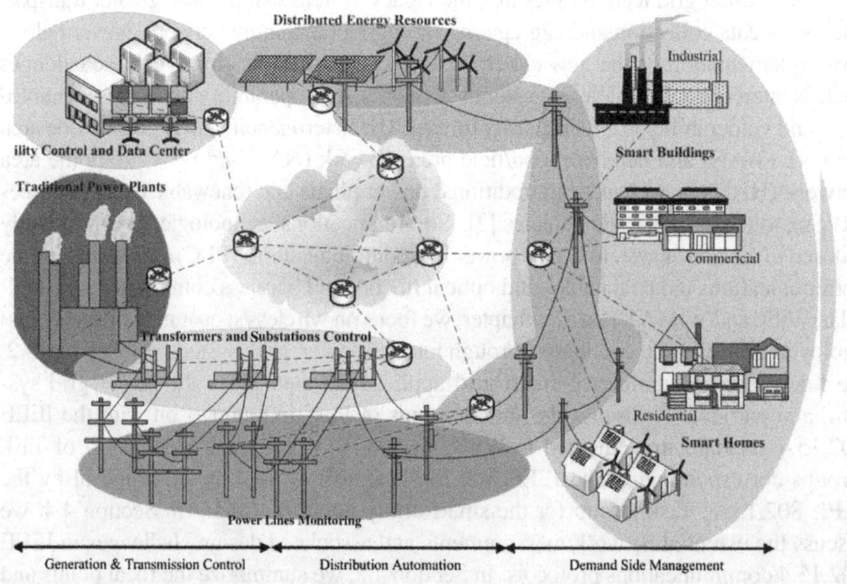

Figure 4.1 Wireless sensor networks deployed in the *five* major domains.

the entire smart grid network at endpoints in which the sheer volume of data sensing and collection are processed in WSNs: traditional power plants, transformers and substations, distributed energy resources (DERs), power lines, and demand-side consumers.

Diverse network load factors are measured in the power system. Some key parameters including voltage and frequency in real power and reactive power are collected from these domains in order to determine the characteristics of specific network buses as well as the system stability [8]. Some data require a magnitude of milliseconds to meet the communication latency requirements. The collected data may be shared among the conventional power plants and renewable power generators. They can also be used for other applications (e.g., demand response for peak reduction) to efficiently control generation and transmission, and at the same time to optimize distribution network operation [9]. Such effective data communications facilitate asset management and outage management in the power system so that load balancing and short response time can be achieved. On the other hand, although some data may be reused for multiple applications, modification to the packet headers is required; moreover, the reused data may not carry sufficient information for some specific applications. Hence, advanced sensors as well as associated sensor data management for the smart grid applications require further development.

In demand-side management, consumers' energy use is monitored and controlled via WSN [4,7]. The detail of energy consumption by households can be obtained through web services on a user interface. In addition to the energy consumption measure, the HAN with an energy management unit is also able to control power generation produced from DERs (e.g., solar panels and wind turbines mounted on rooftops or on the ground) and coordinate the operation of smart appliances and energy storage to achieve high efficiency of energy use. Consumers can make decisions to use certain amounts of energy based on time-of-use price signals received from utilities via smart meters. The intervals of metering data collection can range from a few seconds to hours; different applications can be supported and they are determined by utilities. The so-called demand response can potentially reduce peak loads and minimize the energy cost in favor of both the utilities and consumers. Table 4.1 summarizes fundamental smart grid sensor applications in various deployments.

Popular working groups in WSN applications such as ZigBee, WirelessHART, and ISA100 Alliance have defined their own industrial specifications for different application domains [10]. All of them are based on IEEE 802.15.4, which only specifies PHY and MAC layers and leaves the upper layers to be designed by the application designers. For further details, readers are referred to [10] for the ZigBee stack architecture, [3,11] for the WMSN architectures and testbeds, and [2,8,12,58] for the communications architectures and design issues for the smart grid.

In addition to the IEEE 802.15.4 technology suitable for the networks at endpoints in the power grids (e.g., HAN, smart metering in neighborhoods), communications using power line carriers is another viable approach [2,12,58]. PLC uses power line cables as the communications medium for data transmission. Similar to wireline networks (using coaxial cables and optical fibers), utilizing the existing assets that have been deployed in the current power and communications networks can save the unnecessary investment as well as time for deployment. Despite the attractive advantages,

Table 4.1 Sensor Applications Used in the Smart Grid

Field Categories	Components	Applications	Acceptable Latency
Conventional power stations and nuclear power	Power generation based on coal, oil, natural gas, petroleum, nuclear fusion	Steam turbines, gearboxes, motors, cooling pumps	Low to medium
Distributed energy resources (DERs)	Solar, wind, water power, biofuels, combined heat and power (CHP), microCHP, fuel cells, storage	Power output measure and metering, temperature, pressure, microgrid operation modes	Low to medium
Transformers and substations	SCADA, PMU, RTUs, AMI	Video monitoring, coordination of generation and distribution, Volt/VAR control	Low
Power line monitoring	Transmission and distribution lines	Environmental measures, line sag monitoring, transfer capability measures	Low
Smart utilization	Smart homes and buildings	Energy use and consumption management, customer participation (demand response)	High

the shortcomings of PLC include (1) high bit error rates (due to noisy power lines, e.g., motors, power supplies), (2) limited capacity (attributed to the number of concurrent network users and applications concurrently being used), (3) high signal attenuation (dependent of geographical locations), (4) phase change between indoor and outdoor environments, and (5) disconnected communications due to opened circuits. Readers are referred to [5, 12, 58] for detailed comparisons of different technologies.

While information exchange in the power systems mainly involve metering data and control messages among meters and sensors, the giant grid can be seen as a large-scale sensor network composed of numbers of small WSNs in different domains. In the rest of the chapter, we will introduce the IEEE 802.15.4 standard, and discuss associated communications protocols and specifications of the network designs.

4.3 THE IEEE 802.15 TECHNOLOGIES: WIRELESS PERSONAL AREA NETWORKS

WSN is often deployed under the auspice of the wireless personal area network (WPAN), which was initially designed to have characteristics of low complexity, low data rate, and low energy consumption. WSN facilitates large-scale and fast deployment, as well as low implementation cost. It supports short-range wireless communications and is typically adopted in industry and smart homes. WSNs have been widely used in diverse applications, such as surveillance for crimes, traffic congestion avoidance, telemedicine and e-healthcare, and environmental monitoring. Nevertheless, sensor networks used in power systems are mostly wired-based and are not interconnected [4]. Moreover, WSNs deployed in the smart grid system have more stringent requirements than those in other applications in terms of communications link quality, radio frequency (RF) interference, quality of service (QoS) provisioning, latency, and security [3,4,12,13]. Several challenges have been identified in the studies, among which severe interference in shared ISM bands with the existing communications networks is a main concern.

In order to alleviate the problem, techniques such as multichannel access [14], WiFi features adoption [15], and cognitive radio [16] integrated in WSNs were proposed to enhance the performance of the smart grid communications. Utilizing the channel resource over time (e.g., parallel transmission using dual transceivers) helps address the coexistence issue, as well as reduce traffic loads (including retransmission) and energy consumption to some degree. It may also mitigate the packet loss caused by collision, congestion, and wireless impairments. In other words, spectrum sensing management, data traffic management, and power control management are important elements that predominantly determine how optimized the network performance can be achieved, subject to the constraints of complexity, cost, overhead, and power consumption.

By the second quarter of 2011, the IEEE 802.15 Working Group (WG) for WPAN consisted of *nine* Task Groups (TGs) [17], as described in Table 4.2

Table 4.2 IEEE 802.15 Task Groups

IEEE 802.15 Task Groups
(1) Bluetooth (2005)
(2) Coexistence of WPAN with other wireless devices operating in unlicensed frequency bands (2003)
(3) High rate WPAN (HR-WPAN) (2003)
(b) MAC amendment/enhancement (2006)
(c) Millimeter wave alternative PHY (2009)
(4) Low rate WPAN (LR-WPAN) (2003)
(a) Alternative PHY (2007)
(b) Revision and enhancement for LR-WPAN-2003 (2006)
(c) Alternative PHY to support Chinese frequency bands (2009)
(d) Alternative PHY to support Japanese frequency bands (2009)
(e) MAC amendment and enhancement for LR-WPAN-2006 (2012)
(f) Active RFID system; new PHY and enhancement to LR-WPAN-2006 for RFID (2012)
(g) Smart utility networks/neighborhood SUN (2012)
(5) Mesh topology capability in WPAN (2009)
(6) Body area network (2012)
(7) Visible light communication VLC (2011 to present)
IGthz—TeraHz Interest Group (2008 to present)
WNG—Wireless next generation (2008 to present)

Wireless MAC and PHY specifications have been defined for different WPAN purposes among TGs [18]. IEEE 802.15.1 was originally developed by the Bluetooth Special Interest Group. It defines PHY and MAC for the conventional WPAN. The coverage of wireless connectivity with fixed or portable/handheld digital wireless devices operated by a person or object is up to 10m (in radius) of a personal operating space (POS). Unlike ZigBee, Bluetooth supports much shorter range and coordinates no more than *seven* devices in its network. Besides, it is power-hungry (i.e., the supported lifetime is only a few days) due to the FHSS (frequency-hopping spread spectrum) technology employed in the PHY. Similarly, *Z-Wave* Alliance [19], a proprietary standard designed for home automation operating in around 900MHz, is not as popular as ZigBee.

IEEE 802.15.2 addresses the limitation of coexistence of IEEE 802.15.1-2002 WPANs and IEEE 802.11b-1999 WLANs operated in unlicensed ISM frequency bands. It provides a number of modifications to other standards in IEEE 802.15 for enhancing coexistence with other wireless devices, as well as recommended practices for IEEE 802.11-1999 devices to facilitate coexistence with IEEE 802.15 devices.

IEEE 802.15.3 was meant for wireless multimedia to support high data rates in WPAN required for time dependent and different consumer applications, such as large file transfer in video and digital still imaging. IEEE 802.15.3b adds enhancements to improve the efficiency of IEEE 802.15.3 including the newly defined MLME-SAP (MAC layer management entity-service access point), ACK (acknowledgment) policy and implied-ACK, LLC/SNAP (logical link control/subnetwork access protocol) data frame, and a method for CTA (channel time allocation). IEEE 802.15.3c, namely *mmWave*, enables data rates greater than 5Gbps operating in the 60GHz band and defines a beam-forming negotiation protocol to improve the communications range for transmitters. It also supports aggregation of incoming data and ACKs, respectively, into single packets to improve MAC efficiency by reducing retransmission overhead as well as facilitating coexistence with microwave systems in WPAN. Applications such as real-time video streaming, HDTV, video on demand, and content downloading are supported.

IEEE 802.15.4b, i.e., 802.15.4-2006, the basis for the ZigBee specification, adds enhancements and corrections to IEEE 802.15.4-2003. Major modifications are reducing unnecessary complexities, increasing flexibility in security key usage, and supporting additional frequency bands in various countries. IEEE 802.15.4a [20] provides enhanced resistance to multipath fading with very low transmit power. In order to alleviate the problem, two alternative PHYs were developed. One is to use an ultra wideband (UWB) impulse radio operating in the unlicensed UWB spectrum (i.e., sub-GHz or below 1GHz, 3–5GHz, and 6–10GHz) to increase the precision ranging capability to an accuracy of one meter or better. Another one is to employ chirp spread spectrum (CSS) in the unlicensed 2.4GHz ISM band to support long-range links or links for mobile devices moving at high speed by adopting the unique windowed chirp technique in order to enhance robustness and mobility. The CSS method outperforms 802.15.4b (250Kbps), 802.15.3 (22Mbps), 802.15.1 (1Mbps), and 802.11b (1, 2, 5.5, 11Mbps) operating in the 2.4GHz ISM band. Moreover, IEEE 802.15.4c defines an alternate PHY in addition to those in IEEE 802.15.4b and IEEE 802.15.4a to support one or more of the Chinese 314–316MHz, 430–434MHz, and 779–787MHz bands.

It also provides modifications to MAC needed to support the associated PHY. IEEE 802.15.4d specifies alternate PHYs for the Japanese 950MHz band, and modifies MAC to support the new frequency allocation. By the time of this publication, IEEE 802.15.4-2011 which is a revision of the 2006 version will be published as a single document to consolidate the previous three amendments (i.e., 2 PHYs and 1 MAC) in order to avoid inadequacies or ambiguities discovered in the earlier standards. Table 4.3 provides detailed information on specifications for IEEE 802.15.4b/a/c/d.

IEEE 802.15.5 provides an architectural framework enabling WPAN devices to promote interoperable, stable, and scalable wireless mesh topologies. The features include the extension of network coverage without either increasing the transmit

Table 4.3 PHY Specifications in IEEE 802.15.4a, b, c, and d

Standard	Year	Frequency band (MHz)	Data Rate (kb/s)	Chip Rate (kchip/s)	Bit-Symbol Ratio	Channel Bandwidth (MHz)	Number of Channels	PHY (DSSS) with Modulation Employed
15.4b	2006	868–868.6	20, 100, 250	300, 400, 400	1, 4, 20	<1	1(0)	BPSK, O-QPSK, ASK
		902–928	40, 250, 250	600, 1600, 1000	1, 5, 4		10(1–10)	BPSK, PSSS-ASK, O-QPSK
		2400–2483.5	250	2000	4	2	16(11–26)	O-QPSK
15.4a	2007	250–750 3244–4742 5944–10,234	851 (mandatory); 110, 6810, 27,240 (optional)	–	–	500	1(0) 4(1–4) 11(5–15)	BPM-BPSK
		2400–2483.5	1000 (mandatory), 250 (optional)		6, 1.5		14	CSS-DQCSK
15.4c	2009	779–787	250	1000	4	–	4(0–3)	O-QPSK/MPSK
		868–868.6, 902–928	20, 40	300, 600	1		4(4–7), reserved	BPSK
15.4d	2009	950–956	20/100	300/–	1	–	10(0–9)/12(10–21)	BPSK/GFSK
		2400–2483.5	250	2000	4		Reserved	O-QPSK

power or receiver sensitivity, enhanced reliability via route redundancy, easier network configuration, and longer battery life on devices. Lee et al. [21] discussed issues of addressing and unicast/multicast routing. They further investigated the mesh routing in HR-WPAN supporting QoS by using hierarchically logical tree and address blocks. Solutions of energy saving from asynchronous and synchronous aspects as well as the support of portability for mobile devices in LR-WPAN were further presented. Other WGs in progress [17] include the following:

- *TG4e* enhances and adds functionality to the IEEE 802.15.4-2011 MAC. The improvement will support the industrial markets and permit compatibility with modifications being proposed within the Chinese WPAN. It will further enable various application spaces including factory/process/building automations, asset tracking, home medical health/monitor, and telecommunications applications.
- *TG4f* works on the specifications of an active RFID (RF IDentification) tag device. Such a device is typically attached to an asset or a person with a unique identification. It acquires the ability to produce its own radio signal by employing ambient energy harvested from the surrounding environment.
- *TG4g* is creating a PHY amendment to IEEE 802.15.4-2011 to provide a globally fundamental standard for the smart utility neighborhood (SUN) network operating in the 700 MHz–1 GHz and 2.4 GHz ISM bands with data rates supported in between 40 kbps and 1 Mbps. The associated IEEE 802.11ah developing standard, which will define the use of frequencies below 1 GHz for WiFi networks, has been considered as a direction for the IEEE 802.15.4g participants.
- *TG6* is developing a standard optimized for very-low-power devices worn on/around or implanted in human/animal bodies to serve a variety of applications for medical purposes and others.
- *TG7* introduces a new communications technology that is different from the traditional RF technology and uses visible light having wavelength between ~400 nm (750 THz) and ~700 nm (428 THz). This technology has mainly been tested in restricted areas such as aircraft, spaceships, and hospitals. Moreover, the group is also looking into the future of LED (light-emitting diode) evolution for applications of illumination, display, ITS, and others, in the interest of its attractive potential for environmental protection, energy saving, and efficiency.
- *IGthz* intends to explore the feasibility of the terahertz frequency band roughly from 300 GHz to 3 THz for wireless communications.
- *WNG* is charged by the IEEE 802.15 Wireless Next-Generation standing committee to facilitate and stimulate presentations and discussions on new wireless related technologies within the defined scope.

Notably, some of these standards will be completed by early 2012; meanwhile, it can be foreseen that more Working Group activities will be formed to address related issues with respect to the IEEE 802.15 standard. Readers are referred to Reference [17] for the corresponding updates. Nonetheless, among the aforementioned standards, IEEE 802.15.3 HR-WPAN and IEEE 802.15.4 LR-WPAN are the most promising technologies to support smart grid applications with various bandwidth requirements.

SUN, specified by the IEEE 802.15.4g Task Group (TG4g), has been developed to tackle a number of technical challenges in communications systems for the utility operators: (1) how to manage high volumes of metering data and control messages among a large number of meters/sensors (or nodes) in SUN networks throughout the AMI, and (2) how to establish self-configuring and self-healing utility networks in an efficient and cost-effective manner. The legacy IEEE 802.15.4 has been amended to provision the PHY (by TG4g) and MAC (by the IEEE 802.15.4e Task Group or TG4e) layer requirements in the SUN design. Three modulation formats in the PHY layer proposal are the multirate frequency shift keying (MR-FSK), multirate orthogonal frequency division multiplexing (MR-OFDM), and multirate offset quadrature phase shift keying (MR-OQPSK) [59]. Depending on different regions and network requirements (e.g., dense urban areas versus distant rural locations), various modulation modes, data rates, bandwidths, and channel spacing must be adaptively configured and allocated. The primary issue in SUN is coexistence with homogeneous and heterogeneous systems, especially in sharing the same network resources. Utilizing sub-GHz frequency bands (i.e., license-exempt bands below 1 GHz)[1] [59] as well as facilitating multi-PHY management (MPM) with the common signaling mode (CSM)[2] [60] is a foreseeable solution to signal interference in SUN networks.

In addition to the proposals of the state-of-the-art PHY schemes for SUN, most of the MAC protocols specified in IEEE 802.15.4 are adopted for SUN only with minor changes. Therefore, we will review a number of issues and challenges in the following section that have been addressed based on recent LR-WPAN studies predominantly in MAC designs. The survey on network measurements in the legacy IEEE 802.15.4 protocol will provide useful collation for investigation of SUN networks research.

4.4 RESEARCH EFFORTS ON THE LEGACY IEEE 802.15.4 WIRELESS PERSONAL AREA NETWORK

4.4.1 LR-WPAN Studies and Challenges

IEEE 802.15.4 LR-WPAN essentially employs the TDMA (time division multiple access) method along with the CSMA-CA (carrier sense multiple access with collision avoidance) medium contention mechanism for its network operation. Meanwhile, it mainly adopts DSSS (direct-sequence spread spectrum) rather than FHSS for various modulation schemes in order for battery-powered (or power-constrained) devices to save considerable energy as well as lengthen the network longevity.

Network topology control and traffic engineering involve a number of nodes connected in a network, placement of the nodes, data generation from the nodes, and

[1]The frequency bands allocated in domains/countries for SUN are 470–510 MHz (China), 863–870 MHz (Europe), 902–928 MHz (United States), 950–958 MHz (Japan), and 2.4–2.4835 GHz (worldwide).

[2]The CSM (a PHY signaling specification, proposed by TG4g [18]) is able to bridge the negotiations among networks with different PHY layer designs (MR-FSK, MR-OFDM, and MR-OPQSK) to mitigate the inter-PHY interference.

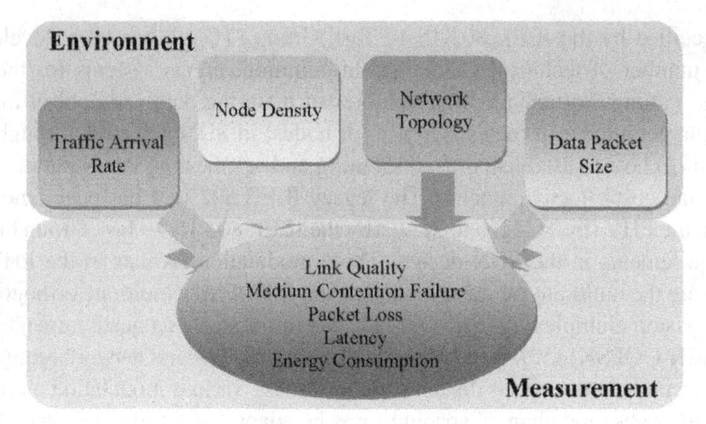

Figure 4.2 The influential network metrics and measurements.

accumulated loads throughout a network. Figure 4.2 depicts the key elements dependent on different scenarios that predominantly determine the network performance. Each of them can have tremendous and consequential impacts on

- frequency of wireless medium contention;
- success ratio of data delivery, which can be degraded by collisions from regular CSMA contention and hidden node transmission, congestions due to heavy traffic loads, as well as losses and drops due to the inherent wireless deterioration and buffer overflow;
- latency induced by unnecessary delayed transmission from the exposed node problem, a clumsy increase in MAC CSMA backoff periods, and inflexible routing design;
- rate of energy depletion affected by the duty-cycle arrangement as well as data aggregation and fusion mechanisms.

4.4.1.1 Wireless Impairments

Background noise and signal attenuation are inevitable in wireless transmission environments. Along the propagation path, losses attributed to refraction, diffraction, and scattering are likely to occur; multipath further incurs fading effects in mobile conditions. Several well-known path-loss models including free space, two-ray ground, and log-distance path have been extensively used to derive radio channel effects in RF-based wireless propagation network testing and experiments [22].

Interference is a critical phenomenon that could also degrade the overall performance in LR-WPANs. It is predominantly dependent upon the power transmission range, distance between transmitters and receivers, as well as orientation of antennae. Specifications in PHY such as ED (receiver energy detection)[3] within the current channel, LQI (link quality indicator) for received packets and channel frequency

[3]ED is similar to the so-called receive signal strength indicator (RSSI) employed in IEEE 802.11.

selection, as well as CCA (clear channel assessment) for CSMA-CA, are principal attributes to tackle the effect of interference. Analyses of multichannel, overlapping channel, and cross- or interchannel interference models and solutions for various scenarios can be found in [23–25].

4.4.1.2 Data Packet Length and Data Delivery

The PHY payload (i.e., PHY service data unit or PSDU) in IEEE 802.15.4 is limited to 127 bytes[4] [18]. Excluding the control bytes, the application payload carrying useful information is approximately 100 bytes. The payload is further reduced to a range of 60 bytes and 80 bytes if routing and security parameters are appended to the PHY overhead [19]. In other words, more than one-third of the space of each transmitted data packet in IEEE 802.15.4 is occupied for the control overhead purpose. Issues addressed in LR-WPAN entail (1) the use of bandwidth efficiently in transmission of small-size data packets to attain high goodput, and (2) the appropriateness of data packet size along with useful information to achieve low delay and low packet-loss rate in HAN and SUN. Meanwhile, security and privacy protection mechanisms supported in the data packets brings further challenges, especially when energy saving is required.

Furthermore, the network topologies supported in LR-WPAN are star, tree, and mesh. The routing schemes may include source routing (up to 5 hops), tree routing (up to 10), mesh routing (up to 30), and broadcasting (up to 30 as well) [19]. Choosing an appropriate scheme or hybrid among these routing protocols is critical for different networking environments. For example, tree routing might surpass mesh routing when a network becomes denser and crowded; source routing may not be valid in multihop transmission once beyond *five* hops when considering energy consumption, routing table size, and additional overhead factors. Apparently, data aggregation performs a key role in LR-WPAN.

4.4.1.3 CSMA-CA Contention Collision (CC)

The unslotted CSMA-CA channel access mechanism in IEEE 802.15.4 usually works well when the node density is sparse and nodes are more uniformly distributed in the network. Once beyond a certain boundary, the overall network performance can be degraded dramatically as more nodes are contending for the same medium. This results in transmission failures, leading to large backoff periods, and finally connection terminations. Each node in LR-WPAN specifies three variables for each transmission attempt:

- *NB:* the number of times that CSMA-CA is required to backoff; it is denoted by *macMaxCSMABackoffs*: 0–5 (default = 4).
- *BE:* a backoff exponent that is used to calculate the backoff period (i.e., $0 \sim (2^{BE} - 1)$) a node shall wait before attempting to access a channel; it is denoted by *macMaxBE*: 3–8 (default = 5); *macMinBE*: 0 ~ *macMaxBE* (default = 3).
- *CW:* the contention window length that represents the number of backoff periods in ensuring that a channel is free; it is denoted by CW: 0–2,

[4]TG4g has proposed the PHY frame size to be a minimum of 1500 bytes for SUN [59].

(default = 2). This variable is only used in the beacon-enabled operational mode (to be discussed in Section 4.2). Two successful clear channel assessments (CCAs) in a row are required before transmission. Otherwise, CW is always reset to 2.

Notably, the CW parameter in LR-WPAN is used differently than that in IEEE 802.11 WLAN. The difference is that CW in IEEE 802.11 can be doubled when network congestion occurs, and may be frozen if a packet loss is detected.

4.4.1.4 Hidden Node Collision (HNC) and Exposed Node Problem (ENP)

The hidden node problem (HNP) [34, 35, 37] is a well-known problem in wireless communications. The issue should be carefully addressed since the probability of having a hidden node in a wireless network can be as high as 41% [26]. The situation happens when two or more sending nodes outside the transmission range of each other transmit data to the same node in the next hop nearly at the same time, thus likely resulting in data collision at the receiving node. These sending nodes are hidden from each other because they are unable to detect the existence of one another. Unlike the RTS/CTS handshake schemes[5] used in IEEE 802.11 networks, LR-WPAN does not support the probing mechanism because of the inherent characteristics of energy saving and small data transmission. Instead, a few grouping techniques[6] proposed to mitigate HNP in LR-WPAN include pulse signal tactic and dynamic channel allocation mechanisms to prevent channels from being wasted [34, 35].

The exposed node problem (ENP) is the opposite of HNP that may also occur in WPAN and degrade the network performance [36]. In short, two or more sending nodes inside the transmission range of each other are actually allowed to transmit nearly at the same time without severely interfering one another, if their destined nodes in the next hop are outside the transmission range of each other. Both the HNP and ENP can be tackled by various power control management, which may, however, induce a trade-off between raising and reducing transmission power (as related to receiver sensitivity) at the transmitters and receivers. They have been challenges in WSN topology control as well as in LR-WPAN. For further studies, readers are referred to [37] for discussions on accumulated overheads attributed to contention collision from HNP, as well as consequent retransmission effects on the system performance in terms of MAC delay. In addition, Zhang and Shu [38] were among the first to study how to optimize the packet size in order to maximize the network resource and energy efficiency. Both research works used a cross-layer approach.

[5]RTS/CTS was originally presented in MACA [27], followed by MACAW (an enhancement to MACA) [28], and FAMA (an improvement over MACAW) [29]. RTS/CTS (as well as associated variants) is a probing mechanism used in IEEE 802.11 to reduce data collisions introduced by the HNP.

[6]In traditional WSNs, the main purpose of clustering is to perform data aggregation and collection of a large amount of correlated raw information collected from sensors [30–33]. The means of choosing a node as a cluster head are mostly based on the least cluster-head change, lowest ID, and highest connectivity. On the other hand, the grouping tactic proposed in WPAN here is to mitigate HNP. Both have the same goal of saving energy.

4.4.2 Demonstration Environments and Related Works

Research in WSN and IEEE 802.15.4 has been approached by the following means:

- Theoretical analysis and derivation based on mathematical methods, e.g., Markov chain, statistical and probabilistic models.
- Emulation devices and modules that produce empirical results, e.g., TinyOS, Tmote Sky, CC2420, MICAz motes.
- Simulation tools that run under both open-free and proprietary simulators and modelers, e.g., NS-2, OPNET, OMNeT++, MATLAB, NetSim.
- A hybrid that combines the theoretical analyses and derivation with validation of either emulation or simulation experiments.

As a result, many tools, simulation codes, and models have been made publicly available to the research community, for instance, NS-2 [39] in 2006, OPNET [40] in 2007, and OMNeT++ [41] in 2007.

Figure 4.3 illustrates an overview of PHY/MAC specifications as well as several principal research issues in IEEE 802.15.4. Variables specified in PHY/MAC LR-WPAN are critical elements for the network design. They determine how well the resources are allocated and utilized, how QoS is supported, and how interference and medium contention can be mitigated under different network scenarios. The issues of routing, fairness, security, and privacy on top of PHY/MAC in LR-WPAN are also addressed. The relevant contents are presented in the following.

4.4.2.1 *Operation Modes (Beacon-Enabled Versus Beaconless)*

The functionality of IEEE 802.15.4 LR-WPAN provides two modes of network configuration: beacon-enabled (B-E) using the slotted CSMA-CA channel access method, and beaconless (BL, i.e., beacon-disabled) adopting the conventional

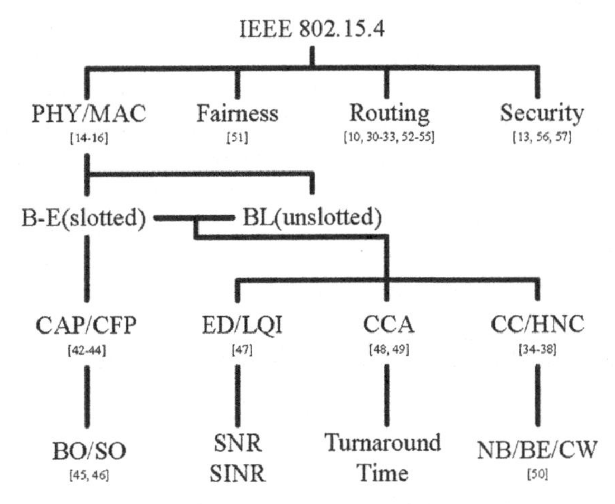

Figure 4.3 IEEE 802.15.4 PHY/MAC measurements and research challenges.

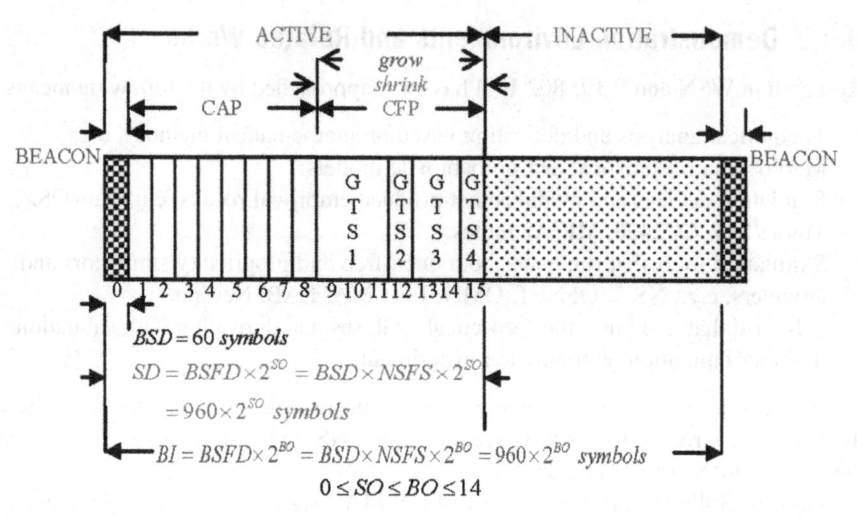

Figure 4.4 The superframe structure of IEEE 802.15.4.

unslotted CSMA-CA contention scheme. Despite two differentiated modes, the use of beacons is still required for network discovery. The B-E operation supports low-latency data transmission and synchronization. The WPAN coordinator regularly transmits periodic beacons so that nodes associated with its network are synchronized in the superframe structure bounded by two consecutive beacons. In other words, each beacon is transmitted in the first slot of each superframe.

As shown in Figure 4.4, the superframe is constructed by active and inactive portions defined by the superframe duration (SD) and beacon interval (BI), which are dependent upon the *aBaseSuperFrameDuration* (BSFD), superframe order (SO), and beacon order (BO).[7] Each node in WPAN enters a low power or sleep mode when the inactive portion is activated. The duty cycle is determined by the ratio of length of the active and inactive periods. The active portion is divided into *sixteen* equal time slots[8] including the first slot (i.e., slot 0) occupied by the network beacon. It comprises contention access period (CAP), which employs the unslotted CSMA-CA mechanism, and contention-free period (CFP), which defines a number of guarantee time slots (GTSs). Each time slot, denoted by *aBaseSlotDuration* (BSD), contains *sixty* symbols as defined in the standard. There are in total up to *seven* GTSs that can be allocated by the WPAN coordinator. Each GTS may occupy more than one slot period. Information such as starting slot, length, direction (i.e., transmit or receive), and associated node address are involved in GTS allocation and management. Each GTS is allocated on a FCFS (first come first serve) basis and released when it is not required. A node that has been allocated a GTS may still operate during CAP. GTS is activated and assigned when requested.

---

[7]BO is set to 15 when the BL mode is operated.

[8]The total number of time slots described in a superframe is denoted by the parameter *aNumSuperFrameSlots* or NSFS = 16, i.e., from slot 0 to slot 15.

In the slotted operation, each portion of a period is contiguous to each other among the beacon, CAP, and CFP (if GTS is required), as well as the inactive section (if the sleep mode is activated). There is always a slot boundary rule for each node to follow: a node begins to transmit on the next available slot boundary when the channel is idle. Otherwise, it allocates the boundary of the next backoff slot before it goes into the backoff stage. If the time between the next available backoff slot boundary and before the end of the active period is not enough for a node to complete its transmission, it may have to wait until the arrival of the next superframe. Hence, this will likely result in a higher delay. Readers are referred to [18] for detailed specifications for IEEE 802.15.4.

4.4.2.2 CAP and CFP (with GTS) Management

The design of CAP and CFP determines the QoS level of data transmission specified in smart grid applications. For example, the mission-critical messages required for small latency have a higher priority to be allocated with GTS resources for transmission than routinely collected noncritical data. Bhatti et al. [42] proposed an idea of swapping the position of CAP and CFP in order to grant the retransmission attempt of GTS to proceed in CAP of the same superframe upon a failed transmission in GTS. The means of shifting with a minor change in the superframe structure has favored GTS transmission. This results in a reduced latency under a small traffic rate while the overall MAC end-to-end delay still remains almost the same, according to their simulation results.

Park et al. [43] focused on time-critical data applications by proposing a GTS allocation mechanism based on the analysis of possible collisions in CAP[9] and interframe spacing (IFS) periods specified in the standard [18]. The authors validated their theoretical model with Monte Carlo simulations for GTS allocation based on Poisson, normal, and gamma distributions. They concluded that the GTS allocation mechanism likely results in lower throughput when BO becomes larger than 3 by reason of an increase in wasted bandwidth. GTS *requests* are likely to drop once the number of requests exceeds 6 and $BO \leq 1$. The overall maximum throughput for different number of GTS requests (i.e., 4–7 in this case) can be obtained when $BO = 2-3$, for a given duty-cycle rate. In their simulations, only uplink GTS transmission was considered.

Similarly, Ndih et al. [44] claimed to be the first to investigate service differentiation and traffic prioritization (i.e., two classes were defined) by analyzing CAP management under an unsaturated traffic condition. The authors have validated their mathematical derivation by using Markov chain analysis with a number of simulations. The idea is to allow nodes with high-priority data to start transmission once $CW = 1$ (i.e., only one successful CCA is required), and $CW = 2$ is still required for the rest of the nodes. However, this might bring up a fairness issue as low-priority data can be deferred forever once the number of competing nodes increases. Modified versions are desired to attain an acceptable boundary to adduce the circumstance. See Section 4.2.7 for discussion to alleviate the fairness problem.

---

[9]GTS *requests* can be collided in CAP with regular contention attempts from other nodes.

4.4.2.3 SO and BO Measurement

The length of SO and BO is considered in relation to the need for power saving on each device at the cost of transmission latency. If nodes in the network are not power-constrained, the SO value can be set equal to BO in order to perform data transmission and reception whenever needed. Chen et al. [45] analyzed several measurements with respect to the mean interarrival time (IAT) considering only a 1-byte data packet. The authors varied the lengths of CAP and CFP (only 1 GTS was assumed here) by changing the BO value while keeping SO = 0, and changing the SO value with BO = 8. Interesting phenomena were discovered from two experiments under a 3-node star topological environment (i.e., one PAN coordinator and two child nodes): (1) a traffic pattern with a larger mean IAT tends to induce higher end-to-end delay than that with smaller mean IAT under different combinations of 50% duty-cycle,[10] and (2) a higher ratio of BO to SO is likely to provoke higher end-to-end delay as well as packet loss. Consequently, it degrades the end-to-end goodput to some extent with varied IAT tested in their simulations.

Huang et al. [46] further emphasized the definition of duty-cycle as follows:

$$\text{duty-cycle} = \left(\frac{1}{2}\right)^{\text{BO}-\text{SO}}, \quad 0 \leqslant (\text{BO}-\text{SO}) \leqslant 14 \tag{4.1}$$

Interestingly, both (3,1) and (12,10) have the same 25% duty-cycle, but are constructed by different combination sets. The authors continued to state that given the same duty-cycle, a combination composed of larger BO and SO values tends to consume more power due to a longer duration of being active, whereas a combination composed of smaller BO and SO values is expected to have higher latency due to smaller time slots allocated in each short superframe. Besides, as the number of beacons increases, it may lead to a rise in power depletion once beyond a certain threshold.

Later, the authors also used Markov chain to analyze their MAC CSMA model. By setting the traffic load exponentially distributed and packet IAT lognormally distributed, they showed that the packet drop rate increases when the BO value is large (i.e., the active duration squeezes while SO is fixed, causing a buffer overflow) due to small buffer size and large variance in traffic load distribution. They concluded that the system goodput and energy consumption are dependent upon the queuing drop rate when the buffer size is limited. Otherwise, the performance outcome is more dominated by the traffic load distribution with a smaller variance while the buffer size is assumed sufficiently large.

4.4.2.4 ED and LQI Assessment

On the PHY layer perspective, determination of ED and LQI is a good approach to identify the radio condition. Gungor et al. [47] performed a number of field tests[11] with Tmote Sky sensor nodes in real-world power delivery and distribution systems by measuring

[10](BO, SO) can be equal to (BO, BO−1)=(1,0),(3,2),...,(11,10), and (13,12), which makes a 50% duty cycle for each combination set. The sets can have different performance outcomes although their duty cycles are the same.

[11]An outdoor 500 kV substation, an indoor industrial power control room, and an underground network transformer vault.

background noises, interference levels, and link quality. By using LQI and RSSI (or ED) metrics, they observed that (1) the background noise, which could be affected by variation in temperature and interference as well as time, turns out to be higher for the indoor scenario than the outdoor environment; (2) channel 26 in IEEE 802.15.4 is not influenced by IEEE 802.11b interference; (3) the interestingly interchangeable use of LQI and RSSI (or ED) is based on the received signal strength.[12] Finally, they stated that LQI is a reliable metric for WSN field operation according to their empirical measurements with respect to the closely correlated packet reception probability and loss rate.

A number of alternative solutions to interference have also been proposed. Sreesha et al. [16] adopted features of cognitive radio (CR) to WSNs by adding a geo-locator database presumably maintained by the FCC (Federal Communications Commission) that maintains a list of white spectrum spaces (the unused spectrums by primary users) in specific locations; meanwhile, modifying RPL (routing protocol in low power and lossy networks) in the network layer to support the proposed CR functionalities. Li et al. [15] suggested that WiFi-based WSNs with features of IEEE 802.11 should be adopted to support video monitoring and large-scale data collection and transmission applications in the smart grid. Moreover, Bilgin and Gungor [14] further proposed a multichannel access scheme in WSN for the wireless automatic meter reading system in the smart grid by using channel 25 and 26 in the IEEE 802.15.4 spectrum. Similar to [47], only the interference from IEEE 802.11b was compared in their simulations.

4.4.2.5 CCA Determination

CCA, as part of CSMA-CA, is performed to determine whether a specific radio channel is busy or idle prior to data transmission. Goyal et al. [48] identified that collision may occur under MAC during the receive-to-transmit (Rx-to-Tx) turnaround time (as well as Tx-to-Rx) even if a channel was initially detected as idle. They proposed a model for CSMA waiting time that is essentially performed before the CCA for every packet transmission attempt by analyzing the probability of CCA failure, collision per transmission, and packet loss as well as latency. Furthermore, Ergen et al. [49] proposed an adaptive MAC engine containing a collection of preset optimal protocols for different network conditions to avoid time spent on restarting the design process each time. In the study, they focused on unslotted random access and TDMA protocols using the two-dimensional Markov chain approach to analyze CCA performance as well as NB and BE variables. The authors substantiated the analyses by studying the healthcare SPINE case as well as the experiments implemented on the TinyOS and Tmote Sky testbed.

4.4.2.6 NB, BE, and CW Examination

The NB, BE, and CW variables are essentially adjusted based on different network environments to optimize network performance. Continuing from the work reported in [48], Rohm et al. [50] further suggested that the CCA duration should be larger

[12] LQI is a good estimator when the signal is found below and close to the sensitivity threshold, i.e., −94 dBm; otherwise, RSSI (or ED) is recommended.

than the Rx-to-Tx (as well as Tx-to-Rx) turnaround time (i.e., 12 symbols) in order to avoid collisions. Since parameters of NB and BE can be directly affected as a consequence of CCA, the authors discovered the following results based on their simulation outcomes with respect to the variation in network traffic loads:

- Adjusting BE becomes insignificant such that the situations of packet loss and consecutively failed CCA seem unimproved when the traffic load grows.
- Increasing BE may also increase the probability of collision due to rising HNP under heavy traffic.
- A value in the range between 5 and 6 for BE is shown to be the best trade-off between the packet-loss ratio and latency (up to 50ms).
- A reasonable packet latency (<30ms) and lower packet-loss rate can be obtained under the condition of 150 packets with 133-byte each per second and NB=7. Since the convention only supports the value of NB up to 5, a modification to the standard is required. The value is set back to 4 when the traffic load exceeds a derived threshold.

Although increasing the BE value seems to bring down the probability of packet loss, the subsequent cost of latency cannot be neglected. A robust and flexible mechanism is required to avoid the consequent collisions and retransmission failures once the backoff period in CSMA is extended, especially for a denser traffic network.

4.4.2.7 Fairness

In the power grid system, the mission-critical data are always superior to the noncritical ones. Fairness may be considered when data have the equal priority level during medium contention. The GTS allocation mechanism should be fairly designed for prioritized data in IEEE 802.15.4. In contrast to the traffic prioritization scheme proposed in [44], Huang et al. [51] addressed fairness by proposing an adaptive GTS allocation (AGA) determined by the success of GTS requests. AGA supports four traffic classes assigned to each node with a priority number. It avoids the starvation of light-traffic nodes where a node assigned with GTS slots may finish its transmission before the allocated GTS expires (and hence, resources are wasted). Basically, a node generating heavy or more recent data traffic is likely to have a higher probability of staying in a higher priority state. Meanwhile, the priority number assignment is determined by the modified multilevel AIMD (additive-increase and multiplicative-decrease), which is dependent upon the present traffic-level state of the node.

Fairness is met such that a node staying in a higher-level state with temporary transmission interruption will slightly be demoted to a lower state. On the other hand, a node in a lower-level state can be promoted to a higher state if a consecutive success of GTS requests is achieved. While this scheme works well particularly under heavy traffic, it is also able to dynamically change the bandwidth allocation by discontinuing the GTS mechanism and adopting the original CAP contention-based access once the traffic load becomes much lighter. The dynamic adoption tactic averts resource wastage and system complexity.

4.4.2.8 Routing Arrangement

While IEEE 802.15.4 does not specify any standards on top of PHY and MAC, the de facto AODV (ad hoc on-demand distance vector) routing algorithm in the network layer of LR-WPAN is essentially adopted in most research works. There have been a number of modified schemes based on AODV including AODV from Uppsala University (AODVUU), ad hoc on-demand multipath distance vector (AOMDV), rumor AODV (RAODV), and multitree-based routing, as well as an enhanced version of AODV using cluster techniques to alleviate the broadcast storm problem (i.e., flooding) [52–54]. Meanwhile, an energy-efficient routing strategy based on OLSR (optimized link state routing) along with the deterministic MAC layer was proposed in [10], which responds to the requirements specified in power generation industry.

With respect to power consumption, a hybrid routing scheme unifying *flat* and *hierarchical* multihop algorithms was proposed in [33]. The authors defined the so-called sink connectivity area (SCA) in which a set of nodes are within the maximum transmission range of the sink node. The nodes located inside the SCA are anticipated to deplete their power quickly because they act as relay nodes most of the time to convey a vast amount of sensed data received from the nodes outside the SCA. In order to minimize power consumption for the neighboring nodes of the sink, a hierarchy method with clustering is adopted in the network outside the SCA to reduce the number of data packets entering the neighboring nodes. In addition to these routing algorithms designed for the homogeneous network, new integrated routing techniques in supporting IPv6 via 6LoWPAN besides the data packet fragmentation and compression specified in RFC4944 need to be developed [55]. Likewise, innovative routing mechanisms for the foreseeable smart home energy implemented with smart meters and energy management units require further investigation in order to accommodate heterogeneity in smart grid HAN.

4.4.2.9 Security and Privacy

Security and privacy have been one of the primary concerns as data communications of the smart grid are open to the public as compared to the legacy power grid. Assurance of secure communications networks with privacy protection for energy profiles of consumers is essential. On account of the low computation capability and high overhead constraints in LR-WPAN, it becomes even more challenging to choose an appropriate suite of security features among a variety of cryptographic techniques. IEEE 802.15.4 only supports the basic 128-bit AES encryption[13] in the MAC layer. In addition to the fundamental CIAA (confidentiality, integrity, authentication, and availability) attributes as well as intrusion detection that must be assured in WSNs, issues of the limited number of access control list (ACL) entries (i.e., only 255 different keys can be stored) and lack of group keying (i.e., current pair-wise schemes may not be adequate) were identified [56]. Notably, security architecture for the smart grid WSNs was proposed in [13], specifying that security standards and testing/

[13]Types of encryption used in IEEE 802.15.4 include AES-CTR (AES-counter mode) encryption only, AES-CBC-MAC (AES-cipher block chaining message authentication code) authentication only, and AES-CCM (AES-counter with CBC-MAC) both encryption and authentication.

evaluation for both hardware and software need to be developed in order to support various smart grid applications from power generation/dispatch to utilization.

Potential attacks to LR-WPAN are similar to those that have been extensively studied in IEEE 802.11 and WSNs. Typical attacks at layers of the OSI model include jamming, tempering, tapping, compromised MAC, data modification, denial of service, man-in-the-middle, routing loops, Sybil/wormhole/sinkhole, and network flooding. They will require much attention once a variety of intelligent electronic devices in HAN integrate with IP architecture as well as with other wireless technologies.

Furthermore, privacy, as a vital part of security matters in smart grid communications from the customer perspective (through HAN), is comparable to patients' medical records in the home and hospital (through body area networks). Misic [57] proposed a modified cryptographic scheme for healthcare WSN by adopting elliptic curve cryptography (ECC). ECC is proven to be lightweight computationally and uses smaller key sizes for obtaining the same security level as compared to RSA (Rivest, Shamir, and Adleman). The mechanism can be further examined and explored for the scenario of the smart grid. In Table 4.4, we summarize the key parameters of the IEEE 802.15.4 protocol as well as issues of fairness, routing, and security/privacy on the performance of LR-WPAN.

Table 4.4 Network Design and Challenges in IEEE802.15.4-Based WSN

Layer	Design Criteria	Network Metrics	Challenges
Physical	Interference from using the 2.4 GHz ISM band; supporting range and number of nodes for acceptable performance	ED and LQI assessments	Coexistence with other existing networks (e.g., WiFi); interoperability and interconnection with heterogeneous networks (e.g., HomePlug, IPv6)
Medium access control	CAP and CFP arrangement based on density of nodes with data generation and QoS requirements	B-E and BL modes, GTS assignment, duty-cycle	Trade-off between hidden node and exposed node problems; power may be a limited resource
Fairness	Aggressiveness of medium contention and allocation request from nodes, and the corresponding compensation schemes for others	CAP and CFP decision, NB/BE/CW design	Adaptation of GTS allocation as well as backoff mechanism for efficient medium utilization
Routing	Data collection and data fusion based on different applications; clustering mechanisms	Decided by designers	Capability of nodes due to constrained resources and functions supported in the sensor devices
Security and Privacy	Lightweight encryption and intrusion detection algorithms; security mechanism for the WSN communications infrastructure	Decided by designers	Limited resources

4.5 CONCLUSION

This chapter drills into an in-depth survey focusing on IEEE 802.15.4 LR-WPAN-based WSNs. WSN is an essential component of the communications infrastructure in supporting a variety of applications and management functions in the smart grid. Typical systems such as SCADA, PMU, RTUs, AMI, and smart metering require intelligent monitoring and control under an advanced infrastructure of WSNs. Smart grid applications characterized by video and audio streams, still images, and scalar data can be provisioned in LR-WPAN (IEEE 802.15.4 based) and HR-WPAN (IEEE 802.15.3 based) in accordance with bandwidth and latency requirements. The utilization of frequency bands below 1 GHz and multi-PHY management proposed by Task Group IEEE 802.15.4g will create a new field of research for the design of smart utility networks. Adopting WiFi features, multichannel access, and cognitive radio (including TV white space) techniques in the legacy IEEE 802.15.4 technology will also enhance the capability of conventional WSNs and the efficiency of spectrum use.

Moreover, innovative mechanisms and models of integrating IP and other technologies with WSNs need to be developed to facilitate smart grid data communications and management. Notably, the WPAN technology might not be a suitable approach to HAN in some countries (e.g., parts of Europe) owing to its applicability and practice issues—this makes the PLC-based Homeplug Standard as well as broadband over PLC an alternative solution. The complementary strategy of combining WPAN with PLC technologies should be considered to provision the sensor applications in the various smart grid domains. Such approach to interoperability in both PHY and MAC layer protocols among different technologies will require extensive studies and investigation.

Future studies on IEEE 802.15.4 LR-WPAN-based WSNs for smart grid communications should focus on the following:

- Design of data prioritization related to specific applications and QoS requirements
- Adequacy of control (i.e., overheads) and data packet size (including commands)
- Schemes for multi-PHY management
- Assessment of communications link quality
- Innovation of MAC medium contention
- Flexibility of routing mechanisms
- Fairness issues upon adopted schemes
- Security/privacy models for protecting data and associated transmission

REFERENCES

[1] The Smart Grid: An Introduction, Book Publication, US Department of Energy, 2008. <http://energy.gov/oe/downloads/smart-grid-introduction-0>.

[2] C. Lo, N. Ansari, The progressive Smart Grid system from both power and communications aspects, IEEE Communications Surveys and Tutorials, in press.

[3] I.F. Akyildiz, T. Melodia, K.R. Chowdhury, A survey on wireless multimedia sensor networks, Computer Network 51 (4) (2007) 921–960.
[4] M. Erol-Kantarci, H.T. Mouftah, Wireless multimedia sensor and actor networks for the next generation power grid, Ad Hoc Network 9 (4) (2011) 542–551.
[5] B.T. Chorpening, D. Tucker, S.M. Maley, Sensors applications in 21st century fossil-fuel based power generation, in: Proceedings of Sensors, 24–27 October 2004, pp. 1153–1156.
[6] C. Saeli, C. Gatti, C. Gemme, How the smart grid and private network get a chance to communicate together? The integrated circuit breaker in line with IEC 61850, in: Proceedings of PCIC EUROPE, 7-9 June 2011, pp. 1–6.
[7] O. Asad, M. Erol-Kantarci, H. Mouftah, Sensor network web services for demand-side energy management applications in the smart grid, in: Proceedings of CCNC, 9–12 January 2011, pp. 1176–1180.
[8] N.A. Hidayatullah, B. Stojcevski, A. Kalam, Analysis of distributed generation systems, smart grid technologies and future motivators influencing change in the electricity sector, Journal of Smart Grid and Renewable Energy (SGRE) 2 (3) (2011) 216–229.
[9] S.T. Mak, Sensor data output requirements for smart grid applications, IEEE Power and Energy Society General Meeting, 25–29 July 2010, pp. 1–3.
[10] A. Agha et al., Which wireless technology for industrial wireless sensor networks? The development of OCARI technology, IEEE Transactions on Industrial Electronics 56 (10) (2009) 4266–4278.
[11] I.F. Akyildiz, T. Melodia, K.R. Chowdhury, Wireless multimedia sensor networks: applications and testbeds, Proceedings of the IEEE 96 (10) (2008) 1588–1605.
[12] W. Wang, Y. Xu, M. Khanna, A survey on the communication architectures in smart grid, Computer Networks 55 (15) (2011) 3604–3629.
[13] Y. Wang, W. Lin, T. Zhang, Study on security of wireless sensor networks in smart grid, in: Proceedings of POWERCON, 24–28 October 2010, pp. 1–7.
[14] B.E. Bilgin, V.C. Gungor, On the performance of multi-channel wireless sensor networks in smart grid environments, in: Proceedings of ICCCN, 31 July–4 August 2011, pp. 1–6.
[15] L. Li, X. Hu, K. Chen, K. He, The applications of WiFi-based wireless sensor network in internet of things and smart grid, in: Proceedings of ICIEA, 21–23 June 2011, pp. 789–793.
[16] A.A. Sreesha, S. Somal, I.T. Lu, Cognitive radio based wireless sensor network architecture for smart grid utility, in: Proceedings of LISAT, 6 May 2011, pp. 1–7.
[17] IEEE 802.15 Working Group for WPAN. <http://ieee802.org/15/index.html>.
[18] IEEE 802.15 WPAN Std. <http://standards.ieee.org/getieee802/802.15.html>.
[19] D. Gislason, ZigBee Wireless Networking, Newnes, 2007.
[20] E. Karapistoli, F.N. Pavlidou, I. Gragopoulos, I. Tsetsinas, An overview of the IEEE 802.15.4a standard, IEEE Communications Magazine 48 (1) (2010) 47–53.
[21] M. Lee et al., Meshing wireless personal area networks: introducing IEEE 802.15.5, IEEE Communications Magazine 48 (1) (2010) 54–61.
[22] T.S. Rappaport, Wireless Communications: Principles and Practice, Prentice Hall.
[23] L.L. Bello, E. Toscano, Coexistence issues of multiple co-located IEEE 802.15.4/ZigBee networks running on adjacent radio channels in industrial environments, IEEE Transactions on Industrial Informatics 5 (2) (2009) 157–167.
[24] G. Xing et al., Multichannel interference measurement and modeling in low-power wireless networks, in: Proceedings of RTSS, 1–4 December 2009, pp. 248–257.
[25] S. Myers, S. Megerian, S. Banerjee, M. Potkonjak, Experimental investigation of IEEE 802.15.4 transmission power control and interference minimization, in: Proceedings of SECON, 18–21 June 2007, pp. 294–303.

[26] Y.C. Tseng, S.Y. Ni, E.Y. Shih, Adaptive approaches to relieving broadcast storms in a wireless multihop mobile ad hoc network, IEEE Transactions on Computers 52 (5) (2003) 545–557.
[27] P. Karn, MACA: a new channel access method for packet radio, in: Proceedings of ARRL/CRRL, 1990, pp. 134–140.
[28] V. Bharghavan, A. Demers, S. Shenker, L. Zhang, MACAW: a media access protocol for wireless LANs, in: Proceedings of ACM SIGCOMM, 1994, pp. 212–225.
[29] C.L. Fullmer, J.J. Garcia-Luna-Aceves, Floor acquisition multiple access (FAMA) for packet-radio networks, in: Proceedings of ACM SIGCOMM, 1995, pp. 262–273.
[30] H. Nakayama, Z.M. Fadlullah, N. Ansari, N. Kato, A novel scheme for WSAN sink mobility based on clustering and set packing techniques, IEEE Transactions on Automatic Control 56 (10) (2011) 2381–2389.
[31] R. Machado, N. Ansari, G. Wang, S. Tekinay, Adaptive density control in heterogeneous wireless sensor networks with and without power management, IET Communications 4 (7) (2010) 758–767 (special issue on vehicular ad hoc and sensor networks).
[32] H. Nakayama, N. Ansari, A. Jamalipour, N. Kato, Fault-resilient sensing in wireless sensor networks, Computer Communications 30 (11–12) (2007) 2375–2384.
[33] H. Nishiyama, A.E.A.A. Abdulla, N. Ansari, Y. Nemoto, N. Kato, HYMN to improve the longevity of wireless sensor networks, in: Proceedings of IEEE GLOBECOM, 6–10 December 2010, pp. 1–5.
[34] C.H. Kwon, R.J. Tek, K.H. Kim, S.H. Yoo, Dynamic group allocation scheme for avoiding hidden node problem in IEEE 802.15.4, in: Proceedings of ISCIT, 28–30 September 2009, pp. 637–638.
[35] J.H. Lee, J.K. Choi, S.J. Yoo, Group node contention algorithm for avoiding continuous collisions in LR-WPAN, in: Proceedings of CIT, 11–14 October 2009, pp. 69–74.
[36] D. Koscielnik, Influence of the hidden stations and the exposed station for the throughput of the LR-WPAN, in: Proceedings of ISIE, 5–8 July 2009, pp. 1190–1195.
[37] H.W. Tseng, S.C. Yang, P.C. Yeh, A.C. Pang, A cross-layer mechanism for solving hidden device problem in IEEE 802.15.4 wireless sensor networks, in: Proceedings of GLOBECOM, 30 November–4 December 2009, pp. 1–6.
[38] Y. Zhang, F. Shu, Packet size optimization for goodput and energy efficiency enhancement in slotted IEEE 802.15.4 networks, in: Proceedings of WCNC, 5–8 April 2009, pp. 1–6.
[39] J. Zheng, M.J. Lee, A comprehensive performance study of IEEE 802.15.4, Sensor Network Operations, Wiley-IEEE Press.
[40] P. Jurcik, A. Koubaa, M. Alves, E. Tovar, Z. Hanzalek, A simulation model for the IEEE 802.15.4 protocol: delay/throughput evaluation of the GTS mechanism, in: Proceedings of MASCOTS, 24–26 October 2007, pp. 109–116.
[41] F. Chen, F. Dressler, A simulation model of IEEE 802.15.4 in OMNeT++, in: Proceedings of FGSN, July 2007, pp. 35–38.
[42] G. Bhatti, A. Mehta, Z. Sahinoglu, J. Zhang, R. Viswanathan, Modified beacon-enabled IEEE 802.15.4 MAC for lower latency, in: Proceedings of GLOBECOM, 30 November–4 December 2008, pp. 1–5.
[43] P. Park, C. Fischione, K. Johansson, Performance analysis of GTS allocation in beacon enabled IEEE 802.15.4, in: Proceedings of SECON, 22–26 June 2009, pp. 1–9.
[44] E. Ndih, N. Khaled, G.D. Micheli, An analytical model for the contention access period of the slotted IEEE 802.15.4 with service differentiation, in: Proceedings of ICC, 14–18 June 2009, pp. 1–6.

[45] F. Chen, N. Wang, R. German, F. Dressler, Simulation study of IEEE 802.15.4 LR-WPAN for industrial applications, Wireless Communications and Mobile Computing 10 (5) (2010) 609–621.
[46] Y.K. Huang, A.C. Pang, H.N. Hung, A comprehensive analysis of low-power operation for beacon-enabled IEEE 802.15.4 wireless networks, IEEE Transactions on Wireless Communications 8 (11) (2009) 5601–5611.
[47] V.C. Gungor, B. Lu, G.P. Hancke, Opportunities and challenges of wireless sensor networks in smart grid, IEEE Transactions on Industrial Electronics 57 (10) (2010) 3557–3564.
[48] M. Goyal et al., A stochastic model for beaconless IEEE 802.15.4 MAC operation, in: Proceedings of SPECTS, 13–16 July 2009, pp. 199–207.
[49] S. Ergen, P.D. Marco, C. Fischione, MAC protocol engine for sensor networks, in: Proceedings of GLOBECOM, 30 November–4 December 2009, pp. 1–8.
[50] D. Rohm, M. Goyal, H. Hosseini, A. Divjak, Y. Bashir, Configuring beaconless IEEE 802.15.4 networks under different traffic loads, in: Proceedings of AINA, 26–29 May 2009, pp. 921–928.
[51] Y.K. Huang, A.C. Pang, H.N. Hung, An adaptive GTS allocation scheme for IEEE 802.15.4, IEEE Transactions on Parallel and Distributed Systems, 19 (5) (2008) 641–651.
[52] A. Shuaib, A. Aghvami, A routing scheme for the IEEE 802.15.4-enabled wireless sensor networks, IEEE Transactions on Vehicular Technology 58 (9) (2009) 5135–5151.
[53] H. Fariborzi, M. Moghavvemi, EAMTR: energy aware multi-tree routing for wireless sensor networks, IET Communications 3 (5) (2009) 733–739.
[54] S. Gowrishankar, S. Sarkar, T. Basavaraju, Performance analysis of AODV, AODVUU, AOMDV and RAODV over IEEE 802.15.4 in wireless sensor networks, in: Proceedings of ICCSIT, 8–11 August 2009, pp. 59–63.
[55] U. Payer, S. Kraxberger, P. Holzer, IPv6 label switching on IEEE 802.15.4, in: Proceedings of SENSORCOMM, 18–23 June 2009, pp. 650–656.
[56] R. Riaz, K.H. Kim, H. Ahmed, Security analysis survey and framework design for IP connected LoWPANs, in: Proceedings of ISADS, 23–25 March 2009, pp. 1–6.
[57] J. Misic, Enforcing patient privacy in healthcare WSNs using ECC implemented on 802.15.4 beacon enabled clusters, in: Proceedings of PerCom, 17–21 March 2008, pp. 686–691.
[58] V.C. Gungor, F.C. Lambert, A survey on communication networks for electric system automation, Computer Networks 50 (7) (2006) 877–897.
[59] C.S. Sum, H. Harada, F. Kojima, L. Zhou, R. Funada, Smart utility networks in TV white space, IEEE Communications Magazine 49 (7) (2011) 132–139.
[60] C.S. Sum, M.A. Rahman, C. Sun, F. Kojima, H. Harada, Performance of common signaling mode for multi-PHY management in smart utility networks, in: Proceedings of ICC, 5–9 June 2011, pp. 1–5.

CHAPTER 5

Smart Grid Communications Networks: Wireless Technologies, Protocols, Issues, and Standards[1]

Quang-Dung Ho, Tho Le-Ngoc

Department of Electrical and Computer Engineering, McGill University, 3480 University Street, Montreal, Quebec, Canada H3A 2A7

5.1 INTRODUCTION

Increased demands on electrical energy, incidences of electricity shortages, power quality problems, rolling blackouts, electricity price spikes, and environmental issues have urged many nations to enhance the efficiency and reliability of their existing power grids as well as to seek alternative sources of high-quality and reliable electricity. Those challenges and issues drive motivations for the modernized electric power network, or smart grid (SG). "Smart Grid is an automated, widely distributed energy delivery network characterized by a two-way flow of electricity and information, capable of monitoring and responding to changes in everything from power plants to customer preferences to individual appliances" [1]. It can monitor, protect, and automatically optimize the operation of its interconnected elements including central and distributed power plants, energy storage stations, transmission and distribution networks, industrial and building automation systems, end-user thermostats, electric vehicles, appliances, and other household devices. Primary objectives of SG are to allow utilities to generate and distribute electricity efficiently and to allow consumers to optimize their energy consumptions.

In addition to the incorporation of information and communications technology (ICT) into the electric power grid, extensive use of distributed energy resources (DER) is another key feature of SG. DER encompass a wide range of distributed generation (DG) and distributed storage (DS) technologies located close to where electricity is used (e.g., a home or business). DG technologies include internal combustion (IC) engines, gas turbines, microturbines, photovoltaic (PV) systems, fuel cells, wind power, and combined cooling heat and power (CCHP) systems. DS technologies include batteries, supercapacitors, and flywheels. The use of DER results in three major benefits: increased power quality and efficiency, enhanced customer

[1] This work was partially supported by the Natural Sciences and Engineering Research Council (NSERC) through a Strategic Grant and the Strategic Research Network Smart Microgrid Network (NSMG-Net).

reliability and energy independence, and significant environmental benefit. First, bringing sources closer to loads contributes to enhancement of the voltage profile, reduction of distribution, and transmission of bottlenecks and losses. Besides, a CCHP system that delivers both electricity and useful heat from an energy source such as natural gas can enhance the use of waste heat. Second, when there are failures or turbulence in the main power grid, customer loads are not affected since they can be automatically islanded and powered by local generators. Third, the use of renewable distributed energy generation technologies and "green power" such as wind, PV, geothermal, biomass, or hydroelectric power can also reduce CO_2 emission.

It is critically important to note that the key to achieve these potential benefits of SG is to successfully build up a smart grid communications network (SGCN) that can support all identified SG functionalities including advanced metering infrastructure (AMI), demand response (DR), electric vehicles, wide-area situational awareness (WASA), distributed energy resources and storage, distribution grid management, etc. [2].

This chapter first presents the layered architecture of SG with the emphasis on its communications layer or SGCN. Typical required bandwidth and latency of numerous traffic types that SGCN is expected to carry are studied. Advantages and disadvantages of wireless communications technologies that could be employed for SGCN are discussed in detail. Since the neighbor area network (NAN) is the most important segment of SGCN, its distinguishing characteristics and requirements are investigated. They help to identify desired features and performance criteria that serve as an indispensable guideline that allows network engineers to develop relevant networking protocols for NAN. Existing protocols that have been suggested for NAN are also surveyed to give motivation for future research issues. Finally, a number of interoperability standards developed for SG are summarized.

5.2 SMART GRID COMMUNICATIONS NETWORK (SGCN)

5.2.1 An Overview of SG

Figure 5.1 shows the overall architecture of SG, which can be decomposed into a power system layer and communications layer. The power system layer, as can be seen in Figure 5.1a, is an integration of various electrical power generation systems, power transmission and distribution grids, substations, microgrids, customers, and control centers. *Power generation* includes the facilities for generating power in central as well as distributed locations. *Power transmission grid* refers to the high-voltage network of electric cables used to take bulk power from generation facilities to power distributions facilities near populated areas. *Power distribution grid* is the process in which the high-voltage power is down-converted and disseminated to the consumers through a mesh network of cables reaching all the way to consumer premises. A *microgrid* (MG) is an integrated energy system consisting of DER and interconnected loads that can operate in parallel with the grid or in an intentional island mode. It is

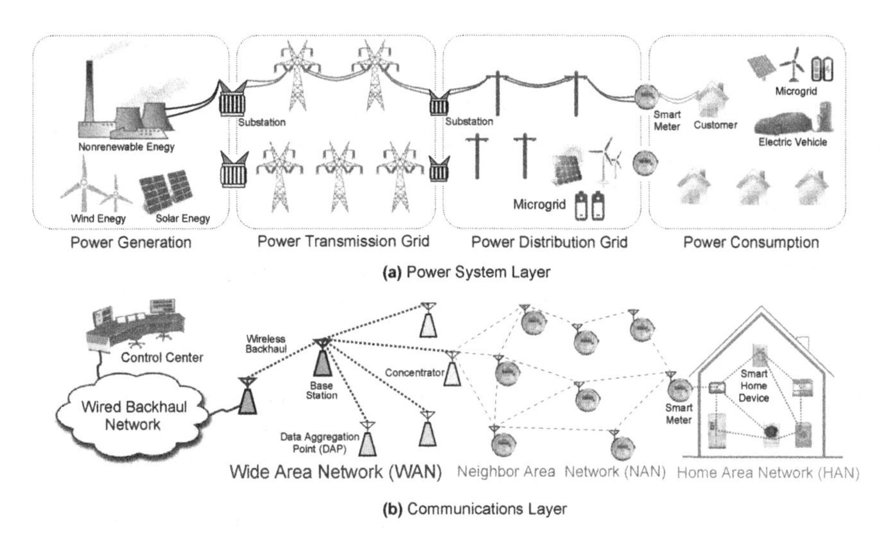

Figure 5.1 The overall layered architecture of SG.

introduced in SG in order to break the distribution system with a high penetration of DER down into small clusters for efficient control and management [3,4]. *Consumption* refers to either a private or commercial entity that consumes power from the distribution network. The *control center* is the central nerve system of the power system. It senses the pulse of the power system, adjusts its condition, coordinates its movement, and provides defense against exogenous events. The distinguishing feature of SG, compared to the existing electrical grid, is that those sub-systems are integrated with the supporting SGCN as shown in Figure 5.1b. SGCN is primarily responsible for monitoring, controlling, and automating the whole power grid. The integration of consumers and DER into the grid requires an automated, distributed, and secure control system of tremendous scale, with reliable, flexible, and cost-effective networking as the fundamental enabling technology.

5.2.2 Overall Architecture of SGCN

SGCN is typically partitioned into various segments including home area network (HAN), building area network (BAN), industrial area network (IAN), neighbor area network (NAN), field area network (FAN), and wide area network (WAN). This chapter however only considers three representative segments of SGCN, i.e., HAN, NAN, and WAN, as can be seen in Figure 5.1b. Those three segments are interconnected through gateways: a smart meter (SM) between HAN and NAN and a concentrator between NAN and WAN. The SM collects the power-usage data of the home by communicating with the home network gateway or functioning as the gateway itself. The concentrator aggregates the data from SMs and relays it to the grid operator's control center. Instructions for optimizing the power grid and user energy consumption can be sent from the control center to intelligent electronic devices (IEDs) and customer devices through WAN, NAN, and HAN in the opposite direction. The characteristics of each communications network segment are summarized as follows.

5.2.2.1 *Home Area Network (HAN)*

A HAN is used to gather sensor information from a variety of devices within the home, and deliver control information to these devices to better control energy consumption. Smart home devices (e.g., heaters, air conditioners, washing machines, dryers, refrigerators, kitchen stoves, dishwashers, chargers for electric car, etc.) can be monitored and controlled by a residential control center or consumers to optimize the power supply and consumption. HAN can support functions such as cycling heaters or air conditioners off during peak load conditions, sharing consumption data with in-home displays, or enabling a card-activated prepayment scheme. SMs installed in each residential unit work as communications gateways that relay information between HANs and NANs. Typically, HAN needs to cover areas up to 1000 sq ft and to support from 1 to 10 kbps.

5.2.2.2 *Neighbor Area Network (NAN)*

A NAN It is the communications network between SMs and concentrators of WAN. It collects information from many households in a neighborhood and relays them to WAN. NAN endpoints are SMs mounted on the outside of single family houses or on the roof of multiple dwelling units. The SM is a device that is at the heart of SG transformation. It records a user's electrical, water, or gas usage at a set interval, and then provides a way for this data to be read electronically. SMs also involve real-time or near-real-time sensing, power outage notification, and power quality monitoring. The number of SMs that one concentrator communicates varies from a few hundreds to a few thousands depends on the grid topology and the communications protocol used by NAN. NAN usually needs to cover areas of 1–10 square miles and to support from 10 to 100 kbps.

5.2.2.3 *Wide Area Network (WAN)*

A *WAN* aggregates data from multiple NANs and conveys it to the utility private network. It also enables long-haul communications among different data aggregation points (DAPs) of power generation plants, DER stations, substations, transmission and distribution grids, control centers, etc. The utility's WAN is responsible for providing the two-way network needed for substation communications, DA, and power quality monitoring while also supporting aggregation and backhaul for AMI and any demand response and demand-side management applications. WANs may cover a very large area, i.e., thousands of square miles, and could aggregate thousands of supported devices that require 10–100 Mbps of data transmission.

Each segment of SGCN may utilize different communication technologies depending on the transmission environments and amount of data being transmitted.

There are various candidate technologies including ZigBee/IEEE 802.15.4, Wi-Fi, 3G/4G Cellular, WiMAX, Z-Wave, HomePlug, Wireless M-Bus, Wavenis, Ethernet, Power Line Communications (PLC), etc. Section 5.3 summarizes a number of candidate wireless technologies that could be used for each segment of SGCN. Criteria used to assess the applicability of each technology include transmission rate, communications range, power consumption/network lifetime, network deployment and maintenance costs, network scalability, and standard availability.

5.2.3 QoS Requirements in SGCN

SGCN is expected to carry various types of traffic that require different quality of services (QoSs) in terms of bandwidth, latency, and reliability as addressed in [2, 5–7]. Table 5.1 summarizes typical bandwidths and delays required by a number of representative traffic types in SGCN.

Table 5.1 Typical QoS Requirements of Some Representative Types of Traffic in SGCN

Traffic Types	Descriptions	Bandwidth	Latency
Meter reads	Meters report energy consumption (e.g., the 15-min interval reads are usually transferred every 4 h)	Up to 10 kbps	2–10 s
Demand response (DR)	Utilities to communicate with customer devices to allow customers to reduce or shift their power use during peak demand periods	Low	500 ms to min
Connects and disconnects	To connect/disconnect customers to/from the grid	Low	A few 100 ms to a few minutes
Synchrophasor	The major primary measurement technologies deployed for WASA	A few 100 kbps	20–200 ms
Substation SCADA	Periodical polling by the master to IEDs inside the substation	10–30 kbps	2–4 s
Inter-substation communications	Emerging applications such as DER might need GOOSE communications outside substation	–	12–20 ms
Substation surveillance	Video site surveillance	A few Mbps	A few seconds
FLIR for distribution grids	To control protection/restoration circuits	10–30 kbps	A few 100 ms

Continued

Table 5.1 Typical QoS Requirements of Some Representative Types of Traffic in SGCN (*Continued*)

Traffic Types	Descriptions	Bandwidth	Latency
Optimization for distribution grids	Volt/VAR optimization and power quality optimization on distribution networks	2–5 Mbps	25–100 ms
Workforce access for distribution grids	To provide video and voice access to field workers	250 kbps	150 ms
Protection for microgrids	To respond to faults, isolate them, and ensure that loads are not affected	-	100 ms–10 s
Optimization for microgrids	To monitor and control the operations of the whole MG in order to optimize the power exchanged between the MG and the main grid	-	100 ms to min

5.2.3.1 Meter Reads

Energy consumption is usually read by SMs on a periodical basis and the associated traffic is predictable and has long latency requirements. For example, the 15-min interval reads are transferred to the utility every 4 h during the day or every 8 h at night. Primary factors that determine the bandwidth for meter reads are the size of reads and the frequency of reads. Each meter may need 10 kbps of data transmission and latency is in the range of 2–10 s.

5.2.3.2 Demand Response (DR)

DR technologies allow utilities to communicate with home devices such as load controllers, smart thermostats, and home energy consoles. DR is essential to allow customers to reduce or shift their power use during peak demand periods. It simply sends a shut-off command to an appliance and thus its bandwidth requirement is quite low. Estimates of the latency requirements of DR fall into a wide range, from as little as 500 ms, to 2 s, up to several minutes. Certain iterations of DR may be considered mission-critical in that failure to reduce energy use will lead to a system overload situation. If DR is used as a load balancing tool, however, the responsiveness of the system may not be critical, and thus latency could be higher.

5.2.3.3 Connects and Disconnects

Connects and disconnects may happen in various cases that require different response times. In the case when the customer moves, this kind of signal can tolerate long latencies. However, in the case when the connect/disconnect operation is used as a response to grid conditions, it may require a few hundreds of milliseconds of latency.

5.2.3.4 Synchrophasor (Synchronized Phasor Measurements)

Synchrophasor is one of the major primary measurement technologies being deployed for WASA. It refers to the implementation of a set of technologies designed to improve the monitoring of the power system across large geographic areas so there can be an efficient response to power system disturbances and cascading blackouts. Synchrophasor traffic has varying levels of latency requirements, ranging from 20 ms to 200 ms depending on the applications. The required bandwidth is a few hundreds of kbps and it is determined by the number of phasor measurement units, word length, number of samples, and frequency.

5.2.3.5 Substation Supervisory Control and Data Acquisition (SCADA)

A SCADA system handles the traffic generated when the master periodically polls IEDs inside the substation. The required bandwidth depends on the number of polled devices and it is forecasted to be around 10–30 kbps. The latency requirement is typically from 2 s to 4 s.

5.2.3.6 Inter-Substation Communications

Communications within a substation are based on Generic Object Oriented Substation Event (GOOSE) and require latencies that could be as low as 4 ms. Emerging applications such as DER may need GOOSE communications between multiple substations. The latency requirement for inter-substation communications ranges from 12 ms to 20 ms.

5.2.3.7 Substation Surveillance

Surveillance applications require high bandwidths up to a few Mbps, especially for video surveillance. The primary factors that determine the required bandwidth are the number of cameras and resolutions of the videos. This kind of traffic can tolerate higher latencies of a few seconds.

5.2.3.8 Fault Location, Isolation, and Restoration (FLIR) for Distribution Grids

For protection circuits required for commercial customers, this service has very low latencies in the range of a few milliseconds. The bandwidth required by FLIR depends on the complexity of the circuits and the number of steps of communications involved before the fault can be isolated. The bandwidth Typically, it varies from 10 kbps to 30 kbps.

5.2.3.9 Operation Optimization for Distribution Grids

Optimization is the service that deals with the Volt/VAR optimization and power quality optimization on the distribution network. It becomes more important in DER scenarios. This kind of service may generate 2–5 Mbps of traffic and request 25–100 ms of delay bound.

5.2.3.10 Workforce Access for Distribution Grids

Workforce access is the service that provides expert video and voice access to field workers. The service also enables the workforce to access local devices. It typically requires 250 kbps of bandwidth and 150 ms of latency.

5.2.3.11 Protection for Microgrids

Protection is a response to faults originating from the main power grid or microgrid. The faults are isolated to ensure that loads are not affected. Typical tolerable latency is from 100 ms to 10 s.

5.2.3.12 Operation Optimization for Microgrids

A MG control center monitors and automates the operations of the whole MG in order to optimize the power exchanged between the MG and the main grid. Latency requirements of this operation vary from 100 ms to a few of minutes.

It is noted that in addition to bandwidth and latency requirements, SGCN will need to be robust and secure. High network availability is absolutely critical along with predictable subsecond convergence for any failures. The network should have a degree of fault tolerance for increased resiliency and should support self-recovery. The estimated reliability varies from 99% to 99.99% [5]. Besides, the network should support a secure end-end transport ensuring confidentiality, integrity, and privacy of the data for meeting North American Electric Reliability Corporation-Critical Infrastructure Protection (NERC-CIP) regulatory requirements.

5.3 WIRELESS COMMUNICATIONS TECHNOLOGIES FOR SGCN

5.3.1 HAN

Many different technologies are considered to be used in HAN technologies to transfer data to the control center for analysis and optimization such as ZigBee/IEEE 802.15.4, Wi-Fi, Ethernet, Z-Wave, HomePlug, Wireless M-Bus, Wavenis, etc. Wireless communications technologies are preferable choices due to their low cost and flexibility of infrastructure.

5.3.1.1 ZigBee/IEEE 802.15.4

ZigBee has emerged as the dominant standard for energy monitoring, control, and management networks. ZigBee is designed specifically for highly reliable, low-power and low-cost control, and monitoring applications [8]. It is built on top of the IEEE 802.15.4 radio standard [9], which defines the PHY and MAC layers. ZigBee can be used to embed wireless communications into virtually any smart energy or HAN device from SMs and climate controls to lighting ballasts and smoke and security alarms without the prohibitive cost and disruption of installing hard wiring. It enables devices to self-assemble into robust wireless mesh networks that automatically configure and heal themselves and enable many individual devices to work for years on battery power alone. Moreover, ZigBee compliance not only offers a wireless control technology for smart energy, home automation, commercial building automation, and a wide variety of other sensing and monitoring applications, but ZigBee has now emerged as the preferred standard for HAN applications as well. The ZigBee standard took a big step forward for HAN energy management with the introduction of the ZigBee Smart Energy Profile (SEP) [10]. The SEP defines the standard behaviors of secure, easy-to-use HAN devices such as programmable thermostats and in-home displays. The SEP offers utilities a true open standard for implementing HAN communications. It also benefits consumers by allowing them to manage their energy consumption wisely using automation and near-real-time information, while having the ability to choose interoperable products from a diverse range of manufacturers.

5.3.1.2 Wi-Fi

Wi-Fi could be also suitable for HAN. It is based on very mature technology and has a large installed base in home networking. Wi-Fi devices are also capable of supporting low-data-rate/low-power applications as well. Wi-Fi is being included in a very wide range of portable and stationary consumer electronics devices, and its home market share will only increase. Wi-Fi operates in unlicensed spectrum, and it is therefore subject to interference. Fortunately, Wi-Fi is resilient to many types of interference and in fact coexists very well with other technologies that share these bands. Besides, Wi-Fi has a mature ecosystem and widely demonstrated interoperability. The Wi-Fi Alliance's certification program is the benchmark for all other wireless technologies. Hundreds of vendors deploy Wi-Fi technology in a wide range of devices. Ongoing technological innovations are bringing tremendous improvements to Wi-Fi power dissipation profiles and significant reduction in chip cost [11]. As a result, Wi-Fi could be a good candidate technology for HAN.

5.3.2 NAN

Typically, a wireless network is established among SMs of a cluster of houses in an area. The transmission range is usually more than 500 m and can potentially incorporate a multihop mesh approach. The protocols and standards considered for NAN are required to be reliable, secure, power efficient, low latency, low cost, diverse path, and scalable technology, and include the ability to support burst, asynchronous upstream traffic. There are quite a few wireless technologies in contention to be used to implement NAN such as IEEE 802.15 Smart Utility Network (SUN), Wi-Fi, 3G/4G cellular, and WiMAX.

5.3.2.1 IEEE 802.15.4g SUN

The IEEE 802.15.4g SUN standard specifies amendments to the IEEE 802.15.4 standard to facilitate very large-scale process control applications such as the utility SG network capable of supporting large, geographically diverse networks with minimal infrastructure with potentially millions of fixed endpoints [12]. Therefore, IEEE 802.15.4g SUN has been considered as one of the candidate technologies for NAN. It can support mesh topology and in fact, typical deployment scenarios for actual SUN implementations are based on this topology due to the fact that mesh networking can offer reliable access to/from the meters at a reasonable deployment cost. However, the detailed specification of the network topologies and routing protocols are out-of-scope of the IEEE 802.15.4g standard.

5.3.2.2 *Wi-Fi*

Municipal-scale Wi-Fi network infrastructure has already been deployed using 802.11 technology. This includes systems, for example, that provide access covering up to 500 m from the access point, interconnected by point-to-point links based on 802.11 technology and using proprietary mesh protocols. Modern municipal Wi-Fi networks typically also support 4.9 GHz access for public safety networks that are also based on 802.11 technology. Newer developments in the 802.11n standard, including support for transmit beam forming, may further enhance the use of Wi-Fi for these outdoor applications. Existing municipal 802.11-based networks are the appropriate scale for NAN. Wi-Fi can connect hundreds of devices on buildings and pole tops in a variety of terrains. The 4.9 GHz public safety application shows that Wi-Fi can be re-banded to support lightly licensed spectrums with different channel sizes. NAN might benefit from operating in lightly licensed spectrums similar to the 4.9 GHz spectrum that has been set aside for public safety applications in the United States [11]. Additional enhancements for NAN can come from work being done within the IEEE 802.11s Task Group to standardize a mesh networking protocol [13].

5.3.2.3 *3G/4G Cellular*

Cellular networks can be a good option for communicating between SMs and the utility. The existing communications infrastructure avoids utilities from spending operational costs and additional time for building a dedicated communications infrastructure. Cellular network solutions also enable smart metering deployments spreading to a wide area environment. 3G and LTE are the cellular communications technologies available to utilities for smart metering deployments. When data exchanges between SMs and the utility becomes more frequent to meet the requirements for emerging SG applications, NAN will need to carry a huge amount of traffic and a high data rate connection would be required.

5.3.2.4 *WiMAX*

WiMAX can transport application data from and to terminal devices that use an intermediary wireless interface, such as ZigBee or Wi-Fi. This is likely to be the case for many smart metering applications, at least initially, with meters transmitting data to concentrators that in turn are connected with WiMAX base stations. As volumes grow and prices decrease, WiMAX will become widely used as a module to connect SMs directly to the WiMAX network. This approach will enable the deployment of more advanced applications that require real-time control and wider bandwidth channels. Initially, SMs with WiMAX modules are more likely to be employed in rural, low-density areas, where WiMAX base stations can cover wide areas and may result in cost savings over the concentrator model. Besides, WiMAX satisfies all utility requirements: advanced metering, demand response, SCADA, distribution automation, voice, real-time outage, fault detection, and real-time service restoration.

5.3.3 WAN

Smart grid WANs may cover a very large, e.g., nationwide, area and could aggregate ten thousand supported devices. Therefore, multimegabit capacity will be required, and the links involved may range from subkilometer to multikilometer distances. WANs can be also implemented over fiber using Ethernet protocol. 3G/4G cellular, WiMAX, and Wi-Fi are most commonly recommended wireless technologies for WAN communications.

5.3.3.1 3G/4G Cellular

3G/4G cellular networks are the fastest and least-expensive way for electric utilities to deploy a WAN to monitor and control SG. Cellular network operators can wirelessly connect important grid assets such as breakers, sensors, remote terminal units (RTUs), transformers, and substations directly to the utility's operation center. This dramatically reduces the utility's up-front deployment costs and timeframes and leverages the cellular carriers' massive investments in network operations and maintenance to reduce the overall cost of ownership. WiMAX currently exhibits a higher bandwidth and a lower latency, compared to 3G cellular communications. However, with the imminent LTE deployment from multiple carriers, WiMAX might lose those advantages. Unlike WiMAX deployments, LTE will mostly reuse existing cellular networks and should be a straightforward evolution of the 3G cellular networks. Due to the popularity of cellular networks and the associated economies of scale that come with a large number of subscribers, 4G cellular is expected to be more economical and more widely available than WiMAX.

5.3.3.2 WiMAX

WiMAX can provide the backhaul link to the network operating center (NOC) or, more commonly, to the nearest or most cost-effective fiber connection. It fulfills the need for a secure, wide area broadband communications network for distribution and substation automation which otherwise is difficult to achieve by using other communications networks. Utilities may adopt proprietary networks or public networks in deploying SG applications. It is observed that utilities prefer to deploy private networks as there is a concern for data control, security, and reliability in public networks. WiMAX can be used as a wireless backhaul network that connects the utility control center with the AMI networks. However, WiMAX might lose its advantages when 4G cellular becomes widely available.

5.3.3.3 Wi-Fi

Existing citywide deployments of Wi-Fi networks demonstrate the clear applicability of Wi-Fi as a SG WAN technology. Minneapolis is just one example of a metropolitan installation in which Wi-Fi is used not only for neighborhood network access but in WAN backhaul portion of the system as well. Today such metropolitan area WANs incorporating standard 802.11 Wi-Fi in point-to-point or point-to-multipoint links embody a variety of proprietary network management approaches, demonstrating that Wi-Fi technology could be similarly incorporated into the future standardized SG management framework for WAN communication. A key advantage of Wi-Fi for WAN is its use of free, unlicensed spectrum. This makes it practical for a city or utility to own and operate a large private wireless network for SG. WiMAX and cellular data networks can provide the required service, but are usually owned and operated by large carriers who pay for the frequency licenses.

5.4 WIRELESS NETWORKING IN NAN

Among all the communications segments of SGCN presented in the previous section, NAN has been attracting the most concern from both academia and industry since it involves gathering a huge volume of various types of data and distributing important control signals from and to millions of devices installed at customer premises. NAN is the most critical segment that connects utilities and customers in order to enable primarily important SG applications including AMI, DR, distributed energy resources, and storage management, etc. Relevant and efficient networking protocols for NAN are the focus points of a large amount of research in the area of SG over the last few years. In this section, important characteristics and requirements of NAN are addressed. They help to pinpoint desired features and performance criteria that networking protocols developed for NAN must have. Those desired features and performance criteria in turn serve as an indispensable guideline that allows network engineers to select potential candidate routing protocols among existing ones that have been proposed for wireless ad hoc and sensor networks, to improve and to adapt those candidate protocols, or to propose new protocols that are most relevant and effective for NAN. Advantages and disadvantages of various wireless ad hoc and sensor network routing protocols in a NAN scenario are also investigated. Existing protocols that have been proposed for NAN are finally surveyed.

5.4.1 Characteristics and Requirements of NAN

NAN is basically a multihop wireless network; it therefore exhibits many characteristics that are commonly found in a general wireless ad hoc mesh network and wireless sensor network. However, due to the nature of its applications and its deployment environment, the design and development of NAN need to consider numerous distinguishing characteristics and requirements as follows.

First, by nature, this type of network needs to support a huge number of devices that distribute over large geographical areas. Current AMI networks have been designed with 1000–50,000 endpoints per DAP. Actual deployment quantity of endpoints per DAP will be based on the endpoint density and design limit thresholds imposed by the network designers to address application latency requirements and providing "headroom" in the network [14]. Applications for NANs should include the scope of future expansion and thus any routing protocol must be scalable to network size. Another desired feature of the routing protocol is self-configurability, i.e., a new node can join the network with minimal human intervention. The new node autonomously advertises its existence to other nodes, acquires information about surrounding nodes, and selects the best routes for data transmission.

Second, this network is heterogeneous and location-aware, i.e., network devices may differ in terms of processing capacity, power supply, transmission rate and power, etc., and their locations are usually fixed and known. NAN endpoints are SMs usually mounted on the outside of single family houses or on the roof of multiple dwelling units. SMs are usually implemented with limited processing capacity and radio transmission power. Routers and DAPs can be placed on top poles along distribution lines or in substation areas. They are mainly designed to have power supply coming from the grid and thus can have higher transmission power levels and communications ranges. However, they will be powered with their own batteries in the case of a power outage. Besides, almost all devices in NANs are mounted to fixed structures and are not relocated during their whole lifetime. As a result, their locations are known and unchanged after being installed. Networking protocols should take full advantage of this valuable location information. It is also important to note that mobility support is not required by NAN. Additionally, routing protocols should minimize energy consumption of nodes that are operating with batteries in order to maximize their lifetime and thus network lifetime.

Third, despite the fact that network nodes in NAN are static in their locations, the wireless links between those nodes dynamically change over time due to various factors. Multipath channel fading is the first and the most important factor because it may introduce significant variations to wireless link conditions. Next, surrounding environments (e.g., temporary structures, trees, moving cars or trucks, etc.) and harsh weather conditions (e.g., heavy rain, snow storms, etc.) may also fluctuate the link quality. A typical example is that a big truck can block a SM from a nearby router for a few hours or even longer. Besides, when there is a power outage, a number of network nodes have to lower their radio transmission power levels after switching their power supply to battery because of limited capacity. As a result, some links may render their connectivity. Therefore, in order to minimize packet loss and delay, NAN needs to have mechanisms to measure or estimate the instantaneous link quality and to be able to adapt well to any link connectivity change.

Fourth, devices of NAN are required to operate in outdoor environments without regular maintenance whereas their desired lifetime is 10 years or even 20 years. The routing protocol therefore must be self-healing. Given that the network is composed of thousands of nodes, it should be robust to the failures of a few links or nodes.

The network must be able to detect any kind of failures and to have a fast response to them. The traffic needs to be automatically rerouted to alternate paths if primary paths fail or detoured around regions having problems in order to minimize transmission corruption and delay fluctuation.

Fifth, NAN carries different types of traffic that require a wide range of QoSs in terms of transmission rate, delivery reliability, delay, and security. For example, energy consumptions at the customer premises are typically read every 15 min and that information is reported back to the utilities every 4 h (during day time) or 8 h (during night time). For each SM, this kind of traffic may need up to 10 kbps and can tolerate long delays of 2–10 s. On the other hand, critical missions of DR applications may only need to transmit short messages occasionally, however those messages require very high reliability (e.g., 99.99%) and very strict delay (e.g., 100 ms). As a result, QoS awareness and provisioning are important features that NAN must support.

Sixth, typical traffic carried in the network includes multi-point-to-point (MP2P) and point-to-multiple-point (P2MP). MP2P is the main traffic volume carried by NAN in which energy consumption, metering profile, customer's energy consumption preferences, power quality, outage detection warning messages, etc., collected by SMs at customer premises flow toward the DAP. P2MP refers to traffic from the DAP towards SMs. This traffic usually conveys real-time pricing information, demand response instructions, and other control messages from utility control centers to customers for optimizing user energy consumption, smoothing out power consumption peaks, and resolving various failures at distribution level. Point-to-point (P2P), which refers to traffic exchanged between any two nodes in the network, might not be very important in NANs.

Finally, since NAN conveys data for millions of customers for energy consumption and cost reduction as well as monitoring/controlling signals for distribution grid stabilization and optimization, it is very vulnerable to privacy and security issues. As the grid incorporates smart metering and load management, user and corporate privacy is becoming more critical. Electricity use patterns could lead to disclosure of not only customers' energy consumption habits but also their personal activities (e.g., staying home versus being away from home, sleeping versus watching television). Increases in power draw might suggest changes in business operations. Such energy-related information could support criminal targeting of homes or provide business intelligence to competitors [15]. Besides, the consequences of a successful denial-of-service (DoS) attack against power control systems could include being unable to control system components, resulting in unstable operations. Unless adequate local controls are in place for each element, local physical damage could take place on a grid-wide scale. This could result in significant financial losses, some cases in which interdependent systems cause cascade effects, and a loss of the very efficiency that SG is supposed to yield [16]. For those reasons, cybersecurity is another critical issue when developing networking protocols for NAN.

5.4.2 A Review on Wireless Routing Protocols

Over the last few decades, several routing protocols have been proposed and studied for wireless ad hoc and sensor networks [17–22]. They can be classified into different protocol families depending on underlying network structure (i.e., flat, hierarchical, and location-based routing) and protocol operation (i.e., multipath-based, query-based, negotiation-based, QoS-based, and coherent-based). Routing protocols can be proactive (each node actively collects current network status and maintains one or more tables containing routing information to every other nodes in the network) or reactive (routes are created as and when required by performing route discovery and selection procedures on-demand). In this section, key features, advantages, and disadvantages of a number of representative routing protocols (including flooding-based, cluster-based, location-based, and self-organizing coordinate protocols) are investigated in the light of requirements for NAN in order to facilitate the selection of candidate protocols for this network.

Flooding-based protocols enable P2P traffic patterns and rely on broadcasting data and control packets by each node into the entire network. In its conventional implementation, a source node sends a packet to all of its neighbors, each of which relays the packet to their neighbors, until all the nodes in the network (including the destination) have received the packet. Despite its simplicity, pure flooding suffers from many disadvantages including implosion, i.e., redundant copies of messages are sent to the same node by different neighbors or through different paths, and resource blindness, i.e., flooding lacks consideration for energy constraints of nodes when transmitting packets [23]. Flooding protocols are only particularly useful for P2P communications among a small number of mobile nodes without the need for any routing algorithm and topology maintenance. Those disadvantages render the application of flooding-based protocols to NAN infeasible.

Location-based protocols route the traffic based on the knowledge of a node's position together with those of its neighbors and the sink node. Greedy geographic routing (GEO) is the simplest form of geographic routing. When a node receives a message, it relays the message to its neighbor geographically closest to the sink [24]. Since geographic distance is not necessarily radio communication distance, the drawback of GEO routing is that the selection of next hop merely based on geographic distance may lead to void areas where the traffic cannot advance further toward the destination. More advanced location-based routing protocols that attempt to improve the delivery rate are proposed in [25–29]. The advantage of this kind of routing protocols is that it can achieve network-wide routing while maintaining only neighborhood information at each node, hence significantly reducing signaling overheads and the complexity of the routing solution. The fact that locations of nodes in NAN are fixed and accurately known promotes location-based protocols as one of promising solutions for this network.

Self-organizing coordinate protocols counteract the biggest drawback of geographic routing protocols by building a viable coordinate system based on communication distance rather than geographic distance. The aim of such coordinate systems in the context routing protocols is not to mimic geographic location but rather to be of use for feasible routing solutions. The Routing Protocol for Low Power and Lossy Networks (RPL) is a representative protocol that captures most of the ideas introduced by self-organizing coordinate protocols [30]. Advantages of RPL can be summarized as follows. First, RPL basically builds up a directed acyclic graph (DAG) whose structure matches the physical structure of NAN. Root nodes represent DAPs, leaf nodes represent SMs, and other nodes inside the DAG represent routers that maintain connectivity between root and leaf nodes. Second, MP2P and P2MP, which are typically required by NAN, are the primary communications supported by RPL over its DAG. Third, by employing different routing metrics and cost functions, RPL can construct multiple instances of DAGs over a given physical network. Each instance can be dedicated for a specific routing objective. This facilitates QoS differentiation and provisioning for different types of traffic that NAN needs to carry. Moreover, with the trickle timer that governs the network state update, RPL requires less signaling overhead and thus it is more energy-efficient. Finally, the roots of DAGs can act as trusted entities that enable security in the network.

Cluster-based protocols are based on a hierarchical network organization. Nodes are grouped into clusters, with a cluster head elected for each one. Data transmission typically goes from cluster members to the cluster head, before going from the cluster head to the sink node. Since cluster heads are responsible for relaying and processing a high volume of data, they typically have higher energy and computation capability. This kind of routing can support MP2P, P2MP, and P2P traffic. Clusters are built and maintained as a function of the various parameters of the nodes and system. Those parameters include node energy, link quality, traffic pattern, data correlations between nodes, etc. [31–37]. The drawbacks of this class of routing protocols are that it cannot capture the link dynamics, and head selection, cluster formation, and maintenance introduce significant signaling overhead. Besides, protocols like LEACH [36] and HEED [34] assume that time division multiple access (TDMA) and code division multiple access (CDMA) are used for intracluster and intercluster communications, respectively, and that nodes can tune their communications range through transmission power. Those assumptions make them nearly impractical for real deployment. Fortunately, splitting the network into smaller clusters efficiently limits the data flooding area. This offers benefits in scalability, lifetime, and energy consumption. Additionally, since nodes physically close to each other are likely to sense correlated events, data can be efficiently aggregated at the cluster head to reduce network load. Implementation of security is also easier since cluster heads can act as trusted entities in the network. Those advantages seem to be very attractive to NAN. First, NAN is naturally organized into multiple clusters. Each cluster serves a few thousands of SMs and data is managed by the DAP acting as the cluster head. Second, SMs that have some underlying correlations can be placed into the same cluster. For example, SMs located in the same distribution feeder may send similar notification messages at the same time when their feeder fails. If those messages are gathered by their cluster head, redundancy can be detected and resolved efficiently to minimize the network traffic volume while still assuring that no important information is lost throughout the network. Finally, the heads of each cluster can offer important security features that are required by NAN.

5.4.3 Existing Routing Protocols Proposed for NAN

Various routing protocols previously proposed for wireless ad hoc and sensor networks have been considered for NAN. They include GEO, RPL, and the IEEE 802.11s Hybrid Wireless Mesh Protocol (HWMP). Relevant modifications and improvements to those protocols have been also suggested to customize them for NAN. Following is a survey of literatures mostly related to those protocols.

Performance of GEO in various realistic smart metering scenarios is presented in [38]. Using simulations, received packet ratios given by the protocol are measured against network scales, offered traffic rates, and placements of routers and DAPs. For the small-scale scenario with 350 SMs, 2 routers, and 1 DAP, the simulations show that the system performs with a received packet ratio of 100% for a message frequency of 1 message per 4 h. However, success rate decreases with increasing message frequency due to collisions in some central nodes. For a large-scale scenario with 17,181 SMs, multiple routers and DAPs, an overall success rate of the system is observed with 99.99% for a message frequency of 1 message per 4 h. In this case, it is noted that there are some isolated zones due to coverage gaps. Geographical distributions of packet success rate and hop count are analyzed in order to determine the suitable number and placement of routers and DAPs that would result in an improved performance.

Performance of RPL implemented in an experimental platform using TinyOS is presented in [39]. The results in [39] indicate that RPL performs similarly to the Collection Tree Protocol (CTP), the *de facto* standard data collection protocol for TinyOS 2.x [40], in terms of packet delivery and protocol overhead. Compared to CTP, RPL can provide additional functionalities, i.e., it is able to establish bidirectional routes and support various types of traffic patterns including MP2P, P2MP, and P2P. Therefore, the authors in [39] conclude that RPL is more attractive for practical wireless sensing systems.

The work in [41] analyzes the stability of RPL whose DAG is built based on link layer delays. It is observed that the fluctuation in the delay introduced by the IEEE 802.15.4 MAC layer negatively influences RPL's stability. That fluctuation forces the nodes to change their best parent along the routing path so frequently that it results in significant end-to-end delay jitters. In order to damp the link layer delay fluctuation, the author proposes to introduce a memory in the delay calculation. Simulations of a small network demonstrate that the proposed solution can reduce the mean and variance of the end-to-end latency and thus improve the protocol stability.

The authors in [42] provide a practical implementation of RPL with a number of proper modifications so as to fit into the AMI structure and meet stringent requirements enforced by the AMI. In particular, the Expected Transmission Time (ETT) link metric and a novel ETX-based rank computation method are used to construct and maintain the DAG. ETX is measured by a low-cost scheme based on a MAC layer feedback mechanism. A reverse path recording mechanism to establish the routes for downlink communications (i.e., from gateways to end-devices) is also proposed. The mechanism is purely based on the processing of uplink unicast data traffic (i.e., from end-devices to gateways), and hence does not produce extra protocol overhead. Extensive simulation results in [42] show that, in a typical NAN with 1000 SMs, and in the presence of shadow fading, the proposed RPL-based routing protocol outperforms some existing routing protocols like Ad hoc On-Demand Vector (AODV) routing, and produces satisfactory performances in terms of packet delivery ratio and end-to-end delay.

Self-organizing and self-healing solutions for RPL are proposed in [43]. SMs are able to automatically discover DAPs in their vicinity and setup a single or multihop link to a selected DAP. A distinguishing feature in [43] is that DAPs may choose to operate at different frequencies in order to accommodate a scalable large network consisting of multiple trees. SMs perform channel scanning to detect DAPs and select the best one. Also, SMs can detect loss of connectivity arising from failed nodes/links/concentrator and automatically recover from such failures by dynamically connecting to an alternative concentrator in their vicinity. Numerous performance parameters of the proposed RPL are studied by simulations. They include DAP discovery latency and effects of DAP failures to packet delivery rate and recovery latency. The results in [43] have demonstrated that the proposed solution exhibits self-organizing properties and it is therefore appealing from a deployment perspective.

In [44], a simulation-based performance evaluation of RPL in real-life topology with empirical link quality data is presented. This study focuses on the mechanisms that RPL employs to repair link or node failures. Global repair is implemented with the help of periodic transmission of a new DAG sequence number by the DAG root. For local repair, upon losing parents, a node will try to quickly and locally find an alternate parent. Results in [44] show that when local repair mechanism is employed, the network fixes the local connection outage to parent much quicker than if using a global repair mechanism only. However, there are a few incidents where the outage time gets large to an order comparable with DAG sequence number period, mainly in cases when packet delivery ratio is low or when DIS or DIO is not heard for a long time. The behaviors and performance of those two mechanisms thus need further study and improvement for outdoor and large-scale networks like NAN.

The authors in [45] promote the use of IEEE 802.11s based wireless mesh network for NAN since that kind of network can provide high scalability and flexibility, along with low installation and management costs. The IEEE 802.11s HWMP combines two types of routing modes: proactive tree building mode and on-demand mode. In proactive mode, Root Announcement (RANN) messages are flooded by the gateway and mesh nodes in order to allow each node to maintain the best route to the gateway. When the path to the root is considered obsolete due to transmission failure, a new path is discovered before the RANN period using an on-demand path discovery algorithm identical to the traditional AODV routing. In [45], the airtime cost metric calculation method defined in the 802.11s standard is modified and the route fluctuation prevention algorithm is proposed in order to stabilize the route selection. Simulations in [45] demonstrate that modified HWMP can offer higher reliability and guarantee better end-to-end delay.

In [46, 47], a tree-based routing protocol that is an extension of the IEEE 802.11s HWMP is proposed for wireless mesh networks with multiple gateways. A heuristic backpressure-based packet-scheduling scheme capable of balancing the traffic load amongst the gateways is developed to maximize the overall throughput. A channel assignment scheme is also included to support multichannel aided routing in a distributed manner. Simulations in [46, 47] show that proposed schemes provide a significant gain in network performance in terms of delay and throughput and can enhance the self-healing ability of the network in case of node failures.

5.4.4 Open Issues in NAN

Despite the fact that there is a large amount of research dealing with networking for NAN over the last few years as presented in the above section, NAN is still facing many technical challenges that need further study. A number of the most important unsolved or not completely solved issues in NAN are addressed in the following.

5.4.4.1 Downlink Communications

It is observed that almost all existing studies only consider uplink communications (i.e., from end-devices to DAPs) while downlink communications (i.e., from DAPs to end-devices) are also important because they are responsible for conveying various types of high-valued data, e.g., real-time pricing information for user energy management, and critical controlling signals, e.g., demand response instructions for distribution grid optimization, from utility control centers to NAN end-devices. Downlink communications enable many advanced SG applications including demand response, distributed energy resources and storage management, etc. System performance in terms of delay and delay jitter of this type of communication needs to be investigated since many services are delay-sensitive and even require near-real-time or real-time responses.

5.4.4.2 QoS Differentiation and Provisioning

Another limitation of existing studies is that they assume a single traffic type in the whole network and no mechanism for QoS differentiation and provisioning has been addressed. This kind of mechanism is very important because in fact the network carries multiple types of traffic that require very much different QoSs. Certain types of traffic, e.g., meter readings, require a high transmission rate but can tolerate considerable delay while others for critical applications, e.g., demand response, do not need to use channel often, however, their latency and jitter requirements are very stringent. Furthermore, traditional QoS differentiation and provisioning approaches may need to be revisited to cater to SG traffic because traffic to be generated by SG applications will likely be quite different from that generated by traditional web browsing, downloading, or streaming applications that are in use today, with a mix of both real-time and non-real-time traffic being generated and distributed across different segments of the communications network.

5.4.4.3 *Network Self-Healing*

NAN is built to optimize the operations of the electric power system (EPS) consisting of SMs, smart electronic devices, substations of the distribution grid, etc. In normal operation modes, NAN only carries out routine procedures such as AMI, DR, etc. However, when one or more failures take place in the electric power system, a NAN must be able to cope with many emerging challenges in order to ensure that the failures can be localized and resolved. For example, when a feeder is cut or a distribution substation fails, a segment of the distribution grid will experience grid power outage. As a result, NAN devices within that affected segment will need to switch to battery power supply and reduce their communications range. This in turn results in abrupt and dramatic changes in the communications network topology. At the same time, those affected nodes may attempt to report to the utility the same warning messages, e.g., reporting the power outage to the control center. Consequently, the control center may wish to receive more detailed and frequent system status updates from end-devices or to send critical control signals to devices. That wave of such kind of incurred traffic may easily overload the network, NAN may thus be unable to work in its stable operating region and consequently the whole system cannot recover from the failures. Even worse, local failures may result in cascaded failures, large-scale fluctuations or catastrophic situations in the whole grid. For those reasons, NAN must have some mechanisms specially designed for handling such situations. Those mechanisms may relate to smart data mining and filtering, traffic prioritization, etc., in order to make sure that redundant information is suppressed and thus does not overload the network, so that critical messages can access the channel and follow the best routes to reach their intended destinations in a timely manner.

5.4.4.4 *Multicasting*

For optimizing user energy consumption and smoothing out power demand peaks, a time-of-use energy pricing policy is applied in SG. To facilitate this, utilities need to inform customers of the energy price at every predetermined period of time. It is important to note that the energy price does not only vary over time of the day but also depends on the geographic or grid management regions that customers belong to. In other words, at a given time, different customer groups may receive different offered prices in order to ensure that energy consumption in the whole grid is balanced and optimized. Obviously the most efficient way to deliver the same message to a group of users is multicasting (rather than unicasting). As a result, multicasting is of vital importance in NAN. Since wireless channel capacity is limited while the number of nodes in NAN is considerable, multicasting has to be done in such a manner that all intended receivers of the same multicast group can receive the message properly and the utilized network resource is minimal. To the best of our knowledge, there has been no study on this interesting research area in NAN.

5.4.4.5 *Cluster-Based Routing*

Despite the fact that there is numerous work dealing with the routing issue for NAN, cluster-based routing protocols have not been taken into account. In fact, NAN in nature has cluster-based network architecture, i.e., it consists of multiple DAPs and each is responsible for the communications with a group of SMs. The problem of how to associate a number of SMs to a given DAP and how traffic is routed in each cluster so that the communication is efficient and reliable is very interesting and practical. Many parameters need to be considered when forming clusters, including geographic distance from SMs to DAP, residue energy of each SM, link quality between each node pair, tolerable end-to-end reliability and latency, total traffic volume being carried by each DAP, etc. Correlation between SMs at electric power and information levels is a unique yet especially important feature in NAN and thus needs to be exploited. For example, SMs that tend to have a power outage at the same time due to the fact that they are in the same feeder need to be attached to the same cluster so that efficient data filtering can be done at the cluster head.

5.4.4.6 Network Design

NAN is an interconnection of a very large number of devices that generate and receive different types of data with various QoS requirements. How to design and provision a scalable and reliable network so that data can be delivered to intended destinations with satisfied QoSs is a challenging task. The work in [38] presents a real-life scenario where NAN is conveying traffic for thousands of SMs. Due to the considerable network scale, multiple routers, and DAPs with higher transmission rates and longer radio ranges compared to those of SMs, are used. Simulation results in [38] show that routers and DAPs can increase the overall packet transmission success rate by bridging traffic and thus reducing average hop count per packet. Unfortunately, systematic approaches used to determine the number of routers and DAPs and the places to install them in order to optimize the network performance while minimizing the cost are not studied in [38]. How to configure the parameters in the system for optimum functionality, and how to determine the optimal number and placement of network devices are among the most important issues to be addressed when designing and deploying NAN, which in nature consists of a very large number of nonhomogenous devices distributed in large geographical areas. Besides, approaches used to forecast network traffic characteristics introduced by advanced SG applications such as demand response, wide-area situational awareness, etc., need extensive study to support optimal network design.

5.5 SMART GRID STANDARDS

Interoperability can be defined as "the ability of two or more systems or components to exchange information and to use the information that has been exchanged" [48]. There have been many technological solutions and commercial products for SG, however, the key challenge is that the overall SG system is lacking widely accepted standards. This limits the interoperability between SMs, smart monitoring/controlling devices, renewable energy sources, and emerging advanced applications, and thus prevents their integration. The adoption of interoperability standards for the overall system is the most critical prerequisite for making an SG system a reality. Seamless interoperability, robust information security, increased safety of new products and systems, a compact set of protocols, and communication exchange are some of the objectives that can be achieved with standardization efforts for SG.

There are many regional, national, and international standards development organizations (SDOs) working toward this goal, e.g., National Institute of Standards and Technology (NIST), American National Standards Institute (ANSI), International Electrotechnical Commission (IEC), Institute of Electrical and Electronics Engineers (IEEE), International Organization for Standardization (ISO), International Telecommunication Union (ITU), etc. In addition to SDOs, there are a number of alliances that recognize the value of a particular technology and attempt to promote specifications as standards for that technology. For example, some well-known alliances related to the utility industry in the HAN market space are ZigBee Alliance, Wi-Fi Alliance, HomePlug Powerline Alliance and, Z-Wave Alliance [49]. The following section review the key roles and activities of NIST in developing standards for inter-operability of SG.

5.5.1 NIST and Its Activities on SG Standards

NIST has been assigned the "primary responsibility to coordinate development of a framework that includes protocols and model standards for information management to achieve interoperability of smart grid devices and systems…" (Energy Independence and Security Act of 2007, Title XIII, Section 1305). Its primary responsibilities include (i) identifying existing applicable standards, (ii) addressing and solving gaps where a standard extension or new standard is needed, and (iii) identifying overlaps where multiple standards address some common information. NIST has developed a three-phase plan to accelerate the identification of an initial set of standards and to establish a robust framework for sustaining development of the many additional standards that will be needed and for setting up a conformity testing and certification infrastructure.

NIST Framework and Roadmap for Smart Grid Interoperability Standards, Release 1.0 [50], is the output of Phase I of the NIST plan. It describes a high-level conceptual reference model for SG, identifies 25 relevant standards (and an additional 50 standards for further review) that are applicable to the ongoing development of SG, and describes the strategy to establish requirements and standards to help ensure SG cybersecurity. Release 2.0 of the NIST Framework and Roadmap for Smart Grid Interoperability Standards [51] details progress made in Phases II and III. Major deliverables have been produced in the areas of SG architecture, cybersecurity, and testing and certification. Release 2.0 [51] presents 34 reviewed standards (and an additional 62 standards for further review). The listed standards have undergone an extensive vetting process and are expected to stand the "test of time" as useful building blocks for firms producing devices and software for SG. Ongoing standards coordination and harmonization process carried out by NIST will ultimately deliver communication protocols, standard interfaces, and other widely accepted and adopted technical specifications necessary to build an advanced, secure electric power grid with two-way communication and control capabilities [51].

In addition to reviewing and selecting applicable standards for SG, NIST has another important contribution in identifying a set of Priority Action Plans (PAPs) for developing and improving standards necessary to build an interoperable SG. Those PAPs arise from the analysis of the applicability of standards to SG use cases and are targeted to resolve specific critical issues. Each PAP addresses one of the following situations: a gap exists, where a standard extension or new standard is needed; or an overlap exists, where two complementary standards address some information that is in common but different for the same scope of application [50, 51]. The organization of the PAPs is summarized in Table 5.2.

As an illustrative example, PAP 02 deals with wireless communications for SG. It provides key tools and method to assist SG system designers in making informed decisions about wireless technologies. An initial set of quantified requirements has

Table 5.2 PAPs Identified by NIST

Supporting	PAPs
Metering	Meter Upgradeability Standard (PAP 00) Standard Meter Data Profiles (PAP 05) Translate ANSI C12.19 to the common semantic model of the Common Information Model (CIM) (PAP 06)
Enhanced customer interactions with SG	Standards for energy usage information (PAP 10) Standard demand response signals (PAP 09) Develop common specification for price and product definition (PAP 03) Develop common scheduling communication for energy transactions (PAP 04)
Smart grid communications	Guidelines for the use of IP protocol suite in SG (PAP 01) Guidelines for the use of wireless communications (PAP 02) Harmonize power line carrier standards for appliance communications in the home (PAP 15)
Distribution and transmission	Develop CIM for distribution grid management (PAP 08) Transmission and distribution power systems model mapping (PAP 14) IEC 61850 objects/distributed network protocol 3 (DNP3) mapping (PAP 12) Harmonization of IEEE C37.118 with IEC 61850 and precision time synchronization (PAP13)
New smart grid technologies	Energy storage interconnection guidelines (PAP 07) Interoperability standards to support plug-in electric vehicles (PAP 11)

been brought together for AMI and initial DA communications. This work area investigates the strengths, weaknesses, capabilities, and constraints of existing and emerging standards-based physical media for wireless communications. The approach is to work with the appropriate SDOs to determine the characteristics of each technology for SG application areas and types. Results are used to assess the appropriateness of wireless communications technologies for meeting SG applications.

Complete lists and details of standards and PAPs that have been addressed by NIST can be found in [50, 51]. In the next section, a number of important and representative standards for SG are summarized.

5.5.2 A Review on Standards for SG

Figure 5.2 shows a number of representative SG standards in SGCN. IEEE P2030 is one of the most important standards due to the fact that it gives the overall guidelines and frameworks for developing interoperability standards for SG. Following is an overview of the IEEE P2030 standard and other SG standards for smart metering, plug-in hybrid electric vehicles (PHEVs), building automation, substation automation, and DER integration.

5.5.2.1 *IEEE P2030*

This is a standard guide for SG interoperability of energy technology and information technology operation with the electric power system (EPS), and end-use applications and loads [1]. This guide provides a knowledge base and reference model

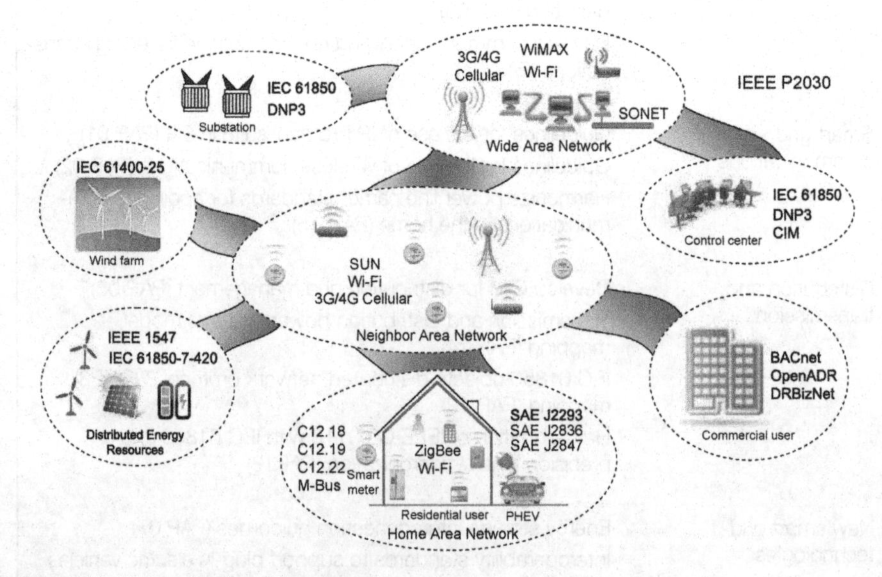

Figure 5.2 Representative SG standards.

toward interoperability of SG. The knowledge base addresses terminology, characteristics, functional performance and evaluation criteria, and the application of engineering principles for SG interoperability of EPS with end-use applications and loads. The P2030 reference model, namely the smart grid interoperability reference model (SGIRM), presents three different architectural perspectives with interoperability tables and charts. The IEEE 2030 series of standards will address more specific technologies and implementation of an SG system (e.g., P2030.1 Electric Vehicle, P2030.2 Storage Energy Systems).

The SGIRM is the central part of the IEEE P2030 standard. It is intended to present interoperable design and implementation alternatives for systems that facilitate data exchange between SG elements, loads, and end-use applications. The IEEE P2030 SGIRM encompasses conceptual architectures of SG from power systems, communications, and information technology perspectives and characteristics of the data that flows between the entities within these perspectives. Each conceptual architecture presents a set of labeled diagrams that offer standards-based architectural direction for the integration of energy systems with ICT infrastructures of the evolving SG. It aims to establish a common language and classification for the SG community to communicate effectively. The interfaces between entities in each architecture will typically contain a wide variety of data. The IEEE SGIRM data classification reference table provides guidance in identifying a set of characteristics for the data at those interfaces. It is a starting point in determining appropriate classifications for the data.

The interoperability architectural perspectives (IAPs) primarily relate to logical, functional considerations of power systems, communications, and information technology interfaces for SG interoperability. The power systems IAP (PS-IAP) mostly represents a traditional view of the EPS, while the communications technology IAP (CT-IAP) provides a means to getting the data from place to place and the information technology IAP (IT-IAP) provides a means to manipulate data to provide useful information. A summary of the three perspectives is as follows:

- *PS-IAP:* The emphasis of the power system perspective is the production, delivery, and consumption of electric energy including apparatus, applications, and operational concepts. This perspective defines seven domains common to all three perspectives: bulk generation, transmission, distribution, service providers, markets, control/operations, and customers.
- *CT-IAP:* The emphasis of the communications technology perspective is communication connectivity among systems, devices, and applications in the context of SG. The perspective includes communications networks, media, performance, and protocols.
- *IT-IAP:* The emphasis of the information technology perspective is the control of processes and data management flow. The perspective includes technologies that store, process, manage, and control the secure information data flow.

The IEEE SGIRM data classification reference table presents various data characteristics (e.g., reach, information transfer time, latency, etc.) and their corresponding value ranges (representative of values that are typically used). The user of the table may need to identify more appropriate data characteristics and values for their specific circumstances.

Besides, the IEEE P2030 SGIRM methodology provides understanding, definitions, and guidance for design and implementation of SG components and end-use applications for both legacy and future infrastructures. The key to using the IEEE P2030 SGIRM is to determine the relevant interfaces, data flows, and data characteristics based on the intended SG application requirements and goals. Once the data requirements of the goals have been defined, the users, based on SGIRM, select a set of interfaces on each IAP of the model that meet the data needs. These interfaces and data flow characteristics are key elements for subsequent SG architectural design and design of implementation operations. The determination of these interfaces is the first step toward determining the implementation of the intended SG application requirements and goals. To assist in this step, the PS-IAP interface tables are provided to identify logical information to be conveyed, the CT-IAP interface tables identify the general communication options of the interface, and the IT-IAP data flow tables identify the general data types.

5.5.2.2 ANSI C12.18

ANSI C12.18 is an ANSI standard that is specifically designed for meter communications and responsible for two-way communications between smart electricity meters (C12.18 device) and a C12.18 client via an optical port.

5.5.2.3 ANSI C12.19

ANSI C12.19 is an ANSI standard for utility industry end-device data tables. This standard defines a table structure for data transmissions between an end-device and a computer for utility applications using binary codes and XML content. ANSI C12.19 is not interested in defining device design criteria or specifying the language or protocol used to transport that data.

5.5.2.4 ANSI C12.22

ANSI C12.22 is an ANSI standard that defines a set of application layer messaging services that are applicable for the enterprise and end-device components of an AMI for SG. The messaging services are tailored for, but not limited to, the exchange of the data table elements defined and published in ANSI C12.19, IEEE P1377/D1, and MC1219. This standard uses AES encryption to enable strong, secure communications, including confidentiality and data integrity. Its security model is extensible to support new security mechanisms. Unlike C12.18 or C12.21 protocols, which only support session-oriented communications, C12.22 provides both session and sessionless communications that help to reduce the complexity of handling communication links on both sides with less signaling overhead. Because of its independence from the underlying network technologies, the ANSI C12.22 open standard enables interoperability between AMI end-devices and the utility data management system across heterogeneous network segments. ANSI C12.22 works with other layers of the IP protocol suite to achieve an end-to-end communication.

5.5.2.5 M-Bus

M-Bus is an European standard for the remote interaction with utility meters and various sensors and actuators. It uses a reduced OSI layer stack that includes physical, link, and application layers. The primary focus of the standard is on simple, low-cost, battery-powered devices. Noteworthy is the support for Device Language Message Specification (DLMS)/Companion Specification for Energy Metering (COSEM) in the lower layers. The Dutch Smart Meter Requirements specifies wired and wireless M-Bus as the means of communication between a metering installation and other (gas, water, etc.) meters, albeit with improved security (AES instead of DES). As this standard is already widely used in meters and reasonably future-proof it is a good contender for local data exchange in SG.

5.5.2.6 SAE J2293

SAE J2293 was developed by the Hybrid Committee that is a part of SAE International and provides requirements for electric vehicles (EV) and electric vehicle supply equipment (EVSE). It standardizes the electrical energy transfer from electric utility to EV.

5.5.2.7 SAE J2836

SAE J2836 supports use cases for communications between plug-in EV and the power grid for energy transfer and other applications.

5.5.2.8 SAE J2847

SAE J2847 supports communication messages between PEVs and grid components.

5.5.2.9 Building Automation Control Network (BACnet)

BACnet is a standard communication protocol that was developed by the American Society of Heating, Refrigerating, and Air-Conditioning Engineers (ASHRAE) for building automation and control networks and supporting the implementation of intelligent buildings with full integration of computer-based building automation and control systems from multiple manufacturers.

5.5.2.10 Open Automated Demand Response (OpenADR)

OpenADR provides communications specifications for fully-automated DR signaling with utility price, reliability, or event signals to trigger customers' preprogrammed energy management strategies. The adoption of OpenADR to SG is very important to provide effective deployment of dynamic pricing, demand response, and grid reliability.

5.5.2.11 Demand Response Business Network (DRBizNet)

DRBizNet is for a highly flexible and automated network of collaboration exchanges and Intelligent Agents (IA) designed on open interoperability standards to support

DR. DRBizNet as a service-oriented architecture, orchestrating standards-based Web services across a business network built on the Internet. The utility industry CIM is leveraged where applicable. It includes distributed registries, DR collaboration exchanges, IA, and standardized web services interface with DR participants.

5.5.2.12 IEC 61850

IEC 61850 is a flexible, open standard that defines the communication between devices in transmission, distribution, and substation automation systems. To enable seamless data communications and information exchange between the overall distribution networks, it is aimed to increase the scope of IEC 61850 to the whole electric network and provide compatibility with the CIM for monitoring, control, and protection applications. This technology is implemented by modern manufacturers in their latest power engineering products like distribution automation nodes/grid measurement and diagnostics devices [52]. IEC 61850 is being extended to cover communications beyond the substation to integration of distributed resources and between substations.

5.5.2.13 DNP3

DNP3 is used for substation and feeder device automation as well as for communications between control centers and substations.

5.5.2.14 IEEE 1547

IEEE 1547 standard establishes technical requirements for EPSs interconnecting with distributed generators such as fuel cells, PV, microturbines, reciprocating engines, wind generators, large turbines, and other local generators. It focuses on the technical specifications for, and the testing of, the interconnection itself. Besides, it provides requirements relevant to the performance, operation, testing, safety, and maintenance of the interconnection.

5.5.2.15 IEC 61850-7-420

IEC 61850-7-420 defines IEC 61850 information models to be used in the exchange of information with DER that comprise dispersed generation devices and dispersed storage devices. It utilizes existing IEC 61850-7-4 logical nodes where possible, but also defines DER-specific logical nodes where needed.

5.5.2.16 IEC 61400-25

IEC 61400-25 is a communications standard that provides uniform information exchange for monitoring and control of wind power plants. This addresses the issue of proprietary communication systems utilizing a wide variety of protocols, labels, semantics, etc., thus enabling one to exchange information with different wind power plants independently of a vendor. It is a subset of IEC 61400, which is a set of standards for designing wind turbines.

5.6 CONCLUSIONS

This chapter gives insight into the smart grid communications network (SGCN) by presenting its layered architecture, typical types of traffic that it may carry, and associated quality of service requirements, as well as candidate wireless communications technologies that can be employed for its implementation. Networking issues that the neighbor area network (NAN) segment of SGCN needs to tackle are highlighted by exploring characteristics and requirements of this network segment and identifying important gaps that existing wireless routing protocols need to cover for their applicability into NAN. It is observed that protocols for downlink communications (from control center to power grid devices and end-user devices) should receive immediate attention due to the fact that this communication is critical for many advanced smart grid applications. Mechanisms for QoS differentiation and provisioning to properly serve a mix of non-real-time and real-time traffic play a very important role in the successful deployment of the smart grid. Additionally, network self-healing capability, multicast routing, cluster-based routing, and optimal network design are among vital considerations for reliable and efficient operations of SGCN. Note that there are other considerations that cannot be mentioned in this work due to space limitations. They include cybersecurity and privacy, machine-to-machine communications platforms and protocols, system design for long-lived network devices, middleware and application programming interfaces, firmware upgradeability, etc. This work also reviews a number of standards for smart grid interoperability.

REFERENCES

[1] Draft Guide for Smart Grid Interoperability of Energy Technology and Information Technology Operation with the Electric Power System (EPS), and End-Use Applications and Loads, IEEE P2030, 2011.

[2] U.S. Department of Energy, Communications Requirements of Smart Grid Technologies, Report, 5 October 2010.

[3] N. Hatziargyriou et al., Microgrids, IEEE Power and Energy Magazine 5 (2007) 78–94.

[4] B. Kroposki et al., Making microgrids work, IEEE Power and Energy Magazine 6 (2008) 40–53.

[5] Y. Gobena et al., Practical architecture considerations for smart grid WAN network, in: Proceedings of Power Systems Conference and Exposition (IEEE/PES), 2011, pp. 1–6.

[6] K. Hopkinson et al., Quality-of-service considerations in utility communication networks, IEEE Transactions on Power Delivery 24 (3) (2009) 1465–1474.

[7] IEEE Standard Communication Delivery Time Performance Requirements for Electric Power Substation Automation, IEEE Std 1646-2004, IEEE Power Engineering Society.

[8] ZigBee: The Choice for Energy Management and Efficiency, White paper, ZigBee Alliance, 2007.

[9] Part 15.4: Wireless Medium Access Control (MAC) and Physical Layer (PHY) Specifications for Low-Rate Wireless Personal Area Networks (WPANs), IEEE Std 802.15.4-2006, IEEE Computer Society.

[10] Smart Energy Profile 2.0 Public Application Protocol Specification, ZigBee Alliance, 12 July 2011.

[11] Wi-Fi for the Smart Grid: Mature, Interoperable, Security-Protected Technology for Advanced Utility Management Communications, Wi-Fi Alliance, September 2009.

[12] P. Beecher, SG Smart Metering Utility Networks Project Draft PAR, IEEE P802.15-08-0705-005-0nan, September 2008.

[13] A. Sgora, D.D. Vergados, P. Chatzimisios, IEEE 802.11s wireless Mesh Networks: challenges and perspectives, Lecture Notes of the Institute for Computer Sciences, Social Informatics and Telecommunications Engineering 13 (2009) 263–271.

[14] David Cypher, NIST Priority Action Plan 2 – Guidelines for Assessing Wireless Standards for Smart Grid Applications. National Institute of Standards and Technology.

[15] K. KhHurana et al., Smart-grid security issues, IEEE Security and Privacy 8 (1) (2010) 81–85.

[16] F. Cohen, The smarter grid, IEEE Security and Privacy 8 (1) (2010) 60–63.

[17] E. Alotaibi, B. Mukherjee, A survey on routing algorithms for wireless ad-hoc and mesh networks, Computer Networks (2011)

[18] T. Watteyne et al., From MANET to IETF ROLL Standardization: a paradigm shift in WSN routing protocols, IEEE Communications Surveys and Tutorials 13 (4) (2011) 688–707.

[19] P. Levis, A. Tavakoli, S. Dawson-Haggerty, Overview of existing routing protocols for low power and lossy networks, IETF ROLL, IETF draft, 15 April 2009, draft-ietf-roll-protocols-survey-07.

[20] K. Akkaya, M. Younis, A survey on routing protocols for wireless sensor networks, Ad Hoc Networks 3 (3) (2005) 325–349.

[21] M. Abolhasan, T.A. Wysocki, E. Dutkiewicz, A review of routing protocols for mobile ad hoc networks, Ad Hoc Networks 2 (1) (2004) 1–22.

[22] J.N. Al-Karaki, A.E. Kamal, Routing techniques in wireless sensor networks: a survey, IEEE Wireless Communication 11 (6) (2004) 6–28.

[23] Y.C. Tseng, S.-Y. Ni, Y.S. Chen, S. Jang-Ping, The broadcast storm problem in a mobile ad hoc network, Wireless Networks 8 (2–3) (2002) 153–167.

[24] I. Stojmenovic, S. Olariu, Handbook of Sensor Networks: Algorithms and Architectures, John Wiley & Sons, Inc., Hoboken, New Jersey, 2005 October

[25] P. Bose, P. Morin, I. Stojmenovic, J. Urrutia, Routing with guaranteed delivery in ad hoc wireless networks, in: Third ACM International Workshop on Discrete Algorithms and Methods for Mobile Computing and Communications (DIAL), Seattle, WA, USA, 20 August 1999, pp. 48–55.

[26] B. Karp, H. Kung, GPSR: Greedy perimeter stateless routing for wireless networks, in: Annual International Conference on Mobile Computing and Networking (MobiCom), Boston, MA, USA, 6–11 August 2000, pp. 243–254.

[27] H. Frey, I. Stojmenovic, On delivery guarantees of face and combined greedy-face routing algorithms in ad hoc and sensor networks, in: 12th ACM Annual International Conference on Mobile Computing and Networking (MobiCom), Los Angeles, CA, USA, 23–29 September 2006, pp. 390–401.

[28] E. Elhafsi, N. Mitton, D. Simplot-Ryl, End-to-end energy efficient geographic path discovery with guaranteed delivery in ad hoc and sensor networks, in: 19th Annual International Symposium on Personal, Indoor and Mobile Radio Communications (PIMRC), Cannes, France, 15–18 September 2008, pp. 1–5.

[29] H. Kalosha, A. Nayak, S. Rhrup, I. Stojmenovic, Select-and-protest-based beaconless georouting with guaranteed delivery in wireless sensor networks, in: 27th Conference on Computer Communications (INFOCOM), Phoenix, AZ, USA, 13–18 April 2008, pp. 346–350.
[30] T. Winter et al., RPL: IPv6 Routing Protocol for Low power and Lossy Networks, ROLL IETF Internet-Draft, 13 March 2011, draft-ietf-roll-rpl-19.
[31] W.B. Heinzelman, A.P. Chandrakasan, H. Balakrishnan, An application-specific protocol architecture for wireless microsensor networks, IEEE Transactions on Wireless Communication 1 (4) (2002) 660–670.
[32] Z. Zhou, S. Zhou, S. Cui, J.H. Cui, Energy-efficient cooperative communication in a clustered wireless sensor network, IEEE Transactions on Vehicular Technology 57 (6) (2008) 3618–3628.
[33] O. Younis, M. Krunz, S. Ramasubramanian, Node clustering in wireless sensor networks: recent developments and deployment challenges, IEEE Network 20 (3) (2006) 20–25.
[34] O. Younis, S. Fahmy, HEED: a hybrid, energy-efficient, distributed clustering approach for ad hoc sensor networks, IEEE Transactions on Mobile Computing 3 (4) (2004) 366–379.
[35] A. Manjeshwarm, D.P. Agrawal, TEEN: a routing protocol for enhanced efficiency in wireless sensor networks, in: Workshop on Parallel and Distributed Computing Issues in Wireless Networks and Mobile Computing, vol. 3, April 2001, pp. 2009–2015.
[36] A. Manjeshwar, D.P. Agrawal, APTEEN: a hybrid protocol for efficient routing and comprehensive information retrieval in wireless sensor networks source, in: International on Parallel and Distributed Processing Symposium (IPDPS), Fort Lauderdale, FL, USA, 15–19 April 2002, pp. 195–202.
[37] A. Boukerche, R.W. Pazzi, R. Araujo, Fault-tolerant wireless sensor network routing protocols for the supervision of context-aware physical environments, Journal of Parallel and Distributed Computing 66 (4) (2006) 586–599.
[38] B. Lichtensteiger et al., RF mesh system for smart metering: system architecture and performance, in: Proceedings of IEEE Smart Grid Communication, Maryland, USA, 1–3 October, 2010, pp. 379–384.
[39] J. Ko et al., Evaluating the performance of RPL and 6LoWPAN in TinyOS, in: Proceedings of the Workshop on Extending the Internet to Low power and Lossy Networks (IP+SN), Chicago, IL, 2011.
[40] O. Gnawali et al., Collection tree protocol, in: Proceedings of the Seventh ACM Conference on Embedded Networked Sensor Systems (SenSys), November 2009, pp. 1–14.
[41] M. Nucolone, Stability analysis of the delays of the routing protocol over low power and lossy networks, Master's thesis, KTH Electrical Engineering, 2010.
[42] D. Wang et al., RPL based routing for advanced metering infrastructure in smart grid, TR2010-053, Mitsubishi Electric Research Laboratories, July 2010.
[43] P. Kulkarni et al., A self-organising mesh networking solution based on enhanced RPL for smart metering communications, in: Proceedings of IEEE International Symposium on a World of Wireless, Mobile and Multimedia Networks (WoWMoM), 20–24 June 2011, pp. 1–6.
[44] J. Tripathi, J.C. de Oliveira, J.P. Vasseur, Applicability study of RPL with local repair in smart grid substation networks, in: Proceedings of the First IEEE International Conference on Smart Grid, Communications (SmartGridComm), 4–6 October 2010, pp. 262–267.

[45] J.S. Jung et al., Improving IEEE 802.11s wireless mesh networks for reliable routing in the smart grid infrastructure, in: Proceedings of IEEE International Conference on Communications Workshops (ICC), 5–9 June 2011, pp. 1–5.
[46] H. Gharavi, B. Hu, Multigate mesh routing for smart grid last mile communications, in: Proceedings of IEEE Wireless Communications and Networking Conference (WCNC), 28–31 March 2011, pp. 275–280.
[47] H. Gharavi, B. Hu, Multigate communication network for smart grid, Proceedings of the IEEE 99 (6) (2011) 1028–1045.
[48] Institute of Electrical and Electronics Engineers, IEEE Standard Computer Dictionary: A Compilation of IEEE Standard Computer Glossaries, New York, NY, 1990.
[49] E.W. Gunther et al., Smart Grid Standards Assessment and Recommendations for Adoption and Development, EnerNex Corporation (2009) February
[50] NIST Framework and Roadmap for Smart Grid Interoperability Standards, Release 1.0, NIST Special, Publication 1108, January 2010.
[51] Draft NIST Framework and Roadmap for Smart Grid Interoperability Standards, Release 2.0, NIST Special, Publication, October 2011.
[52] V.C. Gungor et al., Smart grid technologies: communication technologies and standards, IEEE Transactions on Industrial Informatics 7 (4) (2011) 529–539.

CHAPTER 6

Intercell Interference Coordination: Towards a Greener Cellular Network*

Duy Trong Ngo, Duy Huu Ngoc Nguyen, Tho Le-Ngoc
Department of Electrical and Computer Engineering, McGill University, Canada

6.1 INTRODUCTION

The past decade has witnessed a tremendous growth in the cellular network market with continuous increase in the sales of mobile wireless devices. As the demand for cellular voice and data traffic dramatically escalates, the energy expenditure in wireless networks is pushed to the limit. The study in [1] estimates that the total energy consumed by cellular networks, wireline communication infrastructure, and Internet accounts for more than 3% of worldwide electric energy utilization. The rising carbon footprint and energy costs of operating cellular networks have driven an emerging trend of securing higher energy efficiency among the network operators and regulatory bodies. On the other hand, energy efficiency is also crucial to prolonging the operating time of user terminals in a wireless network. This is because most of these devices are powered by batteries with capacities that are unable to keep pace with the increasing power dissipation of signal processing circuits [2].

In multicellular wireless systems, interference among network devices has always been a critical issue. With the current tendency of adopting universal frequency reuse where all cells have the potential to use all available radio resources, the management of intercell interference (ICI) is of particular importance. As air interface is a shared medium of several wireless users, device energy consumption is not only affected by point-to-point communication links, but also the interaction among different links in the network. Intuitively, each network entity would selfishly try to achieve its highest possible performance without considering the impact caused to others. In an interference-limited setting, it is likely that unnecessary power is expended by the network device, whether it is a base station or a user terminal. In a move towards "greener" cellular networks, coordination among different base stations (BSs) is envisioned to be an effective means to mitigate the adverse consequences of interference, while allowing various energy-efficiency objectives to be realized.

*This work is supported in part by the Natural Science and Engineering Research Council of Canada (NSERC) through Strategic and Discovery Grants, the Alexander Graham Bell Canada Graduate Scholarship for Doctoral Studies (CGS-D), and the McGill Engineering Doctoral Award (MEDA).

In recent 3GPP LTE Releases, several forms of interference avoidance and coordination techniques were proposed with the main objective of efficiently controlling the ICI, especially at the cell-edge user equipment (UE) [3]. In intercell interference coordination (ICIC), such techniques as time-domain solutions (e.g., subframe alignment, OFDM symbol shift) and frequency-domain methods (e.g., channel orthogonalization) explicitly control how the radio resources are utilized to moderate the ICI. In a more advanced coordination technique, namely coordinated multipoint transmission and reception (CoMP), the intercell transmission, instead of being considered as the source of interference, is taken into account as an extra means to enhance the overall system performance. The main concept of CoMP is to coordinate multiple cell sites to jointly transmit data signals to cell-edge UEs or to jointly receive and decode the UEs' uplink transmissions. It is envisioned in [3] that CoMP is able to significantly improve the network-wide performance in both uplink and downlink transmissions. However, this performance enhancement may come at the expense of excessively high complexity in the joint precoding/decoding process and the demand for ideal backhaul transmissions and synchronization among the coordinated BSs.

The above advanced solutions, designed for homogeneous cellular networks, are predicted to soon reach their theoretical limits. Fortunately, it has been shown that an enormous system gain can still be achieved by reducing the cell size and deploying many small cells, called femtocells. With spectral efficiency substantially boosted via spatial reuse, a heterogeneous macrocell/femtocell deployment has the potential to use all available resources taking into account varying traffic load and channel conditions [4,5]. Because of their size, smaller cells (i.e., picocell, femtocell) are also much more power-efficient in extending cellular broadband coverage. As shown in [6], a 7:1 operational energy advantage ratio can be achieved by adopting dense smaller cells over expanding the current macrocell network, for an approximately similar indoor coverage. Moreover, the work of Claussen et al. [7] shows that a joint deployment of macrocell/picocell, where 20% of the customers are served by picocells, can reduce the total network energy consumption by up to 60% compared with a network consisting of macrocells only. Nevertheless, since small cells, e.g., femtocells, operate in the licensed spectrum owned by wireless operators and share this spectrum with macrocell networks, limiting the cross-tier interference from femtocell users at a macrocell BS becomes an indispensable condition. Due to the random deployment of femtocell BSs, users in a femtocell may also suffer severe interference caused by nearby femtocells.

Much of the research effort in the past focuses on exploiting different techniques to achieve spectral efficiency. However, a higher network throughput usually implies a large amount of energy consumed. Because spectral efficiency and energy efficiency can conflict with each other, it is crucial to carefully study how to balance these two design metrics [8]. When energy efficiency is explicitly stated as the design objective, a vital task is to design interference management schemes that reduce energy consumption while satisfying the data-rate requirements. Often, the existing solutions have to be adapted and new solutions are devised to accommodate the trade-off between attaining higher spectral efficiency and realizing better energy efficiency.

This chapter aims to provide an extensive literature review of the current techniques applicable to multicell interference coordination. Specifically, we consider interference management in two important network deployments, namely CoMP and femtocell, which are envisioned to be the future technologies. For energy efficiency, we discuss the current state-of-the-art achievements. The trade-off between spectral and energy efficiencies is also discussed. The rest of this chapter is organized as follows: Section 6.2 presents the issues of interference in multicellular networks. Both traditional homogeneous and small-cell heterogeneous systems are discussed. Section 6.3 reviews the available techniques that deal with the interference issues in homogeneous systems. Recent advances in CoMP are then discussed in Section 6.4. Section 6.5 studies the interference management mechanisms in femtocell networks. Interference coordination with the explicit goal of achieving energy efficiency is presented in Section 6.6. Finally, Section 6.7 concludes this chapter.

6.2 INTERFERENCE ISSUES IN MULTICELL SYSTEMS

6.2.1 Frequency Reuse and the Problem of Interference in Homogeneous Cellular Systems

A cellular network is a radio network distributed over land areas called cells, each of which has one base station (BS) to serve all user equipments (UEs) operating within that cell. The downlink refers to the channel from the BS to the UEs, whereas the uplink refers to the channel from the UEs to the BS. A typical example of cellular networks is depicted in Figure 6.1. In this multiuser multicell setting it is noted that

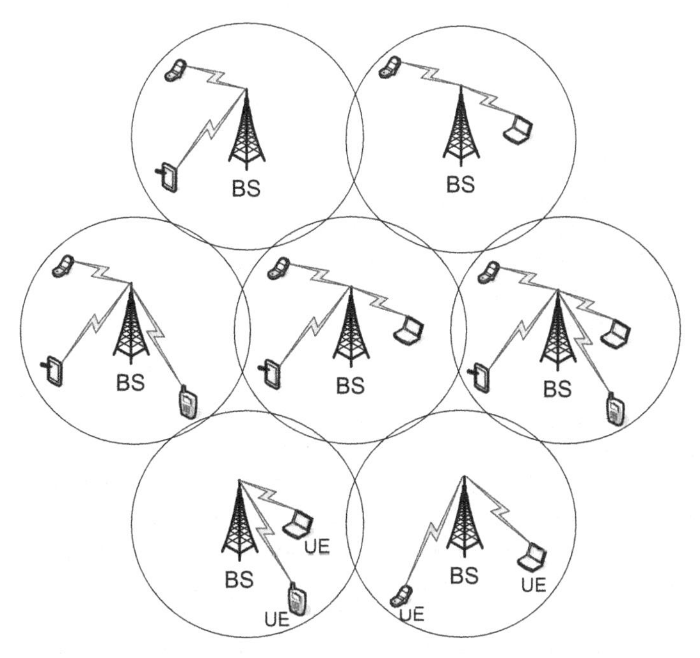

Figure 6.1 An example of multicellular communication systems.

when several transmitters (i.e., the BSs in the downlink and the UEs in the uplink) emit their signals on the same frequency, and within the same geographical location, the receiver (i.e., the UEs in the downlink and the BSs in the uplink) sensing that frequency is likely not able to distinguish which transmitter it is listening to. Essentially, the broadcast nature of the wireless medium results in a fundamental problem of interference in wireless communication.

Conventional cellular systems deploy a fractional frequency reuse scheme where cells located close to one another are allotted with different frequency bands so as to avoid dominant intercell interference (i.e., cochannel interference). Cells that are sufficiently far from each other may reuse the same band. However, segmenting frequency reuse suffers from reduced spectral efficiency. This problem is severe, given that radio spectrum is a scarce resource that is usually expensive for telecommunication operators to afford. In order to support the ever-increasing demand by a huge number of wireless terminals for higher quality-of-service (QoS), next-generation wireless networks are envisioned that employ universal frequency reuse in which all cells operate on the same radio frequency. While offering more efficient spectral utilization, the universal reuse of radio spectrum may degrade the network capacity if the critical issue of ICI is not properly addressed.

In Code Division Multiple Access (CDMA), all UEs in all cells are allowed to simultaneously transmit over all the available frequency bands. With signals differentiated via the use of orthogonal codes, Wideband CDMA (WCDMA) transmission is used in the two popular 3G standards, namely, Universal Mobile Telecommunications System (UMTS) and High Speed Packet Access (HSPA). Thanks to interference averaging, WCDMA receivers are known to be capable of separating signals at very low levels of signal-to-interference-plus-noise ratio (SINR). With unity spectral reuse factor where all UEs either within the same cells or in different cells share the same spectrum, interference remains a critical problem in CDMA systems.

On the other hand, frequency-selective fading is one of the major impairments of a wireless channels, particularly in multipath environments such as indoor and urban areas. As the channel responses differ among different frequencies in these environments, it can be challenging to alleviate the distortion that wideband signals (e.g., those in CDMA-based systems) experience when transmitted over such channels. In this situation, narrow-band signals like those in Orthogonal Frequency Division Multiplexing (OFDM) systems are more preferred as they are more robust to this type of fading. OFDM modulation has also proven to offer higher spectral efficiency and it can be efficiently implemented through fast Fourier transform (FFT) blocks. The IEEE Wireless Interoperability for Microwave Access (WiMAX) standard uses OFDM as the multicarrier technology in the PHY layer, while the 3GPP Long Term Evolution (LTE) standard employs Orthogonal Frequency Division Multiple Access (OFDMA) in the downlink and Single-Carrier FDMA (SC-FDMA) in the uplink [3].

With the use of OFDMA, intracell interference among UEs within the same cell can be suppressed. This is due to the assumption of exclusive subchannel assignment in OFDMA-based systems, i.e., one subchannel can be used by at most one UE at a particular time instant. However, aggressive frequency reuse allows a common spectrum to be shared among the UEs belonging to different cells. While interference averaging helps reduce the effect of interference in CDMA, in OFDMA systems there is not a scenario where one transmitter is enough to completely interfere and jeopardize a given subchannel. As there are multiple subchannels available for use, one has to also resolve the problem of properly assigning subchannels to different UEs within each cell, and power must be optimally allocated to maximize system performance as well as to reduce interference induced to other cells.

6.2.2 Cochannel Deployment of Heterogeneous Small-Cell Networks

It should be pointed out that traditional macrocells are generally designed to provide large coverage and are not efficient in providing high throughput, especially in very dense urban and indoor environments. To meet the demand for ubiquitous wireless coverage and higher throughput, several techniques have been proposed in the literature. Allowing for concurrent use of different frequencies, carrier aggression effectively increases the bandwidth allocated to the UEs. Also, multiple-antenna solutions are attractive in that substantial diversity and multiplexing gains are exploited. Further, multiple cells employing CoMP can coordinate their scheduling to serve UEs with unfavorable link conditions; and thus, this technique is particularly useful to mitigate outage at the cell edges. However, it is noted that these advanced solutions, which are designed for homogeneous cellular networks, are reaching the theoretical limits.

Recent research studies have shown that an enormous system gain can still be achieved by reducing the cell size through the deployment of small cells, called femtocells [4,5]. Femtocell base stations (BSs) are low-power, miniature wireless access points that are set at a home and connected to backhaul networks via residential wireline broadband access links, e.g., digital subscriber lines (DSL), cable broadband connections, or optical fibers. Typically, the range of a femtocell is less than 50 m and it serves a dozen active users [9]. The coexistence of a macrocell and several femtocells in a heterogeneous network is illustrated in Figure 6.2. The benefits of femtocells are summarized in the following [10,11].

- *Higher capacity:* With a large number of small cells, more users can be packed into a given area in the same spectrum, allowing for larger area spectral efficiency (i.e., total number of active users per Hz per unit area). The lower transmit power of femtocells and the signal isolation due to penetration losses by walls may also significantly limit the interference from neighboring femtocells and macrocells. This results in a higher femtocell capacity gain. Because macrocells no longer needs to transmit at higher power to compensate for high penetration loss to cover indoor areas, macrocell capacity is increased as well.
- *Better coverage with lower power consumption:* Thanks to the close proximity between UEs and their home femtocell BSs, femtocells can lower transmit power (to less than 23 dBm, compared with 46 dBm by macrocell) and still be able to

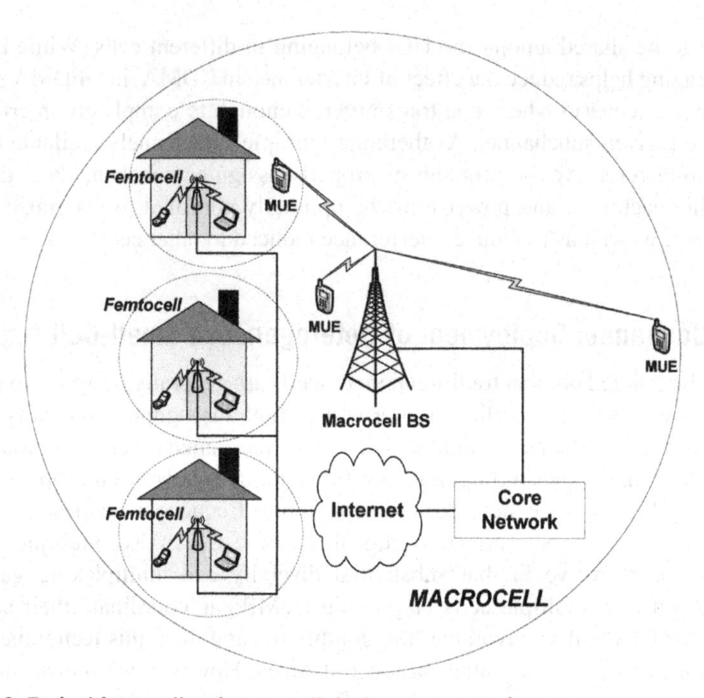

Figure 6.2 Typical femtocell and macrocell deployment scenario.

achieve a high SINR. Substantially improved coverage is particularly important in indoor areas where macrocell signals cannot reach.

- *Macrocell offload:* Traffic originating from indoors can be absorbed by femtocells via the IP backhaul networks, thus offloading a significant portion of traffic that would have otherwise been directed to the macrocell. Consequently, the macrocell network can dedicate its resources to better service its macrocell users.
- *Cost effectiveness:* Compared to the traditional cell-partitioning approach for which a large number of more expensive macrocell BSs are typically required after the extensive site survey and network planning process, femtocells can be integrated into an existing cellular network infrastructure with ease. Mainly deployed by end users, the small-cell solution requires low capital expenditures and operating expenses.

In femtocells, cochannel deployment is attractive where femtocell users (FUEs) share the same frequency bands with macrocell users (MUEs). Indeed, it is the only feasible solution for limited spectrum (i.e., 20 MHz or less) and without impacting the peak data rate of legacy UEs. Moreover, the deployment of low-cost BSs is made possible while the higher-cost carrier aggregation-capable UE is not required. Here again, the critical issue of interference arises as femtocells utilize the spectrum

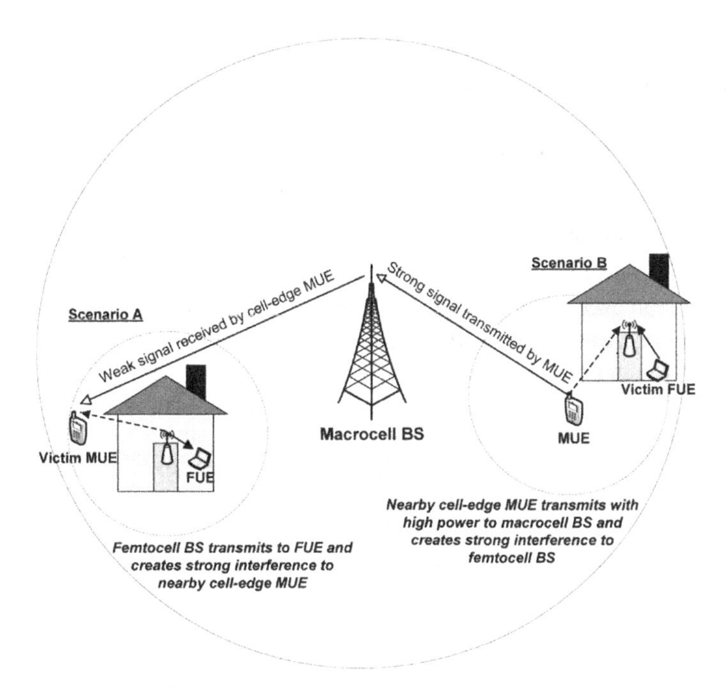

Figure 6.3 Strong cross-tier interference in femtocell/macrocell deployment.

already allocated to the macrocells. In addition to the regular intratier interference within the macrocells and the femtocells, there are now cross-tier interferences from macrocells to femtocells and from femtocells to macrocells. In many instances, the cross-tier interferences can be hard to control and its effects are particularly severe. Figure 6.3 illustrates two typical cross-tier interfering scenarios that occur in a mixed macrocell/femtocell deployment. In Scenario A, a victim cell-edge MUE is strongly interfered with by the downlink transmission of a nearby femtocell BS. In Scenario B, an MUE that is located far away from its serving macrocell BS transmits at high power in the uplink to compensate the path losses. This may jam the transmission of a nearby victim FUE.

6.2.3 BS Coordination for Interference Management in Multicell Systems

6.2.3.1 Homogeneous Network Deployment

In a conventional homogeneous multicell system, interference is usually controlled on a per-cell basis. The ICI is treated as background noise by each cell, the BS of which has no intention to control the interference induced to other cells. BS coordination has recently emerged as an effective means to mitigate cochannel interference in multicell networks. Power control with pricing to enforce cooperation among different cells is particularly applicable to CDMA-based cellular systems. As for multicell networks that use OFDMA, a significant system gain can be realized via coordinating individual BSs and users. Depending on the available bandwidth of the backbone network connecting the BSs, either a low or a high level of coordination is assumed. In certain instances, a locally optimal solution is still acceptable when centralized approaches that require a large amount of communication among cells cannot be afforded.

The latest 3GPP LTE-Advanced Release considers CoMP as an enabling technology to improve coverage, throughput, and efficiency [3]. Actively dealing with the ICI, this solution even takes advantage of the intercell transmissions to enhance the overall system performance. In the downlink, CoMP coordinates the simultaneous transmissions from multiple BSs to the UEs. Such coordination is especially helpful for the cell-edge UEs, whose link conditions are usually unfavorable due to the long distance to their corresponding BSs. In the uplink, CoMP allows for the exploitation of the multiple receptions at multiple cells to jointly decode the uplink signals from the UEs. It is maintained in [3] that CoMP is able to significantly improve the link performance in both uplink and downlink transmissions.

Following the convention in literature, CoMP can be classified into the following three schemes according to the extent of coordination among cells.

- *Joint signal processing (JP):* User data is exchanged among the coordinated BSs such that the multiple BSs are simultaneously transmitting/receiving data signals to/from the UEs within the coordinated areas. This scheme is also referred to as *network multiple-input multiple-output (MIMO)*.
- *Interference coordination (IC):* User data is not exchanged among the BSs. Each UE transmits/receives data signals to/from its (single) serving BS. Nevertheless, control information among the coordinated BSs is exchanged to jointly control the ICI.
- *Interference aware (IA):* There is no information exchange among the transmitting entities. However, the interference is estimated at each receiving entity and fed back to its corresponding transmitter. Acting as a rational entity, each BS selfishly adjusts its transmit/receive strategy according to the knowledge of the interference measured by itself or at its connected UEs. Essentially, the IA scheme represents a strategic noncooperative game (SNG) with the BSs being the players in that game.

It is worth noticing that different CoMP schemes impose different requirements on the data and signaling exchanges, as well as the channel state information (CSI) knowledge needed at the coordinated BSs. In the JP scheme, the antennas of multiple BSs of large single antenna array [12,13]. Data streams intended for all the UEs are jointly processed and transmitted from all the antennas. Apparently, such an approach is the most complex CoMP mode (i.e., with the highest level of coordination) as it requires a significant amount of signaling to be exchanged among the BSs via an ideal backhaul. In the next level of coordination, the IC scheme allows a BS to transmit data only to the UEs within its cell [14,15]. Under both JP and IC schemes, each BS may need to know the CSI of all the UEs in the system, even that of the unconnected UEs in order to fully control the ICI. On the contrary, each BS is only required to know the CSI of its connected UEs under the IA scheme, which corresponds to the lowest level of coordination.

6.2.3.2 Small-Cell Heterogeneous Network Deployment

Femtocell deployment represents a paradigm shift from the traditional centralized macrocell approaches to a more uncoordinated and autonomous solution. The advantages offered by femtocell, however, come with several technical issues, raising various design challenges that need to be overcome. While interference is also a problem within the conventional macrocellular networks, the development and implementation of interference management solutions for femtocell networks are more challenging for the following reasons [9,16,17].

- *Unplanned deployment:* The interference situation is more acute here because femtocells are randomly deployed without the network planning that would normally be undertaken. Since the femtocell BSs and users can be moved or switched on/off at any time, conventional network optimization becomes inefficient as the operators cannot control the number and the location of the newly deployed small cells. The roll-out of unplanned femtocells also creates new cell boundaries and, accordingly, users are more likely to suffer from strong ICI, especially those that happen to be in the cell-edge areas. In such cases, not only are the MUEs badly affected, a poor level of performance is achieved by the FUEs.
- *Access priority:* Since femtocells operate in the licensed spectrum owned the macrocell network, it is imperative to limit the cross-tier interference induced by the FUEs to the MUEs, who have strictly higher priority than the FUEs in accessing the underlying frequency bands [18]. In this way, the performance of the preferential users can effectively be protected. On the other hand, the lower-tier FUEs should configure their transmission parameters to exploit the residual network capacity beyond what is needed to support the QoS requirements of all MUEs.
- *Limited control/signaling:* The residential network infrastructure can only provide limited capacity for the exchange of control and signaling information. Because the wireline backhaul may not belong to the network operators, delay can be a significant issue.

The above issues render centralized interference management difficult. Instead, distributed approaches are always preferable in practical applications, where macrocell BSs and femtocell BSs together with their serviced users coordinate and adapt their transmission parameters (e.g., transmission rate and power) in a decentralized fashion [10,18]. One of the central research themes is to develop autonomous interference management schemes such that (i) the QoS requirements of the existing MUEs with higher access priority always be maintained, and (ii) residual network capacity be effectively exploited by the newly deployed FUEs to optimize their own performance.

6.3 INTERFERENCE MANAGEMENT TECHNIQUES IN HOMOGENEOUS CELLULAR SYSTEMS

6.3.1 Power Control of CDMA-Based Multicell Networks

Let P_i be the transmit power of user i and σ^2 be the power of additive white Gaussian noise (AWGN). Denote the channel gain from the transmitter of user i to its receiver by $g_{i,i}$, and that from the transmitter of user j to the receiver of user $i \neq j$ by $g_{i,j}$. Note that the transmitter of user i in the downlink is indeed the BS that serves user i, whereas in the downlink it is the UE i. The received signal-to-interference-plus-noise ratio (SINR) of user i can be written as

$$\gamma_i = \frac{g_{i,i} P_i}{\sum_{j \neq i} g_{i,j} P_j + \sigma^2}. \tag{6.1}$$

As seen from (6.1), a large unwanted signal power $\sum_{j \neq i} g_{i,j} P_j$ can significantly decrease the SINR, thereby degrading the quality of communication.

Power control has been proven very effective in dealing with interference in traditional CDMA-based wireless networks. One of the most popular solutions is Foschini–Miljanic's algorithm [19], a power control scheme that enables users to eventually achieve their fixed target SINRs by iteratively adapting their transmit power according to

$$P_i[t+1] = \frac{\gamma_i^{\min}}{\hat{\gamma}_i[t]} P_i[t], \tag{6.2}$$

where $\gamma_i^{\min}$ is the target SINR, $P_i[t]$ the transmit power of user i, and $\hat{\gamma}_i[t]$ the measured SINR at the receiver of user i at time t. It is clear from (6.2) that this simple algorithm can be implemented distributively by individual users without the need of cooperation among themselves. As long as the target SINRs are feasible, the iterative algorithm in (6.2) converges to a Pareto-optimal solution at a minimal aggregate transmit power $\sum_i P_i$. However, it should be noted that if there exists an infeasible SINR target, the resulting transmit power will diverge to infinity. This is because each user attempts to meet its own required SINR no matter how high the power consumption can be.

The works in [20–23] investigate several other power control schemes from a game-theoretical point of view. Solutions derived via the application of noncooperative game theory is appealing since they can be determined in a distributive fashion. Here, individual users selfishly optimize their own performance regardless of the actions of other users. Denoting the utility (or payoff) function of user i as $U_i(P_i, \mathbf{P}_{-i})$ where $\mathbf{P}_{-i}$ is the power vector of all users except i, the power control game can be formally expressed as

$$\max_{P_i \geqslant 0} U_i(P_i, \mathbf{P}_{-i}) \tag{6.3}$$

for each user i. Depending on the type of utility function $U_i(\cdot)$, solutions to the individual problem in (6.3) can be found and different games are formulated with various convergence characteristics. In most instances and under certain conditions, the underlying games settle at a Nash equilibrium $\mathbf{P}^* = [P_i^*]$, a stable and predictable state at which no user has any incentive to unilaterally change its power level, i.e.,

$$U_i(P_i^*, \mathbf{P}_{-i}^*) \geqslant U_i(P_i, \mathbf{P}_{-i}^*), \quad \forall P_i \geqslant 0, \tag{6.4}$$

for every user i.

Although the achieved Nash equilibrium (NE) gives a stable operating point, it does not guarantee Pareto efficiency. To alleviate this shortcoming, various pricing schemes are developed in [24] to improve the efficiency of the equilibrium solutions. It is noteworthy that the pricing mechanism can implicitly bring cooperation to the users while maintaining the noncooperative nature of the game. In this case, the total utility of user i is defined as

$$U_{\text{tot},i}(P_i, \mathbf{P}_{-i}) = U_i(P_i, \mathbf{P}_{-i}) - C_i(P_i, \mathbf{P}_{-i}), \tag{6.5}$$

where $C_i(\cdot)$ is the cost function imposed to user i. For each problem $\max_{P_i \geqslant 0} U_{\text{tot},i}$ of user i, different choices of utility and cost functions are possible. Typically, the resulting solution is some modified version of the SINR/power balancing algorithm in (6.2). By selecting proper utilities and a linear cost function $C_i(P_i, \mathbf{P}_{-i}) = \alpha_i P_i$, the studies in [25,26] show that noncooperative games with pricing can substantially enhance the NE if reasonable deviations from the target SINR are allowed. For instance, with $U_i = \beta_i(\gamma_i^{\min} - \gamma_i)^2$ the transmit power in [26] is updated according to

$$P_i[t+1] = \left(\frac{\gamma_i^{\min}}{\hat{\gamma}_i[t]} P_i[t] - \frac{\alpha_i}{\beta_i} \frac{P_i^2[t]}{\hat{\gamma}_i^2[t]} \right)^+, \tag{6.6}$$

where $(\cdot)^+ = \max(\cdot, 0)$. Numerical results show that the Nash solution of [26], via the use of linear price $C_i = 5P_i$, converges faster than the power balancing algorithm in (6.2). In particular, given the SINR target $\gamma_i^{\min} = \gamma^{\min} = 5.0\ \forall i$, an almost 50% savings in power is achieved with only a 5% reduction in the average SINR.

Nevertheless, it is unclear how far the Nash solutions offered by [25,26] are from the actual global optima of the original design problems. Using a different pricing scheme that is linearly proportional to SINR, i.e., $C_i(P_i, \mathbf{P}_{-i}) = \alpha\gamma_i$, the work of [27] shows that the outcome of the noncooperative power control game in single-cell systems is a unique and Pareto-efficient NE. As such, various design goals can be met by setting dynamic prices for individual users. However, in multicell settings where transmit powers of all users need to be jointly optimized across different cells, ICI cannot simply be treated as noise like in [27].

Different from the proposed algorithm in [19] where distributed power control can attain a given fixed and feasible SINR target, the work of [28] considers the distributed Pareto-optimal joint optimization of SINR assignment and power control for multicellular systems. It is argued that a fixed SINR assignment is certainly not suitable in the context of data-service networks. Instead, SINRs should be adjusted to the limit of the system's capacity.A high SINR is translated into better throughput and reliability while a low SINR implies lower data rates. In [28], the feasible SINR region is characterized in terms of the loads at the BSs as well as the indication of potential interference from mobile users. With a reparametrization via the left Perron–Frobenius eigenvectors and a locally computable ascent direction, a distributed solution is derived for the uplink that guarantees Pareto optimality. In implementing the proposed algorithm, a certain level of network cooperation is required. Specifically, pertinent channel information is to be exchanged over the control channel within individual links, and an additional channel is used for broadcasting the load information. Yet, the results obtained here directly apply to homogeneous networks where there exist no differentiated classes of users with distinct access priority and design specifications.

6.3.2 Joint Subchannel and Power Allocation in Multicell OFDMA Systems

Compared to the CDMA approach, OFDMA—the multiuser version of OFDM—provides three dimensions of diversity, i.e., time, frequency, and multiuser, for a more efficient allocation of radio resources. Upon dividing the available spectrum into multiple subchannels, the SINR of UE i in cell m on subchannel k can be expressed as

$$\gamma_{m,i}^{(k)} = \frac{g_{m,i}^{(k)} P_m^{(k)}}{\sum_{n \neq m} g_{n,i}^{(k)} P_n^{(k)} + \sigma^2}, \tag{6.7}$$

where $P_m^{(k)}$ is the transmit power of BS m on subchannel k , and $g_{m,i}^{(k)}$ the channel gain from BS m to user i on subchannel k. As can be seen, the ICI term $\sum_{n \neq m} g_{n,i}^{(k)} P_n^{(k)}$ is the major obstacle that degrades the system performance.

Using noncooperative game theory, the work of [29] solves the for resources competition in a multicellular OFDMA-based network. Assuming the interference from other UEs is fixed, the solution to the pure noncooperative game for individual UEs is indeed of the iterative waterfilling type. It may happen that some undesirable NE with low performance is obtained or, even worse, there exists no NE at all. If the cochannel interference is severe on some subchannel, the NE may not be optimal for the entire system, and there might be several NEs and/or multiple locally optimal solutions. As a result, [29] introduces the concept of "virtual referee." By mandatorily changing the game rules whenever needed, this referee can help improve the outcome of the formulated game. For example, it may reduce the transmit power of UEs whose channel conditions are unfavorable. Those generating significant interference to other UEs may as well be prohibited from using certain subchannels. In doing so, the remaining cochannel UEs can share the corresponding subchannel more effectively. Representing the cooperation among different cells, the virtual referee can be implemented at the BSs without incurring much complexity of the networks.

On the other hand, the study in [30] considers the problem of joint subchannel assignment and power allocation in the downlink of a multicell OFDMA network. Different from [29], the players in the noncooperative game are now the BSs (not the users), who are in charge of allotting subchannels to users within their servicing cells and of deciding how much power to be distributed over those subchannels. Letting c be the price per unit power and σ^2 be the power of AWGN, the utility function of BS m is given by

$$U_m(\mathbf{P}, \mathbf{A}_m) = \sum_i \sum_k a_{m,i}^{(k)} \log\left(1 + \frac{P_m^{(k)} g_{m,i}^{(k)}}{\sum_{n \neq m} P_n^{(k)} g_{n,i}^{(k)} + \sigma^2}\right) - c \sum_k P_m^{(k)}, \tag{6.8}$$

where $\mathbf{P} = [P_m^{(k)}] \succeq \mathbf{0}$ is the network power vector, and $\mathbf{A}_m = [a_{m,i}^{(k)}]$ is the channel assignment matrix of BS m with $a_{m,i}^{(k)} = 1$ if subchannel k is assigned to user i and $a_{m,i}^{(k)} = 0$ otherwise. The utility based on the power cost in (6.8) provides a measure to implicitly enforce the cooperation among the BSs. Given a network power vector $\mathbf{P}$, it is shown that each BS m assigns subchannel k to user i^* if

$$i^* = \arg\max_i \log\left(1 + \frac{P_m^{(k)} g_{m,i}^{(k)}}{\sum_{n \neq m} P_n^{(k)} g_{n,i}^{(k)} + \sigma^2}\right). \tag{6.9}$$

In such a case, $a_{m,i^*}^{(k)}(\mathbf{P}) = 1$. With a fixed optimal subchannel assignment $\mathbf{A}_m^*$ already found, the optimal power allocation is then derived as

$$P_m^{(k)} = \left(\frac{1}{c + \lambda_m^*} - \frac{\sum_{n \neq m} P_n^{(k)} g_{n,i^*}^{(k)} + \sigma^2}{g_{m,i^*}^{(k)}}\right)^+, \tag{6.10}$$

where $\lambda_m^*(\sum_k P_m^{(k)} - P_{\max}) = 0$; $\lambda_m^* \geqslant 0$ is the Lagrangian multiplier for the maximum power constraint $P_{\max}$ at BS m; and i^* denotes the user with $a_{m,i}^{(k)} = 1$. The above iterative process (i.e., finding the optimal subchannel assignment for a fixed power allocation and then determining the optimal power allocation for a given subchannel assignment) is repeated until an equilibrium is finally reached. It is proven that the iterative algorithm proposed by Kwon and Lee [30] converges to a unique NE under certain conditions.

Taking an optimization approach to solve the interference management problem in multicell OFDMA-based networks, the authors of [31] adapt three iterative strategies for the following problem of coordinated scheduling and power allocation:

$$\begin{aligned} \max_{\mathbf{P},\mathbf{i}} \quad & \sum_m \sum_k w_{i(m,k)} R_{m,i(m,k)}^{(k)} \\ \text{subject to} \quad & \sum\nolimits_k \mathbf{P}^{(k)} \preceq \mathbf{P}_{\max}. \end{aligned} \tag{6.11}$$

Here, $i(m,k)$ denotes the user i being served by BS m on subchannel k; $w_{i(m,k)} \geqslant 0$ is the weight of that user $i(m,k)$; $\mathbf{P}^{(k)} \succeq \mathbf{0}$ is the transmit power vector of all users on subchannel k; and $R_{m,i(m,k)}^{(k)} = \log\left[1 + \gamma_{m,i(m,k)}^{(k)}(\mathbf{P}^{(k)})\right]$ is the corresponding achieved throughput. It is noted that all the proposed solutions require a centralized unit to collect and process the complete channel state information. The first algorithm, a multicarrier extension of the SCALE algorithm [32], is proven convergent to a solution that satisfies the necessary optimality conditions of the original nonconvex combinatorial optimization problem. Utilizing Lagrangian dual decomposition, the second algorithm is derived that provides the optimal value in the limit that the number of available subchannels is very large. This result is obtained based mainly on the solutions of [33,34]. Finally, a multicell improved iterative waterfilling algorithm is also developed. Although not provably convergent, this algorithm can offer a solution that satisfies the necessary optimality conditions. Upon noticing that a large bandwidth for channel feedback to the centralized unit is usually unaffordable, several reduced-feedback procedures have also been suggested in [31], which give rise to more practical solutions that require a significantly reduced level of network signaling.

To overcome the low performance and/or the high complexity of the algorithms in [31], Reference [35] proposes a low-complexity and fully distributed scheme called REFIM (REFerence-based Interference Management) to solve the weighted sum rate maximization problem in the downlink of multicell networks, similar to (6.11). Upon noting that (6.11) requires intractable computational complexity to determine the globally optimal subchannel and power allocation, [35] aims at obtaining a practical solution with affordable computational and signaling overhead. Specifically, REFIM decomposes the original problem into several subproblems, each of which is solved by one BS based on the concept of reference user. Also, feedback overhead over backhaul networks is reduced both temporally and spatially.

In a similar study, Wang and Vandendorpe [36] considers the problem of maximizing the weighted sum of the minimal user rates of coordinated cells in the downlink of an OFDMA system. Here, the objective in (6.11) is redefined as

$$\sum_{m} w_m \min_{i \in \mathcal{U}_m} \sum_{k} R_{m,i}^{(k)}, \tag{6.12}$$

where $w_m \geqslant 0$ represents the weight assigned to minimal user rate of cell m, and $\mathcal{U}_m$ denotes the set of UEs belonging to cell m. It is required that multiple BSs be coordinated by a centralized resource allocation algorithm, and that a jointly optimized subchannel and power distribution be determined subject to a total power constraint $\mathbf{P}_{\max}$ at each BS. Similar to [30], the iterative algorithm proposed by Wang and Vandendorpe [36] alternatively optimizes the subcarrier assignment and the power allocation so that (6.12) keeps increasing until convergence. In each iteration, the allotment of subchannels is updated by resolving a mixed integer linear program for each cell, whereas the optimal distribution of power is found via a duality-based numerical algorithm that successively solves a set of convex problems.

For other heuristic approaches to control the interference in OFDMA-based multicell networks, the readers are referred to [37–44] and references therein.

6.4 COORDINATED MULTIPLE POINT: RECENT ADVANCES IN HOMOGENEOUS MULTICELL INTERFERENCE MANAGEMENT

6.4.1 System Model for Downlink CoMP

In order to present a common framework for the downlink CoMP transmission, let us consider a multiuser multicell network with Q coordinated cells concurrently serving K UEs, as illustrated in Figure 6.1. Unlike the multiple access techniques, namely CDMA and OFDMA, considered in Section 6.3, this CoMP system model utilizes the Space Division Multiple Access (SDMA) technique for an additional degree of freedom. Herein, it is assumed that each BS and each UE is equipped with M transmit and 1 receive antennas, respectively. It should be noted that while the discussion in this part is limited to the case of single antenna at the UEs, it can be straightforwardly extended to systems with multiple receive antennas. Denote $\mathcal{Q}$ as the set of BSs and $\mathcal{K}$ as the set of UEs. In addition, denote $\mathcal{Q}_i \subseteq \mathcal{Q}$ as the set of BSs serving UE i and $\mathcal{K}_q \subseteq \mathcal{K}$ as the set of UEs being served by BS q. Note that unlike the system model presented in Section 6.3, the intended data signals at a UE might be sent from multiple BSs, i.e., $|\mathcal{Q}_i| > 1$ under this CoMP framework.

Consider the downlink transmission to a particular UE, say UE i; its received signal y_i can be modeled as

$$y_i = \sum_{q=1}^{Q} \mathbf{h}_{q_i}^H \mathbf{x}_{q_i} + \sum_{j \neq i}^{K} \sum_{q=1}^{Q} \mathbf{h}_{q_i}^H \mathbf{x}_{q_j} + z_i, \tag{6.13}$$

where $\mathbf{x}_{q_i}$ is an $M \times 1$ complex vector representing the transmitted signal at BS q for UE i, $\mathbf{h}_{q_i}^*$ is an $M \times 1$ complex channel vector from BS q to UE i, and z_i is AWGN with power σ^2. Let s_{q_i} be the parameter indicating the assignment of UE i to BS q, where $s_{q_i} = 1$ if BS q transmits data to UE i, and $s_{q_i} = 0$ otherwise. This assignment is assumed to be known systemwide. Also note that $\mathcal{K}_q$ and $\mathcal{Q}_i$ are now defined as $\mathcal{K}_q = \left\{ \{s_{q_i}\}_{i=1}^{Q} | s_{q_i} = 1 \right\}$ and $\mathcal{Q}_i = \left\{ \{s_{q_i}\}_{q=1}^{Q} | s_{q_i} = 1 \right\}$. In the coordinated beamforming design under consideration, the transmitted signal $\mathbf{x}_{q_i}$ can be represented as $\mathbf{x}_{q_i} = s_{q_i} \mathbf{w}_{q_i} u_i$, where u_i is a complex scalar representing the signal intended for UE i, and $\mathbf{w}_{q_i}$ is an $M \times 1$ beamforming vector designed for UE i. Without loss of generality, let $\mathbb{E}[|u_i|] = 1$. It is easy to verify that the SINR at UE i is indeed

$$\gamma_i = \frac{\left| \sum_{q=1}^{Q} s_{q_i} \mathbf{h}_{q_i}^H \mathbf{w}_{q_i} \right|^2}{\sum_{j \neq i}^{K} \left| \sum_{q=1}^{Q} s_{q_j} \mathbf{h}_{q_i}^H \mathbf{w}_{q_j} \right|^2 + \sigma^2}. \tag{6.14}$$

Note that the term $\sum_{j \neq i}^{K} \left| \sum_{q=1}^{Q} s_{q_j} \mathbf{h}_{q_i}^H \mathbf{w}_{q_j} \right|^2$ denotes the interuser interference, including both intracell and intercell interferences, measured at UE i. This combined interference term is simply treated as background noise at the UE. In what follows, the CoMP downlink transmission is investigated under two design criteria: (i) CoMP for power minimization with guaranteed quality-of-service (QoS) requirements expressed in terms of the targeted achievable SINRs at the UEs, and (ii) CoMP for rate maximization under power constraints at the BSs.

6.4.2 CoMP for Power Minimization

This section investigates the CoMP beamforming strategies that minimize the transmit power at the BSs such that the target SINRs are met at all UEs, i.e. $\gamma_i \geqslant \gamma_i^{\min}, \forall i$. Depending on the knowledge of the CSI possessed at each BS and the CoMP mode employed, this section examines how the beamformers at the BSs are adjusted to ultimately maintain the SINR requirements at the UEs.

6.4.2.1 *Interference Aware*

Under the IA scheme, each UE only receives data signals from its connected BS, i.e., $|\mathcal{Q}_i| = 1$. As mentioned before, the CoMP under IA scheme can be considered as a strategic noncooperative game (SNG), where the players are the BSs and the payoff functions are the transmit powers. More specifically, each player competes with other players by choosing the downlink beamformer design that greedily minimizes its own transmit power subject to a given set of target SINRs at the UEs within its cell.

Define the precoding matrix $\mathbf{W}_q = \left[\{\mathbf{w}_{qi}\}_{\forall i \in \mathcal{K}_q}\right]$ as the strategy of BS q, and $\mathbf{W}_{-q}$ the precoding strategy of all the BSs except for BS q. Further define the set of admissible beamforming strategies $\mathbf{W}_q \in \mathcal{S}_q(\mathbf{W}_{-q})$ of cell q as $\mathcal{S}_q(\mathbf{W}_{-q}) = \{\mathbf{W}_q \in \mathbb{C}^{M \times |\mathcal{K}_q|} : \gamma_i \geqslant \gamma_i^{\min}, \forall i \in \mathcal{K}_q\}$.

For UEs connected to BS q, denote the total ICI induced by other BSs plus noise as $r_i(\mathbf{W}_{-q}) = \sum_{m \neq q}^{Q} \left\|\mathbf{W}_m^H \mathbf{h}_{m_i}\right\|^2 + \sigma^2, \forall i \in \mathcal{K}_q$. It is assumed that each UE measures this term and reports back to its connected BS. This facilitates the beamforming strategy at each BS. Furthermore, denote $\mathbf{r}_q = [\{r_i\}_{i \in \mathcal{K}_q}]^T$. Mathematically, the corresponding multicell beamforming game has the following structure:

$$\mathcal{G}_P = \left(\mathcal{Q}, \{\mathcal{S}_q(\mathbf{W}_{-q})\}_{q \in \mathcal{Q}}, \{t_q(\mathbf{W}_q)\}_{q \in \mathcal{Q}}\right), \tag{6.15}$$

where $t_q(\mathbf{W}_q) = \|\mathbf{W}_q)\|_F^2$ is the utility function, defined as the transmit power at BS q. Given the beamforming design of the others, reflected by the background noise r_i, $\forall i \in \mathcal{K}_q$, the optimal or best response strategy of BS q is the solution of the following optimization problem:

$$\begin{aligned} \min_{\mathbf{W}_q} \quad & \sum_{i \in \mathcal{K}_q} \|\mathbf{w}_{qi}\|^2 \\ \text{subject to} \quad & \frac{\left|\mathbf{w}_{qi}^H \mathbf{h}_{qi}\right|^2}{\sum_{j \neq i, j \in \mathcal{K}_q} \left|\mathbf{w}_{qj}^H \mathbf{h}_{qi}\right|^2 + r_i} \geqslant \gamma_i^{\min}, \forall i \in \mathcal{K}_q. \end{aligned} \tag{6.16}$$

Note that this single-cell downlink beamforming problem was optimally solved by several approaches in literature, such as uplink–downlink duality [45–47], and convex second-order cone programming (SOCP) [48,49]. In a multicell configuration, the problem arisen here is that when one player changes its beamforming matrix, the other players also need to change their own beamformers in order to achieve their target SINRs. It is of interest to investigate whether game $\mathcal{G}_P$ eventually converges into a stable point, i.e., an NE; and if an NE exists, does its uniqueness hold? In this case, a feasible strategy profile $\mathbf{W}^\star = \{\mathbf{W}_q^\star\}_{q=1}^{Q}$ is an NE of game $\mathcal{G}_P$ if

$$\|\mathbf{W}_q^\star\|_F^2 \leqslant \|\mathbf{W}_q\|_F^2, \quad \forall \mathbf{W}_q \in \mathcal{S}_q(\mathbf{W}_{-q}^\star), \quad \forall q \in \mathcal{Q}. \tag{6.17}$$

At the NE point, given the beamforming matrices of other cells, a BS does not have the incentive to unilaterally change its own beamforming matrix, i.e., it would consume more power to obtain the same SINR target.

To simplify the analysis of game $\mathcal{G}_P$, it is observed that the norm-1 beam patterns, $\tilde{\mathbf{w}}_{q_i} = \mathbf{w}_{q_i}/|\mathbf{w}_{q_i}|$, $\forall i \in \mathcal{K}_q$, are independent of the ICI plus noise vectors $\mathbf{r}_q$ [50]. Thus, whenever $\mathbf{r}_q$ changes, BS q only needs to adjust the allocated power for its connected UEs, but not the beam patterns. Consequently, once the beam patterns are determined, the multicell beamforming game $\mathcal{G}_P$ can be simplified as a power allocation game.

For the simplest case of $|\mathcal{K}_q| = 1$, let i be the only UE connected to BS q. As the optimal beam pattern for UE i is the maximum ratio transmitting (MRT) beamformer, i.e., $\tilde{\mathbf{w}}_{q_i} = \mathbf{h}_{q_i}/\|\mathbf{h}_{q_i}\|$, its transmit power P_i at time $t+1$ is given by

$$P_i[t+1] = \frac{\gamma_i^{\min} r_i}{\|\mathbf{h}_{q_i}\|^2} = \frac{\gamma_i^{\min}}{\hat{\gamma}_i[t]} P_i[t], \quad i \in \mathcal{K}_q, \tag{6.18}$$

where $\hat{\gamma}_i[t]$ is the measured SINR at UE i at time t. Note that if $|\mathcal{K}_q| = 1$, $\forall q$, the above power update would be exactly the same as the power control for CDMA-based wireless networks in (6.2). However, while the iterative algorithm (6.2) converges to a Pareto-optimal solution, it is not necessarily true for the iteration in (6.18). This is due to the noncooperative design of the MRT beamformers $\tilde{\mathbf{w}}_{q_i}$'s.

For a more complicated case of $|\mathcal{K}_q| > 1$, one may first determine the optimal beam patterns for UEs $i \in \mathcal{K}_q$ by solving the optimization (6.16) in the absence of the ICI. Then, with $\tilde{\mathbf{w}}_{q_i}$, $i \in \mathcal{K}_q$ known, the allocated power vector $\mathbf{p}_q = [\{P_i\}_{i \in \mathcal{K}_q}]$ for the UEs connected to BS q is updated accordingly to the amount of ICI [50]. It is shown in [50] that this power update fits into the framework of standard functions [51]. As a result, the update always converges to a unique fixed point, if such a fixed point exists. This fixed point then corresponds to the unique NE of game $\mathcal{G}_P$. Readers are referred to [50] for the necessary and sufficient conditions on the existence of game $\mathcal{G}_P$'s NE. A key observation from those conditions is that the game's NE always exists if the ICI is sufficiently small.

6.4.2.2 Interference Coordination and Joint Signal Processing

In the previous section, a fully decentralized approach to the CoMP under the IA scheme was studied using a game theory framework where the NE of the system was characterized. However, it is widely accepted that the NE need not be Pareto-efficient. via the coordination among the BSs, a significant power reduction can be obtained by jointly designing all the beamformers at the same time. Nonetheless, this advantage may come with the cost of having to pass messages among the BSs. To this end, a unified framework to analyze CoMP under both IC and JP schemes is provided with the objective of jointly minimizing the transmit power across the BSs.

Denote $\mathbf{w}_i = [\mathbf{w}_{1_i}^T, \ldots, \mathbf{w}_{Q_i}^T]^T$ as the beamformer from Q BSs to UE i. Also if we denote

$$\mathbf{h}_{i,j} = \left[\text{diag}(s_{1_j}, \ldots, s_{Q_j}) \otimes \mathbf{I}_M\right] \left[\mathbf{h}_{1_i}^T, \ldots, \mathbf{h}_{Q_i}^T\right]^T, \tag{6.19}$$

then $|\mathbf{h}_{i,j}^H \mathbf{w}_j|^2$ is effectively the interference caused by UE j's signal at UE i. The joint power minimization at the BSs with guaranteed QoS at the UEs can be stated as

$$\begin{aligned} \min_{\mathbf{w}_i} \quad & \sum_{i=1}^{K} \|\mathbf{w}_i\|^2 \\ \text{subject to} \quad & \frac{|\mathbf{h}_{i,i}^H \mathbf{w}_i|^2}{\sum_{j \neq i}^{K} |\mathbf{h}_{i,j}^H \mathbf{w}_j|^2 + \sigma^2} \geqslant \gamma_i^{\min}, \quad \forall i. \end{aligned} \tag{6.20}$$

Note that the BS assignment variable s_{qi} is not needed in the objective function of (20). This is due to the fact if BS q does not transmit data signal to UE i, i.e., $s_{qi} = 0$, allocating power to make $\mathbf{w}_{qi} \neq \mathbf{0}$ does not affect the SINR of UE i while increasing the objective function. Under this formulation, it is observed that the multicell downlink problem resembles its well-known single-cell counterpart. The difference here is the nominal channel vector $\mathbf{h}_{j,i}$, which carries the information on the assignment of UE i to the coordinated BSs. Thus, existing techniques in [45,47–49] that optimally solve the single-cell beamforming problem can be easily adapted to solve this multicell problem as well. However, because these solution approaches were initially developed for a single-cell networks, their implementations are rather centralized and thus may not be suitable for real-time multicell communications.

Among the first works that directly tackle the CoMP problem under the IC scheme, the work in [15] showed that the optimal beamformers can be found in a distributed manner. Nonetheless, the distributed implementation of the algorithm proposed in [15] comes with several implications, including perfect channel reciprocals, instant signaling exchanges, and synchronization among the coordinated BSs. To alleviate these drawbacks, one may consider a new game that retains the advantages of the beamforming game $\mathcal{G}_P$, i.e., fully distributed implementation, no message passing, and no synchronization. In this new game, instead of selfishly minimizing its transmit power as in game $\mathcal{G}_P$, each BS also attempts to minimize the ICI induced to its unconnected UEs by deploying a pricing mechanism. More specifically, the utility function at a BS, say BS q, is now defined as

$$U_q(\mathbf{W}_q) = \sum\nolimits_{i \in \mathcal{K}_q} \|\mathbf{w}_{qi}\|^2 + \sum_{j \notin \mathcal{K}_q} \pi_{qj} \|\mathbf{W}_q^H \mathbf{h}_{qj}\|^2, \tag{6.21}$$

where $\pi_{qj} \geqslant 0$ is the pricing factor and $\|\mathbf{W}_q^H \mathbf{h}_{qj}\|^2$ is the ICI at user $j \notin \mathcal{K}_q$, caused by BS q. To distinguish with other CoMP schemes, this new game is referred

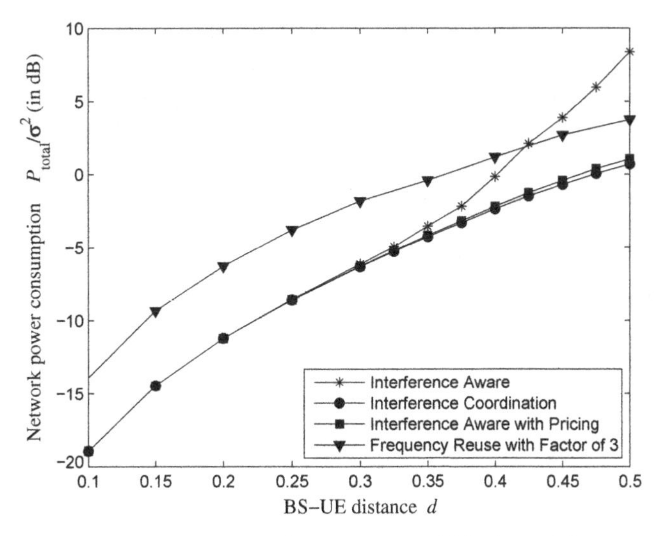

Figure 6.4 Total network power consumption to support 1 bit/s/Hz per UE versus the intercell BS–UE distance.

to as the *Interference Aware with Pricing* scheme hereafter. It is noted that this downlink beamforming game can be implemented at the system where partial channel information is available. More specifically, if the channel to UE $j \notin \mathcal{K}_q$ is known at BS q, a pricing factor $\pi_{qm_j} > 0$ is set to motivate BS q to adopt a more sociable strategy by steering its beamformers to the directions that cause less ICI to UE j. Otherwise, the pricing factor π_{qm_j} is set to 0. Certainly, the game is no longer purely competitive. As each BS adopts a more cooperative strategy, the new game's NE is typically Pareto-dominating the noncooperative game $\mathcal{G}_P$'s NE [50]. Interestingly, under a *right* pricing scheme, the new game's NE is able to approach the same optimal performance established by the IC scheme [50].

To numerically compare the power consumption in a multicell network with and without CoMP, let us consider a 3-cell network with 2 users per cell. It is assumed that the BSs are equidistant, and their distance is normalized to 1. The distance between a UE and its serving BS is also set the same, at d. Of the two UEs in each cell, one is located closely to the borders with other cells, whereas the other far away. Without CoMP, the 3 cells operate on 3 different frequency bands with a frequency reuse factor of 3. On the contrary, each cell utilizes all 3 bands when CoMP schemes (IC, IA, and IA with pricing) are employed. Herein, the IA with pricing scheme is implemented with partial channel information, where each BS also knows the channels to cell-edge UEs in the other two cells. Note that these UEs are much more susceptible to ICI than other UEs.

Figure 6.4 depicts the total network power consumption to support 1 bit/s/Hz at each UE. At small intercell UE–BS distances, the power usages of the 3 CoMP schemes are very similar to each other and about 5 dB better than that of the frequency reuse scheme. This is because the intracell channels are much stronger than the intercell channels. However, as d increases, the effect of ICI is noticeable. The NE point obtained from the IA scheme becomes inefficient compared to those of the coordinated designs and even the frequency reuse. In this case, the IC scheme, which fully controls the ICI, or the IA with pricing scheme, which controls the ICI to the highly interfered UEs, is able to preserve the performance advantages of CoMP over the frequency reuse scheme.

6.4.3 CoMP for Rate Maximization

In the previous sections, the CoMP schemes with guaranteed QoS requirements are examined. However, it may not be possible to meet these requirements at all times or the BSs may have to transmit at an excessive amount of power to maintain them. In this part, the CoMP schemes are investigated to maximize the data rates given to the UEs with strict power constraints imposed at the BSs.

6.4.3.1 *Interference Aware*

Under IA mode, each BS greedily maximizes the sum data rate of its connected UEs only. If we let P_q be the maximum transmit power at BS q, then the CoMP transmission can be mathematically modeled as a noncooperative game

$$\mathcal{G}_R = \left(\mathcal{Q}, \{\mathcal{P}_q\}_{q\in\mathcal{Q}}, \{U_q\}_{q\in\mathcal{Q}}\right), \tag{6.22}$$

where $\mathcal{P}_q = \left\{\mathbf{W}_q \in \mathbb{C}^{M\times|\mathcal{K}_q|} : \sum_{i\in\mathcal{K}_q} \|\mathbf{w}_{q_i}\|^2 \leqslant P_q\right\}$ is the set of admissible strategies at BS q and U_q is the utility function at BS q, defined as the sum rate given to its connected UEs. That is,

$$U_q(\mathbf{W}_q, \mathbf{W}_{-q}) = \sum_{i\in\mathcal{K}_q} \log\left(1 + \frac{|\mathbf{w}_{q_i}^H \mathbf{h}_{q_i}|^2}{\sum_{j\neq i, j\in\mathcal{K}_q} |\mathbf{w}_{q_j}^H \mathbf{h}_{q_i}|^2 + r_i}\right). \tag{6.23}$$

For the case of $|\mathcal{K}_q| = 1$, it is clear that BS q imposes the MRT beamformer to maximize the SINR at its only connected UE. BS q also transmits at its maximum power limit P_q. Thus, if $|\mathcal{K}_q| = 1, \forall q$, the characterization of game $\mathcal{G}_R$'s NE is immediate. There is a unique and pure NE in game $\mathcal{G}_R$, where each BS utilizes the MRT beamformer and transmits at its maximum power limit [52].

For the case of $|\mathcal{K}_q| > 1$, the utility function (6.23) is not concave in $\mathbf{W}_q$. Therefore, it is generally difficult to find the best response strategy for player q. Consequently, it is no longer straightforward to characterize the NE of game $\mathcal{G}_R$. However, should zero-forcing (ZF) precoding be applied on a per-cell basis, the intracell interference can be suppressed and the utility function at each BS then becomes a concave function of the powers allocated to its connected UEs [53].[1]

[1] ZF precoding is only possible if the number of UEs connecting to each BS does not exceed the number of transmit antennas M. We assume that this is the case for the considered system model.

Let $\mathbf{H}_q = \left[\{\mathbf{h}^*_{qi}\}_{\forall i \in \mathcal{K}_q}\right]$ be the downlink channel matrix to the $|\mathcal{K}_q|$ UEs connected to BS q. Under the sum power constraint, the optimal ZF precoding matrix $\mathbf{W}_q$ at BS q is the pseudoinverse of $\mathbf{H}_q$, i.e., $\mathbf{W}_q = \mathbf{H}_q^H(\mathbf{H}_q\mathbf{H}_q^H)^{-1}$. Note that this precoding matrix only depends on $\mathbf{H}_q$, but not on ICI plus noise vector $\mathbf{r}_q$. Thus, under ZF precoding on a per-cell basis, the sum rate maximization game $\mathcal{G}_R$ becomes a power allocation game

$$\bar{\mathcal{G}}_R = \left(\mathcal{Q}, \{\bar{\mathcal{P}}_q\}_{q\in\mathcal{Q}}, \{\bar{U}_q\}_{q\in\mathcal{Q}}\right), \tag{6.24}$$

where the set of admissible strategies is redefined as

$\bar{\mathcal{P}}_q = \{\mathbf{p}_q \in \mathbb{R}^{|\mathcal{K}_q|\times 1} : \sum_{i\in K_q} P_i \leqslant P_q, P_i \geqslant 0\}$ and the utility function is $\bar{U}_q(\mathbf{p}_q, \mathbf{p}_{-q}) = \sum_{i\in\mathcal{K}_q} \log\left(1 + \frac{P_i}{\lambda_i r_i}\right)$ with $\lambda_i = \left[(\mathbf{H}_q\mathbf{H}_q^H)^{-1}\right]_{i,i}$. In this case, the NE is obtained from a feasible strategy profile $(\mathbf{p}^\star_q, \mathbf{p}^\star_{-q})$ if

$$\bar{U}_q(\mathbf{p}^\star_q, \mathbf{p}^\star_{-q}) \geqslant \bar{U}_q(\mathbf{p}_q, \mathbf{p}^\star_{-q}), \quad \forall \mathbf{p}_q \in \bar{\mathcal{P}}_q,\ q \in \mathcal{Q}. \tag{6.25}$$

It is noted that the optimal user power allocation P_i, $i \in \mathcal{K}_q$ at BS q can be obtained by the well-known waterfilling solution $P_i = \left[\mu_q - \lambda_i r_i\right]^+$, where the water level μ_q is chosen to meet the sum power constraint. Thus, game $\bar{\mathcal{G}}_R$ fits naturally into the framework of iterative waterfilling (IWF) games [54]. Since each utility function $\bar{U}_q$ is continuous in $(\mathbf{p}_q, \mathbf{p}_{-q})$ and concave in $\mathbf{p}_q$ and each strategy set $\bar{\mathcal{P}}_q$ is convex and compact, the existence of the game's NE is always guaranteed. Readers are referred to [53] for the sufficient conditions on the uniqueness of game $\bar{\mathcal{G}}_R$'s NE.

6.4.3.2 *Interference Coordination and Joint Signal Processing*

With the IC and JP schemes, the coordinated BSs jointly maximize the network sum rate of all UEs. Under individual power constraint at each BS, the optimization can be generically stated as

$$\max_{\mathbf{w}_{qi}} \sum_{i=1}^{K} \log(1+\gamma_i), \tag{6.26}$$

where γ_i is given in (6.14). Due to the presence of the interference term, the objective function is nonconvex in $\mathbf{w}_{qi}$. Thus, finding its global minimum is highly complex and not suitable for real-time communications. On the contrary, an algorithm that can obtain at least a locally optimal solution is more desirable. If the algorithm can be implemented in a distributed manner across the coordinated BSs with only local channel information needed, it will be even more attractive.

Approaches to solve the nonconvex problem (6.26) are quite diverse in literature. For the JP scheme, ZF-based precoding was examined in [55], where an analytical solution was provided. Using successive convex approximation as a lower bound for the nonconvex part of (6.26), the work in [56] proposed an algorithm that converges monotonically to a locally optimal solution. For the IC scheme, the network sum rate can be locally maximized through an iterative minimization of the weighted mean-square error (WMMSE), as proposed in [57]. Alternatively, through an interference mechanism [58,59], the inherently nonconvex optimization problem can be decomposed and solved on a per-cell basis. This approach is considered next.

Under the IC scheme, the problem of network sum rate maximization can be stated as

$$\max_{\mathbf{w}_{qi}} \quad \sum\nolimits_{q=1}^{Q} \sum_{i \in \mathcal{K}_q} \left(1 + \frac{|\mathbf{w}_{qi}^H \mathbf{h}_{q_i}|^2}{\sum\nolimits_{j \neq i, j \in \mathcal{K}_q} |\mathbf{w}_{qj}^H \mathbf{h}_{q_i}|^2 + r_i} \right)$$
$$amp; \text{subject to } \sum\nolimits_{i \in \mathcal{K}_q} \|\mathbf{w}_{q_i}\|^2 \leqslant P_q. \tag{6.27}$$

To isolate the ICI terms that render problem (6.27) nonconvex, define the sum rate at all other BSs except BS q as $f_q(\mathbf{W}_q, \mathbf{W}_{-q}) = \sum_{m \neq q} U_m(\mathbf{W}_q, \mathbf{W}_{-q})$. At an instance of $(\mathbf{W}_q, \mathbf{W}_{-q})$, evaluated at $(\bar{\mathbf{W}}_q, \bar{\mathbf{W}}_{-q})$, if we upon taking the Taylor expansion of $f_q(\cdot)$ and retaining only the linear term, then (6.27) can be approximated by a set of Q per-cell problems

$$\max_{\mathbf{w}_{qi}} \quad \sum_{i \in \mathcal{K}_q} \log \left(1 + \frac{|\mathbf{w}_{qi}^H \mathbf{h}_{q_i}|^2}{\sum\nolimits_{j \neq i, j \in \mathcal{K}_q} |\mathbf{w}_{qj}^H \mathbf{h}_{q_i}|^2 + r_i} \right) - \sum_{j \neq \mathcal{K}_q} \pi_{qj} \left\| \mathbf{W}_q^H \mathbf{h}_{qj} \right\|^2$$
$$\text{subject to} \quad \sum_{i \in \mathcal{K}_q} \|\mathbf{w}_{q_i}\|^2 \leqslant P_q, \tag{6.28}$$

where π_{qj}, obtained from the negative derivative of $f_q(\cdot)$ with respect to $\mathbf{w}_{qi}$[60], is the pricing factor charged on the ICI caused by BS q. Clearly, (6.28) is the sum rate maximization problem at BS q with an interference pricing component to discourage the selfish behavior of BS q. Thus, this approximation approach can be regarded as the IA with pricing scheme.

It is noted that problem (6.28) is nonconvex for the general case of $|\mathcal{K}_q| > 1$. Similar to the approach in Section 6.4.3.1, ZF precoding might be applied on a per-cell basis to make optimization (6.28) convex. However, unlike the ZF precoder considered in Section 6.4.3.1, the Moore–Penrose pseudoinverse is no longer optimal in this case due to the presence of the interference pricing term. Nonetheless, the optimal solution (including optimal beamfomers and power allocation) to problem (6.28) can be analytically obtained in a closed form [60]. It is shown in [60] that the network sum rate always increases after the optimization as each BS, similar to the case of $|\mathcal{K}_q| = 1$ presented in [58,59]. As a result, the considered IA with pricing scheme always converges to a locally optimal solution under the constraint of ZF precoding on a per-cell basis.

To numerically compare the achievable network sum rates with and without CoMP, let us assume the same network configuration as in Section 6.4.2. We consider three CoMP schemes, namely IA, IA with pricing, and IC obtained from the WMMSE algorithm [57], and the frequency reuse scheme. Note that the IA and IA

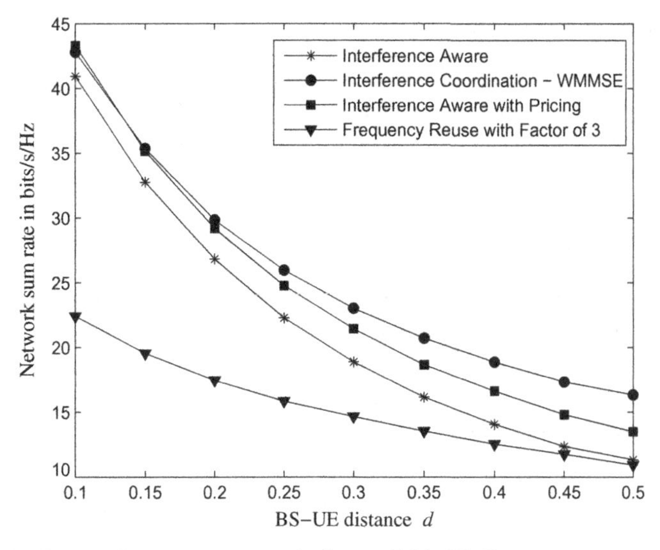

Figure 6.5 Total network sum rate versus the intercell BS–UE distance.

with pricing schemes here are obtained with the per-cell ZF precoding. Figure 6.5 depicts the achievable network sum rate versus the intercell BS–UE distance d with the total network power constraint $P_{\text{total}}/\sigma^2 = 10$ dB. As observed from this figure, at small d, CoMP schemes significantly outperform the frequency reuse scheme due to the relatively small ICI. However, at high d, the ICI becomes more apparent and the IA scheme starts losing its performance advantage. The other two CoMP schemes still perform fairly well due to their active control of the ICI.

6.5 ADVANCED INTERFERENCE COORDINATION TECHNIQUES FOR FEMTOCELL NETWORKS

It is noted that the above discussed interference management approaches might not always work well in the low SINR regimes with the transmit signals significantly attenuated. This is especially true in residential and office settings, where macrocell signals cannot reach indoor users due to the isolation by walls. Small-cell deployment overcomes these issues by bringing the network closer to the end users, through which the radio link quality is substantially enhanced due to the reduced transmitter–receiver distance. Further, a large number of femtocells allow for more efficient utilization of the radio spectrum per unit area. However, there remains the key issue of how to effectively handle the intratier and cross-tier interferences in such heterogeneous scenarios, where the interference is far more unpredictable due to the ad hoc nature of femtocell deployment.

6.5.1 CDMA-Based Femtocells

For CDMA-based femtocell networks, the power control games are formulated and analyzed by Hong et al. [61], Guruacharya et al. [62], and Chandrasekhar et al. [63]. In particular, [63] requires the macrocell user (MUE) to solve the following optimization problem:

$$\max_{0\leqslant P_0\leqslant P_{\max}} \quad U_0(P_0,\gamma_0|\mathbf{P}_{-0}) = -(\gamma_0-\gamma_0^{\min})^2, \tag{6.29}$$

where MUE is denoted as user 0 and $\gamma_i^{\min}$ is the minimum target SINR of user i. It is noteworthy that such a choice of utility function does not always guarantee the minimum required SINRs to be achieved for the prioritized MUE; rather, only a "soft" SINR is provided. On the other hand, femtocell user (FUE) i must solve the following individual problem:

$$\max_{0\leqslant P_i\leqslant P_{\max}} \quad U_i(P_i,\gamma_i|\mathbf{P}_{-i}) = R(\gamma_i,\gamma_i^{\min}) + b_i\frac{C(P_i,\mathbf{P}_{-i})}{I_i(\mathbf{P}_{-i})}, \tag{6.30}$$

where the reward function is $R(\cdot) = 1-\exp[-a_i(\gamma_i-\gamma_i^{\min})]$ and the penalty is $C(\cdot) = -g_{0,i}P_i$. Also, $I_i(\cdot)$ is the interference power at the receiver of user i, and a_i, b_i are constants. Note that because $C(\cdot)$ depends on the actual cross-tier interference $g_{0,i}$, explicit information about the cross-channel gains is required in the power control algorithm proposed by [63]. Indeed, it is quite challenging to estimate these channel values in practice due to the random fluctuations caused by shadowing and short-term fading effects.

The study in [64] presents a joint power and admission control solution for distributed interference management in two-tier CDMA-based femtocell networks. Different from [63], this work proposes an effective dynamic pricing scheme combined with admission control to indirectly manage the cross-tier interference. Together with their distributive nature, the developed schemes are more tractable in view of practical implementation under the limited backhaul networks available for femtocells. Specifically, the following utility and cost functions are used for each MUE i:

$$U_i(\gamma_i) = \frac{1}{1+\exp[-b_i(\gamma_i-c_i)]}, \tag{6.31}$$

$$C_i(P_i) = a_i^{(m)}P_i. \tag{6.32}$$

Here, b_i and c_i respectively control the steepness and the center of the sigmoid function in (6.31), whereas $a_i^{(m)}$ is the pricing coefficient. If we let $f_i(\gamma_i) = U_i'(\gamma_i)$, then it can be shown that the optimal target SINR is indeed $\gamma_i^* = f_i^{-1}\left(\frac{I_i}{g_{ii}}C_i'(P_i)\right)$. The update of power for MUE i is then

$$P_i[t+1] = \max(\gamma_i^*,\gamma_i^{\min})\frac{I_i[t]}{g_{ii}}, \tag{6.33}$$

where $I_i[t]$ is the total received interference power at time t, and g_{ii} the direct link from the transmitter to the receiver of MUE i. The choices of functions in (6.31) and (6.32) are shown capable of robustly protecting the performance of all the active MUEs.

Given that the MUEs' QoS requirements are already supported, if the FUEs also wish to maintain their respective QoS requirements, the operation of the latter may cause network congestion, and hence badly affect the performance of the MUEs. In such cases, the FUEs should be penalized by appropriately regulating their operating parameters. To balance the achieved throughput and the power expenditure for FUEs, Ngo et al. [64] choose a utility function that captures the Shannon capacity for the FUEs and a linear cost function. Altogether, the net utility for FUE j is defined as

$$U_{\text{tot},j}(\gamma_j, P_j) = W \log\left(1+\gamma_j\right) - a_j^{(f)} P_j, \tag{6.34}$$

where W denotes the system bandwidth, and $a_j^{(f)}$ the pricing coefficient. The value of $P_j \geqslant 0$ that globally maximizes $U_{\text{tot},j}$ in (6.34) can be derived as $P_j^* = \max\left(W/a_j^{(f)} - I_j/g_{jj}, 0\right)$. It is apparent that by increasing the pricing coefficient $a_j^{(f)}$, one can decrease the transmit power of FUE j, i.e., can penalize the user who creates undue cross-tier interference. This pricing mechanism can be realized without acquiring the cross-channel gain information, unlike the one in [63], which may not be affordable in two-tier networks due to limited backhaul capacity. It is important to point out that the solution of [64], while corresponding to the Nash equilibrium, is not Pareto-optimal in general. Approaching the interference management problem from the optimization perspective, the work in [65], on the other hand, attempts to come up with distributed solutions in which (i) all users attain their respective SINRs that are always optimal in the Pareto sense, (ii) every MUE i is protected with $\gamma_i \geqslant \gamma_i^{\min}$, and (iii) every FUE j has its utility globally maximized. For this purpose, a complete characterization of the Pareto-optimal boundary of the SINR feasible region is provided. Here, the complex interdependency between macrocell and femtocell networks whose access priorities are intrinsically distinct is explicitly revealed. The findings of [65] confirm that the Pareto-optimal SINRs of FUEs can only be attained conditionally upon the guaranteed performance of the macrocell network. With the introduction of new variable sets and via the load-spillage parametrization [28], it is shown that every SINR point lying on the newly characterized boundary can be realized. Based upon the specific network utility function of FUEs, which is defined as [66]

$$U(\gamma_j) := \begin{cases} \log(\gamma_j), & \text{if } \alpha = 1, \\ (1-\alpha)^{-1}\gamma_j^{1-\alpha}, & \text{if } \alpha \geqslant 0, \end{cases} \tag{6.35}$$

and also the minimum SINR requirements of MUEs, a unique operating SINR point is determined. Finally, transmit power adaptation based on the well-known Foschini–Miljanic's algorithm [19] is performed to attain such a design target. As can be observed from Figure 6.6, the proposed algorithm in [65] converges to the globally optimal solution for different utilities. Notably, while the performance of

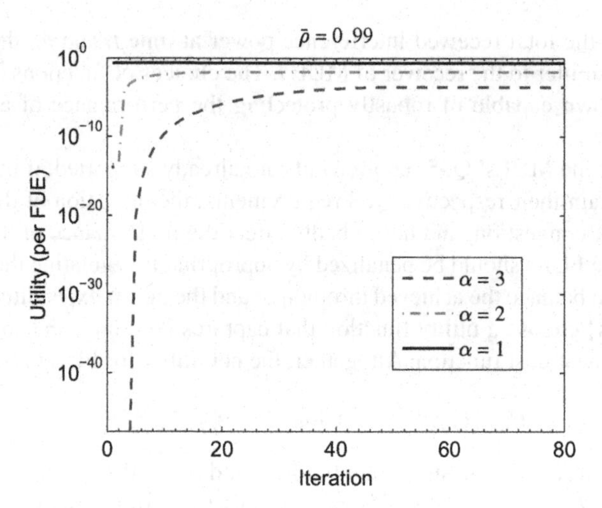

Figure 6.6 Convergence of FUE's utility.

the femtocell network is optimized, the minimum SINRs prescribed for MUEs are always guaranteed (see Table 6.1).

6.5.2 OFDMA-Based Femtocells

In the context of OFDMA-based femtocell networks, the work in [67] investigates the joint allocation of resource blocks and transmit powers. Considering the downlink, the utility of each of the femtocell BSs, which are the players in the underlying game, includes not only the system capacity of the femtocells, but also other sources of interference (i.e., femtocell to macrocell, macrocell to femtocell, and femtocell to femtocell). The formulated game belongs to the class of exact potential game, which has been shown to always converge to an NE when a best response adaptive strategy is applied. Similar to [30], the solution to this game is an iterative process in which (i) the optimal resource block allocation is determined given a transmit power policy, (ii) a waterfilling allocation of power for femtocells is computed for a fixed resource block allocation, and (iii) steps (i) and (ii) are repeated until convergence.

Table 6.1 SINR of MUEs for $(\mathbf{H}_{11}\mathbf{D}(\mathbf{\Gamma}_m^{th})) = 0.9526 < \rho = 0.99$ and $\alpha = 2$.

MUE Index	1	2	3	4	5	6	7	8	9	10
Target SINR	1.578	1.507	1.440	1.376	1.315	1.256	1.200	1.147	1.095	1.047
Achieved SINR	1.578	1.507	1.440	1.376	1.315	1.256	1.200	1.147	1.095	1.047

As discussed earlier, there is an ultimate need to protect the preferential MUEs in heterogeneous macrocell/femtocell network settings. The study of [68] considers the downlink of a spectrum underlay OFDMA-based cognitive radio (CR) network, in which the secondary BSs are allowed to transmit on the same subchannel with the primary users (PUs) as long as the generated interference is limited to an acceptable level. It can be argued that the roles of MUEs and FUEs in macrocell/femtocell settings are very similar to those of primary users (PUs) and secondary users (SUs) in CR networks, respectively. Therefore, the results obtained in [68] can be directly applicable to two-tier networks.

Given a threshold ϵ and a limit $I_{\text{lim}}^{\text{PU}}$, it is required that [68]

$$\Pr\left[I_j^{\text{PU}} > I_{\text{lim}}^{\text{PU}}\right] \leqslant \epsilon, \tag{6.36}$$

where I_j^{PU} is the total interference at the receiver of PU j. To this end, it is recommended that the secondary BSs limit their transmit power on the subchannels over which the PUs are currently active. Subject to this power constraint, Choi et al. [68] proposes a joint subchannel and transmission power allocation scheme that maximizes the capacity of the CR network, through which the ICI among different CR cells is also controlled. Using the Lagrangian dual method, the original design problem is decomposed into multiple subproblems in the dual domain, each of which is solved by an efficient algorithm. Attractively, the developed solution can be implemented in a distributive manner. Although the dual-domain solution does not always coincide with the primal one, Yu and Lui [34] have proven that the duality gap approaches zero when the number of OFDMA subchannels is sufficiently large. Numerical examples have shown the trade-off between the total achieved rates of SU for different limits on the interference imposed to the PUs. In particular, more SUs throughput has to be sacrificed for a lower value of $I_{\text{lim}}^{\text{PU}}$. It is also apparent from those numerical results that the solution proposed by Choi et al. [68] outperforms the fixed subchannel allocation scheme.

6.6 MULTICELL INTERFERENCE COORDINATION WITH EXPLICIT ENERGY-EFFICIENT DESIGN OBJECTIVE

The previous sections discuss interference coordination with the main objective of gaining higher spectral efficiency. As the green evolution becomes a major trend in the wireless communication field, energy-efficient transmission plays a more important role [69]. Unfortunately, achieving higher spectral efficiency and realizing better energy efficiency are not always consistent. As they can sometimes conflict with each other, it is imperative to carefully study how to balance these two design metrics [8].

6.6.1 Trade-off between Spectral Efficiency and Energy Efficiency

It is widely accepted that a higher SINR level at the output of the receiver will result in a lower bit error rate, which in turn translates into a higher throughput.

This can be easily justified from Shannon's formula that defines the achievable transmission rate as

$$R = W \log_2\left(1 + \frac{P}{WN_o}\right) \quad \text{[b/s]}, \tag{6.37}$$

where P is the transmit power, W is the system bandwidth and N_o is the noise power spectral density. The spectral efficiency (SE) can then be expressed as

$$\nu_{SE} = \frac{R}{W} = \log_2\left(1 + \frac{P}{WN_o}\right) \quad \text{[b/s/Hz]}. \tag{6.38}$$

Achieving a high SINR level, however, usually requires a high transmit power, implying low battery life of user terminals. Therefore, transmission schemes that ensure an efficient use of energy are crucial to prolonging the operation time of the user equipment. When energy efficiency (EE) is the main design objective, the following figure-of-merit can be used [70]:

$$\nu_{EE} = \frac{R}{P} = \frac{W \log_2\left(1 + \frac{P}{WN_o}\right)}{P} \quad \text{[b/J]}. \tag{6.39}$$

In fact, ν_{EE} represents the total number of bits that can be transmitted reliably to the receiver per Joule of energy consumed. Capturing very well the trade-off between throughput and energy expenditure,(6.39) is therefore especially appropriate for applications where EE is more important than throughput maximization. From (6.38) and (6.39), it is easy to see that

$$\nu_{EE} = \frac{\nu_{SE}}{(2^{\nu_{SE}} - 1)N_o}. \tag{6.40}$$

Evidently, ν_{EE} decreases monotonically with respect to ν_{SE}. In particular, $\nu_{EE} \to \nu_{EE}^{max} = 1/(N_o \ln 2)$ as $\nu_{SE} \to 0$, whereas $\nu_{EE} \to \nu_{EE}^{min} = 0$ as $\nu_{SE} \to \infty$.

In practical systems, electronic circuit energy consumption needs to be taken into account in the definition of EE as follows [71]:

$$\nu_{EE} = \frac{W \log_2\left(1 + \frac{P}{WN_o}\right)}{P + P_c} \quad \text{[b/J]}. \tag{6.41}$$

With circuit power P_c, the monotonic relation between ν_{EE} and ν_{SE} is no longer applicable. As shown in Figure 6.7, the EE–SE curve changes from the cup shape to the bell shape. Also clear from Figure 6.7 is that a lower energy efficiency level is realized for higher circuit power.

6.6.2 Energy-Efficient Interference Management in Multicell Networks

In [72], an energy-efficient power optimization scheme is proposed for multicarrier multicell systems. It is assumed that there is no cooperation among the users and that all users apply the same policy with their own local information. Specifically,

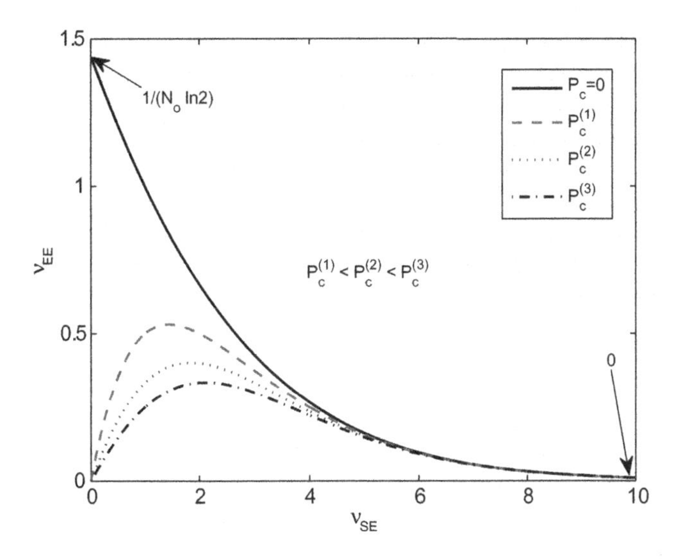

Figure 6.7 Trade-off between energy and spectral efficiency.

the interference on subchannel k (with bandwidth $B_{\rm sc}$) at the receiver of user i can be expressed as $I_i^{(k)} = \sum_{j \neq i} g_{j,i}^{(k)} P_j^{(k)}$. The data rate of user i across all subchannels is then

$$R_i = \sum_k \log_2 \left(1 + \frac{g_{i,i}^{(k)} P_i^{(k)}}{I_i^{(k)} + B_{\rm sc} N_o} \right). \tag{6.42}$$

The total transmit power $P_i = \sum_k P_i^{(k)}$ is used for reliable data transmission, whereas circuit power $P_{\rm c}$ is the average energy consumption of device electronics. The energy efficiency of user i is then simply

$$\nu_i = \frac{R_i}{P_i + P_{\rm c}}. \tag{6.43}$$

Given the power allocation of all other users, $\mathbf{P}_{-i}$, each user i is required to solve the following best-response problem:

$$f_n(\mathbf{P}_{-i}) = \arg\max_{\mathbf{P}_i} \nu_i(\mathbf{P}_i, \mathbf{P}_{-i}). \tag{6.44}$$

As $\nu_i(\mathbf{P}_i, \mathbf{P}_{-i})$ is strictly quasiconcave[2] with respect to $\mathbf{P}_i$, it is shown that there exists at least one equilibrium point in this power control game. In frequency-selective channels, the NE is unique if

- $\| f_n(\mathbf{P}_{-i}) - f_n(\check{\mathbf{P}}_{-i}) \| < \| \mathbf{P}_{-i} - \check{\mathbf{P}}_{-i} \|$ for any different $\mathbf{P}_{-i}$ and $\check{\mathbf{P}}_{-i}$, or
- $\| \frac{\partial \mathbf{I}_i}{\partial \mathbf{P}_{-i}} \| < 1/\sup_{\mathbf{I}_i} \| \frac{\partial f_i}{\partial \mathbf{I}_i} \|$.

[2]A function f defined on a convex subset $\mathcal{S}$ of a real n-dimensional vector space is strictly quasiconcave if $f(\lambda \mathbf{x} + (1-\lambda)\mathbf{y}) > \min\{f(\mathbf{x}), f(\mathbf{y})\}$ for all $\mathbf{x}, \mathbf{y} \in \mathcal{S}$, $\mathbf{x} \neq \mathbf{y}$, and $\lambda \in (0,1)$.

It has been shown that the energy-efficient power allocation developed by Miao et al. [72] can enhance the total network energy efficiency $\nu_{tot} = \sum_i \nu_i$ in interference-limited scenarios. Moreover, it refines the trade-off between energy and spectral efficiencies thanks to the conservativeness of the proposed allocation. As the interference from other cells is restricted, the network throughput is also improved.

On the other hand, the work of [73] devises a distributed energy-efficient joint power control and BS assignment algorithm in multicell CDMA-based wireless systems with the use of pricing. Assuming each user transmits data at the rate of R over the bandwidth W, the utility of each user i received at its assigned BS a_i is defined as

$$u_{a_i,i}(\mathbf{P}) = \frac{RL}{M}\frac{f(\gamma_{a_i,i})}{P_i} \quad \text{[b/J]}, \tag{6.45}$$

where $f(\cdot)$ is the efficiency function that approximates the probability of successful reception, L is the number of information bits in a packet of size M, and $\gamma_{a_i,i}$ is the corresponding received SINR at BS a_i. Because each BS has a different degree of congestion, a local pricing factor c_{a_i} should be determined and enforced based on the amount of traffic offered in its service area. With a linear pricing scheme, the power control game can then be expressed as

$$\max_{P_i,a_i} u^c_{a_i,i}(P_i,\mathbf{P}_{-i}) = u_{a_i,i}(P_i,\mathbf{P}_{-i}) - c_{a_i} P_i \tag{6.46}$$

for all user i. Here, the pricing factor c_{a_i} is broadcast by BS a_i, and each user would then use the pricing factor associated with its BS assignment in computing the optimal power to maximize its net utility in (6.46).

The difficulty in resolving the optimization with two-dimensional space, i.e., transmit power P_i and base station a_i, in (6.46) can be overcome upon proving that [73]

$$\max_{P_i,a_i} u^c_{a_i,i}(P_i,\mathbf{P}_{-i}) \equiv \max_{P_i} [\max_{a_i} u_{a_i,i}(P_i,\mathbf{P}_{-i})] - c_{a_i} P_i. \tag{6.47}$$

Therefore, the joint design problem can be simplified to first solving the BS assignment as a maximum SINR assignment and then selecting transmit power that maximizes user total utility. Specifically, the original problem (6.46) is resorted to $\max_{P_i} u_i(P_i,\mathbf{P}_{-i}) - c_i P_i$ where $u_i(\mathbf{P}_i) = \max_j u_{j,i}(\mathbf{P}) = \max_j \gamma_{j,i}$.Numerical results show that the improvement in energy efficiency with the use of linear pricing is above 25%. Ultimately, this signifies the advantages of incorporating pricing in the design of energy-efficient resource allocation schemes.

6.7 CONCLUSIONS

6.7.1 Chapter Summary

The deployment of any cellular networks often poses two apparently conflicting goals: spectral efficiency and energy efficiency. On one aspect, the demands for higher throughput and higher QoS to mobile end users may drive the deployment of more and more BSs for better spectral efficiency. On the other, there is a major concern about the environmental effects of operating a huge number of cellular sites due to their high energy consumption. With the current trend of promoting universal frequency reuse in cellular networks for better utilization of radio resources, there is a certain need to explore and implement new communication paradigms that proactively deal with the imminent issue of intercell interference. Effective coordination in intercell interference is hence the key to optimizing the two design goals towards a greener cellular network.

In this chapter, several state-of-the-art techniques in controlling ICI under different multicell network settings have been surveyed. For conventional homogeneous networks, this chapter has delineated various CoMP schemes to efficiently coordinate or even take advantage of the ICI. For heterogeneous networks, the implementation of femtocell networks and their interference management mechanisms have been discussed in detail. Finally, this chapter presented many current advances in ICI coordination to improve the energy efficiency of the cellular networks while maintaining a good trade-off with the spectral-efficiency goal.

6.7.2 Some Potential Research Directions

Although current research efforts on CoMP have shown tremendous potential in enhancing both spectral and energy efficiency of coordinated networks, there remain several challenges in implementing CoMP solutions in practice. First, the improvement offered by CoMP relies on the availability of channel state information (CSI) knowledge across all coordinated cells [3]. In practical scenarios, to accurately acquire and quickly exchange such information is not a trivial task at all. Hence, the effects of quantization, fast-varying channels, and CSI feedback delay should be taken into consideration in the design of any CoMP schemes. Second, it has been shown that most resource allocation problems in CoMP are inherently nonconvex. Typically difficult to solve for global optimality, these problems call for suboptimal algorithms that can be efficiently implemented [13]. A potential future research direction is to devise practical algorithms that give a good trade-off between achieving optimal performance and incurring low computational complexity. It is desirable that these algorithms should be robust to CSI uncertainty while allowing for distributed implementation with only local CSI knowledge required. Lastly, as energy efficiency is a rather new design criterion in CoMP, it should be further discussed and addressed in the upcoming CoMP standardization process.

On the other hand, it is believed that small-cell networks have the real potential to offer substantial energy savings. Nonetheless, such a deployment can only be more energy-efficient if properly operated [74]. Further research is needed to determine the optimal cell sizes and the locations to deploy femtocell BSs, along with taking into account the energy expended for the backhaul and signaling overhead. In doing so, the energy consumption of macrocell BSs and user equipments can be further decreased for a given system performance [69]. Moreover, it is predicted that the use of cooperative relaying can help improve the network coverage of femtocells without introducing significant interference to the existing macrocell [75]. For such a network architecture, power-efficient resource allocation techniques (e.g., energy-efficient modulation, selective relaying) should be devised and adapted to better realize the potential energy savings.

REFERENCES

[1] G.P. Fettweis, E. Zimmermann, ICT energy consumption – Trends and challenges, in: Proceedings of the International Symposium on Wireless Personal Multimedia Communications (WPMC), Lapland, Finland, September 2008.

[2] K. Pentikousis, In search of energy-efficient mobile networking, IEEE Communications Magazine 48 (1) (2010) 95–103.

[3] 4G. Americas, 4G Mobile Broadband Evolution: 3GPP Release 10 and Beyond, February 2011.

[4] H. Claussen, L.T.W. Ho, L.G. Samuel, An overview of the femtocell concept, Bell Labs Technical Journal 3 (1) (2008) 221–245.

[5] D. Lopez-Perez, A. Valcarce, G. de la Roche, J. Zhang, OFDMA femtocells: A roadmap on interference avoidance, IEEE Communications Magazine 47 (9) (2000) 41–48.

[6] C. Forster, I. Dickie, G. Maile, H. Smith, M. Crisp, Understanding the environmental impact of communication systems, 2009, Report for Ofcom (online). <http://stakeholders.ofcom.org.uk/binaries/research/technologyresearch/environ.pdf>

[7] H. Claussen, L. Ho, F. Pivit, Effects of joint macrocell and residential picocell deployment on the network energy efficiency, in: Proceedings of the IEEE International Symposium on Personal, Indoor and Mobile Radio Communications (PIMRC), September 2008, pp. 1–6.

[8] G. Miao, N. Himayat, Y.G. Li, A. Swami, Cross-layer optimization for energy-efficient wireless communications: A survey, Wirel. Commun. and Mob. Comp. 9 (4) (2009) 529–542 http://dx.doi.org/10.1002/wcm.698 (online)

[9] D. Lopez-Perez, I. Guvenc, G. de la Roche, M. Kountouris, T. Quek, J. Zhang, Enhanced intercell interference coordination challenges in heterogeneous networks, IEEE Wireless Communications Magazine 18 (3) (2011) 22–30.

[10] V. Chandrasekhar, J.G. Andrews, A. Gatherer, Femtocell networks: A survey, IEEE Communications Magazine 46 (9) (2008) 59–67.

[11] C. Patel, M. Yavuz, S. Nanda, Femtocells [industry perspectives], IEEE Wireless Communications 17 (5) (2010) 6–7.

[12] M. Karakayali, G. Foschini, R. Valenzuela, Network coordination for spectrally efficient communications in cellular systems, IEEE Wireless Communications 13 (4) (2006) 56–61.

[13] D. Gesbert, S. Hanly, H. Huang, S.S. Shitz, O. Simeone, W. Yu, Multi-cell MIMO cooperative networks: A new look at interference, IEEE Journal on Selected Areas in Communications 28 (9) (2010) 1380–1408.

[14] L. Venturino, N. Prasad, X. Wang, Coordinated linear beamforming in downlink multi-cell wireless networks, IEEE Transactions on Wireless Communications 9 (4) (2010) 1451–1461.

[15] H. Dahrouj, W. Yu, Coordinated beamforming for the multicell multi-antenna wireless system, IEEE Transactions on Wireless Communications 9 (5) (2010) 1748–1759.

[16] M. Yavuz, F. Meshkati, S. Nanda, A. Pokhariyal, N. Johnson, B. Raghothaman, A. Richardson, Interference management and performance analysis of UMTS/HSPA+femtocells, IEEE Communications Magazine 47 (9) (2009) 102–109.

[17] H.-S. Jo, C. Mun, J. Moon, J.-G. Yook, Interference mitigation using uplink power control for two-tier femtocell networks, IEEE Transactions on Wireless Communications 8 (10) (2009) 4906–4910.

[18] G.d.l. Roche, A. Valcarce, D. Lopez-Perez, J. Zhang, Access control mechanisms for femtocells, IEEE Communications Magazine 48 (1) (2010) 33–39.

[19] G.J. Foschini, Z. Miljanic, A simple distributed autonomous power control algorithm and its convergence, IEEE Transactions on Vehicular Technology 42 (4) (1993) 641–646.

[20] H. Ji, C.-Y. Huang, Non-cooperative uplink power control in cellular radio systems, Wireless Network 4 (3) (1998) 233–240.

[21] Z. Han, K.J.R. Liu, Noncooperative power-control game and throughput game over wireless networks, IEEE Transactions on Communications 53 (10) (2005) 1625–1629.

[22] J.W. Lee, R.R. Mazumdar, N.B. Shroff, Downlink power allocation for multi-class wireless systems, IEEE/ACM Transactions on Networking 13 (4) (2005) 854–867.

[23] E. Altman, T. Boulogne, R. El-Azouzi, T. Jiminez, L. Wynter, A survey of network games in telecommunications, Computer and Operations Research (2006) 286–311.

[24] C.U. Saraydar, N.B. Mandayam, D.J. Goodman, Efficient power control via pricing in wireless data networks, IEEE Transactions on Communications 50 (2) (2002) 291–303.

[25] M. Xiao, N.B. Shroff, E.K.P. Chong, A utility-based power control scheme in wireless cellular systems, IEEE/ACM Transactions on Networking 11 (2) (2003) 210–221.

[26] S. Koskie, Z. Gajic, A Nash game algorithm for SIR-based power control in 3G wireless CDMA networks, IEEE/ACM Transactions on Networking 13 (5) (2005) 1017–1026.

[27] M. Rasti, A.R. Sharafat, B. Seyfe, Pareto-efficient and goal-driven power control in wireless networks: A game-theoretic approach with a novel pricing scheme, IEEE/ACM Transactions on Networking 17 (2) (2009) 556–569.

[28] P. Hande, S. Rangan, M. Chiang, X. Wu, Distributed uplink power control for optimal SIR assignment in cellular data networks, IEEE/ACM Transactions on Networking 16 (6) (2008) 1420–1433.

[29] Z. Han, Z. Ji, K. Liu, Non-cooperative resource competition game by virtual referee in multi-cell OFDMA networks, IEEE Journal on Selected Areas in Communications 25 (6) (2007) 1079–1090.

[30] H. Kwon, B.G. Lee, Distributed resource allocation through noncooperative game approach in multi-cell OFDMA systems, in: Proceedings of the IEEE International Conference on Communications (ICC), vol. 9, 2006, pp. 4345–4350.

[31] L. Venturino, N. Prasad, X. Wang, Coordinated scheduling and power allocation in downlink multicell OFDMA networks, IEEE Transactions on Vehicular Technology 58 (6) (2009) 2835–2848.

[32] J. Papandriopoulos, J. Evans, SCALE: A low-complexity distributed protocol for spectrum balancing in multiuser DSL networks, IEEE Transactions on Information Theory 55 (8) (2009) 3711–3724.

[33] R. Cendrillon, W. Yu, M. Moonen, J. Verlinden, T. Bostoen, Optimal multiuser spectrum balancing for digital subscriber lines, IEEE Transactions on Communications 54 (5) (2006) 922–933.

[34] W. Yu, R. Lui, Dual methods for nonconvex spectrum optimization of multicarrier systems, IEEE Transactions on Communications 54 (7) (2006) 1310–1322.

[35] K. Son, S. Lee, Y. Yi, S. Chong, REFIM: A practical interference management in heterogeneous wireless access networks, IEEE Journal on Selected Areas in Communications 29 (6) (2011) 1260–1272.

[36] T. Wang, L. Vandendorpe, Iterative resource allocation for maximizing weighted sum min-rate in downlink cellular OFDMA systems, IEEE Transactions on Signal Processing 59 (1) (2011) 223–234.
[37] G. Li, H. Liu, Downlink radio resource allocation for multi-cell OFDMA system, IEEE Transactions on Wireless Communications 5 (12) (2006) 3451–3459.
[38] H. Zhang, L. Venturino, N. Prasad, P. Li, S. Rangarajan, X. Wang, Weighted sum-rate maximization in multi-cell networks via coordinated scheduling and discrete power control, IEEE Journal on Selected Areas in Communications 29 (6) (2011) 1214–1224.
[39] K. Yang, N. Prasad, X. Wang, An auction approach to resource allocation in uplink OFDMA systems, IEEE Transactions on Signal Processing 57 (11) (2009) 4482–4496.
[40] I. Koutsopoulos, L. Tassiulas, Cross-layer adaptive techniques for throughput enhancement in wireless OFDM-based networks, IEEE/ACM Transactions on Networking 14 (5) (2006) 1056–1066.
[41] K. Yang, N. Prasad, X. Wang, A message-passing approach to distributed resource allocation in uplink DFT-spread-OFDMA systems, IEEE Transactions on Communications 59 (4) (2011) 1099–1113.
[42] N. Ksairi, P. Bianchi, P. Ciblat, W. Hachem, Resource allocation for downlink cellular OFDMA systems – Part I: Optimal allocation, IEEE Transactions on Signal Processing 58 (2) (2010) 720–734.
[43] M. Pischella, J.-C. Belfiore, Weighted sum throughput maximization in multicell OFDMA networks, IEEE Transactions on Vehicular Technology 59 (2) (2010) 896–905.
[44] B. Da, R. Zhang, Cooperative interference control for spectrum sharing in OFDMA cellular systems, in: Proceedings of the IEEE International Conference on Communications (ICC), June 2011, pp. 1–5.
[45] F. Rashid-Farrokhi, L. Tassiulas, K.J. Liu, Joint optimal power control and beamforming in wireless networks using antenna arrays, IEEE Transactions on Communications 46 (10) (1998) 1313–1323.
[46] E. Visotsky, U. Madhow, Optimum beamforming using transmit antenna arrays, in: Proceedings of the IEEE Vehicular Technology Conference, vol. 1, May 1999.
[47] M. Schubert, H. Boche, Solution of the multiuser downlink beamforming problem with individual SINR constraints, IEEE Transactions on Vehicular Technology 53 (2004) 18–28.
[48] A. Wiesel, Y.C. Eldar, S. Shamai, Linear precoding via conic optimization for fixed MIMO receivers, IEEE Transactions on Signal Processing 54 (2006) 161–176.
[49] W. Yu, T. Lan, Transmitter optimization for the multi-antenna downlink with per-antenna power constraints, IEEE Transactions on Signal Processing 55 (6) (2007) 2646–2660.
[50] D.H.N. Nguyen, T. Le-Ngoc, Multiuser downlink beamforming in multicell wireless systems: A game theoretical approach, IEEE Transactions on Signal Processing 59 (7) (2011) 3326–3338.
[51] R.D. Yates, A framework for uplink power control in cellular radio systems, IEEE Journal on Selected Areas in Communications 13 (7) (1995) 1341–1347.
[52] E. Larsson, E. Jorswieck, Competition versus cooperation on the MISO interference channel, IEEE Journal on Selected Areas in Communications 26 (7) (2008) 1059–1069.
[53] H. Nguyen-Le, D.H.N. Nguyen, T. Le-Ngoc, Game-based zero-forcing precoding for multicell multiuser transmissions, in: Proceedings of the Vehicular Technology Conference, San Francisco, CA, September 2011.
[54] W. Yu, G. Ginis, J.M. Cioffi, Distributed multiuser power control for digital subscriber lines, IEEE Journal on Selected Areas in Communications 20 (5) (2002) 1105–1115.

[55] R. Zhang, Cooperative multi-cell block diagonalization with per-base-station power constraints, IEEE Journal on Selected Areas in Communications 28 (9) (2010) 1435–1445.
[56] C.T.K. Ng, H. Huang, Linear precoding in cooperative MIMO cellular networks with limited coordination clusters, IEEE Journal on Selected Areas in Communications 28 (9) (2010) 1446–1454.
[57] Q. Shi, M. Razaviyayn, Z.-Q. Luo, C. He, An iteratively weighted MMSE approach to distributed sum-utility maximization for a MIMO interfering broadcast channel, IEEE Transactions on Signal Processing 59 (9) (2011) 4331–4340.
[58] C. Shi, R.A. Berry, M.L. Honig, Monotonic convergence of distributed interference pricing in wireless networks, in: Proceedings of the IEEE International Symposium on Information Theory, Seoul, Republic of Korea, June–July 2009, pp. 1619–1623.
[59] S.-J. Kim, G.B. Giannakis, Optimal resource allocation for MIMO ad hoc cognitive radio networks, IEEE Transactions on Information Theory 57 (5) (2011) 3117–3131.
[60] D.H.N. Nguyen, T. Le-Ngoc, Block Diagonal Precoding in Multicell MIMO Systems: Competition and Coordination, in preparation.
[61] E.J. Hong, S.Y. Yun, D.-H. Cho, Decentralized power control scheme in femtocell networks: A game theoretic approach, in: Proceedings of the IEEE International Symposium on Personal, Indoor and Mobile Radio Communications (PIMRC), September 2009, pp. 415–419.
[62] S. Guruacharya, D. Niyato, E. Hossain, D.I. Kim, Hierarchical competition in femtocell-based cellular networks, in: Proceedings of the IEEE Global Telecommunications Conference (GLOBECOM), December 2010, pp. 1–5.
[63] V. Chandrasekhar, J.G. Andrews, T. Muharemovic, Z. Shen, Power control in two-tier femtocell networks, IEEE Transactions on Wireless Communications 8 (8) (2009) 4316–4328.
[64] D.T. Ngo, L.B Le, T. Le-Ngoc, E. Hossain, D.I. Kim, Distributed interference management in femtocell networks, in: Proceedings of the IEEE Vehicular Technology Conference (VTC-Fall), San Franciso, CA, September 2011.
[65] D.T. Ngo, L.B. Le, T. Le-Ngoc, Distributed Pareto-optimal power control in femtocell networks, in: Proceedings of the IEEE International Symposium on Personal, Indoor and Mobile Radio Communications (PIMRC), Toronto, ON, Canada, September 2011.
[66] J. Mo, J. Walrand, Fair end-to-end window-based congestion control, IEEE/ACM Transactions on Networking 8 (5) (2000) 556–567.
[67] L. Giupponi, C. Ibars, Distributed interference control in OFDMA-based femtocells, in: Proceedings of the IEEE International Symposium on Personal, Indoor and Mobile Radio Communications (PIMRC), September 2010, pp. 1201–1206.
[68] K.W. Choi, E. Hossain, D.I. Kim, Downlink subchannel and power allocation in multi-cell OFDMA cognitive radio networks, IEEE Transactions on Wireless Communications 10 (7) (2011) 2259–2271.
[69] Z. Hasan, H. Boostanimehr, V. Bhargava, Green cellular networks: A survey, some research issues and challenges, IEEE Communications Surveys and Tutorials 13 (4) (2011) 524–540 (fourth quarter)
[70] Y. Chen, S. Zhang, S. Xu, G. Li, Fundamental trade-offs on green wireless networks, IEEE Communications Magazine 49 (6) (2011) 30–37.
[71] G. Miao, N. Himayat, G. Li, Energy-efficient link adaptation in frequency-selective channels, IEEE Transactions on Communications 58 (2) (2010) 545–554.
[72] G. Miao, N. Himayat, G. Li, S. Talwar, Distributed interference-aware energy-efficient power optimization, IEEE Transactions on Wireless Communications 10 (4) (2011) 1323–1333.

[73] C.U. Saraydar, N.B. Mandayam, D.J. Goodman, Pricing and power control in a multicell wireless data network, IEEE Journal on Selected Areas in Communications 19 (10) (2001) 1883–1892.
[74] J. Hoydis, M. Kobayashi, M. Debbah, Green small-cell networks, IEEE Transactions on Vehicular Technology 6 (1) (2011) 37–43.
[75] R. Pabst, B. Walke, D. Schultz, P. Herhold, H. Yanikomeroglu, S. Mukherjee, H. Viswanathan, M. Lott, W. Zirwas, M. Dohler, H. Aghvami, D. Falconer, G. Fettweis, Relay-based deployment concepts for wireless and mobile broadband radio, IEEE Communications Magazine 42 (9) (2004) 80–89.

CHAPTER

Energy-Efficient Green Radio Communications for Delay Tolerant Applications

7

Ashok Karmokar, Alagan Anpalagan

Ryerson University, 350 Victoria Street, Toronto, ON, Canada M5B 2K3

7.1 INTRODUCTION

With the continual increase of dependence on electrical systems in our daily lives, the consumption of electrical energy is increasing at a rapid pace. The demand and the price of electrical energy vary according to time of the day/month/year in order to lower the production cost and stabilize generation systems. The information and communication technology (ICT) industry sector is itself responsible for a significant portion of total global CO_2 emission and global warming. According to a recent study, ICT accounts for about 6% of the energy consumption. To save Mother Earth from greenhouse gas, it is therefore crucial to optimize and schedule energy consumption in every use in our daily life. In recent years, the use of handheld communications devices has increased exponentially due to widespread use of high bandwidth and high speed wireless data applications. This calls for a smart energy-optimized packet scheduler in wireless devices that judiciously schedule packet transmission over fading channels so that minimum energy is used yet other quality of service (QoS) requirements for the traffic are maintained. The associated problem become a dynamic packet transmission policy with the randomly varying channel and the data packet arrivals from the wireless applications.

Energy-efficient green radio is very important in order to reduce carbon emission due to inefficient use and hence extra generation of electrical energy. Currently, many electrical systems are not optimized at the system level. Hence, additional energy is required due to several individual and/or local optimizations in a system. In this chapter, we discuss the energy optimization issue for a generalized case of cross-layer packet adaptations in wireless networks. We study a system where time-varying packets are coming from higher-layer applications whose instantaneous state may be unknown at the transmitter. The transmitter equipped with a finite buffer is transmitting over a wireless fading channel whose gains are correlated, but the exact instantaneous state of the channel is also unknown at the time of transmission. Only the statistical models for the traffic and channel are known. The job of the controller at the transmitter is to make a decision on the number of packets to be transmitted in a particular time-slot and the corresponding transmission power so that the long-term average energy of the transmission is minimized within the allowable quality of service requirements, such as packet buffering delay, packet error rate, and overflow. We show that by judiciously taking action based on cross-layer interaction and information, the scheduler can save a significant amount of energy, thus making the transmitter green. Therefore, it helps to reduce the electrical energy generation and the carbon footprint.

7.1.1 Why Green Design Is Important for Wireless Systems

Wireless devices communicate over air medium where the power gain is randomly varying with time, space, and frequency. Contrary to wired or optical communications media, the channel gain could be too low sometimes. This causes erroneous reception at the receiver and requires higher transmitter power for transmitting the same amount of information data within a given QoS requirement. On the other hand, at other times the gain of the channel could be high in the peak, which permits using lower transmission power for transmitting the same amount of information data while maintaining the same given QoS or permits using higher order modulation and higher error control rate to transmit more information data for a given power with the same QoS requirements. Therefore, intelligent and efficient techniques are especially crucial for transmission over a wireless channel in order to minimize power usage. In order to achieve reliable and energy-efficient transmission over a wireless channel, a plethora of adaptation, scheduling, and radio resource management (RRM) schemes have been proposed and utilized in different layers of the open system interconnection (OSI) protocol stack. In the PHY, for example, the modulation and coding rate can be adapted with the channel condition. In brief, the controller may choose to transmit nothing or a minimum number of packets using the lowest modulation order (e.g., BPSK) and lowest rate coding (using more redundant bits) when the condition of the channel is bad, and may choose to transmit with higher order modulation (e.g., 64-QAM) and higher rate coding (using fewer number of redundant bits) as the channel improves. In the medium access control (MAC) layer, depending upon channel gains, the users can be chosen adaptively for transmission in a given time-slot. In opportunistic scheduling, for example, the base station obtains channel gain for all the users and the users with the best channel gains are chosen for transmission in a particular time-slot in order to maximize system throughput and/or minimize the total power consumption.[1]

7.1.2 Energy-Efficient Cross-Layer Techniques in the Literature

The previous section dicussed why energy-efficient design of protocols and algorithms in different OSI layers are important to achieve green wireless systems. However, in a traditional network, the optimization is usually carried out considering a

[1]In order to address the inherent unfairness issue in the simple form of opportunistic scheduling as described above, various fairness strategies have been included in the literature, where the users that are not among the best set are also scheduled for transmission in order to maintain fairness.

respective layer's objectives based on only local information ignoring other layers' design parameters or information. This fact gives a locally optimal, but globally suboptimal solution. In recent years, significant attention has been placed by the wireless research community on cross-layer optimization for adaptation, scheduling, and resource allocation techniques due to the resulting system level performance improvements. These techniques exploit interdependency and interaction among PHY, MAC, and higher layers in an integrated manner. In cross-layer techniques, a layer interacts and exchanges information with other layers to set up its own strategy. For example, in application layer and PHY interaction, the application layer adapts the QoS and source coding with the channel state information (CSI) from the PHY. The PHY on the other hand adjusts its rate and power according to the application being handled and its traffic characteristics. Cross-layer adaptive congestion control in the transport layer adapts its strategy with the CSI from the PHY utilizing the interaction between the transport layer and the PHY. Using interaction with and information from PHY, the protocols in the network layer may make routing decisions and the scheduling controller in MAC layer may select the best users. Several techniques are studied in the literature based on interaction and information exchange among different layers. Without loss of generality, in this chapter we consider the adaptive cross-layer packet transmission technique that optimizes transmission power dynamically with channel and traffic while maintaining other QoS requirements.

The research on cross-layer energy minimization by adapting with randomly varying system parameters are not new. It has been discussed in the literature in the past. Energy-optimized protocols and techniques have been a major area of research among the wireless research community. Various techniques in the literature are proposed in order to extend the battery life of the handheld mobile devices as well as make the base station energy efficient by maintaining the total power constraint. A comprehensive overview of cross-layer design for energy-efficient wireless communications particularly focusing on system-based approaches toward energy optimal transmission, resource management across time, frequency and spatial domains, and energy-efficient hardware implementations has been presented in [1]. The authors in [2] have surveyed various energy-efficient advanced technologies such as interference mitigation techniques, multiple-input, multiple-output (MIMO), and cooperative communications as well as cross-layer self-organizing networks and the technical challenges of future radio access networks beyond the Long Term Evolution Advanced (LTE-Advanced) standard.

In PHY, various energy-efficient techniques for multiantenna systems have been proposed and evaluated in the recent literature. We discuss a few of them briefly below: using cooperative beamforming framework, a cross-layer optimal weight design for single-beam beamforming and suboptimal weight designs for multibeam beamforming have been discussed in [3] for energy efficiency. In [4], the authors have proposed a cross-layer approach to switch between a MIMO system with two transmit antennas and a single-input multiple-output (SIMO) system to conserve mobile terminals' energy in adaptive MIMO systems. The authors in [5] have studied energy efficiency for Rayleigh fading networks and showed how to map the wireless fading channel to the upper-layer parameters for cross-layer design. PHY energy performance using rate and power adaptation and optimization techniques aided with data-link layer information is improved in the following works. In [6], energy-efficient operation modes in wireless sensor networks (WSNs) have been studied based on cross-layer design techniques over Rayleigh fading channels using a discrete-time queuing model. Using a three-dimensional nonlinear integer programming technique, the authors in [7] have shown that the joint optimization of the PHY and data-link layer parameters (e.g., modulation order, packet size, retransmission limit, etc.) contribute noticeably to the energy saving in energy-constrained wireless networks. The authors in [8] have studied energy-efficient cross-layer design of adaptive modulation and coding using a sleep mode for the system. There are few cross-layer design techniques in the literature [9,10] that are to be considered or even channels to be included while taking a decision on transmission in order to optimize power by looking at the buffer occupancy only when it has insufficient packets. It is assumed that the buffer size is infinite and it does not play any role in making a transmission decision in these works.

Game theory has been used in a lot of literature to design energy-efficient algorithm that consider cross-layer issues. An energy-efficient MAC algorithm has been proposed in [11], where each node sets the contention window size with respect to the residual energy, the harvesting energy, and the transmit power using game-theoretic and cross-layer optimization approach. In [12], the authors have focused on the energy-efficient optimization and cross-layer design problem involving power control and multiuser detection in wireless code division multiple access (CDMA) data networks via a noncooperative game-theoretic approach. The authors in [13] have considered a noncooperative power control game for maximum energy efficiency with a fairness constraint on the maximum received powers in cognitive CDMA wireless networks, where both the primary and secondary users coexist in the same frequency band. In [14], the authors have proposed a game-theoretic energy-efficient model in order to study the cross-layer power and rate control problem with QoS constraints in multiple-access wireless networks.

The following two works deal with the energy-efficient cross-layer design to improve performance across different layers. In [15], the authors proposed a cross-layer window control algorithm in the transport layer to improve the throughput and power control algorithm in the PHY to reduce power consumption for a wireless multihop network. The problem of minimizing the total power consumption in a multihop wireless network has been studied in [16]. A low-complexity, power-efficient, and distributed algorithm is given. Two other works deal with performance evaluation of cross-layer design in wireless standards. In [17], a cross-layer design involving PHY and application layer for mobile WiMAX is given that achieves reduced power consumption and packet loss along with increased throughput. The authors in [18] have presented a cross-layer and energy-efficient mechanism for transmitting voice-over-Internet protocol packets over IEEE 802.11 wireless local area networks (WLANs) and shown that it improves the energy efficiency of a station and WLAN utilization without sacrificing voice qualities. In [19], the authors propose a two-phase energy-efficient resource allocation solution to efficiently solve the sleeping time (by maximizing shutdown time) versus scaling (by leveraging the modulation, code rate, and transmission power) trade-off across the PHY, communications, and link layers for multiple users with MPEG-4 video transmission over a slow-fading channel. The authors in [20] propose a system model using a dynamic frequency selection algorithm for WLANs with the objective of reducing consumed power by introducing a scheme for channel allocation under energy conserving criteria. In [21], the authors propose two sleep modes, namely a real-time dynamic resource activation/deactivation method and a semistatic method where resources are kept unchanged for longer periods, for green base station in 2G and HSPA systems that shut down a number of resources to optimize energy consumption at the network scale while preserving the QoS perceived by users.

7.1.3 Related Work on Cross-Layer Green Adaptive Techniques

Transmit power and rate adaptation can yield significant power savings for delay-tolerant data applications and hence help to reduce the energy footprint and greenhouse emissions when adaptation and optimization are done based on cross-layer interaction and information exchange. There are some works in the literature that deal with energy optimization for delay-tolerant data applications [22,23]. Many cross-layer studies in the literature have investigated the fundamental relationship between average power consumption with other QoS requirements. For example, the authors in [24] have investigated the problem of optimal trade-off between average power and average delay for a single user communicating over a memoryless block-fading channel using information-theoretic concepts. In [25], convexity properties and characterization of the delay-power region of different schedulers are discussed for independent and identically distributed (i.i.d.) traffic in fading channels. In [26,27], the authors have studied the optimal power and rate allocation problem for wireless systems, where the cross-layer adaptation policy is chosen to optimize the transmission power, the buffering delay, the bit error rate (BER), and the packet overflow. The authors formulate the problem as average cost per stage Markov decision process (MDP) problem and sought optimal solutions based on both the unconstrained and constrained formulations. Simple suboptimal heuristic policy is also investigated. An optimization-based suboptimal threshold scheduler is described in [28] for an i.i.d. block-fading channel. The authors optimized the average power subject to the average delay and the packet loss rate constraints. An offline packet scheduling scheme for an additive white Gaussian noise channel model has been analyzed in [29], with the goal of minimizing energy, subject to a deadline constraint. An online lazy packet scheduling algorithm is also devised that varies transmission time according to a backlog. It is shown to be more energy efficient than a deterministic scheduler with the same queue stability region and a similar delay. A study with a similar objective is also carried out in [30], where the authors used cumulative curves methodology and a decomposition approach. A heuristic policy for the case of arbitrary packet arrivals to the queue with individual deadline constraints is presented. Optimal transmission scheduling with energy and deadline constraints for a satellite transmitter is given using a Shannon capacity formula in [31]. Power adaptation strategies for delay-constrained channels are investigated in [32] to maximize expected capacity and minimize outage capacity under both short-term and long-term power constraints over i.i.d. flat block-fading channels. It is assumed that the CSI is fed back to the transmitter in a causal manner. The authors in [33] studied constrained MDP (CMDP)-based techniques for optimal power and rate allocation in MIMO wireless systems. They used the concepts of stochastic dominance, submodularity, and multimodularity to prove the monotonicity property of the optimal randomized policy.

The type-II hybrid automatic repeat request (HARQ) technique employing incremental redundancy has been proposed in different wireless standards for increased reliability and efficiency of packet transmissions. In [34], the authors have proposed a semi-Markov decision process (SMDP)-based cross-layer adaptation technique to optimize transmission power, delay, and overflow. The optimal policy of transmission power and rate is found using the equivalent discrete-time CMDP-based technique. Because of the random variation of the channel, packet errors occur in bursts. Hence, sometimes the perfect state of the current channel cannot be accurately predicted at the receiver and may not be available at the transmitter while transmitting. Rather, the packet error feedback signal is received from the receiver after the packet has been transmitted. In the following literature, no perfect CSI is assumed at the transmitter. In [35], the authors formulate the decision process of determining transmission (when to attempt or suspend) over a Gilbert-Elliott fading channel as a partially observable Markov decision process (POMDP). The optimal policy for the throughput versus energy efficiency trade-off problem is derived for a time horizon less than or equal to 13, and shown to be a threshold rule that varies with the memory present in the channel error process. A suboptimal, limited-lookahead implementation of this policy is also simulated and its performance compared with the persistent retransmission and probing protocols. The work of Choi et al. [35] is extended in [36] by studying the impact of various feedback structures and the effect of channel memory on the performance, design, and structure of this scheme. In both works, the channel state is not directly observable, thus the transmission decisions must be based on positive acknowledgment (ACK) and negative acknowledgment (NAK) information provided over a feedback channel. A formulation of the opportunistic file-transfer problem using a Stop-and-Wait ARQ transmission protocol over a two-state Gilbert-Elliott fading channel is given in [37]. The optimal trade-off between the transmission energy and the latency is formulated as a POMDP and then reformulated as a Markovian search problem. For this simple problem when power is not adapted, an optimal threshold control policy is shown. In [38], a theoretical analysis of a combined optimization of the scheduling layer with the PHY is given. The channel is modeled with a hidden Markov model (HMM), and the solution of the resulting optimization problem is shown to be a POMDP. In [39], a cross-layer adaptation technique is studied when both the states of the traffic and the channel are hidden from the scheduler. The scheduler can only observe them from the observations. For traffic, the observations are the number of packet arrivals, and for the channel, the observations are the ACK/NAK feedback. The authors studied a heuristic-based suboptimal policy to solve the resulting POMDP problem since the optimal solution is computationally infeasible.

A cross-layer energy minimization problem for multiuser settings under an MDP framework is investigated in the following works. In [40], the authors consider several basic cross-layer resource allocation problems (e.g., the transmission rate and the power assigned to each user) for wireless fading channels. The fundamental performance limits with higher-layer QoS (such as delay) are characterized in the survey. An MDP-based selection technique for user, power, and rate is proposed in [41]. In [42], the multiuser cross-layer adaptive transmission problem is decoupled into an equivalent number of single-user problems, so that optimal adaptive single-user policies can be used. The authors in [43] consider a single source media streaming data to multiple users equipped with individual buffers over a shared wireless channel. The problem is formulated as an MDP where the objective of the source transmitter is to dynamically allocate power that minimizes total power consumption and packet holding costs, and at the same time satisfy strict buffer underflow constraints (which prevents the user buffer from emptying so that playout quality is maintained). In [44], an energy-efficient uplink scheduling algorithm for minimizing the average power of each user subject to individual delay constraint in a multiuser wireless system is studied. The authors proposed a learning-based single-user optimal algorithm for the multiuser setting, where each user bids their rate to the base station. The base station schedules the user with highest biding rate user.

MDP-based cross-layer optimization and adaptation techniques are also shown to be energy efficient for cognitive radio, cooperative, and ad hoc networks. In [45], optimal and two-step lookahead suboptimal opportunistic spectrum access strategies for channel probing and transmission scheduling problem in multichannel cognitive radio networks are studied. Assuming that the transmitter does not have complete information on the channel states, the threshold structural properties of the optimal strategies are described by index policies. Using a POMDP framework, the problem of optimal cooperative sensing scheduling for cognitive radio network is considered in [46], where the action is to determine the number of secondary users that should be assigned to sense each channel in order to maximize energy efficiency. A dynamic programming optimal algorithm and polynomial time heuristic algorithm to solve the problem of joint cross-layer optimization of power control, scheduling, and routing subject to stringent packet delay and multiaccess interference constraints in mobile ad hoc networks are presented in [47] to minimize systemwide energy consumption at all the nodes with deterministic mobility pattern, traffic load, and channel conditions. In [48], an energy-efficient random access scheduler is studied for an ALOHA-based network, where each user sends a packet by selecting a power-minimizing slot while considering the delay limit to ensure QoS. With and without the knowledge of the channel distribution to all users, the authors have proposed a dynamic programming-based optimal solution and a heuristic-based suboptimal solution (which learns channel statistics by observing initial time-slots), respectively. In [49], the authors have studied energy harvesting and cooperative communications for wireless sensor networks (WSNs) to develop energy-efficient strategies (whether to transmit directly or via relay). The techniques have utilized both the MDP and the POMDP-based framework for the case when the state information of the relay is completely available and partially available, respectively. In [50], a delay-constrained minimum energy cooperative broadcast problem for a multihop wireless network is formulated and solved by combining dynamic programming and linear programming techniques. Since the problem is NP-complete, an approximate result is given by splitting the problem into three areas (ordering, scheduling, and power control), and an analytical lower bound is also derived. In [51], using a decode-and-forward scheme and dynamic programming techniques, the authors propose an asymptotically optimal policy for relay-assisted wireless systems with a deadline to minimize the sum energy of all nodes.

The energy optimization for wireless sensor networks is of paramount importance due to extending the life of the sensor and hence the networks. The authors in [52] considered a distributed MAC protocol which trades off between Likelihood-Ratio Information (LRI) and CSI to reduce the total transmission energy for efficient detection in wireless sensor networks (WSNs). In [53], the trade-off between tracking performance of an moving object through a wireless sensor network and associated energy consumption is studied by casting the scheduling problem as a POMDP, where the control actions correspond to the set of sensors to activate at each time step. The authors in [54] investigate the scheduling problem in multiaccess wireless sensor networks with the objective of minimizing the transmission energy cost under average packet delay constraint by formulating the problem as a constrained stochastic dynamic programming problem.

7.2 SYSTEM MODELS

Let us consider a communication system over a time-slotted Gilbert-Elliot channel, where a transmitter terminal (TT) with finite buffer is communicating with its receiver terminal (RT). The TT has the capacity to hold B packets in its buffer. We assume that the arrival of additional packets at the buffer will push some packets from the buffer to be dropped on a first-in first-out (FIFO) basis. That is, the packets that arrive first will be dropped first when there is no space in the buffer to make room for the newly arrived packets. For many wireless network applications, the delayed packets are not as useful as new packets. Assume that the TT has no deadline to finish transmission. However, it must maintain the packet delay and the overflow requirement demanded by the handled traffic. Let T_s denote the length of a time-slot in seconds. The time-slot is the basic unit of the time that we follow in the remaining part of the chapter. In each time-slot, one modulated symbol is transmitted over the fading channel. The transmission power of the modulated symbol depends on the modulation format (i.e., number of bits the symbol contains), BER requirement, and the channel gain. Several time-slots constitute a radio frame, also called a block. Suppose each packet coming from the upper-layer application consists of N_p bits and one radio frame has N_f time-slots (in other words, N_f symbols are transmitted in a radio frame). Now, in a particular time-slot if a modulation scheme with X_{m} bits/symbol is selected, the corresponding number of packets X_{p} that can be transmitted is given by following relationship:

$$X_{\mathrm{p}} = X_{\mathrm{m}} \frac{N_f}{N_{\mathrm{p}}}. \tag{7.1}$$

In the above expression, we assume that control bits are negligible as compared to information data bits. Usually in the literature, when computing throughput, N_f is assumed to be equal to the number of time-slots associated with transmitting data information in the radio frame.

7.2.1 Traffic Model

The nature of the real traffic for modern wireless networks is bursty and correlated. The incorporation of traffic models into the analysis is important for dynamic packet adaptations in wireless networks supporting a wide variety of incoming traffic having different QoS requirements, such as constant-bit-rate (CBR) real-time traffic (e.g., voice), variable-bit-rate (VBR) real-time traffic (e.g., video streaming, teleconferencing, games, etc.) as well as CBR or VBR best-effort data traffic (e.g., file transfer, web browsing, messaging, etc.). Different traffic types exhibit different burstiness and different correlation. While some traffic, such as interactive data and compressed video, is highly bursty, other traffic, such as large files, is continuous (less bursty). Some traffic is highly correlated, while some is independent and identically distributed (i.i.d.).

A Markov model is usually used to capture both the memory and burstiness of the network traffic. In this model, the states of the traffic states are governed by an underlying Markov chain. In any state, the incoming packet arrival may follow uniform, Poisson, Bernoulli, etc., distribution. Let $\mathcal{F} = \{f_1, f_2, \ldots, f_F\}$ denote the state space of the traffic, where $f_i, i = 1, 2, \ldots, F$ denotes the ith state and F is the total number of traffic states. The traffic transition matrix can be written as

$$\mathcal{P}_f = \begin{bmatrix} P_{f_1,f_1} & P_{f_1,f_2} & \cdots & P_{f_1,f_F} \\ P_{f_2,f_1} & P_{f_2,f_2} & \cdots & P_{f_2,f_F} \\ \vdots & \vdots & \ddots & \vdots \\ P_{f_F,f_1} & P_{f_F,f_2} & \cdots & P_{f_F,f_F} \end{bmatrix}, \tag{7.2}$$

where P_{f_i,f_j} represents the traffic state transition probability from state f_i to state f_j.

A more generalized way for modeling a wide range of traffic is to use the hidden Markov model (HMM), where the states of the incoming traffic are assumed to be unknown to the controller of the transmitter. However, these states are estimated from the observations $\mathcal{O}_{\mathrm{a}} = \{a_0, a_1, \ldots, a_A\}$. In some practical situations, the states are not known exactly, but the packet arrival in a particular time-slot is known. An HMM model is usually used for these situations.

7.2.2 Channel Model

The gains of the wireless channels vary randomly over time. However, the gains over two consecutive time-slots are correlated. This memory of the wireless channels is usually captured using a Markov channel model. Perhaps the most used Markov model in the literature is the Gilbert-Elliot channel model, where the gain is partitioned into two regions. When the gain falls in first region, the channel is in a bad state and otherwise it is in good state. Certainly, the average gain of the good channel state is higher than the bad channel state, hence it permits transmission with less transmission power using the same modulation scheme keeping the same BER requirement or with more bits/symbol using the higher order modulation keeping the same BER and power requirement.

Assume $\mathcal{C} = \{c_1, c_2\}$ denotes the state space of the channel, where c_1 corresponds to a bad channel and c_2 corresponds to a good channel. Let P_{c_1,c_2} and P_{c_2,c_1} denote the cross-transition probabilities from state c_1 to state c_2, and state c_2 to state c_1 respectively. We can write the channel transition matrix as follows:

$$\mathcal{P}_\mathrm{c} = \begin{bmatrix} P_{c_1,c_1} & P_{c_1,c_2} \\ P_{c_2,c_1} & P_{c_2,c_2} \end{bmatrix}. \tag{7.3}$$

Note that since the sum of all outbound transition probabilities from a particular state is 1, we can write the self-transition probabilities as follows: $P_{c_1,c_1} = 1 - P_{c_1,c_2}$ and $P_{c_2,c_2} = 1 - P_{c_2,c_1}$. Let the row vector $\Pi_\mathrm{c} = [\pi_{c_1}, \pi_{c_2}]$ represent the stationary distribution of the channel state, hence it satisfies

$$\Pi_\mathrm{c} = \Pi_\mathrm{c}\mathcal{P}_\mathrm{c}. \tag{7.4}$$

With normalizing condition $\pi_{c_1} + \pi_{c_2} = 1$, stationary distribution Π_c corresponds to the normalized left eigenvalue of the transition matrix $\mathcal{P}_\mathrm{c}$ associated with the eigen value 1. For the two-state channel, we can write it as $\Pi_\mathrm{c} = \left[\frac{P_{c_2,c_1}}{P_{c_1,c_2}+P_{c_2,c_1}}, \frac{P_{c_1,c_2}}{P_{c_1,c_2}+P_{c_2,c_1}}\right]$. The Gilbert-Elliot model can be generalized by partitioning the gain into more than two regions and the resulting channel model is know as a finite state Markov channel (FSMC). Without loss of generality, we carry out our discussion for the Gilbert-Elliot model.

Let α and $\bar{\alpha}$ denote the instantaneous channel gain and the average channel gain respectively. If $f_\alpha(\alpha)$ is the probability density function (pdf) of the wireless channel gain, we can write $\bar{\alpha} = \mathbb{E}\{\alpha\} = \int_0^\infty \alpha f_\alpha(\alpha)d\alpha$. We assume that the channel gain is block-fading so that it remains the same for whole duration of the time-slot. The relation between the Doppler frequency f_m and the time-slot T_s should satisfy the condition $T_\mathrm{s} f_m \ll 0.423$ in order for the gain α to remain constant for at least a time-slot. Although any wireless channel can be incorporated in the study, let us discuss a MIMO system that has n_T transmit antennas and n_R receive antennas with diversity order of $L \triangleq n_\mathrm{T} n_\mathrm{R}$. The MIMO system can be represented by the following channel matrix:

$$\mathbf{H} = \left[\alpha_{jk} \exp(\mathrm{i}\phi_{jk})\right]_{j,k=1}^{n_\mathrm{R},n_\mathrm{T}}, \tag{7.5}$$

where i is an unit imaginary number, i.e., $\mathrm{i}^2 = -1$, α_{jk} is the path gain between the kth transmit and jth receive antennas, and ϕ_{jk} is the corresponding phase, which is uniformly distributed across $[0, 2\pi]$. The output matrix $\mathbf{Y}_{n_{\mathrm{R}} \times N_f}$ of the received signal is related to input matrix $\mathbf{X}_{n_{\mathrm{T}} \times N_f}$ of the transmitted symbols by $\mathbf{Y} = \mathbf{HX} + \mathbf{V}$, where $\mathbf{V}_{n_{\mathrm{R}} \times N_f}$ is the receiver noise matrix with elements i.i.d. complex circular Gaussian random variables, each with a $\mathcal{CN}(0, \sigma^2)$ distribution. When the fading follows a Nakagami-m distribution, the pdf of the path gain α_{jk} can be written as

$$f_\alpha(\alpha_{jk}) = 2\left(\frac{m}{\bar{\alpha}_{jk}}\right)^m \frac{\alpha_{jk}^{2m-1}}{\Gamma(m)} \exp\left(-\frac{m\alpha_{jk}^2}{\bar{\alpha}_{jk}}\right), \quad \alpha_{jk} \geqslant 0. \tag{7.6}$$

In a transmit diversity system through space–time block coding (STBC), each of the $N_i \leqslant N_f$ input symbols is mapped to n_{T} orthogonal sequences of length N_f and transmitted simultaneously with n_{T} transmit antennas. The input–output relationship of each subchannel before maximum-likelihood detection can be described by the following relationship [10]:

$$y_{\mathrm{s}} = c\|\mathbf{H}\|_F^2 x_{\mathrm{s}} + v_{\mathrm{s}}, \tag{7.7}$$

where c is a code-dependent constant that satisfies $c = \frac{P_{\mathrm{t}}}{P_{\mathrm{s}} n_{\mathrm{T}} R_{\mathrm{c}}}$, P_{s} is the average energy per transmitted symbol, P_{t} is the total transmit power per symbol time, and $R_{\mathrm{c}} = \frac{N_i}{N_f}$ is the information code rate of the STBC. In (7.7), $\|\cdot\|_F$ denotes the matrix Frobenius norm, x_{s} corresponds to either the real or imaginary part of a transmitted symbol with power $P_{\mathrm{s}}/2$, and v_{s} is the noise term after STBC decoding with distribution $\mathcal{N}(0, c\|\mathbf{H}\|_F^2\sigma^2/2)$. The effective received signal-to-noise ratio (SNR) per symbol at the output of the decoder is therefore given by

$$\gamma = \frac{\bar{\gamma}}{m n_{\mathrm{T}} R_{\mathrm{c}}} \|\mathbf{H}\|_F^2, \tag{7.8}$$

where $\bar{\gamma} \triangleq m P_{\mathrm{t}}/\sigma^2$ is the average SNR per receive antenna, and $\|\mathbf{H}\|_F^2 \triangleq \sum_{j,k} \alpha_{jk}^2$ is the sum of L independent gamma random variables, each with parameter m and unit mean. Using random variable transformation techniques, the SNR at the output of the receiver can be given by the gamma distribution with parameter $m_{\mathrm{T}} = mL$ and mean $\bar{\gamma}_{\mathrm{T}} = \langle\gamma\rangle \triangleq mL\bar{\gamma}/mn_{\mathrm{T}}R_{\mathrm{c}} = n_{\mathrm{R}}\bar{\gamma}/R_{\mathrm{c}}$. Its pdf is given by [10]

$$f_\gamma(\gamma) = \frac{\gamma^{m_{\mathrm{T}}-1}}{\Gamma(m_{\mathrm{T}})} \left(\frac{m_{\mathrm{T}}}{\bar{\gamma}_{\mathrm{T}}}\right)^{m_{\mathrm{T}}} \exp\left(-\frac{m_{\mathrm{T}}\gamma}{\bar{\gamma}_{\mathrm{T}}}\right), \quad \text{for } \gamma \geqslant 0. \tag{7.9}$$

If we suppose γ_1 represents the value of the partitioning channel gain, then the average channel gains for bad and good channels can be written respectively as $\bar{\gamma}_1 = \int_0^{\gamma_1} \gamma f_\gamma(\gamma) \mathrm{d}\gamma$ and $\bar{\gamma}_2 = \int_{\gamma_1}^{\infty} \gamma f_\gamma(\gamma) \mathrm{d}\gamma$. The channel will be in state c_1 and c_2 when the gains are in the interval $[0, \gamma_1)$ and $[\gamma_1, \infty)$, respectively. Among the different channel partitioning methods, the equal probability method (EPM) maintains good balance between accuracy and complexity. Let γ_{i-1} and γ_i denote the lower and upper received SNR thresholds associated with channel state c_i. Thus, the probability of staying in channel state $c_i \in \mathcal{C}$ can be given by

$$\pi_{c_i} = \int_{\gamma_{i-1}}^{\gamma_i} f_\gamma(\gamma)\mathrm{d}\gamma = F_\gamma(\gamma_i) - F_\gamma(\gamma_{i-1}) = \frac{\gamma\left(m_\mathrm{T}, \frac{m_\mathrm{T}}{\bar{\gamma}_\mathrm{T}}\gamma_i\right) - \gamma\left(m_\mathrm{T}, \frac{m_\mathrm{T}}{\bar{\gamma}_\mathrm{T}}\gamma_{i-1}\right)}{\Gamma(m_\mathrm{T})}, \tag{7.10}$$

where the cumulative distribution function (cdf) of the received SNR is given as

$$F_\gamma(\gamma_i) = \frac{\gamma\left(m_\mathrm{T}, \frac{m_\mathrm{T}}{\bar{\gamma}_\mathrm{T}}\gamma_i\right)}{\Gamma(m_\mathrm{T})}. \tag{7.11}$$

In EPM, γ_1 is determined so that the stationary probabilities are equal in all states, i.e., $\pi_{c_1} = \pi_{c_2}$. The transition probability P_{c_i,c_j} from state $c_i \in \mathcal{C}$ to state $c_j \in \mathcal{C}$ can be approximated by the ratio of the expected number of level crossings at the received SNR γ_k and the steady-state probability of channel state c_i as

$$P_{c_i,c_j} \approx \frac{N_{\gamma_k} T_\mathrm{s}}{\pi_{c_i}}, \quad k = i \text{ for } j = i+1, \; k = j \text{ for } j = i-1. \tag{7.12}$$

The level crossing rate N_{γ_k} depends on the maximum Doppler frequency f_m, and π_{c_i} is the steady-state probability of channel state c_i. The level crossing rate at the corresponding received SNR threshold γ_i can be deduced in a similar way as [55]

$$N_{\gamma_i} = \frac{\sqrt{2\pi} f_m}{\Gamma(m_\mathrm{T})} \left(\frac{m_\mathrm{T}\gamma_i}{\bar{\gamma}_\mathrm{T}}\right)^{m_\mathrm{T}-\frac{1}{2}} \exp\left(-\frac{m_\mathrm{T}\gamma_i}{\bar{\gamma}_\mathrm{T}}\right). \tag{7.13}$$

When the states of the channel are not known, but only the observations are known, the channel can be modeled as HMM. In an HMM model, the above-discussed channel model serves as an underlying channel, which is unobservable. However, its states can be estimated via the received ACK/NAK observation from the receiver. Let $\mathcal{O}_\mathrm{c} = \{\omega_1, \omega_2\} = \{\mathrm{ACK}, \mathrm{NAK}\}$ denote the channel observation vector.

7.3 A DECISION THEORETIC FORMULATION

The partially observable Markov decision process (POMDP) is a generalized framework for formulating problems where a system makes dynamic decisions based on the observed or the belief of the hidden state parameters of the system. When the state components are perfectly observed at the beginning of the decision epoch,[2] the

[2]For DT-MDP, the decision epoch is equal to the time-slot. We use both the time-slot and the decision epoch interchangeably.

formulations can be done with the discrete-time Markov decision process (DT-MDP) or continuous-time Markov decision process (CT-MDP). The optimal decision rule, μ^*, which gives the rule for choosing an action in each state, for DT-MDP is relatively easier to compute using either an unconstrained MDP (UMDP) formation or a constrained MDP formulation (CMDP). While a UMDP problem is solved via dynamic programming algorithm, a CMDP problem is solved via linear programming techniques. When the system state is unknown, the computation of the optimal decision rule becomes infeasible for anything other than a simple problem with less than 10 states. Therefore, the system state for both such cases is usually estimated using previously taken action and corresponding obtained observation feedback. From prior action–observation sets, a probability distribution can be maintained in each time-slot using a forward filtering formula. This probability distribution is known as a *belief state* or *information state*. In this chapter, we discuss a policy iteration (PI) algorithm to solve the UMDP problem using a dynamic programming technique. As the name implies, a UMDP problem can accommodate only one objective to find the optimal decision rule. When the problem has more than one objective, a composite objective function is formulated by combining all the objective functions with appropriate weighting factors. The weighting factors have the role of doing trade-offs among the objectives.

Generalized POMDP formulations have been used to solve various wireless networking decision making problems. In this section, our challenge is to find a policy given the state information of the buffer and observations of the traffic and channel available as shown in Figure 7.1a. A POMDP problem can be defined by a tuple $(\mathcal{S},\mathcal{U},\mathcal{P}_s,\mathcal{G},\mathcal{O},\mathcal{P}_o)$, with $\mathcal{S}$ as its system state space, $\mathcal{U}$ as its action space, $\mathcal{P}_s$ as its state transition probability matrix, $\mathcal{G}$ as its cost matrix, $\mathcal{O}$ as its observation space, and finally $\mathcal{P}_o$ as its observation probability matrix. We discuss each ingredient of the POMDP problem in the following.

7.3.1 States

The state of the considered system is composite and can be expressed by its nature of traffic, channel condition, and buffer occupancy status. We can write the system state space as $\mathcal{S} = \mathcal{F} \times \mathcal{C} \times \mathcal{B} = \{s_1, s_2, \ldots, s_S\}$ with the total number of states being $S = F \times C \times (B+1)$, where s_l, $l = 1, 2, \ldots, S$ denotes S system states. If we suppose that the traffic state, channel state, and buffer state at time-slot n are f^n, c^n, and b^n respectively,[3] then the corresponding system state can be given by $s^n = C(B+1)(f^n - 1) + (B+1)(c^n - 1) + b^n$.

7.3.2 Actions

Actions describe the task of the scheduler to be performed at a particular state. The scheduler may have different choices and also the choices may be different

[3]Unless specified otherwise, superscript n denotes the value of a variable at time-slot n.

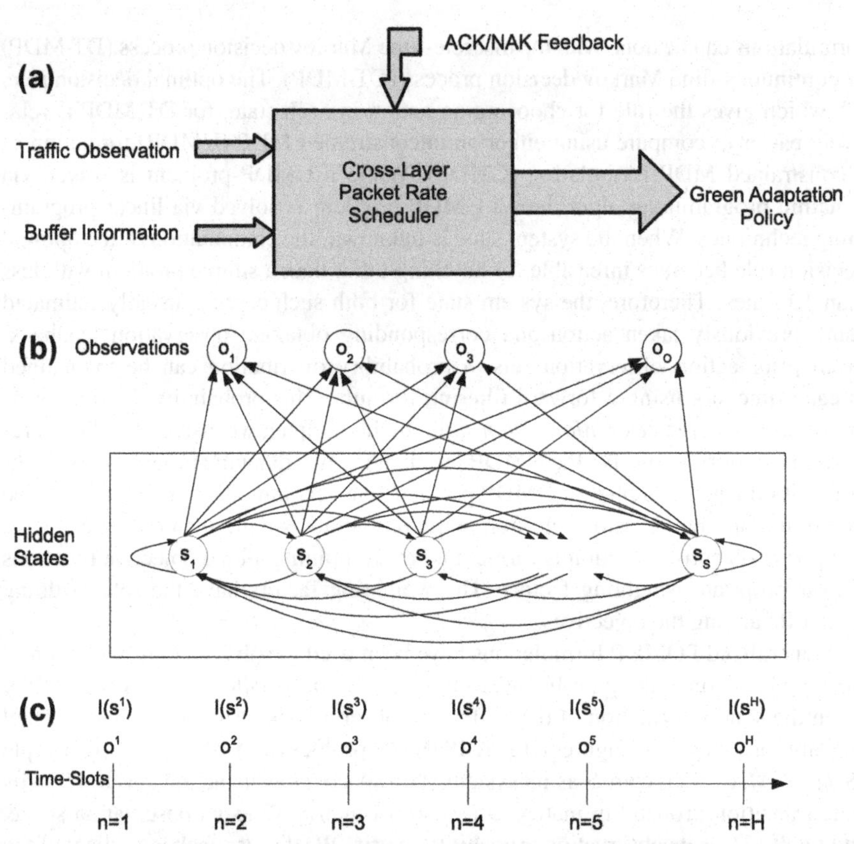

Figure 7.1 (a) Block diagram of the cross-layer green adaptation policy, which is provided with buffer state, channel feedback, and traffic information; (b) state space and its transitions and observations; (c) time-slot sequence, and corresponding observation sequence and information state sequence over all the time-slots.

in different states. We denote the set of actions by $\mathcal{U} = \{u_1, u_2, \ldots, u_U\}$, where U is the total number of possible actions available to the scheduler. Although the scheduler can perform many different tasks at a particular state, in this chapter we consider the transmission rate and corresponding power as the action for the problem. Let $\mathcal{X} = \{X_1, X_2, \ldots, X_U\}$ denote the set of transmission rate in bits/symbol, where rate X_i corresponds to action u_i. There is a one-to-one mapping between the transmission rate and the number of packets taken from the buffer. Also suppose $\mathcal{W} = \{w_1, w_2, \ldots, w_U\}$ denotes the set of packets corresponding to the action set $\mathcal{U}$ and transmission set $\mathcal{X}$, hence we can write $w^n = \Psi(u^n)$, where Ψ is a mapping function that maps the action to the corresponding number of packets to be taken from the buffer for transmission.

7.3.3 Transition Probabilities

The transitions among the states are governed by the state transition probabilities. In a POMDP framework, the transition probability for an action u^n expresses the probability of switching from a particular hidden system state $s^n = s_i$ at time-slot n to another hidden system state $s^{n+1} = s_j$ at the next time-slot $n+1$. For our problem, it is composite and depends on the individual transition probabilities for traffic arrivals, channel transition, and buffer transition as follows:

$$\mathbf{P}_{\mathrm{s}}(u_i) = \mathbf{P}_{\mathrm{a}}(u_i) \otimes \mathbf{P}_{\mathrm{c}}(u_i) \otimes \mathbf{P}_{\mathrm{b}}(u_i) = \begin{bmatrix} P_{s_1,s_1}(u_i) & P_{s_1,s_2}(u_i) & \cdots & P_{s_1,s_S}(u_i) \\ P_{s_2,s_1}(u_i) & P_{s_2,s_2}(u_i) & \cdots & P_{s_2,s_S}(u_i) \\ \vdots & \vdots & \ddots & \vdots \\ P_{s_S,s_1}(u_i) & P_{s_S,s_2}(u_i) & \cdots & P_{s_S,s_S}(u_i) \end{bmatrix},$$

where the transition probability P_{s_q,s_r} for action $u^n = u_i$ can be given by

$$P_{s_q,s_r}(u_i) = P_{a_j,a_x} P_{c_k,c_y} P_{b_l,b_z}, \tag{7.14}$$

where $s_q = [a_j, c_k, b_l]$ and $s_r = [a_x, c_y, b_z]$. For a particular action u_i and traffic state f_j, the probability of occupying buffer state $b^{n+1} = b_z$ from state $b^n = b_l$ is given by

$$P_{b_l,b_z} = \sum_{x=0}^{x=A} \delta(b_z - b_l - a_x + \Psi(u_i)) P(a_x | f_j), \quad \forall c_k \in \mathcal{C}, \tag{7.15}$$

where function $\delta(x)$ returns 1 when $x = 0$ and returns 0 otherwise, and $P(a_x|f_j)$ is the probability of a_x arrivals in state f_j. The relations between system states, and traffic, channel, and buffer states are as follows: $q = (j-1)(B+1)C + (k-1)(B+1) + l$ and $r = (x-1)(B+1)C + (y-1)(B+1) + z$.

7.3.4 Costs

The choice for an action in a state is driven by associated costs. The scheduler chooses the action that incurs the lowest cost. An action in a particular time-slot causes a specific cost for each objective that we like to attain. Suppose that our prime objective is to minimize power cost for the green radio systems, and at the same time our goals are to minimize delay and overflow as well, so that they are within tolerable limits. A cost function $G(s_i, u_j)$ describes the relationship between the state–action pair (s_i, u_j) and the cost. We describe three cost functions associated with the considered objectives as follows:

(1) *Power:* The transmitter power in a particular slot determines the power cost. We can write the power cost function $G_{\mathrm{P}}(s_i, u_j) = P_{\mathrm{t}}$, where P_{t} is the instantaneous power for action u_j in state s_i to transmit w_j number of packets. Although our discussed model is general enough to accommodate any modulation and channel coding scheme, for brevity we concentrate on modulation rate only. Let $P_{\mathrm{b}}(\gamma)$ denotes the instantaneous BER with received SNR γ. An approximate expression of BER for M-QAM is given in [56] and it can be expressed by

$$P_b(\gamma) = \frac{2}{v_i}\left(1 - \frac{1}{\sqrt{M}}\right)\sum_{j=1}^{\frac{\sqrt{M}}{2}} \operatorname{erfc}\left((2j-1)\sqrt{\frac{3v_i\gamma P_t}{2(M-1)P}}\right), \qquad (7.16)$$

where $v_i = \log_2(M)$ is the number of bits that modulates a 2^{v_i}-QAM symbol. The sum of the first two terms in (7.16) has been found to give the best approximation to the system BER for all values of M and SNR [56]. Let $\overline{P}_b(c_j, u_i)$ denote the average BER for channel state c_j when the scheduler chooses action u_i in a particular time-slot n. Therefore, $\overline{P}_b(c_j, u_i)$ can be obtained as follows:

$$\overline{P}_b(c_j, u_i) = \frac{1}{\pi_{c_j}} \int_{\gamma_{j-1}}^{\gamma_j} P_b(\gamma) f_\gamma(\gamma) d\gamma. \qquad (7.17)$$

Given a particular channel state c_j and action u_i, the transmitter power for a chosen rate and average BER can be calculated numerically from (7.16) and (7.17).

(2) *Delay:* The delay cost is determined by the number of slots of the packet in the buffer before transmission. We can write the delay cost function as $G_D(s_i, u_j) = \frac{b_k - 1}{A_l}$, where b_k is the corresponding buffer state for system state s_i and A_l is the average packet arrival in corresponding traffic state.

(3) *Overflow:* The overflow cost is equal to the number of packets dropped from the buffer as a result of insufficient storage. We follow a first-in first-out (FIFO) service strategy at the buffer. The packet overflow cost in a time-slot can be determined as $G_O(s^n, u^n) = b^n - w^n + a^n$.

7.3.5 Observations

We assume that the exact states of the traffic as well as the channel are not available to the scheduler. However, they are observable through respective observations. The observation for the problem hence consists of traffic observation a^n and channel feedback observation ω^n and the set of observations can be denoted as $\mathcal{O} = \mathcal{O}_a \times \mathcal{O}_c = \{o_1, o_2, \ldots, o_O\} = \{(a_0, \omega_1), (a_0, \omega_2), (a_1, \omega_1), \ldots, (a_A, \omega_2), \}$, where $O = (A+1) \times 2$ is the total number of observations in the set.

7.3.6 Observation Probabilities

Let $P(a_l | f_i, u_k)$ and $P(\omega_m | c_j, u_k)$ denote the observation probability for a_l, $l = 0, 1, \ldots, A$ packet arrivals in traffic state f_i and for receiving feedback ω_m in channel state c_j when action u_k is chosen. Note that the packet arrival does not depend on the action u_k, so we can write $P(a_l|f_i, u_k) = P(a_l|f_i)$, $\forall u_k$. The observation probability for an action u_k can be written as $P_o = P(a_l|f_i) \times P(\omega_m|c_j, u_k)$.

The packet arrivals can be uniformly, Bernoulli, Poisson, etc., distributed. For uniformly distributed traffic, $P(a_l|f_i) = 1/\overline{A}_i$. If we let $\overline{A}_i$ denote the average arrival rate in traffic state $f_i \in \mathcal{F}$, then the average arrival $\overline{A} = \sum_{i=1}^{F} \pi_{f_i} \overline{A}_i$, where π_{f_i} is the stationary probability of state f_i.

The positive acknowledgment (ACK) probability for channel state–action pair (c_i, u_j) can be written as

$$P_A(c_i, u_j) = (1 - \overline{P}_b(c_i, u_j))^{X_j N_f}. \tag{7.18}$$

The negative acknowledgment (NAK) probability is $P_N = 1 - P_A$. The states of the system and their transitions, and the observations from different states are shown in Figure 7.1b pictorially. The system in POMDP formulation works as follows: in each time-slot n, an action u^n is chosen from the admissible set of actions based on the estimated state s^n and the system moves to a new state s^{n+1} according to the system transition probability matrix $\mathcal{P}_s(u^n)$ for that action u^n. For the state–action pair (s^n, u^n), the system receives an observation o^n according to the observation probability matrix $\mathcal{P}_o$ and incurs costs according to cost functions $G_P(s^n, u^n)$, $G_D(s^n, u^n)$, and $G_O(s^n, u^n)$. Next time-slot $n + 1$ begins and all the steps are repeated. In the next section, we discuss two policies for evaluating the performance of the POMDP systems. The evolution of time-slots and the observations received in each time-slot are shown in Figure 7.1c. The updated information states are also shown in the same figure.

7.4 SOLUTION TECHNIQUES AND POLICY

In this section, we discuss the determination of the decision rule, μ, which is the rule that determines the action for specific state in a specific time-slot, i.e., $u^n = \mu(s^n)$. We call a set of decision rules for all time-slots a policy, $\pi = \{\mu^1, \mu^2, \ldots, \mu^H\}$, where H is the horizon of the problem. For a stationary system, the states and transition probabilities do not change over time and hence decision rules remain the same. Therefore, in the sequel, we use μ to denote the decision rule for any time-slot and for brevity we use μ and π interchangeably. Let Π denote the set of all stationary policies π. As discussed before, the calculation of optimal policy μ^* for the POMDP problem is infeasible for practical applications. Therefore, we consider two policies, namely Fully Observable Optimal Policy (FOOP) and Maximum-Likelihood Heuristic Policy (MLHP), for the problem as discussed below.

7.4.1 Fully Observable Optimal Policy (FOOP)

Since the optimal policy of the POMDP problem is infeasible due to not having perfect information of the states, we first consider the case when the states are assumed to be completely observable. This case gives us the optimal performance for the scheduler and serves as a benchmark for comparison with the heuristic policy discussed in the next section. The optimal policy for the completely observable Markov decision process (COMDP) problem gives us the upper bound of the achievable performance. We formulate the problem as an infinite horizon average cost UMDP problem, where our objective is to minimize a weighted sum of the three discussed cost functions, $G_{\mathrm{T}}(s^n, u^n) = G_{\mathrm{P}}(s^n, u^n) + \beta_1 G_{\mathrm{D}}(s^n, u^n) + \beta_2 G_{\mathrm{O}}(s^n, u^n)$, where β_1 and β_2 are weighting factors. It can seen that there could be different total costs for different values of β_1 and β_2. Each combination of the weighting factors determines a unique optimal policy for the adaptation problem because of the unique cost function. The average expected weighted total cost over a long horizon for stationary policy μ can be given by

$$G_{\mathrm{T}}^{\mu} = \lim_{H \mapsto \infty} \frac{1}{H} \mathbb{E}_{\mu} \left[\sum_{n=1}^{H} G_{\mathrm{T}}(s^n, \mu(s^n)) \right]. \tag{7.19}$$

Therefore, our objective is to find optimal stationary policy $\mu^* \in \Pi$ so that G_{T}^{μ} is minimized. The problem can be solved using a dynamic programming algorithm, e.g., the relative value iteration algorithm (RVI), policy iteration (PI) algorithm, etc. It is noted that infinite-horizon formulation does not necessarily mean the communications need to be carried out over an infinitely long time, rather it gives a stationary policy that does not change from one time-slot to another [57].

The heart of the dynamic programming algorithm lies in the Bellman equation (also called the dynamic programming equation). It can be written as

$$\lambda + h(s_i) = \min_{u \in \mathcal{U}_{s_i}} \left[G_{\mathrm{T}}(s_i, u) + \sum_{s_j \in \mathcal{S}} P_{s_i, s_j}(u) h(s_j) \right], \tag{7.20}$$

where λ is the optimal average cost, $h(s_i)$ is the differential cost for each state $s_i \in \mathcal{S}$ with respect to a chosen reference state s_r, and $\mathcal{U}_{s_i}$ is the set of admissible actions in state s_i. It can be noted that not all the actions may be available in all the states. Therefore, the set of admissible (or allowable) actions can be different in different states.

In the relative value iteration algorithm, the relative values of the expected cost for all the states $s_i \in \mathcal{S}$ with respect to reference state s_r are updated iteratively until an equilibrium is reached [58]. The policy under equilibrium condition is the optimal policy. In the policy iteration algorithm, on the other hand, the expected relative costs $h(s_i)$, $s_i \in \mathcal{S}$ are calculated for a given policy $\mu^{(k)}$, where superscript k is an iteration index. Then based on these calculated expected costs, a new improved policy $\mu^{(k+1)}$ is determined that yields maximum expected relative cost using one-step lookahead. The algorithm terminates when the policy improvement step yields no change in the expected costs.

The detailed computational steps for the PI algorithm are given below [59]:

Initialization: Set $k = 0$ and select an arbitrary initial policy $\mu^{(0)}$.

Policy evaluation: Select a state s_r as a reference state and put $h^{(k)}(s_r) = 0$. Find $\lambda^{(k)}$, $h^{(k)}(s_i)$, $i = 1, 2, \ldots, S$ by solving the following set of equations:

$$\lambda^{(k)} + h^{(k)}(s_i) = G_{\mathrm{T}}(s_i, \mu^{(k)}(s_i)) + \sum_{j=1}^{S} P_{s_i, s_j}(\mu^{(k)}(s_i)) h^{(k)}(s_j), \quad \forall s_i \in \mathcal{S}. \tag{7.21}$$

Policy improvement: Find new improved policy using

$$G_{\mathrm{T}}(s_i, \mu^{(k+1)}(s_i)) + \sum_{j=1}^{S} P_{s_i, s_j}(\mu^{(k+1)}(s_i)) h^{(k)}(s_j) = \min_{u \in \mathcal{U}_{s_i}} \left[G_{\mathrm{T}}(s_i, u) + \sum_{j=1}^{S} P_{s_i, s_j}(u) h^{(k)}(s_j) \right]. \tag{7.22}$$

Termination: If $\mu^{(k+1)} = \mu^{(k)}$, the algorithm terminates; otherwise, the process is repeated with $\mu^{(k+1)}$ replacing $\mu^{(k)}$.

In a particular time-slot n, the state of the system is found from the traffic, channel, and buffer states. The action for that system state is the FOOP policy. The resulting power, delay, overflow, and throughput are computed.

7.4.2 Maximum-Likelihood Heuristic Policy (MLHP)

FOOP, discussed above, gives maximum achievable performance and optimal policy assuming the states are available completely. However, for the general problem at hand, the states may not be available in all practical situations. Therefore, the COMDP algorithm cannot be used for solving a POMDP problem. In this section, we discuss a heuristic policy based on the most probable state, called MLHP. Let $\mathcal{Z} = \{z_1, z_2, \ldots, z_S\}$ denote the belief of the states, where $z^n = z_i = I(s_i) = P(s_i)$ is the belief state (or information state) of a physical system state $s^n = s_i \in \mathcal{S}$ in time-slot n. As discussed before, belief is the probability distribution over the states.

In MLHP, at time-slot n for a particular belief $z^n = z_i \in \mathcal{Z}$, the policy can be represented as

$$\mu_{\mathrm{ML}}(z^n) = \mu^*_{\mathrm{MDP}}\left(\arg\max_{s^n \in \mathcal{S}} I(s^n)\right), \tag{7.23}$$

where $\mu^*_{\mathrm{MDP}}(s_i)$ is the optimal policy for state s_i of the system as computed in the earlier section. The maximum-likelihood policy given by (7.23) is stationary since the optimal policy of the underlying perfectly observable MDP is stationary.

The MLHP works as follows: first MLHP finds the most probable state of the system, $s^n = s_i$, from the current belief, $I(s^n)$, $s^n = s_i \in \mathcal{S}$. If two or more states are equally likely, it chooses one arbitrarily. Then the action $u^n = \mu^*(s^n)$ for that particular state in time-slot n is found using the optimal policy μ^* that has been computed using the dynamic programming algorithm for the underlying MDP.[4]

[4]It is assumed that the transition probability and cost models are known, so that optimal policy can be computed using dynamic programming techniques. However, the exact states are unknown in any time-slot.

The new belief, $I(s^{n+1})$, for state $s^{n+1} = s_j$ at the next time-slot $n+1$ is updated using the following filtering formula:

$$I(s^{n+1} = s_j) = \alpha P(o^n | s^{n+1} = s_j, u^n) \sum_{s^n = s_i} P_{s_i,s_j}(u^n) I(s^n = s_i), \quad \forall s_j \in \mathcal{S}, \tag{7.24}$$

where α is a normalizing constant that makes the belief sum to 1. We can show that, $\forall s_j \in \mathcal{S}$,

$$I(s_j) = \frac{P(o^n | s_j, u^n) \sum_{s_i} P_{s_i,s_j}(u^n) I(s_i)}{\sum_{s_j} P(o^n | s_j, u^n) \sum_{s_i} P_{s_i,s_j}(u^n) I(s_i)}. \tag{7.25}$$

That is, a full belief is maintained during execution, but the scheduler uses the maximum-likelihood state for determining the next action (using the underlying MDP). MLHP has performed very well for many real-world POMDP problems in the area of robotic controls. Note that when the observation probability $P(o^n | s^n, u^n)$ is known, we can calculate $P(o^n | s^{n+1}, u^n)$ using the following relationship:

$$P(o^n | s^{n+1} = s_j, u^n) = \sum_{s^n = s_i} P(o^n | s^n = s_i, u^n) P_{s_i,s_j}(u^n). \tag{7.26}$$

7.5 RESULTS

In this section, we present Monte Carlo simulation results for $H = 10^5$ time-slots to show the trade-off between the power and delay for the scheduler. We also discuss the throughput performance under both the FOOP and MLHP policies. Unless specified otherwise, we use the following data for the simulations: number of antenna in the transmitter and receiver, $n_\text{T} = 2$ and $n_\text{R} = 1$, the fading parameter of the Nakagami-m channel, $m = 1$, number of channel states, $C = 2$, normalized Doppler frequency, $f_m T_\text{s} = 0.1$, normalized average channel gain, $\overline{\gamma} = 1$, information code rate of STBC, $R_\text{c} = 1$, average BER in all channel states, $\overline{P}_\text{b} = 10^{-4}$, buffer size $B = 50$, number of actions including no transmission action, $U = 4$, the set of transmission rates, $\mathcal{X} = \{0, 2, 4, 6\}$ bits/symbol (corresponding to QPSK, 16-QAM, and 64-QAM modulations), number of traffic states, $F = 2$, average arrival rate in state f_1 and f_2, $\overline{A}_1 = \overline{A}_2 = 1.0$, the number of symbols in a block, $N_f = 1000$, and the packet size in bits/packet, $N_\text{p} = 1000$. The distribution of the traffic is assumed to be Poisson with probability of $a^n = a_k$, where $k = 0, 1, \ldots, A$, packets arrival is $P(a_k | f_i) = \exp(-\overline{A}_i) \frac{(\overline{A}_i)^k}{k!}$. The traffic states are equally likely. Without loss of generality, we carry out simulation assuming that only channel state may be hidden, but the traffic state and the buffer state are known at the transmitter at the time a decision is made.

The optimal policies of the underlying fully observable MDP are found using weighting factors, β_1 varied from 0.1 to 100, and $\beta_2 = 0$. As discussed before, for each value of the weighting factor, we get a unique policy. We use the policy iteration algorithm as described in Section 7.4.1 to compute optimal policies for each pair of weighting factors. For FOOP Monte Carlo simulation, the samples of the traffic state and the channel state for the whole horizon are generated using their respective transition matrix and uniform initial state probabilities. In the first time-slot, the buffer state is assumed to be b_0 although any starting state will give the same long-term average power, delay, throughput, or overflow. Then it is updated in each time-slot using the packet arrival and packet transmission information. The optimal policy for the system state is applied in this policy in a particular time-slot. For MLHP Monte Carlo simulation, the channel state is assumed to be unknown and it is estimated and updated using belief update formula (7.25). The packets received in error are dropped and not retransmitted, however ACK/NAK feedback is sent to the transmitter to update belief on the channel. The channel state with maximum probability in a given time-slot is assumed to be the underlying channel. Using the generated traffic state and updated buffer state as before, and the estimated channel state from the information state, the system state is found and the optimal action for the system state is applied in a particular slot. Using the policies determined for FOOP and MLHP, the average power, average delay, average overflow, and average throughput are determined for both cases.

In Figure 7.2, normalized average power as a function of normalized average delay for different numbers of transmit antennas and fading change speed is shown.

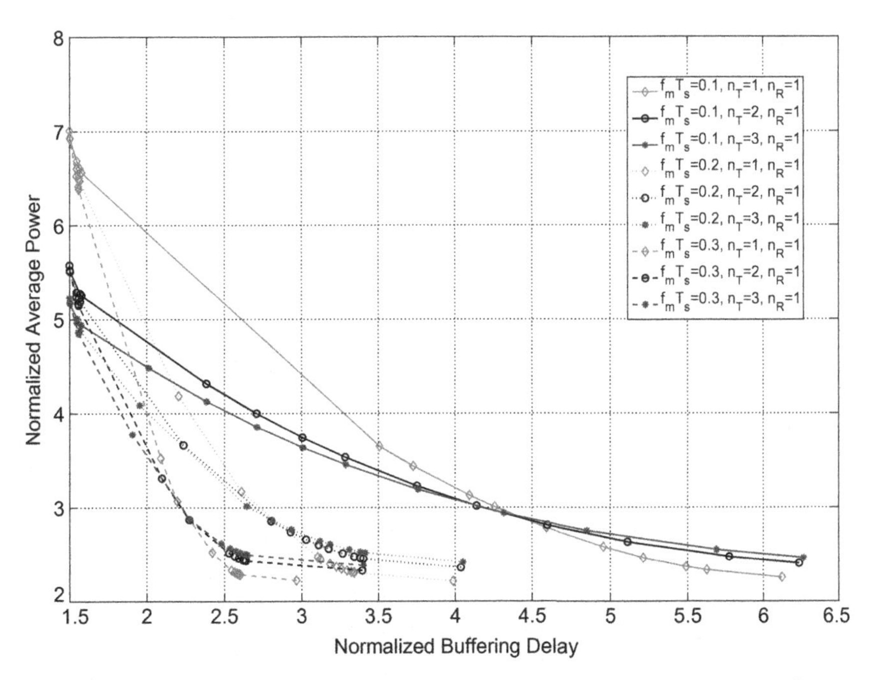

Figure 7.2 The power versus delay trade-off with speed of fading state change and dimension of transmitter diversity. Each point on the curve corresponds to a pair of weighting factor values.

The main characteristic of these curves is a trade-off between the power and the delay. In other words, in order to keep delay low, we need to use more power and vice versa. The engineering implication in the area of green energy is that when the delay limit is increased, the power consumption can be decreased. That makes the device green! A majority of the high bandwidth wireless traffic is relatively delay insensitive, which means it need not be sent immediately. Rather, depending on channel state, buffer state, and traffic state, it can wait in the buffer for a few time-slots before transmission. For example, when the buffer occupancy is relatively low, the packets will be immediately sent when the channel is good, and the packets will wait until the channel changes to a good state if it is in a bad state now. When the buffer occupancy is relatively high, the scheduler need to consider the overflow and delay limits and make an intelligent decision. A good channel state in this situation is not a problem, but in a bad channel state, the scheduler may need to use higher power in order to avoid overflow and maintain the delay limit. Therefore, when the delay limit is larger, the scheduler has more flexibility especially when more power is necessary to achieve a given BER. That is, it can wait more time for a better channel before using more power. This fact is revealed from the curves: the average transmitter power decreases as the average delay limit increases. When the fading rate increases, the fall of power with increased delay limit is faster. This is due to the fact that the

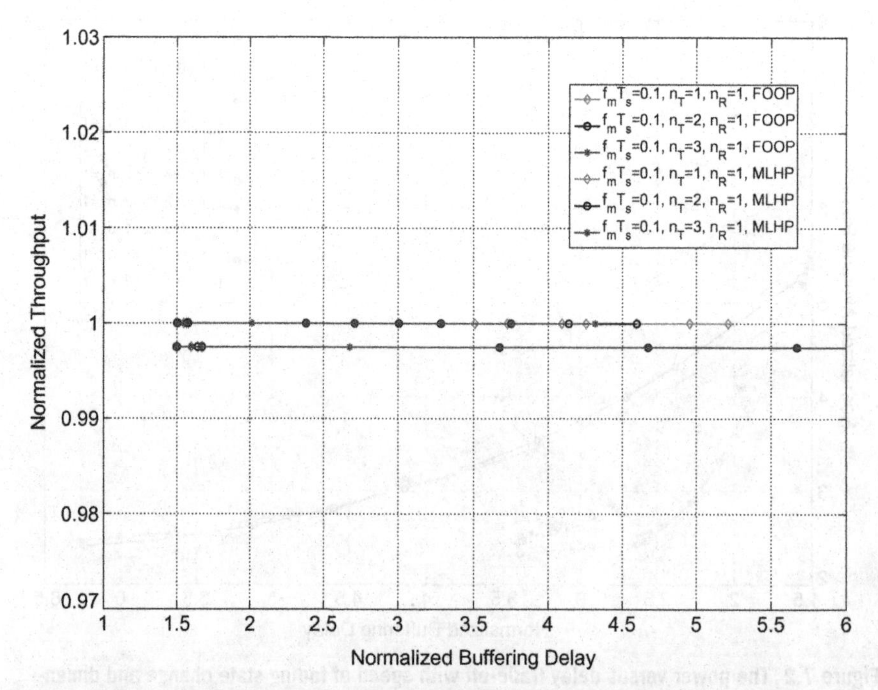

Figure 7.3 Normalized average throughput for FOOP and MLHP algorithms for two situations when the states are known and partially known though observations.

faster channel permits us to get a better channel at a faster rate compared to the worse channel. Therefore, the scheduler needs to wait less for sending the same number of packets with lower power. When the number of transmitter antennas increases for the same fading speed, the transmitter needs less power in lower delay limit region. However, the reverse phenomenon is seen in the higher delay limit regions.

Figure 7.3 shows the average throughput as a function of average delay for both FOOP and MLHP with the same parameters as the previous figure. It is seen that in terms of throughput, the MLHP performs almost the same as the FOOP. Slightly less throughput is observed due to the suboptimal nature of the heuristic policy, which makes decisions based on information state, not on the actual state. Note that in our investigated schedulers the throughput remains the same irrespective of fading rate and/or transmitter antenna diversity. The scheduler is smart enough to take care of maintaining the overflow, BER, and delay, and hence the normalized throughput remains the same.

7.6 CONCLUSION

In this chapter, we discussed two packet schedulers over wireless channels, which may possess memory. The first scheduler deals with the situation when all the states, namely, traffic state, buffer state, and channel state, are known at the transmitter. In some practical situations, the exact traffic and channel states may not be known at the time of transmission. To handle these situations, we propose a heuristic policy for the resulting partially observable Markov decision process. The optimal policies for the underlying Markov decision process problem are determined using a policy iteration algorithm. The formulation of the problem is described through its objectives and associated components. We discussed the traffic and channel models for both cases taking real-world scenarios into consideration. We discussed associated literature that deals with energy-aware design. We provided simulation results to show the energy optimization and saving features of the green schedulers. Future work in this area may consider multiuser, multiantenna, multicarrier, and multiple access scenarios. Also, other upper-layer information may be combined at the physical layer transmissions.

REFERENCES

[1] G. Miao, N. Himayat, Y.G. Li, A. Swami, Cross-layer optimization for energy-efficient wireless communications: a survey, Wiley Wireless Communications and Mobile Computing 9 (4) (2009) 529–542.

[2] S. Liu, J. Wu, C.H. Koh, V. Lau, A 25 $\text{Gb/s}(/\text{km}^2)$ urban wireless network beyond IMT-advanced, IEEE Communications Magazine 49 (2) (2011) 122–129.

[3] L. Dong, A.P. Petropulu, H.V. Poor, Weighted cross-layer cooperative beamforming for wireless networks, IEEE Transactions on Signal Processing 57 (8) (2010) 3240–3252.

[4] H. Kim, C.-B. Chae, G. de Veciana, R.W. Heath, A cross-layer approach to energy efficiency for adaptive MIMO systems exploiting spare capacity, IEEE Transactions on Wireless Communications 8 (8) (2009) 4264–4275.

[5] G. Li, P. Fan, K.B. Letaief, Rayleigh fading networks: a cross-layer way, IEEE Transactions on Communications 57 (2) (2009) 520–529.
[6] X.-H. Lin, Y.-K. Kwok, H. Wang, Cross-layer design for energy efficient communication in wireless sensor networks, Wiley Wireless Communications and Mobile Computing 9 (2) (2009) 251–268.
[7] H. Cheng, Y.-D. Yao, Link optimization for energy-constrained wireless networks with packet retransmissions, Wireless Communications and Mobile Computing, vol. 12, no. 6, pp. 553–566, Apr. 2012.http://dx.doi.org/10.1002/wcm.996.
[8] J. Gong, S. Zhou, Z. Niu, Queuing on energy-efficient wireless transmissions with adaptive modulation and coding, in: Proceedings of IEEE ICC'11, Kyoto, Japan, June 2011, pp. 1–5.
[9] Q. Liu, S. Zhou, G.B. Giannakis, Cross-layer combining of adaptive modulation and coding with truncated ARQ over wireless links, IEEE Transactions on Wireless Communications 3 (2004) 1746–1755.
[10] A. Maaref, S. Aı¨ssa, Combined adaptive modulation and transacted ARQ for packet data transmission in MIMO systems, in: Proceedings of IEEE Globecom'04, Dallas, TX, vol. 6, November 29–December 3, 2004, pp. 3818–3822.
[11] H. Kim, H. Lee, S. Lee, A cross-layer optimization for energy-efficient MAC protocol with delay and rate constraints, in: Proceedings of the IEEE International Conference on Acoustics, Speech and Signal Processing (ICASSP), 2011, pp. 2336–2339.
[12] S. Buzzi, V. Massaro, H.V. Poor, Energy-efficient resource allocation in multipath CDMA channels with band-limited waveforms, IEEE Transactions on Signal Processing 57 (4) (2009) 1494–1510.
[13] S. Buzzi, D. Saturnino, A game-theoretic approach to energy-efficient power control and receiver design in cognitive CDMA wireless networks, IEEE Journal of Selected Topics in Signal Processing 5 (1) (2011) 137–150.
[14] F. Meshkati, H.V. Poor, S.C. Schwartz, R.V. Balan, Energy-efficient resource allocation in wireless networks with quality-of-service constraints, IEEE Transactions on Communications 57 (11) (2009) 3406–3414.
[15] H.-J. Lee, J.-T. Lim, Cross-layer congestion control for power efficiency over wireless multihop networks, IEEE Transactions on Vehicular Technology 58 (9) (2009) 5274–5278.
[16] L. Lin, X. Lin, N.B. Shroff, Low-complexity and distributed energy minimization in multihop wireless networks, IEEE/ACM Transactions on Networks 18 (2) (2010) 501–514.
[17] D. Triantafyllopoulou, N. Passas, L. Merakos, N.C. Sagias, P.T. Mathiopoulos, E-CLEMA: a cross-layer design for improved quality of service in mobile WiMAX networks, Wiley Wireless Communications and Mobile Computing 9 (9) (2009) 1274–1286.
[18] S.-L. Tsao, C.-H. Huang, An energy-efficient transmission mechanism for VoIP over IEEE 802.11 WLAN, Wiley Wireless Communications and Mobile Computing 9 (12) (2009) 1629–1644.
[19] S. Pollin, R. Mangharam, B. Bougard, L.V. der Perre, I. Moerman, R. Rajkumar, F. Catthoor, MEERA: cross-layer methodology for energy efficient resource allocation in wireless networks, IEEE Transactions on Wireless Communications 6 (2) (2007) 617–628.
[20] V. Miliotis, A. Apostolaras, T. Korakis, Z. Tao, L. Tassiulas, New channel allocation techniques for power efficient WiFi networks, in: Proceedings of IEEE PIMRC'2010.
[21] L. Saker, S.-E. Elayoubi, T. Chahed, Minimizing energy consumption via sleep mode in green base station, in: Proceedings of IEEE WCNC'2010.
[22] C. Comaniciu, N.B. Mandayam, H.V. Poor, Radio resource management for green wireless networks, in: Proceedings of IEEE VTC'09-Fall, Anchorage, AK, September 2009, pp. 1–5.

[23] A. Ephremides, Energy concerns in wireless networks, IEEE Wireless Communications Magazine 9 (4) (2002) 48–59.
[24] R.A. Berry, R.G. Gallager, Communication over fading channels with delay constraints, IEEE Transactions on Information Theory 48 (2002) 1135–1149.
[25] D. Rajan, A. Sabharwal, B. Aazhang, Outage behavior with delay and CSIT, in: Proceedings of IEEE ICC'04, Paris, France, vol. 1, June 20–24, 2004, pp. 578–582.
[26] A.K. Karmokar, D.V. Djonin, V.K. Bhargava, Optimal and suboptimal packet scheduling over time-varying flat fading channels, IEEE Transactions on Wireless Communications 5 (2006) 446–457.
[27] D.V. Djonin, A.K. Karmokar, V.K. Bhargava, Joint rate and power adaptation for type-I hybrid ARQ systems over correlated fading channels under different buffer-cost constraints, IEEE Transactions on Vehicular Technology 57 (2008) 421–435.
[28] H. Wang, N.B. Mandayam, A simple packet-transmission scheme for wireless data over fading channels, IEEE Transactions on Communications 52 (2004) 1055–1059.
[29] E. Uysal-Biyikoglu, A.E. Gamal, B. Prabhakar, Energy-efficient packet transmission over a wireless link, IEEE/ACM Transactions on Networks 10 (2002) 487–499.
[30] M. Zafer, E. Modiano, Optimal rate control for delay-constrained data transmission over a wireless channel, IEEE Transactions on Information Theory 54 (9) (2008) 4020–4039.
[31] A. Fu, E. Modiano, J.N. Tsitsiklis, Optimal transmission scheduling over a fading channel with energy and deadline constraints, IEEE Transactions on Wireless Communications 5 (2006) 630–641.
[32] R. Negi, J.M. Cioffi, Delay-constrained capacity with causal feedback, IEEE Transactions on Information Theory 48 (2002) 2478–2494.
[33] D.V. Djonin, V. Krishnamurthy, MIMO transmission control in fading channels—a constrained Markov decision process formulation with monotone randomized policies, IEEE Transactions on Signal Processing 55 (2007) 5069–5083.
[34] A.K. Karmokar, D.V. Djonin, V.K. Bhargava, Cross-layer rate and power adaptation strategies for IR-HARQ systems over fading channels with memory: a SMDP-based approach, IEEE Transactions on Communications 56 (2008) 1352–1365.
[35] J.D. Choi, K.M. Wasserman, W.E. Stark, Effect of channel memory on retransmission protocols for low energy wireless data communications, in: Proceedings of IEEE ICC'99, Vancouver, BC, vol. 3, June 6–10, 1999, pp. 1552–1556.
[36] D. Zhang, K.M. Wasserman, Energy efficient data communication over fading channels, in: Proceedings of IEEE WCNC'00, Chicago, IL, vol. 3, September 23–28, 2000, pp. 986–991.
[37] L.A. Johnston, S. Krishnamurthy, Opportunistic file transfer over a fading channel: a POMDP search theory formulation with optimal threshold policies, IEEE Transactions on Wireless Communications 5 (2006) 394–405.
[38] A. Ekbal, K.-B. Song, J.M. Cioffi, QoS-constrained physical layer optimization for correlated flat-fading wireless channels, in: Proceedings of IEEE ICC'04, Paris, France, vol. 7, June 20–24, 2004, pp. 4211–4215.
[39] A.K. Karmokar, D.V. Djonin, V.K. Bhargava, POMDP-based coding rate adaptation for type-I hybrid ARQ systems over fading channels with memory, IEEE Transactions on Wireless Communications 5 (2006) 3512–3523.
[40] R.A. Berry, E.M. Yeh, Fundamental performance limits for wireless fading channels—cross-layer wireless resource allocation, IEEE Signal Processing Magazine 21 (2004) 59–68.
[41] Z.K.M. Ho, V.K.N. Lau, R.S.-K. Cheng, Cross-layer design of FDD-OFDM systems based on ACK/NAK feedbacks, IEEE Transactions on Information Theory 55 (10) (2009) 4568–4584.

[42] A.T. Hoang, M. Motani, Decoupling multiuser cross-layer adaptive transmission, in: Proceedings of IEEE ICC'04, Paris, France, vol. 5, June 20–24, 2004, pp. 3061–3065.
[43] D.I. Shuman, M. Liu, O.Q. Wu, Energy-efficient transmission scheduling with strict underflow constraints, IEEE Transactions on Information Theory 57 (3) (2011) 1344–1367.
[44] N. Salodkar, A. Karandikar, V.S. Borkar, A stable online algorithm for energy-efficient multiuser scheduling, IEEE Transactions on Mobile Computating 9 (10) (2010) 1391–1406.
[45] N.B. Chang, M. Liu, Optimal channel probing and transmission scheduling for opportunistic spectrum access, IEEE/ACM Transactions on Networks 17 (6) (2009) 1805–1818.
[46] T. Zhang, D.H.K. Tsang, Optimal cooperative sensing scheduling for energy-efficient cognitive radio networks, in: Proceedings of IEEE INFOCOM'2011.
[47] J. Zhang, Q. Zhang, B. Li, X. Luo, W. Zhu, Energy-efficient routing in mobile ad hoc networks: mobility-assisted case, IEEE Transactions on Vehicular Technology 55 (1) (2006) 369–379.
[48] W. Lee, C.-S. Hwang, J.M. Cioffi, Energy-efficient random-access scheduler for delay-limited traffic in wireless networks, in: Proceedings of IEEE GLOBECOM 2007, pp. 5048-5052, Washington, DC, Dec. 2007.
[49] H. Li, N. Jaggi, B. Sikdar, Relay scheduling for cooperative communications in sensor networks with energy harvesting, IEEE Transactions on Communications 10 (9) (2011) 2918–2928.
[50] M. Baghaie, B. Krishnamachari, Delay constrained minimum energy broadcast in cooperative wireless networks, in: Proceedings of IEEE INFOCOM'2011.
[51] C.K. Ho, P.H. Tan, S. Sun, Relaying for energy-efficient scheduling with deadline, in: Proceedings of IEEE ICC'2010.
[52] K. Cohen, A. Leshem, Energy-efficient detection in wireless sensor networks using likelihood ratio and channel state information, IEEE Journal on Selected Areas in Communications 29 (8) (2011) 1671–1683.
[53] G.K. Atia, V.V. Veeravalli, J.A. Fuemmeler, Sensor scheduling for energy-efficient target tracking in sensor networks, IEEE Transactions on Signal Processing 59 (10) (2011) 4923–4937.
[54] X. Mao, P. Qiu, Energy efficient scheduling in multi-access wireless sensor networks: a stochastic dynamic programming method, in: Proceedings of the European Wireless Conference, 2010, pp. 1–5.
[55] C.-D. Iskander, P.T. Mathiopoulos, Fast simulation of diversity Nakagami fading channels using finite-state Markov models, IEEE Transactions on Broadcasting 49 (2003) 269–277.
[56] J. Lu, K.B. Letaief, J.C.-I. Chuang, M.L. Liou, M-PSK and M-QAM BER computation using signal-space concepts, IEEE Transactions on Communications 47 (2) (1999) 181–184.
[57] S. Russel, P. Norvig, Artificial Intelligence: A Modern Approach. second ed., Prentice Hall, Upper Saddle River, NJ, 2003
[58] M.L. Puterman, Markov Decision Processes: Discrete Stochastic Dynamic Programming, John Wiley & Sons, New York, NY, 1994
[59] D.P. Bertsekas, Dynamic Programming and Optimal Control. second ed., Athena Scientific, Belmont, MA, 2001

CHAPTER

Green Computing and Communication Architecture

Tarik Guelzim, Mohammad S. Obaidat

Department of Computer Science and Software Engineering, Monmouth University, West Long Branch, NJ 07764, USA

8.1 INTRODUCTION

Enterprises,[1] governments, and institutions at large have a new important agenda: tackle the most challenging environmental issues and adopt environmentally sound practices. We are very passionate about advancing information technology (IT), nevertheless, IT has been contributing to environmental problems, which most people do not realize. Computers and IT infrastructure consume significant amounts of electricity, placing a big burden on power grids. Besides, in order to cope with the increasing demand for content, every year close to 150,000 new cellular towers are deployed across the world to serve around 400 million new subscribers worldwide. The downlink per subscriber is also increasing, going from an average of 2 MB per day to 16 MB. Besides the technical complexity behind developing and maintaining such a network, meeting this demand alone is becoming a burden on energy consumption, especially if we take into consideration the energy cost, which is skyrocketing in the middle of ever-changing geopolitics and economics around the world. Today, it is clear that it is no longer an option to focus *only* on the technological development of broadband Internet to solve the capacity problem nor it is an option to hold on to the current networking communication protocols that have reached a road block with regards to their initial design and features [1–2].

The impressive economic growth of China and India has led their citizens to use more and more information and communication devices. Given the fact these two countries account for about 40% of the world's population, this means that the amount of energy consumed and amount of carbon dioxide emitted by such The information and communication technology (ICT) devices are increasing at a high rate.

In the telecommunication markets, energy consumption accounts for more than half (50%) of the operating expenses of major Internet service providers (ISPs) and causes 1% of the entire world's carbon footprint. To give the reader a relative view of this problem, each base station antenna along with the operating equipment requires

[1]Regardless of internet speed or physical media.

on average 8800kWh per year, which is the average energy required to run a two bedroom house in the USA or a little under 200,000 USA households if we take into account an average ISP network of 15,000 cell towers [3].

In other specialized fields such as sensor networks, it is clear today that energy preservation is primordial to the wide spread of this technology as well as widespread adoption by groups other than the military. Much research in this field has focused on innovative ways to propose secure, energy-efficient dynamic routing protocols for heterogeneous wireless sensor networks in urban environments [4].

Combining the growth in energy demand by major communication network infrastructures and the cost that takes on a positive slope, we cannot neglect the environmental and societal fallouts which will affect societies across the world if not addressed properly.

To this day, there is more hype than real science behind the term green computing and communications, which does not have a clear scientific definition other than its commercial and marketing connotation. What is needed is rather a standard to adopt that considers efficiency in communication network equipment along with the use of alternative sources of energy [5–8].

8.2 GREEN COMPUTING AND GOVERNMENTAL EFFORT

In a nutshell, the study of green communications requires investigation in many areas such as radio frequency (RF) hardware, media access control (MAC) protocols, general networking, material design, and integration of renewable energy. In this sense, metrics are key elements to properly measure and provide indicators to allow making better judgments and to strategically plan the growth of the different communication network infrastructures. Organizations today are realizing that greenhouse gas (GHG) emissions are directly impacted by the energy source as well as the consumption levels in the field [9].

The following equation summarizes the above statement:

$$\begin{aligned}\text{reduced energy consumption} &= \text{reduced greenhouse emissions}\\ &= \text{reduced operational cost for business.}\end{aligned}$$

In an attempt to jump-start this field, governments have started many programs, each of which takes a different approach at lowering energy consumption in a specific subdomain. The following section names a few [10,11].

8.2.1 Energy Star

In a programmatic effort by the U.S. government, the Energy Star program has provided administrative guidance in reducing energy and power consumption in appliances and other devices, both civil and military. Energy Star is one of the first technologies that was built in and used inherently in data centers. This is because virtually every computer device today has this technology incorporated natively within it. In an attempt to further the reach of this technology on a global distributed scale, programs such as the *eSustainability* initiative [12] have called for broad directives such as smart motors, smart logistics, smart grids, and smart networks as a way of standardizing architectures by taking into energy use from the ground up [13].

The new Energy Star 4.0 standard regulates energy performance of external and internal power supplies. It also gives power consumption specifications for different hardware states: idle, sleep, and standby modes. Computers meeting these requirements will save energy in all modes of operation. This new standard also requires OEMs to educate end users about efficient computer use.

8.2.2 Electronics Energy Efficiency Effort

The electronics industry has put a lot of effort into improving component efficiency because they are the basic building blocks of any system. The Silicon Integration Initiative (S2I) and the Power Forward Initiative (PFI) [14] are creating conventions for low power design by standardizing computer hardware interfaces. These flexible models are deployed in real-field experiments by large telecommunication providers, such as Vodafone. Initial forecasts have shown to be promising; it is possible to reduce carbon emission by 50% over a decade. Reducing carbon emission translates to the use of the alternative energy sources complemented by energy efficient components [15].

8.2.3 The TIP Program

In another governmental effort, the TIP managed green communication program proposes a new solution that addresses the inherent limitations of the vertically separated design of OSI layers. TIP puts the emphasis on combining layer 1 and layer 2 (physical and MAC) to promote factoring research effort in one area. It also promotes using scientifically rigorous definitions and metrics to allow agencies to quantify and qualify green communication infrastructures. More of a framework than a specific technology, TIP managed programs are being experimented with in fields such as transportation, medical systems, and finance [16, 17].

The TIP framework [18] mapping to green communication is summarized in Table 8.1.

8.2.4 EPEAT Program

The EPEAT program [19] is a comprehensive environmental rating that helps promote greener computer and electronic equipment. It is the result of a collaboration between businesses, advocacy, government, and academia. EPEAT evaluates electronic products on 23 required criteria and 28 optional ones, which are grouped into eight main categories:

1. Reducing and eliminating environmentally sensitive materials
2. Selecting materials

Table 8.1 TIP Program Guidance Summary.

	Mapping to Administrative Guidance	Justification of Government Awareness	TIP Funding
Metrics and measurement development	Synergy with emphasis on energy initiatives (PCAST) [18] and the reliance on several NIST current programs.	Current research is still blurred by the limited design of layered architecture, mainly OSI.	TIP is said to be transformational because it questions the very basics of the communications architecture in use today. It promotes finding ways other than layer components such as distributed multiagent networks. Overcoming the current vertical separation between layers will result in a radical environmental and infrastructure impact but shall allow ISPs to meet the high demand for broadband communication networks.

3. Designing for the product's end of life (recycling)
4. Product longevity
5. Energy conservation
6. End of life management
7. Corporate performance
8. Packaging

EPEAT identifies its registered products as bronze, silver, or gold. Bronze products meet all 23 required criteria; silver meets all 23 criteria in addition to at least 14 optional criteria; and the gold class meet all 23 required criteria in addition to at least 21 optional ones. Optional criteria are freely selected by manufacturers.

Today, all registered computers have reduced levels of cadmium, lead, and mercury to better protect human health.

8.3 ENERGY-BASED MONITORING ARCHITECTURE FOR GREEN COMPUTING

8.3.1 Study of Power Consumption in Traditional PC Architecture

The following graph depicts the energy consumption [20] by different parts of a traditional computer architecture under normal usage conditions (see Figure 8.1).

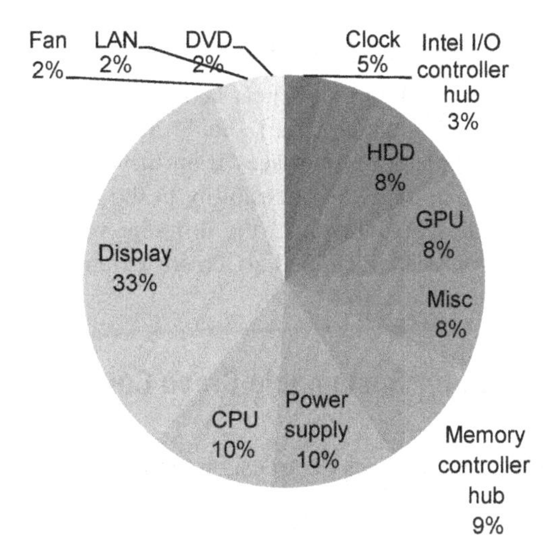

Figure 8.1 Conventional PC energy consumption.

8.3.2 Energy Monitoring

More than ever, enterprise architects and decision makers need tools such as indicators and dashboards to allow them to successfully forecast the energy consumption of their future system (SCADA system, data center, etc.). Based on these forecasts, existing systems can be optimized. Besides, technologies and design decisions can also be impacted by those metrics when planning for future systems [15,16]. Table 8.2 summarizes a few energy monitoring tools [21,22] that are in use today.

Table 8.2 Energy Monitoring Tools [31]

Tools/Technique	Usage
Circuit meters	Used in data centers especially in a "rack group" setup. Circuit meters allow the precise measurement of the energy consumption of a group of server racks.
Power strip meters	Also used in data centers. Power strip meters allow for the monitoring of a group of interconnected systems
Plug meter	The plug meter allows for the monitoring of one physical system
External thermal sensor meter	The thermal sensor meter allows a more grained sensing capability at a floor level.
Internal server thermal meter	Used inside a physical system.

8.3.3 Extensible Computing Architecture

Any green computing architecture today must take into consideration the environmental impact it incurs. IT architects [23] should aggregate the communication models with the different metering interfaces from the ground up to allow better synergy and effectiveness in terms of extensibility in the long run. Because of the lack of standards and the segmentation of the different vendors, several reusable architectural blocks have been identified to create extensible green computing architecture systems (see Table 8.3).

8.3.4 Best Practices for Sustainable Green Computing System Design

The following are a few best practices when designing a green computing architecture.

8.3.4.1 Understanding Business Objectives

Understanding the business objectives in an environment driven company can be a daunting task. Nevertheless, it is critical to be aware of the regulations, entities, and norms that are applicable to a company before starting any design blueprints.

Table 8.3 Extensible Green Computing Architecture Basic Elements

Issue	Description
Proprietary energy APIs	Most vendors have developed their own APIs to interface and communicate with metering devices. Especially in large IT structures, we can find more than one interfacing API each corresponding to a different system. This fragmentation phenomenon is similar to having to use many cell phones and using a different battery charger for each one of them. Inevitably, this incurs and multiplies additional maintenance cost for the different architectures.
Environmental communication bus	Because there is a heterogeneous metering environment, system architects attempt to design architectures that comprise a common communication bus that acts as a façade when communicating with the different subsystems. Consolidating the communication bus makes it possible to create better monitoring and reporting tools [24].
Configuration management	Configuration is an important aspect of any system. In large systems, configuration changes can be costly (e.g., number of backups, number of allowed concurrent sessions, disk quota, redundancy, etc.) and can add work load on the system. Configuration management should allow correlating energy consumption to configured system profiles and also allow adapting the system dynamically based on its GHG measurement [24].

8.3.4.2 Understanding Power Consumption and Impact

Knowing how the energy is consumed and in what parts of the system is as important as developing the system itself. By using the metrics and indicators given by the different deployed sensors, we can for instance apply dynamically different configurations to different situations.

8.3.4.3 Developing a Monitoring Strategy

This step is directly linked to the business strategy of the company and can be summarized in defining metrics that are relevant to the system being designed.

8.3.4.4 Build Environment Sustainability into Configuration Management

When it comes to large systems, configuring system execution profiles is an essential step. It is important that the configuration management system is built around a common green computing strategy in order to allow enabling or disabling certain features, components, hardware, among others, either statically or dynamically based on received metrics from the different sensors.

8.3.5 Incorporating Alternative Energy in the Basic BTS Architecture

Several cellular operators are experimenting with the use of alternative energy such as wind and solar in order to operate cellular base stations. Currently, there are still many base stations in developing countries that run on diesel fuel. In Namibia for instance, Mobile Telecommunication Limited (MTC) and Motorola initiated a 90-day trial in 2007 to study the feasibility of using solar panels to fuel cellular base stations [25]. This trial used a 6 kW wind turbine and 28 kW solar panels in addition to small batteries to power critical equipment. Results have suggested that approximately 659 kg CO_2 could be saved by eliminating the diesel generator [26]. Many countries have launched major projects to produce clean energy from solar panels. Examples include Germany, Spain, USA, Japan, Italy, Czech Republic, Belgium, China, France, and India. As for wind energy projects, there are many that have been launched all over the world. Major countries that have wind energy projects include China, USA, Germany, Spain, India, Italy, France, UK, Canada, and Denmark. Energy produced by wind is increasing by a rate of 30% annually.

The increase in oil prices and climate change concerns have led to growth in renewable energy projects, legislation, incentives by governments and power energy providers, and commercialization of such technology.

8.4 GREEN COMMUNICATION PROTOCOLS AND MODELS

The following section details a few *green* computing protocol models. Because of the natural need for energy-efficient models and protocols by sensor networks, much of the research has been done in this area and it is also a focus area of this chapter.

8.4.1 Cognitive Radio

Recent advances and breakthroughs in the area of cognitive radio [27,28] have significant potential in green communication. Cognitive radio, as its name implies, relies fundamentally on sensing the surrounding environment and readapting its behavior on the different elements collected. This technology also uses the information gathered to learn future patterns. The integration of contextual "awareness" in the wireless spectrum opens a wide range of power control possibilities. This is because the energy produced or made available by the generators is scaled in relation to the traffic requirements and also by taking into consideration the current state of the entire power grid in which the infrastructure making use of this technology resides. As an example, controlling the energy is dissipated by consuming devices weight in the load balancing with the energy supply in the grid thus contributing to the overall performance of the grid.

This technology does not, however, come without drawbacks, in reality, in today's setups; using less power at the cell level must be compensated for by more power consumption at the device level resulting in a zero sum gain. It is clear that an interdisciplinary method in the telecom field between cell tower and cell phone handheld manufacturers must be put into place to push forward to an interlayer approach.

8.4.2 First Order Radio Model (FORM)

The first order radio model (FORM) has been developed in an attempt to solve energy inefficiency issues with sensor networks. Sensor networks are usually distributed in different geographical areas to monitor predefined phenomena. Sensor nodes are usually very small with limited capabilities in terms of their computation power, memory, and sometimes power. Main applications of sensor networks include environmental sensing, air pollution monitoring, nuclear, biological, and chemical attack detection, reconnaissance of enemy forces, battlefield surveillance, forest fire detection/monitoring, structural monitoring, health applications, and digital and smart home applications, among others [29]. Due to the resource constraints of the sensor nodes, redundancy of coverage area must be reduced for effective utilization of the available resources. If two nodes have the same coverage area in their active state, and if both nodes are activated simultaneously, it leads to redundancy in the network and waste of precious sensor resources [30].

Because of the inherent limitations both in design and in security, there is a necessity to deploy these sensor devices in large numbers (1000s in areas such military intelligence) in order to obtain reliable results [31,32]. This is done by combining several unreliable data signals to produce an accurate one. This technique is known as data fusion.

Under normal conditions, FORM [33] diminishes energy dissipation between the transmitter and the receiver nodes. To send a message the radio dissipation is

$$\epsilon_{elec} = 50\ \text{nJ/bit}.$$

and

$$\epsilon_{amp} = 100\ \text{pJ/bit/m}^2.$$

Transmitting a message of $k$bits a distance d radio expends

$$E_{\tau x}(k,d) = E_{\tau x-\mathrm{elec}}(k) + E_{\tau x\text{-amp}}(k,d) = E_{\tau x\text{-elec}} * k + \epsilon * k * d^2.$$

The receiving end radio expands as such:

$$E_{Rx}(k) = E_{Rx\text{-elec}}(k).$$

8.4.3 Direct Communication Protocol (DCP)

This protocol is used for routing in combination with a low energy dissipation model such as FORM. Using the direct communication protocol (DCP), each sensor sends data directly to the base station. Only the base station receives messages. Although this might seem like a design limitation for some, it is acceptable. This is because if the base station is distant away from the sending nodes and the computed energy required to receive the data (based on the formula by FORM) is large, this will drain the node's battery and the entire network will become irresponsive and inaccurate progressively.

8.4.4 Minimum Transmission Energy (MTE) Protocol

Using the Minimum Transmission Energy (MTE) protocol, nodes employ intermediate nodes for routing to send data to the base station. The intermediate node's path is chosen such that it minimizes the energy amplifier E_{amp} equation (using the FORM model). This protocol has a few drawbacks however. In fact, if relied on to minimize the amplifier equation, all nodes close to the base station will be chosen as intermediate nodes since they do not require a large amount of signal amplification to send data to the base station. This implies that these nodes will run out of battery quickly and thus the section in which they belong will be out of order quickly as well.

8.4.5 Clustering

The clustering technique tries to carry the benefits of both DCP and MTP. Under clustering, nodes are organized into clusters with a "delegate" base station. These base stations then transmit consolidated data to the main base. This technique works best if the base station is not an energy-constrained node.

8.4.6 System Power Consumption (SPC)

System Power Consumption (SPC) is yet another communication model that aims at lowering energy consumption. Recent research efforts [34] have proposed to limit energy consumption by adapting modulation techniques, coding, and radiated power. The research work in [34] is on the same energy optimization track as the power amplifier (PA) model described in a previous section.

8.4.7 Network Topology and Operation

One of the leading telecom companies, Huawai, is leading the way to providing an energy efficient base transceiver station (BTS) by including PA improvement along with operational strategies as well as an advanced cooling architecture [35]. At the core of Huawai's technology is the Doherty-enhanced technology that is combined with a custom power amplification chip to elevate amplifier efficiency to 45% instead of the current 33%. Recent advanced firmware upgrades address shutdown and power-off technologies like the time-slot and channel shutdown to reduce static power consumption by as much as 60%. In addition, the relative location between an antenna and the radio has been retaught because some field studies have shown that having the radio equipment at the ground level connected to antennas by means of feeder cables contributes to a lot of energy waste. Instead, eliminating many cable meters by redesigning the base station in a modular way and bringing the equipment closer to the antenna can save energy.

8.4.8 Low Energy Adaptive Clustering Hierarchy (LEACH) Protocol

The Low Energy Adaptive Clustering Hierarchy (LEACH) protocol [34,35] improves on the previously described clustering algorithms by distributing cluster information and by adding local processing to reduce global communication. In addition, LEACH adds random rotation of the cluster heads to maximize and extend the system lifetime. This protocol is broken into the setup and steady-state phase.

The setup phase starts with advertisement in which each node decides whether to become a cluster head by choosing a random number between 0 and 1. The first node to choose a value below a threshold is denoted by $T(n)$ where

$$T(n) = \begin{cases} \dfrac{P}{1-P*\left(r*\text{mod}\frac{1}{P}\right)} & \text{if } n \in G, \\ 0 & \text{otherwise,} \end{cases}$$

where P is the desired percentage of cluster header, R is the current round, and G is the set of nodes that were not cluster heads in any of the $\frac{1}{P}$ rounds.

All "elected" cluster heads broadcast an advertisement message to all nodes in the network using the carrier sense multiple access (CSMA) MAC protocol employing the same transmit energy.

Once the broadcast is successful, the clustering phase starts and each node is required to inform the cluster head that it will join its cluster. As in the advertisement phase, the CSMA MAC protocol is used to transmit the clustering information. Based on the clustering requests received from the different nodes, each cluster head creates a time division multiple access (TDMA) schedule and informs each node which frame it can transmit within. The radio of each nonclustering node can be turned off until the node reaches its allocation time, thus minimizing the energy dissipation of each one of them. Cluster heads use data compression to create a composite signal to the master cluster base station with all of the received messages.

Experimental results in [36] have shown that the LEACH protocol reduces energy consumption between seven and eight times when compared to DCP or MTE alone. Another important advantage that LEACH provides is that nodes in a mesh sensor network die out in a random fashion rather than a predictable fashion such as in DCP or MTE. Ideas for improving LEACH include features such as hierarchical clusters and an energy-based threshold function that incorporates MTE.

8.4.9 Network Elements—Power Amplifier (PA)

A classical cell base station is made up of the following main parts:

- The feeder network
- The radio
- The baseband unit

According to recent studies [36], these units consume energy in the following manner (see Table 8.4).

Out of the 80% energy consumed by the base station's radio, over 50% is consumed by the power amplifier. Advanced simulation models have shown that the deployment of larger base stations, also called macrocells, in combination with smaller cells, picocells, can potentially give a 60% efficiency gain over the current modes of deployment. As a side note, besides the energy efficiency this solution presents, it also gives higher and better capacity than the traditional setup. Some problems are still yet to be solved with regard to efficient handover, security, and load balancing before this technology can be deployed at a larger scale.

Another track of research [37] in the power amplifier (PA) area is a multicarrier base station technology such as GSM Quadruple Transceiver. This latter device uses six carriers to bring down the energy consumption to 30%.

8.4.10 Wireless Network Distributed Computing (WNDC)

The current traditional wireless model empowers a "selfish" node behavior that is solely based on improving performance, quality of service (QoS), or capacity in the surrounding areas that are covered by its cell. The requirements of each base station must rather be coordinated within the overall network. Advances in wired distributed computing have defined initial models from which wireless distributed communication can be inspired. The ability to apply such concepts in the wireless

Table 8.4 Energy Consumption of Base Station Network Elements

Element	Consumption (%)
Feeder network	15
The radio	80
The baseband unit	5

domain remains a challenge given the disruptive characteristics of the wireless channel. Distributed design trade-offs involve the following:

- A cross-system interaction between the application layer and the communication layer
- Simplifying the interaction complexity in the underlying networking, radio, and physical layers

These trade-offs make it possible to lower the power consumption per node and by consequence to the whole network. In addition it leads to demand-supply matching as well as simplifying small node form factor design with lower processor clock rates. Each of these small nodes contributes to parts of the communicated data and thus processes the communication in a distributed and load-balanced manner. This solution can scale out linearly by adding more nodes and potentially lower the amount of processing each node has to do, which again translates into a small energy consumption footprint [38].

8.4.11 Dynamic Spectrum Access (DSA)

Dynamic Spectrum Access (DSA) [39] provides the sensing and dynamic reconfiguration capabilities that leverage the use of spectrum white holes; this allows supporting opportunistic transmission without requiring extra spectrum bandwidth. To build such systems, a power efficient communication processor is necessary to support the high speed operation of this model, especially on low power mobile devices [39]. DSA requires satisfying two contradicting needs: high processing speed and low power consumption. The current general purpose processors or digital signal processors are not suited for such a task because simply their processing power efficiency (GOPS/mW) cannot meet the growth of the DSA requirement, especially in mobile devices. Researchers have exploited similarities between several communication algorithms (synchronization, channel estimation, etc.) in order to reduce the algorithm search space and identify the fundamental mathematical properties. The result of this research led to the development of a green communication processor. It offered solutions to many issues:

1. Offering real-time processing using low clock frequency operation
2. Reducing RF accessing needs since this is a power-consuming factor
3. Processing multiple instructions with each clock cycle

8.5 DESIGNING GREEN COMPUTING MODELS

8.5.1 Green Processor Design

Green computer design aims at reducing the environmental impact of computing by adopting new technologies and using new techniques and materials while balancing performance with the economic viability of the green solution.

Moving from single core to multicore processors saves power while increasing power performance. This is because the race now has shifted from increasing the processor frequency to chip layout optimization in order to increase the number of cores, usually with lower frequency, into one single chip. Some research suggests that 50% energy consumption gain can be achieved by only decreasing the chip frequency by 15%. Other initiatives concentrate on dividing the cache into segments and only enabling, i.e., powering them when needed.

Another track of bleeding edge research has been experimenting with the replacement of the silicon transistor by nanocarbon tubes in our computer CPU and memory chips. Although the results are encouraging, it is still not at the commercial phase [40].

8.5.2 Greening Data Centers

The continued rise of the Internet and Web applications is driving the rapid growth of data centers. Enterprises are installing more servers and expanding their capacity as well. With operational cost of servers that increase year after year, the cost of running data centers continues to increase steadily. We can improve data centers efficiency by using new energy-efficient equipment, improving air flow management to reduce cooling requirements, investing in energy management software, and adopting environmentally friendly designs for data centers. Some of the currently deployed measures in the industry are the following:

1. *Energy conservation:* Companies like HP, IBM, SprayCool, and Cooligy are developing technologies such as liquid cooling and nanofluid cooling into servers and racks. Others use green energy sources such as hydrogen fuel cells as alternative green power sources.
2. *Virtualization:* This is a key strategy to reduce data center power consumption. With virtualization, one server can host multiple virtual servers that use the computing resources efficiently because computing resources (CPU clock cycles) are often not used at their maximum capacity.

8.5.3 Green Supercomputing

Until recently, supercomputers have enjoyed a "free ride" on institutional infrastructure. Data center supercomputers provide unparallel computational horsepower for solving scientific and engineering problems. This horsepower comes at the expense of power consumption not only to run the super computer, but also to cool it down. By nature, supercomputing focuses on performance and "occasionally" cost/performance where performance is defined as speed. As with consumer-end PCs, supercomputer speed has increased tremendously. However, this focus on performance/speed has let other evaluation metrics go unchecked. Supercomputers consume significant amounts of energy and generate a lot of heat. Consequently, in order to keep these super computers running, huge cooling systems must be put in place in order to prevent these systems from crashing or returning unreliable results. Green Destiny is the first major instantiation of the Supercomputing in Small Spaces project. It is arguably the first supercomputer that is energy efficient in an attempt to harness the power of going "green." Green Destiny has put into perspective two types of supercomputer architectures:

1. A low-power supercomputer that balances performance and power at system integration time
2. A power-aware supercomputer that adapts power to performance needs when the system is running

Both of these approaches aim at reducing power consumption and improving energy efficiency.

8.5.4 Web Services Based SaaS and the Cloud

Software as a Service (SaaS) is a buzzword in the realm of cloud computing. The basic idea is that we can reduce the number of applications deployed in data centers by consolidating similar applications by SaaS providers. One can easily see the benefits of such a model because it significantly reduces the need for building data centers by companies and organizations. There are many major players in the market today such as salesforce.com and amazon.com who offer SaaS services at a low cost relative to the real cost it would incur if it was built, managed, and maintained by individual companies.

The most common implementation of SaaS is in the form of generic Web Services [41], where a Web service is an implementation of the functionality of a business enterprise, which can be utilized by users of different applications. The major key benefits of Web Service based SaaS can be summarized as follows:

1. Cost (money) effectiveness
2. Time effectiveness
3. Focus on business needs rather than the construction and maintenance of IT infrastructure
4. Gain immediate access to state-of-the-art innovations rather than maintaining legacy infrastructure
5. Gaining instant benchmarking data based on community feedback of SaaS
6. And most importantly, save energy by reducing storage, computational, and processing power at the servers by making clients light and servers thick

8.5.5 Green Operating Systems (OS)

There are many ways to optimize a client environment to save energy and reduce environmental impact. Recent operating systems (OS) have made a clear design cut between the different layers that make up a general purpose operating system (see Figure 8.2).

Each layer has its own services and provides services to the above layer. Each layer has to work in coordination with other layers. This layering model allows dividing

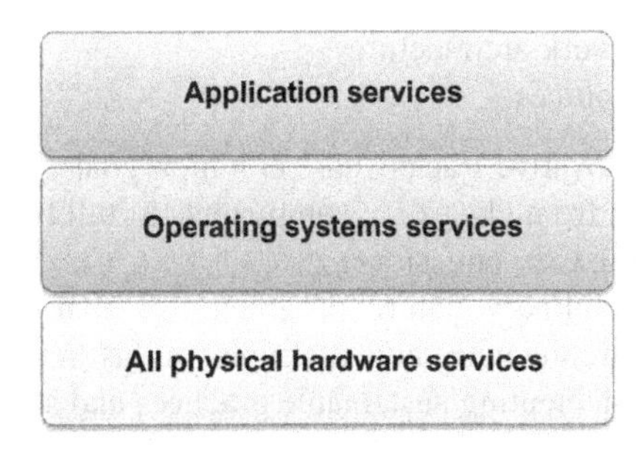

Figure 8.2 Three-tier architecture for green operating systems [40].

the "green" requirement into three scopes; each one is handled by the contributor of the subpart in the final OS [42]. These scopes are

1. *Physical services environment:* Acquiring an Energy Star 4.0 system that recognizes ACPI 3.0 power management capabilities in an OS such as Windows 7 allows the OS to manage power for processors, including dual- and quad-core types, and attached devices, and to utilize advanced hibernation and sleep capabilities.
2. *Operating system services environment:* To leverage advanced energy efficient computing hardware, the operating environment must deliver advanced performances and power management capabilities to the end user as well as advanced users such as system administrators. To reach this goal, it is crucial to standardize configurations, and minimize the number of running system services to reduce energy consumption. Standardized configuration allows for the streamlining of energy consumption policies in corporate IT.
3. *Application services environment:* In order to reduce the amount of resources a client must use to run a fully installed application, architect teams can leverage client application virtualization. This is similar to SaaS with the difference that it is deployed at a "smaller" scale; the client-end PC. Microsoft is working on many projects in the application layer that lower power consumption by optimizing and unifying the presentation layer (colors, themes, etc.). The power-aware Windows Presentation Foundation (WPF) application can use less power by dynamically readapting the application to the power state of the client [43].

8.6 CONCLUSIONS

In this chapter, we have reviewed the major research and technical directions in the field of green communications and networks. These orientations focus on

1. Energy-efficient protocols
2. Energy-efficient signal processing methods
3. Energy-efficient adaptive algorithms
4. Energy dissipation in computation

5. Energy-efficient network architectures
6. Alternative energy sources

All of the above innovative tracks share common goals, which are minimizing the energy consumption from electronic components to full blown systems and as a consequence diminishing CO_2 emissions.

Given the multidisciplinary nature of green ICT that ranges from material engineering to system architecture, breakthroughs in this field are arriving at a fast pace. Nevertheless, implementing sustainable practices and strategies in businesses, products, and services is as important as the work of their engineering counterparts in order to achieve this milestone.

Ignoring power consumption as a design constraint of IT systems will result in supercomputing systems with very high operational cost and low reliability. This is to say that future petaflop machine will require 75–100 MW to power up and later cool down, which can be estimated to cost several million U.S. dollars in a span of a year.

In addition to moving itself in a greener direction and leveraging other environmental initiatives, IT can help create greener awareness among IT professionals, businesses, and the general public by assisting in building communities, engaging groups in participatory decisions, and supporting education and green advocacy campaigns. Besides IT itself being green, it can support and assist other initiatives by offering innovative models, simulations, and decisional tools such as

1. Software tools for analyzing and modeling the environmental impact of certain decisions
2. Platforms for eco-management and emission trading
3. Tools for monitoring and reporting carbon emissions of data centers
4. Engineering and integrating an environmentally green network that may replace the current Internet

Green ICT is a hot topic today and will continue to be an important issue for several years to come. To foster green ICT, we need to understand the key environmental impacts arising from IT as well as the major issues that need to be addressed. ICT infrastructure, products and services, operations, and applications must turn into "green" sustainable forms.

The challenges of green ICT are immense. Nevertheless, recent development indicates that the IT industry is aware and has the "will" to tackle these issues given that the enterprise business orientations are not impacted by such measures, at least not at the same instance in time.

REFERENCES

[1] The World Telecommunication/ICT Indicators Database, International Telecommunications Union, 2007.

[2] R. Singh, Wireless Technology: To Connect the Unconnected, World Bank, Unpublished work.

[3] M. Etoh, T. Ohya, Y. Nakayama, Energy consumption on mobile networks, in: Proceedings of SAINT 2008, vol. 1, 2008, pp. 365–368.

[4] M. Obaidat, S. Dhurandher, D. Gupta, N. Gupta, A. Asthana, DEESR: dynamic energy efficient and secure routing protocol for wireless sensor networks in urban environments, Journal of Information Processing Systems 6 (3) (2010) 269–294.

[5] Ericsson, Green power to bring mobile telephony to billions of people, 2008. <http://www.ericsson.com/ericsson/press/videos/2008/081215-green-power.shtml>.

[6] S. Lee, N. Golmie, Power-efficient interface selection scheme using paging of WWAN for WLAN in heterogeneous wireless networks, in: Proceedings of the IEEE International Radio Communications (PIMRC'06), vol. 4, Helsinki, Finland, 2006, pp. 1742–1747.

[7] "Green" Telecom: More than Just Marketing, in GLG News: Gerson Lehrman Group. <https://www.gplus.com/hardware/insight/-green-telecom-more-than-just-marketing-21960>.

[8] S. Arbanowski, eMobility stimulates "green research," 2009. <http://www.emobility.eu.org/documents/Newsletter/Newsletter102008_web.pdf>.

[9] K. Roth, F. Goldstein, J. Kleinman, Energy consumption by commercial office and telecommunications equipment, in: Energy Information Administration: Promises and Pitfalls, 2002, pp. 2.20–7.21.

[10] The Obama–Biden New Energy for America Plan, The White House, 2009. <http://www.cfr.org/united-states/obama-biden-new-energy-america-plan-january-2009/p18306>.

[11] The Energy Imperative: Report Update, President's Council of Advisors on Science and Technology, November 2008. <http://www.whitehouse.gov/files/documents/ostp/PCAST/PCAST%20Energy%20update-final.pdf>.

[12] SMART 2020: Enabling the low carbon economy in the information age, The Climate Group/the Global eSustainability Initiative (GeSI), 2008. <http://www.smart2020.org/_assets/files/02_Smart2020Report.pdf>.

[13] Report to Congress on Server and Data Center Energy Efficiency: Public Law 109-431, U.S. Environmental Protection Agency, ENERGY STAR Program, 2007.

[14] Johan Vounckx, Nadine Azemard, Philippe Maurine, The power forward initiative: charting the industry's course to achieve enhance power management solutions for advanced process geometries, Integrated Circuit and System Design: Power and Timing Modeling, Optimization and Simulation, vol. 4148, 2006, p. 673.

[15] Advanced Energy Initiative: Research and Development in the President's 2009 Budget, Office of Science and Technology Policy – Executive Office of the President, 2008. <http://www.whitehouse.gov/files/documents/ostp/Budget09/AdvancedEnergyInitiative1pager.pdf>.

[16] AAAS R&D Funding Update on the 2009 Stimulus Appropriations Bill, AAAS, 2009. <http://www.aaas.org/spp/rd/stim09c.htm>.

[17] K. Sriram, Y.-T. Kim, D. Montgomery, Architectural considerations for the mapping distribution protocols, in: Proceedings of the 72nd IETF (in the RRG meeting), Dublin, Ireland, July 27–August 1, 2008.

[18] R. Min, A. Chandrakasan, A framework for energy-scalable communication in high-density wireless networks, in: Proceedings of ISLPED International Symposium on Low Power Electronics and Design, Monterey, CA, 2002, pp. 36–41.

[19] J. Omelchuck, V. Salazar, J. Katz, H. Elwood, W. Rifer, The implementation of EPEAT: electronic product environmental assessment tool – the implementation of an environmental rating system of electronic products for governmental/institutional

procurement, in: Proceedings of the 2006 IEEE International Symposium on Electronics and the Environment, Portland, May 2006, pp. 100–105.
[20] R. Cheda, D. Shookowsky, S. Stefanovich, J. Toscano, Profiling energy usage for efficient consumption, The Architecture Journal: Green Computing Issue 18 (2008) 24–27.
[21] F. Hunt, V. Marbukh, Measuring the utility/path diversity tradeoff in multipath protocols, in: Proceedings of the Fourth International Conference on Performance Evaluation Methodologies and Tools, Pisa, Italy, October 20–22, 2009.
[22] A. Amanna, J. Reed, T. Bose, T. Newman, Metrics and measurement technologies for green communications, National Institute of Standards and Technology, Technology Innovation program, March 2009.
[23] Low Power Coalition: Request for Technology to Support Low Power Design, Silicon Integration Initiative, 2008.
[24] HP, Intel, Microsoft, Phoenix, and Toshiba, The Advanced Configuration and Power Interface (ACPI) 3.0b Specification, ACPI, Ed., 2008.
[25] V. Marbukh, K. Mills, Demand pricing and resource allocation in market-based compute grids: a model and initial results, in: Proceedings of the Seventh IEEE International Conference on Networking, April 2008, pp. 752–757.
[26] Green Base Station – The Benefits of Going Green, Mobile Europe, 2008.
[27] B. Le, T.W. Rondeau, C.W. Bostian, Cognitive radio realities, Wireless Communication and Mobile Computing 7 (9) (2007) 1037–1048.
[28] H. Zhang, Cognitive radio for green communications and green spectrum, in: Proceedings of 2008 CHINACOM, Hangzhou, China, 2008.
[29] M.S. Obaidat, S. Misra, Fundamentals of Wireless Sensor Networks, Cambridge University Press, in press, 2012.
[30] S. Misra, M.P. Kumar, M.S. Obaidat, Connectivity preserving localized coverage algorithm for area monitoring using wireless sensor networks, Computer Communications 34 (12) (2011) 1484–1496.
[31] K. Mills, A brief survey of self-organization in wireless sensor networks, Wireless Communications and Mobile Computing 7 (9) (2007) 823–834.
[32] H. Gharavi, Multichannel mobile ad hoc links for multimedia communications, Proceedings of the IEEE 96 (1) (2008) 77–96.
[33] W. Heinzelman, A. Chandrakasan, H. Balakrishnan, Energy-efficient communication protocols for wireless microsensor networks, in: Proceedings of the 33rd Hawaii International Conference on Systems Science, vol. 2, January 2000, p. 10.
[34] B. Bako, F. Kargl, E. Schoch, M. Weber, Advanced adaptive gossiping using 2-hop neighborhood information, in: IEEE Global Telecommunications Conference, IEEE GLOBECOM, December 2008, pp. 1–6.
[35] S. Sasanus, D. Tipper, Y. Qian, Impact of signaling load on the UMTS call blocking/dropping, in: Proceedings of IEEE VTC'2008-Spring, Singapore, May 11–14, 2008.
[36] Program Solicitation: NSF 08-611-Cyber-Physical Systems (CPS), National Science Foundation: Directorate for Computer and Information Science and Engineering, Directorate for Engineering, 2008.
[37] J. Rowley, D. Haig-Thomas, Unwiring the planet – wireless communications and climate change, in: ITU International Symposiums ICTs and Climate Change, London, UK, 2008.
[38] C.-K. Liang, K.-C. Chen, A green software-defined communication processor for dynamic spectrum access, in: Proceedings of the 21st Annual IEEE International

Symposium on Personal, Indoor and Mobile Communications, September 2010, pp. 774–779.
[39] K.C. Chen, R. Prasad, Cognitive Radio Networks, Wiley.
[40] M.S. Obaidat, Trends and challenges in green informational and communication systems, in: IEEE ICCIT 2011, Aqaba, Jordan, March 2011. <http://www.iccit-conf.org/archive/2011/?page_id=124>.
[41] Mydhili K. Nair, V. Gopalakrishna, Generic web services: a step towards green computing, International Science and Engineering Journal 1 (3) (2009) 248–253.
[42] G. Appasami, K. Suresh Joseph, Optimization of operating systems towards green computing, International Journal of Combinatorial Optimization Problems and Informatics 2 (3) (2011) 39–51.
[43] R.A. Sheikh, U.A. Lanjewar, Green computing – embrace a secure future, International Journal of Computer Applications 10 (4) (2010) 22–27.

CHAPTER

Green Computing Platforms for Biomedical Systems 9

Vinay Vijendra Kumar Lakshmi, Ashish Panday, Arindam Mukherjee, Bharat S. Joshi

Department of Electrical and Computer Engineering, University of North Carolina at Charlotte, USA

9.1 INTRODUCTION TO GREEN COMPUTING IN THE BIOMEDICAL FIELD

Computing in biomedical systems can be classified into three categories:

1. Implantable device
2. Portable/embedded devices
3. Server

Application of green computing in any of the categories requires knowledge of the attributes of each of them. For instance, an implantable device requires a continual source of energy, efficient thermal management, and high reliable. The battery life for portable devices is critical. However, their performance cannot be compromised. In the case of biomedical servers, application of renewable energy sources and performance hold higher significance. Overall, we can ascribe one or more of the following attributes when analyzing the application of green computing in any of the categories of biomedical systems:

1. Power consumption
2. Renewable energy resource—energy harvesting
3. Heat dissipation
4. Minimizing area
5. Cost
6. Performance
7. Reliability

9.1.1 Implantable Devices

A wide variety of electronic devices are being used to monitor the physiological parameters of the human body. In some cases, they are completely replacing complex biological functions. Pacemakers, cardioverter-defibrillators, and cochlear implants are a few such medical devices. Because of extreme financial and ethical liabilities, careful considerations have to be made about testing implantable devices and minimizing the associated risk. The advancements in implantable devices and the concerns to be addressed while designing them are summarized in [60].

Most of the implantable devices are inactive most of the time and activate based on a stimulus from the body. However, there are many devices that stimulate the human body and potentially process a lot of data. Such devices have to be highly reliable (refer to [51, 59, 61]) as replacing defective devices frequently is laborious, costly, and in the worst case life threatening. Abouei et al. propose a protocol in [50] that uses frequency-shift keying to overcome the reliability and power cost issues in an implantable wireless body area network (WBAN).

As mentioned before, implantable devices also need a continual source of power and efficient thermal management techniques. Probably the most important attribute that needs monitoring in biomedical implantable devices is heat dissipation and the generated electromagnetic fields. Lazzi documents the causes and effects of heat and electromagnetic waves generated in implantable devices in [32]. Lazzi uses the example of a dual-unit retina prosthetic used to restore partial vision to the blind. Tang et al. present a communication scheduling scheme in [33] to reduce the thermal effect of an implantable biosensor network. An alternative approach to thermal management would be to design the algorithm and the microchip of the implantable device efficiently in order to reduce the amount of power consumed by the device.

Suhail and Oweiss propose a novel design of a wavelet-based module for real-time processing in implantable neural interface devices in [34]. They aim at reducing the area and power consumed by the implanted hardware by performing computations using fixed-point representations instead of traditional floating-point computations. They claim that the integrity of the outputs is still maintained. They implement their circuit design in 18 μm CMOS, which occupies only 0.22 mm^2 and consumes 76 μW [35]. Zhang et al. implemented a low power micro control unit (MCU) specifically designed for implantable medical devices used for monitoring and stimulus in [49]. They propose that with more compact code, efficient operations from instruction improvement, and DMA channels, the power consumed was reduced significantly. Gong et al. included energy harvesting for wirelessly powered medical electronic devices by including a self-powered integrated rectifier [54].

9.1.2 Embedded Platforms

The most important criteria for an embedded platform are battery life, area, and cost. There is always a compromise between these factors and performance. With the advancement in technology, biomedical devices have evolved into smaller and modular compositions while maintaining their functionality. However due to limitations of the human body, many of these devices are not approved for medical practice. In order to make these systems more accessible, one can increase the performance area power-consumed ratio of the systems.

From physiological monitoring systems to recognition systems, the majority of biomedical systems are being implemented on embedded platforms. Many research groups have presented their approaches of moving the majority of the computations involved with each of these systems onto embedded platforms rather than servers. Obviously, there has to be communication of data with the data centers to possibly run complex algorithms along with maintenance of existing databases to store(retrieve) data.

Yang et al. present a unique fingerprint matching technique in [36]. The method involves dividing the algorithm into secure and nonsecure parts. The secure part of the algorithm is relatively small and comprises sensitive biometric template information. The rest of the algorithm runs on LEON, an embedded platform. Some other research on real-time personal identification systems on embedded platforms include those by Lee et al. [37] and Alt and Jaekel [63]. Maltoni et al. describes many techniques for fingerprint identification in [38].

Penhaker et al. describe a smart embedded biotelemetry system in [39]. Their model, called HomeCare, monitors the users' movements and location along with physiological parameters like blood pressure, weight, EKG, etc. The data is transmitted from the person to the embedded devices via Bluetooth or ZigBee technology. The embedded processor does all basic preprocessing and evaluation of life threatening states. This data is then transmitted to the centralized processing unit. Some other physiological monitoring systems include those by Kunze et al. [40], where they use ubiquitous computing to run their algorithms, and Lin et al.'s wireless PDA-based physiological monitoring system for patient transport [41].

Recent advances in wireless technologies have opened up a new generation of WBANs for providing health care services [52]. Wearable WBANs provide RF communication between on-body sensors and an external control unit. Given their limited power sources, sensors in body area networks (BANs) have to be energy-efficient to ensure longevity and safety of the network. Several research groups have proposed ways to increase energy efficiency to keep up with the power demands [56–58]. However, increasing the reliability of BANs increases the amount of power consumed. The relationship of power management and reliability of medical devices is discussed in [53] and [55].

9.1.3 Servers

Typically servers are responsible for maintaining and communicating large amounts of data with remote devices like implanted/wearable sensor networks, or embedded devices. They are also responsible for potentially running complex kernels that cannot run on these remote devices due to limitations on the resources available at the device end. Since multiple complex kernels run to maintain, communicate, or compute in tandem, they have to be implemented efficiently to optimize the amount of time, power, and resources utilized. These kernels can, potentially, tend to be bottlenecks for real-time analysis of data and are mostly be analyzed offline. Consequently, unique challenges and opportunities are emerging for the involvement of high performance and green computing in this field. Furthermore, according to [42], healthcare facilities consume 4% of the total energy consumed in the United States. So there is a great opportunity to involve renewable resources and energy harvesting in this category of biomedical systems.

A very prominent field in biomedical studies involves analyzing and classifying DNA structures and identification systems. These systems tend to take a long time to classify a DNA strand and compare with the data stored previously in databases. A large number of organizations also rely on personal identification systems like fingerprints, iris recognition, and voice recognition.

Some research groups are also interested in understanding the basic building blocks of the several kinds of cells in a living being. Sanbonmatsu and Tung present their multimillion atom simulation results of nanoscale systems in [43]. The simulations were run on the Pittsburgh supercomputing center Lemieux machine over 1024 CPUs. Such systems consume high power even though 85% parallel efficiency is observed. Therefore power management is another critical component of green computing in biomedical servers.

By efficiently organizing the power management in computers, and their components like hard disks, RAM, etc., one can preset the time after which the computer is turned off if not in use. Ultra-low power electronic design plays a major role in this movement [2]. Proper coding and using optimized algorithms that are written proficiently with fewer lines of code can reduce the load on the servers, thus reducing energy consumption. Another approach can involve usage of thin terminal devices, which are used for connecting to servers. These devices are very thin and use one-eighth of the total power used in normal desktops.

Arguably, in servers with high power requirements, there is a great opportunity to involve renewable resources and energy harvesting as mentioned above. With such high power demands, engaging alternate green and sustainable energy resources in the biomedical field can result in substantial reduction in the demand from the network.

9.2 SURVEY OF GREEN COMPUTING PLATFORMS

In this section we survey the use of existing platforms to achieve efficiency in terms of area, power consumption, thermal management, and other attributes of implantable devices, embedded platforms, and servers.

9.2.1 Implantable Devices

The examples of [17] (Figures 9.1 and 9.2) represent two biomedical application contexts where several ultra-low power RF, sensor, analog processing, electrode recording, and electrode-stimulation building-block circuits are useful. Figure 9.1 shows the mechanical topology of a cochlear implant or bionic ear whose configuration bears similarity to that in many biomedical implants today. Such cochlear

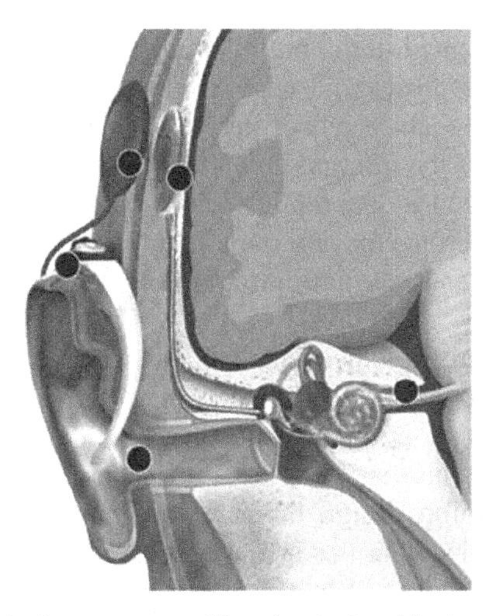

Figure 9.1 An example of a low-power cochlear implant or bionic ear.

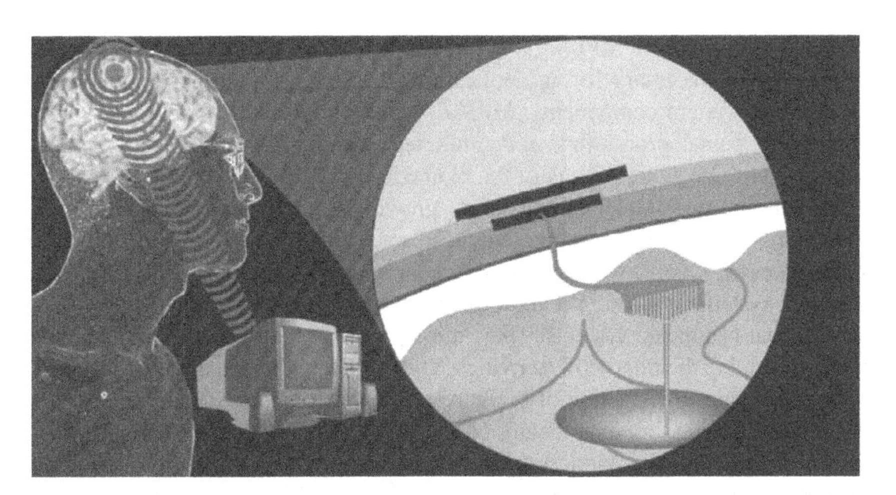

Figure 9.2 Configuration of a brain implant or brain machine interface (BMI).

implants enable profoundly deaf people, for whom a hearing aid is not beneficial, to hear. In [17], Sarpeshkar explains that a cochlear implant mimics the frequency-to-place stimulation of the healthy auditory nerve and the compression or gain-control properties of the healthy ear. Cochlear implants today have the speech processor in a behind-the-ear (BTE) unit as shown in Figure 9.1. They are evolving towards fully implanted systems, where all components will be implanted inside the body.

Figure 9.2 shows the mechanical topology in a brain implant or in a brain machine interface (BMI) that could be used to treat blindness in the future.

Electrodes that are implanted in the brain stimulate one of its visual regions. The internal unit can receive power and data wirelessly from an external unit as in cochlear implants. The received power periodically recharges an implanted battery or constantly powers the internal unit wirelessly. The external unit is capable of wireless communication to a computer via a standard ultra-wideband (UWB), ZigBee, or Bluetooth interface for programming, monitoring, and debugging functions. The external unit and internal unit are mechanically aligned via electronic magnets.

Implants for several other biomedical applications, e.g., cardiac pacemakers, spinal cord stimulators, deep-brain stimulators for the treatment of Parkinson's disease, vagal-nerve stimulators, etc., utilize several similar building-block circuits, to architect slightly different systems that all operate by and large with the same technology base. In fact, biocompatibility design, hermetic design, mechanical design, and electrode design share several similarities in all of these applications.

9.2.2 Embedded Platforms

Multiprocessor system-on-chips (MPSoCs) play a major role in such biomedical embedded platforms. An MPSoC is a VLSI system that incorporates most or all of the components necessary for an application. It also uses multiple programmable processors as system components. MPSoCs are widely used in networking, communications, signal processing, and multimedia among other applications. Thus MPSoCs constitute a huge chunk of the biomedical systems.

BAN design and implementations are growing for wireless health monitoring applications. NXP Semiconductor announced a versatile ultra-low power biomedical signal processor, CoolBio™, meeting the requirements of future wearable biomedical sensor systems promising a solution for more comfortable, cost-effective, and time-efficient healthcare systems. They allow people to be monitored and followed up at home, doing their daily life activities. The CoolBio biomedical signal processor allows drastic power reduction of the wireless BAN biomedical sensor nodes. Processing and compressing data locally on the BAN node limits power hungry transmission of data over the wireless link, while adding motion artifact reduction and smart diagnosis at the same time.

The NXP CoolBio biomedical signal processor (Figure 9.3) [62] consumes only 13 pJ/cycle when running a complex ECG (electrocardiogram) algorithm at 1 MHz and 0.4 V operating voltage. This biomedical signal processor is programmable using the C language. The CoolBio biomedical signal processor is voltage and performance scalable supporting a frequency range of 1 MHz up to 100 MHz with an operating voltage from 0.4 to 1.2 V.

9.2.2.1 Intel Atom

The Intel Atom is an ultra-low voltage x86 processor designed with 45 nm CMOS technology. It is mainly used in mobile platforms. The Atom has an integrated power management chip and clock companion IC as stated in [18]. ICs like DA6011 and

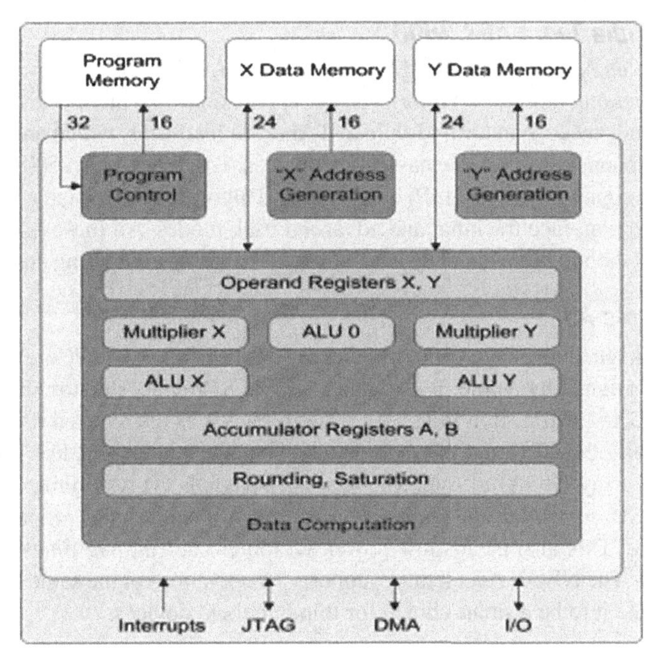

Figure 9.3 Architecture of CoolBio biomedical signal processor.

ROHM usually consist of a power management control block, a power management block, a clock synthesizer, and a few programmable registers which work on reducing the noise, achieving low quiescent current, real-time dynamic switching of voltage and frequency between multiple performance modes, varying core operation voltage and processor speeds to save on Atom's power, and improving its performance. Figure 9.4, taken from [18], shows the block diagram of the power management of Intel ATOM.

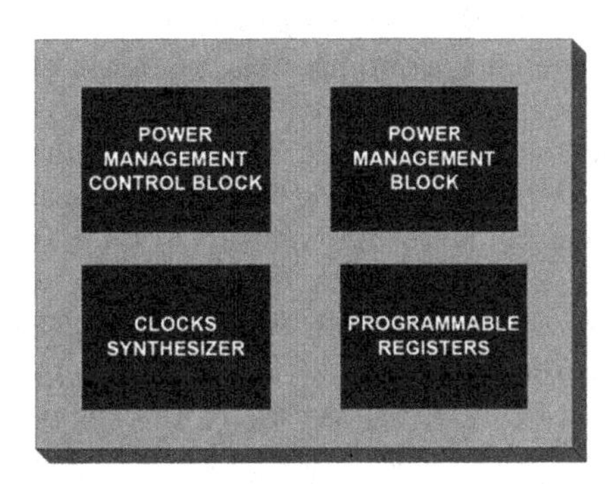

Figure 9.4 Power management in the Intel ATOM.

9.2.2.2 NVidia Tegra APX 2500

This chip has an ARM11 600MHz MPCore processor, and ultra-low power NVIDIA graphics processing unit (GPU) for low power applications like mobile devices [19]. It supports the advanced Windows Mobile and Android platforms, which are now being used in smartphones and portable navigation devices. The Tegra APX 2500 has an integrated image signal processor (ISP) with high-end algorithms that enables image and video stabilization, face tracking, and advanced trick modes. All these features make it suitable for mobile biomedical devices which need visual computing and imaging.

9.2.2.3 Apple A5

This is a package on package (PoP), dual core ARM Cortex-A9 MPCore system-on-chip (SoC) designed by Apple. It also has a NEON SIMD accelerator and dual core PowerVR SGX543MP2. In a PoP, two or more packages are stacked on top of one another, usually the CPU and the memory with standard interfaces like low density I/O ball grid array (BGA) to route the necessary signals. This minimizes the track length between controller and memory, resulting in faster signal propagation and reduced noise. This also means low power. Graphics can run five times faster than the Apple A4. The NEON Accelerator supports 16 operations at the same time. These features enable it to be a smart choice for thin handheld devices.

9.2.2.4 Samsung Exynos 4210

This 32-bit RISC SoC is designed by Samsung for handheld devices. It has a dual core ARM-Cortex-A9. With eight channels of I2C, GPS baseband, USB, and HSIC interfaces it provides good support for wired and wireless communications. It is also available in PoP. It promises to give the best mobile 3D performance and native triple display capability, while achieving the lowest power consumption in its class. The company boasts of ultimate performance with green technology in [8].

9.2.2.5 AMD Fusion

The key aspect to note is that all the major system elements–x86 cores, vector (SIMD) engines, and a unified video decoder (UVD) for HD decoding tasks—attach directly to the same high speed bus, and thus to the main system memory. According to N. Brookwood of AMD in [20], this design concept eliminates one of the fundamental constraints that limit the performance of traditional integrated graphics controllers/integrated graphics processors (IGPs). Fusion's ability to fit CPU cores, GPU cores, and NorthBridge (for communication with the rest of the chipset) onto a single piece of silicon proves to be a holistic approach towards management. Depending on the workloads, Fusion can power various parts of the chip up and down reducing the power consumption by a few milliwatts every time, which in the aggregate amounts to significant power savings. AMD Fusion has *CoolCore Technology,* which can reduce processor power consumption by dynamically turning off sections of the processor when inactive. Figure 9.5 shows the block diagram of the AMD Fusion processing platform [20], which integrates a CPU–GPU system via an accelerated processing unit (APU).

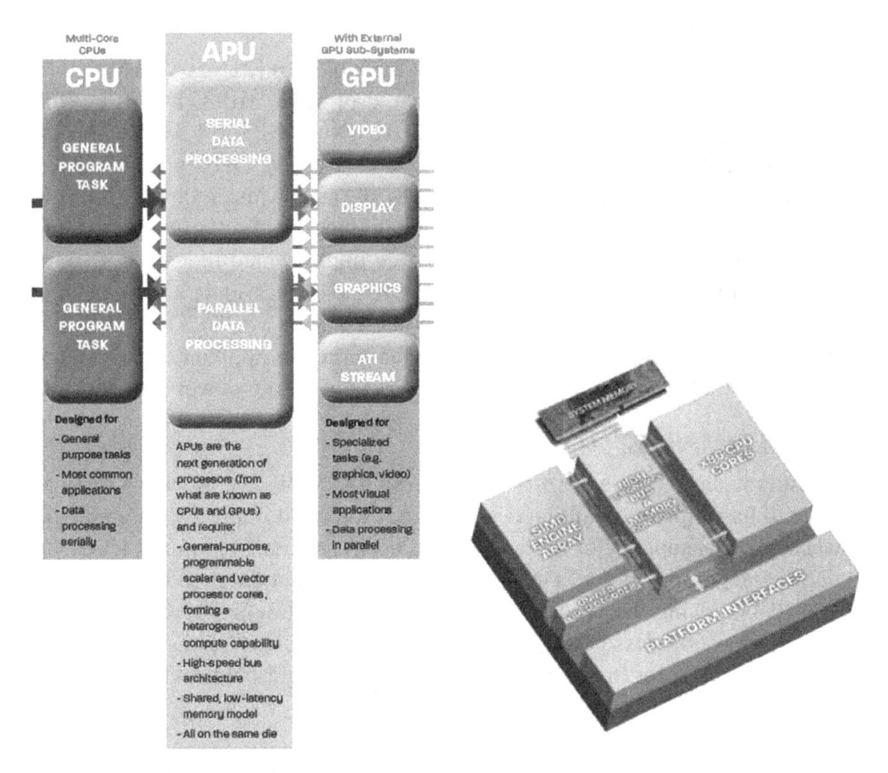

Figure 9.5 APU in AMD Fusion.

9.2.2.6 IBM PowerPC 476FP

This is an embedded processor core that is a 4-issue, 5-pipeline superscalar RISC. It has a tightly integrated floating-point unit that exceeds three gigaflops at 1.6 GHz. The core supports both symmetric multiprocessing (SMP) and asymmetric multiprocessing (AMP). The performance, evaluated with the Dhrystone 2.1 benchmark, shows 2.71 DMIPS per MHZ and a dynamic power dissipation of 1.2 W at 1.0 V. The graph (Figure 9.6) given in [9] summarizes the performance. It is very attractive for the biomedical processor market.

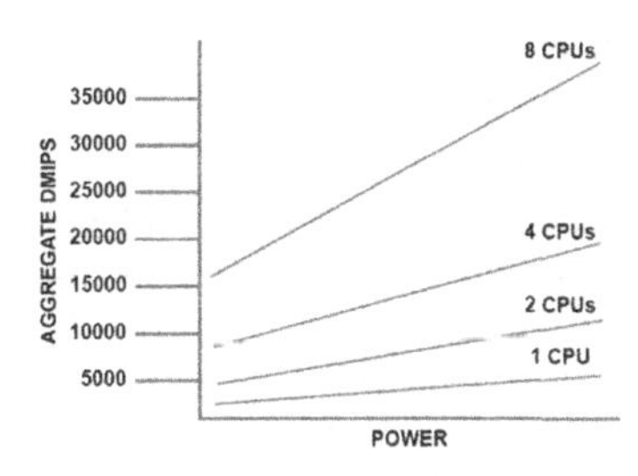

Figure 9.6 Performance of IBM PowerPC.

9.2.2.7 MIPS32 1074K

This is a coherent multiprocessor from MIPS Technologies [21], i.e., it has two different kinds of processor cores—superscalar, out-of-order MIPS32 74 K in 40 nm technology. It is able to achieve a maximum of 12,000 DMIPS at 1.5 GHz. The lowest power dissipation is less than 1 W dynamic power at 1.25 GHz for a greater than 6300 total Coremark, with 5000 total DMIPS. It has a Cluster Power Controller (CPC) which consists of multicore power gating, clock gating, and reset management, which work in conjunction with each core implemented in a separate power domain. It is currently used in digital televisions, set-top boxes, and a variety of home-networking applications. One of the green applications of MIPS32 1074K is home automation, controlling and switching power for particular appliances and modules within the home/office building and thereby helping in saving power automatically.

9.2.2.8 TI OMAP5432

Texas Instruments has made use of an ARM Cortex-A15 MPCore processor in its new SoC [11]. It has improved multiprocessing bandwidth, and improves streaming performance, with its support for more out-of-order instructions, optimized L1 caches, and tighter integration with NEON/VFP. TI has proved that the mobile devices with TI OMAP5432, using the processing abilities of 28 nm low-power Cortex-A15, at 2 GHz consumes 60% less power than the devices which use a Cortex-A9 processor. Hence it can be used for future biomedical mobile and hand-held devices.

9.2.3 Servers

MPSoCs have been the basis of modern day servers. These processors are specifically designed to impart high performance in low power modes. Let us take a look at a few of the most efficient servers in the market, which are aimed at saving power without compromising much on performance metrics like CPI.

9.2.3.1 Sun Niagara 3

Sun's 1 billion-transistor, 16-core Niagara 3 processor is a great example of modern multiprocessor-turned-SoC (system on a chip). Everything about this design is focused on pushing large numbers of parallel instruction streams and data streams through the processor socket at once. The shared cache is small, the shared pipes are wide, and the end result is a chip that is all about maintaining a high rate of flow, and not one that is aimed at collecting a large pile of data and chipping away at it with heavy equipment. Figure 9.7, taken from [22], shows the layout of Sun's Niagara.

The above features have made it possible for Niagara 3, often called the SPARC T3, to achieve high throughput and power efficiency compared to its predecessor SPARC T2 when it comes to servers which offer green computing and occupying the least area as specified in [22].

The Oracle WebLogic Server 11*g* software was used to demonstrate the performance of the Avitek Medical Records sample application. A configuration using SPARC T3-1B and SPARC Enterprise M5000 servers from Oracle was used and

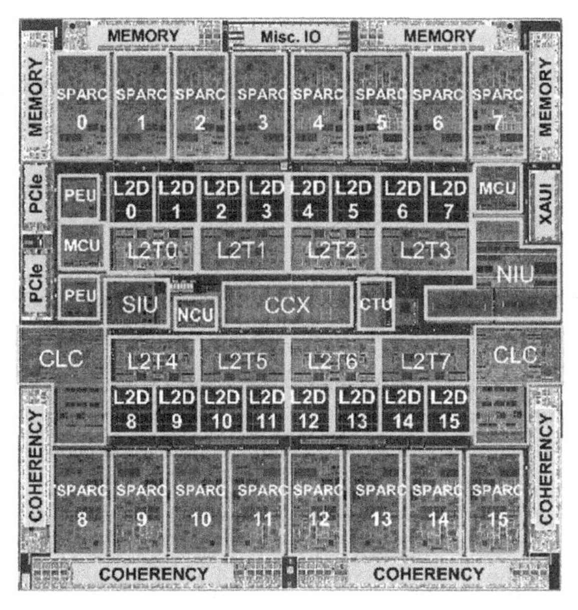

Figure 9.7 Sun's Niagara 3: die micrograph.

showed excellent scaling of different configurations as well as doubling of previous generation SPARC blade performance. Avitek Medical Records (or MedRec) is an Oracle WebLogic Server 11*g* sample application suite that demonstrates all aspects of the J2EE platform [23]. MedRec showcases the use of each J2EE component, and illustrates best practice design patterns for component interaction and client development. Oracle WebLogic Server 11*g* is a key component of Oracle Fusion Middleware 11*g*.

The MedRec application provides a framework for patients, doctors, and administrators to manage patient data using a variety of different clients. Patient data includes

- *Patient profile information:* a patient's name, address, social security number, and log-in information
- *Patient medical records:* details about a patient's visit with a physician, such as the patient's vital signs and symptoms as well as the physician's diagnosis and prescriptions

MedRec comprises two main Java EE applications supporting different user scenarios:

MedRecEar—Patients log in to the Web application (PatientWebApp) to register their profile or edit. Patients can also view medical records or their prior visits. Administrators use the Web application (AdminWebApp) to approve or deny new patient profile requests. MedRecEar also provides all of the controller and business logic used by the MedRec application suite, as well as the Web Service used by different clients.

Table 9.1 Performance Landscape

Server	Processor	Memory	Maximum TPS
SPARC T3-1B	1 × SPARC T3, 1.65 GHz, 16 cores	128 GB	28,156
SPARC T3-1B	1 × SPARC T3, 1.65 GHz, 8 cores	128 GB	14,030
Sun Blade T6320	1 × UltraSPARC T2, 1.4 GHz, 8 cores	64 GB	13,386

PhysicianEar—Physicians and nurses login to the Web application (PhysicianWebApp) to search and access patient profiles, create and review medical records, and prescribe medicine to patients. The physician application is designed to communicate using the Web Service provided in MedRecEar.

The MedRecEar and PhysicianEar applications are deployed on an Oracle WebLogic Server 11g instance called MedRecServer. The PhysicianEar application communicates with the controller components of MedRecEar using Web Services. Performance for the application tier is presented in Table 9.1 [23]. Results are the maximum transactions per second (TPS).

9.2.3.2 IBM POWER 7

At 1.2 billion transistors, IBM's new 45 nm POWER7 processor [5] (Figure 9.8) is only a little bigger than Niagara 3, but it couldn't be more different. Where Niagara 3 keeps a large number of relatively weak cores busy by moving data onto and off of the chip using ample I/O resources, POWER7's approach to feeding a smaller number of much more robust cores is to cache large amounts of data on-chip so that

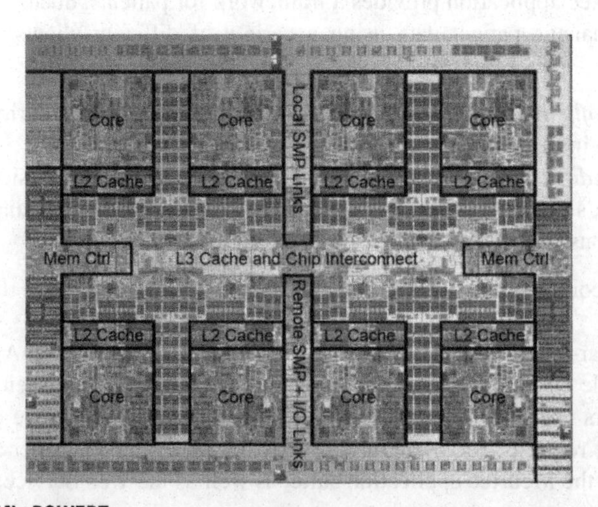

Figure 9.8 IBM's POWER7.

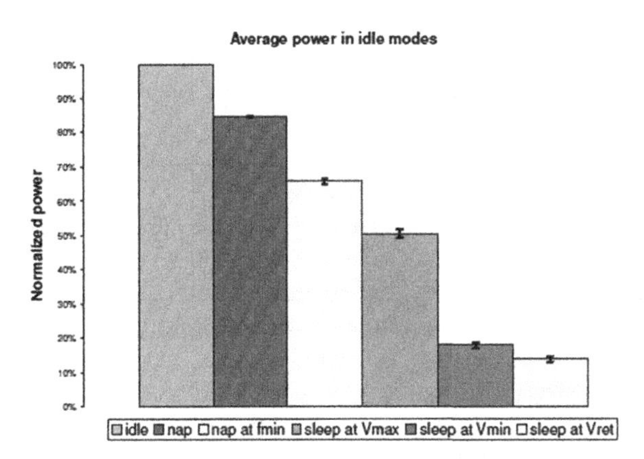

Figure 9.9 Comparison of processor idle modes.

the cores can grind through it in batches. This being the case, POWER7 has the most remarkable on-chip cache hardware of any processor on the market.

Sun's Niagara is aimed at networked server operations where lots of simultaneous, lightweight requests have to be serviced—databases, Web servers, and the like. In contrast, POWER7 has the capability to process a smaller number of more compute-intensive tasks at a high rate of speed. Power7's ability to manage millions of transactions in real time is necessary for applications such as smart electrical grids and biomedical applications like wireless health monitoring.

The new architectural feature of POWER 7 that makes it green is the Nap and Sleep modes. Both Nap and Sleep are hypervisor-privileged modes that maintain a few of the architected processor resources; exit from a power-saving instruction is similar to a thread-level reset. Power-saving instructions that trigger the entry to these modes also cause dynamic SMT mode switching [5]. The idle modes that lead to power saving are compared as shown in Figure 9.9.

9.2.3.3 *The Cell Processor*

This is the first implementation of the Cell Broadband Engine Architecture (CBEA), which is a fully compatible extension of the 64-bit PowerPC Architecture. Its initial target is the PlayStation 3 game console, but its capabilities also make it well suited for other applications such as visualization, image and signal processing, and various scientific and technical workloads [3].

Figure 9.10 shows a high-level view of the first implementation of Cell BE. It includes a general-purpose 64-bit Power Processing Element (PPE). In addition, the Cell BE incorporates eight Synergistic Processing Elements (SPEs) interconnected by a high-speed, memory-coherent Element Interconnect Bus (EIB). This initial implementation of Cell BE is targeted to run at 3.2 GHz. The Cell BE is a best-of-breed design that delivers significant computational power, high bandwidth, excellent thermal management and performance, and leading area efficiency within the constraints of a process technology.

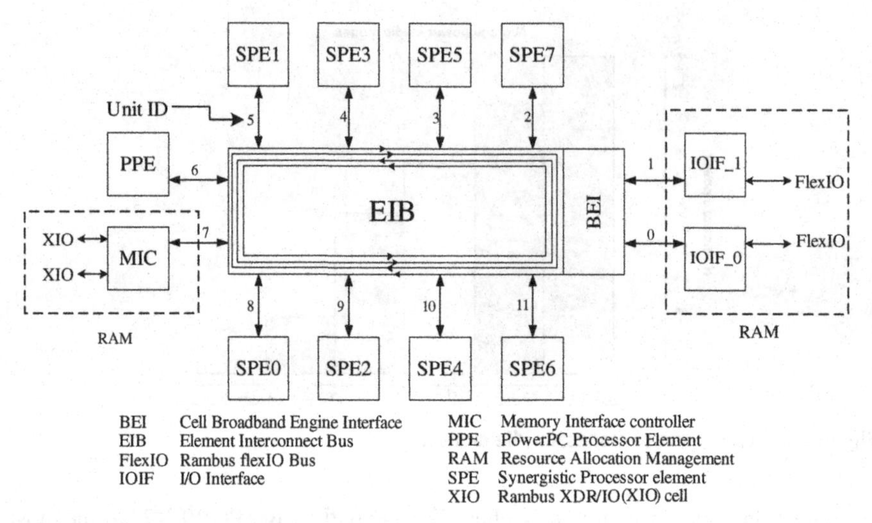

Figure 9.10 Cell processor block diagram.

Even though all the processors have unique features to save power, they also have unique designs which sometimes favor some applications and sometimes can dampen the performance of the application. Therefore a careful selection of algorithm and a suitable architecture is important to extract maximum benefits, both from performance and power points of view, from these processors. In the next section, we will see an example of a popular biomedical application, pairwise correlation(PWC), and realize an algorithm and architecture pair to the best of our ability so that maximum performance (and lowest power consumption) can be extracted from one of the architectures mentioned above.

9.3 ANALYSIS OF BIOMEDICAL APPLICATIONS

In this section we propose a methodology for optimizing the implementation of biomedical kernels with the example of pairwise correlation (PWC), a very common compute-intensive kernel that is used in the biomedical field. We will show that with an appropriate algorithm and architecture selection, PWC can be implemented efficiently, with high performance and maximum resource utilization. Figure 9.11 shows a standard flowchart that can be followed for any biomedical algorithm. We will realize each section for PWC and try to optimize the implementation of PWC.

A promising line of work in EEG, and more specifically intracranial EEG (icEEG), is the study of task-related neural activity or an expression of aberrant connectivity or seizure spread [25]. Other encouraging domains of research are the recent development with fMRI studies and the development of the method of connectivity analysis [44–47]. As with the connectivity analysis performed with fMRI

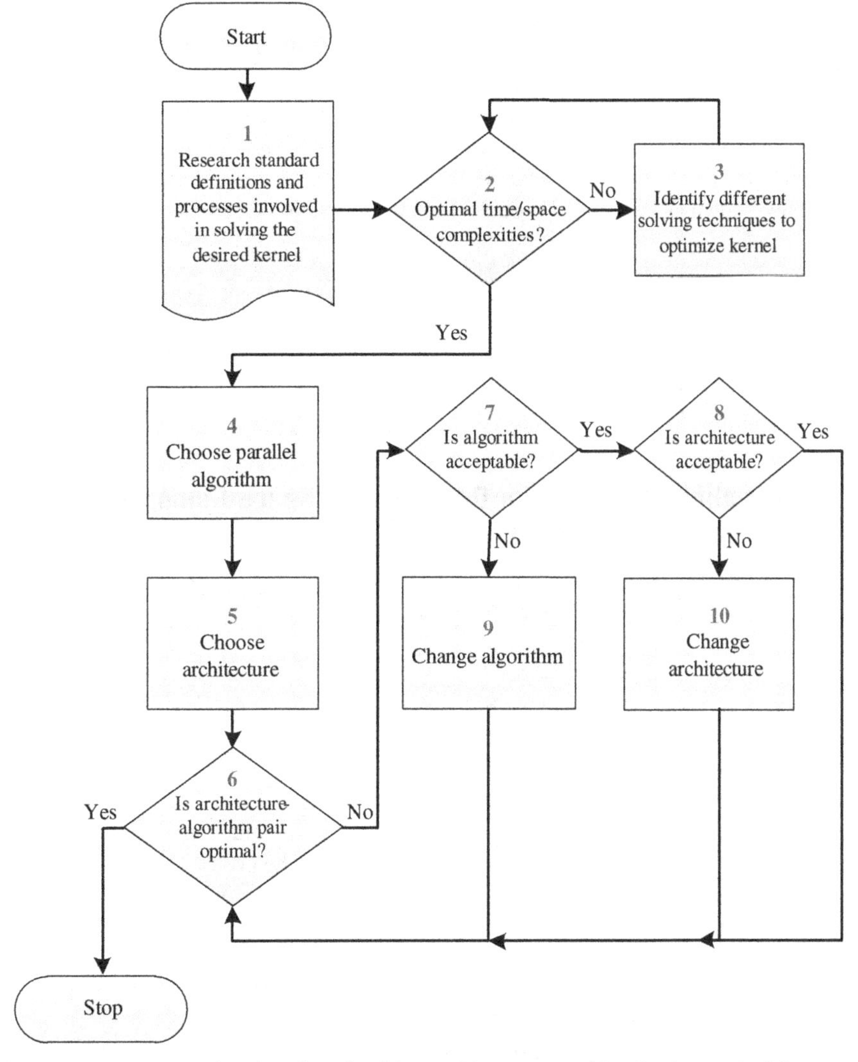

Figure 9.11 Flowchart for choosing algorithm-architecture combination best suited for an application.

data, the correlation of time series of each channel with all other channels (PWC) is a common time-series analysis measure in epilepsy and neuroscience EEG studies. These studies are valuable because approximately 0.4–1% of the world's population suffers from epilepsy [48] and nearly 36% of epilepsy patients have seizures that cannot be controlled by any available antiepileptic drug [1]. The alternative for patients with uncontrolled seizures is epilepsy surgery, which is appropriate for a small fraction of patients, is expensive, and requires specialized teams. In recent years brain implantable devices have emerged as an alternative to pharmacotherapy and surgery for the control of seizures.

Correlation describes the degree of relationship between two variables. The coefficient of correlation, r, of two sets of data X and Y, with n elements, is calculated by dividing the covariance of X and Y by the product of their standard deviations given by Eq. (9.1). Equation (9.1) can also be related to block 1 in Figure 9.11. The measure, r, is more commonly known as Pearson's Product-Moment Correlation Coefficient (PPMCC)

$$r_{XY} = \frac{\mathrm{cov}(X,Y)}{\sigma_x \sigma_y} = \frac{E[(X-\mu_X)(Y-\mu_Y)]}{\sigma_X \sigma_Y}, \tag{9.1}$$

where X: $\{x_1, x_2, x_3, \ldots, x_n\}$; Y: $\{y_1, y_2, y_3, \ldots, y_n\}$; r: coefficient of correlation $-1 \leqslant r \leqslant 1$; cov($X$, Y): covariance of X and Y; σ_X: standard deviation of X; σ_Y: standard deviation of Y; μ_X: expectation of X; and μ_Y: expectation of Y.

9.3.1 Simplifying PPMCC for Optimal Time/Space Complexities (Blocks 2 and 3 of Figure 9.11)

Since X and Y are sets with cardinality n, PPMCC can be estimated by using a sample correlation coefficient (SCC) as shown in (9.2):

$$r_{XY} = \frac{\mathrm{cov}(X,Y)}{\sigma_x \sigma_y} = \frac{\sum_{i=1}^{n}(x_i-\mu_x)(y_i-\mu_y)}{\sqrt{\sum_{i=1}^{n}(x_i-\mu_x)^2}\sqrt{\sum_{i=1}^{n}(y_i-\mu_y)^2}}. \tag{9.2}$$

In Eq. (9.2) cov(X, Y) is the covariance of X and Y, σ_x and σ_y are the standard deviations of X and Y and μ_x and μ_y are the expectations of X and Y. To solve PPMCC using SCC, μ_x and μ_y have to be computed first. These values are then subtracted from each sample in the respective set, multiplied, and the products added to calculate the covariance and standard deviations. Table 9.2 consolidates the operations required to compute PPMCC in this fashion.

Table 9.2 Number of Operations of Eq. (9.2)

	Addition	Multiplication	Subtraction	Division	Square Root
Expectation $X(\mu_x)$	n	0	0	1	0
Expectation $Y(\mu_y)$	n	0	0	1	0
cov(X, Y)	n	N	$2n$	0	0
Standard deviation $X(\sigma_x)$	n	N	n	0	1
Standard deviation $Y(\sigma_y)$	n	N	n	0	1
r_{XY}	0	1	0	1	0
Total	$5n$	$3n+1$	$4n$	3	2

Another way to interpret PPMCC by simplifying μ_x and μ_y is given by (9.3):

$$r = \frac{n\sum_{i=1}^{n} x_i y_i - \sum_{i=1}^{n} x_i \sum_{i=1}^{n} y_i}{\sqrt{n\sum_{i=1}^{n} x_i^2 - (\sum_{i=1}^{n} x_i)^2} - \sqrt{n\sum_{i=1}^{n} y_i^2 - (\sum_{i=1}^{n} y_i)^2}}. \quad (9.3)$$

Table 9.3 aggregates the operations involved in solving Eq. (9.3).

Obviously Eq. (9.3) has a lower number of computations than Eq. (9.2). Therefore, to optimize the performance of the PWC kernel, in order to save energy without compromising the resolution of the results, Eq. (9.3) should be adopted. The following observations can be made by recognizing PPMCC in this manner. First, the computations of $\sum x_i$, $\sum y_i$, $\sum x_i^2$, $\sum y_i^2$, and $\sum x_i y_i$ are linear directed acyclic graphs (DAG) and computationally intensive. Second, $\sum x_i$, $\sum y_i$, $\sum x_i^2$, $\sum y_i^2$, and $\sum x_i y_i$ are iterative in nature and do not require costly functions like divide and square root. Third, $\sum x_i$, $\sum y_i$, $\sum x_i^2$, $\sum y_i^2$, and $\sum x_i y_i$ can be computed separately and the results used to compute PPMCC. Fourth, the number of instances of the costly operations in Eq. (9.3), divide and square root, are 1 and 2, respectively. Finally, the iterative nature of Eq. (9.3) can be manipulated to exploit temporal and spatial localities.

Extending correlation computation to multiple channels, it can be defined as the set of correlation values for all possible pairs that can be formed from m channels. This set is called PWC. A more literal way of representing the elements of PWC for multiple channels is given by Eq. (9.4):

$$r_{(i,j)} = \frac{n\sum_{k=1}^{n} x_{(i,k)} x_{(j,k)} - \sum_{k=1}^{n} x_{(i,k)} \sum_{k=1}^{n} x_{(j,k)}}{\sqrt{n\sum_{k=1}^{n} x_{(i,k)}^2 - (\sum_{k=1}^{n} x_{(i,k)})^2} - \sqrt{n\sum_{k=1}^{n} x_{(j,k)}^2 - (\sum_{k=1}^{n} x_{(j,k)})^2}}, \quad (9.4)$$

where i, j: ith, jth channel where $1 \leqslant i, j \leqslant m$; $x_{(i,k)}$, $x_{(j,k)}$: kth sample from ith, jth channel where $1 \leqslant i, j \leqslant m, i \neq j$, and $1 \leqslant k \leqslant n$; $r(i, j)$: correlation coefficient between ith, jth channel where $1 \leqslant ij \leqslant m$. The ith and the jth channel will be represented as X_i and X_j in the rest of the chapter.

Table 9.3 Number of Operations of Eq. (9.3)

	Addition	Multiplication	Subtraction	Division	Square Root
$\sum_{i=1}^{n} x_i$	n	0	0	0	0
$\sum_{i=1}^{n} y_i$	n	0	0	0	0
$\sum_{i=1}^{n} x_i y_i$	n	n	0	0	0
$\sum_{i=1}^{n} x_i^2$	n	n	0	0	0
$\sum_{i=1}^{n} y_i^2$	n	n	0	0	0
Numerator	0	2	1	0	0
Denominator	0	4	3	0	2
r_{XY}	0	0	0	1	0
Total	$5n$	$3n+6$	4	1	2

9.3.2 Choosing the Initial Algorithm and Architecture (Blocks 4 and 5 of Figure 9.11)

To test the working of the kernel proposed by Eq. (9.4) let us start by implementing PWC on a single core processor. The Intel Xeon is chosen as the initial platform as it was easily accessible. It is a quad-core processor with 64KB L1 cache and 256KB L2 cache. VTune, a performance analysis tool available for Intel processors, is used to profile the execution kernel. The performance of the serial implementation with O3 optimization is tabulated in Table 9.4.

This data shows that the implementation experiences a substantial number of cache misses. Since the implementation experiences misses, the processor remains idle (due to lack of data) without imparting any throughput. Consequently, power consumption is high and performance per unit power is low. Another reason for power consumption to be high is that only one core is participating in the implementation whereas the other three cores are idle but consuming power. Parallelizing the algorithm can potentially decrease this metric. This can be achieved by using OpenMP, a powerful open-source parallelizing tool. The performance of the parallelized implementation using OpenMP is tabulated in Table 9.5.

Clocks per instruction (CPI) have improved from 0.84 to 0.67 in the parallel code. L2_MISS% has drastically reduced to 25.67. Since only L1 cache is private, the hit rate in L1 increases on an Intel quad-core Xeon processor. Thereby parallelizing the kernel reduces the power consumption.

9.3.3 Identifying the Nature of the Parallelized PWC (Block 6 of Figure 9.11)

To analyze an algorithm is to determine the amount of resources (such as time and storage) necessary to execute it. Usually the efficiency of an algorithm is stated as a function relating the input length to the number of steps (time complexity) and storage space (space complexity) required to run the algorithm and is usually represented with an *O* (pronounced as *Big O*).

Table 9.4 Performance of Serial Code on Intel Xeon Dual-Core Processor

	CPI	L1I_MISS%	L1D_MISS%	L2_MISS%
Serial code	0.84	22.98	91.54	60.77

Table 9.5 Performance of OpenMP Code on Intel Xeon Dual-Core Processor

	CPI	L1I_MISS%	L1D_MISS%	L2_MISS%
Parallel code (OMP)	0.67	27.84	89.23	25.67

By analyzing Eq. (9.4) one can realize that the cardinality of PWC is always $\binom{m}{2}$ and time and space complexity is proportional to m^2. Each pair involves computation of $\sum x_i$, $\sum y_i$, $\sum x_i^2$, $\sum y_i^2$, and $\sum x_i y_i$. Therefore for larger values of m, the PWC calculations become a bottleneck for real-time diagnostic algorithms. It should also be noted at this point that the complexities of computations of each element of PWC is also directly proportional to n.

9.3.4 Efficiently Parallelizing PWC (Blocks 7–10 of Figure 9.11)

There are three aspects of parallelizing any kernel that are important to consider when designing the implementation scheme, i.e., computation, communication between cores/memory, and space utilization. We will now see how by carefully studying these segments of an implementation one can efficiently parallelize a kernel.

Since PWC involves iterative computations of $\sum x_{(i,k)}$, $\sum x^2_{(i,k)}$, and $\sum x_{(i,k)} x_{(j,k)}$ (see Eq. (9.4)) the ideal approach is to partition the sampled data and stream it to multiple SIMD processing elements. Ideally, the data set for PWC can be represented as a 2D matrix $m \times n$ where m represents the number of channels and n represents the number of samples per channel. The number of operations for computing all elements of PWC is presented in Table 9.6.

Therefore the computation time complexity is $O(mn^2)$. Regarding communication between cores, Jimenez-Gonzalez and Martor [26,24] propose that a ring algorithm is the most efficient way to communicate data between multiple processing elements when the compute kernel consists of a pairwise type of computation. The ring algorithm considers a network of p processors connected by a ring network. In the first step of the ring algorithm, the processor i sends its data to processor $((i+1)$

Table 9.6 Number of Operations of Eq. (9.4)

	Addition	Multiplication	Subtraction	Division	Square Root
$\sum_{i=1}^{n} x_i$	mn	0	0	0	0
$\sum_{i=1}^{n} y_i$	mn	0	0	0	0
$\sum_{i=1}^{n} x_i y_i$	$\binom{m}{2} n_{\text{n\textbackslash)}}$	$\binom{m}{2} n$	0	0	0
$\sum_{i=1}^{n} x_i^2$	mn	mn	0	0	0
$\sum_{i=1}^{n} y_i^2$	mn	mn	0	0	0
Numerator	0	$2m$	m	0	0
Denominator	0	$4m$	$3m$	0	$2m$
r_{XY}	0	0	0	m	0
Total	$4mn + \binom{m}{2} n$	$m(2n+6) + \binom{m}{2} n$	$4m$	m	$2m$

mod p) and receives data from processor (($i-1$)mod p). In the consecutive steps of the Ring Algorithm, processor i forwards the data it received in the immediate previous step from processor (($i-1$)mod p) to processor (($i+1$)mod p). This process continues till all possible combinations are exhausted.

The current architecture allows us to communicate between cores through lower levels of memory. With the communication scheme modified, the Intel Xeon might not give the best performance over communication between cores. As suggested in block 9 of Figure 9.11, other architectures should be explored. The IBM Cell Broadband Engine (CBE) has an Elemental Interconnect Bus (EIB) for communication between the cores, with a ring topology. It utilizes a message-passing form of communication that is ideal for the ring algorithm. This architectural feature can be exploited to reduce the intercore communication time thereby saving power by reducing processor idling time. Apart from the ring topology there are some other features of CBE that can help us improve performance and save power like software controlled storage organization, asynchronous DMA transfer, and vectorization of computations.

According to our calculations, a speedup of more than 56 is observed in the implementation on Cell over the serial implementation on Intel Xeon with both the processors running at 3.2 GHz. But is the architecture-algorithm pair optimal? A thorough design space exploration can potentially answer this question.

9.4 DESIGN FRAMEWORK FOR BIOMEDICAL EMBEDDED PROCESSORS AND SURVEY OF SIMULATOR TOOLS

Owing to the ongoing research efforts in the field of green computing, the technological evolution process is perpetual. In the quest for designing an optimum architecture, simulator tools play a pivotal role. Simulator tools are used to perform design space exploration and assist researchers in arriving at innovative platforms. A standard flow for exploring the architecture of any kind of application-specific embedded platform is shown in Figure 9.12.

To arrive at an apropos embedded architecture for a specific application, a suitable benchmark suite has to be established which caters to all the requirements of the biomedical domain. For performing processor architecture exploration with the benchmark suite, simulation is one of the most important techniques used for evaluation, analysis, and estimation of performance, power dissipation, and energy consumption of selected candidate designs. Numerous simulator tools are available in the research community, which help in arriving at a green computing platform in the field of biomedical systems. In this section we discuss three such tools that help in design space exploration.

9.4.1 CASPER

CASPER is an execution 0driven cycle 0accurate chip multithreading (CMT) simulator used to evaluate performance, power, and energy consumption of heterogeneous multicore processors. The decision made about each micro-architectural feature of a

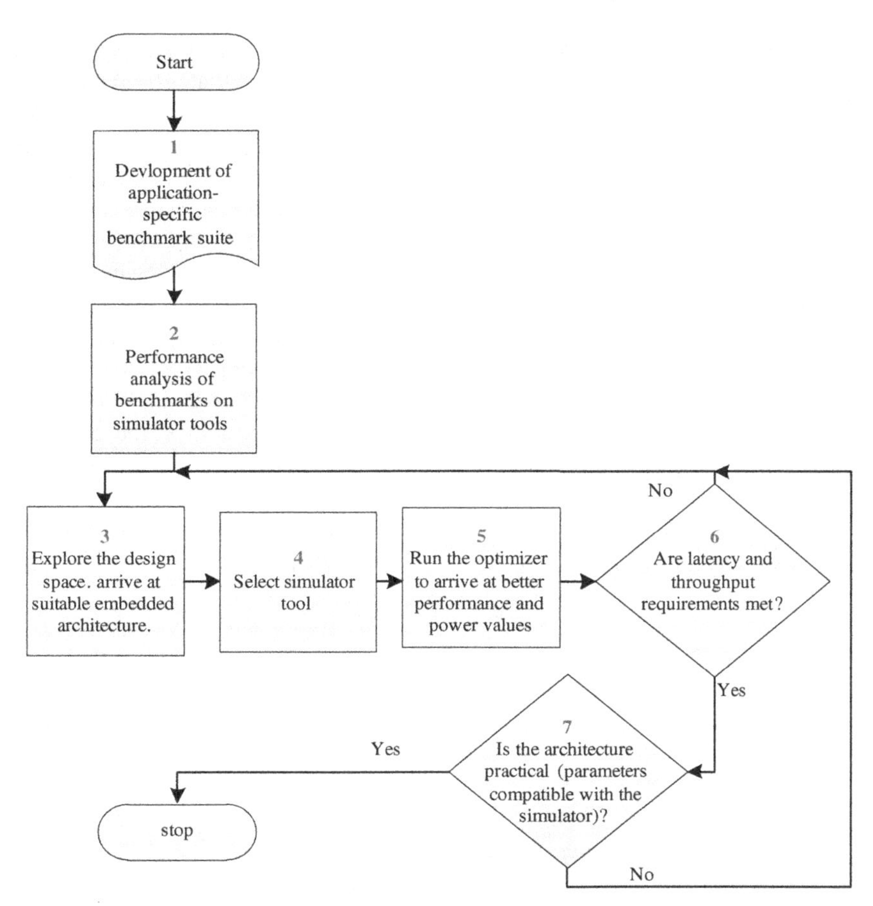

Figure 9.12 Design flow for biomedical embedded processors.

chip has implications in terms of area, power, performance, and energy consumption of the chip. There are simulation tools available to estimate performance in terms of cycles per instruction (CPI) at the front end of the design flow, but the tools that evaluate the power/energy consumption are towards the end of the design process and drastically increase the redesign effort. CASPER has deep chip vision technology that helps in calculating an accurate estimate of area, delay, power dissipation, and energy consumption during simulation. This technology uses switching factors of micro-architectural components that are found from cycle-accurate simulation and as per characterized HDL libraries of the micro-architectural components, that can be scaled over a range of different architectural parameters. CASPER targets an open-sourced ISA and processor architecture, under OpenSPARC of Sun MicroSystems Inc.

CASPER can simulate any number of cores with any number of hardware threads per core. The shared memory subsystem is flexible to accommodate L2 and L3 unified caches, although the first release includes the L2 cache model only. The cave at here is that once any of the cache functional strategies such as coherence protocol or an inclusive or exclusive strategy are modified such that it deviates from the original Ultra SPARC T1 architecture, there is no way to verify the system behavior. All the L1 private caches in the core subsystem are also parametrized as well. The inter connection network can be either a cross bar switch similar to T1, or a ring topology or a hypercube. In [10] Datta et al. a crossbar interconnect the between the core and memory subsystem is chosen (see Figure 9.13).

The CASPER instruction pipeline (per core) consists of the following stages: Fetch, Thread Schedule, Decode, Execute, Memory, WriteBack. Each thread possesses separate integer and float register files. Traps, exceptions, and interrupt control, detection, and processing are all identical to the UltraSPARC T1 architecture. An accurate SPARC V9 implementation helps the user to run any instruction traces run on Solaris 10. These traces can be dumped from SAM T1 and SAM T2 simulators. Once the trace files are generated, the configuration file is changed to define a set of parameters like number of cores, number of threads, L1 cache size, etc., as needed for a particular CMT architecture. CASPER uses a counter to calculate the CPI, Cacti to measure the cache hit latencies, and deep chip vision technology to measure power and area.

9.4.2 M5

Developed specifically to enable research in TCP/IP networking, the M5 simulator provides features necessary for simulating networked hosts, including full-system capability, a detailed I/O subsystem, and the ability to simulate multiple networked

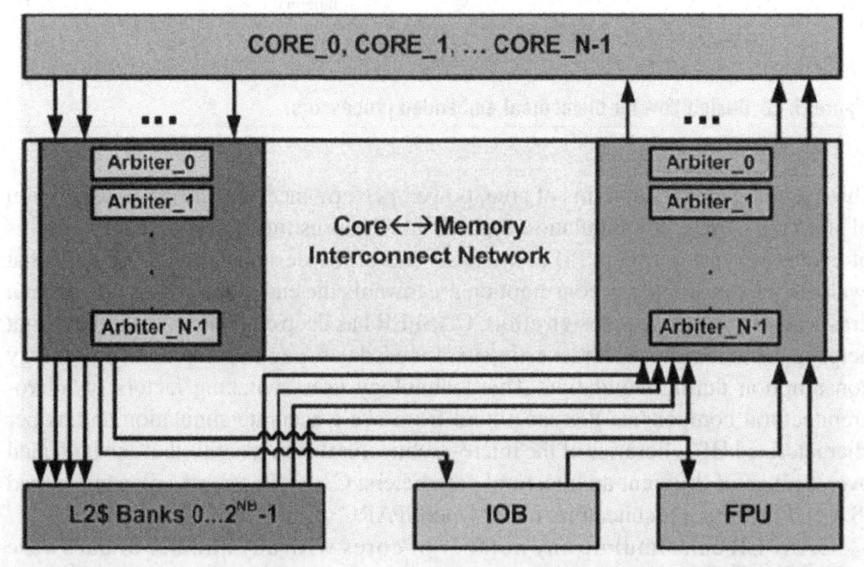

Figure 9.13 Processor model showing crossbar interconnect between core and memory subsystem.

systems deterministically [6]. A body sensor network design and implementation for wireless health monitoring applications can be simulated in order to develop a green computing platform.

A simulated system is merely a collection of objects (CPUs, memory, devices, and so on), so it is possible to model a client-server network by instantiating two of these collections and interconnecting them with a network link object. All of these objects are in the same simulation process and share the same global event queue, satisfying our requirement for deterministic multisystem simulation.M5 is implemented using two object-oriented languages: Python for high-level object configuration and simulation scripting (when flexibility and ease of programming are important) and C++ for low-level object implementation (when performance is key). M5 represents all simulation objects (CPUs, buses, caches, and so on) as objects in both Python and C++. Using Python objects for configuration allows flexible script-based object composition to describe complex simulation targets. Once the configuration is constructed in Python, M5 instantiates the corresponding C++ objects, which provide good runtime performance for detailed modeling.

The key requirements for performance modeling of networked systems are the following:

- the ability to execute operating-system as well as application code (known as *fullsystem simulation*), as this is where the bulk of the network protocol implementation resides
- Detailed performance models of the memory and I/O subsystems, including network interface devices
- The ability to model multiple networked systems (for example, a server and one or more clients) in a deterministic fashion

Traditional CPU-centric simulators, such as SimpleScalar, lack nearly all of these features. Among the simulators available, only SimOS provides the majority of the necessary features, missing only the ability to deterministically model multiple systems. Fullsystem simulation is one of the features which CASPER lacks.

9.4.3 MV5

With an increase in demand for green computing platforms to host modern applications that exhibit diverging characteristics, the need for heterogeneous systems is growing. While many latency-oriented workloads such as compilers (e.g., GCC) and compressors (e.g., bzip2) remain with sequential executions, many throughput-oriented workloads are transitioning to parallel executions. They include media processing, scientific computing, physics simulation, and data mining. Workloads can also be compute-intensive or data-intensive. Moreover, different types of workloads can execute simultaneously over the same system. Each workload may demonstrate different kinds of parallelism, including instruction-level (ILP), data-level (DLP), thread-level (TLP), and memory-level parallelism (MLP), which challenges future architectures to optimize for less execution time and more power- and area-efficiency. In order to accommodate both latency-oriented and throughput-oriented workloads, the system is likely to present a heterogeneous setting of cores. In particular, sequential code can achieve peak performance with an out-of-order core while parallel code achieves peak throughput over a set of simple, in-order cores. Several architecture designs have been proposed to address the challenges posed by heterogeneous workloads.

The developers of MV5 [14] envision that future CPU–GPU integration, as a representative heterogeneous multicore architecture, will provide both CPU-like, heavy weighted out-of-order (OOO) cores, and GPU-like, light-weighted in-order (IO) cores on the same chip. A shared memory will be provided using coherent caches connected through an on-chip network (OCN). This leads to a large design space: cores may have a different number of hardware thread contexts and different, SIMD (single instruction, multiple data) widths; caches can present different hierarchies, and one cache can be private or shared among multiple cores; the OCN may exhibit different topologies; and the coherence protocol has several variations as well. The modularity of M5 makes it easy for the developers of MV5 to extend it by means of additional components. The cache latency of MV5 is modeled using Cacti. It is assumed the instructions per cycle (IPC) for an in-order core is one, except for memory references. The IPC for an OOO core is measured separately. Energy/power consumed is measured using Cacti and Wattch. Cacti 4.2 is used to measure dynamic energy for reads and writes, and the leakage power of the caches. Dynamic energy of each core, the energy consumed for each time a unit of the pipeline, fetch and decode, integer ALU, floating-point ALU, or register file is accessed, is measured using Wattch. Area estimates of different functional units in an AMD Opteron processor (130 nm technology) are obtained from a publicly available die photo and they are scaled down to 65 nm technology with a scaling factor of 0.7. The heterogeneity that can be achieved using MV5 can be shown by looking at a few of the system configurations (Figure 9.14) achieved by changing the Python script.

The simulator features are compared in Table 9.7.

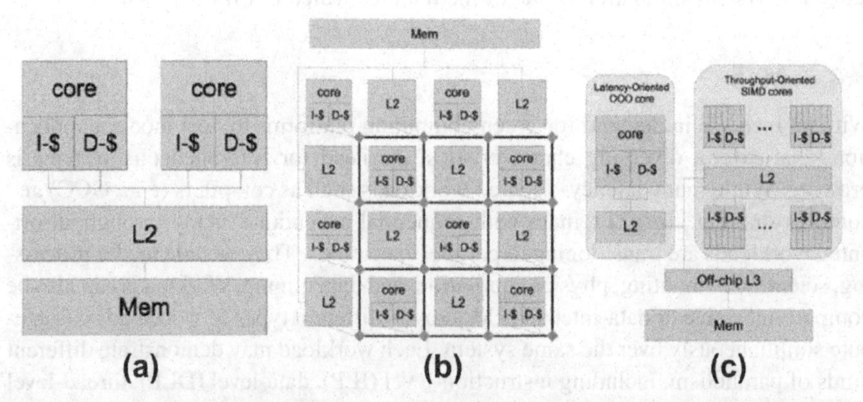

Figure 9.14 Various system configurations: (a) dual core; (b) tiled cores; (c) heterogeneous cores.

Table 9.7 Comparison of Simulator Features

Features	MV5	M5	CASPER
Full-system simulation	×	✓	×
System-call emulation	✓	✓	×
I/O disk	✓	✓	×
ISA	Alpha	Various	Sparc
Emulated thread API	✓	✓	✓
Category	Event driven	Cycle driven	Trace driven
IO core	✓	✓	✓
Multithreaded core	✓	✓	×
OOO core	✓	✓	✓
SIMD core	✓	✓	×

9.5 DEVELOPMENT AND CHARACTERIZATION OF BENCHMARK SUITE

For selecting a good embedded multicore processor we need good benchmarks. A good multicore benchmark will identify bottlenecks in the multicore system design including memory and I/O bottlenecks, computational bottlenecks, and real-time bottlenecks [7]. In addition, a good multicore benchmark will identify synchronization problems where code and data blocks are split, distributed to various compute engines for processing, and then the results are reassembled. The best multicore benchmark for biomedical application code is usually available for a homogeneous architecture. One of the codes is PWC. Many more chip vendors build multicore ASSPs with heterogeneous architectures. Figure 9.15 compares SMP and heterogeneous architectures. Multicore ASSPs and similarly architected SoCs can be simple two-core designs that combine a general-purpose processor and a DSP core, or

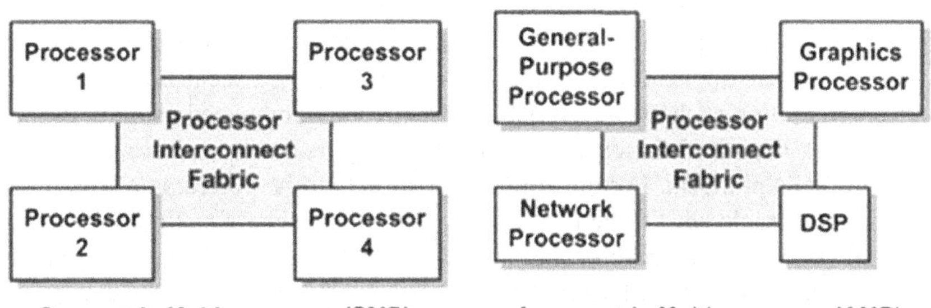

Figure 9.15 Comparison of SMP and heterogeneous multicore architectures.

complex many-core devices that incorporate tens or hundreds of processor cores. Multicore chips employ a variety of interconnect technologies including buses, point-to-point connections, and NoCs (networks on chip). Consequently, multicore architectures are highly differentiated and multicore benchmarks must be differentiated as well.

9.5.1 Requirements for Multicore Benchmark

Can any sequential code be used for the purpose of benchmarking a green computing platform consisting of multicore architectures? To answer this question, we need to understand that benchmarks for multicore architectures should be either compute- or memory-intensive, or some combination of both.

Distributed memory systems, which are more commonly used for SoC designs, give each processor its own local memory. When one core requires data from another core or when processor cores must synchronize data, the data must be moved from one local memory to the other or the control code must switch from one processor core to the other (which is nearly impossible for heterogeneous systems). Even when each processor core has its own local memory, there can still be memory bottlenecks depending on how data moves around and onto or off of the chip. A good multicore benchmark will detect such memory bottlenecks.

Scalability is another important benchmark criterion. It's not uncommon to have hundreds of threads in a relatively complex program. If the number of threads exactly matches the number of processor cores, performance could scale linearly assuming no memory-bandwidth limitations. Realistically the number of threads often exceeds the number of cores and performance depends on other factors such as cache utilization, memory and I/O bandwidth, intercore communications, OS scheduling support, and synchronization efficiency. A processor incurs performance penalties when it oversubscribes computing resources and a good multicore benchmark will highlight such situations.

Heterogeneous multicore architectures require an entirely different benchmarking strategy. Unlike the SMP benchmarks, heterogeneous cores require careful workload partitioning because that's exactly the way they're used in practice. Benchmarking heterogeneous devices in a relevant manner with portable code is a huge challenge because each part of the system likely uses a different development-tool chain and communication among the different processor cores is not standardized.

The biomedical application kernels such as PWC can be executed on a homogeneous configuration such as the SPARC-v9 architecture and tested using CASPER. We are trying to extend the existing biomedical application kernels to execute on a heterogeneous platform. For the application kernels to be ported as a benchmark on MV5, the kernels need to be rewritten by defining various functionalities of the sequential code as specific kernels using Fractal APIs and executed in parallel as suggested in [15]. Benchmarks have been implemented in two different ways to test the heterogeneous architecture. In the first method, a single kernel is split into multiple threads to be executed on multiple cores of an SoC architecture. In the second method a heterogeneous benchmark is generated by combining several kernels. For example in an architecture with both IO and OOO cores, suitable kernels are chosen for each type of core—a compute-intensive benchmark can be scheduled to work on the in-order cores, and the memory-intensive benchmark can be executed on the OOO cores.

9.5.2 Performance Analysis of PWC

According to [4], the above discussed Cell processor is power efficient when it comes to biomedical applications of the streaming data type. Diagnosis for epilepsy using electroencephalogram (EEG) signals is one such scenario as mentioned in [25]. Typically, the number of EEG signals/channels, collected from grids of electrodes, are of the order of hundreds and they are expected to approach the thousand sensors mark in the near future. Furthermore, the sampling frequency (f) of the channels is at least 1 KHz as mentioned in [12]. These computations are currently done offline, but with a few strategies in implementation of the algorithm, an optimized approach on a fast multicore processor like CBE can solve the problem in real time, bound by a few constraints. There is a growing need to perform these diagnoses in real time to aid health care providers, including surgeons, in decision-making processes that will lead to improved quality of life and prevent undesirable consequences, such as readmission to hospitals resulting in prolonged suffering and higher health care costs. In [4], a parallel scheme for real-time computations of PWC of EEG signals has been proposed.

9.5.2.1 CASPER Simulation

Performance of the PWC code in a Sun machine, with architectural features of Ultra-Sparc T1 is evaluated through simulations run on CASPER

The serial kernel was run through Simulator Architecture Model (SAM) to generate a trace file. The trace file was run through CASPER, with one core one thread configuration. The architectural parameters, like L1 data and instruction cache size, associativity, and bit length were varied to arrive at the optimum performance of an instruction per cycle (IPC = 1/CPI) to power consumption ratio of 0.00995. The optimum processor's architecture parameter set was found to have an L1 data and instruction cache of 1024 kb each and associativity of 4 (see Figures 9.16 and 9.17).

9.5.2.2 MV5 Simulation

The same kernel was rewritten in a parallel fashion using Fractal APIs. Sections of the code were split into three different kernels for calculating $\sum x_{(i,k)}$, $\sum x^2_{(i,k)}$, and $\sum x_{(i,k)}x_{(j,k)}$, and the results were taken into account in the main kernel for calculation of pairwise correlation in the main program. All three kernels were made to run parallel, since it was compiled in simdOMP format[16], i.e.,

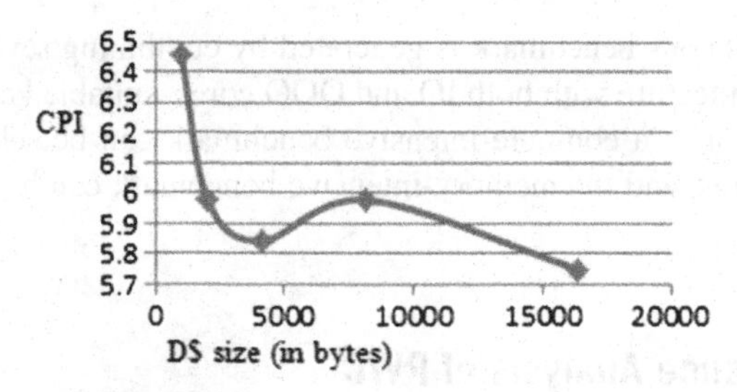

Figure 9.16 CPI per core on CASPER.

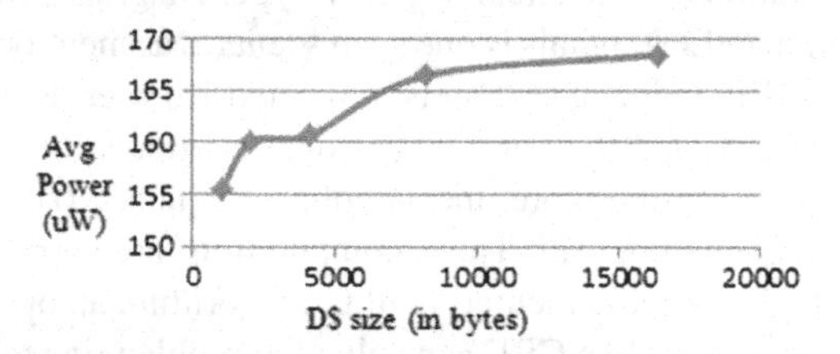

Figure 9.17 Average power per core on CASPER.

compiled to run on the simulated system, with partitioning that resembles OpenMP (segment parallel for loops into chunks, each assigned to a thread). Another benchmark, which is part of image processing in various biomedical applications, is also used for the simulation. A filter kernel is used for edge detection of an image. It does a convolution on a 3×3 neighborhood. Here the simulated system has ALPHA architecture, which has both in-order (SIMD) and out-of-order (OOO) cores as indicated in Table 9.8.

The heterogeneous configuration is then changed to verify the proper combinations of SIMD and OOO cores. In Table 9.9, the third configuration has PWC running on SIMD cores and FILTER on OOO cores and vice versa in the fourth configuration.

Table 9.8 Analysis of Parallel Version of the Code on MV5

	Frequency (GHz)	Number of SIMD CPUs	Number of OOO CPUs	No. of HW+SW Threads	Benchmark Used	Host Memory Usage (MB)	Simulation Time (s)
Fractal_smp	1	4	0	64+2	FILTER	1.217	0.019065
Fractal_smp	1	4	0	64+2	PWC	1.207	0.001364
Config_hetero	1	2	2	32+2	FILTER PWC	2.234	0.070888
Config_hetero	1	2	2	32+2	FILTER PWC	2.255	1050.42

Table 9.9 Analysis of Parallel Version of the Code (per CPU Results) on MV5 with Various Configurations

	Total Energy of CPU (mJ)	Total Leakage Energy of CPU (mJ)	Clock Active Energy (μJ)	Total Cache Energy (mJ)	D$ Miss Rate	I$ Miss Rate	Floating ALU Active Energy (mJ)	Integer ALU Active Energy (mJ)
Fractal_smp on FILTER	26.358100	1.713209	0.000956	2.186035	0.257	0.162	1.0877987	1.785665
Fractal_smp on PWC	0.010644	0.002118	0.000188	0.003291	2.195	0.078	0.0004001	0.000844
Config_hetero FILTER+PWC	29.918695	1.543702	0.000292	12.097728	2.876	0.018	2.241064	4.005570
Config_hetero FILTER+PWC	32.2689733	4.1982526	0.000182	0.0081944	1.747	0.000001	1.0877992	1.7854455

As we can see from the above table, the third combination consumes the least compared to other heterogeneous configurations. Hence it can be deduced that PWC, which has three kernels running in simdOMP fashion, performs better on a SIMD core and FILTER, which has a single kernel and is compute-intensive, runs better on a OOO core.

9.6 DESIGN SPACE EXPLORATION AND OPTIMIZATION OF EMBEDDED MICRO ARCHITECTURES

In this section we survey the different approaches used for design space exploration for multicore processor architecture. In [10], CASPER is used for cycle-accurate multicore simulations to arrive at a power and performance measure for a particular benchmark optimized for network processing. The benchmark suite used to evaluate the performance and power dissipation of candidate designs in CASPER is *Embedded Network Packet Processing Benchmark* (ENePBench), which emulates the IP packet processing tasks executed in a network router. To arrive at suitable micro-architectural configurations, the in-loop simulations, i.e., running CASPER for various permutations and combinations of varying parameters, is replaced with statistical regression models in order to speed up the estimation of power-performance values. In CASPER the core-level optimization with the chip-level shared memory micro-architecture optimization are integrated. The idea is to capture the diminishing effects of shared secondary cache contention and interconnection on single thread performance in the cores, which prohibits the scaling of cores as well as the number of threads per core in the design. In this step a macro-simulator is used, which accounts for the simultaneous accesses from different cores, and hence it is possible to estimate the L2 cache access latency, which changes due to contention.

Randomly chosen core-level parameters are used to run the simulations to arrive at the regression models for power and performance. The models are then passed to a genetic algorithm based optimization engine to explore the design space. An algorithm-based optimization engine called the Fast Genetic Algorithm (FGA) [13] is used to generate a set of 10 best optimized core micro-architectures with minimal power dissipation. The small hill-climbing technique, embedded in the simulated annealing (SA) method, enables us to quickly converge to an optimal design thus giving us a shared memory heterogeneous many-core micro-architecture. The cost function in the SA method is the average power dissipation per cycle and constraints are the real-time throughput boundaries of the packets. The L2 macro-simulator values are then integrated. For every chosen L2 cache configuration and a set of optimized micro-architectures, macro-simulation is performed, and a optimal many-core design which dissipates minimum power is obtained.

The advantages of the exploration tool designed in [10] are that it is integrated with CASPER, and follows the same command line conventions. The disadvantage is that only in-order many-core architecture can be explored. However for further power reduction and greener computing, even OOO cores have to be taken into consideration. For this purpose we make use of MV5, rather than CASPER. In MV5 a broader parameter set has to be explored, since more variations of configurations can be achieved. MV5 comes with a driver tool for this purpose. The advantage is that, various configurations can be specified easily with a simple Python script and the combinations can be made to be executed simultaneously on a cluster of machines.

The driver tool helps users create, manage, and analyze experiments. This tool works for a cluster of machines with a shared file system. The tool is used to manage jobs in batches by specifying the set of system parameters to explore. The cluster is divided into two types of machines. The monitor is the system that creates and manages the batches and warriors are the systems in which the batches are executing in parallel. First a server script is run on the main system. A monitor is created in one of the systems to monitor and control the execution of batches in various warriors. The following diagram represents a scenario where two monitors are created to handle three warriors (see Figure 9.18).

As the architectural design complexity increases the time taken to explore it also increases exponentially, irrespective of the platform. Both in CASPER and MV5, it would be hard to evaluate the complete design space to arrive at an optimum architecture which is in line with the green computing philosophy, by consuming the least amount of power in spite of having good performance numbers in terms of CPI, memory access times, and the like.

There are many ways of arriving at approximation models. Ipek et al. [28] use artificial neural networks (ANNs) to automatically create global approximation models of the architecture design space. They randomly sample points throughout the design space and train the neural networks, and these trained sets are used to predict

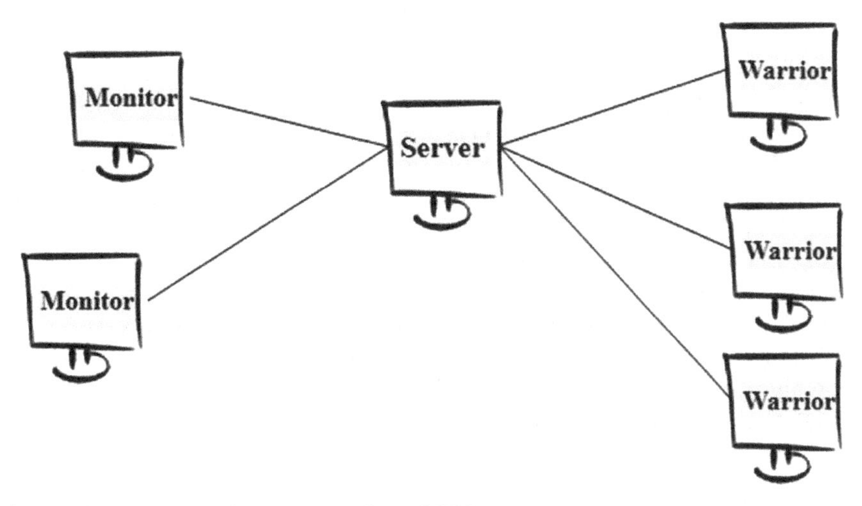

Figure 9.18 Server, monitors, and warriors of MV5.

over the rest of the design space. With the advantage of achieving 98–99% prediction accuracy after only training 1–2% of the design space, it has the disadvantage of being in the form of a "black box" to the designer, since it does not give relationships between the input parameters. Lee and Brooks [29] perform linear regression modeling in order to generate approximation functions for power and performance. This is the method adopted by Datta et al. [10] for micro-architectural design space exploration. This method is a statistical process that has to be carried out by the architect, rather than an automated algorithm, and requires a prior understanding of relationships among the parameter set.

Cook and Skadron [27] propose a solution for finding the right set of design points for users who do not have an in-depth knowledge of the unexplored architecture. He uses Audze-Eglais uniform latin hypercube design of experiments (DOE) to select design points included in the sample set. DOE is described in detail in Bates et al. [30]. A genetic algorithm is used to implement DOE, and to arrive at sample points which are evenly distributed.

MV5 can be used to execute the PWC and FILTER kernels (discussed in Section 9.5) at these design points and evaluate the performance and power for a particular set of IO and OOO cores. The resulting full-scale simulation data is then used by the genetic programming algorithm to construct the response surface, which is a nonlinear approximation of the data. The specific details of genetic programming are described in [31].

Genetically programmed response surfaces (GPRS) avoid coupling the search time of the algorithm to the running time of the simulation. In other words, the GPRS process costs O(simulationTime_sampleSize+evaluationTime_stepsToConverge) whereas a heuristic search costs O(simulationTime_evaluationTime_stepsToConverge). For GPRS, stepsToConverge is relatively agnostic to design space dimensionality and sampleSize grows only linearly with design space dimensionality. In order to offer comparable scalability, the time to convergence of a heuristic search could not increase by more than a few evaluations per dimension added to the design space. Another fundamental difference between predictive model generation and heuristic searching is the level of insight the end result provides into the nature of the design space. GPRS provides a global map of the entire design space in the form of an analytic function, making explicit to the architect which variables have a significant impact on the target performance measure and what interrelationships exist among them. A global view of performance behavior across the entire design space is a superset of the guidance an architect could expect from a heuristic search based approach.

We find that the models generated by GPRS accurately and robustly depict the true behavior of power and performance measures on MV5. GPRS tells the user the significance of specific architecture parameters by including them explicitly in an analytical equation, and thus guiding a user's insight into how the parameters have to be optimized and which subspaces have to be explored to arrive at better green computing biomedical embedded processors. The disadvantage of MV5 is that it does not support full system simulation; future trends can achieve the right nonpredictive design space of exploration tool. One such method would be porting a real-time operating system to MV5 and achieving scheduling policies to select the right kind of compute-intensive or compute-intensive application code on the right set of IO or OOO cores within the specified real-time constraints of biomedical applications. Also predictive tools like GPRS can be integrated with the nonpredictive tool set to arrive at a self-testing environment, where once the best suitable green computing embedded processor architecture is predicted, the power and performance aspects of it should be verified by the simulator tool automatically.

9.7 CONCLUSION AND FUTURE WORK

In this chapter we have discussed the issues involved in designing "green" processing platforms for biomedical systems including the following:

(i) Methodologies for the characterization of biomedical applications for ultra-low energy and low heat producing embedded implantable devices, as well as for low power dissipation but high performance embedded computing platforms. For a compute-intensive PWC benchmark with complexity $O(mn^2)$, the above stated methodology has achieved a CPI of 0.67 and L2 cache miss percentage of 25.67 on the Intel Xeon dual-core processor, and given a speedup of 56 when compared to the serial version of the PWC kernel. These efforts might lead to initiatives to development of appropriate biomedical benchmark suites.

(ii) Outlines of the procedure to be followed for the design space exploration of processor micro-architectures using existing simulation tools and optimizers. We have discussed deficiencies of existing embedded processing platforms as used in biomedical systems, and identified future research directions for developing more efficient domain-specific embedded processors. Our preliminary results as seen in Section 9.5 show that a heterogeneous configuration with two IO and two OOO consumes less energy per CPU (29.918 mJ) compared to a homogenous configuration on MV5's alpha architecture simulation. Further research that will cater to optimizing the architecture for both performance and power will prove to be significant.

Other research areas of immediate interest include development of better instruction set architectures (ISAs) and the corresponding cross-compilers. These can be used to generate optimized executables for platforms and simulators, upgrading existing simulation platforms to support full system mode with real-time kernel libraries to account for the latency and throughput of the real-life applications. These can also be used to develop advanced real-time operating systems and scheduling algorithms to schedule the various applications on different heterogeneous cores to meet the hard real-time constraints. The ideas presented in this chapter are expected to stimulate new research directions for greener computing platforms in biomedical systems.

REFERENCES

[1] P. Kwan, M.J. Brodie, Early identification of refractory epilepsy, The New England Journal of Medicine 342 (2000) 314–319.

[2] B. Ackland, A. Anesko, D. Brinthaupt, S.J. Daubert, A. Kalavade, J. Knobloch, E. Micca, M. Moturi, C.J. Nicol, J.H. O'Neill, J. Othmer, E. Sackinger, K.J. Singh, J. Sweet, C.J. Terman, J. Williams, A single-chip, 1.6-billion, 16-b MAC/s multiprocessor DSP, IEEE Journal of Solid-State Circuits 35 (3) (2000) 412–424.

[3] M. Kistler, M. Perrone, F. Petrini, Cell multiprocessor communication network: built for speed, IEEE Micro 26 (3) (2006) 10–23.

[4] A. Panday, B. Joshi, A. Ravindran, J. Byun, H. Zaveri, Study of data locality for real-time biomedical signal processing of streaming data on cell broadband engine, in: IEEE SoutheastCon 2010, 18–21 March 2010.

[5] M. Ware, K. Rajamani, M. Floyd, B. Brock, J.C. Rubio, F. Rawson, J. B. Carter, Architecting for power management: the IBM® POWER7™ approach, in: IEEE High Performance Computer Architecture (HPCA), 9–14 January 2010.

[6] N. Binkert, R. Drenslinski, L. Hsu, K. Lim, A. Saidi, S. Reinhardt, The M5 Simulator: Modeling Networked Systems, IEEE Computer Society.

[7] S. Gal-On, M. Levy, S. Leibson, How to survive the quest for a useful multicore benchmark, ECN Magazine, December 2009.

[8] Ultimate Performance with Green Technology. <http://www.samsung.com/global/business/semiconductor/support/brochures/downloads/systemlsi/Exynos4210.pdf>.

[9] Introduction to IBM PowerPC 476FP Embedded Processor Core, April 2011.

[10] K. Datta, A. Mukherjee, G. Cao, J. Byun, A. Ravindran, B. Joshi, D. Vahia, CASPER: A SPARCV9-based Cycle-Accurate Chip Multi-threaded Architecture Simulator for Performance, Power, Energy and Area Analysis, IPDPDS.

[11] B. Carlson, Going "Beyond a Faster Horse" to Transform Mobile Devices, May 2011.

[12] H.P. Zaveri, W. Williams, J. Sackellares, A. Beydoun, R. Duckrow, S. Spencer, Measuring the coherence of intracranial electroencephalograms, Clinical Neurophysiology 110 (10) (Dec 1999) 1717–1725.

[13] A. Presta, Fast Genetic Algorithm, Book, 2007.

[14] J. Meng, K. Skadron, A Reconfigurable Simulator for Large-Scale Heterogeneous Multicore Architectures, University of Virginia, Performance Analysis of Systems and Software, IEEE.

[15] J. Meng, J. Sheaffer, K. Skadron, Exploiting inter-thread temporal locality for chip multithreading, in: Proceedings of the 24rd International Parallel and Distributed Processing Symposium, 2010.

[16] J. Meng, S.R. Tarapore, S. Che, J. Huang, J.W. Sheaffer, K. Skadron, Programming with Relaxed Streams, UVA Tech, Report CS-2007-17.

[17] R. Sarpeshkar, Ultra Low Power Bioelectronics: Fundamentals, Biomedical Applications, and Bio-Inspired Systems, 2010.

[18] C.C. Yew, The Power Management IC for the Intel® Atom™ Processor E6xx Series and Intel® Platform Controller Hub EG20T, January 2011.

[19] NVIDIA Tegra APX Series. <http://www.nvidia.com/object/product_tegra_apx_us.html>.

[20] N. Brookwood, AMD Fusion™ Family of APUs: Enabling a Superior, Immersive PC Experience, March 2010.

[21] MIPS32 1074K: High-Performance Multiprocessor IP for Web-Connected Digital Home SOCs.

[22] X. Chen, J. Hu, N. Xu, Regularity-constrained floorplanning for multi-core processors, in: International Symposium on Physical Design, March 2011.
[23] Overview of the Avitek Medical Records Development Tutorials. <http://download.oracle.com/docs/cd/E13222_01/wls/docs81/medrec_tutorials/overview.html>.
[24] Programming the Cell Broadband Engine Architecture. <www.redbooks.ibm.com/abstracts>.
[25] H.P. Zaveri, S.M. Pincus, I.I. Goncharova, R.B. Duckrow, D.D. Spencer, S.S. Spencer, Localization-related epilepsy exhibits significant connectivity away from the seizure-onset area, NeuroReport 20 (9) (2009) 891–895.
[26] D. Jimenez-Gonzalez, X. Martor, Analysis of cell broadband engine applications, in: 2007 IEEE Symposium on Performance Analysis of System and Architecture (ISPASS), pp. 210–219.
[27] H. Cook, K. Skadron, Predictive Design Space Exploration Using Genetically Programmed Response Surfaces, DAC 2008.
[28] E. Ipek, S.A. McKee, B.R. de Supinski, M. Schulz, R. Caruana, Efficiently exploring architectural design spaces via predictive modeling. In: The 12th ACM International Conference on Architectural Support for Programming Languages and Operating Systems (ASPLOS XII), San Jose, CA, October 2006.
[29] B. Lee, D. Brooks, Accurate and efficient regression modeling for microarchitectural performance and power prediction, in: The 12th ACM International Conference on Architectural Support for Programming Languages and Operating Systems (ASPLOS XII), San Jose, CA, October 2006.
[30] S.J. Bates, J. Sienz, D.S. Langley, Formulation of the Audze Eglais uniform Latin hypercube design of experiments, Advances in Engineering Software 34 (8) (2003) 493–506.
[31] L. Alvarez, Design optimization based on genetic programming, PhD Thesis, University of Bradford, 2000.
[32] G. Lazzi, Thermal effects of bioimplants, IEEE Engineering in Medicine and Biology Magazine 24 (5) (2005) 75–81.
[33] Q. Tang, N. Tummala, S.K.S. Gupta, L. Schwiebert, Communication scheduling to minimize thermal effects of implanted biosensor networks in homogeneous tissue, IEEE Transactions on Biomedical Engineering 52 (7) (2005) 1285–1294.
[34] Y. Suhail, K.G. Oweiss, A reduced complexity integer lifting wavelet-based module for real-time processing in implantable neural interface devices, in: Engineering in Medicine and Biology Society, 2004, IEMBS '04, 26th Annual International Conference of the IEEE, vol. 2, 1–5 September 2004, pp. 4552–4555.
[35] A.M. Kamboh, M. Raetz, K.G. Oweiss, A. Mason, Area-power efficient VLSI implementation of multichannel DWT for data compression in implantable neuroprosthetics, IEEE Transactions on Biomedical Circuits and Systems 1 (2) (2007) 128–135.
[36] S. Yang, I.M. Verbauwhede, A secure fingerprint matching technique, in: Proceedings of the Workshop on Biometrics Applications and Methods, November 2003, pp. 89–94.
[37] J.K. Lee, S.R. Ryu, K.Y. Yoo, Fingerprint-based remote user authentication scheme using smart cards, Electronics Letters 38 (12) (2002) 554–555.
[38] D. Maltoni, D. Maio, A.K. Jain, S. Prabhakar, Handbook of Fingerprint Recognition, Springer, New York, 2003
[39] M. Penhaker, M. Černý, L. Martinák, J. Spišák, A. Válková, HomeCare—Smart embedded biotelemetry system, in: World Congress on Medical Physics and Biomedical Engineering 2006, IFMBE Proceedings, vol. 14, Part 7, 2007, pp. 711–714.

[40] C. Kunze, U. Grobmann, W. Stork, K.D. Müller-Glaser, Application of ubiquitous computing in personal health monitoring systems, Biomedizinische Technik/Biomedical Engineering 47 (s1a) (2002) 360–362.
[41] Y. Lin, I. Jan, P.C. Ko, Y. Chen, J. Wong, G. Jan, A wireless PDA-based physiological monitoring system for patient transport, IEEE Transactions on Information Technology in Biomedicine 8 (4) (2004) 439–447.
[42] EIA, 2006 Energy Information Administration (EIA), Commercial Buildings Energy Consumption Survey (CBECS): Consumption and Expenditures Tables, Table C3A, U.S. Department of Energy, 2006.
[43] K.Y. Sanbonmatsu, C.S. Tung, High performance computing in biology: multimillion atom simulations of nanoscale systems, Journal of Structural Biology 157 (3) (2007) 470–480.
[44] M. Greicius, B. Krasnow, A. Reiss, V. Menon, Functional connectivity in the resting brain: a network analysis of the default mode hypothesis, PNAS 100 (2002) 253–258.
[45] M. Fox, M. Raichle, Spontaneous fluctuations in brain activity observed with functional magnetic resonance imaging, Nature Reviews Neuroscience 8 (9) (2007) 700–711.
[46] N. Weiskopf, R. Sitaram, O. Josephs, R. Veit, F. Scharnowski, R. Goebel, N. Birbaumer, R. Deichmann, K. Mathiak, Real-time functional magnetic resonance imaging: methods and applications, in: Proceedings of the International School on Magnetic Resonance and Brain Function, 2007, pp. 989–1003.
[47] R. deCharms, Applications of real-time fMRI, Nature Reviews Neuroscience 9 (9) (2008) 720–729.
[48] ILAE Commission Report. The epidemiology of the epilepsies: future directions, International League Against Epilepsy, Epilepsia 38 (5) (1997) 614–618.
[49] X. Zhang, H. Jiang, B. Zhu, X. Chen, C. Zhang, Z. Wang, A low-power remotely-programmable MCU for implantable medical devices, in: 2010 IEEE Asia Pacific Conference on Circuits and Systems (APCCAS), 6–9 December 2010, pp. 28–31.
[50] J. Abouei, J.D. Brown, K.N. Plataniotis, S. Pasupathy, Energy efficiency and reliability in wireless biomedical implant systems, IEEE Transactions on Information Technology in Biomedicine 15 (3) (2011) 456–466.
[51] P. Borgesen, E. Cotts, Implantable medical electronics assembly – quality and reliability considerations, in: Proceedings from the 2004 SMTA Medical Electronics Symposium, Minneapolis, Minnesota, April 2004.
[52] R. Schmidt, T. Norgall, J. Mörsdorf, J. Bernhard, T. von der Grün, Body area network BAN: a key infrastructure element for patient-centered medical applications, Biomedizinische Technik (Berlin) 47 (2002) 365–368.
[53] A. Nahapetian, F. Dabiri, M. Sarrafzadeh, Energy minimization and reliability for wearable medical applications, in: 2006 International Conference on Parallel Processing Workshops, 2006, ICPP 2006 Workshops, pp. 8–318.
[54] C.-S.A. Gong, M.-T. Shiue, Y.-P. Lee, K.-W. Yao, Self-powered integrated rectifier for wireless medical energy harvesting applications, in: 2011 International Symposium on VLSI Design, Automation and Test (VLSI-DAT), 25–28 April 2011, pp. 1–4.
[55] B. Otal, C. Verikoukis, L. Alonso, Fuzzy-logic scheduling for highly reliable and energy-efficient medical body sensor networks, in: IEEE International Conference on Communications Workshops, 2009, ICC Workshops 2009, 14–18 June 2009, pp. 1–5.
[56] K.K. Venkatasubramanian, A. Banerjee, S.K.S. Gupta, Green and sustainable cyber-physical security solutions for body area networks, in: Sixth International Workshop on Wearable and Implantable Body Sensor Networks, 2009, BSN 2009, 3–5 June 2009, pp. 240–245.

[57] Z. Besic, D. Zrno, D. Simunic, Energy efficient sensing in wireless sensor networks in body area networks, in: MIPRO, 2011 Proceedings of the 34th International Convention, 23–27 May 2011, pp. 1703–1706.

[58] X. Zhang, H. Jiang, F. Li, S. Cheng, C. Zhang, Z. Wang, An energy-efficient SoC for closed-loop medical monitoring and intervention, in: Custom Integrated Circuits Conference (CICC), 2010 IEEE, 19–22 September 2010, pp. 1–4.

[59] M. Porter, P. Gerrish, L. Tyler, S. Murray, R.Mauriello, F. Soto, G. Phetteplace, S. Hareland, Reliability considerations for implantable medical IC's, in: IEEE International Reliability Physics Symposium Proceedings 2008, April 2008, pp. 516–523.

[60] S. Oesterle, P. Gerrish, Peng Cong, New interfaces to the body through implantable-system integration, in: Solid-State Circuits Conference Digest of Technical Papers (ISSCC), 2011 IEEE International, 20–24 February 2011, pp. 9–14.

[61] M. Soma, Reliability of implantable electronic devices: two case studies, IEEE Transactions on Reliability 35 (5) (1986) 483–487.

[62] <http://www.coolflux.com/nxp-coolflux-dsp>.

[63] E. Alt, J. Jaekel, Advanced personal identification systems and techniques, US Patent 6580356, issued date June 17, 2003.

CHAPTER

Green Data center Infrastructures in the Cloud Computing Era

10

Sergio Ricciardi[a], Francesco Palmieri[b], Jordi Torres-Viñals[c], Beniamino Di Martino[b], Germán Santos-Boada[a], Josep Solé-Pareta[a]

[a]*Technical University of Catalonia (UPC) – BarcelonaTech, Department of Computer Architecture, 08034 Barcelona, Spain,*
[b]*Second University of Naples (SUN), Department of Information Engineering, I-81031 Aversa, Italy,*
[c]*Barcelona Supercomputing Center (BSC), Autonomic Systems and eBusiness, 08034 Barcelona, Spain*

10.1 INTRODUCTION

In the past few decades, society has experienced drastic changes in the way information is accessed, stored, transmitted, and processed. Data has been digitized to allow electronic processing, by migrating from physical to digital supports, and the Internet has made it accessible from whatever device is connected to the global network. This has produced deep changes in society, industries, and communications, giving rise to the so-called *Information and Communication Society* (ICS).

Several milestones can be identified in the history of the ICS, approximately every 15 years [1]. At the beginning of the ICS, there were the mainframes: big machines that processed jobs in batch, i.e., sequentially and offline. With the advent of personal computers (PCs), the users could have their own small computation and storage resources on their desktops. This was a significant change in the paradigm, since each user could process its jobs on its own. The interconnection of the PCs was possible with the advent of the Internet, mainly based on the client–server model. The servers grew in size and functionality, and federated data centers and farms distributed throughout the world started cooperating to offer more and more services to the users, evolving into the actual Internet-scale computing paradigm and service facilities commonly known as grid and cloud infrastructures.

Such a paradigm represents a technological revolution in the way that information is accessed, stored, and shared from whatever device is connected to the Internet which, in turn, evolves from a "simple" connection infrastructure to an integrated platform offering services to millions of smart, always-connected terminal devices. The fundamental advantage of the aforementioned cloud paradigm is the abstraction between the physical and logical resources, needed to provide services, and the users, which can simply *use* the services they need (according to the so-called *software, platform, and infrastructure as a service* models) without having to worry about *how* they are actually implemented.

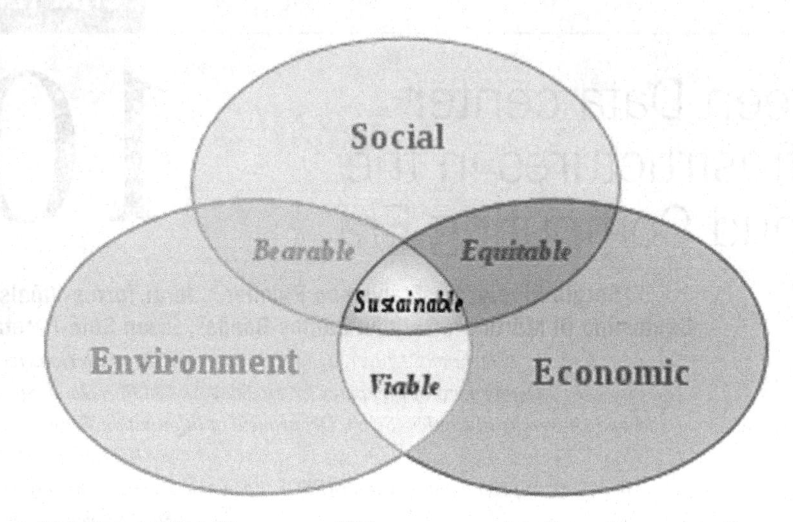

Figure 10.1 Eco-sustainability as an equilibrium among society, the environment, and economics.

In addition to the large number of technological and architectural issues originated by the above paradigm change, the ICS is now facing another major challenge: *eco-sustainability*. Eco-sustainability has become a major issue in the ICS since resource exploitation, climate change, and pollution are strongly affecting our planet. In particular, eco-sustainability refers to the equilibrium among society, the environment, and economics (Figure 10.1), by meeting the needs of the present generation without compromising the ability of future ones to meet their own needs [2].

Today, the ICS has the opportunity to put together the above evolutionary distributed computing technologies and eco-sustainability into an integrated ICT paradigm, encompassing both the ever-increasing need of development and the preservation of the world's natural resources. The ICT sector has in fact a unique ability to reduce its ecological footprint [basically, energy consumption and greenhouse gas (GHG) emissions] as well as the ones of other industry sectors through the use of technologically innovative solutions [3]. The first step is therefore the reduction of the ICT ecological footprint, which will drive the change towards societal growth and prosperity.

The key factors of such an energy-oriented paradigm are (1) *energy-efficiency*, which lowers the energy requirement of the involved service provisioning devices without affecting their performance and (2) *energy-awareness*, which adapts the overall energy consumption to the current load, introducing energy-proportionality, and exploits the use of renewable energy sources and emerging *smart grid* power distribution networks. Energy-efficiency and energy-awareness goals can be achieved together into an (3) *energy-oriented* systemic approach that considers the whole lifecycle assessment (LCA) of any new solution to ensure that it will be effective and avoid a rebound effect.[1]

[1]There is a logical problem related to having more efficient components: higher efficiency may lead, in fact, to decreased costs and to increased demand, possibly overcoming the gains obtained with efficiency, a phenomenon known as *rebound effect* or, in other contexts, as *Jevons Paradox* or the *Kazzom-Brokes postulate*[4].

10.2 BACKGROUND

10.2.1 The Energy Problem

Humans' demand on the biosphere, known as the human ecological footprint,[2] has been estimated to be equivalent to 1.5 planet Earths [5], meaning that humanity uses ecological resources 1.5 times faster than the capacity of Earth to regenerate them. Simply stated, humanity's demands exceed the planet's capacity to sustain us. The scarcity of the traditional fossil energy sources with the consequent rising energy costs have become one of the major challenges for the ICS. Therefore, as part of the ecological footprint, the energy consumption and the concomitant GHG emissions of cloud computing infrastructures (data centers and ultra-high speed interconnection networks) have imposed new constraints for ICT.

The ever-increasing data volumes to be processed, stored, and accessed every day through the modern Internet infrastructure result in the data centers' energy demand growing a faster and faster pace. For this reason, energy-oriented data centers are being investigated in order to lower their ecological footprint. However, since the electrical energy needed to power the ICT is not directly present in nature, it has to be derived from primary energy sources, i.e., from sources directly available in nature, such as oil, sun, and nuclear, that may be renewable or nonrenewable. Nonrenewable energy sources, like fossil fuels, are burned emitting large quantities of GHG in the atmosphere (usually measured in carbon dioxide equivalent,[3] CO_2e), thus contributing to global warming and pollution. Renewable sources may be also exploited to generate electricity. Most of the renewable energy sources are clean (usually referred to as *green*) as they do not emit GHG during their use (with the exception of biomasses[4]) although some other drawbacks may be still present, like visual and noise impact in the case of wind turbines or large surfaces covered in the case of solar panels. Nevertheless, from an ecological point-of-view, renewable energy sources are beneficial over their entire lifecycle [6], even if their cost and efficiency may still be low when compared to fossil-based energy sources [7], which is the main reason why their employment has to be carefully considered with a assessment of advantages and drawbacks. Besides being clean, renewable energy sources—as the name suggests—are virtually *inexhaustible*, thus they are the perfect candidate to support eco-sustainable growth. Nonetheless, renewable energy sources may be not always available; sun, wind, and tide are cyclic or even almost unpredictable phenomena, though some inertia is guaranteed by battery packs and potential energy-storing techniques.

---

[2]Resource exploitation, pollution, climate change, global warming, and global dimming all form part of the human ecological footprint.

[3]Carbon dioxide equivalent is a quantity that describes, for a given mixture and amount of greenhouse gas, the equivalent amount of CO_2 that would have the same global warming potential (GWP), when measured over a specified timescale (generally, 100 years).

[4]When burned to generate electrical energy, biomasses emit large quantity of CO_2, but the growing of the plants partially compensates the emissions. Biomasses, however, have other impacts on crops since they take away soil for agriculture.

10.2.2 Smart Grids

Both fossil-based fuels and renewable energy sources can be employed together in a dynamic adaptive fashion within the intelligent power distribution system provided by the smart grids. Smart grids are therefore emerging as a promising solution both to achieve drastic reduction in GHG emissions and to cope with the growing power requirements of ICT network infrastructures. They promise to change the traditional energy production/consumption paradigm in which one large (legacy) energy plant provides the whole region with energy, towards a configuration in which many small renewable energy plants (e.g., solar panels placed on the top of the buildings, wind turbines in the courtyards, etc.) interchange the energy with the power distribution grid, by producing their own energy and releasing the excesses to the smart grid. The smart grid then redistributes it together with the energy produced by the legacy power plants to the sites where the energy is needed or the renewable energy is not currently available. Smart grids open a new scenario in which the energy production and consumption can be closely matched, avoiding peak power production, and in which the energy quantity, quality, and cost vary as a function of the power plant producing it. Therefore, smart grids are foreseen to play a fundamental role in reducing GHG emissions and energy costs since they allows premises, data centers, storage, and computational power to be interconnected to different energy sources and possibly relocated near renewable energy plants or where the environmental conditions are favorable (e.g., a cold climate can be exploited to efficiently cool down machines).

10.2.3 Follow-the-Energy and Follow-the-Data

Therefore, an energy-aware paradigm relying on smart grid infrastructure will be able to either choose to direct the computing tasks or the data towards a site which is currently green powered (thus, in a *follow-the-energy* manner), or to request of the smart grid a quantity and quality of energy (e.g., from an available renewable energy production site) for the facility (in a *follow-the-data* manner). Such an energy-aware paradigm unveils totally new potential for the ICT that have not been explored before, not only for the cloud computing infrastructure but for the entire ICT, industrial, and transportation sectors.

In this sense, a follow-the-energy (e.g., follow-the-sun, follow-the-wind, follow-the-tide, etc.) approach and the knowledge of the current power consumption of the devices may be taken into account into an integrated approach to optimize the overall energy consumption, GHG emissions, energy costs, and performance. In the follow-the-energy approach, the preferred sites to which data or tasks are retrieved, stored, or transmitted are the ones currently powered by green renewable energy (e.g., the current energy source is solar and it is daytime). In this case, it is possible to "make light from light," in the sense that the energy coming from the solar panels can power the fiber optics and transmit the required data. This approach will be commonly used to load the facilities that are already powered by green energy sources.

In the follow-the-data approach, instead, the smart grid will be able to fulfill an energy provisioning request specifying both the quantity and quality of the required energy. The smart grid will thus reply in the same way as a typical telecommunication network, operating under the control of an engineering-capable control plane, i.e., by fulfilling the request and establishing the appropriate energy path or rejecting the request according to profitability/availability criteria. If the smart grid control plane decides to fulfill the request, an appropriate energy path has to be established between one of the energy sources available that can satisfy the request. The path will be created, like in the telecom network domain, by establishing the correct circuits in the smart energy switches distributing the power from the energy sources to the grid. This way, the energy is driven from a preferably renewable energy source towards the desired site, and the information of the current energy source will be forwarded to the site through the deployed smart meters. This approach will be commonly employed to switch the energy source powering the facility with a greener source when available.

The two approaches are not exclusive and can both be used at the same time. Anyway, in some cases one approach will be preferable to the other. For example, when it would be profitable to use some specific data center sites, the follow-the-data approach will be preferable (i.e., forward an energy provisioning request to the smart grid), whereas, if the data or the computing power can be obtained by different equally cost sites (e.g., as in the case of content distribution networks with replicated data in all the sites), the follow-the-energy approach will be preferable (i.e., routing the data request thorough the greenest sites).

10.2.4 Energy Containment Strategies

In such a dynamic and heterogeneous context, it is essential for the cloud computing infrastructure (several data centers widespread around the world) to be *aware* of the current energy source that is powering its equipment and possibly request the smart grid (energy provider) a specific power provisioning (quantity and quality of energy) in order to exploit the energy sources and lower its ecological footprint. Such information is critical to manage and operate the cloud in the greenest way and it will be a requirement in the CO_2 containment strategies that are being approved by the governments, such as cap and trade, carbon offset, carbon taxes, and green incentive frameworks [4]. In this scenario, two main approaches have been developed to reduce the carbon footprint: carbon neutrality and zero carbon. In the carbon neutrality approach, the industry's GHG emissions are partially or totally compensated (hence, *neutrality*) by a credit system (e.g., cap and trade or carbon offset); besides, incentive or tax models are also possible to encourage the use of green sources and limit industry carbon footprints. In the zero carbon approach, green renewable energy sources are employed and no GHG emissions are released at all. Zero carbon is considered to be the only long-term viable solution as it does not suffer from the rebound effect: even with increased demand no GHG are emitted at all. Thus, to achieve eco-sustainability, energy-efficiency and energy-awareness should both be exploited in a systemic energy-oriented approach leveraging smart grids employing green renewable energy sources and techniques that range from high-level policies to low-level technological improvements cooperating with each other. This is a complex task that represents one fundamental step of the challenge that humanity has to face in the 21st century: not only inverting the global warming trend but also achieving sustainable solutions for the decades to come. Here, *sustainability* represents the key word in order to successfully address all these problems.

10.2.5 ICT Energy-Efficiency Metrics

The huge energy demand originated by data centers worldwide (approximately quantified in about 1.5–2% of global electricity, growing at a rate of 12% annually [8,9]) is strongly conditioned not only by the power required by the individual runtime, storage, and networking facilities that constitute their basic building blocks, but also by cooling (referred to as HVAC, heat, ventilation, and air conditioning), uninterruptible power supply systems (UPS), lighting, and other auxiliary facilities. The *power usage effectiveness* (PUE) index, proposed together with the *data centers infrastructure efficiency* (DCiE) by the GreenGrid [10], is a universally recognized metric, used to estimate the energy-efficiency of a data center by considering the impact of the auxiliary components with respect to the basic ones. PUE is defined as the ratio between the total amount of power required by the whole infrastructure and the power directly delivered to the ICT (computing, storage, and networking) facilities:

$$\text{PUE} = \frac{\text{total data center power}}{\text{ICT equipment power}}. \tag{10.1}$$

DCiE is expressed as the percentage of the ICT equipment power by total facility power:

$$\text{DCiE} = \frac{\text{ICT equipment power}}{\text{total data center power}} \cdot 100\%. \tag{10.2}$$

A PUE value of 2 or, equivalently, a DCiE of 50%, can be typically observed by examining most of the current installations [11], demonstrating that HVAC and UPS systems approximately double the data center energy needs. Furthermore, the need for redundancy and the use of more sophisticated and expensive energy supply systems make things worse in the largest and mission critical installations. In this scenario, the cooling facilities represent the most significant *collateral energy drain*. However, their energy-efficiency is improving thanks to new cost-containing cooling strategies based on the use of computational fluid dynamics and air flow reuse concepts (free-cooling, cold aisles ducted cooling, etc.).

10.2.6 Energy and the Cloud

As for the ICT equipment, the computing and storage facilities can be considered the most energy-hungry components. As an example, the Barcelona Supercomputing Center (a medium-size data center, hosting about 10,000 processors) has the same yearly energy demand of a small (1200 houses) town, with a power absorption of 1.2 MW, resulting in an energy bill of more than 1 million Euros [12,13].

A significant variety of computing equipment populates the state-of-the-art data centers, ranging from small-sized high-density server systems (in single unit arrangements or blade enclosures) with computational capabilities limited to 2–4 multicore CPUs, to large supercomputers with hundreds of symmetric CPUs/thousands of cores or more complex parallel (e.g., systolic array or vector-based) computing architectures.

Different server architectures can be properly customized for specific computing or service provisioning tasks such as deploying network info-services (HTTP, FTP, DNS, or e-mail servers) or managing large databases. In addition, servers may assume specialized roles within a cloud computing organization by behaving as general-purpose "worker" nodes or as control devices running specialized resource brokering/scheduling systems that manage the dynamic allocation of jobs/applications or virtual machines on the available worker nodes and/or assign, with the role of storage pool managers, the required storage space to them.

Considering that, also within a fairly dimensioned farm, with a limited degree of over-provisioning to handle peak loads, most of the servers operate far below their maximum capacity most of the time [14,15], a lot of energy is usually wasted, leaving great space for potential savings with energy-proportional architectures [14] and strategies consolidating underutilized servers. The devices belonging to a server farm are, in fact, usually always powered on also when the farm is currently solicited by a very limited computational/service burden or is totally idle. This consideration can be immediately exploited by a service-demand matching approach, consolidating the current load on a minimum size subset of the available resources, and putting into sleep mode all the remaining ones, with the immediate effect of greatly reducing the energy consumption, as will be presented in Section 10.7. Such a subset can dynamically expand or shrink its dimensions by powering up or down some servers, when necessary, to provide more computing or storage capacity, or reduce the current energy consumption when the load falls under a specific threshold.

10.2.7 Energy-Saving Approaches

The energy-saving approaches currently available can be described by the three different "do less work," "slow down," and "turn off idle elements" strategies. In the first one, the applications/jobs requiring services to the cloud are optimized in time and space in order to keep the execution load at a minimum level, resulting in reduced power consumption. The second strategy starts from the consideration that the faster a process runs the more resource intensive it gets. In a complex runtime application, the speed of some component processes does not perfectly match and thus several resources remain locked for some time without being really useful. By lowering the speed of the faster activities such unnecessary waiting or resource locks can be avoided by not affecting the overall application completion time. Processes can be slowed down in two ways: they can be run with adaptive speeds, by selecting the minimal required speed to complete the process in time or, alternatively, intermediate buffering techniques can be introduced so that instead of running a process immediately upon arrival, one can collect new tasks until the buffer is full and then execute them in bulk. This allows for some runtime components to be temporarily switched off resulting in significantly lower power consumption. Finally, the last strategy refers to the opportunity of exploiting the availability of a low-power consumption status, the *sleep mode*, of the involved device. That is, the "turn off idle elements" approach aims at switching the devices into sleep mode during their inactivity periods and restoring them to their fully operational status when more power is needed. If properly used, all the strategies based on the use of sleep mode may represent a very useful means for achieving great power savings in large data centers, when they are lightly loaded. Such infrastructures are usually designed according to a modular approach, since they are built up by a number of logical-equivalent elements (bulks of servers or computing aisles), so that unloaded bulks can be dynamically put into sleep mode during low-load periods. Such approaches may be employed in orthogonal dimensions in the sense that they may and should act in concert and simultaneously, multiplying their benefits with respect to a one-dimensional optimization.

10.3 STATE OF THE ART

Several approaches have been proposed in literature in order to contain the energy demand of modern data centers.

The works presented in [8,11,16,17] focus on the use of sleep mode to attain energy-efficiency in different ways. In particular, Koomey [8] asserts that a really effective strategy to achieve significant power savings can be based on switching off most of the available servers during the night or in the presence of a limited load and using the full capacity of the servers only during peak hours. On the other hand, between the other approaches proposed, the work in [11] presents a power containment technology based on the use of sleep mode at the single component level where specific technological features can be exploited to achieve a significant degree of optimization. Another perspective is discussed in [16] analyzing the impact on network protocols for saving energy by putting network interfaces and other components to sleep. The tutorial [17] explains several network-driven power management strategies to dynamically switch computers back and forth from sleep/power-idle mode. In [18], an energy manager for data center network infrastructure is presented, dynamically adjusting the set of active network elements (interfaces and/or switches) to accommodate new capacity to support increased data center traffic loads. In [19] several ideas are presented: legacy equipment may undergo hardware upgrades (such as modified power supply modules) and their network presence may be transferred to a proxy or agent allowing the end device to be put in low-power mode during inactive periods while being virtually connected to the Internet. The authors also promote the use of renewable energy sources, such as solar, wind, or hydropower, as a valid alternative for supplying power to ICT equipment. Such a strategy seems to be well suited for data centers, which can be located near to renewable energy production sites. However, since renewable energy sources tend to be unpredictable and not always available (e.g., wind), or may present significant variations during day and night (e.g., sun), to really benefit from their usage the involved jobs/applications and the associated data, in case of supply variations, should be migrated from one data center to the other, where the energy is currently available, according to the follow-the-sun or chase-the-wind scenario [4]. This implies the presence of an energy-efficient and high capacity communication network always available to support the above facility. Based on similar concepts, a study presented in [20] investigates cost and energy-aware load distribution strategies across multiple data centers, using a high performance underlying network. The study evaluates the potential energy cost and carbon savings for data centers located in different time zones and partially powered by green energy and determined that, when striving at optimizing the overall green energy usage, green data centers can decrease CO_2 emission by 35% by leveraging on the green energy sources at only a 3% cost increase.

This chapter focuses on improving the operating energy-efficiency of the computing resources (mainly the available servers) within a complex distributed infrastructure, which are responsible for the greatest part of data centers, energy consumption.

10.4 ENERGY-AWARE DATA CENTER MODEL

A distributed service-provisioning infrastructure, such as a cloud, is composed by data centers spread throughout the world, eventually belonging to different facilities, acting together as a federated entity.

Each of these cooperating data centers contains a large number of servers whose main task is running applications, virtual machines, or processes submitted by the cloud clients, typically by using the Internet. Each individual server is characterized by a processing capacity, depending essentially on the number of CPUs/cores and on the quantity of random access memory (RAM) available and by a storage capacity dynamically assigned to it by some centralized or distributed storage management system. Thus, the data center workload at the time is measured by considering the applications/jobs that need to be processed or are still running at this time. Usually, data centers are designed with a significant degree of resource over-provisioning to always reserve some residual capacity needed to operate under peak workload conditions [14,15] and ensure a certain scalability margin over time. However, in the presence of a limited workload, the servers that are totally idle, since they have no processes to run, are normally kept turned on, thus adversely affecting the overall power consumption and introducing additional and unnecessary costs in the energy bill.

In such a context, most of the effort should be oriented to the reduction of the active/running servers to a minimal subset and turning off the idle ones, according to the previously presented "turn off idle elements" strategy. For this sake, a high-level energy-aware control logic is needed to dynamically control the data center power distribution by exploiting load fluctuations and turning off inactive servers to save energy. Thus, as the current load decreases under a specific attention threshold, a properly designed policy should clearly identify (1) *how many* servers and (2) *which* of them have to be powered down, and the correct procedures to perform this task have to be implemented, ensuring that the physical and logical dependencies among the devices within the data center will be always respected (as described in Section 10.4.1).

In particular, such an operating policy has to be implemented by using a proactive algorithm, running within the data center resource broker/scheduling logic, that constantly monitors the current runtime load and adaptively determines the subset of servers that can be turned on or off, by using properly crafted actuator functions with the specific task of correctly switching the operating status between on and off and vice versa.

10.4.1 Physical and Logical Dependencies

In modern data center farms participating in grid or cloud-based distributed computing infrastructures, a large number of devices are interconnected together into cooperating clusters, which are made up of heterogeneous computing, storage or communication nodes, each with its own role in the farm. Each of these nodes, often known as a computing resource broker or control element (CE), storage manager/element (SE), disk server (DS), gateway, router, or whatever else, has its own hardware and software features that must be considered when operating in the data center. Furthermore, nodes interact among them according to their logical role in the farm and, secondary but not least, to their physical placing, as depicted in Figure 10.2.

Such logical and physical dependencies must be evaluated, especially in power management operations where devices are switched on and off.

An example of a *power-on* procedure executed on the node SE_1 is illustrated in Figure 10.3, in which the highlighted nodes are turned on, starting from the node UPS_1 and going down to nodes $RACK_1$, $FARM_1$, and eventually SE_1.

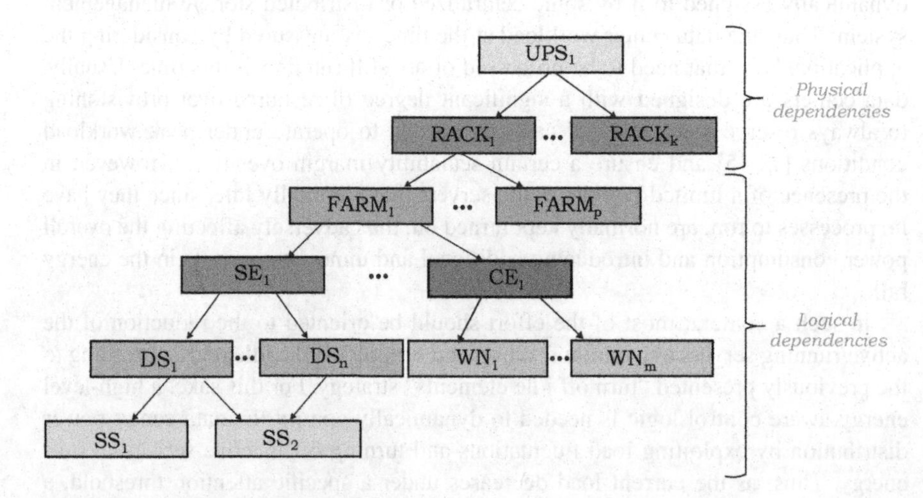

Figure 10.2 Dependency graph for a grid farm.

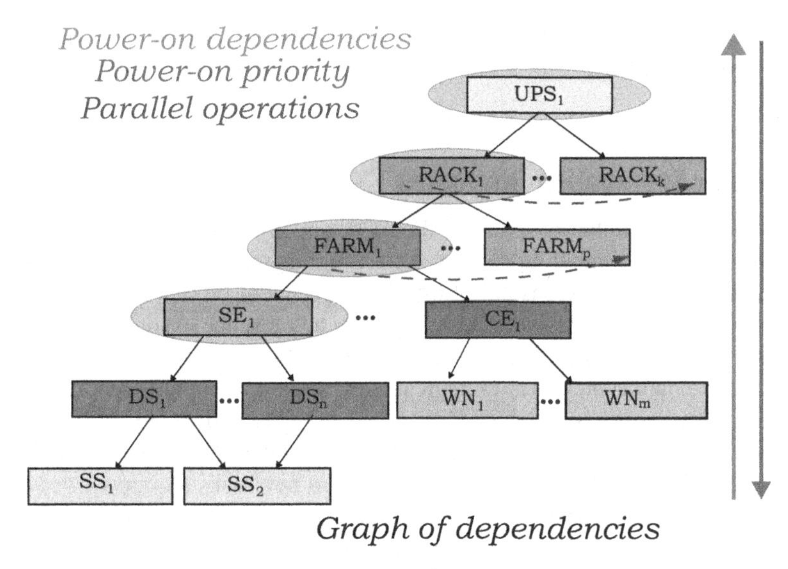

Figure 10.3 Power-on procedure executed on device SE_1**.**

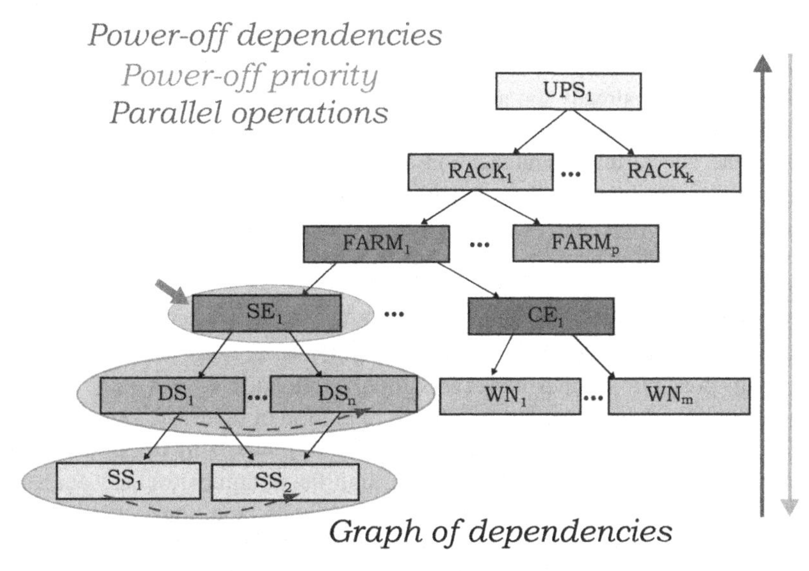

Figure 10.4 Power-off procedure executed on device SE_1**.**

Analogously, a *power-off* procedure executed on the node SE_1 is depicted in Figure 10.4, in which the highlighted nodes are turned off, starting from the nodes SS_1 and SS_2 and going up to the nodes $\mathrm{DS}_1, \ldots, \mathrm{DS}_n$ and eventually to the node SE_1. In this way, the portion of the cluster related to the storage subsystem will be completely turned off while the remaining part of the cluster will be still working.

10.4.2 Job Aggregation Strategies

The above model implicitly assumes that each job/process or virtual machine is assigned to a single computing core, in such a way that each multiprocessor server, with n total cores available, is able to run n independent tasks without experiencing any performance slowdown. Accordingly, in the presence of multicore systems, a runtime task consolidation strategy can be applied, by moving the current tasks throughout the data center farm to aggregate them on a minimal subset of servers, in order to allow a greater number of idle servers to be turned off. In the presence of multicore CPUs with n cores per server, several consolidation strategies are possible. Such strategies can also be progressively applied as new tasks require service to the data center, to avoid expensive recompaction procedures and complex job migration processes affecting already running applications. For example, in a data center using a traditional *first-fit* scheduling strategy, a new task is assigned to the first server, among the available ones, with at least a single core available. As a valid alternative, a *best-fit* strategy aims at compacting the available tasks as much as possible; thus each task is assigned to a server with just one core free (and, thus, $n - 1$ already busy) if any such server exists. Otherwise, it looks for a server with only two free cores, then with three, and so on, up to n, according to the principle of distributing the load on the servers that are already the most loaded ones in an effort to totally saturate their available capacity. Note that, in this aspect, the optimization of data centers differs from the one of telecommunication networks, in which the load balancing criteria has to try to *not* saturate the available resources (e.g., fiber links) in order to leave enough "space" (e.g., bandwidth) for future connection requests to come.

Clearly, the first-fit scheme is faster, but it has the disadvantage of leaving a large number of servers only partially loaded. On the other hand, best-fit gives the more satisfactory results according to the aforementioned consolidation strategy, since it compacts the jobs as much as possible on a few servers and leaves the maximum possible number of servers totally unloaded so that they can be immediately powered down, with a significant reduction of the wasted energy (see Figure 10.5). Besides achieving optimal compaction, the best-fit strategy is also more profitable since a multicore server with a significant number of busy cores is statistically less likely to get free of all its runtime duties (and, thus, of being put into sleep mode) than a server with a low number of jobs. The inherent computational complexity of the best-fit strategy may be improved to work in a constant amortized time by implementing per server priority queues with Fibonacci heaps, so that it will not introduce additional burden to the overall computing facility. Unfortunately, in the presence of single-core devices, no further aggregation is possible and hence energy savings can be achieved only by powering down the idle servers.

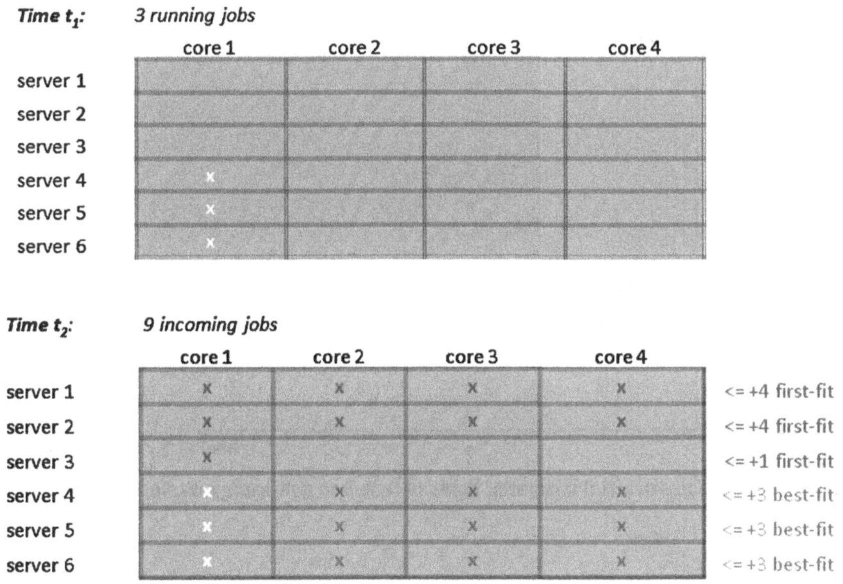

Time t_1: *3 running jobs*

	core 1	core 2	core 3	core 4
server 1				
server 2				
server 3				
server 4	x			
server 5	x			
server 6	x			

Time t_2: *9 incoming jobs*

	core 1	core 2	core 3	core 4	
server 1	x	x	x	x	<= +4 first-fit
server 2	x	x	x	x	<= +4 first-fit
server 3	x				<= +1 first-fit
server 4	x	x	x	x	<= +3 best-fit
server 5	x	x	x	x	<= +3 best-fit
server 6	x	x	x	x	<= +3 best-fit

Figure 10.5 Visualization of first-fit and best-fit allocation strategies for a subset of six servers with four cores each. At time t_1, there are three running jobs arranged (as a result of previous allocations) as in (a); at time t_2, nine new jobs arrive (b). First-fit will use up to 2× more servers than best-fit.

10.5 TRAFFIC FLUCTUATION

The data center workload is usually characterized by recurring fluctuation phenomena where higher utilization periods (e.g., during some hours of the day) are followed by lower utilization ones (e.g., during the night) and so on. Due to the regularity of these recurrence phenomena, driven by the 24 h or weekly rhythm of human activities, the aforementioned fluctuations are typically predictable within certain fixed time periods (e.g., day/night or working day/weekend cycles, months of the year, etc.) and they can be described by a pseudo-sinusoidal trend [18,21]. The theoretical daily workload variation for a typical energy-unaware data center [18] is shown in Figure 10.6.

It can be immediately observed that, while the demand load follows a pseudo-sinusoidal trend, the power drained remains almost constant during both the high and low usage periods. This is essentially due to the impact of computing resources that are always kept up and running when they are underutilized or not utilized at all, thus wasting a large amount of energy during the low-load periods. The fundamental idea behind more energy-conscious data center resource management is introducing *elasticity* in computing power provisioning under the effect of a variable demand, by adaptively changing the data center capacity so to follow the current demand/load, as shown in Figure 10.7. This can be accomplished by dynamically managing, at the

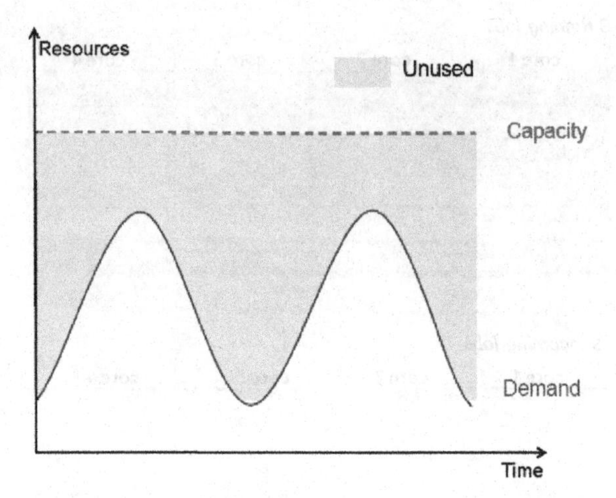

Figure 10.6 Capacity-demand mismatch leads to resource and energy wastes.

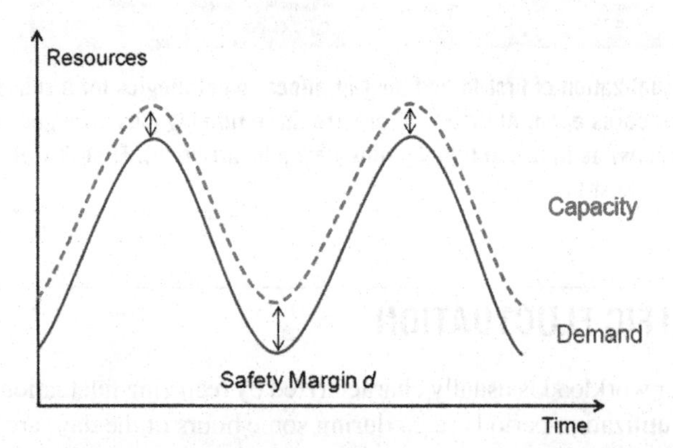

Figure 10.7 Theoretical provisioning elasticity concept. The capacity curve should resemble the demand curve as close as possible, leaving a safety margin to serve possible peak loads.

resource scheduling level, the allocation of tasks to the available servers, and putting the unused ones in low-power sleep mode and rapidly resuming a specific block of sleeping servers when more capacity is required and the currently operating server pool is not able to accommodate it.

10.6 ENERGY-ORIENTED OPTIMIZATION

Since a distributed computing infrastructure, such as a cloud, can be seen both as a single federated entity but also as a set of autonomous data centers working together, energy-oriented optimization can be introduced at different levels. Energy-efficiency

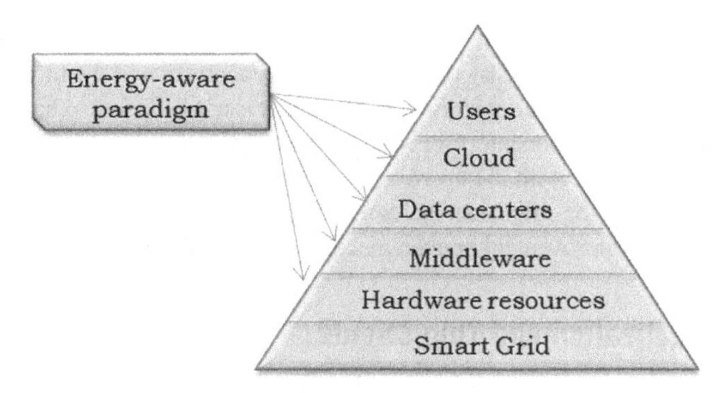

Figure 10.8 Energy-awareness multilevel approach overview.

is the first step towards reducing the ecological footprint of cloud infrastructures. It is related to the "do more for less" paradigm, and aims at using more efficient computing or manufacturing techniques that lower the components, power requirements. At a higher level of abstraction, it is possible to introduce the energy-awareness at all levels, meaning that the device (from the hardware resources to individual servers up to the data center and the whole grid/cloud) is *aware* of its power requirements, and adapts its behavior to its current load and even to the current energy source that is feeding it with energy (as sketched in Figure 10.8). Energy-awareness can be applied levelwise, from a local data center basis (*intrasite optimization*) to a wider cloud scope (*intersite optimization*).

10.6.1 Energy-Efficiency

From the lowest granularity, energy-efficiency can be implemented at the devices level, designing and building more energy-efficient components that decrease the power consumption while not disrupting the offered service level, or even increasing it. As an example of technological innovation and energy-efficiency, consider the ASCI Red supercomputer of Sandia National Laboratories (in Albuquerque, New Mexico, USA), built in 1997. With its peak performance of 1.8 teraflops, it was the most powerful supercomputer in the world (according to TOP500) between June 1997 and June 2000, and occupied a physical size of 150 m^2, with a power absorption of 800,000 W. Just 9 years later, in 2006, the Sony Playstation 3 achieved the same peak performance with a power absorption lower than 200 W and a physical size of 0.08 m^2[13].

As another example, consider the Apple iMac in the year 2000: it had the dimensions of a CRT monitor, a weight of 15.8 kg and its technical specifications were 500 MHz CPU, 128 MB of RAM, and 30 GB of storage, with a power consumption of about 150 W. In 2010, 10 years after, Apple released the iPhone 4, which, despite its handy-size and 137 g of weight, had a 1 GHz CPU, 512 MB of RAM, and 32 GB of storage capacity, with a power absorption of just 2 W. Furthermore, it also added advanced functionalities, like GPS, Wi-Fi, compass, mobile phone, etc., in line also with the"do more for less" paradigm.

Looking for energy-efficiency in the design of the new hardware has more or less been always present in the industry processes, and it remains one of the most important issues in reducing energy requirements. Nonetheless, it is not sufficient to have more and more energy-efficient components, for two main reasons. From one side, the energy-efficiency alone cannot cope with the ever-increasing needs of energy and the consequent carbon footprint. From the other side, increased efficiency can lead to the rebound effect, which can limit any improvement or even make the ecological footprint worse.

10.6.2 Virtualization and Thin Clients

From the user point-of-view, the ICT market is evolving towards a new network-centric model, where the traditional energy-hungry end-user devices (e.g., PCs, workstations, etc.) are being progressively replaced by mobile lightweight clients characterized by moderate processing capabilities, limited energy requirements, and high-speed ubiquitous network connectivity (e.g., smart phones, handhelds, netbooks, etc.), significantly incrementing the strategic role of the network for connecting them to the cloud infrastructures. These devices typically do not directly run complex and expensive applications but rely on *virtual machines* and remote storage resources residing on distributed cloud organizations, accessible from the Internet, to accomplish their more challenging computing tasks, thus limiting their roles to very flexible and high-level interfaces to the cloud services. Such an evolution in users' habits, if properly managed, can be exploited to significantly reduce the power consumption of both data centers and user-level computing equipment.

Modern virtualization [16, 19] technologies may play a fundamental role by dynamically moving, in a seamless way, entire virtual machines and hence their associated computations, between different data centers across the network, in such a way that the data centers placed near renewable energy plants may execute the involved computational demands with a lower energy cost and a reduced carbon footprint with respect to traditionally power consuming (and almost idle) PCs statically located in the users' premises. Increasing the computing/runtime density and power in sites where green and/or less expensive energy is available will be the upcoming challenge for next generation data centers.

Virtualization can be applied on two levels. First, multiple virtual machine instances can be shared on a single physical server, by reducing the number of servers and thus the power consumption. Second scalable systems using multiple physical servers can be built, allowing most servers to be switched off during low usage periods, and only using the full capacity of the computing farm during peak hours.

Resource sharing may also play an important role both for data centers and network equipment. In data centers, virtualization may be exploited by an energy-aware middleware that dynamically moves individual tasks or virtual machines to the most energy-convenient high-density sites (those characterized by the lowest carbon footprint or energy costs) in order to increase, as much as possible, their computational burden. From the networking perspective, an energy-aware control plane may properly route the traffic/connections associated to the above task/VMs by privileging intermediate nodes currently fed with renewable energy while simultaneously grouping the above connections on the same path instead of spreading them over the whole network. This management practice will maximize the usage of the already active devices/paths and consequently save the energy resulting from temporarily powering down the networking devices that should serve the alternate/secondary and no longer useful paths.

10.7 INTRASITE OPTIMIZATION

In principle, the instantaneously available capacity in a data center should closely follow the current load (Figure 10.7). However, the instantaneous capacity function cannot be described by a *continuous* curve, but is instead a step function in which each step is associated to a variation in the quantity of *discrete* resources (e.g., single servers or blocks of servers in the farm) being turned on or off. Hence, the demand curve must be approximated with a step-shaped service curve that is always able to support the current demand/load while minimizing the overall energy consumption (Figure 10.9). Since the power drained is directly proportional to the number of active servers, the closer the service curve approximates the demand one, the lower will be the power wasted.

Furthermore, the instantaneously available capacity should be dimensioned by ensuring the presence of a safety margin (i.e., a minimal distance d to be always maintained between the values assumed by the demand and the capacity curves, expressed as number of servers or cores) to properly handle bursty traffic phenomena or peak loads. This margin is directly associated to a certain number of servers that are preventively kept turned on to erve new incoming tasks whose number cannot be easily foreseen by using statistical observations. Reducing the distance margin d directly implies decreasing the energy consumption, but also introduces, as

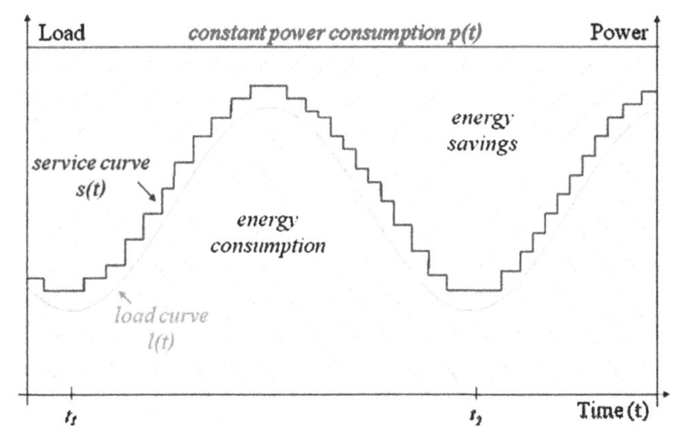

Figure 10.9 Service-demand matching.

Server type	*Power on (hardware)*	*Power off (software)*	*Power off (hardware)*
Computing Element (CE)	120	20	5
Storage Element (SE)	180	10	5
Home Location Register (HLR)	120	60	5
Pizza Box Form Factor Servers	120	10	5
Blade Servers (Dell® DRAC)	160	45	45
Storage Server (IBM® DS400 Storage System)	60	10	10

Figure 10.10 Complete power ON/FF times (s) for different legacy devices.

a side effect, an increment in the average service delay for incoming tasks since the number of new jobs that can be immediately served without reactivating previously sleeping devices decreases proportionally with the safety margin. Conversely, larger values for the distance *d*, introduce more tolerance to extemporaneous load variations, and thus a greater number of tasks can be immediately served as they arrive (with no delay), but, obviously, the energy consumption increases proportionally (since more servers have to be kept up and running). Thus, the safety margin *d* has to be large enough to avoid unnecessary oscillations of a certain group of servers back and forth from the sleep-mode, in correspondence to limited variations of the load. This is also required to avoid, as possible, the peaks in the power absorption that are usually experienced during the start-up phase or when switching from the sleep-mode status to the fully operating one. Therefore, the safety margin *d* can be considered as *upper-bounded* by the energy consumption and *lower-bounded* by the peak load absorption capacity together with the minimization of the aforementioned power fluctuation requirement.

Choosing a safety margin value of *d*, ensures that a bulk of $k \leqslant d$ incoming tasks can be immediately served without waiting. Thus, the *d* parameter also identifies the size of the zero-waiting queue of the tasks that are served as they arrive. If $k > d$, there will be $k - d$ jobs that will have to wait a time before they can get served, where is the start-up time of the servers (obviously, if the load reaches the site maximum capacity, all the new tasks will have to wait for new resources to become available). The start-up time may sensibly vary with the available technologies. In the presence of *agile* servers equipped with enhanced sleep-mode capabilities, the value of may range in the order of a few ms, while for less sophisticated legacy equipment a complete bootstrap procedure will be required and the start-up time may grow up to some minutes. As a general rule, the higher is the value, the higher has to be the safety margin *d*, and consequently the lower will be the energy saving, whereas lower values of *t*, allow significantly greater energy-savings margins. In Figure 10.10 we reported the (software and hardware) power off and power on times measured in the

INFN[5] Tier2 site of the CERN[6] LHC[7] experiment. When enhanced sleep-mode is not employed, servers need several tens or even hundreds of seconds to switch their state from turned down to up and running, under the control of specific *wake-on-LAN* or external power management facilities (e.g., "intelligent" power distribution units, PDU). Such high times clearly indicate that the enhanced sleep-mode feature is required to make modern data centers agile and may bring great benefits in terms of reduced energy waste and consequent electrical bills.

10.7.1 Analytically Evaluating the Energy-Saving Potential

In order to evaluate, from the analytical point-of-view, the upper-bound for the energy-saving potential of the discussed service-demand elasticity approach, we can consider instantaneous transitions among the sleep and the active states ($t = 0$) and theoretical sinusoidal traffic, like the one depicted in Figure 6. The demand curve represents the request for service load experienced during the day, while the service curve represents the servers that need to be fully operating to process the job requests. Without the introduction of any energy-saving policy, the power consumption of the data center remains constant [18] over the entire observation interval, and the energy required is represented by the integral of the power drained over time:

$$\int_{t_1}^{t_2} p(t)\mathrm{d}t, \tag{10.3}$$

where $p(t)$ is the power consumption function and t_1 and t_2 are considered the extremes of the observation time interval. Ideally, the lower bound for the data center energy consumption is defined as

$$\int_{t_1}^{t_2} l(t)\mathrm{d}t, \tag{10.4}$$

where the $l(t)$ function describes the demand/load curve. Such a curve can be closely approximated by the adaptive service curve $s(t)$, which is the step function that establishes the minimum set of runtime resources that have to be powered on to serve the current demand. Therefore, with the proposed energy-saving schema, the theoretical energy consumption is defined as

$$\int_{t_1}^{t_2} s(t)\mathrm{d}t. \tag{10.5}$$

[5]Italian National Institute for Nuclear Physics.

[6]European Organization for Nuclear Research.

[7]Large Hadron Collider.

Clearly, between the above equations it holds that (10.4) < (10.5) << (10.3), and the bigger the difference between the values assumed by Eqs. (10.3) and (10.5), the greater the energy saving. Theoretically, the energy saving is upper-bounded by

$$\int_{t_1}^{t_2} (p(t) - l(t))\, dt, \tag{10.6}$$

while the actual energy saving is defined as

$$\sum_{i=1}^{n} (p(i) - s(i)) \cdot \Delta_i, \tag{10.7}$$

where n is the number of intervals in which the time interval $[t_1, t_2]$ is partitioned and Δ_i is the duration of the ith time interval; note that the value n defines the time basis on which the optimization process is executed; thus Eq. (10.7) represents the potential energy savings that can be achieved.

10.7.2 The Service-Demand Matching Algorithm

Given a specific demand/load curve, the service-demand matching algorithm determines the service curve that is able to always satisfy the current demand while minimizing the number of active servers and hence the power needed. As an example, consider a scenario in which the demand curve increases between the times t_i and t_{i+1}. As a consequence, the distance from the service curve decreases from d_i to d_{i+1}. Since $d_{i+1} < d$, the algorithm detects the increment in the demand (totally absorbed by the safety margin d, thus no service delay occurs in this case) and consequently increases the number of active servers by turning on $s_{i+1} - s_i$ servers.

In the opposite situation, when a decrement $d_{i+1} > d$ of the demand curve occurs, it causes the algorithm to decrease the service curve from s_{i+1} to s_i, thus allowing more devices to be put into sleep mode.

10.7.3 Experimental Evaluation

To evaluate the effectiveness of the service-demand matching algorithm (best-fit aggregation with sleep mode during low-load periods) in terms of energy-saving potential, a large data center simulation model, composed of more than 5000 heterogeneous single-core and multicore servers, has been built starting from data available in literature [14, 15] concerning the observation over a six-month period of a large number of Google servers.

The servers' power consumption was modeled as depicted in Figure 10.11, in which the server always consumes a fixed power needed for the device to stay on, and a load-dependent variable power, equal at maximum to the fixed part, scaling linearly with the CPU load [15].

The jobs arrive with a pseudo-sinusoidal trend as illustrated in Figure 10.6. The duration of the jobs was taken to be exponentially distributed [22], extrapolated from a duration of 1 to a maximum of 24 h, according to the distribution reported in Figure 10.12.

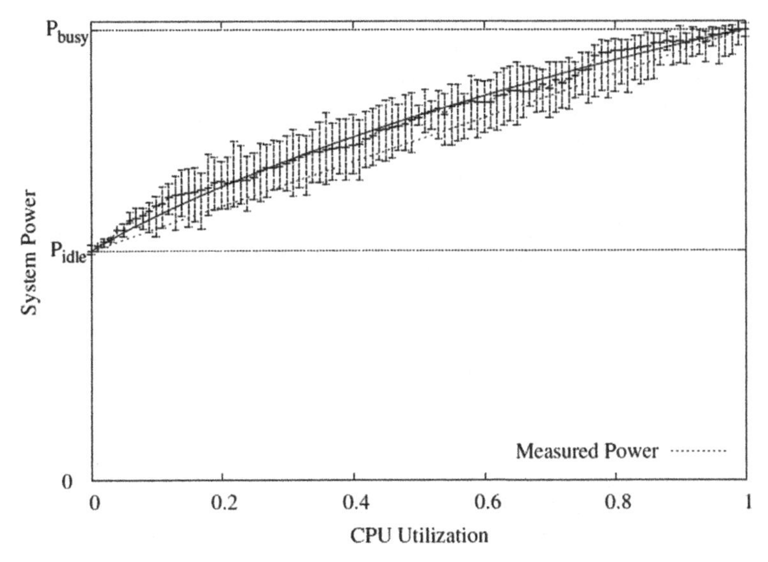

Figure 10.11 Server energy model: the power consumption varies linearly with the CPU load.

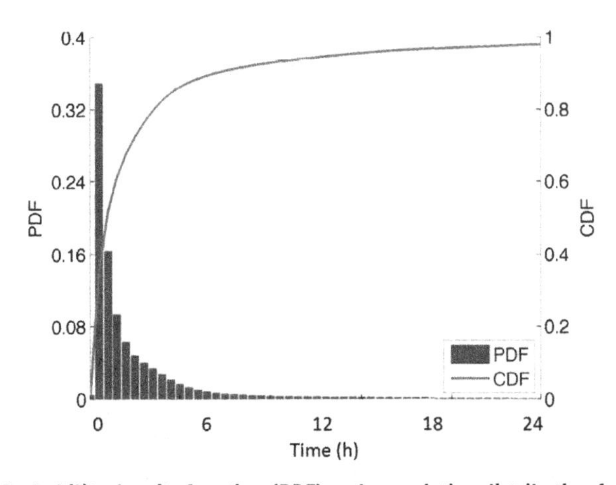

Figure 10.12 Probability density function (PDF) and cumulative distribution function (CDF) of job duration.

First, we evaluated the effectiveness of the approach considering nonzero transition times (t) between the on and the sleep-mode states (taking as a reference the value of Figure 10.10). Day 1 of the simulation is depicted in Figure 10.13. At the beginning, no job is present in the data center. As time passes, we monitor the energy consumption with all devices on (i.e., without our approach, red area) with the energy consumption of the approach (blue area), and plot the obtained energy savings (green

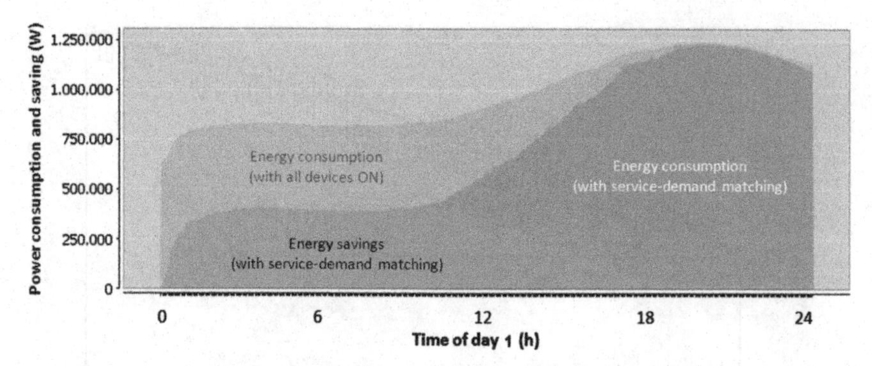

Figure 10.13 Energy consumption of day 1 with and without the service-demand matching algorithm.

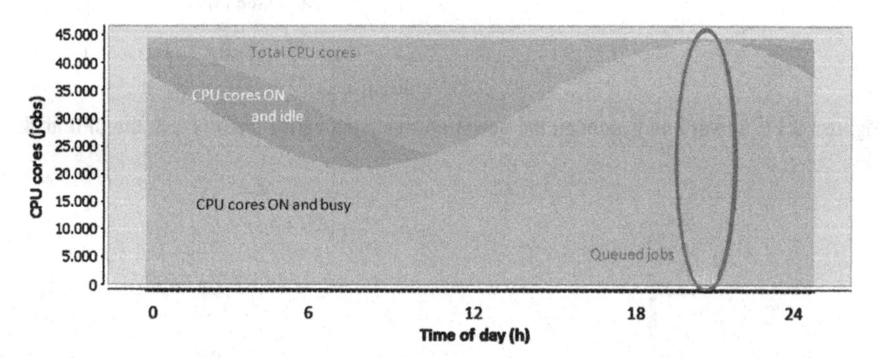

Figure 10.14 Energy consumption of day *n* with and without the service-demand matching algorithm.

area). Note that, when the servers are turned on but idle (during the first 12 h approximately), they consume about half of their maximum power consumption, which is in turn achieved when the CPU works at 100%, according to the energy model of Figure 10.11.

In Figure 10.14 we reported the generic day n of simulation. We can observe that a number of CPUs (set by the safety margin d) is kept on even if idle to serve future jobs that may arrive at the data center. However, a peak of traffic can still cause a scenario where a number of jobs will have to wait, as observed in Figure 10.15.

Therefore, we evaluated the impact of variations in the safety margin d against the potential savings, in terms of reduced energy consumption (MWh), GHG emissions (tons of CO_2) and economical cost (Euros).

For a commercial/industrial facility like a data center, we assumed that the average cost of energy is about 0.12 Euros/kWh [23], and we considered fossil-fueled energy plants powering the data centers, which emit 890 g of CO_2 per kWh [9].

Several simulation experiments have been run with different values of the safety margin d. The observed results show that the maximum achievable cost savings may

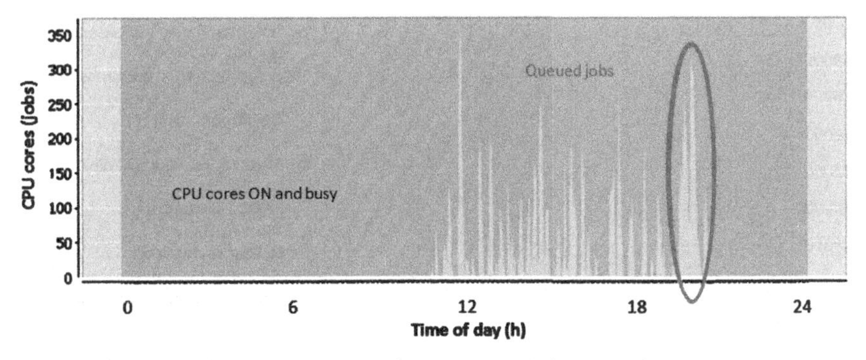

Figure 10.15 Queued jobs that have to wait due to a peak in the traffic load.

reach about 1.5 millions Euros, with a reduction of more than 13 GWh in the energy consumption and more than 11 ktons of CO_2 in GHG emissions. These results are not surprising since the servers are rarely utilized at their maximum capacity and operate, for most of their time, at an average load ranging between 10% and 50% of their maximum utilization levels [14].

As expected, the *d* value significantly affects the overall energy savings along with the consequent CO_2 emissions and electric bill costs. The best results have been achieved with lower values of the safety margin. When evaluating the impact of the safety margin, since the basic goal of the experiments was to provide a lower bound for the energy savings of modern and future data centers, the transition time between the powered on and off states has been set to 0, so that switching the servers between the sleep and operating mode introduces no delay (i.e., all the considered servers are agile). As a consequence, the frequency of the load variations (i.e., how and how often the traffic load varies in time) only affects the number of transitions between the on and off (or sleeping and operating) states, but it does not influence the energy savings at all, as each variation is immediately followed by the corresponding power mode switching action on the involved servers.

The efficiency that can be achieved in resource utilization may reach values between 20% and 68%, meaning that a high percentage of the servers can be put into sleep mode for a considerable time. The saving margins decrease almost linearly as the *d* values increase (see Figure 10.16). In fact, while the load is far from the actual data center capacity, both the achieved energy savings and the safety margin *d* vary linearly but, as the load approaches higher values, the *d* threshold will exclude a higher number of devices from being switched down, leading to relatively lower savings. When considering multicore devices, job aggregation is possible. When comparing the two aggregation/consolidation strategies based on first-fit and best-fit scheduling of new tasks in Section 10.4, we observed that first-fit performed significantly worse than best-fit (up to 50%), so that in evaluating the consolidation strategy we only focused on the best-fit task scheduling scheme.

Finally, varying the number of cores per servers shows a common behavior: the more cores available in the data center, the higher the energy consumption. This is due to the

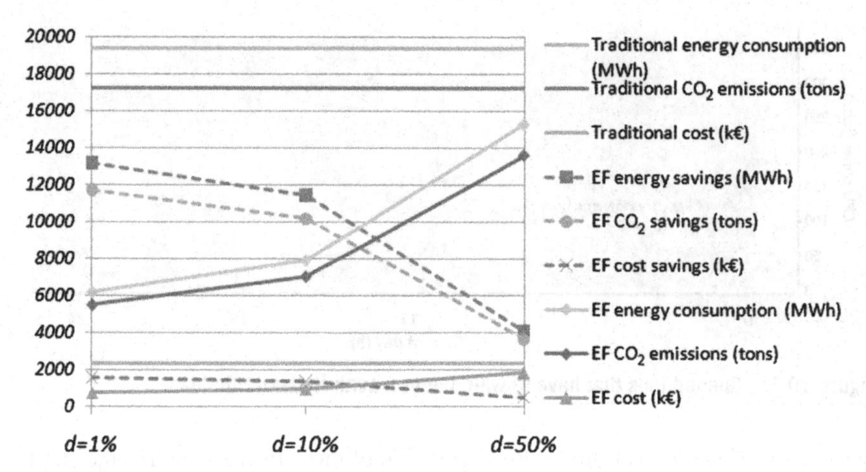

Figure 10.16 Energy, CO_2 and costs with varying safety margins (d) for a large data center (5000 servers).

fact that the data centers usually operate very far from their actual maximum utilization capacity and this causes multicore servers to run with only a few jobs even when using the best-fit scheduling strategy. That is, multicore servers are more affected by internal fragmentation phenomena (not all the cores are always busy). Thus, in the presence of a limited load, assigning a task to a single-core server costs less than executing it on a multicore server (due to the greater energy consumption of the latter device), while, at higher loads, the greater computing density of multicore servers may be exploited by the best-fit strategy to lower the overall data center energy consumption.

10.8 INTERSITE OPTIMIZATION

When dealing with intersite optimization, new opportunities are added to the discussed intrasite optimization. With the cloud distributed over the territory, even with large distances (two sites of the same confederation may be in different continents), new perspectives are possible for reducing energy consumption, GHG emissions, and energy costs, especially in conjunction with the discussed smart grid infrastructure. In this sense, a *follow-the-energy* approach [4] and the knowledge of the *current* electricity prices in the different regions on which the cloud infrastructure spans may be taken into account and exploited by an energy-aware task scheduling paradigm to optimize the overall energy consumption (see Figure 10.17) and/or the use of renewable energy sources. This requires the presence of a high performance communication infrastructure, joining the cloud sites and supporting the capability of dynamically moving the cloud tasks from one site to another one with minimum latency, according to a fully distributed scheduling paradigm.

For example, since electricity prices may present significant geographic and temporal variations, due to differences in the actual regional demand or in the use of a

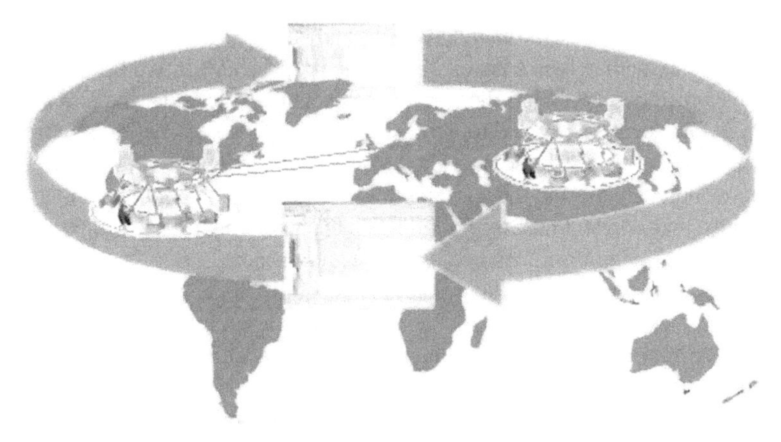

Figure 10.17 Follow-the-energy paradigm and intersite optimization schema in the cloud infrastructure.

cheaper energy source, by working on the energy cost dimension, significant reductions can be achieved by adaptively moving and rescheduling the running tasks from a cloud site to other locations, offering runtime services to the cloud, where the electricity prices and associated taxes are the lowest during particular hours of the day.

Energy-oriented scheduling criteria may also be introduced in multisite clouds by optimizing the choice of energy sources in such a way that runtime resources located within sites powered by renewable sources are always preferred when available. In fact, data centers may be equipped with a dual power supply: the always-available power coming from dirty energy sources and the not-always-available power coming from renewable energy sources. Consider, for instance, the availability of energy produced by solar panels; it is strongly correlated with the time of the day, since it is known that no energy will be produced during the night and that some energy is expected to be produced during the day. Such knowledge should be included in the distributed scheduling logic, by implementing automatic follow-the-sun or chase-the-wind paradigms. Accordingly, the aforementioned scheduling system should have the additional goal of distributing the available power to the incoming new tasks or to the already running ones to dynamically follow the 24-h daylight cycle. Thus, when the sun is shining in a specific geographic area, all the incoming new jobs should be scheduled on the sites located in areas that are entirely powered by solar energy. Analogously, the tasks already running in sites that are powered by dirty energy sources should be eventually forced to pass through the underlying network infrastructure to be relocated on sits powered by the solar energy. Of course, when the daylight is no longer available, the above tasks should be dynamically moved back elsewhere, e.g., where the sun is now shining, in such a way that the use of green energy sources is maximized and the carbon footprint minimized.

As another example, we can imagine some sites powered by wind energy where power supply is a pseudorandom process depending on the availability of wind. Due to the inertia of the power generating mechanisms and batteries, a drop in the wind power does not result immediately in a power generation drop. Hence, if wind stops, it is possible to dynamically reconfigure the load over the cloud, to consider the new distribution of available clean energy and reoptimize its carbon footprint. Differently from the case of the daylight, whose duration is known in advance, a decrease in wind strength is much more unpredictable and the warning time is shorter. This should be only handled with adaptive and efficient task migration mechanisms implemented within the cloud middleware. For this reason, it is necessary to develop novel migrating schemes and resource allocation mechanisms that take advantage of the early notification of the forecast power variation of clean sources with time-varying power output. Furthermore, another interesting perspective in energy-aware job allocation comes from linking job scheduling/dispatching to the different available electricity prices, dynamically and continuously moving data to areas/devices where electricity costs are lower.

10.9 CONCLUSIONS

In order to support all the above adaptive behaviors, energy-related information associated to the data centers belonging to the cloud need to be introduced as new constraints (in addition to the traditional ones, e.g., computing and storage resources details) in the formulations of dynamic job allocation algorithms. Down-clocking or sleep mode should be handled as new capabilities of the data center equipment that need to be considered at the cloud traffic engineering layer, and the associated information must be conveyed to the various devices within the same energy-management domain. This clearly requires modifications to the current protocols and middleware architecture. However, in many cases, the carbon footprint improvements will be achieved at the expense of the overall performance (e.g., survivability, level of service, stability, etc.), which can in turn be compensated through over-designing (increase of CAPEX) or over-provisioning (increase of OPEX). In fact, by putting energy-hungry equipment or components into low-power mode, or creating traffic diversions driven by reasons different from the traditional load balancing ones, we implicitly reduce the available capacity of the cloud and hence the experienced delays tend to be longer and/or data centers more congested, decreasing the overall service quality. This implies that the new algorithms empowering the energy-aware middleware should be driven by smart heuristics that always take into account the trade-off between performance and energy savings.

REFERENCES

[1] J. Torres, Empreses en el nuvol, Libros de cabecera, 2011, ISBN: 978-84-938303-9-7.
[2] United Nations General Assembly Report of the World Commission on Environment and Development: Our Common Future. Transmitted to the General Assembly as an Annex to Document A/42/427 – Development and International Co-operation, 1987.
[3] SMART 2020: Enabling the Low Carbon Economy in the Information Age, The Climate Group, 2008.

[4] B. St. Arnaud, ICT and global warming: opportunities for innovation and economic growth (online). <http://docs.google.com/Doc?id=dgbgjrct_2767dxpbdvcf>.
[5] Living Planet Report 2010, The biennial report, WWF, Global Footprint Network, Zoological Society of London, 2010.
[6] C.J. Koroneos, Y. Koroneos, Renewable energy systems: the environmental impact approach, International Journal of Global Energy Issues 27 (4) (2007) 425–441.
[7] J.O. Blackburn, S. Cunningham, Solar and nuclear costs – the historic crossover, NC WARN: Waste Awareness and Reduction Network, July 2010. <www.ncwarn.org>.
[8] J. Koomey, Estimating total power consumption by servers in the US and the world (online). <http://enterprise.amd.com/Downloads/svrpwrusecompletefinal.pdf>.
[9] BONE project, WP 21 Topical Project Green Optical Networks: Report on year 1 and updated plan for activities, NoE, FP7-ICT-2007-1 216863 BONE project, December 2009.
[10] The Green Grid, The Green Grid Data Center Power Efficiency Metrics: PUE and DCiE, Technical Committee White Paper, 2008.
[11] W. Vereecken, W. Van Heddeghem, D. Colle, M. Pickavet, P. Demeester, Overall ICT footprint and green communication technologies, in: Proceedings of ISCCSP 2010, Limassol, Cyprus, March 2010.
[12] J. Torres, Green computing: the next wave in computing, Ed. UPCommons, Technical University of Catalonia (UPC), February 2010.
[13] P. Kogge, The tops in flops, IEEE Spectrum (2011) 49–54.
[14] L.A. Barroso, U. Hlzle, The case for energy-proportional computing, IEEE Computer 40 (2007) 33–37.
[15] X. Fan, W. Dietrich Weber, L.A. Barroso, Power provisioning for a warehouse-sized computer, in: Proceedings of ISCA, 2007.
[16] M. Gupta, S. Singh, Greening of the Internet, in: Proceedings of the ACM SIGCOMM, Karlsruhe, Germany, 2003.
[17] K. Christensen, B. Nordman, Reducing the energy consumption of networked devices, in: IEEE 802.3 Tutorial, 2005.
[18] B. Heller, S. Seetharaman, P. Mahadevan, Y. Yiakoumis, P. Sharma, S. Banerjee, N. McKeown, Elastictree: saving energy in data center networks, Proceedings of the Seventh USENIX Symposium on Networked System Design and Implementation (NSDI), San Jose, California, USA, ACM., pp. 249–264.
[19] W. Van Heddeghem, W. Vereecken, M. Pickavet, P. Demeester, Energy in ICT – trends and research directions, in: Proceedings of IEEE ANTS 2009, New Delhi, India, December 2010.
[20] K. Ley, R. Bianchiniy, M. Martonosiz, T.D. Nguyen, Cost- and energy-aware load distribution across data centers, in: SOSP Workshop on Power Aware Computing and Systems (HotPower'09), Big Sky Montana, USA, 2009.
[21] M. Armbrust, A. Fox, R. Griffith, A. Joseph, R. Katz, A. Konwinski, G. Lee, D. Patterson, A. Rabkin, I. Stoica, M. Zaharia, Above the clouds: a Berkeley view of cloud computing, Technical report No. UCB/EECS-2009-28, University of California at Berkley, USA, February 2009.
[22] D. Meisner, T.F. Wenisch, Stochastic queuing simulation for data center workloads, in: EXERT: Exascale Evaluation and Research Techniques, workshop held in conjunction with ASPLOS, March 2010.
[23] U.S. Energy Information Administration, State electricity profiles (online). <http://www.eia.gov/electricity/state/>.

CHAPTER

11 Energy-Efficient Cloud Computing: A Green Migration of Traditional IT

Hussein T. Mouftah, Burak Kantarci

School of Electrical Engineering and Computer Science,
University of Ottawa, Ottawa, ON, Canada

11.1 INTRODUCTION

Over the last decade, information technology (IT) has introduced tremendous advances for the purpose of high performance data processing, data storage, and high speed wired/wireless communications in the Internet. Due to the limitations of local resources and power-efficiency concerns, distributed systems have emerged as feasible solutions each of which aims at a specific target. For instance, IBM's *autonomic computing* aims at building self-configuring, self-healing, self-optimizing, and self-protecting distributed computer systems for complex and unpredictable computing environments [1]. As another distributed system, *grid computing* comprises a group of computers located at physically distant locations and sharing computing and storage resources to accomplish a specific task on an on-demand basis [2]. Client–server and peer-to-peer networking are other distributed systems where the former denotes a distributed system consisting of a high performance server and lower performance clients, and the latter defines a distributed system consisting of peers that share hardware resources such as storage, processing power, and network link capacity [3]. In parallel to the distributed computing services *utility computing* has been proposed as another business model that delivers the computing resources among several computers based on an on-demand basis and bills the computing facilities such as in electricity and Internet billing [4].

The advent of distributed systems have eventually evolved to the *cloud computing* concept where almost infinite computing resources are available on-demand as well as the dynamic provisioning, release, and billing of computing facilities. Furthermore, capital/operational expenditures of the service providers are being reduced [5]. In fact, as stated by Zhang et al., cloud computing is not a new technology but it is a novel business model that brings several distributed system concepts together [6]. Therefore, several definitions of cloud computing exist [7] while most of the literature relies on the definition of the National Institute of Standards Technology (NIST). According to the NIST, cloud computing denotes a shared pool of resources available to the users, which can dynamically be provisioned and released without interacting with the service provider [8].

Grid computing seems to be the closest paradigm to cloud computing since it offers a distributed powerful and cost-efficient computing platform over the Internet. However, cloud computing offers a homogeneous pool of resources to the users as opposed to the heterogeneity of grid computing. Furthermore, cloud computing also differs from grid computing by the high reliability assurance of the service providers [9]. For instance, Amazon-EC2 assures 99.995% reliability, which leads to 27 min or less downtime per year [10].

Figure 11.1 illustrates an overview of the cloud computing architecture. As seen in the figure, four layers, namely the application, platform, infrastructure, and hardware layers form the *cloud.* The above two layers provide Software as a Service (SaaS) and Platform as a Service (PaaS), respectively, while the two bottom layers provide

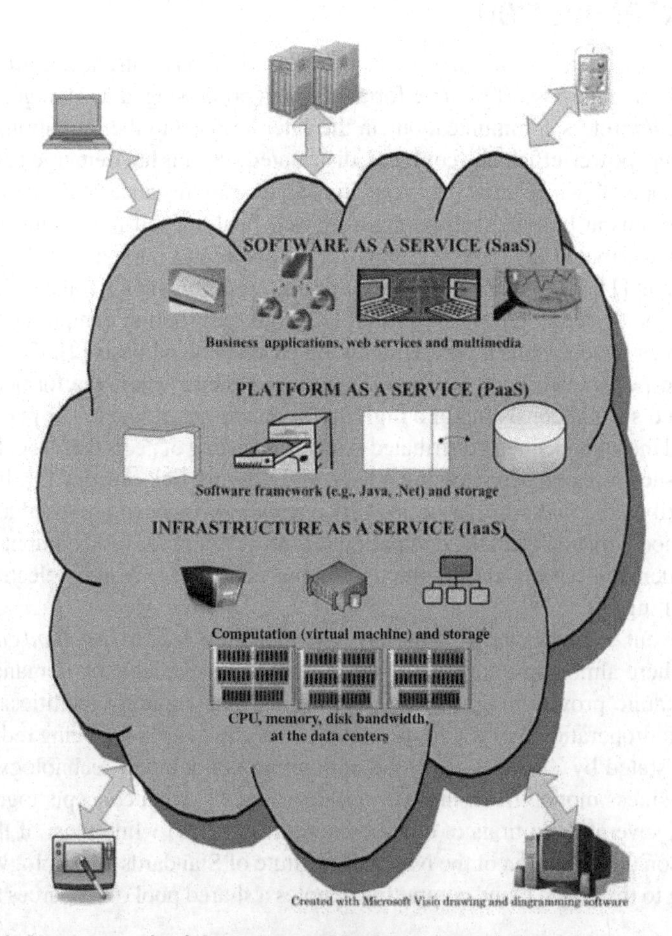

Figure 11.1 Cloud computing infrastructure of business enterprises.

Infrastructure as a Service (IaaS) to the end users. Hayes defines cloud computing as basically a shift in the geography of computation [11] and reports the attractive features of cloud computing along with the challenges. For instance, users are able to use the word processors offered by Google Docs or the Adobe's Buzzword without installing any word processors in their local computers. Besides, enterprise computing applications such as marketing and customer relation services can run on corporate servers without installation of any software on the local computers. Storage in the cloud is also another service offered to the users in terms of IaaS, e.g., Amazon Web Services, Google App Engine, IBM Smart Business Storage Cloud, etc.

Four deployment models of cloud computing exists as follows: (i) *Public clouds* serve through the Internet backbone and operate in a pay-as-you-go fashion (see Figure 11.2a). (ii) *Private clouds* are dedicated to an organization where the files and tasks are hosted within the corresponding organization. Thus, the users from various departments of the organization can access the cloud through a corporate network which is connected to the data center network through a gateway router (see Figure 11.2b). (iii) *Community clouds* enable several organizations to access a shared pool of cloud services forming a community of a special interest, and (iv) *hybrid clouds* are a combination of the public, private, and the community clouds with the objective of overcoming the limitations of each model [12].

In [6,11] several challenges in cloud computing have are discussed including security, privacy, reliability, virtual machine migration, automated service provisioning, and energy management. For the last decade, information and communication

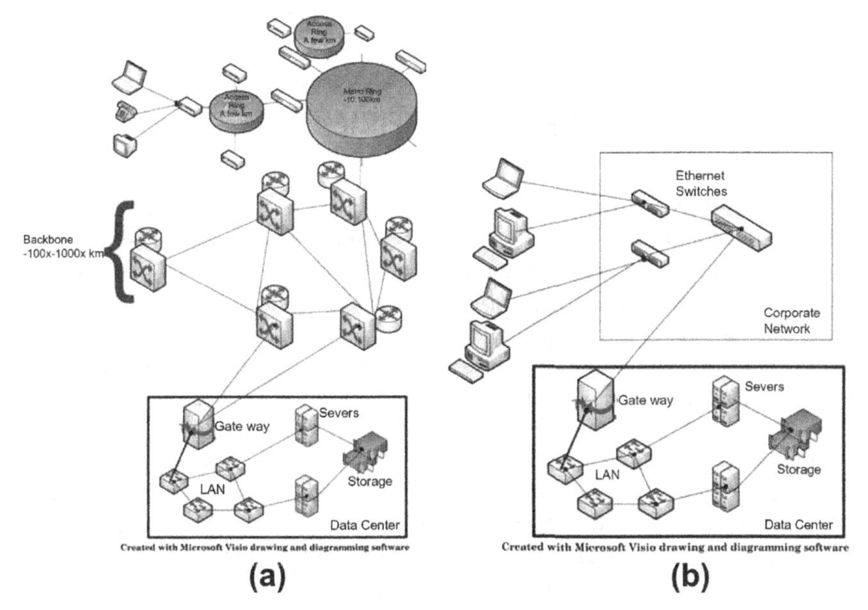

Figure 11.2 Fundamental deployment models of cloud computing: (a) public cloud, (b) private cloud.

technologies (ICTs) have been experiencing the energy bottleneck problem due to the exponential growth of the power consumption in the Internet [13], and improving the energy-efficiency in the Internet has been an emerging challenge [14]. As a part of the ICTs, processing, storage, and transmission of the cloud services contribute to the energy consumption of the Internet significantly, and energy-efficient design with respect to power consumption of the network equipment, service types, demographics, and service level agreements is emergent. For instance, power consumption of an IPTV service has three components such as storage power of the video, processing power consumed by the video servers, and the transmission power consumed by the transport network. Storage power is independent of the download demands however as the number of download demands per hour increases, power consumption of the video servers and the transport medium increases exponentially where the dominating factor is the transmission power. In [15], it has been shown that, for a single standard definition video replicated in 20 data centers, the total power consumption moves towards hundreds of kilowatts as the download request rate approaches 100 demands per hour.

This chapter focuses on the energy management challenges in cloud computing, which mainly have three aspects, namely processing, storage, and transport [16]. The chapter is organized as follows: in Section 11.2, we explain the relation between green ICT and energy-efficient cloud computing. We present the energy management in the data centers in terms of energy-efficient processing and storage in Section 11.3. Section 11.4 studies the optimal data center placement problem in order to ensure efficient utilization of renewable energy resources. Section 11.5 presents the studies to minimize the power consumption of transport services in cloud computing. Finally, the conclusion, research challenges, and further discussions are presented in Section 11.6.

11.2 GREEN ICT AND ENERGY-EFFICIENT CLOUD COMPUTING

Since the mid-2000s, energy consumption of the Internet-based applications has become a concern due to the rise of greenhouse gas (GHG) emissions; hence power-saving architectures and protocols have started being considered [17]. As of 2009, ICTs consume 4% of the global electricity consumption, and it is expected to double in the next few years [18]. Zhang et al. have presented a comprehensive survey on energy-efficiency schemes in various parts of the telecommunication networks [19]. Energy-management is mainly based on turning off the unused components such as line cards, network interfaces, router, and switch ports or putting these components in *low-power* (*sleep*) mode when they are idle. However, energy-efficient design and planning is also emergent for energy-efficient ICTs and it must be complemented by power-saving hardware and architectures [20].

According to a report in 2009, contributions of IT equipment to global GHG emissions are as follows: PCs and monitors contribute 40%, servers and their cooling equipment 23%, fixed line telecommunication equipments 15%, mobile telecommunications equipment 9%, local area and office network devices 7%, and printers 6% [18]. Based on this distribution, moving the IT services to distant servers can lead to significant energy savings in PCs and monitors. In this sense, cloud computing seems to provide a solution to the green IT objective. However, migration of services to distant servers in remote data centers will have two main impacts on global energy consumption and consequently GHG emissions: (i) utilization of the data center servers will increase, and (ii) migrating the resources over the Internet will overload the Internet backbone traffic leading to increased power consumption of routing and switching equipments. Hence, as Baliga et al. have stated, energy-efficiency in cloud computing needs to maintain a balance between processing, storage, and transport energy [16].

11.2.1 Motivation for Green Data Centers

Figure 11.3 illustrates a sample data center layout and the associated thermal activity. As seen in the figure, a data center consists of several rack towers where each rack contains servers, switches, or storage devices [21]. In order to enable the transport of cool air supplied by the computer room air conditioner (CRAC) into the aisles of the data center, the rack towers are placed on a raised floor in either a back-facing or front-facing manner. Cool air is supplied into the aisles through the vents towards the front faces of the assets in the racks forming cool aisles. Assets in the racks emit the hot air through back faces, which is channeled to the CRAC. As stated in [21], this layout design is cost-efficient however it may lead to hot spots in the data center since hot and cold air is mixed, and cooling is not uniformly distributed in the data center. Hence, cooling power may even exceed the IT equipment power. In the absence of an energy-efficiency design, a typical data center consumes 2000–3000 kWh/(m^2year) leading to high operational and capital expenditures [22]. According to the report of the Energy Star program in 2007 [23], data centers consumed 61 billion kWh of energy in 2006, which contributed to 1.7–2.2% of the total electricity consumption in the U.S., and it was expected to double in the next year. The worldwide picture is similar to that in the U.S., where 1.1–1.5% of the global electricity consumption is contributed by the worldwide data centers. A significant portion of this energy consumption is due to the supporting systems such as lighting, uninterrupted power

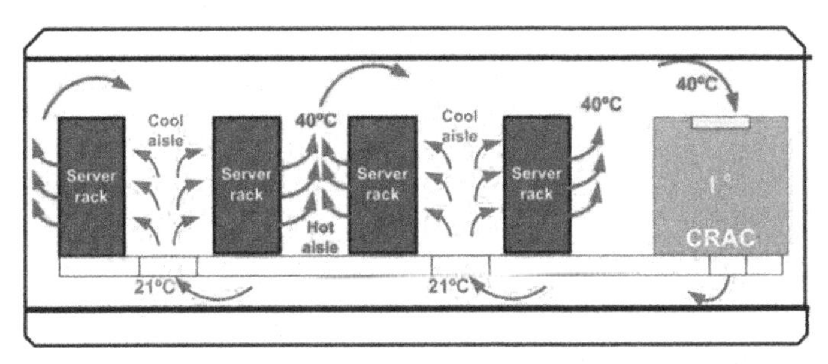

Figure 11.3 Layout design and the associated thermal activity in a typical data center.

supply (UPS), and heating, ventilation, and air conditioning (HVAC) units in the data centers. Thus, energy savings are possible if the power consumption of the supporting systems can be reduced.

Two main metrics are used to evaluate the energy-efficiency of a data center. The first metric is the ratio of the power consumption of the HVAC (i.e., CRAC) equipment to the power consumption of the IT equipment as shown in Eq. (11.1); thus, it denotes the cooling efficiency in the data center, i.e., *data center efficiency (DCE)*. A reasonable DCE value for a data center is 0.5 while a good target is reported to be above 0.625 [24]. The second metric evaluates the ratio of the power consumed by the IT equipment to the total power consumption in the data center as shown in Eq. (11.2); thus, it denotes the overall energy-efficiency in the data center, i.e., *power usage efficiency (PUE)*[25]. A typical PUE value for a data center is 2 however, today companies like Google, Amazon, and Microsoft have been able to build data centers with PUE values below 1.3 [26]:

$$\mathrm{DCE} = \frac{P_{\mathrm{HVAC}}}{P_{\mathrm{IT\ Equipment}}} \tag{11.1}$$

$$\mathrm{PUE} = \frac{P_{\mathrm{IT\ Equipment}}}{P_{\mathrm{total}}} \tag{11.2}$$

Figure 11.4 illustrates the data center efficiency report of Google where quarterly and 12-months average PUE of all of its data centers are shown. According to the power efficiency report of Google, by the end of the first half of 2011, 12-month and quarter-based average PUE values have been reduced to 1.15 and 1.14, respectively. It has been reported that reduced PUE can be achieved by joint use of the following three methods: thermal management, power-efficient

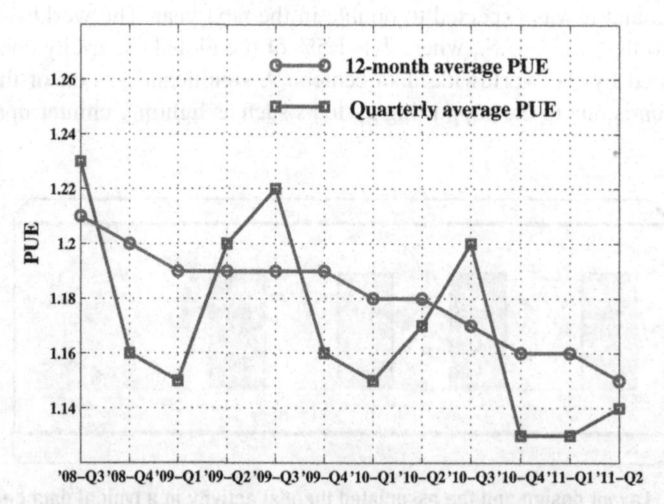

Figure 11.4 Quarterly and 12-month average PUE of all Google data centers [27].

hardware in servers together with optimized power distribution among the servers, and deployment of free (without chillers) cooling methods such as evaporating water [27].

11.2.2 Motivation for Green ICT

Baliga et al. have formulated the energy to transmit one bit from a data center to the user through a corporate network (i.e., private clouds) and energy to transmit one bit from a data center to a user through the Internet (i.e., public clouds) [16]. The notation used in the formulation is as follows:

F_{overhead} Overhead factor for redundancy, cooling, and other factors
F_{util} Overhead for underutilization of the core routers
P_{ses} Power consumption of small Ethernet switches
C_{es} Capacity (bps) of small Ethernet switches
P_{es} Power consumption of Ethernet switches
C_{es} Capacity (bps) of Ethernet switches
P_{bg} Power consumption of border gateway routers
C_{bg} Capacity (bps) of border gateway routers
P_{g} Power consumption of data center gateway routers
C_{g} Capacity (bps) of data center gateway routers
P_{ed} Power consumption of edge routers
C_{ed} Capacity (bps) of edge routers
P_{c} Power consumption of core routers
C_{c} Capacity (bps) of core routers
O_{c} Overprovisioning factor for core routers
H_{c} Hop count factor for the core traffic
P_{w} Power consumption of wavelength division multiplexing (WDM) equipment
C_{w} Capacity (bps) of WDM equipment

$$E_{\text{public}} = F_{\text{overhead}} \cdot F_{\text{util}} \cdot \left(\frac{3 \cdot P_{\text{es}}}{C_{\text{es}}} + \frac{P_{\text{bg}}}{C_{\text{bg}}} + \frac{P_{\text{g}}}{C_{\text{g}}} + \frac{2 \cdot P_{\text{ed}}}{C_{\text{ed}}} + \frac{O_{\text{c}} \cdot H_{\text{c}} \cdot P_{\text{c}}}{C_{\text{c}}} + \frac{(H_{c} - 1) \cdot P_{\text{w}}}{2 \cdot C_{\text{w}}} \right). \tag{11.3}$$

Equation (11.3) uses the layout in Figure 11.2a to formulate the energy to transmit one bit from a data center to a user over the Internet (E_{public}). In the equation, power consumption of Ethernet switches (P_{es}) is multiplied by a factor of three denoting the Ethernet switches in the metro network and in the data center network. Similarly, edge routers are multiplied by a factor of two considering the edge routers in the metro edge and in the data center. Considering the fact that core routers are overprovisioned for future growth of the demand, power consumption of core routers is multiplied by an overprovisioning factor $O_{\text{c}} > 1$, which is taken as 2 [16]. Furthermore, power consumption of the core routers is multiplied by a factor (H_{c}) denoting the core hops, which also contributes to the power consumption of the WDM equipment. Since the WDM equipment is reused while routing within the core, the number of hops in the WDM network is degraded by a factor, which is 2. The whole summation is multiplied by two overhead factors, which are $F_{overhead}$ denoting the redundancy and cooling overheads and F_{util} denoting the underutilization of the core routers. Baliga et al. have reported that setting $F_{overhead}$ and F_{util} to 3 and 2, respectively, is feasible for energy per bit calculation in the public cloud [16]:

$$E_{private} = F_{overhead} \cdot F_{util} \cdot \left(\frac{P_{ses}}{C_{ses}} + \frac{3 \cdot P_{es}}{C_{es}} + \frac{P_g}{C_g} \right) \tag{11.4}$$

In Eq. (11.4), energy consumption per bit for a private cloud is shown. In the equation, small Ethernet switches are the ones connecting the users to the corporate network while those connecting the corporate network to the data centers are the regular Ethernet switches. Since the formulation includes the Ethernet switches in both the corporate network and the data center, power consumption of the Ethernet switches is multiplied by 3. Similar to the public cloud energy consumption, the whole summation is multiplied by the two overhead factors. $F_{overhead}$ is proposed to be set to 3. However, as opposed to the public clouds, utilization is expected to be significantly less than that in the public cloud; hence underutilization overhead is set to 3 assuming 33% of network utilization [16].

Indeed, power consumption in watts can be derived as the product of energy per bit and the traffic intensity in terms of bits per second, and as stated in [14], transport energy starts dominating the energy consumption as user demands get heavier. Furthermore, in [28], the emergence of energy savings in transport services is explained by the following analogy: An average passenger car consumes 150 g CO_2/km while an IP router port consuming 1 kWh energy emits 228 g CO_2 if renewable resources are not used leading to 2 tons of CO_2 emission in a year. Hence, saving an energy consumption (or GHG emission) of one IP router port can save 13,000 km of the journey of a passenger vehicle.

11.3 ENERGY-EFFICIENT PROCESSING AND STORAGE IN CLOUD COMPUTING

11.3.1 Energy-Efficient Processing in Data Centers

Energy management in data centers has various aspects as specified in Section 11.2. These aspects are thermal management, energy-efficient server hardware, optimal power distribution among the servers, and deployment of chiller-less cooling techniques in the data centers. As stated in Section 11.2.1, water evaporation is one of the free-cooling techniques. Cooling towers are used in the data centers where hot water from the data center is carried to the top of the cooling tower and while flowing down, some portion of the water evaporates letting the remaining water cool, and the cooled water is supplied back to the data center from the ground. In [29], four main power saving techniques are surveyed: dynamic component deactivation, workload consolidation, resource throttling, and dynamic voltage and frequency sampling. There are other cooling techniques that eliminate chiller-based cooling, however, this chapter studies the energy-efficient computing perspective of cloud services; hence we will focus on thermal management and optimal power distribution among the servers in the data centers.

11.3.1.1 Thermal-Aware Workload Placement

Optimal power distribution among the servers aims at minimizing heat recirculation in the data center in order to avoid hot spots. Minimizing heat recirculation requires thermal-aware scheduling of the jobs. The total heat that recirculates in the data center (δQ) with n servers is formulated as shown in Eq. (11.5). In the equation, T_i^{in} and T_{sup} denote the inlet temperature of the server-i and the temperature supplied by the CRAC, respectively. Air flow rate at the server-i is denoted by m_i (in kg/s) while C_p is the specific heat of air (in W s/kg K):

$$\delta Q = \sum_{i=1}^{n} C_p \cdot m_i \cdot (T_i^{\text{in}} - T_{\text{sup}}). \tag{11.5}$$

The Minimum Heat Recirculation (MinHR) algorithm has been proposed in [30]. The MinHR algorithm aims at distributing the power (P_i) proportional to the ratio of the heat produced (Q_i) to the heat recirculated (δQ_i) at server i. A reference heat value (Q_{ref}) is set whenever an *event* such as a new CRAC or a new group of servers is installed, and a reference heat recirculation value (δQ_{ref}) is calculated accordingly. In the second step of MinHR, a group of adjacent servers are considered to be a pod; thus, the data center network consists of P pods. The power level of the servers in each pod is determined by P iterations. In each iteration (say iteration j), the power level of the servers in pod-j is set to maximum; then the amount of generated and recirculated heat is calculated denoting the Heat Recirculation Factor for pod-j (HRF_j) as shown in Eq. (11.6). Once all HRF values are obtained, the total power level of each pod is multiplied by the normalized sum of HRFs as seen in Eq. (11.6) thus, power levels ($P_{\text{pod-}j}$) are set as seen in Eq. (11.7):

$$\text{HRF}_j = \frac{Q_j - Q_{\text{ref}}}{\delta Q_j - \delta Q_{\text{ref}}}, \tag{11.6}$$

$$P_{\text{pod-}j} = \frac{\text{HRF}_j}{\sum_{i=1}^{P} \text{HRF}_i} \cdot P_j. \tag{11.7}$$

At the end of each placement, the supply temperature of the CRAC (T_{sup}) is adjusted by T_{adj} by considering the maximum values of the server inlet (T_{in}) and redline temperatures. The redline temperature denotes the safe server inlet temperature. As seen in Eq. (11.8), T_{adj} can have a negative value if the maximum observed inlet temperature exceeds the maximum redline temperature. In such a case, T_{sup} is decreased in order to supply cooler air into the data center:

$$T_{\text{adj}} = T_{\text{redline}}^{\max} - T_{\text{in}}^{\max}. \tag{11.8}$$

MinHR can jointly guarantee minimum heat recirculation and maximum server utilization. However, thermal-aware workload placement lacks cooling awareness since dynamic cooling behaviors of the CRACs are not considered. Thus, minimum heat recirculation cannot always lead to minimum PUE [31].

11.3.1.2 Thermal and Cooling-Aware Workload Placement

Challenges faced by thermal-aware workload placement are being addressed by introducing cooling awareness [31]. Cooling awareness denotes consideration of the cooling behavior of CRAC so that cooling costs are minimized. Figure 11.5 illustrates two scenarios to explain the motivation behind cooling and thermal awareness. Two jobs arrive at the data center where three servers are available. In the first scenario (Figure 11.5a), the jobs are placed in a first-come first-serve (FCFS) fashion, i.e., no cooling nor thermal awareness. Thus, job 1 is scheduled on server 1, and job 2 is scheduled on server 2. Server 3 has the lightest load; hence it requires the thermostat to be 22°C. Scheduling job 2 on server 2 leads to a thermostat setting of 20°C in order to let server 2 work properly. On the other hand, scheduling job 1 on server 1 generates a hot spot in the data center; hence server 1 requires the thermostat setting to be 18°C. In the second scenario, job placement is done by considering thermal activity along with the cooling behavior in the data center. Scheduling job 2 on server 3 does not lead to a thermostat setting lower than 22°C while scheduling job 1 on server 2 requires the thermostat setting to be 20°C for server 2 to work properly. While adjusting the thermostat supply temperature, the lowest temperature requirement has to be considered to ensure that all servers are working properly. Hence in the second scenario, the thermostat supply temperature has to be set to 20°C where in the first scenario, the thermostat setting should be adjusted to 18°C leading to higher cooling costs.

In order to manage this scheduling process, Banerjee et al. have proposed a spatial scheduling algorithm, namely the *Highest Thermostat Setting (HTS)*. The proposed method differs from conventional solutions due to considering multimode operation of the CRAC unit rather than supplying cool air at a constant temperature. Indeed, each operation mode of the CRAC extracts a certain level of heat which

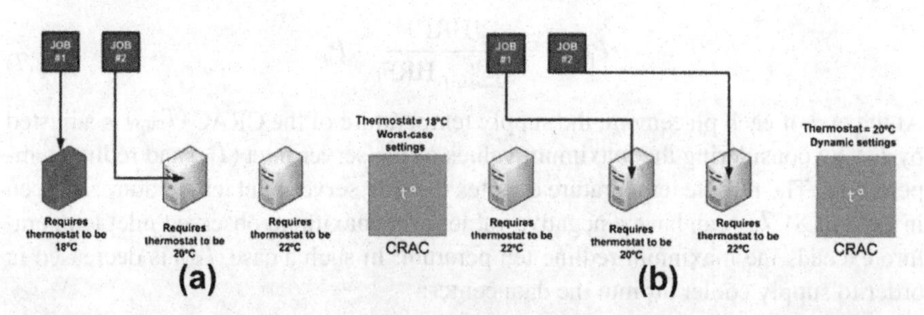

Figure 11.5 Workload placement: (a) first-come first-serve; (b) thermal and cooling-aware [31].

is specific to the corresponding mode. The HTS algorithm initially schedules the arriving jobs in the time domain using either FCFS or earliest deadline first (EDF). Cooling-thermal-aware workload placement is based on coordination between the three steps in (i)–(iii) as shown in Figure 11.6. These three steps can be explained as follows:

(i) *Static ranking of servers:* This step assigns ranks to servers based on their CRAC thermostat setting requirements to meet their redline inlet temperatures for 100% utilization. The server with the highest thermostat temperature requirement is assigned the lowest rank so that the incoming jobs are more likely to be scheduled on the servers with the highest thermostat setting requirement leading to reduced cooling costs. In [31], CRAC high threshold requirements for a server ($T^{\text{th}}_{\text{high}_i}$) are formulated as shown in Eq. (11.9):

$$T^{\text{th}}_{\text{high}_i} = \frac{T^{\text{redline}} - \sum_j d_{ij} \cdot P^{\text{full}}_j}{\sum_j f_{ij}} + \frac{p^{\text{low}}_{\text{ex}}}{r_{\text{ac}}} - \frac{(P^{\text{comp}}_h)^{\text{full}} - p^{\text{low}}_{\text{ex}}}{r_{\text{room}}} \cdot t_{\text{sw}} \quad (11.9)$$

According to the static server ranking function, high threshold requirements of a server are a function of the maximum allowed inlet temperature for the

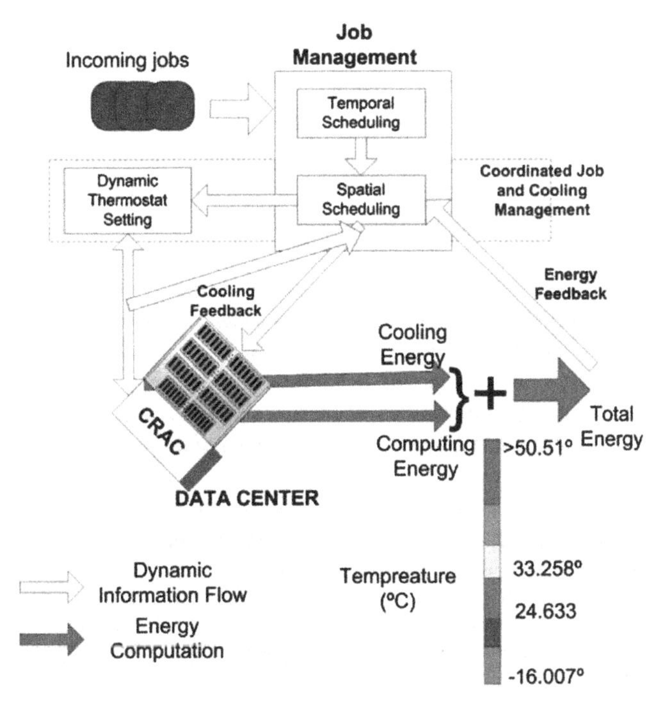

Figure 11.6 Coordinated workload placement proposed in [31].

corresponding server (i.e., the first summation term), temperature change due to CRAC power extraction (i.e., the second summation term), and the increase in temperature due to switching time of CRAC (i.e., the third summation term). In the first term denoting the maximum inlet temperature for server i, T^{redline} denotes the maximum inlet temperature that is specified by the manufacturer, d_{ij} is the heat recirculation coefficient between server i and server j, P_j^{full} is the power consumption of the server when fully utilized, and f_{ij} denotes the fraction of cold air supply flowing from CRAC to server-j. In the second term, $P_{\text{ex}}^{\text{low}}$ denotes the power extracted by CRAC in low-power mode while r_{ac} is the thermal capacity of air supplied by CRAC per unit time. In the third term, where the increase in temperature due to switching time of the CRAC unit is formulated, $(P_h^{\text{comp}})^{\text{full}}$ denotes the total computation power in the data center for a period of h while r_{room} and t_{sw} are the thermal capacity of the air in the data center and the switching time of CRAC between its operation modes, respectively.

(ii) *Job placement:* Upon scheduling the jobs in the time domain and determining the server ranks, servers are sorted in decreasing order with respect to their $T_{\text{high}_i}^{\text{th}}$ values. Jobs are popped from the virtual queue where they are temporally scheduled and each job is aimed to be assigned to the server with the lowest rank value.

(iii) *CRAC thermostat setting determination:* As the jobs are spatially scheduled, the power distribution vector is obtained denoting the power consumption throughout the data center. As opposed to step (ii) where 100% utilization of the servers is considered, here, actual server utilization levels are used to compute the highest possible thermostat value ($T_{\text{high}}^{\text{th}}$). In order to derive the equation for $T_{\text{th}}^{\text{high}}$, heat input to server i and heat output from server i during a short time of dt are formulated. The following notation is used in the equation sets:

l_i	Fraction of air flowing from CRAC to server i
r_{ac}	Thermal capacity of air flowing out of CRAC
a_{ji}	Fraction of air recirculating from server j to server i
r	Thermal capacity of air flowing out of a server
$T_i^{\text{in}}(t)$	Inlet temperature of server i at time t
$T^{\text{sup}}(t)$	Temperature supply of CRAC at time t
$T_i^{\text{out}}(t)$	Outlet temperature of server i at time t
$P_i(t)$	Computation power generated by server i at time t

Equation (11.10) formulates the input heat to server server i in a time interval of dt as the sum of the heat supplied by CRAC to server server i and the heat recirculates from other servers to server i in the time interval, dt.

$$\left(l_i \cdot r_{\text{ac}} + \sum_j a_{ji} \cdot r\right) \cdot T_i^{\text{in}}(t) \cdot \text{d}t = l_i \cdot r_{\text{ac}} \cdot T^{\text{sup}}(t) \cdot \text{d}t + \sum_j a_{ji} \cdot r \cdot T_j^{\text{out}}(t) \cdot \text{d}t. \tag{11.10}$$

Equation (11.11) formulates the outlet heat of server i in a time interval of $\mathrm{d}t$ as the sum of the inlet heat and the heat generated by server i during the time $\mathrm{d}t$:

$$\left(l_i \cdot r_{\mathrm{ac}} + \sum_j a_{ji} \cdot r\right) \cdot T_i^{\mathrm{in}}(t) \cdot \mathrm{d}t + P_i(t) \cdot \mathrm{d}t = r \cdot T_j^{\mathrm{out}}(t) \cdot \mathrm{d}t. \quad (11.11)$$

Since Eqs. (11.10) and (11.11) are derived for each server in the data center, the equation set formed by applying these two equations to each server in the data center defines a vectored operation, which is formulated in Eq. (11.12). Since the maximum inlet temperature for a server should not exceed the redline temperature, Eq. (11.12) can be reformulated as in Eq. (11.13):

$$[T_{\mathrm{in}}(t)] = [F] \cdot [T^{\mathrm{sup}}(t)] + [D] \cdot [P(t)], \quad (11.12)$$

$$[F^{-1}] \cdot [T_{\mathrm{redline}}(t)] - [F^{-1}] \cdot [D] \cdot [P(t)] = [T^{\mathrm{sup}}(t)]. \quad (11.13)$$

The maximum value of the supply temperature is constrained to the high thermostat setting ($T^{\mathrm{th}}_{\mathrm{high}}$), total computing power (P_h^{comp}), power extraction of CRAC in the low mode, the thermal capacity of air supplied by CRAC per unit time (r_{ac}), and the switching time of CRAC from low mode to high mode (t_{sw}). In [31], the maximum value of the supply temperature of CRAC is formulated as shown in Eq. (11.14). Thus, substitution of Eq. (11.14) into Eq. (11.13) yields Eq. (11.15), which computes the highest thermostat setting value for the algorithm:

$$T^{\mathrm{sup}}_{\mathrm{max}} = T^{\mathrm{th}}_{\mathrm{high}} + \frac{P_h^{\mathrm{comp}} - P^{\mathrm{low}}_{\mathrm{ex}}}{r_{\mathrm{room}}} \cdot t_{\mathrm{sw}} - \frac{P^{\mathrm{low}}_{\mathrm{ex}}}{r_{\mathrm{ac}}}, \quad (11.14)$$

$$T^{\mathrm{th}}_{\mathrm{high}^{\mathrm{max}}} = [F^{-1}] \cdot [T_{\mathrm{red}}] - [F^{-1}] \cdot [D] \cdot [P_h] - \frac{P_h^{\mathrm{comp}} - P^{\mathrm{low}}_{\mathrm{ex}}}{r_{\mathrm{room}}} \cdot t_{\mathrm{sw}} + \frac{P^{\mathrm{low}}_{\mathrm{ex}}}{r_{\mathrm{ac}}}. \quad (11.15)$$

In [31], the authors have shown that HST leads to significant energy savings (up to 15%) over the conventional thermal-aware workload placement algorithms. Moreover, by turning off the idle servers, further energy savings (up to 9%) are possible under HST. In cloud computing, requests usually last short; hence rather than setting the thermostat value dynamically, integration of decision making schemes with HST would assure both energy savings and service quality requirements. Furthermore, thermal- and cooling-aware workload placement can further be enhanced to address the trade-off between quality of service and energy savings.

11.3.2 Energy-Efficient Storage in Data Centers

Besides processing and short-term transactions, data centers also serve as long-term storage. Furthermore, data center storage demand is increasing by 50–60%

per year [32]. Hence deployment of low-power storage equipment and energy-efficient storage techniques can save significant energy in the data centers.

11.3.2.1 Solid State Disks (SSDs)

A solid state disk is a storage device consisting of NAND flash memory and a controller. Recently, SSDs have appeared as an alternative to conventional hard disk drives (HDDs) due to being lightweight, having a small form factor, having no moving mechanical parts, and lower power consumption [33]. For the sake of energy-efficiency, deployment of SSDs in data centers is advantageous however there are several issues that have to be addressed prior to integrating SSDs in data centers. In [32], write reliability is pointed to as one of these challenges since a single level SSD cell bit introduces a write penalty after 100,000 writes; hence this introduces a drawback when compared to conventional hard disk drives. Another challenge introduced by SSDs is the cost/GB. As of the first quarter of 2011, an SSD costs around $1.80 per GB while an HDD costs approximately $0.11 per GB. Although it is not likely that the SSD costs per GB can be reduced dramatically in short term, as reported in [32], enhancements in high performance and low power can let SSDs be integrated in the data center storage systems.

11.3.2.2 Massive Arrays of Idle Disks (MAIDs)

MAIDs consist of a large amount of hard disk drives that are used for nearline storage. Thus, a hard disk drive [34] spins up whenever an access request arrives for the data stored in it and the rest of the storage consists of a large number of spun down disks. Although MAID can lead to significant power savings due to nearline storage, it also introduces the trade-off between energy-efficiency and performance since spinning up takes more time than data access does. Hence, vendors such as DataDirect Networks Inc., EMC Corp., Fujitsu, Hitachi Data Systems, NEC Corp., and Nexsan Technologies Inc. have introduced multiple levels of power savings to MAID in order to overcome this trade-off. Furthermore, sleep mode support has also been introduced to the MAID system in order to avoid spinning down during peak load hours [35]. Thus, storage systems supporting spin-down and nearline storage seem promising to ensure energy-efficient storage in the data centers [36].

11.3.2.3 Storage Virtualization

Virtualized storage denotes a logical storage pool that is independent of the physical location of the disks [37]. In virtualized storage, unused storage segments can be consolidated in logical storage units increasing the storage efficiency. Enhanced management of the storage pool leads to reduced storage energy since idle physical resources can be spun down and/or put in the standby mode.

Figure 11.7 illustrates a storage area network (SAN), which enables virtualization of storage units. Servers are connected to the physical resources through SAN switches; hence a global storage pool is available to each server. Whenever a storage block requires allocation, a logical unit number is assigned to the allocated virtual space, which acts as a pointer between the physical storage resource and the logical

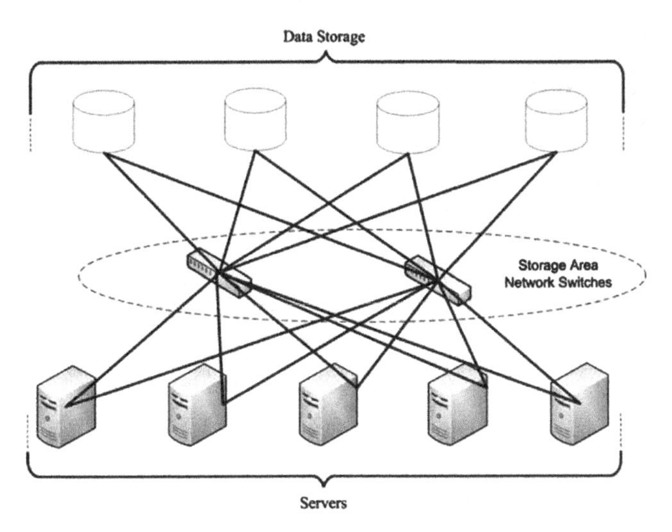

Figure 11.7 Storage area network [37].

storage block. Microsoft's Cluster Server, VMWare's File Server, and IBM's General File System are well-known examples of virtualized storage systems. Besides, IBM System Storage SAN Volume Controller (SVC) is another storage virtualization solution where more active data are moved to SSDs to ensure further energy-efficiency [38].

11.3.3 Monitoring Thermal Activity in Data Centers

As we have seen in the previous subsection, thermal- and cooling-aware workload placement is crucial to meet the desired PUE (and DCE as well) levels. In order to make decisions to increase/decrease the heat supply of the CRAC and place arriving jobs to the servers by avoiding hot spots in the data center, data center operators require online thermal monitoring of the data center. Knowingly, heat distribution in the data center is a function of many parameters including the CRAC temperature setting, workload placement, rack layout design, server types, and so on; hence such a monitoring system would further eliminate the need for on-demand air flow and thermodynamics analysis for dynamic energy management in a data center. Moreover, having the opportunity of thermal monitoring of data centers can also enable the operators to maintain the rack layout of the data centers more efficiently in long term.

The Data Center Genome Project of Microsoft Research and MS Global Foundation Services has proposed using wireless sensor networks (WSNs) for thermal monitoring of the data center [39]. Figure 11.8 illustrates the architecture of a data center monitoring system that mainly consists of three main blocks: data center, data collection/processing/analysis, and the decision-making blocks. Rack layout, cooling and power systems, heat distribution, server performance, and load variations are

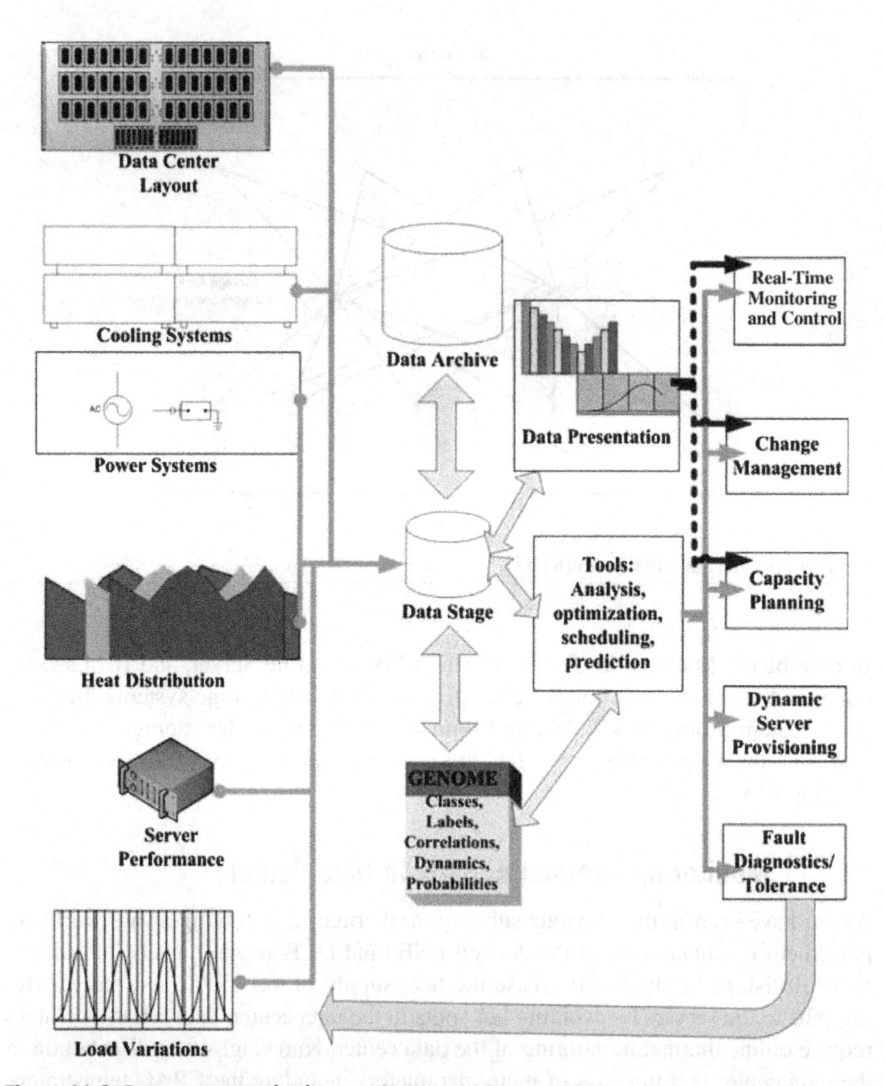

Figure 11.8 Data center monitoring architecture of Genome [39].

the phenomena that are monitored in the data center. The collected data are stored, archived, and used for analysis, optimization, and scheduling purposes. The outputs of the storage/processing block are used for real-time monitoring and control, change management in the data center, capacity planning, dynamic server provisioning, and fault diagnostics or tolerance.

The motivation behind deploying WSNs is explained as WSNs offering a low-cost and nonintrusive technology that has wide coverage and can be repurposed. The Genome Project deploys a sensor network called RACNet in a multimegawatt

(MMW) data center. RACNet consists of approximately 700 sensors. The sensor nodes form a hierarchical topology where master and slave sensor nodes coexist and cooperate. Slave sensors are mounted on the server racks, and they do not have a radio interface. The main function of the slave sensors is collection of the temperature and humidity data and reporting the collected data to the master sensors via serial wired interfaces forming a daisy chain circuit. Each master sensor node has a serial data communication interface and an IEEE 802.15.4 radio interface, which provides the WSN connectivity between the master nodes and the base station.

On the other hand, WSNs face several challenges when deployed in data centers. First of all, the data center introduces a tough radio frequency (RF) environment due to high metal contents of racks and servers, cables, railings, and so on. Besides, since the wireless sensor nodes need to be densely deployed for fast and reliable data collection, packet collision probability is expected to be high. Furthermore, IEEE 802.15.4 radio in RACNet introduces high bit error rate (BER).

In order to overcome the challenges, the Genome project has proposed a reliable Data Collection Protocol (rDCP). The following three key solutions are used by rDCP:

(i) *Channel diversity:* rDCP coordinates multiple base stations among 16 concurrent channels in the 2.4 GHz ISM band without collision. The number of master nodes on a channel is dynamically determined based on link quality measurements.

(ii) *Adaptive bidirectional collection tree:* On each wireless channel, rDCP adaptively builds a bidirectional collection tree. Hence, hop-by-hop connectivity can be met by selecting the high quality links when communicating with the master nodes.

(iii) *Coordinated data retrieval:* Each wireless sensor node (master node) has flash memory on board. Once the master node receives data through the daisy chain of slave nodes, it does not immediately transmit to the base station; instead it stores in the flash memory. Once the base station polls the master node by sending a retrieval request message, the data collected are retrieved over the collection tree in the opposite direction. It is worth noting that rDCP ensures that there is at most one retrieval data stream at a time on a radio channel.

Scalability of the data center monitoring system is a big challenge. For instance, consider that a data center consisting of 1000 racks with 50 servers in each rack required at least 50,000 sensor nodes in order to collect temperature and humidity data. Furthermore, if the humidity and thermal properties collected in different parts of an aisle are highly correlated, some of the sensor nodes will provide redundant data although they can be temporarily put into sleep mode to prolong their battery life. Furthermore, at the sink, processing highly correlated data will consume more processing power than is likely desired. In [40], a self-organizing monitoring system has been proposed for data centers, which consists of scalar sensors mounted on the server racks and thermal cameras mounted on the walls. In the self-organizing data center monitoring system, sensors are grouped as representative (REP) and associate (ASSOC) nodes in order to eliminate the data redundancy. A sensor node i is said to be a potential ASSOC of a sensor node j if their sensed data is similar and correlated. Sensed data of two sensors are said to be similar if the difference between their mean values is below a certain threshold. The sensed data are said to be correlated if their correlation coefficient is greater than a prespecified threshold value. The correlation coefficient is calculated by using a small finite number of samples taken from the data sensed by corresponding sensor nodes. Similarity and correlation thresholds are set by the data center operator in advance, and the values of these parameters may vary between different parts of the data center. A REP node is selected out of a set of ASSOC nodes.

In order to reduce costs, in [40], thermal cameras are also used with the scalar sensor nodes since thermal cameras have larger fields of view. A thermal camera uses infrared radiation from the objects and generates a 2D thermal image of the data center. Localization of hot spots is possible by exchanging the images between the thermal cameras. However, transmission of raw images between thermal cameras requires high bandwidth; hence compression at the sender and decompression at the receiver enhances bandwidth efficiency. The thermal signature of the data center (i.e., expected thermal map) is computed in advance by considering temperature, air flow rate, and workload placement factors. Thermal cameras and scalar sensors collaborate to obtain the actual thermal map. If the observed thermal map is different than the expected thermal map, the observed data are said to be an *anomaly*. A thermal map that is detected as an anomaly is classified by a supervised neural network since an anomaly can be due to the several reasons such as problems due to the cooling system, increase in cold water temperature, misconfiguration of servers, CPU fan failures, security attacks that overload the servers with illegitimate jobs, and so on. In the case of an anomaly detection, the base station can request the sensed data at the anomaly location in finer granularity, i.e., more sensed data are needed from the anomaly region.

11.4 OPTIMAL DATA CENTER PLACEMENT

The location of data centers has a significant impact on the transport energy of cloud services in the Internet backbone. Furthermore, varying load profile on the backbone nodes due to different time zones (such as the nationwide backbone network in the United States, known as NSFNET) affect the energy consumption throughout the day. Besides, while aiming at minimum power consumption in data center placement, taking advantage of availability of renewable resources such as solar panels and wind farms can lead to further reduction in CO_2 emissions. In [28], a mathematical model has been introduced with the objective of minimizing nonrenewable energy consumption at data centers. Wind farms are assumed to be used to power the data centers while solar panels are employed to power the nodes of the transport network. In this section, we briefly summarize the corresponding optimization model for data center placement.

It is assumed that the backbone employs an IP over WDM network with the optical bypass technology to transport the data center and regular Internet traffic as shown in Figure 11.2a for a public cloud. Figure 11.9 focuses on a backbone link in

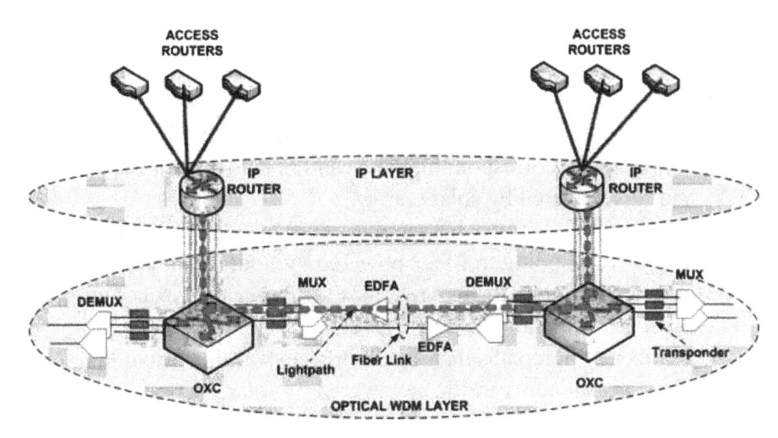

Figure 11.9 Components of an IP over WDM network.

Figure 11.2a. As seen in the figure, the traffic arriving from the access routers are aggregated at the IP routers. The IP layer is bypassed, and the traffic is routed at the optical WDM layer as much as possible. A virtual topology is formed from the physical topology to provision the demands, where a virtual link between two nodes denotes a lightpath between the corresponding nodes in the physical topology. It is not always possible to arrive at the destination node with one virtual hop; hence the traffic is terminated at the optical layer and transmitted to the IP layer where it is routed to the destination over the next virtual hop. This principle is known as *multihop bypass* technology [41]. The notation used in the linear programming (LP) formulation in [28] is shown below:

P_{r} Power consumption of an IP router port
P_{t} Power consumption of a transponder
P_{MD} Power consumption of a (de) multiplexer
DM_i Number of (de) multiplexers at node i
N_i Set of neighboring nodes of node i
P_{e} Power consumption of an erbium-doped fiber amplifier (EDFA)
NEDFA_{mn} Number of EDFAs deployed in the physical link mn
f_{mn} Number of fibers deployed in the physical link mn
P_i^{OXC} Power consumption of the optical switch at node i
$(C_{ij}^t)^{\mathrm{DC}}$ Number of lightpaths that carry data center traffic over the virtual link ij at time t. The corresponding traffic either starts or ends at a data center and it is powered by nonrenewable resources.
$(C_{ij}^t)^{\mathrm{DCW}}$ Number of lightpaths that carry data center traffic over the virtual link ij at time t. The corresponding traffic either starts or ends at a data center and it is powered by wind energy.
$(C_{ij}^t)^{\mathrm{NDC}}$ Number of lightpaths that carry data center traffic over the virtual link ij at time t. The corresponding traffic either starts or ends at a regular node and it is powered by nonrenewable resources.

$(C_{ij}^t)^{\mathrm{NDCW}}$ Number of lightpaths that carry data center traffic over the virtual link ij at time t. The corresponding traffic either starts or ends at a regular node and it is powered by wind energy.

$(C_{ij}^t)^{\mathrm{NDCS}}$ Number of lightpaths that carry data center traffic over the virtual link ij at time t. The corresponding traffic either starts or ends at a regular node and it is powered by solar energy.

$(C_{ij}^t)^{\mathrm{R}}$ Number of lightpaths that regular traffic over the virtual link ij at time t. The corresponding traffic is powered by nonrenewable resources.

$(C_{ij}^t)^{\mathrm{RW}}$ Number of lightpaths that regular traffic over the virtual link ij at time t. The corresponding traffic is powered by wind energy.

$(C_{ij}^t)^{\mathrm{RS}}$ Number of lightpaths that regular traffic over the virtual link ij at time t. The corresponding traffic is powered by solar energy.

Q_i^t Number of aggregation ports at node i at time t that are powered by nonrenewable resources.

$Q_i^{t\mathrm{W}}$ Number of aggregation ports at node i at time t that are powered by wind energy.

$Q_i^{t\mathrm{S}}$ Number of aggregation ports at node i at time t that are powered by solar energy.

$W_{mn}^{\mathrm{DC}^t}$ Number of wavelength channels on physical link mn, carrying the data center traffic. The corresponding traffic either starts or ends at a data center, and it is powered by the nonrenewable resources.

$W_{mn}^{\mathrm{DCW}^t}$ Number of wavelength channels on physical link mn, carrying the data center traffic. The corresponding traffic either starts or ends at a data center, and it is powered by wind energy.

$W_{mn}^{\mathrm{NDC}^t}$ Number of wavelength channels on physical link mn, carrying the data center traffic. The corresponding traffic either starts or ends at a regular node, and it is powered by the nonrenewable resources.

$W_{mn}^{\mathrm{NDCW}^t}$ Number of wavelength channels on physical link mn, carrying the data center traffic. The corresponding traffic either starts or ends at a regular node, and it is powered by wind energy.

$W_{mn}^{\mathrm{NDCS}^t}$ Number of wavelength channels on physical link mn, carrying the data center traffic. The corresponding traffic either starts or ends at a regular node, and it is powered by solar energy.

$W_{mn}^{\mathrm{R}\ t}$ Number of wavelength channels on physical link mn, carrying the regular traffic. The corresponding traffic is powered by the nonrenewable resources.

$W_{mn}^{\mathrm{RW}^t}$ Number of wavelength channels on physical link mn, carrying the regular traffic. The corresponding traffic is powered by wind energy.

$W_{mn}^{\mathrm{RS}^t}$ Number of wavelength channels on physical link mn, carrying the regular traffic. The corresponding traffic is powered by solar energy.

PWF_k^t Output power of wind farm k at time

S_i^t Output power of solar cells of node i at time t

U_k Available portion of the wind farm k output power to supply data centers

ℓ_k^i Loss of output power of wind farm k when transporting to data center i

δ_k^i Binary variable is 1 if data center i can be powered by wind farm k

The objective of the data center placement problem is to minimize the utilization of nonrenewable resources to power the transport medium. Here, this function is presented by the sum of five components: power consumption at the IP routers (P_{IP}), power consumption by the transponders at the optical layer (P_{optical}), power consumption due to amplification through EDFAs (P_{amp}), and power consumption due to optical switching (P_{sw}) as shown in Eq. (11.16). In the equation, T is the set of time points, and multiplying the power consumptions by time gives the energy consumption of the nonrenewable resources. Each component of the summation in the objective function is explained below:

$$\text{minimize} \sum_{t \in T} P_{\mathrm{IP}} + P_{\mathrm{optical}} + P_{\mathrm{amp}} + P_{\mathrm{sw}}. \tag{11.16}$$

Equation (11.17) formulates the power consumption at the IP router ports. The corresponding IP routers are powered by nonrenewable resources and they carry three types of demands as follows: (i) data center traffic that is either originating or arriving at a data center, (ii) data center traffic that either originates or arrives at a regular node, and (iii) regular Internet traffic:

$$P_{\mathrm{IP}} = \sum_{i \in N} \sum_{j \in N, j \neq i} P_{\mathrm{r}} \cdot \left((C_{ij}^{t})^{\mathrm{DC}} + (C_{ij}^{t})^{\mathrm{NDC}} + (C_{ij}^{t})^{\mathrm{R}} \right) + \sum_{i \in N} P_{\mathrm{r}} \cdot Q_{i}^{t}. \tag{11.17}$$

Equation (11.18) formulates the power consumption at the transponders. The corresponding optical nodes are powered by nonrenewable resources and they carry the data center traffic that is either originating or arriving at a data center, data center traffic that either originates or arrives at a regular node, and the regular Internet traffic:

$$P_{\mathrm{optical}} = \sum_{m \in N} \sum_{n \in N_m} P_{\mathrm{t}} \cdot \left(W_{mn}^{\mathrm{DC}^t} + W_{mn}^{\mathrm{NDC}^t} + W_{mn}^{\mathrm{R}\ t} \right). \tag{11.18}$$

Equation (11.19) formulates the power consumption of EDFAs that are powered by nonrenewable resources. Here, it is worth noting that the number of EDFAs deployed in a fiber (NEDFA_{mn}) is calculated as $\lfloor L_{mn}/L_{\mathrm{span}} \rfloor + 1$ where L_{mn} is the length of the fiber between node m and node n and L_{span} is the minimum distance between two EDFAs, which is often taken as 80 km:

$$P_{\mathrm{amp}} = \sum_{m \in N} \sum_{n \in N_m} P_{\mathrm{e}} \cdot \mathrm{NEDFA}_{mn} \cdot f_{mn}. \tag{11.19}$$

The last two components of total transport power consumption are the power consumed by optical switches and the power consumption of (de) multiplexers that are powered by nonrenewable resources and formulated by Eqs. (11.20) and (11.21), respectively:

$$P_{\mathrm{SW}} = \sum_{i \in N} P_{i}^{\mathrm{OXC}}, \tag{11.20}$$

$$\sum\nolimits_{i \in N} P_{\mathrm{MD}} \cdot \mathrm{DM}_{i}. \tag{11.21}$$

The constraints mainly denote the link and power capacity, flow conservation, and wavelength continuity constraints as the rest of the optimization model inherits the energy-minimized design of IP over WDM networks [41], and its details can be found in [28]. Here, we focus on the constraints related to renewable energy consumption as seen in Eqs. (11.22)–(11.24). The constraint in Eq. (11.22) ensures that renewable energy consumption at the router ports and transponders as well as the cooling and computing equipments in a data center cannot exceed the power supplied by a single wind farm. In the equation, power losses due to transportation of the wind energy to the data center are also taken into consideration:

$$\sum\nolimits_{j\in N,i\neq j} P_{\mathrm{r}}^{\mathrm{w}}\cdot\left((C_{ij}^{t})^{\mathrm{DCW}}+(C_{ij}^{t})^{\mathrm{NDCW}}+(C_{ij}^{t})^{\mathrm{RW}}\right)$$
$$+P_{\mathrm{r}}^{\mathrm{w}}\cdot Q_i^{t^w}+\sum_{m\in N_i} P_t^{\mathrm{w}}\cdot\left(W_{im}^{\mathrm{DCW}^t}+W_{im}^{\mathrm{RW}^t}+W_{im}^{\mathrm{NDCW}^t}\right)+P_i^{\mathrm{cool}}+P_i^{\mathrm{compute}}$$
$$\leqslant \mathrm{sum}_{k\in K}\delta_k^i\cdot \mathrm{PWF}_k^t\cdot(1-\ell_i)\cdot U_k,\quad \forall t\in T,\ i\in N. \tag{11.22}$$

By Eq. (11.23), the solar power that is available to a backbone (regular) node sets an upper bound for the renewable energy consumed by the router ports and the transponders of the corresponding node:

$$\sum_{j\in N,i\neq j} P_{\mathrm{r}}^{\mathrm{s}}\cdot\left((C_{ij}^{t})^{\mathrm{NDCS}}+(C_{ij}^{t})^{\mathrm{RS}}\right)$$
$$+P_{\mathrm{r}}^{\mathrm{s}}\cdot Q_i^{t^s}+\sum_{m\in N_i} P_t^{\mathrm{s}}\cdot\left(W_{im}^{\mathrm{RS}^t}+W_{im}^{\mathrm{NDCS}^t}\right) \tag{11.23}$$
$$\leqslant(1-\mathrm{sum}_{k\in K}\delta_k^i)\cdot S_i^t,\quad \forall t\in T,\ i\in N.$$

The left-hand side of Eq. (11.24) formulates the total renewable power consumption by all data centers. Thus, total renewable power consumption of all data centers is constrained to the total power supply of the wind farms:

$$\sum_{i\in N}\left\{(1+\ell_i^k)\cdot\left[\sum_{j\in N,j\neq i} P_{\mathrm{r}}^{\mathrm{w}}\cdot\left((C_{ij}^{t})^{\mathrm{DCW}}+(C_{ij}^{t})^{\mathrm{NDCW}}+(C_{ij}^{t})^{\mathrm{RW}}\right)+P_{\mathrm{r}}^{\mathrm{w}}\cdot Q_i^{t\mathrm{w}}\right.\right.$$
$$\left.\left.+\sum_{m\in N_i} P_t^{\mathrm{w}}\cdot\left(W_{im}^{\mathrm{DCW}^t}+W_{im}^{\mathrm{RW}^t}+W_{im}^{\mathrm{NDCW}^t}\right)\right]+P_i^{\mathrm{cool}}+P_i^{\mathrm{compute}}\right\}$$
$$\leq\sum_{k\in K} PWF_k^t\cdot U_k,\quad \forall t\in T,\ i\in N. \tag{11.24}$$

In [28], optimal data center placement over the NSFNET topology has been tested by using the above LP model. The following three wind farms have been considered: Cedar Creek Wind Farm, Capricorn Ridge Wind Farm, and Twin Groves Wind Farm.

Solar power available to the backbone nodes is affected by the time of the day; e.g., at 23:00 (EST), no solar power is available for the nodes in the Eastern Time Zone while the nodes in the Pacific Time Zone can still have solar power. Without loss of generality, the data centers are assumed to operate with a PUE of 2. According to the LP formulation results, locating the data centers at the center of the NSFNET topology can lead to minimization of nonrenewable power utilization, i.e., minimum CO_2 emissions, as more wind power is available if the data centers are built at the center of the network. Furthermore, the authors have shown that nonrenewable power consumption of the network can be reduced by 20% when compared to a scenario where data centers are randomly located over the NSFNET backbone. Another important result that has been reported is that the optimal placement of data centers by considering renewable resources, and incorporation of multihop bypass technology when routing the demands can lead to reduction in nonrenewable power consumption of up to 77%. Besides, when dynamically provisioning data center requests, in order to ensure fewer hops, replication of data among the data centers based on popularity can also lead to further savings in nonrenewable power consumption as stated by [28].

11.5 ENERGY-EFFICIENT TRANSPORT OF CLOUD SERVICES

11.5.1 From Unicast/Multicast to Anycast/Manycast

Traditionally, connection demands are provisioned with respect to unicast or multicast. Unicast provisioning is denoted by (s, d) where s and d denote the source and destination nodes of the incoming demand, respectively. Multicast provisioning is denoted by (s, D) where s is the source node and D is the set of destination nodes that the messages are required to be delivered to. In distributed computing applications such as cluster computing, grid computing, or cloud computing, one or more destinations are selected out of a group of candidate destinations. Selecting a destination out of a candidate destination set is referred to as *anycast*, and it is denoted as $(s, d_s \in D)$ where d_s is the selected destination out of the set of candidate destinations, D. Selection of a group of destinations out of a candidate destination set is referred as *manycast* provisioning, and it is denoted by $(s, D_s \subseteq D)$ where D_s is a subset of the candidate destinations set, D. Obviously, if $D_s = D$, then, manycast is equivalent to multicast, and if $|D_s| = 1$, then, manycast and anycast are identical.

In this section, we study energy-efficient anycast and manycast provisioning of user demands over the Internet since cloud computing services are provisioned with respect to either an anycast or manycast paradigm.

11.5.2 Energy-Efficient Anycast

11.5.2.1 Anycast with Dynamic Sleep Cycles

In [42], Bathula and Elmirghani have proposed an energy-efficient anycast algorithm for the provisioning of jobs submitted to a computational grid. Despite the differences in programming, business, computation, and data models, cloud and grid computing have common features in architecture, vision, and technology [43]. Hence, anycasting can also be considered for demand provisioning for cloud computing.

In the referred study, the network is represented by a clustered architecture where these computing clusters are interconnected through their boundary nodes. Dynamic sleep cycles are proposed in the clusters in order to save energy. Thus, based on the traffic load profile, the nodes in a cluster are switched to the *off* state, and then, switched back to the *on* state. Routing the jobs submitted to the grid is realized by energy-efficient anycast, i.e., anycast among the *on* clusters without leading to Quality of service (QoS) degradation. For each link j, a network element vector (NEV_i) is defined as follows:

$$NEV_i = [\eta_i \quad \tau_i]^T, \tag{11.25}$$

where η_i and τ_i are the noise factor and the propagation delay of link i, respectively. For a set of links denoting a route from node s to node d, end-to-end (E2E) noise factor would be the product of noise factors of the links along the path while the E2E propagation delay is the sum of the propagation delays of all links along the path. Thus, NEV for a route R can be obtained as shown in Eq. (11.26):

$$NEV_R = [\eta_R \quad \tau_R]^T = \left[\prod_{k=s}^{d} \eta_k \quad \sum_{k=s}^{d} \tau_k\right]^T. \tag{11.26}$$

Since jobs are submitted with their service level agreements (SLAs), the provisioning solution must not violate the SLA. Hence, for each job, θ, a threshold NEV is predetermined as shown in Eq. (11.27), and $NEV_R \leqslant NEV_{TH}^{\theta}$ is the aim:

$$NEV_{TH}^{\theta} = [\eta_{th} \quad \tau_{th}]^T. \tag{11.27}$$

Anycast provisioning of a submitted job ($\theta = (n, D_n)$) is an iterative procedure where an iteration step is illustrated in the flowchart in Figure 11.10. An iteration step works as follows. The initial value of NEV is set to $[1 \quad 0]^T$. The iterations continue until a destination out of the candidate destinations set, D_n is reached. The destinations are sorted in increasing order with respect to their shortest path distances to the source node, n. The destination with the minimum hop distance is selected as the potential destination node (d'). The next hop node to node d' is looked up in the routing table of node n. n_k is the next hop to the potential destination node d'. If node n_k is the boundary node of a cluster which is in the *off* state, then, the candidate destinations set is updated by removing the nodes that are in the same cluster with n_k. Otherwise, NEV is updated as follows:

$$NEV[n-1 \quad n_k] \leftarrow NEV[n-1 \quad n] \circ NEV[n \quad n_k], \tag{11.28}$$

where the Boolean operator $\circ$ denotes multiplication on the noise factor and summation on the propagation delay. If the SLA requirements are met, i.e.,

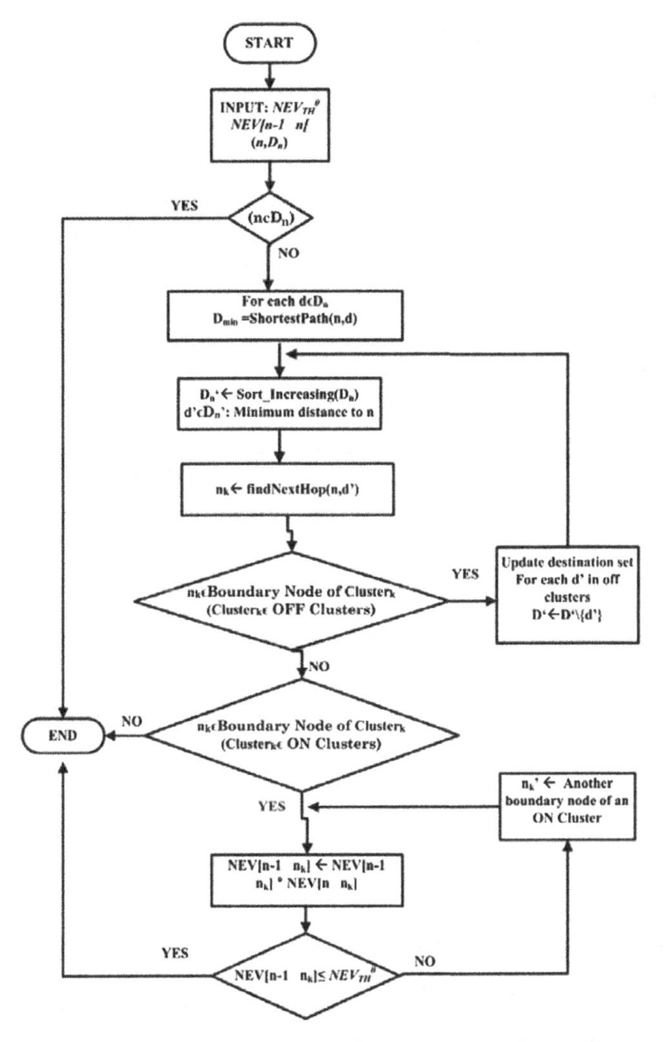

Figure 11.10 An iteration step of an energy-efficient anycast algorithm [42].

NEV $[n-1 \quad n_k] \leqslant \text{NEV}^{\theta}_{\text{TH}}$, the anycast function is called for n_k and the candidate destination set. Otherwise, another candidate destination is selected from another cluster and the above steps are run for the next hop node to the selected cluster.

The proposed anycast scheme is promising in terms of energy-efficiency and SLA guarantee however, dynamic sleep cycle management is crucial to gain maximum benefit in terms of energy-efficiency without SLA violation. Furthermore, the proposed algorithm is based on the assumption that there are at least three clusters in the *on* state in order to provision an incoming request which introduces further requirements in dynamic management of sleep cycles of the clusters.

11.5.2.2 Anycast among Data Center Replicates

In [28], Dong et al. propose the Energy-Delay Optimal Routing (EDOR) algorithm in a network where a few data centers are placed at the backbone nodes and data are replicated in those data centers based on popularity. The algorithm runs on an IP over WDM network, and aims at mapping the physical topology onto a virtual topology by the employment of optical bypass technology as discussed in Section 11.4. Static upstream data center demands are provisioned as follows: The demands are sorted in decreasing order. The first demand in the sorted list is retrieved along with the list of data centers where the desired data are replicated. The algorithm computes all available paths to the retrieved data centers on the virtual topology. If there is sufficient capacity on the virtual paths, the demand is routed based on the shortest path paradigm, and the virtual link capacities are updated to continue with the next demand in the list. Otherwise, a new virtual link is built between the source node and one of the data centers leading to the minimum number of physical hops. Upon updating the virtual topology, the algorithm continues with the next demand in the sorted list.

In [28], the authors show that under the optimal placement of data centers in the NSFNET topology, the EDOR algorithm with multihop bypass technology can ensure an average of 4.5% power savings in the IP over WDM network over non-bypass provisioning. Furthermore, it has also been shown that if the demands are provisioned with respect to the unicast paradigm and with the objective of minimized energy, power savings of the IP over WDM network are limited to 3.7% on average. Thus, anycasting among the replicated data centers lets more demands share bandwidth on the virtual links leading to more power savings. The EDOR algorithm leads to an increase in the propagation delay up to 8% when compared to the shortest distance routing. The EDOR algorithm can be extended by considering SLA requirements such as noise factors and propagation delay. Furthermore, it can be easily modified to provision the demands with respect to the manycast paradigm as well.

11.5.3 Energy-Efficient Manycast

The manycast problem aims to select to reach at a subset of the candidate destinations set originating at the source node where the demand is requested. Considering the employment of optical WDM networks in the Internet backbone, the manycast problem can be defined as the light-tree selection problem [44]. In [45], Kantarci and Mouftah formulated an energy-efficient light-tree selection problem to provision cloud services over optical transport networks and proposed a heuristic for the solution of the corresponding problem.

In Figure 11.11, a cloud over a wavelength routed network is illustrated over the NSFNET topology. Cloud resources are distributed and interconnected through the wavelength routing (WR) backbone nodes. The architecture of a WR node is illustrated at node 3. The main components in a WR switch are the transmitters and receivers where the add and drop traffic is managed, the wavelength switches where the pass-through traffic is transported, and (de) multiplexers where incoming/outgoing traffic is routed among the wavelength switch inlets/outlets or the transceivers. The partial sleep cycle

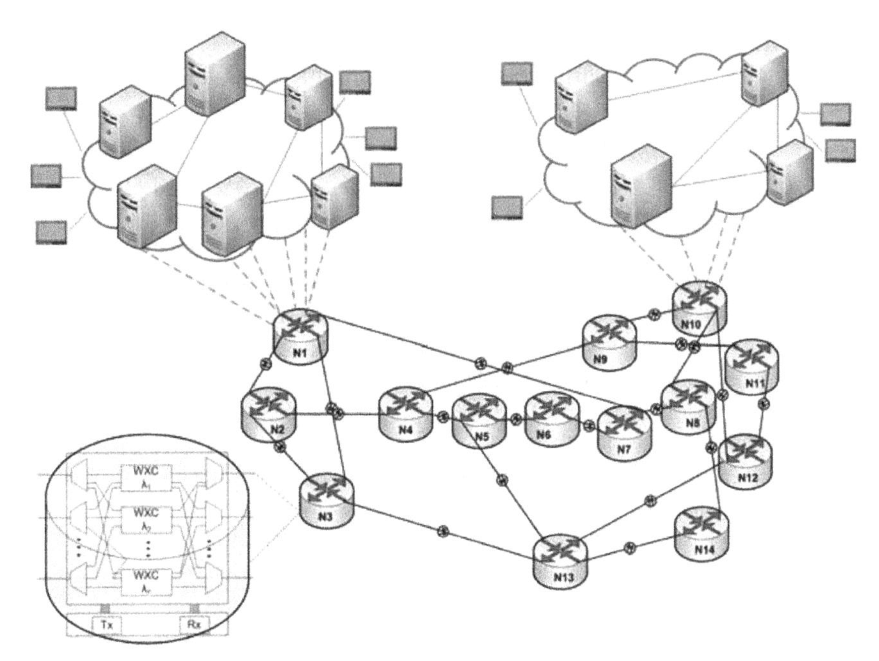

Figure 11.11 Cloud over WR optical network in the NSFNET topology.

idea in [46, 47] is adopted here as follows: A WR node enters the power saving (*off*) mode and the traffic routed via the corresponding node is cut and routed through other nodes that are in the *on* state. To add or drop connections is possible in the *off* mode but pass-through traffic is not routed via a node in the *off* mode. The notation used for optimized provisioning of manycast demands over the wavelength routed network is as follows:

L_{mn} Length of fiber between node m and node n

λ_w^{ij} Number of active wavelength channels on the link ij

χ_i A binary variable, and it is 1 if node is in the power saving mode ($\overline{\chi_i}$ is the one's complement of χ_i)

$\{i, j\}$ Directional link from node i to node j

$u_{i,j}^{x,w}$ A binary variable, and it is 1 if demand x is utilizing the wavelength w on link i, j

$\Phi_{i,j}$ A binary variable, and it is 1 if neither node i nor node j is in the power saving mode

δ_x Lower bound for the number of manycast destinations that have to be reached by demand x

N Number of nodes in the network

O_i^x Addition order of node i to the manycast tree of demand x

Λ_w^x A binary variable, and it is 1 if demand x is utilizing wavelength w on the links of its manycast tree

D_x Set of candidate manycast destinations of demand x

s_x Source node of demand x

Three main components contribute to the power consumption of the optical transport network as follows:

(i) *Power consumption of erbium-doped fiber amplifiers* (E_{EDFA})*:* The number of EDFAs in link ij is a function of the fiber length between node i and node j (dist (i, j)) and the length of a fiber span ($\lfloor L_{mn}/L_{\text{span}} \rfloor + 1$).

(ii) *Power consumption due to switching equipment* (E_{MEMS})*:* MEMS equipment and the wavelength converters in the WR nodes are the major contributors of the switching energy consumption (E_{MEMS}).

(iii) *Idle power consumption*(E_{ON})*:* This denotes the power consumption of a WR node when it is in the *ON* state. Based on the factors mentioned above, total energy consumption in the network can be formulated as shown in Eq. (11.29) where β denotes the average ratio of the time that a WR node spends in the *on* mode. Thus, $(1 - \beta)$ of a WR node's idle power can be saved if the corresponding node spends β of its time in the *on* mode. If the partial sleep cycle idea is adopted, the ratio between the pass-through traffic and the add/drop traffic can be used to determine the value of β. Research by Pramod and Mouftah reports that the add/drop traffic and the pass-through traffic demonstrate a proportion of 3:7 [48]; hence setting β to 0.3 is reasonable as proposed in [45]:

$$\text{Energy} = \sum_i \sum_j (\lfloor L_{mn}/L_{\text{span}} \rfloor + 1) \cdot E_{\text{EDFA}} + \sum_i \sum_j \sum_w \lambda_w^{i,j} \cdot E_{\text{MEMS}} + \sum_i \beta_i \cdot E_{\text{ON}}. \tag{11.29}$$

An optimization model for energy-efficient manycast provisioning is presented below in Eqs. (11.30)–(11.47). The objective function in Eq. (11.30) aims at maximizing the number of nodes that are in the *off* mode. Hereafter, we use the *sleep* and *off* modes interchangeably. The constraint set consists of two parts where the former [Eqs. (11.31)–(11.37)] denotes the energy-efficiency assurance while the latter [Eqs. (11.38)–(11.47)] is the manycast constraint set. Equation (11.31) ensures that a demand x can utilize a channel on link ij if neither node i nor node j is in sleep mode. It is worth noting that it is assumed that the upstream data center traffic traverses at least two hops in the backbone. The next constraint in Eq. (11.32) denotes that the next hop after the source node of a demand has to be in the *on* mode in order to utilize the link between the source node and the corresponding node. In the next constraint [Eq. (11.33)], it is ensured that if node i is one hop before one of the destinations, the light-tree can traverse the corresponding node if and only if it is in the *on* mode. Equation (11.34) ensures that a node can be in either one of the *on* or *off* modes. The next three constraints in Eqs. (11.35)–(11.37) formulates a linear approximation of the product of χ_i and χ_j denoting if two nodes are concurrently in the *off* mode.

$$\text{Maximize} \sum_i \chi_i, \tag{11.30}$$

$$\sum_w u_{i,j}^{x,w} \leqslant \Phi_{i,j}, \quad \forall x, i, j \ (i \neq s_x, \{j\} \nsubseteq D_x), \tag{11.31}$$

$$\sum_w u_{s_x,j}^{x,w} \leqslant \overline{\chi}_j, \quad \forall x, j\ (\{j\} \nsubseteq D_x), \tag{11.32}$$

$$\sum_w u_{i,j}^{x,w} \leqslant \overline{\chi}_i, \quad \forall x, i, j\ (i \neq s_x, \{j\} \subseteq D_x), \tag{11.33}$$

$$\chi_i + \overline{\chi}_i = 1, \quad \forall i, \tag{11.34}$$

$$\Phi_{i,j} - \overline{\chi}_i \leqslant 0, \quad \forall i, j, \tag{11.35}$$

$$\Phi_{i,j} - \overline{\chi}_j \leqslant 0, \quad \forall i, j, \tag{11.36}$$

$$\overline{\chi}_i + \overline{\chi}_j - \Phi_{i,j} \leqslant 1, \quad \forall i, j. \tag{11.37}$$

The remaining of the mathematical model given in Eqs. (11.38)–(11.47) represents the manycast constraints that have initially been presented in [44]. Equation (11.38) denotes that a wavelength on a bidirectional link can be used at most in one direction. Reaching a satisfying number of destinations is guaranteed by Eq. (11.39). The source node must have at least one outgoing wavelength but no incoming wavelengths as formulated by Eqs. (11.40) and (11.41). By Eqs. (11.42)and (11.43), demand x is allowed to utilize at most one wavelength on a bidirectional link. Equations (11.44) and (11.45) stand for the flow conservation constraints while Eq. (11.46) guarantees that loops are avoided. Demand x can utilize one and only one wavelength in its manycast tree as shown in Eq. (11.47).

$$\sum_x (u_{i,j}^{x,w} + u_{j,i}^{x,w}) \leqslant 1, \quad \forall i, j, w, \tag{11.38}$$

$$\sum_i \sum_{j \in D_x} \sum_w u_{i,j}^{x,w} \geqslant \delta_x, \quad \forall x, \tag{11.39}$$

$$\sum_j \sum_w u_{s_x,j}^{x,w} \geqslant 1, \quad \forall x, \tag{11.40}$$

$$\sum_i \sum_w u_{i,s_x}^{x,w} = 0, \quad \forall x, \tag{11.41}$$

$$\sum_i \sum_w u_{i,j}^{x,w} \leqslant 1, \quad \forall x, j \neq s_x, \tag{11.42}$$

$$u_{i,j}^{x,w} + u_{j,i}^{x,w} \leqslant \Lambda^{x,w}, \quad \forall x, w, j, i\ (i < j), \tag{11.43}$$

$$\sum_j (u_{i,j}^{x,w} - N \cdot u_{j,i}^{x,w}) \leqslant 0, \quad \forall x, w, i \neq s_x, \tag{11.44}$$

$$\sum_i (u_{i,j}^{x,w} - u_{j,i}^{x,w}) \leqslant 0, \quad \forall x, w, \{j\} \nsubseteq D_x, \tag{11.45}$$

$$O_i^x - O_j^x + N \cdot u_{i,j}^{x,w} \leqslant N - 1, \quad \forall i, j, x, w, \tag{11.46}$$

$$\sum_w \Lambda^{x,w} = 1, \quad \forall x. \tag{11.47}$$

Solution of an optimization model can take a long time, which may not be feasible to make decisions, even if the demand profile can be predicted a few hours

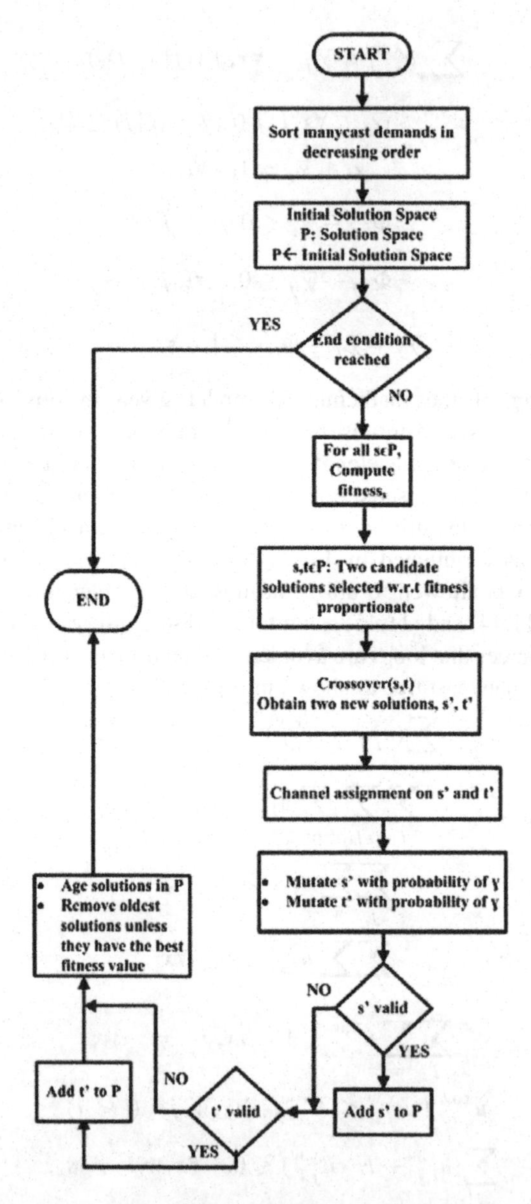

Figure 11.12 Flowchart of the Evolutionary Algorithm for Green Light-tree Establishment (EAGLE) [45].

ahead. Hence, Kantarci, and Mouftah have proposed a heuristic called *Evolutionary Algorithm for Green Light-tree Establishment (EAGLE)*[45]. Figure 11.12 illustrates the flowchart of EAGLE.

As seen in Figure 11.12, EAGLE starts with an initial solution pool which is updated by a number of iterations. The algorithm works as follows: At the beginning of each iteration, a fitness value is calculated for each individual solution. The fitness function can be defined in various ways. In [45], the authors define three different fitness functions as follows:

(i) $F_{\text{Max-Sleep}}$: aims at maximizing the number of nodes in the *off* mode [Eq. (11.48)].

(ii) $F_{\text{Min-Energy}}$: aims at minimizing the total power consumption in the network [Eq. (11.49)].

(iii) $F_{\text{Min-Channel}}$: aims at minimizing the wavelength channel consumption on each link [Eq. (11.50)].

$$F_{\text{Max-Sleep}} = \sum_i \chi_i, \tag{11.48}$$

$$F_{\text{Min-Energy}} = \frac{1}{\text{Energy}}, \tag{11.49}$$

$$F_{\text{Min-Channel}} = \frac{1}{\max\{\sum_w \lambda_w^{i,j}\}_{i,j}}. \tag{11.50}$$

Upon obtaining the fitness values of the individuals, two individual solutions are selected with respect to the fitness proportionate fashion, i.e., each candidate individual can be selected with the probability P, which is the ratio of its fitness to the sum of the fitness values of all individuals. An individual denotes a solution (tree) prior to wavelength assignment. Two selected individuals are used as the inputs of a crossover function in order to obtain two new solutions with a probability of Θ. The crossover function works as follows: upon finding a common link between the branches of two trees, the rest of the branches are exchanged between the new solutions. The crossover is followed by wavelength assignment to the new solutions in first-fit fashion. For each solution, EAGLE employs an aging counter that is incremented by one at the end of each iteration. During the iteration steps, if the number of the solutions in the solution space exceeds the population size (P), the m oldest solutions are removed from the solution space unless any of them has the highest fitness value in the current solution space.

In [45], the performance of EAGLE has been evaluated by assuming each node has access to at least one data center, and considering the upstream data center demand profile as in Figure11.13a. Four time zones, EST, CST, MST, and PST, are assumed as in [28] to assure heterogeneity of the network traffic, and one day is represented by eight slots where each slot corresponds to a three-hour duration. Energy consumption of EDFAs, MEMs equipment, and the idle power consumption are taken as 8, 1.8, and 150 Wh, respectively. The WR transport network offers 10 Gbps capacity on each wavelength channel while at least two destinations are required to be reached out of either three or four destination candidates. EAGLE has been run with a crossover probability of 0.20 and a mutation probability of 0.01. Figure 11.13 illustrates the cumulative energy consumption of the transport network throughout the day when EAGLE runs under the three fitness functions. As seen in the figure,

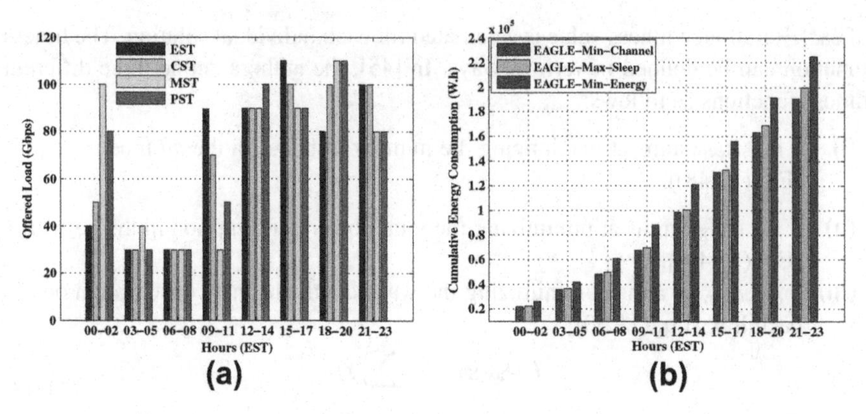

Figure 11.13 (a) Offered load profile in four time zones during the day. (b) Cumulative energy consumption of EAGLE throughout the day with the three fitness functions.

$F_{\text{Min-Channel}}$ (EAGLE-Min-Channel) leads to the highest energy consumption since setting the objective as minimizing the maximum number of wavelength channels per fiber can lead to longer routes during the evolution process. On the other hand, the other fitness functions, namely $F_{\text{Max-Sleep}}$ (EAGLE-Max-Sleep) and $F_{\text{Min-Energy}}$ (EAGLE-Min-Energy) introduce significant energy savings. Moreover, instead of only aiming to avoid the maximum number of pass-through nodes, provisioning the demands by considering the overall energy consumption in the transport network leads to more energy savings at the end of the day.

Since it is based on an evolutionary heuristic, EAGLE provides a suboptimal solution for the energy-efficient provisioning of the cloud services. Hence, further improvements are possible in manycast provisioning by considering more efficient fitness functions. Furthermore, faster heuristics running in dynamic environments are emergent to eliminate the need for lookahead demand profiles since demands are provisioned and released dynamically in cloud computing.

11.6 SUMMARY AND CHALLENGES

By bringing many parallel and distributed system concepts together, cloud computing is expected to serve many business areas such as health [49], education [50], scientific computation [51], multimedia service delivery [52], and so on. Migration of local resources to a shared pool of resources distributed over the Internet introduces many advantages but at the same time increases the communication energy, and consequently increases GHG emissions. Furthermore, data centers that are the main hosts of cloud computing environment emit significantly large amount of GHGs. Hence, energy-efficient and green migration of IT services by cloud computing is emergent.

Table 11.1 Energy Management Solutions in Cloud Computing

Energy Management Solution	Processing	Storage	Transport
Thermal-aware workload placement in data centers	j		
Cooling management in data centers	j		
Dynamic component deactivation	j		
Dynamic voltage and frequency sampling	j		
Data center monitoring	j	j	
Massive array of idle disks (MAIDs)		j	
Solid-state disks (SSDs)		j	
Virtualization by storage area networks		j	
Optimal data center placement	j	j	j
Energy-efficient anycast			j
Energy-efficient manycast			j

This chapter has provided an overview of energy-efficiency issues and the existing solutions. Table 11.1 summarizes the energy management techniques in cloud computing. Energy-efficiency of processing and storage in cloud computing has been studied in the sense of energy management in data centers. Thermal- and cooling-aware solutions for efficient workload placement problems are required to improve data center PUE values. Furthermore, thermal monitoring of data centers is crucial to assist cooling management and workload placement processes. The effect of data center locations on the energy-efficiency of cloud computing has also been reviewed with consideration of the associated transport network energy. Building the data centers close to renewable resources such as wind farms and use of solar panels at the transport network nodes can minimize the dependency on nonrenewable resources and lead to further reduction in GHG emissions of processing, storage, and transport energies of cloud computing. Energy-efficient provisioning of the cloud services over the transport network has been studied by introducing energy-efficient anycast and manycast techniques. Sleep-mode support at the backbone nodes can introduce significant savings.

Despite the existing techniques, there are still open issues for the researchers in this field. Although the existing cloud service providers offer 99.995% availability, this value has to be enhanced for the mission-critical applications of large enterprizes [53]. Enhancement of reliability is possible by replication of content and redundancy of transport resources, which introduce an increase in network energy consumption. Efficient solutions are needed to address the trade-off between service level agreement (SLA) assurance and energy-efficiency. Furthermore, reliability assurance of data center monitoring systems is another emergent subject in energy-efficient cloud computing. The more reliable the monitoring system, the more precise the decisions made for the workload consolidation among the servers. Dynamic, fast, and scalable anycast and manycast algorithms are required for efficient utilization of the network resources while provisioning the cloud services. Last but not least, energy-efficient design of storage area networks (SANs) is another direction for researchers in this field.

REFERENCES

[1] IBM Autonomic Computing Manifesto, 2001 (online). <http://www.research.ibm.com/autonomic.manifesto/>.

[2] U. Schwiegelshohn et al., Perspectives on grid computing, Future Generation Computer Systems 26 (8) (2010) 1104–1115.

[3] R. Schollmeier, A definition of peer-to-peer networking for the classification of peer-to-peer architectures and applications, in: Proceedings of First International Conference on Peer-to-Peer Computing, August 2001, pp. 101–102.

[4] J.W. Ross, G. Westerman, Preparing for utility computing: the role of IT architecture and relationship management, IBM Systems Journal 43 (1) (2004) 5–19.

[5] M. Armbrust et al., Above the Clouds: A Berkeley View of Cloud Computing, UCB/EECS-2009-28, EECS Department, University of California, Berkeley, February 2009 (online). <http://www.eecs.berkeley.edu/Pubs/TechRpts/2009/EECS-2009-28.html>.

[6] Q. Zhang, L. Cheng, R. Boutaba, Cloud computing: state-of-the-art and research challenges, Journal of Internet Services and Applications 1, 7–18.

[7] L.M. Vaquero et al., A break in the clouds: towards a cloud definition, ACM SIGCOMM Computer Communication Review 39 (1) (2009) 50–55.

[8] P. Mell, T. Grance, The NIST Definition of Cloud Computing, 800–145, National Institute of Standards and Technology, January 2011 (online). <http://csrc.nist.gov/publications/>.

[9] D. Kondo et al., Cost-benefit analysis of cloud computing versus desktop grids, in: IEEE International Symposium on Parallel Distributed Processing, May 2009, pp. 1–12.

[10] Amazon EC2 Service Level Agreement, 2008 (online). <http://aws.amazon.com/ec2-sla/>.

[11] B. Hayes, Cloud computing, Communications of the ACM 51 (2008) 9–11.

[12] Vic J.R. Winkler, Chapter 2—cloud computing architecture, in: Securing the Cloud, Syngress, Boston, 2011, pp. 29–53 (online). <http://www.sciencedirect.com/science/article/pii/B9781597495929000026>.

[13] J. Baliga et al., Photonics switching and the energy bottleneck, in: OSA Conference on Photonics in Switching, 2007, pp. 125–126.

[14] K. Hinton et al., Power consumption and energy efficiency in the Internet, IEEE Network 25 (2) (2009)

[15] J. Baliga et al., Architectures for energy-efficient IPTV networks, in: Conference on Optical Fiber Communication (OFC), March 2009, pp. 1–3.

[16] J. Baliga et al., Green cloud computing: balancing energy in processing, storage, and transport, Proceedings of the IEEE 99 (1) (2011) 149–167.
[17] M. Gupta, S. Singh, Greening of the Internet, in: ACM SIGCOMM'03, August 2003, pp. 19–26.
[18] P. Leisching, P. Pickavet, Energy footprint of ICTs: forecasts and network solutions, in: OFC/NFOEC, Workshop on Energy Footprint of ICT: Forecast and Network Solutions, 2009.
[19] Y. Zhang et al., Energy efficiency in telecom optical networks, IEEE Communications Surveys and Tutorials 12 (4) (2010) 441–458.
[20] H.T. Mouftah and B. Kantarci, Energy-Aware Systems and Networking for Sustainable Initiatives, in: W-C. Hu, N. Kaabouch (Eds), Greening the Survivable Optical Networks: Solutions and Challenges for the Backbone and Access, IGI Global, pp. 256–286, 2012.
[21] K. Kant, Data center evolution: A tutorial on state of the art, issues, and challenges, Computer Networks 53 (2009) 2939–2965.
[22] H.S. Sun, S.E. Lee, Case study of data centers energy performance, Energy and Buildings 38 (5) (2006) 522–533.
[23] US Environmental Protection Agency ENERGY STAR Program, Report to Congress on Server and Data Center Energy Efficiency, ENERGY STAR, August 2007 (online). <http://www.energystar.gov>.
[24] C. Belady, The green grid data center power efficiency metrics: PUE and DCiE, Whitepaper, 2008 (online). <http://www.thegreengrid.org/>.
[25] S. Greenberg, W. Tshudi, J. Weale, Self-Benchmarking Guide for Data Center Energy Performance, Lawrence Berkley National Laboratory, (Online) <http://www.lbnl.gov/>, 2006.
[26] Data Center Knowledge (online). <http://www.datacenterknowledge.com>, accessed in July 2012.
[27] Google Data Center Efficiency (online). <http://www.google.com/about/datacenters/inside/efficiency/powerusage.html> (accessed October 2011).
[28] X. Dong, T. El-Gorashi, J.M.H. Elmirghani, Green IP over WDM with data centers, IEEE/OSA Journal of Lightwave Technology 29 (12) (2011) 1861–1880.
[29] A. Beloglazov, A taxonomy and survey of energy-efficient data centers and cloud computing systems, Advance in Computers vol. 82, Elsevier.
[30] J. Moore et al., Making scheduling cool: temperature-aware workload placement in data centers, in: Usenix Ann. Technical Conf., 2005, pp. 61–74.
[31] A. Banerjee et al., Sustainable Computing: Informatics and Systems 1 (2) (2011) 134–150.
[32] D. Reinsel and J. Janukowicz, Data center SSDs: solid footing for growth, Whitepaper, 2008 (online). <http://www.samsung.com/global/business/semiconductor/products/SSD/downloads/datacenter ssds.pdf>.
[33] D. Kim et al., Architecture exploration of high-performance PCs with a solid-state disk, IEEE Transactions on Computers 59 (2010) 878–890.
[34] D. Colarelli, D. Grunwald, Massive arrays of idle disks for storage archives, in: SC Conference, 2002, pp. 47–52.
[35] The State of MAID in Data Centers, April 2009 (online). <http://searchstorage.techtarget.com/report/The-state-of-MAID-in-data-centers>.
[36] X. Zhang et al., Key technologies for green data center, in: Third International Symposium on Information Processing (ISIP), October 2010, pp. 477–480.

[37] D. Barrett, G. Kipper, Virtualization challenges, in: Virtualization and Forensics, Syngress, Boston, 2010, pp. 175–195.

[38] IBM System Storage SAN Volume Controller (online). <http://www-03.ibm.com/systems/storage/software/virtualization/svc/index.html>.

[39] J. Liu et al., Project genome: wireless sensor network for data center cooling, The Architecture Journal Microsoft 18 (2008) 28–34.

[40] H. Viswanathan, E.K. Lee, D. Pompili, Self-organizing sensing infrastructure for autonomic management of green datacenters, IEEE Network 25 (4) (2011) 34–40.

[41] G. Shen, R.S. Tucker, Energy-minimized design for IP over WDM networks, IEEE/OSA Journal of Optical Communications and Networking 1 (2009) 176–186.

[42] B.G. Bathula, J.M.H. Elmirghani, Green networks: energy efficient design for optical networks, in: International Conference on Wireless and Optical Communications Networks (WOCN), April 2009.

[43] I. Foster et al., Cloud computing and grid computing 360-degree compared, in: Grid Computing Environments Workshop (GCE), November 2008, pp. 1–10.

[44] N. Charbonneau, V.M. Vokkarane, Routing and wavelength assignment of static manycast demands over all-optical wavelength-routed WDM networks, Journal of Optical Communications and Networking 2 (7) (2010) 442–455.

[45] B. Kantarci, H.T. Mouftah, Energy-efficient cloud services over wavelength-routed optical transport networks, in: Proceedings of IEEE GLOBECOM, December 2011, pp. SAC06.6.1–SAC06.6.5.

[46] B.G. Bathula, M. Alresheedi, J.M.H. Elmirghani, Energy efficient architectures for optical networks, in: London Communications Symposium (LCS), 2009.

[47] B.G. Bathula, J.M.H. Elmirghani, Energy efficient optical burst switched (OBS) networks, in: IEEE Globecom Workshops, 2009.

[48] S.R. Pramod, H.T. Mouftah, A sharable architecture for efficient utilization of transmitter–receiver ports on an optical switch, in: SPIE Photonics, December 2004.

[49] A. Rosenthal et al., Cloud computing: a new business paradigm for biomedical information sharing, Journal of Biomedical Informatics 43 (2) (2010) 342–353.

[50] N. Sultan, Cloud computing for education: a new dawn? International Journal of Information Management 30 (2) (2010) 109–116.

[51] S.N. Srirama, P. Jakovits, E. Vainikko, Adapting scientific computing problems to clouds using MapReduce, Future Generation Computer Systems 28 (1) (2012) 184–192.

[52] W. Shi et al., Sharc: a scalable 3d graphics virtual appliance delivery framework in cloud, Journal of Network and Computer Applications 34 (4) (2011) 1078–1087.

[53] S. Marston et al., Cloud computing the business perspective, Decision Support Systems 51 (1) (2011) 176–189.

CHAPTER

Green Data Centers

12

Yan Zhang, Nirwan Ansari

Advanced Networking Laboratory, Department of Electrical and Computer Engineering, New Jersey Institute of Technology, Newark, NJ 07102, USA

12.1 INTRODUCTION

In order to provide reliable and scalable computing infrastructure, the high network capacity of data centers is especially provisioned for worst-case or busy-hour load, and thus data centers consume a huge amount of energy. As reported in 2005, the electricity usage of data centers has been almost doubled from 2000 to 2005 [1], and it was predicted to double again in 2011 [1]. The electricity cost accounts for about 20% of the total cost of data centers [2]. However, numerous studies have shown that the average server utilization is often below 30% of the maximum utilization in data centers [3,4], and much of the electrical power usage in data centers is wasted [5]. At low server utilization levels, servers are highly energy-inefficient [6,7]. Therefore, further investigation of energy efficiency of data centers is critical. To quantify the energy efficiency of data centers, several energy-efficiency metrics have been proposed recently, such as Power Usage Effectiveness (PUE) and its reciprocal Data Center infrastructure Efficiency (DCiE) [8], Data Center energy Productivity (DCeP) [9], Datacenter Performance Per Energy (DPPE) [10], and the Green Grid Productivity Indicator [11].

Techniques to improve energy efficiency of data centers have been developed and verified recently. According to a benchmark study of data center power usage [5], servers, storage and communication equipment, and power distribution infrastructure and cooling infrastructure are the three main contributors to power consumption of data centers, and therefore energy-efficient information technology (IT) infrastructure [12–27], and smart power delivery [28–33] and cooling [34–48] are effective solutions to improve the energy efficiency of data centers. Moreover, power management techniques [49–62], such as provisioning, virtualization, and consolidation, can also decrease power usage of data centers significantly. Also, some other energy saving techniques, e.g., power-aware routing [63] and architecture designs [64] for data centers, should be included in the game plan.

In this chapter, the power consumption and energy efficiency of data centers are reviewed first in Section 12.2. In order to quantify the energy efficiency of data

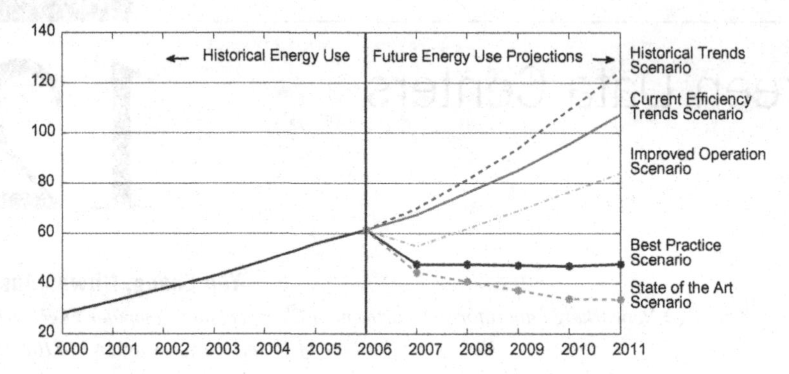

Figure 12.1 EPA report to Congress on Server and Data Center Energy Efficiency in 2007 (adapted from [1]).

centers, several energy-efficiency metrics have been proposed recently. The energy-efficiency metrics for data centers are presented in Section 12.3. The techniques to improve energy efficiency of data centers are discussed in Section 12.4. Finally, Section 12.5 concludes the chapter.

12.2 POWER CONSUMPTION AND ENERGY EFFICIENCY OF DATA CENTERS

The rapid escalating power consumption of data centers has become a central critical issue, and this problem has attracted great attention from government, industry, and academics recently. However, numerous studies have shown that much of the electrical power usage in data centers is wasted. In this section, power consumption of data centers is reviewed first, and energy efficiency of data centers is discussed later in detail.

12.2.1 Power Consumption of Data Centers

In 2007, the Environmental Protection Agency (EPA) Report to Congress on Server and Data Center Energy Efficiency [1] assessed trends in the energy usage and energy costs of data centers and servers in the United States, and outlined existing and emerging opportunities for improved energy efficiency. Based on the power consumption of data centers in the United States from the year 2000 to 2006, the power consumption of data centers is predicted as shown in Figure 12.1 under five different scenarios, namely, historical trends scenario, current efficiency trends scenario, improved operation scenario, best practice scenario, as well as state-of-the-art scenario. This prediction was performed with the total power consumption of the installed base of servers, external disk drivers, and network ports in data centers multiplied by a power overhead factor caused by the power usage of power distribution and cooling infrastructure in data centers. It can be observed that data centers and servers in the U.S. consumed about 61 billion kilowatt-hours (kWh) in 2006 (1.5% of the total U.S. electricity consumption) for a total electricity cost of about $4.5 billion. The energy use of data centers and servers in 2006 is more than double the electricity that was consumed by data centers in 2000. As analyzed in [65], this growth was mostly driven by the increase of the number of "volume servers," which increase the power usage per server and consequently increase the overall power usage of data centers.

The "historical trends" and the "current efficiency trends" are two baseline prediction scenarios to estimate the power consumption of data centers in the absence of expanded energy-efficiency efforts. The historical trends scenario simply estimated the power consumption trends based on the observed power usage from year 2000 to 2006. The "current efficiency trends" scenario projected the power usage trajectory of U.S. servers and data centers by considering the observed efficiency trends for IT equipment and site infrastructure systems. It is estimated the energy usage of data centers could nearly double again in 2011 to more than 100 billion kWh, representing a $7.4 billion annual electricity cost with historical trends and "current efficiency trends."

Three other energy-efficiency scenarios were also explored in the EPA Report to Congress. The "improved operation" scenario utilizes any essentially operational technologies requiring little or no capital investment to improve energy efficiency beyond "current efficiency trends." The "best practice" scenario adopts more widespread technologies and practices in the most energy-efficient facilities in operation today. The "state-of-the-art" scenario maximizes the energy efficiency of data centers using the most energy-efficient technologies and best management practices available today.

The trends in data center electricity usage from 2005 to 2010 were described in [65]. As reported in [65], the rapid growth in data center electricity usage was slowed significantly from 2005 to 2010 due to the economic slowdown since the 2008 financial crisis, and a significant reduction in the actual server installed base with improved virtualization techniques. The total power consumption of data centers increased about 56% from 2005 to 2010 for worldwide data centers (about 1.3% of all electricity usage for the world), and increased only 36% for U.S. data centers (about 2% of all electricity usage for the U.S.) instead of doubling, which is significantly lower than that was predicted by the EPA Report to Congress on data centers [1].

12.2.2 Energy Efficiency in Data Centers

APC White Paper #6 [2] investigated the total cost of ownership (TCO) of physical data center infrastructure, and found that the cost of electrical power consumption contributed to about 20% of the total cost. Raised by the high operational cost that caused power consumption and low utilization of data centers (lower than 30%), academics and industries have started to address various issues to improve the energy efficiency of data centers.

Numerous studies have shown that data center servers rarely operate at full utilization, and it has been well established in the research literature that the average server utilization is often below 30% of the maximum utilization in data centers [3,4] and a great number of servers work in the idle state in these richly connected data centers. At low levels of workload, servers are highly energy inefficient. As shown in [6,7], the power consumption of current commodity servers can be approximated by a model with a large always-present constant power and a dynamic power linear to server performance. The amount of dynamic power of typical servers today is small, at most 25% of the total dissipated power. At the idle state, the power consumed is over 50% of its peak power for an energy-efficient server [4] and often over 80% for a commodity server [6].

Barroso and Hölzle [4] studied the energy efficiency of servers and showed a mismatch between common server workload profiles and server energy efficiency. The power consumption of a typical energy-efficient server as a function of workload utilization is depicted in Figure 12.2 labeled with "Power (typical)." Notably, it consumes about half of its full power at the idle state, and the energy efficiency, defined as the ratio of the server utilization over the uniformed power usage to

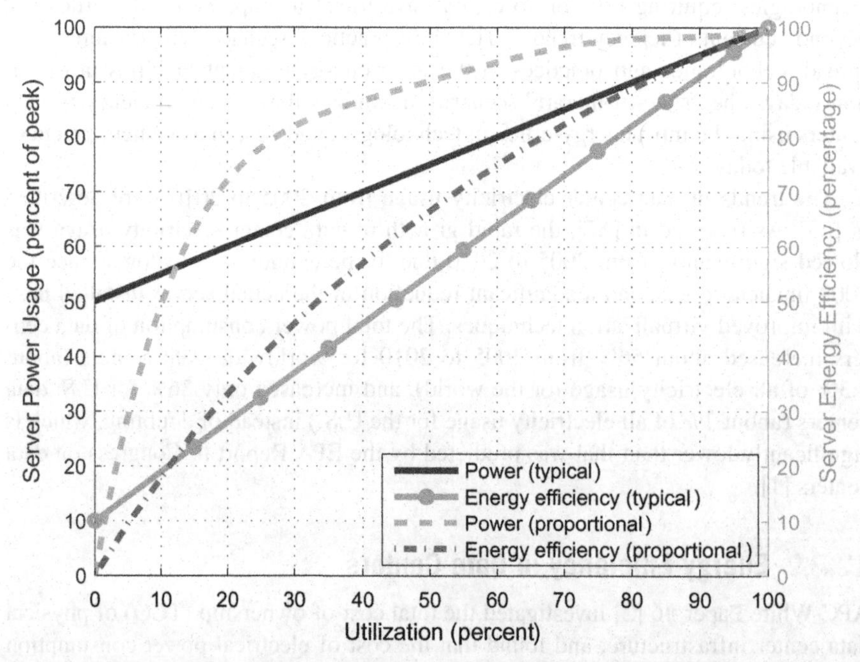

Figure 12.2 Server power usage and energy efficiency at varying utilization levels, from idle to peak performance (adapted from [4]).

that at the maximum server utilization, is quite low at light workload. The servers' energy efficiency below 30% utilization, which is the average server utilization of modern data centers, is less than half of that at the maximum utilization. Thus, it is not power-proportional, as the amount of power used is proportional to the provisioned capacity, not to the workload. They proposed the concept of energy-proportional computing systems that ideally consume almost no power when idle and gradually consume more power as the activity level increases. Figure 12.2 also shows the power usage and energy efficiency of energy-proportional servers with a dynamic power range of 90%. The servers' energy efficiency can be improved greatly at low utilization levels by increasing the dynamic power range from 50% to 90% of the server's peak power usage. As compared to today's typical servers, energy-proportional servers can reduce almost half of the power usage at low utilization.

The energy saving opportunities for data centers [5] have been studied to gain a solid understanding of data center power consumption. IT infrastructure, including servers, storage, and communication equipment, as well as power distribution infrastructure and cooling infrastructure are the three main contributors to power consumption of data centers. Based on a benchmark study of data center power consumption [5], the total power consumption of IT equipment accounts for approximately 40–60% of the power delivered to a data center, and the rest is utilized by power distribution, cooling, and lighting that support IT infrastructure. Figure 12.3 shows the power usage of a typical data center [32].

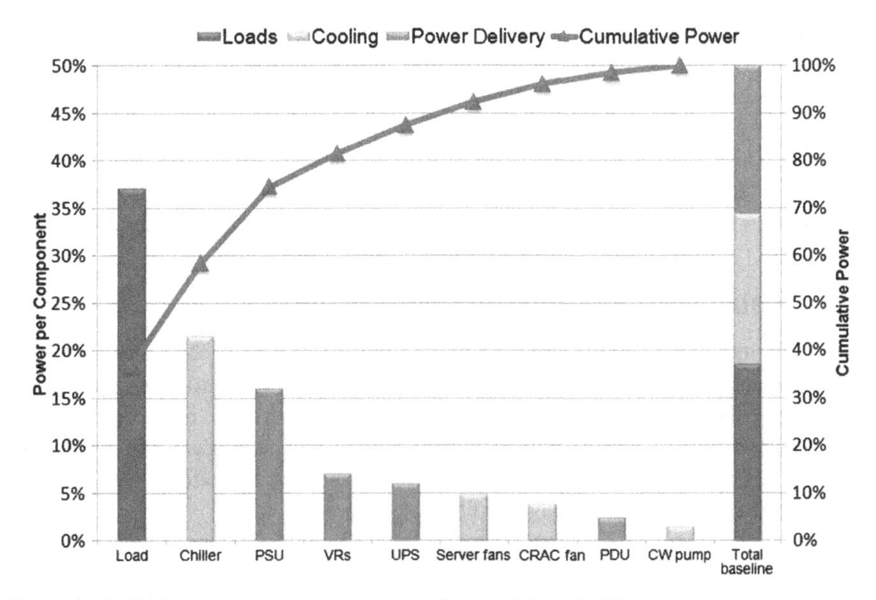

Figure 12.3 Typical data center power usage (adapted from [32]).

12.3 ENERGY-EFFICIENCY METRICS FOR DATA CENTERS

In order to quantify the energy efficiency of data centers, several energy-efficiency metrics such as Power Usage Effectiveness (PUE) and its reciprocal Data Center infrastructure Efficiency (DCiE) [8], Data Center energy Productivity (DCeP) [9], Datacenter Performance Per Energy (DPPE) [10], and Green Grid Productivity Indicator [11] have been proposed to help data center operators improve the energy efficiency and reduce operation costs of data centers.

12.3.1 PUE and DCiE

The most commonly used metric to indicate the energy efficiency of a data center is PUE and its reciprocal DCiE [8]. PUE is defined as the ratio of the total power consumption of W data center to the total power consumption of IT equipment as follows:

$$\text{PUE} = \frac{\text{Total Power Consumption of a Data Center}}{\text{Total Power Consumption of IT Equipment}}, \tag{12.1}$$

$$\text{DCiE} = \frac{1}{\text{PUE}} = \frac{\text{Total Power Consumption of IT Equipment}}{\text{Total Power Consumption of a Data Center}}. \tag{12.2}$$

The PUE metric measures the total power consumption overhead caused by the data center facility support equipment, including the cooling systems, power delivery, and other facility infrastructure like lighting. PUE=1.0 implies that there is no power overhead load at all and all power consumption of the data center goes to the IT equipment. According to the Report to Congress on Server and Data Center Energy Efficiency [1], the average data center in the U.S. in 2006 has a PUE of 2.0, implying that 1 W of overhead power is used to cool and deliver every watt to IT equipment. It also predicts that "state-of-the-art" data center energy efficiency could reach a PUE of 1.2 [66]. Google publishes quarterly the PUE results from data centers with an IT load of at least 5 MW and time-in-operation of at least 6 months [67]. The latest trailing 12-month, energy-weighted average PUE result obtained in the first quarter of this year (2011) is 1.16, which exceeds the EPA's goal for state-of-the-art data center efficiency.

12.3.2 Data Center Energy Productivity (DCeP)

Energy efficiency and energy productivity are closely related to each other. Energy efficiency focuses on reducing unnecessary power consumption to produce a work output, while the energy productivity of a data center measures the quantity of useful work done relative to the amount of power consumption of a data center in producing this work. DCeP [9] allows the continuous monitoring of the productivity of a data center as a function of power consumed by a data center:

$$\text{DCeP} = \frac{\text{Useful Work Produced}}{\text{Total Data Center Power Consumed Producing This Work}}. \tag{12.3}$$

From the above equation, it is easy to see that the DCeP metric tracks the overall work product of a data center per unit of power consumption expended to produce this work.

12.3.3 Datacenter Performance Per Energy (DPPE)

The PUE metric only evaluates the data center facility power consumption. DPPE [10] is proposed as a new metric to evaluate the energy efficiency of data centers as a whole. The DPPE metric indicates data center productivity per unit energy, and it defines four submetrics, namely IT Equipment Utilization (ITEU), ITEE (IT Equipment Energy Efficiency), PUE, and GEC (Green Energy Coefficient). These four submetrics reflect four kinds of independent energy-saving efforts, and are designed to prevent one kind of energy-saving effort from affecting others:

(1) ITEU is essentially the average utilization factor of all IT equipment hosted in a data center. ITEU measures the degree of energy saving by efficient operation of IT equipment through virtual techniques and other operational techniques:

$$\text{ITEU} = \frac{\text{Total Measured Power of IT Equipment}}{\text{Total Rated Power of IT Equipment}}. \tag{12.4}$$

(2) ITEE is defined as the ratio of the total capacity of IT equipment to the total rated power of IT equipment. This metric aims to encourage the installation of equipment with high processing capacity per unit electric power in data centers to promote energy savings:

$$\text{ITEE} = \frac{\text{Total IT Equipment Capacity}}{\text{Rated Power of IT Equipment}}. \tag{12.5}$$

(3) PUE is already defined in Eq. (12.1). PUE indicates the power savings for data center facilities. The lower the power consumption of the facility infrastructure, the smaller the value of PUE.

(4) GEC is defined as the ratio of the green energy produced and used in a data center to its total power consumption. The value of GEC becomes larger if the production of non-CO_2 energy (i.e., photovoltaic power generation) is increased in a data center:

$$\text{GEC} = \frac{\text{Green Energy}}{\text{Total Power Consumption of a Data Center}}. \tag{12.6}$$

Considering the definitions of the above four submetrics, DPPE incorporates these four submetrics and can be expressed as a function of them as follows:

$$\text{DPPE} = \text{ITEU} \times \text{ITEE} \times \frac{1}{\text{PUE}} \times \frac{1}{1 - \text{GEC}}. \tag{12.7}$$

12.3.4 Green Grid Productivity Indicator

In order to understand and compare data centers in more than one dimension, the Green Grid Productivity Indicator [11] has been proposed as a multiparameter framework to evaluate overall data center efficiency. Through the use of a radial graph, relevant indicators such as DCiE, data center utilization, server utilization, storage utilization, and network utilization can quickly, concisely, and flexibly emerge to provide organizational awareness. Each of these indicators has values ranging from 0% to 100% where 100% is the theoretical maximum. Data center operators can set up the target value for each indicator. By plotting the peak and average values of each indicator during the period of monitoring, together with their target and theoretical maximum values on a radial graph, the data center operators are able to assess how well the data center resources are utilized easily, check if the business targets are achieved visually and quickly, and figure out how to spend their efforts to maximize the benefits.

12.4 TECHNIQUES TO IMPROVE ENERGY EFFICIENCY OF DATA CENTERS

The high operational cost and the mismatch between data center utilization and power consumption have spurred interest in improving data center energy efficiency. Energy savings in data centers can come from more energy-efficient hardware infrastructure, including servers, storage, and network equipment. Moreover, some smart power delivery and cooling technologies have been investigated and verified to be effective solutions to save energy. In contrast, tremendous efforts have been made to address power efficiency in data centers with power management in server and storage clusters. Some other techniques have also been proposed to save energy in data centers including power-aware routing and energy-efficient data center architectures. In this section, energy-saving approaches for data centers are reviewed and discussed.

12.4.1 IT Infrastructure Improvements

Approximately 40–60% of power consumption of a data center is devoted to IT infrastructure, which consists of servers, storage, and network equipment, and therefore energy efficiency of IT equipment would be very important to energy savings in data centers.

12.4.1.1 Servers and Storages

High power waste at low workload has prompted a fundamental redesign of each computer system component to exemplify the energy-proportional concept, especially processors since in the past they consumed more power than any other component (a 2005 study found that a processor was responsible for 55% of the power usage of a server [4]). Much research has been done to explore processor designs to reduce CPU power with dynamic voltage and frequency scaling (DVFS) [12,13]. DVFS can be deployed to save energy at the cost of slower program execution by reducing the voltage and frequency. The effectiveness of DVFS in saving energy with moderately intense Web workloads was examined in [3], and the results show that DVFS can save from 23% to 36% of the CPU energy while keeping server responsiveness within reasonable limits. Unfortunately, processors no longer dominate power consumption in modern servers. Processors currently contribute around 25% of the total system power consumption [4]. According to a study shown in [6], the chipset is the dominant constant power consumption in modern commodity servers.

Several techniques can be used to reduce power consumption of memory and disk subsystems [14–17]. A novel, system-level power management technique, power shifting, for increasing performance under constrained power budgets was proposed to rebudget the available power between processor and memory to maintain a server budget [14]. Power shifting is a threshold-based throttling scheme to limit the number of operations performed by each subsystem during an interval of time, but power budget violations and unnecessary performance degradation may be caused by improper interval length. Diniz et al. [15] proposed and evaluated four techniques, called Knapsack, LRU-Greedy, LRU-Smooth, and LRU-Ordered, to dynamically limit memory power consumption by adjusting the power states of the memory devices as a function of the load on the memory subsystem. They further proposed energy- and performance-aware versions of these techniques to trade off between power consumption and performance. Zheng et al. [16] proposed mini-rank, an adaptive DRAM architecture, to limit power consumption of DRAM by breaking a conventional DRAM rank into multiple smaller mini-ranks with a small bridge chip. Dynamic rotations per minute (DRPM) [17] was proposed as a low-level hardware-based technique to dynamically modulate disk speed to save power in disk drives since the slower the disk drive spins the less power it consumes.

In order to design an energy-proportional computer system, each system component needs to consume energy in proportion to utilization. However, many components incur fixed power overheads when active, such as the clock, and thus designing an energy-proportional computer system still remains a research challenge. Several system-level power management schemes have been proposed [18,19] to reduce power consumption by putting idle servers to sleep. Given the state of the art of energy efficiency of today's hardware, energy-proportional systems can be approximated with off-the-shelf non-energy-proportional hardware at the ensemble layer through dynamic virtual machine consolidation [18]. It is also believed that new alternative energy-efficient hardware designs will help design energy-proportional systems at both the single server and ensemble layer. However, this method works at the coarse time scale (minutes) and cannot address the performance isolation concerns of dynamic consolidation.

An energy-conservation approach, called PowerNap [19], was proposed to attune the server power consumptions to server utilization patterns. With PowerNap, the entire system transits rapidly between a high-performance active state and a minimal-power nap state in response to instantaneous load. PowerNap can be modeled

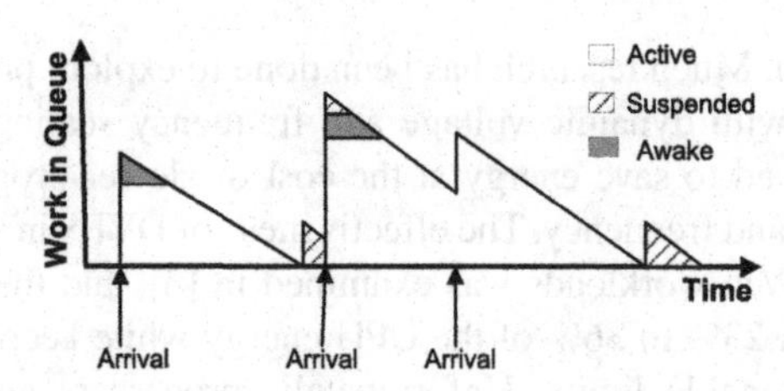

Figure 12.4 The PowerNap analytic model.

as an M/G/1 queuing system with arrival rate λ, and a generalized service time distribution with known first and second moments $E[s]$ and $E[s^2]$. Figure 12.4 shows the analytic models for PowerNap with three arrival jobs. Symmetric transmission latency T_t in and out of the minimal-power nap state is assumed. The service of the first job in each busy period is delayed by an initial setup time I, which includes the wake transition and the remaining portion of a suspend transition delay; it can be expressed as follows:

$$I = \begin{cases} 2T_t - x & \text{if } 0 \leq x < T_t, \\ T_t & \text{if } x \geqslant T_t. \end{cases} \tag{12.8}$$

The first and second moments $E[I]$ and $E[I^2]$ are given by

$$E[I] = \int_0^{\infty} I\lambda \, e^{-\lambda x} \, dx = 2T_t + \frac{1}{\lambda} e^{-\lambda T_t} - \frac{1}{\lambda},$$

$$E[I^2] = \int_0^{\infty} I^2 \lambda \, e^{-\lambda x} \, dx = 4T_t^2 - 2T_t^2 \, e^{-\lambda T_t} - \left(\frac{4T_t}{\lambda} + \frac{2}{\lambda^2}\right)\left[1 - (1 + \lambda T_t)\, e^{-\lambda T_t}\right]. \tag{12.9}$$

The average power with PowerNap can be expressed as

$$P_{avg} = P_{nap} \times F_{nap} + P_{max} \times (1 - F_{nap}), \tag{12.10}$$

where P_{nap} and P_{max} denote the server power consumption when it is in the nap state and active state, respectively. F_{nap} is the fraction of time spent napping, which is defined as the ratio of the expected length of each nap period to the expected busy-idle cycle length:

$$F_{nap} = \frac{\int_0^{T_t} t\lambda \, e^{-\lambda t} \, dt + \int_{T_t}^{\infty} (t - T_t)\lambda \, e^{-\lambda t} \, dt}{\frac{E[S]+E[I]}{1-\lambda E[S]} + \frac{1}{\lambda}} = \frac{e^{-\lambda T_t}(1 - \lambda E[S])}{1 + \lambda E[I]}. \tag{12.11}$$

The response time for an M/G/1 server with exceptional first service is

$$E[R] = \frac{\lambda E[S^2]}{2(1 - \lambda E[S])} + \frac{2E[I] + \lambda E[I^2]}{2(1 + \lambda E[I])} + E[S] \tag{12.12}$$

The first term of $E[R]$ is the Pollaczek–Khinchin formula for the expected queuing delay in a standard M/G/1 queue, the second term is additional residual delay caused by the initial setup time I, and the final term is the expected service time $E[R]$. The second term vanishes when $T_t = 0$.

Rather than requiring fine-grained power-performance states and complex load proportional operation from each system component, PowerNap minimizes idle power and transition time. According to the evaluation reported in [19], PowerNap can save more power with better response time than DVFS. However, a transition time of under 10ms is required for significant power savings; unfortunately, sleep times on current servers are two orders of magnitude larger.

12.4.1.2 Network Equipment

A power measurement study of a variety of networking gears, e.g., switches, routers, wireless access points, was performed in [68] to quantify power saving schemes. A typical networking router power consumption can be divided among four main components: chassis, switching fabric, line cards, and ports. The chassis alone consumes 25–50% of the total router power in a typical configuration [20]. Furthermore, power characteristics of current routers are not energy proportional; even worse, they consume around 90% of their maximum power consumption [20]. Prompted by the poor energy characteristics of modern routers, active research is being conducted in reducing power consumption of networking equipment, and researchers have suggested a few techniques to save energy [21,22], e.g., sleeping and rate-adaption. With a low power "sleep" mode, network equipment can stay in two states, a sleeping state and an active state. Network equipment transits into the low-power "sleep" mode when no transmission is needed, and returns back to the active mode when transmission is requested. However, the transition time overhead of putting a device into and out of the sleep mode may reduce energy efficiency significantly [23], and thus Reviriego et al. [24] explored the use of burst transmission to improve energy efficiency. Another technique to save power is to adapt the transmission rate of network operation to the offered workload [22], based on the fact that the lower the line speed is, the less power the devices consume. Speed negotiation is required in the rate-adaption scheme for both of the transmission ends. Speed negotiation requires from a few hundred milliseconds to a few seconds; this is excessive for many applications. The effectiveness of power management schemes to reduce power consumption of networks based on sleeping and rate-adaption was evaluated in [25]. It was shown that sleeping or rate-adaption can offer substantial savings.

Several schemes have been proposed to reduce power consumption of network switches in [26], including Time Window Prediction (TWP), Power Save Mode (PSM), and Lightweight Alternative. The theme of these schemes is to trade off some performance, latency, and packet loss to reduce power consumption. TWP and PSM schemes concentrate on intelligently putting ports to sleep during idle periods. In TWP, switches monitor the number of packets crossing a port in a sliding time window and predict the traffic in the next sliding window. If the number of packets in the current sliding time window is below a predefined threshold, the switch powers off the port for some time. Packets that arrive at the port in the low power state are buffered. Adaptive sleep time is used in TWP based on the prediction of the traffic for the next sliding window and latency requirements. The performance of TWP relies on the accuracy of the prediction function. The more accurate the prediction function is, the less latency it may cause. PSM is a special case of the TWP where the sleep happens with regularity instead of depending on the traffic flow. PSM is more of a policy-based scheme to power off ports and naturally causes more latency than TWP does. By observing a clear diurnal variation in the traffic patterns, lightweight alternative switches have been proposed. Lightweight alternative switches are low-power integrated switches with lower packet processing speed and line speeds. At low traffic load, only the lightweight alternative switches are powered on to provide the connection. An analytical framework, which can effectively be adopted to dynamically optimize power consumption of a network device while maintaining an expected forwarding performance level, was proposed in [27].

12.4.2 Power Distribution

Power distribution infrastructure is another major power consumer in data centers [5]. Current typical power delivery systems for data centers still use alternating current (AC) power [32], which is distributed from the utility to the facility, and is then stepped down via transformers and delivered to uninterruptible power supplies (UPS). The central UPS generally involves power conversion from AC power to DC power for energy storage to isolate equipment from power interruptions or other disturbances, and power reconversion from DC power back to AC power for facility power distribution to servers and other IT equipment in racks through power distribution units (PDUs). The AC power received at the server or other IT equipment is converted again to DC power within its power supply unit (PSU), which is stepped down again and distributed to different electronics.

It can be observed that several levels of power conversion exist in both data center facilities and within IT equipment that results in significant electrical power losses for a data center, including power losses in UPS, transformers, and power line losses. UPS efficiency varies greatly between various UPS loads. According to a benchmark study of UPS efficiency in data centers performed by Ecos Consulting and EPRI Solutions [28,29], the average load factor of UPSs tested in the field is 37.8%, and the corresponding average UPS efficiency tested in the field is 85.2%. Transformer efficiency varies between 85% and 98% [30]. Major improvements in power technologies have been made in the past few years to reduce power losses at power infrastructure equipment [31]. In order to reduce power conversion stages as those in AC power distribution systems, a DC power distribution system has been demonstrated and evaluated for data centers [32]. With a DC power distribution system, UPSs can obtain an extra 5–10% of energy efficiency. Many materials for DC power delivery for data centers are available at the website [33].

12.4.3 Smart Cooling and Thermal Management

The state of air and liquid cooling solutions in data centers has been reviewed in [34]. Currently, most data centers use liquid cooling for computer room air conditioning (CRAC). Recently, several types of rack-level liquid-cooling solutions have been developed to bring chilled water or liquid refrigerant closer to the servers. One is rear-door liquid cooling with sealed tubes filled with chilled water that cools the air leaving the server, essentially making the rack thermally neutral to the room. Another strategy is a sealed rack connecting to a closed-liquid cooling system. At the sealed rack, the contained airflow within the rack is cooled through a heat exchanger at the rack bottom and the heat is removed in the connected liquid-cooling system. Moreover, in-row liquid-coolers are embedded in rows of data center racks to provide localized air distribution and management. With in-row liquid-coolers, the chilled water or refrigerant piping is run overhead or under the floor to each individual cooler. Suspending from the ceiling, overhead liquid-coolers captures the hot air arising from the hot aisle, cools it, and releases it back to the cold aisle. Data center liquid cooling techniques tend to use naturally cooled water [35], like lake or sea water, which provides electrical savings by eliminating or reducing the need for water chillers in data centers.

The predominant air cooling scheme for current data centers is to use the CRAC units and an under-floor cool air distribution system as shown in Figure 12.5. Racks are typically arranged in rows with alternating airflow directions, and thus "hot" and "cold" aisles are formed that separate the cold air supply from the exhaust air and increase the overall efficiency of the cold air delivery and exhaust air collection from each rack in data centers. The CRAC units supply cold air to a raised floor plenum

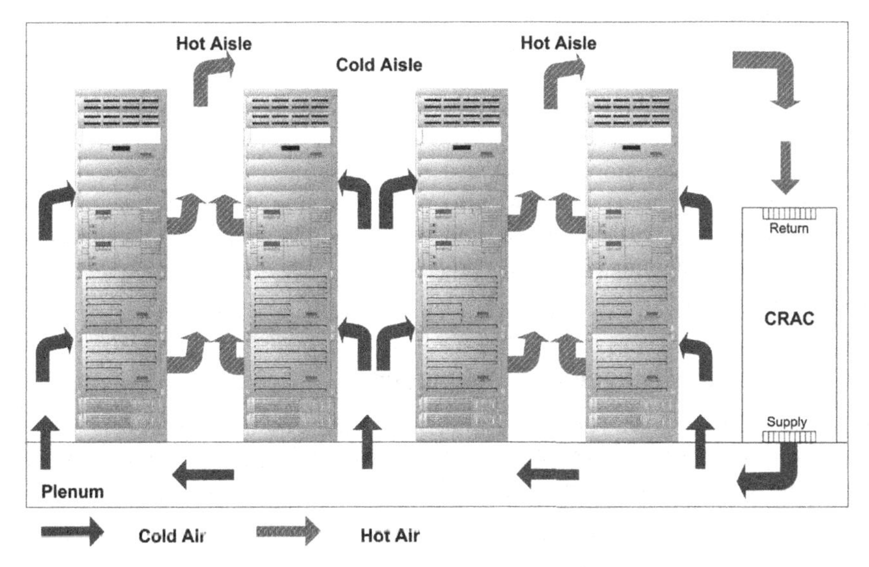

Figure 12.5 The airflow schematics for a typical raised floor plenum and room return air cooling solution.

underneath the racks, and the perforated tiles located near the racks deliver the cool air to the front of the rack servers. The hot exhaust air from the racks is then collected from the upper portion of the facility by the CRAC units, where the hot air is then cooled to complete the airflow loop. Alternative cool air supply and hot exhaust return configurations are also possible to improve the thermal efficiency [36,37]. Green data center air cooling has been proposed by using outdoor air to cool data centers. Hewlett–Packard has officially opened its green data center cooled entirely by cold wind [38]. A hybrid air and liquid data center cooling scheme is described in [39].

Early works on cooling techniques for data centers focused on computational fluid dynamic models to analyze and design server racks and data center configurations to optimize the delivery of cold air/liquid to minimize cooling costs [40,41]. In order to improve cooling efficiency, thermal modeling methodologies have been developed for data centers, and an assessment of these models have been made in [36] with computational fluid dynamics and heat transfer tools. A proactive thermal management approach for green data centers is presented in [37]. Some smart cooling technologies have been investigated and verified to be effective in saving energy. Distributing the workload onto those servers that are more efficient to cool than others was proposed and verified to reduce power consumption of data centers [42,43]. Parolini et al. [44] proposed a coordinated cooling and load management strategy to reduce data center power consumption, and formulated the energy management problem as a constrained Markov decision process, which can be solved by linear programming to generate the optimal strategies for power management. Vasic et al. [45] proposed a thermodynamic model of a data center and designed a novel model-based temperature control strategy, by combining air flow control for long time-scale decisions and thermal-aware scheduling for short time-scale workload fluctuations. Load distribution in a thermal-aware manner incurs the development of fast and accurate estimation techniques for temperature distribution of a data center [46,47]. Several proposals have been made to reduce power consumption in data centers by keeping as many servers as necessary active and putting the rest into low-power sleep mode. However, these proposals may cause hot spots in data centers, thus increasing the cooling power and degrading response time due to sleep-to-active transition delays. In order to solve these problems, PowerTrade [48], a novel joint optimization of idle power and cooling power, was proposed to reduce the total power consumption of data centers.

12.4.4 Power Management Techniques

Tremendous efforts have been made to reduce power consumption in data centers with power management techniques, such as server provisioning, application consolidation, and server virtualization.

12.4.4.1 *Provisioning*

In telecommunications, provisioning is the process of preparing and initializing a network or equipment for application services. In data centers, provisioning is an effective solution to reduce power consumption by turning off the idle servers, storage, and network equipment, or by putting them into a lower power mode.

Chen et al. [49] proposed to reduce power usage in data centers by adopting dynamic server provisioning techniques, which are effective to dynamically turn on a minimum number of servers required to satisfy application-specific quality of service, and load dispatching, which distributes the current load among the running machines. This work specially deals with long-lived connection-intensive Internet services. A power-proportional cluster [50], consisting of a power-aware cluster manager and a set of heterogeneous machines, was designed to make use of currently available energy-efficient hardware, mechanisms for transiting in and out of low-power sleep states, and dynamic provisioning and scheduling to minimize power consumption. This work specially deals with short lived request-response type of workloads. Based on queuing theory and service differentiation, an energy proportional model was proposed for server clusters in [51], which can achieve theoretically guaranteed service performance with accurate, controllable, and predictable quantitative control over power consumption.

Some other techniques benefit from inactive low energy modes in disks. A log-structured file system solution was proposed in [52] to reduce disk array power consumption by powering off a fraction of disks without incurring unacceptable performance penalties because a stripping system could perfectly predict disks for write-access. Based on the observation of significant diurnal variation of data center traffic, Sierra [53], a power-proportional, distributed storage system, was proposed to turn off a fraction of storage servers during tough traffic period. Sierra utilizes a set of techniques including power-aware layout, predictive gear scheduling, and a replicated short-term store, to maintain the consistency and fault-tolerance of the data, as well as good system performance. A power-proportional distributed file system, called Rabbit, was proposed in [54] to provide ideal power-proportionality for large-scale cluster-based storage and data-intensive computing systems by using a new cluster-based storage data layout. Rabbit can maintain near ideal power proportionality even with node failures.

With servers becoming more energy proportional, the network power cannot be ignored. In a recent work, a network-wide power manager, ElasticTree [55], was proposed to dynamically adjust the set of active network elements, links, and switches, to satisfy changing data center traffic loads. ElasticTree optimizes a data center by finding the power-optimal network subset to route the traffic. It essentially extends the idea of power proportionality into the network domain, as described in [4]. Given the data center network topology, the traffic demand matrix, and the power consumption of each link and node, ElasticTree minimizes the total power consumption of a data center by solving a capacitated multicommodity cost flow (CMCF) optimization problem. The data center network architecture can be modeled by a graph $G = (V, E)$, where V and E are the sets of vertices and edges, respectively. There are k traffic flows $K_1, K_2, \ldots, K_k$, defined by $K_i = (s_i, t_i, d_i)$, where, for flow i, s_i is the source, t_i is the destination, and d_i is the traffic demand. Based on the above definitions, the CMCF problem can be formulated as follows:

The objective function is to minimize

$$P_{\text{total}} = \sum_{(u,v)\in E} x_{u,v} \times P^L_{u,v} + \sum_{u\in V} y_u \times P^N_u, \tag{12.13}$$

where $P^L_{u,v}$ represents the power consumption of link $(u,v) \in E$, P^N_u represents the power consumption of node $u \in V$, and $x_{u,v}$ and y_u denote the power status of link $(u,v) \in E$ and node $u \in V$, respectively.

The minimization problem is constrained by the following:

(1) *Flow conservation:* Commodities are neither created nor destroyed at intermediate nodes:

$$\sum_{w\in V} f_i(u,w) = 0 \quad (\forall u \neq s_i, u \neq t_i, i \in [1,k]), \tag{12.14}$$

where $f_i(u,v)$ denotes the traffic flow i along edge $(u,v) \in E$.

(2) *Demand satisfaction:* For each flow i, source s_i sends and sink t_i receives an amount of traffic equal to the traffic demand d_i:

$$\sum_{w\in V} f_i(s_i,w) = \sum_{w\in V} f_i(w,t_i) = d_i \quad (\forall i \in [1,k]). \tag{12.15}$$

(3) *Capacity and utilization constraint:* The total flow along each link $f(u,\ v)$ ($\forall (u,v) \in E$) must be smaller than the link capacity weighed by the link utilization requirement factor α:

$$f(u,v) = \sum_{i=1}^{k} f_i(u,v) \leqslant \alpha C_{uv} x_{uv}, \quad \forall (u,v) \in E, \tag{12.16}$$

where $C_{u,v}$ denotes the capacity of the link $(u,v) \in E$ and $\alpha \in [0, 1]$ represents the maximum link utilization.

(4) *Link power is bidirectional:* A link must be powered on if traffic is flowing in either direction:

$$x_{u,v} = x_{v,u} \quad \forall (u,v) \in E. \tag{12.17}$$

(5) *Switch turn off rule:* A node can be turned off only if all links connected to it are actually turned off:

$$\sum_{w\in V} x_{w,u} \leqslant M y_u, \quad \forall u \in V, \tag{12.18}$$

where M is the total number of links connected to swith $u \in V$.

Mahadevan et al. [56] proposed a network-wide energy monitoring tool, Urja, which can be integrated into network management operations to collect configuration and traffic information from live network switches and accurately predict their power consumption. By analyzing real network traces, several techniques, including disabling unused ports, port rate-adaption, maximizing active ports on a line card, and using fewer switches, were proposed to achieve significantly closer to power-proportional behavior with non-power-proportional devices.

12.4.4.2 Consolidation

Significant opportunities of power savings exist at the application layer, which has the most information on performance degradation and energy trade-offs. As verified in [57], consolidation of applications in cloud computing environments can present a significant opportunity for energy optimization. Their study reveals the energy performance trade-offs for consolidation and shows that optimal operating points exist and the application consolidation problem can be modeled as a modified bin-packing problem. Kansal and Zhao [58] enabled generic application-layer energy optimization, which guides the design choices using energy profiles of various resource components of an application. Intelligent data placement and/or data migration can be used to save energy in storage systems. Hibernator [59], a disk array energy management system, uses several techniques to reduce power consumption while maintaining performance goals, including disk drives that rotate at different speeds and migration of data to an appropriate-speed disk drive.

12.4.4.3 Virtualization

Server virtualization is another effective way to enhance server utilization, consolidate servers and reduce the total number of physical servers, especially when the compatibility or security issues prevent service consolidation. The effects of the reduced number of servers on power consumption of a data center are twofold. First, the power consumption caused by servers is reduced. Second, the cooling requirement should also be reduced comparably. According to [60], data center power consumption can be reduced by up to 50%. Virtualization can also be helpful to build energy proportional storage systems. Sample-Replicate-Consolidate Mapping (SRC-Map) [61], a storage virtualization solution for energy-proportional storage, was proposed by consolidating the cumulative workload on a minimal subset of physical volumes proportional to the I/O workload intensity. GreenCloud [62] enables comprehensive online monitoring, live virtual machine migration, and virtual machine placement optimization to reduce data center power consumption while guaranteeing the performance goals.

12.4.5 Others

Some other techniques have been proposed to reduce the power consumption of data centers from the perspectives of routing [63] and data center architecture [64]. The energy-saving problem in data centers was solved from a routing perspective in [63]. They established the model of energy-aware routing, in which the objective is to find a route for a given traffic matrix that minimizes the total number of switches. They proved that the proposed energy-aware routing model is NP-hard, and designed a heuristic algorithm to solve the energy-aware routing problem. Abts et al. [64]

proposed a new high-performance data center architecture, called energy proportional data center networks, and showed that a flattened butterfly data center topology is inherently more power efficient than the other commonly proposed topology for high-performance data centers.

12.5 CONCLUSIONS

Power consumption is a central critical issue for data centers. In this chapter, we have reviewed the power consumption and energy efficiency of data centers. The electricity usage of data centers has almost doubled from 2000 to 2005, but the rapid growth of data center power consumption has slowed down significantly from 2005 to 2010 owing to the economic slowdown of the 2008 financial crisis, and a significant reduction in actual server installed base with improved virtualization techniques. There was a 56% increase in power consumption from 2005 to 2010 for worldwide data centers, but only about a 36% increase in U.S. data centers. In order to quantify the energy efficiency of data centers, several energy-efficiency metrics for data centers have also been presented such as PUE, DCiE, DCeP, and DPPE. Finally, techniques to improve energy efficiency of data centers are discussed. Numerous studies have shown that IT infrastructure, power distribution infrastructure, and cooling infrastructure are the three main contributors to power consumption of data centers. Therefore, techniques to improve energy efficiency of these hardware infrastructures can decrease the power usage of data centers. The power overhead in data centers caused by power distribution and cooling can be reduced by smart power delivery and cooling technologies. Furthermore, power management techniques, such as provisioning, virtualization, and consolidation, is provided as an effective solution for energy savings in data centers. Some other techniques, such as power-aware routing and energy-efficient data center architecture design, are also candidates to improve energy efficiency of data centers.

REFERENCES

[1] U.S. Environmental Protection Agency, Environmental Protection Agency, Report to Congress on Server and Data Center Energy Efficiency Public Law 109-431, ENERGY STAR Program, August 2007.

[2] APC White Paper #6, Determining total cost of ownership for data center and network room infrastructure, 2005. <http://www.linuxlabs.com/PDF/Data%20Center%20Cost%20of%20Ownership.pdf>.

[3] P. Bohrer, E.N. Elnozahy, T. Keller, M. Kistler, C. Lefurgy, C. McDowell, R. Rajamony, The case for power management in web servers, Power Aware Computing, 2002, pp. 261–289.

[4] L.A. Barroso, U. Hölzle, The case for energy-proportional computing, Computer 40 (12) (2007) 33–37.

[5] N. Rasmussen, APC White Paper #113, Electrical efficiency modeling for data centers, 2007. <http://www.apcmedia.com/salestools/SADE-5TNRLG_R5_EN.pdf>.

[6] S. Dawson-Haggerty, A. Krioukov, D. Culler, Power optimization – a reality check, Technical report UCB/EECS-2009-140, EECS Department, University of California, Berkeley, October 2009.

[7] S. Rivoire, P. Ranganathan, C. Kozyrakis, A comparison of high-level full-system power models, in: Proceedings of the 2008 Conference on Power aware Computing and Systems, San Diego, California, 2008.

[8] The Green Grid, The green grid data center power efficiency metrics: PUE and DCiE, December 2007. <http://www.thegreengrid.org/gg_content/TGG_Data_Center_Power_Efficiency_Metrics_PUE_and_DCiE.pdf>.

[9] The Green Grid, A framework for data center energy productivity, May 2008. <http://www.thegreengrid.org/~/media/WhitePapers/WhitePaper13FrameworkforDataCenterEnergyProductivity5908.ashx?lang=en>.

[10] Green IT Promotion Council, Concept of new metrics for data center energy efficiency: introduction of datacenter performance per energy (DPPE), February 2010. <http://www.greenit-pc.jp/topics/release/pdf/dppe_e_20100315.pdf>.

[11] The Green Grid, The green grid productivity indicator. <http://www.thegreengrid.org/~/media/WhitePapers/White_Paper_15_-_TGG_Productivity_Indicator_063008.pdf?lang=en>.

[12] T. Pering, T. Burd, R. Brodersen, Dynamic voltage scaling and the design of a low-power microprocessor system, in: Proceedings of Power Driven Microarchitecture, Workshop, June 1998.

[13] G. Semeraro, G. Magklis, R. Balasubramonian, D.H. Albonesi, S. Dwarkadas, M.L. Scott, Energy-efficient processor design using multiple clock domains with dynamic voltage and frequency scaling, in: Proceedings of the Eighth International Symposium on High-Performance Computer, Architecture, February 2002, pp. 29–40.

[14] W. Felter, K. Rajamani, T. Keller, C. Rusu, A performance-conserving approach for reducing peak power consumption in server systems, in: Proceedings of the 19th International Conference on Supercomputing, Cambridge, Massachusetts, 2005, pp. 293–302.

[15] B. Diniz, D. Guedes, W. Meira, R. Bianchini, Limiting the power consumption of main memory, in: Proceedings of the 34th Annual Symposium on Computer Architecture, San Diego, 2007, pp. 290–301.

[16] H. Zheng, J. Lin, Z. Zhang, E. Gorbatov, H. David, Z. Zhu, Mini-rank: adaptive dram architecture for improving memory power efficiency, in: Proceedings of the 41st Annual IEEE/ACM International Symposium on Microarchitecture, Washington, 2008, pp. 210–221.

[17] S. Gurumurthi, A. Sivasubramaniam, M. Kandemir, H. Franke, DRPM: dynamic speed control for power management in server class disks, in: Proceedings of the 30th Annual International Symposium on Computer Architecture, San Diego, California, June 2003, pp. 169–181.

[18] N. Tolia, Z. Wang, M. Marwah, C. Bash, P. Ranganathan, X. Zhu, Delivering energy proportionality with non energy-proportional systems-optimizing the ensemble, in: First Workshop on Power Aware Computing and Systems, San Diego, CA, December 7, 2008.

[19] D. Meisner, B.T. Gold, T.F. Wenisch, PowerNap: eliminating server idle power, in: Proceedings of the 14th International Conference on Architectural Support for Programming Languages and Operating Systems, Washington, 2009, pp. 205–216.

[20] J. Chabarek, J. Sommers, P. Barford, C. Estan, D. Tsiang, S. Wright, Power awareness in network design and routing, in: Proceedings of INFOCOM, Phoenix, USA, April 13–18, 2008, pp. 457–465.
[21] IEEE P802.3az Energy Efficient Ethernet Task Force. <http://grouper.ieee.org/groups/802/3/az/public/index.html>.
[22] C. Gunaratne, K. Christensen, B. Nordman, S. Suen, Reducing the energy consumption of ethernet with adaptive link rate (ALR), IEEE Transactions on Computers 57 (4) (2008) 448–461.
[23] P. Reviriego, J.A. Hernandez, D. Larrabeiti, J.A. Maestro, Performance evaluation of energy efficient Ethernet, IEEE Communications Letters 13 (9) (2009) 697–699.
[24] P. Reviriego, J.A. Maestro, J.A. Hernándndez, D. Larrabeiti, Burst transmission for energy-efficient Ethernet, IEEE Internet Computing 14 (4) (2010) 50–57.
[25] S. Nedevschi, L. Popa, G. Iannaccone, S. Ratnasamy, D. Wetherall, Reducing network energy consumption via sleeping and rate-adaption, in: Proceedings of USENIX Symposium on Networked Systems Design and Implementation, San Francisco, USA, 2008, pp. 323–336.
[26] G. Ananthanarayanan, R.H. Katz, Greening the switch, in: Proceedings of Power Aware Computing and Systems Conference, San Diego, California, December 2008.
[27] R. Bolla, R. Bruschi, F. Davoli, A. Ranieri, Performance constrained power consumption optimization in distributed network equipment, in: Proceedings of International Conference on Communications, Dresden, Germany, June 14–18, 2009.
[28] M. Ton, B. Fortenbury, High performance buildings: data centers uninterruptible power supplies (UPS), December 2005. <http://hightech.lbl.gov/documents/UPS/Final_UPS_Report.pdf>.
[29] N. Rasmussen, J. Spitaels, APC White Paper #127, A quantitative comparison of high efficiency AC vs. DC power distribution for data centers, 2007. <http://www.apcmedia.com/salestools/NRAN-76TTJY_R2_EN.pdf>.
[30] TEAM Companies, Enterprise data center: a look at power efficiencies, March 2009. <http://www.team-companies.com/wp-content/uploads/2010/05/PowerEfficiencies2.pdf>.
[31] M. Brown, Utility infrastructure improvements for energy efficiency understanding the supply side opportunity, November 2010. <http://www.state.mn.us/mn/externalDocs/Commerce/CARD_Utility_Infrastructure_Improvements_for_Energy_Efficiency_030311035704_UtilityInfrastructureImprovementsEE.pdf>.
[32] M. Ton, B. Fortenbery, W. Tschudi, DC power for improved data center efficiency, March 2008. <http://hightech.lbl.gov/documents/DATA_CENTERS/DCDemoFinalReport.pdf>.
[33] DC Power for Data Centers of the Future. <http://hightech.lbl.gov/dc-powering/documents>.
[34] M.K. Patterson, D. Fenwick, The state of data center cooling: a review of current air and liquid cooling solutions, White Paper, Digital Enterprise Group, March 2008. <http://download.intel.com/technology/eep/data-center-efficiency/state-of-date-center-cooling.pdf>.
[35] R. Miller, Google Using Sea Water to Cool Finland Project, September 2010. <http://www.datacenterknowledge.com/archives/2010/09/15/google-using-sea-water-to-cool-finland-project/>.
[36] J. Rambo, Y. Joshi, Modeling of data center airflow and heat transfer: state of the art and future trends, Distributed and Parallel Databases 21 (2007) 193–225.

[37] E.K. Lee, I. Kulkarni, D. Pompili, M. Parashar, Proactive thermal management in green datacenters, Journal of Supercomputing (2010) 1–31.

[38] A. Nusca, HP opens first wind-cooled green data center, February 2010. <http://www.smartplanet.com/blog/smart-takes/hp-opens-first-wind-cooled-green-data-center-most-efficient-to-date/4191>.

[39] B.A. Rubenstein, R. Zeighami, R. Lankston, E. Peterson, Hybrid cooled data center using above ambient liquid cooling, in: Proceedings of the 12th IEEE Intersociety Conference on Thermal and Thermomechanical Phenomena in Electronic Systems, June 2010, pp. 1–10.

[40] C.D. Patel, R. Sharma, C.E. Bash, A. Beitelmal, Thermal considerations in cooling large scale high compute density data centers, in: Proceedings of Conference on Thermal and Thermomechanical Phenomena in Electronic Systems, 2002, pp. 767–776.

[41] C. Patel, C. Bash, R. Sharma, M. Beitelmam, R. Friedrich, Smart cooling of data centers, in: Proceedings of InterPack, July 2003.

[42] C. Bash, G. Forman, Cool job allocation: measuring the power savings of placing jobs at cooling-efficient locations in the data center, Technical report HPL-2007-62, HP Laboratories, 2007.

[43] J. Moore, J. Chase, P. Ranganathan, R. Sharma, Making scheduling "cool": temperature-aware workload placement in data centers, in: Proceedings of USENIX Annual Technical Conference, 2005, pp. 61–75.

[44] L. Parolini, B. Sinopoli, B.H. Krogh, Reducing data center energy consumption via coordinated cooling and load management, in: Proceedings of Power Aware Computing and Systems, San Diego, 2008.

[45] N. Vasic, T. Scherer, W. Schott, Thermal-aware workload scheduling for energy efficient data centers, in: Proceedings of the Seventh International Conference on Autonomic Computing, Washington, DC, USA, 2010, pp. 169–174.

[46] Q.H. Tang, T. Mukherjee, S.K.S. Gupta, P. Cayton, Sensor-based fast thermal evaluation model for energy efficient high-performance datacenters, in: Proceedings of the Fourth International Conference on Intelligent Sensing and Information Processing, October 2006, pp. 203–208.

[47] C.M. Liang, J. Liu, L. Luo, A. Terzis, F. Zhao, RACNet: a high-fidelity data center sensing network, in: Proceedings of the Seventh ACM Conference on Embedded Networked Sensor Systems, Berkeley, California, 2009, pp. 15–28.

[48] F. Ahmad, T.N. Vijaykumar, Joint optimization of idle and cooling power in data centers while maintaining response time, in: Proceedings of ASPLOS on Architectural Support for Programming Languages and Operating Systems, Pittsburgh, USA, 2010, pp. 243–256.

[49] G. Chen, W. He, J. Liu, S. Nath, L. Rigas, L. Xiao, F. Zhao, Energy-aware server provisioning and load dispatching for connection-intensive internet services, in: Proceedings of USENIX Symposium on Networked Systems Design and Implementation, San Francisco, 2008, pp. 337–350.

[50] A. Krioukov, P. Mohan, S. Alspaugh, L. Keys, D. Culler, R.H. Katz, NapSAC: design and implementation of a power-proportional web cluster, in: Proceedings of the First ACM SIGCOMM Workshop on Green Networking, New Delhi, India, August 30–September 3, 2010.

[51] X. Zheng, Y. Cai, Achieving energy proportionality in server clusters, International Journal of Computer Networks (IJCN) 1 (2) (2010) 21–35.

[52] L. Ganesh, H. Weatherspoon, M. Balakrishnan, K. Birman, Optimizing power consumption in large scale storage systems, in: Proceedings of the 11th USENIX Workshop on Hot Topics in Operating Systems, San Diego, CA, 2007, pp. 1–6.

[53] E. Thereska, A. Donnelly, D. Narayanan, Sierra: a power-proportional, distributed storage system, Technical report MSR-TR-2009-153, Microsoft Research Ltd., November 2009.
[54] H. Amur, J. Cipar, V. Gupta, G.R. Ganger, M.A. Kozuch, K. Schwan, Robust and flexible power-proportional storage, in: ACM Symposium on Cloud Computing, Indianapolis, Indiana, June 2010.
[55] B. Heller, S. Seetharaman, P. Mahadevan, Y. Yiakoumis, P. Sharma, S. Banerjee, N. McKeown, ElasticTree: saving energy in data center networks, in: Proceedings of the Seventh ACM/USENIX Symposium on Networked Systems Design and Implementation, San Jose, CA, April 2010, pp. 249–264.
[56] P. Mahadevan, S. Banerjee, P. Sharma, Energy proportionality of an enterprise network, in: Proceedings of ACM SIGCOMM, New Delhi, India, August 30–September 3, 2010.
[57] S. Srikantaiah, A. Kansal, F. Zhao, Energy aware consolidation for cloud computing, in: Proceedings of Conference on Power Aware Computing and Systems, San Diego, California, December 2008.
[58] A. Kansal, F. Zhao, Fine-grained energy profiling for power-aware application design, SIGMETRICS Performance Evaluation Review 36 (2) (2008) 26–31.
[59] Q.B. Zhu, Z.F. Chen, L. Tan, Y.Y. Zhou, K. Keeton, J. Wilkes, Hibernator: helping disk arrays sleep through the winter, in: Proceedings of the 20th ACM Symposium on Operating Systems Principles, Brighton, UK, 2005, pp. 177–190.
[60] Fujitsu Siemens Computers, White Paper, Green data center and virtualization – reducing power by up to 50%, August 2007. <http://webobjects.cdw.com/webobjects/media/PDF/Virtualization-Green-Data-Center.pdf>.
[61] A. Verma, R. Koller, L. Useche, R. Rangaswami, SRCMap: energy proportional storage using dynamic consolidation, in: Proceedings of the Eighth USENIX Conference on File and Storage Technologies, San Jose, California, 2010.
[62] L. Liu, H. Wang, X. Liu, X. Jin, W.B. He, Q.B. Wang, Y. Chen, GreenCloud: a new architecture for green data center, in: Proceedings of the Sixth International Conference on Autonomic Computing and Communications Industry Session, Barcelona, Spain, 2009, pp. 29–38.
[63] Y.F. Shang, D. Li, M.W. Xu, Energy-aware routing in data center network, in: Proceedings of the First ACM SIGCOMM Workshop on Green Networking, New Delhi, India, August 30–September 3, 2010, pp. 1–8.
[64] D. Abts, M.R. Marty, P.M. Wells, P. Klausler, H. Liu, Energy proportional datacenter networks, in: Proceedings of the Fifth International Conference on Emerging Networking Experiments and Technologies, Saint-Malo, France, June 19–23, 2010, pp. 338–347.
[65] Jonathan G. Koomey, Growth in Data Center Electricity Use 2005 to 2010, Analytics press, Oakland, CA, August 2011.
[66] Silicon Valley Leadership Group, Data center energy forecast report, July 2008. <https://microsite.accenture.com/svlgreport/Documents/pdf/SVLG_Report.pdf>.
[67] Google Data Centers. <http://www.google.com/corporate/datacenter/efficient-computing/measurement.html>.
[68] P. Mahadevan, P. Sharma, S. Banerjee, P. Ranganathan, A power benchmarking framework for network devices, in: Proceedings of the Eighth International IFIP-TC Networking Conference, Aachen, Germany, 2009, pp. 795–808.

CHAPTER

Energy-Efficient Sensor Networks

13

Megha Gupta[a], Mohammad S. Obaidat[b,1], Sanjay K. Dhurandher[c]

[a]Division of Computer Engineering, Netaji Subhas Institute of Technology, University of Delhi, Azad Hing Fauj Marg Sector 3, Dwarka (Pappankalan), New Delhi 110 078, India,
[b]Department of Computer Science and Software Engineering, Monmouth University, West Long Branch, NJ 07764, USA,
[c]Division of Information Technology, Netaji Subhas Institute of Technology, University of Delhi, Azad Hing Fauj Marg Sector 3, Dwarka (Pappankalan), New Delhi 110 078, India

13.1 INTRODUCTION

A *sensor network* [1–4] is a collection of sensors that can communicate between themselves. Sensors are tiny devices that can sense their residing environment's various activities due to their special manufacturing features. Advanced technologies, such as micro-electro-mechanical systems (MEMS)/wireless communications and digital electronics, have led to the design and manufacture of low-cost, low-power, multifunctional sensor nodes that are small in size and can provide easy communication for short distances. These tiny sensor nodes make up a large network through their cooperative work.

A sensor node consists of the following special features:

- *Sensing device:* A device to capture the activity taking place around it.
- *Data processing device:* A device to process the captured data, analyze it, and determine what information is relevant.
- *Communicating device:* A device to share the data/information with the other neighboring sensors in the network.

A sensor node can perform certain special types of tasks irrespective of the type of application: a sensing task for detection of any event, a task of collection for gathering and analyzing of the information, and a dissemination task for distributing or sharing of the information.

A sensor node can be deployed easily to build a network and inspect a particular phenomenon in the targeted environment. There can be different types of sensors such as thermal, visual, infrared, humidity, pressure, temperature, acoustic,

[1]Fellow of IEEE and Fellow of SCS.

magnetic, seismic, radar, vehicular, light intensity, sound intensity, chemical concentration, water current direction, speed, etc.

With the help of the above types of sensors, a wide range of sensor applications can be designed. Applications can be classified as *underwater* (UW) *applications* and *nonunderwater* (NUW) *applications*. UW applications perform beneath the water, whereas NUW applications are used on land. Underwater static and mobile sensor nodes examples are shown in Figures 13.1–13.3.

Some of the applications of sensor networks [3, 4] include environmental monitoring (weather change/indoor climate control); habitat monitoring of animals and plants (on land/underwater) [5] such as tracking the movements of birds, small animals, water animals, and insects; surveillance and verification of agreement between two nations; structural monitoring (e.g., buildings/bridges) performed on the basis of the data collected periodically; condition-based equipment maintenance; agriculture monitoring and soil makeup; medical applications [6] that include integrated patient monitoring, diagnostics, and tracking and monitoring doctors and patients inside a hospital; urban terrain mapping; natural disaster monitoring (forest fires/earthquakes/volcanoes/flood, etc.); traffic monitoring; smart home appliances; and inventory management and military applications that include battlefield surveillance, friendly/hostile forces tracking, monitoring of equipment, and biological attack detection, among others.

Figure 13.1 An underwater sensor node chip. A small chip comprising all the components: (1) terminal block for solar panel or external 12 V supply; (2) Molex connector for battery (paralleled with connector 1); (3) debugging interface that can be used to monitor phone communications using a PC serial port; (4) ICD2 interface for programming the PIC; (5) Molex connector to mobile phone; and (6) Molex connector to underwater sensor (*courtesy* http://pei.ucc.ie/daithi/construction.html).

Major limiting factors in sensor networks applications are the sensors. Sensors are the main culprit behind the failure of any application due to their following features:

- Limited in computation
- Low memory
- Low power resources
- Slow communication speed

Figure 13.2 Static underwater sensor node [41]; AQUAFLECKS and a mobile node Amour AUV(Autonomous Underwater Vehicle).

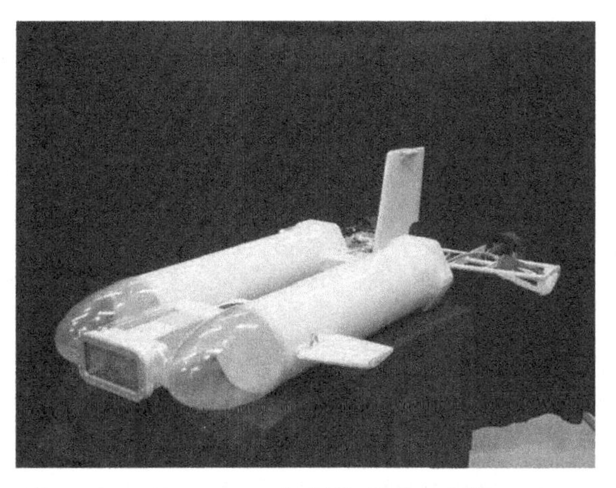

Figure 13.3 A mobile underwater sensor node [41]; Starbug AUV.

- Small bandwidth
- May not have global identification
- More prone to failures due to harsh deployment environments and energy constraints
- Need to be densely deployed in most environments

Due to their small size, sensors are deficient in almost all resources. Energy is the key player out of all the resources, and its shortage creates lots of problems in the network.

The next section describes wireless sensor networks (WSNs), underwater sensor networks (UWSNs), and the importance of energy resources in the life of the network. Section 13.3 reviews the related works and briefly describes some of the energy-efficient routing techniques for terrestrial and UWSNs. Section 13.4 contains the conclusion.

13.2 WIRELESS SENSOR NETWORKS (WSNS)

Wireless sensor networks (*WSNs*) [2, 7, 8] are emerging as a new and hot research technology in the field of computer networking. Researchers are modeling the innovating applications of WSNs. WSNs have captured attention due to the recent advances in micro-electro-mechanical systems [9–12]. It is now practical and economical to build small and low-cost sensors; hence researchers can easily develop and test their WSN applications.

Figure 13.4 shows a WSN. As shown in the figure, the network between the sensor devices is established through the radio component of the sensors. A wireless sensor network can consist of hundreds of sensor nodes. During the communication process, the sensor nodes exchange information and discover the neighboring nodes

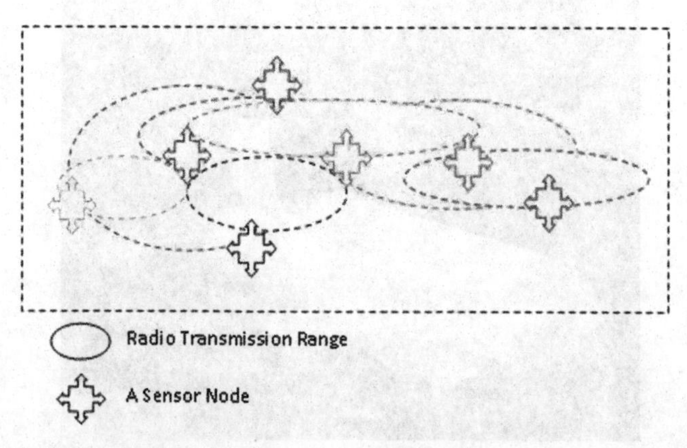

Figure 13.4 A wireless sensor network.

easily. As per the application requirement the sensor nodes determine the route for data forwarding in the wireless sensor network.

13.2.1 Wireless Underwater Sensor Network (WUWSN)

In recent years, underwater sensor networks (UWSNs) [13–23] have been widely used in underwater environments for a wide range of applications, such as pollution monitoring, health monitoring of marine creatures, sensing naval activities, and information gathering. Monitoring an area includes detection, identification, localization, and tracking objects, which are the main objectives of the sensor-based applications. A wireless underwater sensor network (WUWSN) consists of a number of sensor nodes, which can perform various functions such as data sensing, data processing, and communication with the other nodes. WUWSNs are different from the ground-based wireless sensor networks in terms of the communication methods and the mobility of the nodes. For communication, WUWSNs use acoustic signals instead of radio signals. Acoustic signals are used due to their lower attenuation in an underwater environment.

Sensor networks are on their revolutionizing way. However, due to certain issues advances in wireless underwater sensor networks have not matched those in the terrestrial domain.

Some of the major issues that arise in WUWSNs are listed below [24–27]:

- *Low bandwidth:* Available bandwidth is limited and it depends on frequency and transmission range. Hence the size of data packets for transmission needs to be as small as possible.
- *Water pressure:* Different at various levels under the sea, which affects communication.
- *Noise interference:* Interference arises due to the noise produced by sea animals, movements of ships, etc.
- *Salinity:* As the salt content increases in water, the water density also increases and causes delays in communication.
- *Temperature:* Sea temperature fluctuates with tides, which affects acoustic communication.
- *Rayleigh fading:* Causes reduction of signal strength and the receiver receives weak signals.
- *High transmission power:* Water currents act as a major resistive force for wave propagation and increase wave attenuation. Hence for good efficient communication, it becomes necessary to transmit the signals with higher transmission power.
- *Propagation delay:* Propagation delay in underwater is high, which results in propagation speed reduction.
- *Memory limitation:* Sensor nodes are small in size and hence limited memory is available with them.
- *Energy limitations:* Underwater sensor nodes are battery operated and hence the sensor network operation needs to be energy efficient in terms of power usage.

13.2.2 Energy Constraints

Energy is an important factor used in every minor or major operation of any type of application in different environments or scenarios. In an underwater environment, we cannot have plug-in sockets to provide power as per requirements; hence the sensors have limited energy. Sensors are equipped with batteries, but these batteries do have a limited lifetime. To replace or recharge a battery in an underwater environment is a difficult task. Hence as a better alternative, the battery is brought to land for recharging.

Moreover when underwater sensors collect data as per requirements, the collected data needs to be sent back to land, where there are analysis centers to study that data and to bring out productive conclusions. This movement of data from underwater to the land analysis center also becomes very costly in terms of energy consumption of the sensors.

In a sensor network, during wireless communication each node operates in three major modes viz. *transmitting*, *receiving*, *and listening*. In the listening mode the energy expenditure is minimal; this is also called promiscuous mode. However, if the node spends most of the time in promiscuous mode only, it consumes a large portion of the energy. However, in an underwater environment transmitting a packet requires more energy than receiving the packet. Hence, in an underwater environment the receiving and listening mode's energy requirements are negligible as compared to the transmitting mode.

Energy-efficient sensor means sensors having long-lasting batteries to work with. The battery technology is still lagging behind the microprocessor technology. Energy-efficient networking protocols are required nowadays. These energy constraints are keeping researchers busy as they look for innovative protocols in their respective fields. Research is going on to determine ways to save energy and enhance the efficiency of operation, which we will discuss in later sections.

13.2.3 Energy Conservation

In a sensor network, hundreds/thousands of sensors need to be densely deployed; due to this the neighboring communicating nodes come very close to each other. In this network scenario, a multihop data transfer is the best option for power saving in the sensor network. Multihop communication can also reduce signal propagation delays that may arise in long-distance communication. In multihop communication, energy could be saved more if we choose long paths along with a series of short hops rather than short paths along with a line of long hops. Selection of the long-path/short-hops combination saves energy, however, only to a certain degree as it can increase processing overheads/delays/control overheads. Hence, it becomes the requirements of the user that bends the flow of the sensor network viz. more energy saving or less delays.

Another option could be an in-built mechanism to control the usage of power for several operations, and as per the user requirements this option could be availed at the cost of lower productivity/higher delays, but with an increase in the network lifetime.

Other small steps that could save a large amount of energy include the following: (a) turn-off the transceiver when it is not required, (b) use shorter data packets for communication, (c) derive and use multiple paths to travel from source to destination in order to increase the network lifeline, (d) transmit data from the source node only when the destination node is ready, so that data can be transmitted without error, (e) avoid having two or more nodes send data simultaneously to avoid collisions, and (f) highly discourage node idle-listening and overhearing in the operation of the network.

13.3 LITERATURE SURVEY

Sensor nodes lack energy and bandwidth resources. Moreover, to make a sensor network, sensors need to be deployed in large numbers. These necessities require saving energy at all the layers of the networking protocol stack from application layer to physical layer. Most of the research is focused on system-level power awareness such as dynamic voltage alteration, hardware needed for communication, issues of low duty cycle, energy-aware MAC protocols, and energy-efficient routing at the network layer.

At the MAC layer, energy usage can be minimized by:

Avoiding collisions: Collisions can take place during data transmission between sensor nodes and may require retransmission of the data. Data transmission requires a specific amount of energy and retransmission will consume more energy in addition to data transmit energy.

Avoiding overhearing: Overhearing occurs when a sensor node receives data that is destined for its neighbors. A node can overhear only when it remains active in the network and this activeness consumes energy.

Avoiding idle-listening: A node can go into an idle state to save energy. But in the idle state, a node can put its transceivers on to listen to the communication taking place in its neighborhood; this is idle-listening by a node. As the transceivers are in an *on* state, energy consumption takes place.

Avoiding control packets overrun: Control packets are used to carry the relevant information (information other than the data) needed for communication. These packet transmissions consumes energy and if more control packets flow in the network, a proportionate amount of energy will be required to manage them.

Avoiding unnecessary transitions: Avoiding unnecessary transitions between various modes viz. sleep, idle, transmit, and receive.

In this section, we focus on the network layer and discuss energy-efficient routing protocols. At the network layer, researchers are trying to minimize energy usage and maximize network lifetime by:

Efficient routing: Routing is the process of finding the path from the source node to the destination node. It consists of two main phases: *route setup* and *route maintenance*. Route setup is meant to find an efficient path between the source node and destination node. An established path should be reliable and provide guaranteed delivery in all types of scenarios. Route maintenance manages the established path or rebuilds it if required between the source node and the destination node. Rebuilding of a path may be required in case a path is broken due to mobility of sensor nodes, failure of intermediate sensor nodes, or arising of some obstacles between the sensor nodes. An efficient established path could save a large amount of network energy and increase its productivity.

Reliable communication among sensor nodes: In a network, when sensor nodes collect the data, the collected data needs to be sent to a master collector. The source node sends the data to the master collector acting as the destination node either directly or through relay. In data relaying, data are passed through some intermediate nodes, hence the reliability of data becomes a concern. Data should reach the destination without any modification, error, and duplication. Reliable communication will save the energy that can be consumed in resending and checking data, either in the form of complex calculations, large algorithm processing, memory storage overhead, or control packets overrun.

For terrestrial sensor networks, the following are some of the existing energy-efficient routing protocols:

- Directed diffusion [28]
- Rumor routing [29]
- LEACH (Low-Energy Adaptive Clustering Hierarchy) [30]
- TTDD (Two-Tier Data Dissemination) [31]
- GEAR (Geographic and Energy Aware Routing) [32]

13.3.1 Directed Diffusion

The directed diffusion protocol is a data-centric protocol. In this algorithm data is diffused through sensor nodes by using a naming scheme for data. *A naming scheme is used to save energy as it avoids unnecessary operations by the network layer.* Under a naming scheme, it uses attribute–value pairs for the data. By using these pairs sensors are queried on an on-demand basis. An interest is defined with the attribute–value pairs such as time duration, geographical location, etc. An interest entry also contains several gradient fields. A gradient is a reply link with a neighbor from which the interest was received. By interest and gradients, paths are established between the source and data collector nodes. Multiple paths have been established and out of them one is selected by the source node for information passing.

The main point of concern for this protocol is that data is sent through multiple paths and hence creates redundancy in the network. When multiple paths compete with each other, one of the paths is forced to send the data at higher rates, which may result in either loss of data or congestion of the network.

13.3.2 Rumor

Rumor routing is a variation of the directed diffusion protocol. It is applicable where geographic routing cannot be used. Directed diffusion can flood the query in the network when no geographic boundary is defined to diffuse the task. This flooding creates unnecessary overheads in the network and leads to more energy consumption. Rumor routing creates the concept of flooding, i.e. between the event and the query. The main idea is to route the queries to the node that has observed a particular event. This will save the entire network from flooding. When a node detects any event, it generates an agent. The agent's task is to communicate the information about the event. When a node is queried for an event, another node that knows about the route responds to the query by referring to its event table. This saves the cost of flooding the entire network. Rumor routing maintains only one path between source and destination, while in directed diffusion multiple paths exist for data passing between source and destination.

13.3.3 LEACH (Low-Energy Adaptive Clustering Hierarchy)

The LEACH protocol is a form of cluster-based routing. It forms *clusters to minimize the energy dissipation*. The operations of this protocol is divided into two phases: the setup phase and the steady-state phase. The steady phase is of longer duration to minimize the overhead.

The setup phase consists of the following steps:

- The sensor node chooses a random number between 0 and 1.
- If this chosen random number is less than the threshold $T(n)$, the sensor node is a cluster-head (CH):

$$T(n) = \begin{cases} \left\{ \frac{P}{1-P\left[r \bmod \left(\frac{1}{P}\right)\right]} \right\} & \text{if } n \in G, \\ 0 & \text{otherwise,} \end{cases} \tag{13.1}$$

  where P = percentage chance of becoming a cluster-head, r = current round, and G = set of nodes that have not been selected as a cluster-head in the last $1/P$ rounds.
- The cluster-head then advertises its selection.
- After advertisement, the other sensor nodes decide whether they want to be part of this cluster-head's cluster or not, based on the signal strength of the advertisement.

- The cluster-head assigns a timetable to the sensor nodes of its cluster based on the TDMA approach. At the indicated time the nodes can send data to the cluster-head.

The steady phase consists of the following steps:

- Sensor nodes start sensing and transmitting data to cluster-heads.
- The cluster-head aggregates all the data and sends it to the base station.
- After a certain period of time, the network goes again to the setup phase and again starts a new round of cluster-head selection.

This protocol does not address the network partition due to the nodes' failure. Another main point of concern is the selection of the cluster-head in each round. A sensor node becomes the cluster-head with only a certain probability; hence there are more chances that part of the network is left without a cluster-head, leading to inefficient routing.

Various versions of LEACH protocol had been proposed in the literature, and they are are discussed below.

E-LEACH: The *Energy-LEACH* [36] protocol is an improvement over the LEACH protocol. This protocol changes the cluster-head selection procedure. In the first round, a cluster-head is chosen. All the nodes have the same probability of being a cluster-head. After the first round, nodes' energy is also considered in cluster-head selection. A node with high residual energy is chosen as a cluster-head. The cluster-head is the main focused point for communication in the network and hence its working life will affect the communication between the sensor nodes. This protocol selects the cluster-head with maximum remaining energy or maximum working capacity; this provides stability in the network.

TL-LEACH: Two-Level LEACH [37] sends data to the base station in two hops. The cluster-head collects data from the other nodes. It then sends the collected data to the base station through another cluster-head that lies in between it and base station.

M-LEACH: In the *Multihop LEACH* [38] protocol, data is relayed to the base station in multiple hops. This protocol addresses the problem of data transmission from the far clusters to the base station. The cluster-head sends the collected data to the base station through another cluster-head that lies in between it and the base station. Due to multihop communication, a lot of energy is saved at the cluster-head node.

LEACH-C: The *Centralized LEACH* [30] protocol introduces the centralized cluster formation algorithm. During the setup phase, nodes send their remaining energy and location to the sink. After that the sink runs a centralized cluster formation algorithm and forms the clusters for that phase. In each round, new clusters are formed by the sink. This protocol distributes the cluster-heads throughout the network based on the nodes energy and location; hence it may produce better results.

VLEACH: *This is a new version of the LEACH* [39] protocol. In this new version of the protocol, a cluster will have the cluster-head as well as a vice-cluster-head (CH and vice-CH). The vice-cluster-head will take the authority of the cluster when the existing cluster-head dies. This concept saves the energy of the cluster's members, which they use in data collection. Remember if the cluster-head dies, the collected information could not reach the sink and the result is energy wastage by the nodes. With the help of the vice-CH, the collected information can now reach the sink even if the CH dies.

13.3.4 GEAR (Geographical and Energy Aware Routing)

The GEAR protocol uses geographical information for distributing the queries to the appropriate regions. The neighbor selection is done on the basis of energy and the location to route the packet. It conserves more energy than directed diffusion as the forwarding region is restricted. Each node keeps account of two costs for reaching the destination: estimated cost and learned cost.

Hole conditions arise when a node does not have any neighboring node to further forward the packet. In this condition the estimated cost is equal to the learned cost.

The algorithm consists of two phases:

- Routing towards destination region:
  - Nearest neighbor node to the destination region is selected as the next forwarding node.
  - In the hold scenario, the neighbor node is selected on the basis of the learning cost function.
- Data dissemination inside the destination region:
  - Uses restricted flooding or recursive geographic forwarding.

13.3.5 Multihop Routing Protocols

In [33], the *Geographic Adaptive Fidelity scheme*, an energy-aware and location-based routing algorithm, is proposed. It is designed keeping in mind mobile ad hoc networks, but can also work for sensor networks. This algorithm *turns off the unnecessary nodes of the network to save energy*. A virtual grid is created for the network. Nodes use their GPS location to associate themselves with a point in the virtual grid. Multiple nodes on the same point of the grid are considered to cost the same for packet routing. Out of these multiple nodes with the same cost, some nodes are made to sleep in order to save energy. There are three states for a node: discovery (to find the neighbors in the grid), active (nodes taking part in routing), and sleep (node is inactive; the node's sleeping duration is application dependent). Each node communicates the estimated time it will leave the grid. When an active node is going to leave the grid, another sleeping node wakes up and becomes active.

In [34], another energy-aware routing algorithm is proposed that uses a set of suboptimal paths to increase the network life. Paths are chosen by a probability function that depends on the energy consumption of each path. Each node is identified on the basis of class-based addressing that includes the location and the type of the node. The protocol consists of three phases: The setup phase, forwarding phase, and route maintenance phase:

- Setup phase:
  - Routes are found through restricted localized flooding.
  - Each node calculates its total energy cost as follows:
    - If the request is going from node A to node B then node B calculates the cost of the path as

$$C_{B,A} = \text{Cost}(A) + \text{Metric}(B, A). \tag{13.2}$$

    - The metric here is the energy metric that uses transmission and reception costs along with the residual energy of the nodes.
  - Paths having high cost are discarded from use.
  - Node selection is done as per the distance to the destination; nodes nearer to the destination are selected first.
  - The node assigns a probability to each of its neighboring nodes in its routing table (RT) corresponding to the formed paths.
  - The probability is inversely proportional to the cost as defined below:

$$P_{B,A} = \frac{1/C_{B,A}}{\sum_{k\in RT_B} 1/C_{B,\text{Node}\,k}}. \tag{13.3}$$

Node B then calculates the average cost for reaching the destination using the neighbors in its routing table (RT) by the below formula:

$$\text{Cost } B - \sum_{i\in RT_B} P_{B,A} C_{B,A}. \tag{13.4}$$

This average cost for node B is set in the cost field of the request and forwarded.

In the forwarding phase, the node then selects a next forwarding node from its routing table using the probabilities and forwards the data to it. The route maintenance phase uses localized flooding.

In this protocol, a single path is randomly chosen from the multiple paths to save energy. This makes it an improvement over the directed diffusion protocol.

In [35], the BAR protocol is proposed, which works with the explanation that the battery can work longer if it is not used continuously. The battery should be given recovery time for strengthening the network working capacity. But the BAR scheme works with the assumption that each node in the network knows its geographical location.

13.3.6 Underwater Energy-Efficient Protocols

This section reviews some of the energy-efficient routing protocols for underwater sensor networks.

Vector-Based Forwarding: This protocol [13] is an energy efficient and robust algorithm. A routing forwarding vector is defined between the source and the destination. A forwarding region is defined around the routing vector that consists of a predefined radius. Only a set of nodes that are in forwarding region takes part in routing. An intermediate node will be the candidate for the next relay node if the distance between itself and the routing vector is less as compared to the other nodes.

Cluster-Based Protocol: The energy-efficient cluster-based [40] protocol utilizes the direction (up–down transmission) characteristic of an underwater environment and has been shown to be a better performer in terms of whole network operation. It forms clusters that are direction dependent. The cluster-head is chosen in the direction of transmission only. The cluster-head collects the data from its cluster member and sends the collected data to the sink via other cluster-heads on the way.

Distributed Underwater Clustering Scheme (*DUCS*): The DUCS protocol is an energy-efficient and GPS-free routing protocol. Clusters are formed inside the network and a cluster-head is chosen. The cluster-head collects the data from its cluster's members in a single hop. Multihop routing is used to transmit the data to the sink from the cluster-head [41, 42]. The cluster-head a data aggregation technique to remove the redundant data from the collected information. This algorithm uses a TDMA/CDMA schedule to communicate with cluster members and to improve communication as well. It also uses a continuous adjusted timer along with guard time values to prevent data loss.

Reference [43] deals with energy-efficient routing schemes and the paper presents different energy-efficient routing methods. It compares various protocols including new ones. The authors consider all the major parameters that characterize underwater communications such as attenuation, absorption, propagation delay, bandwidth-distance, power-distance relationships, and modem energy consumption profiles. In the developed schemes, the next neighbor is chosen in such a way that the relay length between two nodes could be nearer to the optimum. Energy efficiency is achieved by choosing the next relay node based on the local positioning algorithm.

In Reference [44], the authors present a scheme to reduce energy consumption and delay for underwater communications. The communication architecture is based on a tree structure. Underwater sensors and the underwater sink make the branches of the tree. Underwater sensors are connected to the underwater sink. A routing protocol based on the tree structure needs to use the data aggregation technique in order to reduce energy consumption. The proposed routing protocol uses an energy-aware data aggregation method to create an energy-efficient and delay-decreasing protocol. It reconfigures the aggregation tree via a dynamic pruning and grafting function to find an optimum path from the source node to the sink node.

The authors in [45] introduce the E-PULRP (Energy optimized Path Unaware Layered Routing Protocol) for dense underwater 3D sensor networks. In this work an uplink transmission is considered. Underwater sensor nodes collect and send the information to the stationary sink node. E-PURLP consists of two phases: a layering phase and a communication phase. In the first phase a layering structure is developed around the sink node, which is a set of concentric spheres. The radii of the concentric spheres as well as the transmission energy of the nodes in each layer are chosen considering the probability of successful packet transmissions and minimum overall energy expenditure. In the second phase an intermediate relay node is selected and an on-the-fly routing algorithm is used for packet delivery from the source node to sink node across the identified relay nodes.

13.4 CONCLUSION

Sensor networks are different from the other networks in terms of their design and functioning. This makes working in this field very difficult and challenging. In this chapter, wireless sensor networks/underwater sensor networks along with their applications and issues have been discussed. Energy is an essential and important factor in the lifetime of a sensor network. Energy needs to be used as a precious gem that should not be wasted inefficiently. The main purpose of network establishment is sharing of information through communication and energy is the key requirement for this communication. At the network layer, many routing protocols have been developed to this point for saving energy. Each protocol addresses the energy issue in its own way and provides the solution accordingly. In this chapter, an overview of the energy-efficient routing protocols for terrestrial and underwater sensor networks has been provided. Energy-aware protocols for sensor networks contribute to saving energy and hence help to create more and greener communication networks and systems.

REFERENCES

[1] Jie Liu, Handbook on Theoretical and Algorithmic Aspects of Sensor, Ad-Hoc Wireless, and Peer-to-Peer Networks, Auerbach Publications, July 2005.
[2] M. Ilyas, I. Mahgoub, Handbook of Sensor Networks: Compact Wireless and Wired Sensing Systems, CRC Press, Boca Raton, Florida, 2004.
[3] D. Culler, D. Estrin, M. Srivastava, Overview of sensor networks, IEEE Computer Magazine 37 (8) (2004) 41–49.
[4] I.F. Akyildiz, W. Su, Y. Sankarasubramaniam, E. Cayici, A survey on sensor networks, IEEE Communication Magazine 40 (8) (2002) 102–114.
[5] A. Mainwaring, J. Polastre, R. Szewczyk, D. Culler, J. Anderson, Wireless sensor networks for habitat monitoring, in: Proceedings of ACM International Workshop on Wireless Sensor Networks and Applications, September 2002, pp. 88–97.

[6] L. Schwiebert, S.K.S. Gupta, J. Weinmann, Research challenges in wireless networks of biomedical sensors, in: Proceedings of International Conference on Mobile Computing and Networking, 2001, pp. 151–165.

[7] M. Lee, H. Yoe, Comparative analysis and design of wired and wireless integrated networks for wireless sensor networks, in: IEEE Fifth International Conference on Software Engineering Research, Management and Applications, 2007, pp. 518–522.

[8] J. Pottie, W.J. Kaiser, Wireless Integrated Network Sensors, Proceedings of ACM Communications 43 (5) (2000) 51–58.

[9] D. Estrin, R. Govindan, J. Heidemann, S. Kumar, Next century challenges: scalable coordination in sensor networks, in: Proceedings of International Conference on Mobile Computing Networking, 1999, pp. 263–270.

[10] J.W. Gardner, V. Varadan, O. Awadelkarim, Microsensors, MEMS and Smart Devices, Wiley, New York, 2001

[11] J. Hill, R. Szewczyk, A. Woo, S. Hollar, D. Culler, K. Pister, System architecture directions for network sensors, in: Proceedings of ASPLOS-IX, 2000.

[12] J.M. Kahn, R.H. Katz, K.S.J. Pister, Next century challenges: mobile networking for "smart dust", in: Proceedings of International Conference on Mobile Computing Networking, 1999, pp. 271–278.

[13] S.K. Dhurandher, S. Misra, M.S. Obaidat, S. Khairwal, UWSim: a simulator for underwater wireless sensor networks, Simulation: Transaction of the Society for Modeling and Simulation International 84 (7) (2008) 327–338.

[14] I. Vasilescu, C. Detweiler, D. Rus, AquaNodes: an underwater sensor network, in: International Conference on Mobile Computing and Networking, WUWNet'07, September 2007, pp. 85–88.

[15] P. Xie, J.-H. Cui, L. Lao, VBF: vector-based forwarding protocol for underwater sensor networks, in: Proceedings of IFIP Networking, Coimbra, Portugal, May 2006, pp. 1216-1221.

[16] N. Nicolaou, A. See, P. Xie, J.H. Cui, D. Maggiorini, Improving the robustness of location-based routing for underwater sensor networks, in: Proceedings of Oceana 2007, Europe, 2007, pp. 1–6.

[17] J.M. Jornet, M. Stojanovic, M. Zorzi, Focused beam routing protocol for underwater acoustic networks, in: Proceedings of Mobicam WUWNet, San Francisco, California, USA, September 2008, pp. 75–82.

[18] N. Chirdchoo, W.-S. Soh, K.-C. Chua, Sector-based routing with destination location prediction for underwater mobile networks, Proceedings of International Conference on Advanced Information Networking and Applications Workshops, IEEE.

[19] D. Hwang, D. Kim, DFR: directional flooding-based routing protocol for underwater sensor networks, in: MTS/IEEE, Oceans 2008, September 2008.

[20] S. Gopi, G. Kannan, D. Chander, U.B. Desai, S.N. Merchant, PULRP: path unaware layered routing protocol for underwater sensor networks, in: IEEE International Conference on Communication, 2008, pp. 3141–3145.

[21] E. Cheng, Y. Qi, B. Sun, Z. Zhuang, J. Deng, Research on routing protocol for shallow underwater acoustic ad hoc network, in: ISECS International Colloquium on Computing, Communication, Control, and Management, vol. 2, August 2008, pp. 533–537.

[22] T. Li, Multi-sink opportunistic routing protocol for underwater mesh network, in: IEEE/ICCCAS 2008, May 2008, pp. 405–409.

[23] A. Nimbalkar, Reliable unicast and geocast protocols for underwater inter-vehicle communications, Master of Science thesis, The State University of New Jersey, 2008.

[24] J.-H. Cui, J. Kong, M. Gerla, S. Zhou, Challenges: building scalable mobile underwater wireless sensor networks for aquatic applications, University of Connecticut, USA, UCONN CSE Technical report: UbiNET-TR05-02, 2005.
[25] J. Partan, J. Kurose, B.N. Levine, A survey of practical issues in underwater networks, in: International Conference on Mobile Computing and Networking WUWNet'06, 2006, pp. 17–24.
[26] I.F. Akyildiz, D. Pompili, T. Melodia, Challenges for efficient communication in underwater acoustic sensor networks, ACM SIGBED Review 1 (2) (2004) 3–8.
[27] L. Liu, S. Zhou, J.-H. Cui, Prospects and problems of wireless communication for underwater sensor networks, Wiley WCMC issue on Underwater Sensor Networks 8 (8) (2008) 977–994.
[28] C. Intanagonwiwat, R. Govindan, D. Estrin, Directed diffusion: a scalable and robust communication paradigm for sensor networks, in: Proceedings of ACM MobiCom 2000, Boston, MA, 2000, pp. 56-67.
[29] D. Braginsky, D. Estrin, Rumor routing algorithm for sensor networks, in: International Conference on Distributed Computing Systems, November 2001.
[30] W.B Heinzelman, A.P. Chandrakasan, H. Balakrishnan, An application-specific protocol architecture for wireless microsensor networks, IEEE Transactions on Wireless Communications 1 (4) (2002) 660–670.
[31] F. Ye, H. Luo, J. Cheng, S. Lu, L. Zhang, A two-tier data dissemination model for large-scale wireless sensor networks, in: Proceedings of ACM/IEEE MOBICOM, 2002.
[32] Y. Yu, D. Estrin, R. Govindan, Geographical and energy-aware routing: a recursive data dissemination protocol for wireless sensor networks, UCLA Comp. Sci. Dept. tech. rep., UCLA-CSD TR-010023, May 2001.
[33] Y. Xu, J. Heidemann, D. Estrin, Geography-informed energy conservation for ad hoc routing, in: Proceedings of Annual International Conference on Mobile Computing and Networking (MobiCom), Rome, Italy, July 2001.
[34] R.C. Shah, J.M. Rabaey, Energy aware routing for low energy ad hoc sensor networks, in: Proceedings of IEEE Wireless Communications of Networking Conference (WCNC), Orlando, FL, USA, March 2002.
[35] C. Ma, Y. Yang, Battery-aware routing for streaming data transmissions in wireless sensor networks, Mobile Networks and Applications 11 (5) (2006) 757–767.
[36] F. Xiangning, S. Yulin, Improvement on LEACH protocol of wireless sensor network, in: 2007 International Conference on Sensor Technologies and Applications, pp. 260–264, doi:10.1109/SENSORCOMM.2007.60.
[37] V. Loscrì, G. Morabito, S. Marano, A two-levels hierarchy for low-energy adaptive clustering hierarchy, in: 2005 Vehicular Technology Conference, pp. 1809-1813.
[38] H. Zhou, Z. Jiang, M. Xiaoyan, Study and design on cluster routing protocols of wireless sensor networks, Dissertation, 2006.
[39] M.B. Yassein, A. Al-zou'bi, Y. Khamayseh, W. Mardini, Improvement on LEACH protocol of wireless sensor network (VLEACH), International Journal of Digital Content Technology and its Applications 3 (2) (2009) 132–136.
[40] C.J. Huang, Y.J. Chen, I.-Fan Chen, K.W. Hu, J.J. Liao, D.X. Yang, A clustering head selection algorithm for underwater sensor networks, in: Proceedings of FGCN'08, vol. 1, December 2008, pp. 21–24.
[41] M.C. Domingo, R. Prior, A distributed clustering scheme for underwater wireless sensor networks, in: Proceedings of 18th Annual IEEE International Symposium on Personal,

Indoor and Mobile Radio Communications (PIMRC 2007), Athens, Greece, September 2007.

[42] M.C. Domingo, R. Prior, Design and analysis of a GPS-free routing protocol for underwater wireless sensor networks in deep water, in: Proceedings of UNWAT 2007, Valencia, Spain, October 2007.

[43] M. Zorzi, P. Casari, N. Baldo, A.F. HarrisIII , Energy-efficient routing schemes for underwater acoustic networks, IEEE Journal on Selected Areas in Communications 26 (9) (2008) 1754–1766.

[44] H.Nam, S. An, Energy-efficient routing protocol in underwater acoustic sensor networks, in: IEEE/IFIP International Conference on Embedded and Ubiquitous Computing, 2008, pp. 663–669.

[45] S. Gopi, G. Kannan, U.B. Desai, S.N. Merchant, Energy optimized path unaware layered routing protocol for underwater sensor networks, IEEE "GLOBECOM" 2008.

CHAPTER

Energy-Efficient Next-Generation Wireless Communications

14

Jason B. Ernst

School of Computer Science, University of Guelph, 50 Stone Rd. East, Guelph, Ontario, Canada N1G 2W1

14.1 INTRODUCTION

Many next generation network technologies have been focusing on implementing new standards to deal with low power and energy efficiency since this is an important challenge across all forms of networking in the future. For example, Singh et al. [1] proposes an enhancement to IEEE 802.11n networks that allows mobile stations to operate in a power saving polling mode. In wired networks energy-efficient protocols and equipment lead to reduced energy costs for service providers, data centers, and large institutions that rely heavily on communication technology to provide service and connectivity to users. In wireless networks, there are many unique challenges related to reducing energy requirements. In wireless sensor networks, an energy-efficient protocol may lead to longer lifetimes for the sensor network, which is one of the big challenges in the area.

This chapter is structured as follows: First we discuss the motivation for energy-efficient communications. Next we introduce applications and particular next generation networks where green communication may be applied. After that we provide an overview of energy-efficient architectures, techniques, and protocols for green communication for next generation networks. This is followed by a discussion of the trends, limitations, challenges and open problems in green communication. Finally we finish the chapter with conclusions and future directions.

14.2 MOTIVATION FOR ENERGY-EFFICIENT COMMUNICATIONS

This chapter is broken down into several subsections that discuss various motivations for studying energy-efficient communications in next generation wireless networks.

14.2.1 Energy as a Resource

Around the world, people are beginning to realize that energy is a limited resource similar to oil, gas, or other environmental resources. While there are sources which have not been fully realized due to technical limitations, it is still important for us to develop technologies that use energy efficiently. In the developing world particularly, "energy is a precious resource whose scarcity hinders widespread Internet deployment" [6]. By increasing efficiency in communications it becomes possible to deploy networks more widely. Lower power communications may enable renewable energy technology such as solar and wind to supply power to network equipment that typically could not operate under such power constraints previously.

One common technique that making more efficient use of the spectrum is using cognitive radio [43] to allow for communications into bands which are normally reserved for other uses by detecting when they are being used by the owners of the spectrum and backing off to allow them priority while making use of the frequency when the spectrum is idle. Another benefit to using alternative frequencies is that some frequencies are better suited for longer range applications such as rural access. For instance, using the 700 MHz band that is becoming unused in some countries due to the transition to digital television allows for wireless access technology that offers coverage an order of magnitude higher than traditional Wi-Fi or WiMAX [10].

14.2.2 Handheld and Battery Powered Devices

Since handheld devices are battery powered, there is strong motivation to study how to make these devices communicate more efficiently. In this case, the battery is one of the heaviest, largest, and most expensive parts of the device. Any increase in efficiency may correspond to cheaper and lighter devices, which makes the devices more attractive to consumers. According to [8,9] the annual carbon footprint of an average mobile subscriber is about 25 kg, which is equivalent to operating a 5 W lamp for a year. This also includes the supply chain, vendor, usage, and end-of-life phases in the lifecycle of the handheld, however the part we are most interested in for this context is usage. Furthermore, it is important to make future networks and existing mobile networks more efficient since mobile providers are often required to maintain older technology in parallel with newer technology. For example, as 4G and LTE networks are deployed, there may still be 2G and 3G networks around since many users still use the older devices and networks [7].

There are special challenges when the devices in the network are battery powered, particularly when the devices are not able to be recharged easily. In this case, the challenge becomes twofold. First, the devices should communicate as efficiently as possible. Second, the lifetime of the network should be extended as long as possible. The best method for achieving this is to strategically select routes or cluster-heads such that the overall lifetime of the network is extended. The goal is to avoid partitioning the network so that portions of it are unable to communicate. In some cases, the network may be temporarily partitioned as the network reconfigures cluster-heads or routes, so other techniques such as delay tolerance may be used in conjunction to further improve the efficiency. Instead of simply dropping and retransmitting packets while the network is in the reconfiguration stage, it may be worthwhile to take a store-and-forward approach used in delay tolerant networks to cut down on costly retransmissions. More details on these techniques are provided later in the chapter.

14.2.3 Potential Health Effects

Recently, the health effects of operating wireless radio devices have come into question around the world. There are several studies that claim there are health effects such as headaches, changes in brain physiology, alpha-band activity during sleep, and other problems, however there are many studies that refute and/or dispute those claims [17]. In [41] the authors note several studies that indicate that the power and duration of the exposure may be linked to negative effects in animals and more recently guidelines and standards have been established to avoid these effects. The authors also provide a brief history regarding testing, guidelines, and standard setting that helps in understanding the assumptions and limitations in existing models. Whether or not the effects are true despite modern standards, there is still a perception among many people in the population that wireless communication causes health problems. Any efforts to improve the efficiency means that any potential health effects of the operation of the wireless devices can be reduced. Technologies that can operate with lower power levels, reduced overhead, or avoid interference mean less exposure and likely decreased effects on people and animals nearby.

14.2.4 Reduced Interference and Contention

The wireless spectrum itself has recently been considered as a natural resource since there is only a limited amount of spectrum available for everyone. Because of this, misuse or inefficient use of the spectrum can be thought of as pollution towards the other users of the spectrum [2]. It is also important to transmit data efficiently purely from a performance point of view as well. According to the first International Workshop on Green Wireless, "The transmitted data volume increases approximately by a factor of 10 every 5 years, which corresponds to an increase of the associated energy consumption by approximately 16–20% per year. Currently, 3% of the world-wide energy is consumed by the ICT infrastructure which causes about 2% of the world-wide CO_2 emissions (which is comparable to the world-wide CO_2 emissions by airplanes or one quarter of the world-wide CO_2 emissions by cars). If this energy consumption is doubled every 5 years, serious problems will arise. Therefore, lowering energy consumption of future wireless radio systems is demanding greater attention [3,5]."

14.3 APPLICATIONS AND NEXT-GENERATION NETWORKS WHERE GREEN COMMUNICATIONS CAN BE APPLIED

In this section particular network technologies are discussed along with the unique challenges and an overview on solutions related to each technology. Further details about individual techniques are presented in the next section on protocols and architectures. A recent list of recent real world projects targeting green mobile communications is presented in [42].

14.3.1 Wireless Sensor and Ad Hoc Networks

Since wireless sensor networks are often battery powered, one of the biggest problems is keeping the network as a whole alive as long as possible. If the network becomes partitioned it becomes impossible for portions of the network to continue communicating since there is often no infrastructure to support communications in these networks. Typically, sensor nodes are cheap and are dropped from planes or scattered about the environment randomly. This leaves the network to establish the best communication paths in an ad hoc manner, which may lead to partitioning earlier than if efforts are not made to avoid this problem. One solution is to divide the network into clusters to reduce the average distance, and transmission power, for a single hop communication. There are many other problems to study related to this including cluster-head election, rotating cluster-heads, and other techniques to further improve the lifetime of the network as well. More detail about clustering will be provided in sections "Topology Control" and "Clustering and Caching." It has been suggested in [6] that techniques already used in wireless sensor and ad hoc networks can also be applied to make traditional wired technologies more efficient. Moreover, it has even been suggested that IEEE 802.11b wireless networks are more efficient than some wired transmission technologies in the Internet [6] using data provided by Lucent and studies on the efficiency of wired communications. The possibility that these wireless technologies may be more efficient than wired equivalents does not mean that there is no room for improvement. Instead, it is motivation to further improve the efficiency of both technologies together. Furthermore, the advances found in wireless sensor and ad hoc networks can be adapted and generalized to many current and next generation wireless technologies.

14.3.2 Heterogeneous Wireless Networks

Heterogeneous wireless networks are networks that use multiple radio access technologies (RATs) together in one network. For instance it may be made up of IEEE 802.11 Wi-Fi, Bluetooth, 4G cellular, and other access technologies. Since many devices such as mobile phones, tablets, and laptops come packaged with multiple RATs, it is now possible for devices to remain connected using a choice of different technologies, or multiple technologies at once. In the case of green communication, it is possible to choose an optimal technology for reducing energy consumption. However, it is not a cut and dry as it would seem. While it may be obvious that one technology usually uses more energy than other, there are several factors that may contribute to selecting another technology which may often use more energy. Some of the factors to consider are:

1. Distance to the access point for all available technologies
2. Quality of the signal between the access point
3. Load of the access point
4. Number of users in the immediate region using the same technology

The essence of all of these considerations is to avoid congested regions in the network, or regions where there are interference. This is because retransmissions are very expensive, both in terms of performance, and in terms of energy. It is also important to avoid communication over a large distance if a close access point is available. This is related to the relay nodes, as it is usually cheaper in terms of energy to communicate over a short range with multiple hops than it is to communicate over a large distance in one hop. There are emerging standards that try to address interoperability between networks such as IEEE 802.21, however energy awareness is often not a priority.

One technique proposed by Smaoui et al. [34] tries to encapsulate several of these considerations in conjunction with a power consumption model so that the best network is selected from a variety of characteristics combined with the least power. This is achieved by using the TOPSIS method for multicriteria optimization. While the approach by Smaoui et al. explicitly includes a power consumption model, the majority of other approaches do not consider this, but leave room for it as a potential parameter in their generalized algorithms such as in the proposal by Fiterau et al. [35] or the network selection algorithm by Bakmaz et al. [37]. There are also approaches that are optimal in terms of the best performance, but make no consideration with respect to energy efficiency or power consumption of the individual network types such as the approach by Si et al. [36].

14.3.3 Delay Tolerant Networks

Delay tolerant networks are a research subject on their own; however, they are quite important in many of challenges related to energy efficiency and green communication. Since many of the networks are power constrained, rely on batteries, and often involve mobility, there is a high chance that at some point the network will become partitioned. When part of the network cannot communicate with the rest, a subset of the nodes may be cut off from communicating with the Internet or other intended destinations within a subnet. In Figure 14.1, consider a network where sensor nodes were dropped randomly from a plane. Eventually the nodes on the routing path will use their battery faster than the rest of the nodes and a partition in the network may occur if nodes on each side of the path are outside of communicating range of one another.

Another problem delay tolerance tries to address is high error rates within communication networks. It has typically been taken for granted that communication will occur without errors, however with wireless networks and increasing use of the spectrum, this will become more of a problem all the time.

The last problem that delay tolerance can address is long or variable delay within networks. It has traditionally been assumed that propagation delays are relatively homogeneous within networks, however with long haul wireless links, intermittent connections, and high error rates this is not always the case. It may take some time for a successful communication to occur.

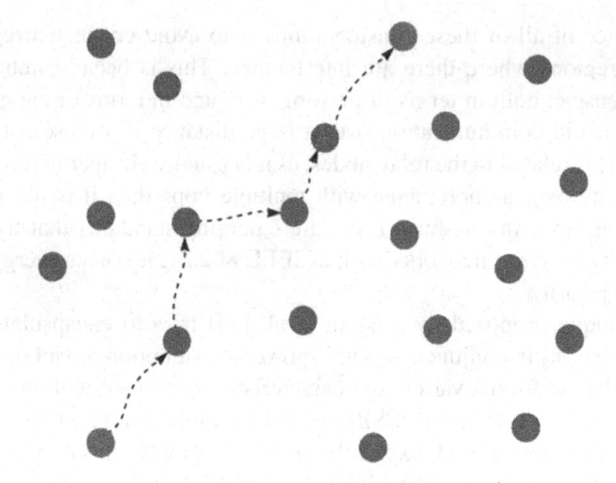

Figure 14.1 Example network vulnerable to partitioning.

There are several ways in which delay tolerant networks can overcome these problems. First, store-and-forward protocols can be used so that eventually the data reaches the intended destination [38]. This takes much longer than traditional protocols and requires that routers have some type of persistent storage such as hard drives or flash memory. Furthermore, data may need to be stored even after transmission until it can ensure the transmission to the next hop was successful. The second way in which delay tolerant networks can solve these problems is by using opportunistic contact to the network [38]. There are two types of opportunistic contact. The first is unscheduled. This may be due to unpredictable movement of nodes that move in and out of communication range or line of sight. The second type is unscheduled contact, which occurs with some type of predictable motion, such as the orbit of a satellite or planet. This also requires time synchronization between the nodes in the network, which is often a difficult problem.

14.3.4 Interplanetary and Intergalactic Networks

Interplanetary and intergalactic networks have unique challenges. In addition to requiring delay tolerance, there are special challenges and unique characteristics that can be taken advantage of. For instance consider a satellite orbiting a planet. Part of the orbit, no communication can occur with Earth because the planet blocks the path of the communication. However, the communication window is predictable because we are able to compute the orbit path. This information can be used for scheduled transmissions so that energy is not wasted trying to send and receive data when the satellite has no chance of successful communication. This type of technique can be extended to other more terrestrial networks such as those on aircraft, cars, and other mobile networks where line of sight may be able to be predicted.

While the scheduled communication techniques follow concepts from delay tolerant networks, there is a protocol that is unique to long haul intergalactic networks. The protocol is called the Licklider Transmission Protocol (LTP), and was named after James Licklider who is considered one of the first people to convince DARPA to fund communications networks like ARPAnet, the precursor to the Internet. This protocol operates over a point-to-point link and handles delay tolerance over this single link. This differs from other delay tolerance approaches that operate across multiple hops and use a store-and-forward approach. The LTP protocol is used because it tries to address some of the problems with TCP. Particularly, TCP is poor at interpreting interference and link errors. Often TCP implementations confuse this with link congestion since a timeout would occur in the case of temporary interference or pool link quality disrupting a transmission. A comparison table between TCP and LTP is presented in [38], but the main differences are that LTP is not connection-oriented, LTP uses selective NACKs on ranges of data in blocks, and there is no congestion control in LTP. Bundle protocols running on top of LTP are able to implement congestion control separately if they wish. Finally, LTP does not use the round trip time (RTT) to determine timeout intervals, it uses a one-way technique that avoids the large RTT that would occur over long distances.

Another competing protocol in this area is called NAK-Oriented Multicast Protocol (NORM) [39], which is also used for delay tolerant networking. It is proposed in [38] that this approach be compared against LTP-based approaches to see which is best, however interplanetary networks are still a young research area since there is limited ability to experiment in the real environment and common tools do not often simulate this environment.

14.3.5 Alternative Energy Networks

Recently, alternative energy sources have been proposed to power everything from cars to everyday devices in the home. It is also possible to take advantage of alternative energy such as wind and solar to power the infrastructure of communications networks. This is an especially high impact method to green communication networks since the infrastructure represents a high proportion of the total amount of energy used. One of the problems with alternative energy networks using sources such as wind or solar is that the wind does not always blow and the sun does not always shine. These sources must be combined with batteries so that they are able to be continuously powered [26]. However, with this type of system in place, the network may actually become more robust, withstanding disruptions to the power grid itself. In developing countries, this type of approach is especially useful since costs of deploying electrical infrastructure may be prohibitive in certain regions. This model has already proven effective in both Kenya [27] and Ethiopia [28]. In the case of Safaricom in Kenya, it is estimated that 25% of the provider's base stations are powered solely by diesel generators due to lack to electrical infrastructure. Additional base stations exist in regions where the power infrastructure is so unreliable that the majority of the time they are also powered by generators [27]. The first trials into green energy at a site that previously used solely diesel energy shows encouraging results.

The generators run just 1.32 h per day resulting in diesel consumption reduction of over 95% [27]. This also contributed to reduced fuel transportation and maintenance costs, so in addition to being "greener," this solution is also more economical. In a white paper by Nokia Siemens, it is noted that while it is more expensive initially to deploy solar panels compared to a diesel generator, the payback period is only 2–4 years and the only maintenance is keeping the panels clean [28]. A more conservative estimate on the payback time is presented in [29], which estimates the payback period to be 5–8 years, while also noting the solar panels typically have a lifespan of 20–25 years. Another model that may prove effective, especially in developed countries is a hybrid energy model where multiple green sources are used in conjunction with the existing electrical grid—including solar, wind, geothermal, hydro, or micro hydro [29]. Furthermore, in disaster scenarios, there may be no electrical infrastructure left to use, so an alternative energy network is appealing here as well.

14.3.6 Military, Emergency, and Disaster Scenarios

In emergencies and disasters, infrastructure is sometimes destroyed. For example, the tsunami in Japan in 2010 destroyed the infrastructure over large areas while the subway bombings in London and Madrid destroyed localized infrastructure. In these cases, portable battery powered equipment can be set up to provide emergency communications to the people who need help and for the rescuers to communicate with them. Furthermore, the rescuers can use the communication technology to co-ordinate with each other.

Military applications are similar as networks may need to be portable since volatile regions often have unpredictable infrastructure. As missions change in physical region, the equipment may need to move with the soldiers. Vehicles such as tanks, light armored vehicles, and unmanned aerial vehicles (UAVs) may act as base stations for communications between soldiers. There are several unique characteristics in military networks that have special challenges associated with them. First, these networks must be extremely reliable. This may mean higher redundancy than traditional networks. It is far more likely that equipment will be destroyed in this scenario, so more extreme techniques will need to be applied to ensure the network does not partition. In some scenarios, the networks may be highly mobile, so there are greater challenges in this case as well. For instance, even the base stations themselves may be mobile in addition to client nodes. On the other hand, missions often have a plan, and trajectories of nodes such as UAVs may be used in conjunction with scheduled transmissions to reduce energy, similar to delay tolerant networks.

14.3.7 Developing Countries and Rural Access Networks

As mentioned previously, there have been several examples of green communications technologies applied in developing countries successfully. These types of models could be applied in places in the developed world where infrastructure is expensive such as rural areas with low population density. Other remote places such as mountain ranges, Northern Canada, desert environments, and other extreme places could also benefit from this type of technology. Combined with techniques such as delay tolerant networking, the networks could be green and robust and further the goal of pervasive communication.

14.4 ENERGY-EFFICIENT NETWORK ARCHITECTURES, TECHNIQUES, AND PROTOCOLS

In this section, protocols, architectures, and techniques are discussed together because often with these approaches the architectures and techniques are tied to the protocols. For instance, caching and clustering have routing protocols that may be uniquely designed to work within only that particular architecture, whereas other networks may make use of more generalized networking protocols. Particularly in the case of energy-efficient communications, one of the methods to improve efficiency is to exploit unique architectural aspects such as repeaters, caches, cluster-heads, and other equipment within the network with less restrictive constraints on their operation such as mesh routers with more CPU and energy availability in comparison with handheld devices.

14.4.1 Topology Control

One unique method of topology control is an aggressive form of topology control proposed in [7]. In this approach, the authors propose shutting some access points off completely when traffic demand is low while still maintaining QoS and coverage levels that are appropriate in the given area. This means that during off-peak hours of use, the capacity of the network is temporarily lowered until the demand requires that additional access points are brought back online. A similar approach is presented in [33], along with some interesting variations including mobility prediction and combining delay tolerance to allow a portion of the access points to be shut off completely.

Another form of topology control is the use of varying sizes of cells in a mobile network. This serves two purposes. First, it allows better use of the frequency in dense areas. Second, since the smaller cells cover smaller areas, the transmission power of devices is lower, resulting in less energy being used, and even less interference and contention with devices nearby. It is suggested in [5,26] that the use of femto cells can reduce the overall network power consumption by a factor of 7 compared with macrocell deployments. Furthermore, in a network that supports a mixture of femto and macrocells, techniques such as [7] could be applied where the femto cells are turned off when the network is not under heavy demand, further saving energy.

Another technique related to topology control is the use of clustering. In this case, certain nodes are elected as cluster-heads and forward traffic on behalf of nodes within the cluster. This reduces the distance that a single hop must traverse, thereby reducing the power required to transmit. There is a trade-off between power and delay however, because more hops generally increases the delay with the increased propagation and transmission delays introduced. More details and specific techniques are discussed in the subsection on clustering and caching.

14.4.2 Power Control

Power control is related to topology control in some applications. For instance, power control can be used to dynamically tune how large an area an access point covers, or how far a host node can communicate. This contrasts the previously discussed topology control technique of varied cell sizes because in that case it was more of a static approach that was designed as an architectural trait. Power control can be used to serve two goals. First it can be used to avoid interference between neighboring nodes. This may be done in conjunction with transmission schedules and compatibility matrices.

Second, power control can be used when the network is densely populated. In this case, each node has many neighbors, and many routing choices. By lowering the transmission power of the nodes strategically, the number of neighbors can be reduced while still leaving a connected network. This lowers the overall power consumption strictly by using less power for communication, but also reduces it in two other ways. First, power is reduced by the lowered probability of interference. Second, since there are fewer neighbor choices the routing becomes simpler, and likely converges faster in the case of on-demand protocols, or uses less intensive computations in global protocols. On the downside, however, there are fewer paths maintained, which may affect the ability for the network to handle poor conditions such as congested links or interference from external sources.

14.4.3 Repeater and Relay Nodes

Relay nodes may make a particularly large improvement in the energy efficiency of a wireless network. Instead of one long hop from one node to another, relays may be strategically deployed to turn the single long hop into two shorter hops. Although this technique depends on the technology, path loss models, and environment, in some cases it has shown to be extremely effective at reducing the path loss. For instance, Harrod et al. [4] and Vadgama [5] claims a path loss reduction from 21 dB to 3–7 dB by switching from cellular access to a two-hop system. According to [5], more than 50% of energy consumption at one prominent UK mobile network provider is attributed to base station transmissions. This is compared with 30% from the mobile switching and core transmission within the network. So improving the efficiency of the base stations is one way to significantly reduce the amount of energy used in mobile networks.

Another unique technique related to repeater and relay nodes is using "Green Antennas" within cellular networks. In this proposal, the normal cellular architecture is used outdoors, but indoors additional green antennas are used only for the uplink from the device to the cellular network. The reasoning is that the mobile device has to use very high power to penetrate buildings while the outdoor towers are usually linked to the grid and not battery powered so they can afford to operator on higher power [26].

14.4.4 Caching, Clustering, and Data Aggregation

Caching and clustering are grouped together because often clustering goes hand in hand with caching. Caching is commonly used to store commonly and frequently requested data at an intermediate point in the network between the hosts and the content providers. This has several advantages to the noncaching approach. First, the round-trip time for requesting the data is reduced since the request must only travel to the cache location. Similarly, the delay is likely reduced since many hops are also reduced. Third, the load on the content provider is reduced because the content is distributed to the caches. Finally, the number of retransmissions is reduced and localized more since disruptions in external networks are isolated. The reason why caching and clustering go together is now clearer. Since a cluster-head is a common point of data aggregation for the other nodes within the cluster for upstream traffic, it makes sense that a cache is located at this point to store commonly requested content in the downstream. Of course, this technique is most useful in networks where there is enough downstream traffic, so networks like wireless sensor networks may see little benefit. Figure 14.2 shows an example of caching, clustering, and aggregation for wireless networks. The figure illustrates the direction of the data in aggregation and caching and where the cluster-head nodes are located with respect to other nodes in the network. It is also possible to see how the network is organized into regions around the cluster-heads.

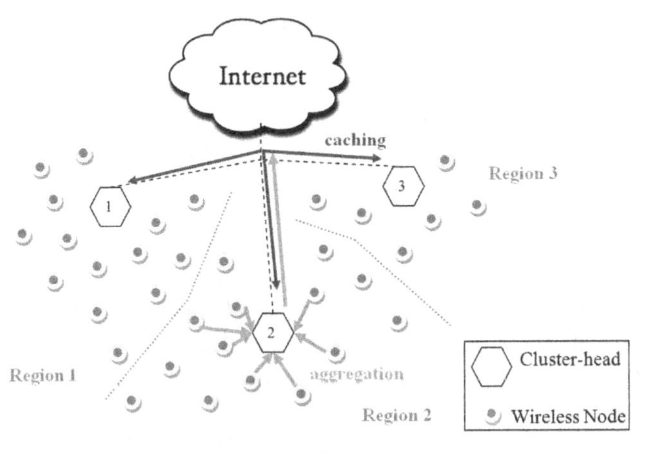

Figure 14.2 Caching, clustering, and aggregation architectures.

Networks such as wireless mesh networks or cellular networks may see improvements, particularly in applications with large numbers of users requesting similar content from the same geographic area. For example a live sporting event may have people checking scores, watching video streams, or listening to audio telecasts live. At a school campus students may check common web pages for course content. In a regional ISP people are likely to visit similar websites, such as the local news websites or regional content.

There have been many proposed techniques for clustering in sensor and ad hoc networks including LEACH [18], HEED [11], SEP [13], ALEACH [12], TEEN [14], APTEEN [15], DEEC [20], SDEEC [16], and TDEEC [19]. Compared with other approaches such as direct transmission, minimum-transmission-energy, multihop routing, and static clustering, approaches such as LEACH can reduce the energy dissipation by as much as a factor of 8 [18]. LEACH tries to reduce energy consumption by randomly rotating the cluster-head so that the energy load is evenly distributed around the network. LEACH was one of the first clustering protocols developed for wireless sensor networks and has problems that have been addressed more recently. For instance, LEACH is a centralized protocol that suffers from problems such as single point of failure. Another improvement on LEACH is selecting the cluster-heads in a more intelligent manner. The cluster-head selection problem is one of the most studied problems in clustering research. Many schemes try to achieve better energy efficiency solely by improving cluster-head selection [13,16,19,20]. For example, SEP [13] proposes a method that takes into account the energy levels relative to other nodes, using the nodes with the most energy as the cluster-heads. Similarly, DEEC [20] and TDEEC [19] use a measurement based on the ratio of residual energy and average energy of the network to select cluster-heads. There are also approaches that aim to extend the overall network lifetime, as opposed to just reducing the energy dissipation. While these problems are related, the network lifetime issue is also important because in some cases, solutions that are energy efficient at particular nodes may leave some strategically located nodes dead sooner, which results in a segmented or partitioned network that is useless for some applications [22].

An alternative approach to the LEACH style of clustering is the TEEN approach [14,15], which combines clustering with data-centric routing. This type of approach lends itself best to networks such as sensor networks where the primary goal of routing is to retrieve specific information, rather than two points of the network communicating with each other. For example, one may wish to ask "What is the average temperature across all of the sensors in region 1?" In this case we are more interested in the result of query than where in particular the data comes from. This lets the network take advantage of on-the-fly aggregation techniques that return only the average rather than all of the individual temperatures directly to the node that made the query. There is also the possibility of time-based requests, for example "What is the temperature every 10 min?" While these types of clustering protocols are good for time-sensitive data and data-centric queries, there is higher overhead compared with protocols such as LEACH [19].

It is also possible to combine clustering with relaying. Both solve similar problems by reducing the distance of a wireless transmission by making smaller, but more hops along towards the destination. In [23], Choi et al. propose a hybrid clustering-relay architecture that aims to reduce energy consumption and prolong network lifetime simultaneously.

Another approach that is related to clustering is simple chaining. This technique avoids the overhead of forming clusters and instead forms an entire chain of nodes from one end of the network to the sink. This type of technique is often used in wireless sensor networks. Instead of each sensor sending data unicast towards the sink in several flows, traffic is combined into larger and larger bundles at each hop and sent together as one larger flow. This technique was proposed in PEGASIS [21].

14.4.5 Energy-Aware Handover Techniques

Handover is used when there are mobile devices in a wireless network. There are two types of handover, vertical and horizontal. Horizontal handover occurs when a device moves within one type of network technology under one operator. For instance a mobile phone user moving within their provider's network must switch cellular towers as they move in and out of range of different towers. Vertical handover occurs when the device must switch technologies or operators. For instance, a device switching from IEEE 802.11 Wi-Fi to 4G cellular coverage illustrates a switch between technologies. A device roaming between operators on a 4G network, for example from Rogers to Telus, illustrates a vertical handover within one technology but across operators. In all types of handover, there are methods to reduce the energy used during this process. Traditional solutions do not typically take into account energy efficiency, but instead focus on making the handover smooth and seamless by preserving connections.

14.4.6 Access Point Selection

In infrastructure networks, access point selection is the problem where a particular node has the choice between accessing more than one infrastructure node for its access to the network. Typically this choice is made based on the signal strength of the candidate access points, however depending on the underlying infrastructure, routing algorithms, and other factors, there may be other choices available. For instance if a particular infrastructure node is congested, it may make more sense to choose a less loaded, but poorer signal quality access point in favor of reducing retransmissions.

Another form of access point selection is proposed in [33]. In this case, there are several variations on traditional cellular and mesh architectures where some base stations or access points are shut off, while others are turned higher. This is similar to approaches for power control and topology control. This could be useful in a scenario where the number of mobile nodes is spare. Another variation presented by Al-Hazmi et al. [33] is to avoid using the infrastructure in some cases and instead revert to ad hoc networking, particularly when nodes are traveling in uniform predictable patterns, such as cars on a highway. Lastly, it was also proposed that some base stations in highly mobile environments are shut off, while delay tolerant approaches are used to skip several stations at once so that losses are avoided. This technique must be combined with mobility prediction so that the correct base station/access point is sent the correct data for the intended mobile node.

14.4.7 Energy-Efficient Routing Algorithms

There are several challenges that must be addressed by good routing algorithms. The schemes should scale so that large numbers of nodes are supported. This type of scalability has contributed to the success of the Internet and should be a goal of all network routing protocols. Secondly, the protocol should be robust and able to support unexpected failures in the network. This is particularly important in low energy networks where nodes may be asleep or may die resulting in paths that are no longer usable. Lastly, since the focus is green communication, the algorithms should try to reduce energy dissipation.

As mentioned previously, one way to achieve energy-efficient routing is to use clustering and data-centric routing like the LEACH [18] or TEEN [14] methods. These data-centric approaches are based on a concept called "directed diffusion," which was proposed first for wireless sensor networks by Intanagonwiwat et al. [24]. These techniques use attribute naming and requests for interest as building blocks for the routing algorithm. For instance a user may request "Give me periodic reports about animal location in region A every *t* seconds." The user is located at a sink node that floods its interest to the entire network. As nodes around the network hear the interest flood, they build a path between the source and the sink using "local gradients." There are various mechanisms to rank and select the best path using these gradients. A competing method to this approach is called "GRAdient Broadcasting" (GRAB) by Ye et al. [25]. In this approach, credits are assigned at traffic sources. As the packets are broadcast, the credits associated with the packet are decremented. This allows for a high level of flexibility near the source and destination in terms of which route to take. As the credits become lower, there is less flexibility to explore and paths that explore too much die off because a lack of credits before they reach the destination.

Another method related to this is chaining, using a technique like the PEGASIS [21] approach. These approaches all apply to wireless sensor networks and are most useful when traffic is data centric rather than source–destination centric. Many of the techniques that do not follow the data-centric paradigm are cross-layer energy-aware routing protocols.

It is also possible to make smarter and more energy-efficient information using geographic routing. Geographic routing requires node positions to be known. When this is the case it is possible to route using the closest physical nodes to the destination or to route around trouble areas in the network. It is also possible to combine this information with information known about the environment to influence routing decisions in a way that may improve energy efficiency. One limitation of geographic routing is in cases where greedy approaches are used to select the next hop. If the routing algorithm simply chooses the physically closest nodes all along the way, it may encounter a situation where the next hop choice is physically out of range of the radios while still being the physically closest hop to the destination. The only path to the destination in this case may be a more roundabout path following the periphery of the network. This problem is solved using "perimeter routing" methods. Another shortcoming of geographic approaches is the requirement of GPS in the devices in the network; however, the cost of GPS has dropped drastically in recent years and is a standard feature in most mobile phones for example. It is also possible to determine relative positioning using various techniques such as signal strength triangulation if at least a subset of the nodes in the network has known positions.

Similar to geographic routing is trajectory-based routing. In this approach, the "general direction" of the traffic is encoded into the packets. So if it is known that the packets should head "south" the routing algorithm tries to take any route it can that will generally take it in this direction. Again these approaches require at least some of the nodes in the network to have a known position for this type of routing to work.

14.4.8 Energy-Efficient Transport Protocols

To start a discussion of energy-efficient transport protocols, it is helpful to consider what is expected in a wireless transport protocol. First the protocol should introduce reliability into the network. Consider TCP as opposed to UDP. TCP uses a system of acknowledgements to notify the sender that packets have been received while UDP makes no such effort. It is important to consider reliable transport in wireless networks because retransmissions are so expensive, particularly when they are end-to-end retransmission. In many more constrained networks such as delay tolerant networks, reliability is link-to-link, so that the progress a packet has made is not lost, requiring retransmission right from the first hop. This model is the store-and-forward model, and may be applicable in some cases to save energy. It is noted in [32] that in order for a transport layer protocol to be energy efficient "poor paths should not be artificially bolstered via mechanisms such as MAC layer ARQ during route discovery and path selection" but at the same time, "recovery should not be costly, since many applications are impervious to occasional packet loss." So there must be a trade-off between energy efficiency and performance in some respects that may also depend on the application of the network. If the network is used to transport data that can tolerate some loss, it may be worthwhile to use recovery mechanisms that result in poorer performance but good energy efficiency otherwise.

The second expectation of a transport protocol is quality of service. This is important because it allows traffic of high priority to be given priority over other traffic. One way to incorporate this concept with green communications is to give traffic that is known to come from other "green networks" a higher priority than traffic coming from less efficient networks. In this way, it is possible not only to improve the efficiency of networks under a single operator, but also to influence other operators to become more efficient themselves in order to secure higher priority traffic. While some may argue this goes against the neutral nature of the Internet, this model is already seen with traffic priorities being biased against users of unwanted applications like peer-to-peer (p2p), torrent, and streaming.

The third expectation of a transport protocol is congestion and flow control. This is particularly important when it comes to green communications. If this stage is neglected and a model similar to UDP is used in the network, without flow or congestion control, many retransmissions will occur. The farther away a node is in the network from a destination node, the less likely it will have a successful communication because the more hops it must traverse, the more likely it is that one of the queues will be full or one of the links along the way will be congested.

Another expectation in communications protocols is fairness. Traditionally, fairness is thought of in terms of performance. This performance is usually measured in terms of throughput or delay. All users paying equal amounts to access a network receive similar delay and throughput within the network. However, this can be extended to resources in general, and since energy is a resource in communication networks, an effort should be made so that all users experience fairness in how much energy they must use to access the network.

The techniques used to achieve some of these goals or expectations may vary drastically depending on the type of network. For instance, the communication model of the network may be many to one in the case of a wireless sensor network. The model may be one to many in the case of a multicast enabled network where many users are streaming content from a common provider. Finally it may be one to one, or peer-to-peer (p2p) in the typical network models. These communication models greatly affect the transport strategy used to enable energy-efficient communication. In the case of sensor networks, as mentioned in the previous section on routing, communication is data-centric, but the nodes are storage limited, so store-and-forward models are also limited. In an extremely unreliable network, the limited storage of the nodes is not much help in reducing retransmissions. Furthermore, in networks where information is not always data-centric, and there are QoS constraints such as low delay, it does not make sense to follow a store-and-forward model. Due to this, the remainder of the section is split into two subsections—"Store-and-Forward Energy-Efficient Transport Protocols" and "End-to-End Energy-Efficient Transport Protocols."

14.4.8.1 Store-and-Forward Energy-Efficient Transport Protocols

The main benefit from store-and-forward as opposed to end-to-end protocols is the ability to retransmit at only the link that has the problem, rather than across the whole path. Figure 14.3 illustrates this difference. Furthermore, depending on the implementation of the end-to-end mechanism, there may or may not be caching at the destination. If there is not, a scenario like Figure 14.3 would result in the resending of packet 3, even though it was successfully received at the destination because it was received out of order. In contrast, the store-and-forward model caches received packets so that once a packet has been successfully received at any router, it can be used regardless of the order it is received. For instance, in Figure 14.3, packet 3 arrives at

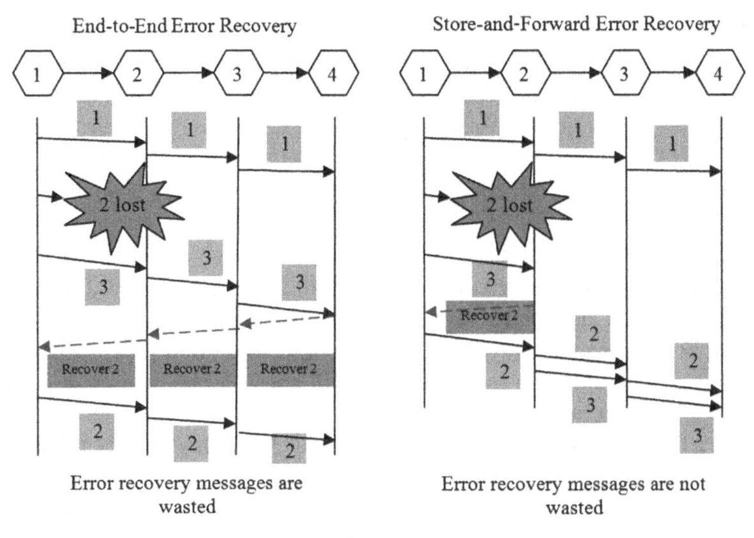

Figure 14.3 End-to-end vs. store-and-forward error recovery.

router 2, and after the loss of packet 2 can be sent without receiving it directly from router 1 again.

One example of a store-and-forward protocol that was designed for wireless sensor networks is called "Pump Slowly, Fetch Quickly" by Wan et al. [30]. This protocol is scalable and energy efficient because it supports minimal signaling to achieve reliable transportation. The basic idea is to distribute data from a source slowly. Any neighbors that experience a loss can then fetch the missing segments quickly.

14.4.8.2 End-to-End Energy-Efficient Transport Protocols

A competing protocol to the store-and-forward approach for sensor networks is called "Simple Wireless Sensor Protocol (SWSP)," which was proposed by Agrawal et al. [31]. SWSP is an event-driven protocol that prepares a "register" message and broadcasts it to the network. If an ACK is not received after 3 "ready to receive" messages from the destination, data is retransmitted. The goal of this protocol is to "inherit the basic characteristics of TCP while remaining less complex and less resource demanding" [31]. One of the limitations with this protocol is the amount of broadcasting used. This may prove to make the protocol less energy efficient than others, however this may be offset by the low resources used in other aspects of the protocol.

Reliable Multi-Segment Transport (RMST) proposed by Stann and Heidermann is another reliable transport protocol for sensor networks [32].

14.4.9 Energy-Efficient Medium Access Protocols

There are several design choices when it comes to medium access in wireless networks. In this approach taken by IEEE 802.11 and other related technologies, there is a reliability mechanism that uses RTS/CTS, ACK, and randomized slot times. RTS/CTS guarantee only a single node will access the medium at a single time within a certain region. The ACK is used as a stop-and-wait protocol whereby a timeout occurs if an ACK has not been received from the destination in a certain period of time. On top of these, there is the notion of Automatic Repeat reQuest (ARQ), which determines how the network behaves in hop-to-hop recovery of lost frames or frames that arrive with errors.

The first choice, which is to have no ARQ, assumes that all reliability is handled via the transport or higher protocol layers. There is some benefit to this approach. If used with routing algorithms like directed diffusion [24], links that are poor will actually appear poor and be avoided by the routing algorithm [32]. If on the other hand, ARQ is used, at higher layers the link will appear to be high quality and may be used when in fact another link should have been chosen. So either no ARQ can be used, or the routing metric should take into account delay since the reliability mechanism at the MAC layer will introduce significantly higher delay, especially on poor links.

The second choice is to always enable ARQ. This will cause stop-and-wait reliability at the MAC layer resulting in links that appear to be reliable, but also have higher delay in the case where they are actually reliable. This may be beneficial in the case where links are only periodically poor, since it may be beneficial to continue using a link rather than initiating the process at the routing level to determine a new route. On the other hand, if the routing algorithm supports maintaining multiple paths, this point may be invalid.

Finally, the last choice is to use "selective ARQ." Packets sent to a single neighbor are sent using ARQ while those that are not use no ARQ. Data that is traveling on an established unicast path is biased toward continuing to use the same path, while packets that are discovering new routes do not use ARQ so that paths that are actually unreliable are not selected [32].

Another consideration, particularly in heterogeneous wireless networks or those that use cognitive radio techniques, is the order in which the medium is sensed. Due to the nature and timing of the various access schemes within individual radio technologies, it is possible to design a schedule for sensing that minimizes the total energy spent listening on each radio [43].

14.5 TRENDS, LIMITATIONS, CHALLENGES, AND OPEN PROBLEMS IN GREEN COMMUNICATIONS

A difficult challenge with green communications is the complexity of the interactions of the different protocol layers. In order for a complete green solution to be found, studies must be undertaken to determine how to best select design and protocol aspects such that energy use is minimized. While one protocol may work well in its own particular layer, when combined with the actions of other layers it may turn out to be very inefficient. Thus, in order to evaluate protocols, one must take care to design experiments that aim to study this complexity between these interactions. Cross-layer protocols may be a method with which to tackle this problem, however care must also be taken to avoid convoluted and unmaintainable solutions. One of the greatest successes of the Internet was the software engineering principles governing the layered methodology that made the technology easily maintainable and easy to understand. While this model may not always be applicable now, particularly with wireless communications, it is worthwhile to carefully consider extensible models for exchanging information from nonadjacent layers.

Another weakness in green communication is the lack of standardization or methods to quantify how energy efficient devices are. For instance in general there are bodies such as ENERGY STAR in North America that try to rate how efficient devices like televisions, computer monitors, and other devices are, however, nothing yet exists for communication devices. It may be possible that some devices exist for the general operation of themselves, but do not quantify the communication of the network, or networking protocols used themselves.

14.6 CONCLUSIONS AND FUTURE DIRECTIONS

It is clear that there are already many energy-efficient techniques and solutions that exist within wireless communications. The problem with current solutions is that they are generally applied only to very particular network types or very restrictive applications. Since next-generation wireless communication is heading towards convergence of technologies, it is logical that many innovations in other areas will be applied wherever possible. This is the future of green communication. Techniques such as data-centric routing, which has been successful in wireless sensor networks, may also be applicable in more generalized wireless networks for applications such as streaming, where the user may be interested in queries like "Stream the movie X" or "Find prices of Y product over the next week." In these cases, the users are more interested in the data, and may not be as concerned about where it is coming from. In applications where data-centric routing is useful, it is also possible to see further improvements in efficiency if ARQs are disabled at the link layer since poor links can then be avoided by routing. Another approach could be a dual strategy so that when links are periodically poor ARQs are still used in some cases until a threshold is reached and ARQs become disabled for the link.

Another area that requires more research and study is applying energy-efficient horizontal handover techniques to vertical handover. While there are some solutions that exist for handing over due to mobility, there are more difficult challenges in making similar decisions when handing over across technologies. In these cases, the device must consider the price of the access network, the capacity, congestion level, and other factors, making it a very complex decision.

This chapter has presented an overview of some of the challenges in green wireless communications. It has also presented several of the most popular and emerging techniques to address these problems. Unique applications, networks, and architectures were also outlined along with challenges and solutions. While energy efficiency has been studied for some time in particular wireless technologies such as sensor networks, there are many others that have not yet taken advantage of this research. However, this area is quickly becoming a hot topic of study because the benefits to the environment, performance, and profitability are difficult to ignore.

REFERENCES

[1] H. Singh, H.-R. Shao, C. Ngo, Enhanced power saving in next generation wireless LANs, in: Proceedings of 64th Vehicular Technology Conference (VTC), 2006, pp. 1–5.

[2] J. Palicot, Cognitive radio: an enabling technology for green radio communications concept, in: Proceedings of 2009 International Conference on Wireless Communications and Mobile Computing, June 2009, pp. 489–494.

[3] <http://www.cwc.oulu.fi/workshops/W-Green2008.pdf>.

[4] T.J. Harrod et al., Intelligent relaying for future personal communications system, in: IEEE Colloquium on Capacity and Range Enhancement Techniques for the Third Generation Mobile Communications and Beyond, February 2000, pp. 1/9–5/9.

[5] S. Vadgama, Trends in green wireless access, FUJITSU Scientific Technical Journal 45 (4) (2009) 404–408.

[6] M. Gupta, S. Singh, Greening of the Internet, in: Proceedings of the 2003 Conference on Applications, Technologies, Architectures and Protocols for Computer Communications (SIGCOMM), 2003, pp. 19–26.

[7] J. Lorincz, A. Capone, D. Begusic, Optimized network management for energy savings of wireless access networks, Computer Networks 55 (3) (2011) 514–540.

[8] E.H. Ong, K. Mahata, J.Y. Khan, Energy efficient architecture for green handsets in next generation IP-based wireless networks, in: Proceedings of IEEE International Conference on Communications (ICC 2011), June 2011, pp. 1–6.

[9] A.B. Ericsson, Sustainable energy use in mobile communications, White paper, EAB-07:021801 Uen Rev B:1–23, 2007.

[10] V. Pejovic, E. Belding, Energy efficient communication in next generation rural-area wireless networks, in: Proceedings of 2010 ACM Workshop on Cognitive Radio Networks (CoRoNet'10), 2010, pp. 19–24.

[11] O. Younis, S. Fahmy, HEED: a hybrid, energy efficient, distributed clustering approach for ad hoc sensor networks, IEEE Transactions on Mobile Computing 3 (4) (2004) 660–669.

[12] M.S. Ali, T. Dey, R. Biswas, ALEACH: advanced LEACH routing protocol for wireless microsensor networks, in: Proceedings of International Conference on Electrical and Control Engineering (ICECE 2008), 2008, pp. 20–22.

[13] G. Smaragdakis, I. Matta, A. Bestavros, SEP: a stable election protocol for clustered heterogeneous wireless sensor networks, in: Proceedings of Second International Workshop on Sensor and Actor Network Protocols and Applications (SAPNA 2004), 2004, pp. 1–11.

[14] A. Manjeshwar, D.P. Agrawal, TEEN: a routing protocol for enhanced efficiency in wireless sensor networks, in: First International Workshop on Parallel and Distributed Computing Issues in Wireless Networks and Mobile Computing, April 2001, pp. 2009–2015.

[15] A. Manjeshwar, D.P. Agrawal, APTEEN: a hybrid protocol for efficient routing and comprehensive information retrieval in wireless sensor networks, in: Proceedings of

International Parallel and Distributed Processing Symposium (IPDPS 2002), 2002, pp. 195–202.

[16] E. Brahim, S. Rachid, A. Pages-Zamora, D. Aboutajdine, Stochastic distributed energy-efficient clustering (SDEEC) for heterogeneous wireless sensor networks, International Congress for Global Science and Technology, Journal of Computer Networks and Internet Research (ICGST-CNIR) 9 (2) (2009) 11–17.

[17] M. Röösli, P. Frei, E. Mohler, K. Hug, Systematic reviews on the health effects of exposure to radiofrequency electromagnetic fields from mobile phone base stations, Bulletin of World Health Organization (2010) 887–896.

[18] W.H. Heinzelman, A. Chandrakasan, H. Balakrishnan, Energy-efficient communication protocol for wireless microsensor networks, in: Proceedings of 33rd Hawaii International Conference on System Sciences, 2000, pp. 1–10.

[19] P. Saini, A.K. Sharma, Energy efficient scheme for clustering protocol prolonging the lifetime of heterogeneous wireless sensor networks, International Journal of Computer Applications 6 (2) (2010) 30–36.

[20] L. Qing, Q. Zhu, M. Wang, Design of a distributed energy-efficient clustering algorithm for heterogeneous wireless sensor networks, Journal of Computer Communications 29 (12) (2006) 2230–2237.

[21] S. Lindsay, C.S. Raghavendra, PEGASIS: power efficient gathering in sensor information systems, in: Proceedings of IEEE Aerospace Conference, Big Sky, Montana, 2002, pp. 1125–1130.

[22] M.J. Handy, M. Haase, D. Timmermann, Low energy adaptive clustering hierarchy with deterministic cluster-head selection, in: Fourth International Workshop on Mobile and Wireless Communications Networks, 2002, pp. 368-372.

[23] D. Choi, S. Moh, I. Chung, ARCS: an energy-efficient clustering scheme for sensor network monitoring systems, International Scholarly Research Journal on Communications and Networking 2011 (2011) 1–10.

[24] C. Intanagonwiwat, R. Govindan, D. Estrin, J. Heidemann, F. Silva, Directed diffusion for wireless sensor networks, IEEE/ACM Transactions on Networking 11 (1) (2003) 2–16.

[25] F. Ye, G. Zhong, S. Lu, L. Zhang, GRAdient Broadcast: a robust data delivery protocol for large scale sensor networks, Journal on Wireless Networks 11 (3) (2005) 285–298.

[26] A. Kumar, Y. Liu, T. Singh, S.S. Khurmi, Sustainable energy optimization techniques in wireless mobile communication networks, in: First International Conference on Interdisciplinary Research and, Development, June 2011, pp. 24.1–24.6.

[27] X. Lixing, P. Tao, Safaricom: Kenya gets greener with alternative energy, Huawei white paper: huawei.com/en/static/hw-079527.pdf, 2009, pp. 38–40.

[28] Nokia Siemens, Ethiopia Telecommunications Corporation (ETC): a success story, white paper: http://www.nokiasiemensnetworks.com/system/files/document/ETC_Ethiopian_communications_network_powered_by_renewable_energy_solutions.pdf.

[29] A. Kumar, Y. Liu, T. Singh, S.G. Singh, Sustainability in wireless mobile networks through alternative energy resources, International Journal of Computer Science and Telecommunications (IJST) 1 (2) (2010) 196–200.

[30] C.-Y. Wan, A.T. Campbell, L. Krishnamurthy, Pump-slowly, fetch-quickly (PSFQ): a reliable transport protocol for sensor networks, IEEE Journal on Selected Areas in Communications 23 (4) (2005) 862–872.

[31] P. Agrawal, T.S. Teck, A.L. Ananda, A lightweight protocol for wireless sensor networks, in: IEEE Wireless Communications and Networking (WCNC 2003), March 2003, pp. 1280–1285.

[32] F. Stann, J. Heidermann, RMST: reliable data transport in sensor networks, in: Proceedings of IEEE Sensor Network Protocols and Applications, May 2003, pp. 102–112.

[33] Y. Al-Hazmi, H. de Meer, K.A. Hummel, H. Meyer, M. Meo, D. Remondo, Energy-efficient wireless mesh infrastructures, IEEE Network 25 (2) (2011) 32–38.

[34] I. Smaoui, F Zarai, R Bouallegue, L Kamoun, Multi-criteria dynamic access selection in heterogeneous wireless networks, in: Sixth International Symposium on Wireless Communication Systems (ISWCS 2009), September 2009, pp. 338–342.

[35] M. Fiterau, O. Ormond, G.-M. Muntean, Performance of handover for multiple users in heterogeneous wireless networks, in: IEEE 34th Conference on Local Computer Networks (LCN 2009), October 2009, pp. 257–260.

[36] P. Si, R. Yu, H. Ji, V.C.M. Leung, Optimal network selection in heterogeneous wireless multimedia networks, in: IEEE International Conference on Communications (ICC 2009), 2009, pp. 1–5.

[37] B. Bakmaz, Z. Bojkovic, M. Bakmaz, Network selection algorithm for heterogeneous wireless environment, in: 18th IEEE Symposium on Personal, Indoor and Mobile Radio Communications (PIMRC 2007), 2007, pp. 1–4.

[38] F. Warthman, Delay-tolerant networks (DTNs): a tutorial, white paper: http://www.dtnrg.org/docs/tutorials/warthman-1.1.pdf, Version 1.1 (accessed December 2011).

[39] R. Wang, S.C. Burleigh, P. Parikh, C.-J. Lin, B. Sun, Licklider transmission protocol (LTP)-based DTN for cislunar communications, IEEE/ACM Transactions on Networking 19 (2) (2011) 359–368.

[40] B. Admanson, C. Bormann, M. Handley, J. Macker, NACK-oriented reliable multicast (NORM) transport protocol, IETF, RFC 5740, November 2009 <http://tools.ietf.org/html/rfc5740> (accessed December 2011).

[41] O.P. Gandi, L.L. Morgan, A.A. de Salles, Y.-Y. Han, R.B. Herberman, D.L. Davis, Exposure limits: the underestimation of absorbed cell phone radiation, especially in children, Electromagnetic Biology and Medicine 31 (1) (2012) 34–51.

[42] X. Wang, A.V. Vasilakos, M. Chen, Y. Liu, T.T. Kwon, A survey of green mobile networks: opportunities and challenges, Mobile Network Applications 17 (1) (2012) 4–20.

[43] Jose Marinho, Edmundo Monteiro, Cognitive radio: survey on communication protocols, spectrum decision issues, and future research directions, Wireless Networks 18 (2) (2012) 147–164.

CHAPTER

Energy-Efficient MIMO–OFDM Systems

15

Zimran Rafique, Boon-Chong Seet

Department of Electrical and Electronic Engineering, School of Engineering, Auckland University of Technology, Private Bag 92006, Auckland 1142, New Zealand

15.1 INTRODUCTION

Due to the advent of mobile multimedia applications such as mobile video on-demand, wireless communication systems that can cope with high data rate demands are required. However, higher data rates necessitate more energy per bit for a given bit error rate (BER), which in turn increases the overall energy consumption of the system and the production of CO_2 emissions that are a threat to global warming. Thus, researching energy-efficient designs for high data rate wireless communication systems is a burning issue. Due to escalating expansion of wireless network infrastructures and exponential growth in traffic rate, a considerable amount of worldwide energy is consumed by information and communication technology (ICT) of which more than 70% is being used by the radio access part/ radio frequency (RF) section [1].

In the late 1990s, multiple-input multiple-output (MIMO) techniques were proposed to achieve higher data rates and smaller BER with the same transmit power and bandwidth required by a single antenna system. Diagonal Bell Labs Layered Space–Time (D-BLAST), a spatial multiplexing technique, was proposed in [2] to transmit independent information sequences using multiple antennas to increase the overall data rate as compared to single antenna system without extra bandwidth or power. Due to the implementation complexity of D-BLAST, a simplified version known as Vertical BLAST (V-BLAST) [3] was proposed by Wolniansky et al., and a prototype implementation of V-BLAST architecture with the Zero Forcing detection algorithm was discussed. The results showed high spectral efficiency as compared to a traditional single-input single-output (SISO) communication system. D-BLAST and V-BLAST systems work efficiently if the number of receiving antennas N_r is greater than or equal to transmitting antennas N_t ($N_t \leq N_r$). A Turbo-BLAST [4] technique was later proposed by Sellathurai and Haykin, which can perform well even if N_r is less than N_t ($N_r < N_t$).

A well-known space–time coding (STC) technique was devised by Alamouti to improve BER performance of the multiantenna setup by transmitting and/or receiving redundant copies of the information signals without decreasing the data rate as compared to a single antenna system [5]. The space–time trellis coding (STTC) technique was invented by Tarokh et al. and performed better than the Alamouti coding technique but with more decoding complexity [6]. Orthogonal space–time block codes (OSTBC) were proposed to further maximize the possible spatial diversity and allow simple decoding algorithms based on linear processing [7]. A spatio-temporal vector coding (STVC) communication structure was suggested by Raleigh and Cioffi as a means for achieving high MIMO channel capacity using singular value decomposition (SVD) and eigenvalue decomposition. However, for this technique, channel state information (CSI) is required at the transmitter as well as receiver side [8].

Multiantenna systems can also be used for beamforming to increase signal- to-noise ratio (SNR) at the receiver and to suppress co-channel interference in multiuser networks [9]. A multifunctional MIMO system design was proposed in [10] to attain multiplexing gain, diversity gain,and beamforming gain. Due to the small form factor and limited energy of some wireless nodes, it is often not realistic to equip each node with multiple antennas to implement MIMO. Instead, a cluster of single-antenna nodes can cooperate to form a virtual antenna array (VAA) to achieve virtual MIMO communication. In virtual MIMO systems, nodes are physically located in different places, and thus timing and frequency asynchronism can be a major problem for such systems [11].

MIMO schemes have good spectral efficiency but with more circuit complexity that consumes energy [12]. In long-distance transmission, circuit energy consumption is much less than transmission energy consumption, but in short distance transmission circuit energy consumption is comparable with transmission energy. MIMO techniques have been proposed for wireless sensor networks (WSNs) where nodes mostly operate on batteries and are expected to work for a long period of time. To evaluate the performance of MIMO techniques in energy-limited WSNs, one must take into account both the circuit and transmission power consumption. In [12], the Alamouti coding technique was used for a MIMO WSN and the performance was compared with SISO for the same throughput and BER. The energy efficiency was also compared over different transmission distances with consideration of circuit and transmission energy consumption. A V-BLAST based WSN was proposed in [13] with the concept of virtual MIMO without spatial encoding on transmitting side nodes, thus eradicating local communication on the transmitting side as in the Alamouti coding technique.

Orthogonal frequency division multiplexing (OFDM) is a multicarrier modulation technique that has the capability to mitigate the effect of intersymbol interference (ISI) at the receiver side. In most of the multiantenna systems, CSI is required for effectual signal processing at the receiver side so that effective channel estimation can be done using OFDM with an optimal training sequence. MIMO techniques are used with OFDM (MIMO–OFDM) to further enhance system performance. MIMO–OFDM systems are capable of increasing the channel capacity even under severe channel conditions [14]. The combination of MIMO and OFDM can also provide two-dimensional space–frequency coding (SFC) in space and frequency using individual subcarriers of an OFDM symbol [15], or three-dimensional coding called space–time–frequency coding (STFC) to achieve larger diversity and coding gains [16, 17]. OFDM can also be used in a multiuser cooperative communication system by assigning a subcarrier to different users for overall transmit power reduction [18].

WOFDM is another multicarrier modulation technique and its properties are being explored in the field of high data rate communication systems [19]. In this modulation technique, the orthogonal subcarriers are generated using a symmetric/asymmetric multilevel quadrature mirror filter (QMF) bank structure. WOFDM as an alternative to Fourier-based OFDM (FOFDM) was first proposed by Lindsey [20] and theoretical discussions were carried out regarding the power spectral density and bandwidth efficiency of WOFDM. The ISI and intercarrier interference (ICI) analysis of WOFDM and FOFDM has been carried out by Negash and Nikookar [21] with an asymmetric multilevel QMF structure. The ISI and ICI of WOFDM were less as compared to FOFDM. These results were further verified by Jamin and Mahonen [22] who showed that WOFDM is a feasible alternative to FOFDM. The BER performance of WOFDM with standard wavelets and FOFDM were also investigated in the presence of frequency offset and phase noise. Both techniques were found to be equally affected by frequency offset and phase noise [23]. The BER performance of WOFDM and FOFDM with V-BLAST system architecture was explored in [24] and the performance of WOFDM-based V-BLAST system architecture was also investigated in [25] using different detection algorithms. It was shown that new wavelet bases can be designed to utilize unoccupied time-frequency gaps of licensed users [26], and to perform even better in the presence of a timing offset [27].

The remainder of the chapter is organized as follows. In Section 15.2, multiantenna systems and their advantages (high data rate, low BER, energy efficiency, beamforming) and disadvantages (circuit complexity, physical size of nodes) are presented along with cooperative (virtual MIMO) and noncooperative (true MIMO) communication. FOFDM and WOFDM are further discussed in Section 15.3. Multiantenna systems with FOFDM and WOFDM are presented in Section 15.4. Finally, concluding remarks are given in Section 15.5.

15.2 MULTIPLE ANTENNA SYSTEMS

In multiple antenna systems, more than one antenna are used on the transmitting and/or receiving side as shown in Figure 15.1. MIMO techniques can be used to increase

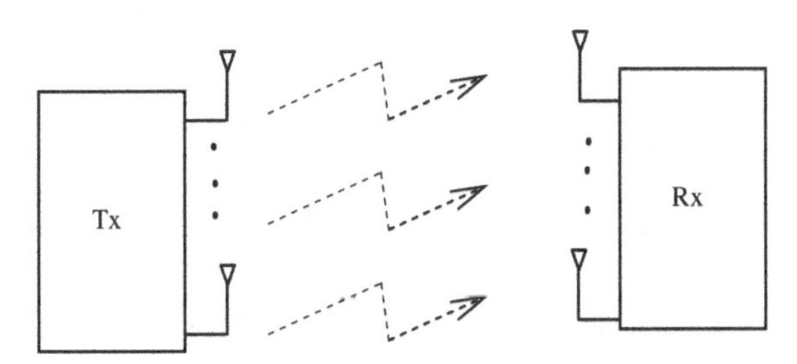

Figure 15.1 MIMO wireless communication system.

the data rate using spatial multiplexing and BER can be improved by using spatial diversity. MIMO techniques can also be used to improve SNR at the receiver and to mitigate co-channel interference (CCI) along with beamforming techniques.

15.2.1 Spatial Multiplexing Techniques

By using spatial multiplexing techniques, the number of users or data rate of a single user can be increased by the factor of the number of transmitting antennas (N_t) for the same transmission power and bandwidth. Since the user's data is being sent simultaneously in parallel over multiple antennas, the effective transmission rate is increased roughly in proportion to the number of transmit antennas used. All the substreams are transmitted in the same frequency band, thus the spectrum is used very efficiently. The transmitted substreams are also independent of one another. Individual transmitter antenna power is scaled by $1/N_t$, thus the total power remains constant and independent of number of the transmitter side antennas (N_t). At the receiver, the transmitted signals are retrieved from received sequences (layers) by using detection algorithms. The basic spatial multiplexing system architecture is shown in Figure 15.2.

15.2.1.1 D-BLAST

D-BLAST is a spatial multiplexing technique to achieve spectral efficiency for a given bit rate and transmission power [2]. For this technique, CSI is required at the receiver side but not at transmitter side. For N_t transmitting antennas and N_r receiving antennas, the *i*th codeword is made up "diagonally" with N_t transmitted signals.

Assume that at time $t = t_1$, only antenna a_1 transmits its data and all other transmitting antenna remain silent. At time $t = t_2$, antennas a_1 and a_2 transmit their data. At time $t = t_{N_t}$, all transmitting antennas (a_1, a_2, ..., a_{N_t}) transmit their data. At time $t = t_{N_t} + 1$, again all transmitting antennas transmit their data, and so on. The code block in time–space dimension for $N_t = 4$ is shown below:

$$S = \begin{bmatrix} s_{11}^{1} & s_{12}^{2} & s_{13}^{3} & s_{14}^{4} & \cdots \\ 0 & s_{22}^{1} & s_{23}^{2} & s_{24}^{3} & \cdots \\ 0 & 0 & s_{33}^{1} & s_{34}^{2} & \cdots \\ 0 & 0 & 0 & s_{44}^{4} & \cdots \end{bmatrix}, \tag{15.1}$$

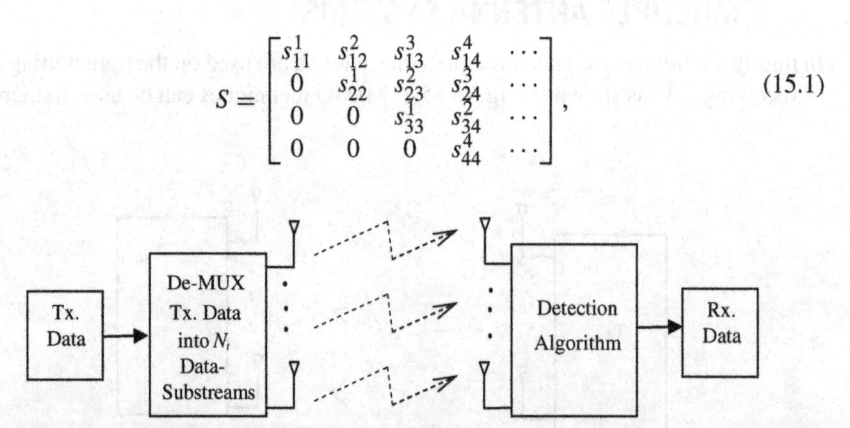

Figure 15.2 Spatial multiplexing system architecture with N_t transmitting and N_r receiving antennas.

where s^{i}_{jk} is the symbol transmitted from the jth antenna at the kth time slot of ith codeword. The received signal at the receiver with N_r receiving antennas, when s^1_{11} was transmitted, can be expressed as follows:

$$\text{Rec} = Hs^1_{11} + \zeta \tag{15.2}$$

where Rec is an Nr × 1 vector, ζ is an Nr × 1 noise vector whose elements are complex Gaussian random variables with zero mean and variance N_o, and H is an Nr × 1 channel matrix. At the receiver side, s^1_{11} can be estimated using maximal-ratio combining (MRC) with the help of following equation:

$$s^{\hat{1}}_{11} = \frac{H^*\text{Rec}}{H^*H}, \tag{15.3}$$

where * represents the hermitian of H (transpose and complex conjugate). Symbol s^1_{22} can be calculated using minimum mean square error (MMSE). The received signal vector can be expressed as

$$\text{Rec} = HX + \zeta, \tag{15.4}$$

where $X = \begin{bmatrix} s^2_{12} \\ s^1_{22} \end{bmatrix}$ and $H = \begin{bmatrix} h_{11} & h_{12} \\ h_{21} & h_{22} \\ h_{31} & h_{32} \\ \vdots & \vdots \\ h_{N_r1} & h_{N_r2} \end{bmatrix}$ and ζ is an $N_r \times 1$ noise vector whose elements are complex Gaussian random variables with zero mean and variance N_o. During this time slot, only antennas a_1 and a_2 transmitted the signals.

The symbol s^1_{22} transmitted at the second time slot from antenna a_2 (i.e., n_t=2) is estimated by using the following steps [28]:

- Replace the first column (all $n_t - 1$ columns) of H by a null column so that

$$H_{\text{new}} = \begin{bmatrix} 0 & h_{12} \\ 0 & h_{22} \\ 0 & h_{32} \\ \vdots & \vdots \\ 0 & h_{N_r2} \end{bmatrix}.$$

- Calculate $G_{\text{MMSE}} = H^*_{\text{new}}\left[H_{\text{new}}H^*_{\text{new}} + \frac{1}{\text{SNR}}I_{N_r}\right]^{-1}$ where $\frac{1}{\text{SNR}}$ is the noise power to signal power ratio and I_{N_r} is the identity matrix of size N_r.
- Select the second (n_t=2) row of G_{MMSE}.
- Estimate $s^1_{22} : s^{\hat{1}}_{22} = G_{\text{MMSE}_2}\text{Rec}$.
- Subtract the effect of $s^{\hat{1}}_{22}\left(\text{Rec} - \begin{bmatrix} h_{12} \\ h_{22} \end{bmatrix} s^{\hat{1}}_{22}\right)$ such that s^2_{12} sees again an interference-free channel and MRC can be used to detect s^2_{12}.

By using D-BLAST, there is wastage of space and time in the initial phase of transmission and error will propagate in the event of detection of a wrong signal.

15.2.1.2 V-BLAST

V-BLAST is a spatial multiplexing technique to achieve spectral efficiency for a given bit rate and transmission power. This technique enables simpler encoding on the transmitter side than D-BLAST [3] and there is no resource (space and time) wastage during the initialization phase of transmission. In this technique, CSI is required at the receiver side but not at the transmitter side. For N_t transmitting antennas and N_r receiving antennas, the ith codeword is made up "vertically" with N_t transmitted signals.

Assume that at time $t = t_1$, all transmitting antennas transmit their data, and at time $t = t_2$, again all transmitting antennas transmit their data, and so on. The code block in time–space dimension for $N_t = 4$ is shown below:

$$S = \begin{bmatrix} s_{11}^1 & s_{12}^2 & s_{13}^3 & s_{14}^4 & \cdots \\ s_{21}^1 & s_{22}^2 & s_{23}^3 & s_{24}^4 & \cdots \\ s_{31}^1 & s_{32}^2 & s_{33}^3 & s_{34}^4 & \cdots \\ s_{41}^1 & s_{42}^2 & s_{43}^3 & s_{44}^4 & \cdots \end{bmatrix}, \tag{15.5}$$

where s_{jk}^i is the symbol transmitted from the jth antenna at the kth time slot of the ith codeword. The received signal at the receiver with N_r receiving antennas, when the ith codeword was transmitted from N_t transmitting antennas can be expressed as follows:

$$\text{Rec} = HX + \zeta, \tag{15.6}$$

where X is an Nt×1 vector, ζ is an Nr×1 noise vector whose elements are complex Gaussian random variables with zero mean and variance No, and H is an Nr×Nt channel matrix. Different detection algorithms have been proposed for the V-BLAST system architecture and in this chapter, we discuss the following three well-known detection algorithms.

15.2.1.2.1 Zero-Forcing Detection Algorithm

The Zero-Forcing (ZF) detection algorithm is a low complexity suboptimal detection algorithm presented in [3]. At each symbol time, it first detects the "strongest" layer and then cancels the effect of this strongest layer from each of the received signals, and proceeds to detect the "strongest" of the remaining layers, and so on. This algorithm consists of the following steps with the assumption that channel response "H" is known at the receiver side.

- Determine the optimal detection order that corresponds to choosing the row of G (also referred as nulling matrix) with minimum Euclidian norm, where P is a column matrix whose entities correspond to Euclidian norm of each row of the G matrix:

$$G = (H^* H)^{-1} H^*,$$

$$P = \begin{bmatrix} \|G_k = 1\|^2 \\ \vdots \\ \|G_k = N_t\|^2 \end{bmatrix}.$$

- Choose the row $\underline{(G)}_i$ and multiply it with Rec to obtain a "strongest" signal y_i where i is the index of row column vector P with the minimum value:
$$y_i = \underline{(G)}_i \text{Rec}.$$
- The estimated value of the strongest transmit signal is detected by slicing to the nearest value in the signal constellation:
$$\widehat{S}_i = Q(y_i).$$
- Since the strongest transmit signal has been detected, its effect should be cancelled from the received signal vector to reduce the detection complexity for the remaining transmit signals:
$$\text{Rec} = \text{Rec} - \widehat{S}_i(H)_i.$$
- Replace $\underline{(P)}_i$ by ∞, so that in the next cycle, the row $\underline{(G)}_i$ with minimum value is chosen.

15.2.1.2.2 MMSE Detection Algorithm

The minimum mean square error (MMSE) detection algorithm has almost the same computational complexity as the ZF detection algorithm but gives more reliable results even with a noisy estimation of the channel due to the introduction of diagonal matrix $\left(\frac{1}{\text{SNR}} I_{N_r}\right)$ in the MMSE filter (G_{MMSE}) equation [29]. This algorithm consists of following steps with the assumption that channel response "H" is known at the receiver side:

- Calculate $G_{\text{MMSE}} = H^*[HH^* + \frac{1}{\text{SNR}} I_{N_r}]^{-1}$ where $\frac{1}{\text{SNR}}$ is the noise power to signal power ratio and I_{N_r} is the identity matrix of size N_r.
- Select the ith row of G_{MMSE} and multiply it with Rec to detect the signal y_i:
$$y_i = \underline{(G_{\text{MMSE}})}_i \text{Rec}.$$
- The estimated value of the signal is detected by slicing to the nearest value in the signal constellation:
$$\widehat{S}_i = Q(y_i).$$

15.2.1.2.3 QR Decomposition Detection Algorithm

The computational complexity of the QR Decomposition detection algorithm is less as compared to the above-mentioned detection algorithms [30] but with a slight performance degradation. This algorithm consists of the following steps with the assumption that channel response "H" is known at the receiver side:

- Denoting channel response matrix $H = QR$, where Q is an $N_r \times N_t$ unitary matrix composed of orthonormal columns with unit norm and R is $N_t \times N_t$, an upper triangular matrix.
- The received signal expression in Eq. (15.6) can be modified to detect the transmitted signals by multiplying it with Q^t (transpose of Q) as follows:
$$\widetilde{\text{Rec}} = Q^t\text{Rec} = Q^t(HS + \zeta) = Q^t QRS + \eta = RS + \eta,$$

where $Q^{t}Q = I$ (identity matrix), and $\eta = Q^{t}\zeta$ is statistically identical to ζ.

- Due to the upper triangular structure of R, the ith element of $\widetilde{\text{Rec}}$ is given by

$$\tilde{\text{rec}}_i = r_{ii} + d_i + \eta_i,$$

where $d_i = \sum_{j=i+1}^{N_t} r_{ij}s_i$ is the interference term.

- The interference free signal element is given by

$$z_i = \tilde{\text{rec}}_i - d_i.$$

- The detected signal corresponding to each receiving antenna can be calculated using $\widehat{y}_i = \frac{z_i}{r_{ii}}$ and sliced to the nearest value in the signal constellation to estimate the symbol $\widehat{S}_i$.

15.2.1.3 Turbo-BLAST

As mentioned in Section 15.1, this technique can performed well even if the number of receiving antennas N_r is less than transmitting antennas N_t ($N_r < N_t$) but with more computational complexity on receiver side. In this technique, data substreams are interleaved using time interleavers and diagonal layering space interleavers. Similar to V-BLAST, there is no resource (space and time) wastage during the initialization phase of transmission as in D-BLAST, and similarly CSI is required at the receiver side but not at transmitter side.

For N_t transmitting antennas and N_r receiving antennas, the ith codeword is made up "diagonally" (as opposed to "vertically" in V-BLAST) with N_t transmitted signals. Assume that at time $t = t_1$, all transmitting antennas transmit their data, and at time $t = t_2$, again all transmitting antennas transmit their data, and so on. The code block in time–space dimension for $N_t = 4$ is shown below:

$$S = \begin{bmatrix} s_{11}^1 & s_{12}^4 & s_{13}^3 & s_{14}^2 & \cdots \\ s_{21}^2 & s_{22}^1 & s_{23}^4 & s_{24}^3 & \cdots \\ s_{31}^3 & s_{32}^2 & s_{33}^1 & s_{34}^4 & \cdots \\ s_{41}^4 & s_{42}^3 & a_{43}^2 & s_{44}^1 & \cdots \end{bmatrix}, \tag{15.7}$$

where s_{jk}^{i} is the symbol transmitted from the jth antenna at the kth time slot of the ith codeword. The received signal at the receiver with Nr receiving and Nt transmitted antennas can be expressed as follows:

$$\text{Rec} = HX + \zeta, \tag{15.8}$$

where X is an Nt×1 vector, ζ is an Nr×1 noise vector whose elements are complex Gaussian random variables with zero mean and variance N_c, and H is an Nr×Nt channel matrix. If x_i is the desired signal to be detected on the receiver side, the above equation can be rewritten as

$$\text{Rec} = \boldsymbol{h}_i x_i + H_i X_i + \zeta, \tag{15.9}$$

where $\boldsymbol{h}_i$ is the ith column of channel matrix H, $H_i = [\boldsymbol{h}_1\ \boldsymbol{h}_2\ \cdots\ \boldsymbol{h}_{i-1}\ \boldsymbol{h}_{i+1}\ \cdots\ \boldsymbol{h}_{N_t}]$ is an $N_r \times N_t - 1$ matrix and $X_i = [x_1\ x_2\ \cdots\ x_{i-1}\ x_{i+1}\ \cdots\ x_{N_t}]$. The following steps of the optimal MMSE algorithm [4] can be used to detect the signals:

- Calculate $G_{\mathrm{MMSE}_i} = \boldsymbol{h}_i^*[\boldsymbol{h}_i^*\boldsymbol{h}_i + \frac{1}{\mathrm{SNR}}]^{-1}$,where $\frac{1}{\mathrm{SNR}}$ is the noise power to signal power ratio.
- $y_i = G_{\mathrm{MMSE}_i}(\mathrm{Rec} - H_i E[X_i])$,where is the expected value of . For the first iteration, it can be assumed that , and with the increase in the number of iterations, .
- The estimated value of the signal is detected by slicing to the nearest value in the signal constellation: .

15.2.2 Space–Time Coding (STC) Techniques

By using space and time (two-dimensional coding), multiple antenna setups can be used to attain coding gain and diversity gain for the same bit rate, transmission power, and bandwidth as compared to a single antenna system. In STC techniques, information bits are transmitted according to some predefined transmission sequence. At the receiver, the received signals are combined by using an optimal combining scheme followed by a decision rule for maximum likelihood detection. The basic Space-time coding system architecture is shown in Figure 15.3.

15.2.2.1 Alamouti Space–Time Coding Technique

As mentioned, STC improves the BER performance of the multiantenna setup by transmitting and/or receiving redundant copies of the information signals without decreasing the data rate as compared to a single antenna system [5]. We consider a multiple antenna setup with two transmitting antennas and one receiving antenna. At time , antenna a_1 transmits symbol and antenna a_2 transmits symbol . At time , antenna a_1 transmits symbol and antenna a_2 transmits symbol , where *

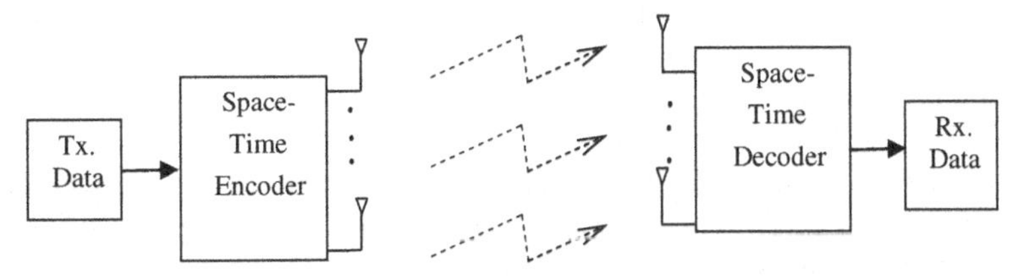

Figure 15.3 Space–time coding system architecture with N_t transmitting and N_r receiving antennas.

denotes the complex conjugate. The received signals (corresponding to time) and (corresponding to time) can be expressed using the following equation:

where is the channel response between the *m*th transmitting and *n*th receiving antenna and it is assumed that fading is constant across consecutive transmitting symbols. The received signals can be combined by using the following equation:

which is then divided by such that

Finally, the signals are estimated by slicing to the nearest value in the signal constellation:

15.2.2.2 Space–Time Trellis Coding (STTC) Technique

This technique was first proposed by Tarokh et al. [6], in which space and time are used to transmit encoded symbols (codewords) to achieve coding gain and also diversity gain. Trellis coded modulation (TCM) [31] is used at the transmitter side and the Viterbi algorithm is used for decoding at the receiver side. The computational complexity of this technique is higher as compared to the Alamouti-based technique but STTC also provides more coding gain and diversity gain [32]. At time *t*, the signal received at the *n*th receiving antenna is given by

(15.10)

where is the channel response between the *m*th transmitting and *n*th receiving antenna, is the signal transmitted at time *t* from the *m*th transmitting antenna and is a complex Gaussian random noise variable with zero mean and variance N_0.

At the transmitter side, the encoder can produce different encoded signals depending on the encoder coefficient set [32]. 4-phase shift keying (PSK) 4-state STT code with two transmitting antennas is shown in Figure 15.4. As an example, Table 15.1 shows the input signal sequence and the signal that are transmitted using a two transmitting antennas system with time-delay diversity. It is clear from Table 15.1 that the

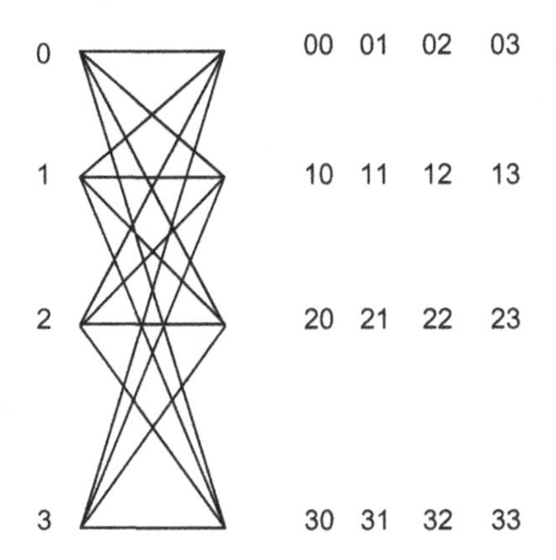

Figure 15.4 4-PSK 4-state space–time code with two transmitting antennas.

Table 15.1 Time-Delay Diversity with Two Antennas

Input	0	1	2	1	1	3	2	3
Tx 1	0	0	1	2	1	2	3	2
Tx 2	0	1	2	1	2	3	2	3

first antenna (Tx 1) transmits the previous value of the second antenna (Tx 2), which in turn transmits the present value, and so on.

At time t, the branch metric at the receiver side is given by

(15.11)

The Viterbi algorithm is used to calculate the trellis path with the lowest metric.

15.2.2.3 Orthogonal Space–Time Block Coding (OSTBC) Technique

The concept of OSTBC was introduced by Tarokh et al. [7]. In this technique, it can be observed that the columns of matrix "O" are pair-wise orthogonal:

- For a real-valued orthogonal set: .
- For a complex-valued orthogonal set: .

Again we consider a multiple antenna setup with two transmitting antennas and one receiving antenna. At time , antenna a_1 transmits symbol and antenna a_2

transmits symbol . At time , a_1 transmits symbol and a_2 transmits symbol , where * represents the complex conjugate. The received signals (corresponding to time and (corresponding to time) can be expressed using the following equation:

where is the channel response between the *m*th transmitting and *n*th receiving antenna , and it is assumed that fading is constant across consecutive transmitting symbols. It can be seen that the columns of matrix "*S*" are pair-wise orthogonal. The received signals can be modified by using the following equation:

where is also orthogonal, and .

The above equation is then divided by such that

Finally, the signals are estimated by slicing to the nearest value in the signal constellation:

It is obvious from Section 15.2.2.1 that the Alamouti coding scheme is OSTBC with complex encoded transmitted blocks. Generalized real and complex orthogonal designs are discussed in detail in [7]. For OSTBCs, simple linear processing is required at the receiver side, thus the computational complexity is less as compared to STTCs.

15.2.2.4 Space–Time Vector Coding (STVC) Technique

In this technique, CSI is required at both the transmitter and receiver side to achieve maximum channel capacity [8]. The SVD method is used to convert the MIMO channel matrix into parallel SISO channels with nonequal gains [33]. Consider a MIMO system with N_t transmitting and N_r receiving antennas, where H is an $N_r \times N_t$ channel matrix. By using SVD, the MIMO channel matrix H can be decomposed as follows:

(15.12)

where U and V are unitary matrices with $N_r \times N_r$ and $N_t \times N_t$ matrix order respectively, and X is an $N_r \times N_t$ diagonal matrix. Before transmitting the encoded signal vector S of order $N_t \times 1$, it is multiplied by matrix V such that . At the receiver, the received signal is first multiplied by as follows:

H can be replaced by and Z by VS; the above equation then becomes

where is statistically identical to . By using the Zero-Forcing method

The ith transmitted signal can be estimated as follows:

Finally, the estimated value of the signal is detected by slicing to the nearest value in the signal constellation:

15.2.3 Beamforming

Multiple antennas are capable of steering lobes and nulls of an antenna beam for co-channel interference cancellation in a multiuser setup to improve SNR and to reduce delay spread of the channel [9]. The basic principle of beamforming (BF) is shown in Figure 15.5.

There are different types of BFs depending upon the system requirements. A simple delay-sum beam-former is shown in Figure 15.6.

By estimating the direction of arrival (DOA) of waves and by using BF techniques, high directivity can be obtained to improve SNR at the receiver, and in the

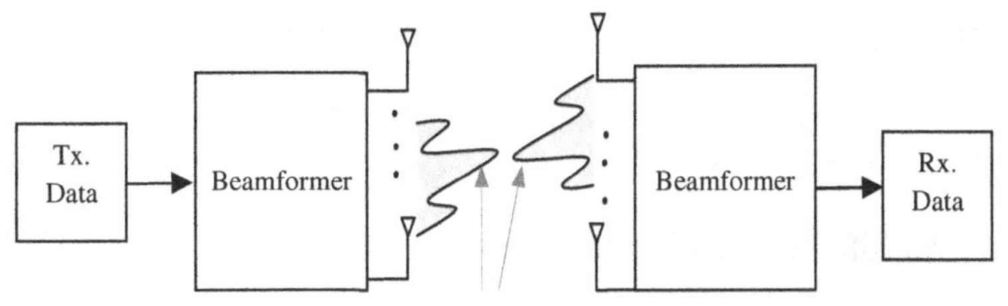

Figure 15.5 A beam-former with N_t transmitting and N_r receiving antennas.

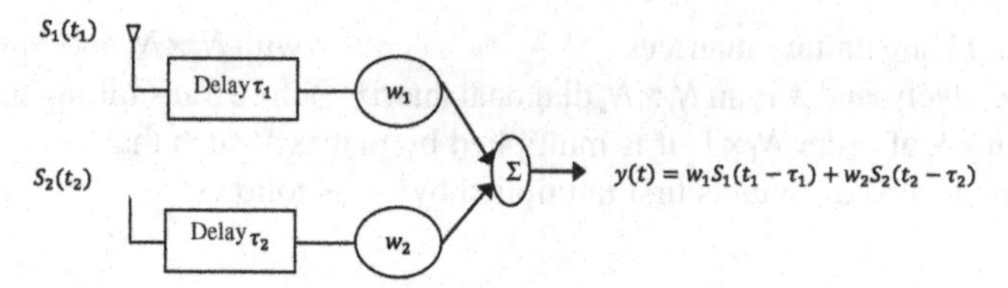

Figure 15.6 A simple delay-sum beam-former.

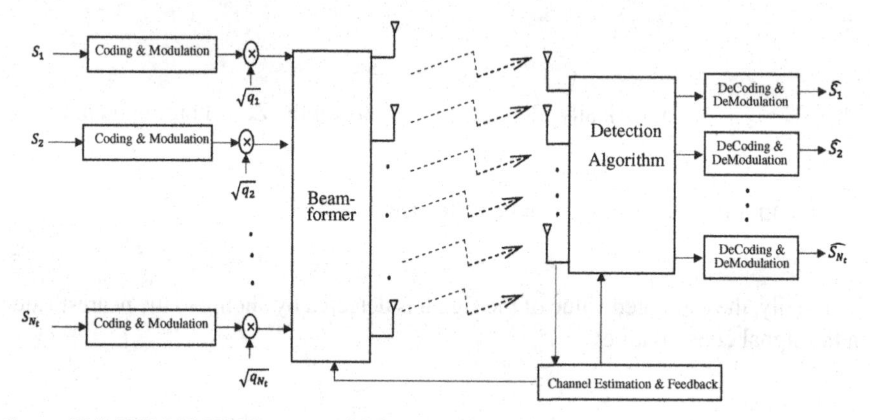

Figure 15.7 V-BLAST MIMO system with beam-former.

same way the transmitter power can be concentrated within the desired region. The delay spread of the channel can also be reduced by steering nulls towards dominant reflectors in the signal propagation path.

A linear precoder with V-BLAST system architecture is used in [34] to achieve maximum spectrum efficiency. In the proposed scheme, the linear precoder is the combination of the BF matrix and power allocation matrix. For the BF matrix, CSI is required at the transmitter side, which is sent to the transmitter using the finite rate channel feedback method as shown in Figure 15.7.

A two-direction Eigen BF along with OSTBC was proposed by Zhou and Giannakis [35] to improve BER performance of the system without rate reduction. To reduce co-channel interference cancellation in multiuser scenario where each user is equipped with multiple antennas, the Alamouti coding scheme along with BF at the receiver side is discussed in [36].

15.2.4 Multifunctional MIMO Systems

As discussed in the above sections, MIMO techniques can be used for achieving multiplexing gain, diversity gain, and beamforming gain. A technique combining STBC and V-BLAST was proposed by Tarokh et al. [37] to obtain diversity gain and multiplexing gain. Another technique combining STBC and beamforming is

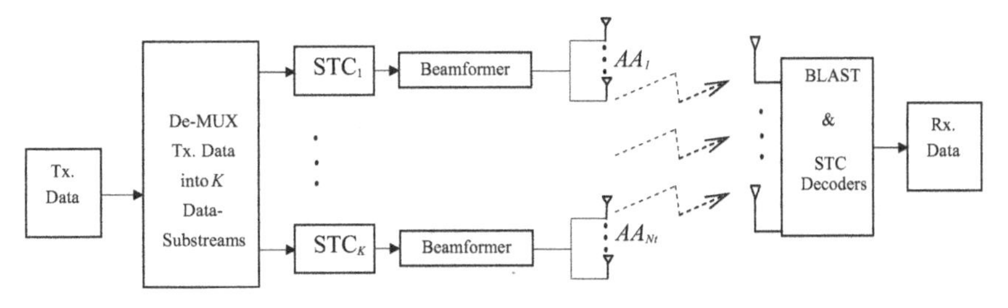

Figure 15.8 Multifunctional MIMO system.

discussed in [38] to achieve diversity gain and SNR gain due to beamforming. In [10], a multifunctional MIMO system is proposed to improve BER performance, throughput, and beamforming (SNR) gain of conventional MIMO systems.

A general multifunctional MIMO system is shown in Figure 15.8 that is capable of combining the advantages of STC, BLAST, and beamforming. The system has N_t transmit antenna arrays (AAs) that are sufficiently apart to experience independent fading. The L_{AA} numbers of elements AA each are spaced at a distance of for achieving beamforming gain. The receiver is equipped with N_r receiving antennas.

The data is transmitted independently from K transmitting nodes to obtain multiplexing gain (K). Before transmission, the data of each node is encoded by an STC encoder to improve BER performance. The data is transmitted using the antenna arrays to achieve beamforming gain. At the receiver, the combination of BLAST and STC decoders (depending on encoders at transmitter side) can be used to retrieve the transmitted signals.

15.2.5 Virtual MIMO Systems

MIMO techniques are capable of providing high system performance without additional transmission power and bandwidth. However, due to the small form factor of some wireless nodes (e.g., mobile phones, sensor nodes), less energy availability, and the need to maintain a minimum distance among the antennas (to avoid fading), it can be difficult to realize the advantages of MIMO techniques for such wireless nodes [11]. Thus, the concept of virtual MIMO (V-MIMO) or cooperative MIMO was proposed for energy and physically constrained wireless nodes [12]. In Figure 15.9, four different V-MIMO models are shown.

In *Model-a*, each source node is equipped with one antenna and transmits data simultaneously without sharing any information among themselves. The destination node is equipped with multiple antennas with the assumption that it has no energy and physical dimension constraints. This model can be utilized with the V-BLAST technique.

In *Model-b*, there are N_t source nodes, one destination node, and relay nodes (R_D) at the destination side each with one antenna. The source nodes transmit their

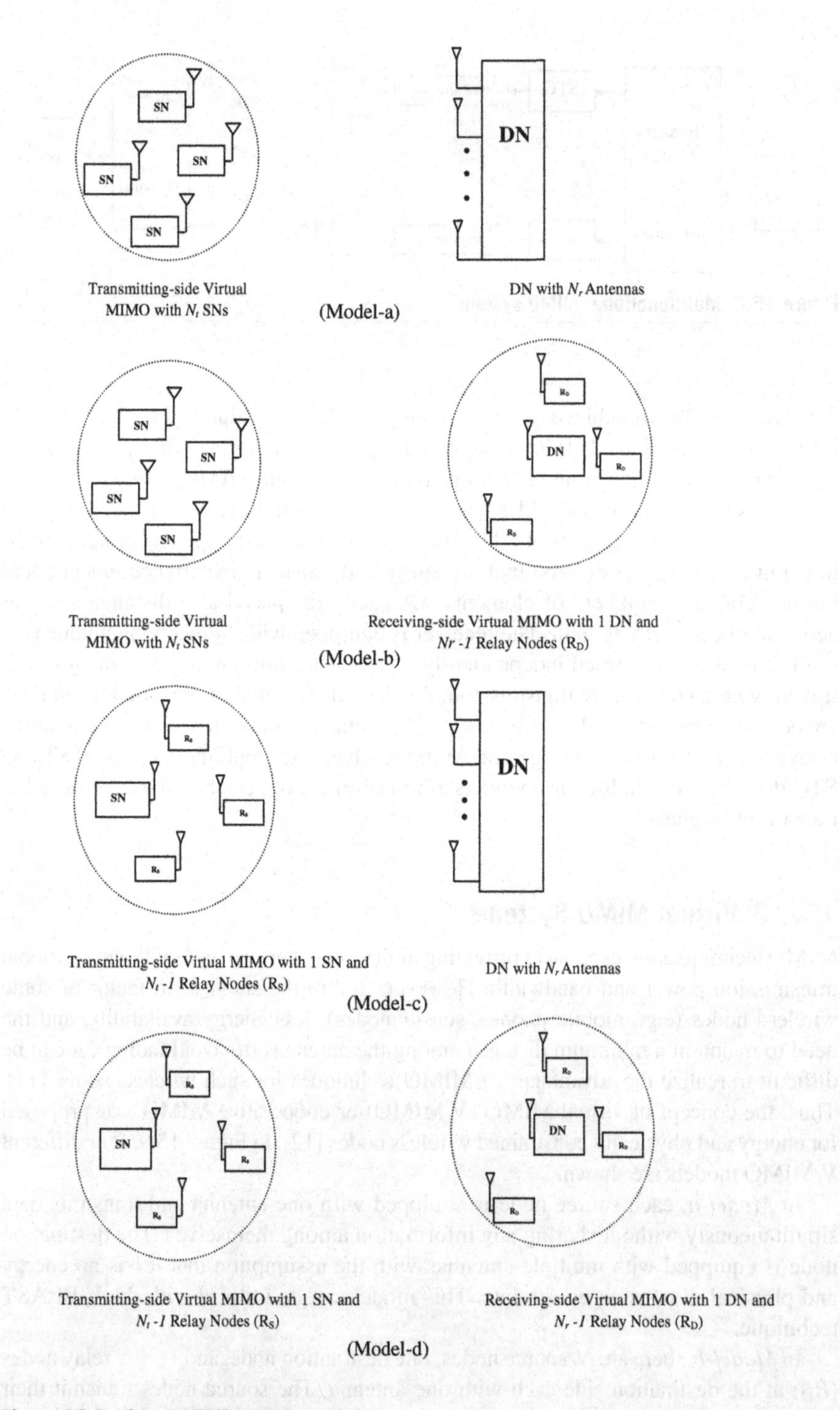

Figure 15.9 Virtual-MIMO system models.

data simultaneously without any local communication and at the destination side, local communication is involved between the destination node and the set of relay nodes. The transmission delay of *Model-b* is greater than transmission delay of *Model-a*.

In *Model-c*, there is one source node, relay nodes (R_S) each with one antenna, and one destination node with N_r receiving antennas. The local communication is between the source node and the set of relay nodes at the transmitter side. This model can be utilized for STC techniques.

In *Model-d*, there is one source node, one destination node, a set of relay nodes (R_S) at the transmitter as well as receiver side (R_D), and each node has a single antenna. The transmission delay of this model is greater than other models. The transmission delay for *Model-d* can be calculated by using the following equation [12]:

(15.13)

where is the number of bits transmitted by each node i, is the total number of symbols received, is the constellation size (bits per symbol) used in the MIMO technique, and are the constellation sizes used during local communication at the transmitter and receiver side, respectively, m_r represents the number of bits after quantization of each symbol received at the receiver side relay nodes, is the symbol duration, and B is the transmission bandwidth.

V-MIMO systems are distributed in nature because multiple nodes are placed at different physical locations to cooperate with each other. V-MIMO systems may also have problems such as time and frequency asynchronism. Each transmitting node is placed at a different distance from receiving nodes, thus the cooperative transmission may not be time synchronized. Every node also has its own oscillator with different frequency error, thus there are also chances of frequency asynchronism [11].

15.2.6 Energy Efficiency of MIMO Systems

MIMO systems are considered bandwidth and energy efficient as compared to SISO for the same throughput and BER. Any SDM and/or STC scheme can be explored with true and/or virtual-MIMO architecture depending upon resource constraints and system performance requirements.

Here in this chapter, energy consumption analysis of *Model-d* is discussed with the assumption that the energy consumed in baseband signal processing blocks may be small enough to be neglected to keep the energy consumption model simple [12, 13]. The energy consumed by MIMO signal processing at the source node (SN) and destination node (DN) is also neglected due to the assumption of the model that SN and DN have no energy constraints, unlike relay nodes R_S and R_D at the transmitter and receiver side. The total energy consumption is due to transmitter side local communication (from SN to R_S nodes), long-haul communication (from SN to receiving side R_D nodes and DN itself) and receiver side local communication (from R_D nodes to DN). The total average power consumption along the signal path for long-haul can be divided into two main components: power consumption of all power amplifiers , and power consumption of all other circuit blocks [39]. As in [40], we assume that the power consumed by power amplifiers is linearly dependent on the transmit power :

(15.14)

where $\alpha = \mu/\varsigma$ with ς being the drain efficiency of the RF power amplifier, and μ being the PAPR [41], which depends on the modulation scheme and associated constellation size [12]. can be calculated according to the link budget relationship [42] as follows:

(15.15)

where is the required energy per bit for a given BER at the receiver side, Rb is the bit rate of the system, dL is the distance between the transmitting and receiving side cluster, Gt and G_r are the transmitter and receiver antenna gains, respectively, λ is the carrier wavelength, M_l is the link margin for compensating the hardware process variations and other additive background noise or interference, and N_f is the receiver noise figure.

Total power consumption in all circuit blocks for long-haul communication with N_t transmitter circuits and N_r receiver circuits using In-Phase/Quadrature-Phase (FOFDM and QAM) transmitter and receiver architecture, as shown in Figure 15.10, can be calculated as

(15.16)

where PDAC, Pmix, Pfil, PLO, PLNA, PIFA, PADC, PPS, and PAdd are the power consumption values for the digital-to-analog convertor (DAC), mixer, filter, local oscillator, low-noise amplifier (LNA), amplifier, analog-to-digital convertor (ADC), phase shifter, and adder, respectively. In addition, the energy models developed in

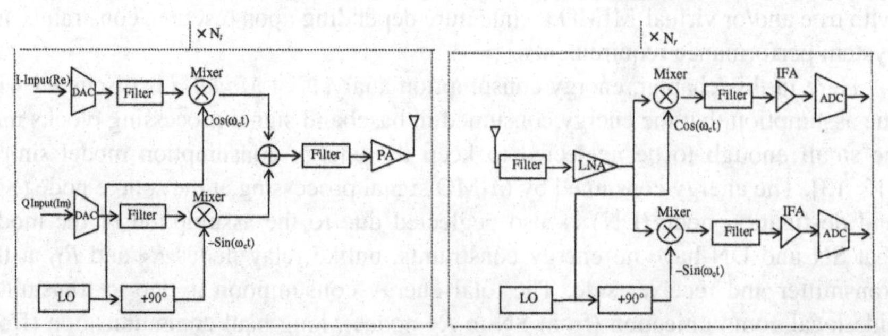

Figure 15.10 Transmitter and receiver architecture (In-Phase/Quadrature-Phase) for FOFDM and QAM (analog).

[39] can be used to estimate the values for P_{DAC} and P_{ADC}. The total energy consumption per bit for long-haul communication can then be obtained as follows:

(15.17)

where R_b is the data rate in bits per second (bps). The total energy consumption per bit for local communication can be obtained as follows:

(15.18)

is the amplifier power of each relay node (RS or RD) during local communication and its value can be obtained by using Eqs. (15.14) and (15.15) and substituting the parameters , R_b, G_t, G_r, d^L with , R_{bi}, G_{ti}, G_{ri}, d^L, where is the required energy per bit for a given BER at the DGN side, R_{bi} is the bit rate of each individual node i, d^l ($\ll d^L$) is the distance between SN and R_S nodes at the transmitter side, and between DN and R_D at the receiver side, G_{ti} is the antenna gain of each relay node, and G_{ri} is the antenna gain of DN. Circuit power consumption for local communication can be calculated using Eq. (15.16) by replacing $N_r = N_t = 1$. The total energy per bit required for communication from all nodes to DN can be calculated using the following equation:

(15.19)

15.3 OFDM AND WOFDM

15.3.1 OFDM

OFDM is an attractive multicarrier technique for mitigating the effects of the multipath delay spread of a radio channel. OFDM is a well-proven technique in applications such as Digital Audio Broadcasting (DAB) and Asymmetric Digital Subscriber Line (ADSL). Furthermore, OFDM is a candidate for the Digital Video Broadcasting (DVB) and high data rate bandwidth-efficient wireless networks [14]. In conventional OFDM, complex exponential Fourier bases are used to generate orthogonal subcarriers consisting of a series of orthogonal sine/cosine functions. In wavelet-based OFDM (WOFDM), wavelet bases are used to generate orthogonal carriers. These bases are generated using symmetric or asymmetric QMF structure of the delay or delay-free type [24].

OFDM exhibits robustness against various kinds of interference and also enables multiple access. In OFDM, the available spectrum B is divided into numerous narrowband subchannels. A data stream is transmitted by frequency division multiplexing (FDM) using N carriers with the frequencies f_1, f_2, f_1, ..., f_N in parallel. Each of the subchannels has the bandwidth $\Delta f = B/N$. The narrowband property of the subchannel justifies the assumption that attenuation and group delay are constant within each channel, allowing equalization to become an easy task.

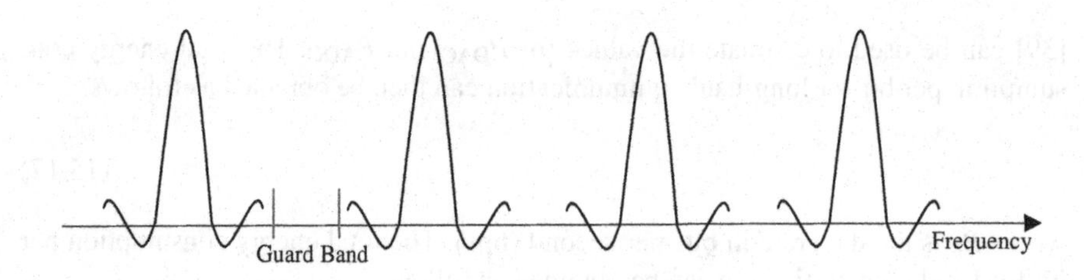

(a) Conventional Frequency Division Multiplex (FDM) Multicarrier Modulation Technique

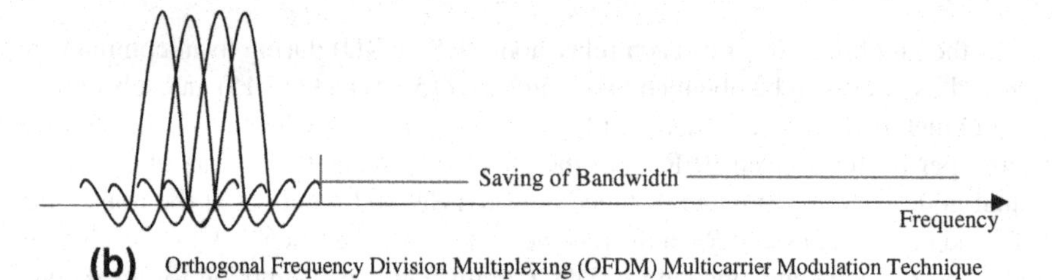

(b) Orthogonal Frequency Division Multiplexing (OFDM) Multicarrier Modulation Technique

Figure 15.11 Comparison of the bandwidth utilization for (a) FDM and (b) OFDM.

15.3.2 OFDM and the Orthogonality Principle

In a typical FDM system, many carriers are spaced apart in such a way that the signals can be received using conventional filters and demodulators. In such receivers, guard bands have to be introduced between the different carriers (Figure 15.11), and the introduction of these guard bands in the frequency domain results in a lower spectrum efficiency.

It is possible, however, to arrange the carriers in an OFDM signal so that the sidebands of the individual carriers overlap (Figure 15.11b) while the signals can still be received without adjacent carrier interference. In order to do this, the carriers must be mathematically orthogonal, i.e., if given a set of signals ψ, where ψ_p is the pth element in the set, the signals are orthogonal if

(15.20)

where * indicates the complex conjugate and interval $[a,\ b]$ is the symbol period.

15.3.3 Fourier-Based OFDM (FOFDM)

In FOFDM, a high data rate substream is demultiplexed into lower data rate substreams to increase the duration of each substream so that ISI can be reduced. The orthogonal subcarriers are generated using sine/cosine bases and the orthogonality is achieved in a time window of width equal to the duration of the symbol. Therefore FOFDM is not band limited. Each subcarrier produces side lobes that in turn create

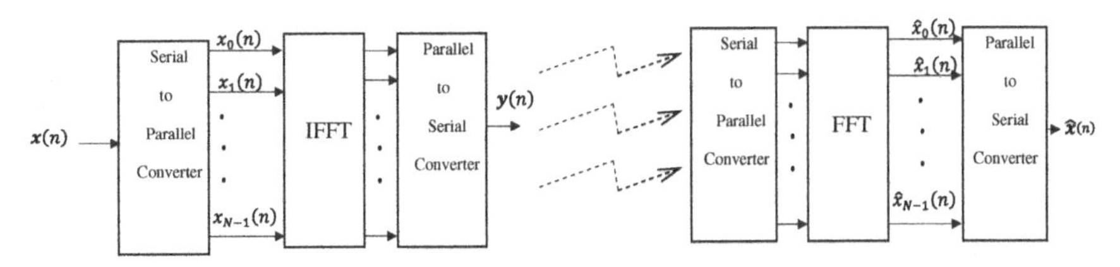

Figure 15.12 A basic FOFDM-based communication system.

ICI, which can be increased due to the multipath channel effect that also causes an increase in ISI. A cyclic prefix (CP)/guard interval (GI) is added to each FOFDM symbol to avoid this problem at the cost of transmission efficiency degradation.

A basic FOFDM-based communication system is shown in Figure 15.12. The FOFDM symbol before transmission without CP can be obtained by taking the inverse fast Fourier transform (IFFT) of the input signal $x_i(n)$:

(15.21)

The serial data stream $x(n)$ is converted into N parallel substreams by using a serial-to-parallel converter. After passing through the channel, FFT is performed at the receiver side to retrieve the signal.

The equivalent structure of the communication system of Figure 15.12 can be represented by the filter bank structure [43] as shown in Figure 15.13.

The filter impulse can be expressed as

(15.22)

where n = 0, 1, 2, 3, …, N − 1, and i = 0, 1, 2, 3, …, N − 1.

Each substream is first upsampled and then passes through the filter with impulse response $f_i(n)$. At the receiver side, the signal passes through the filter with impulse

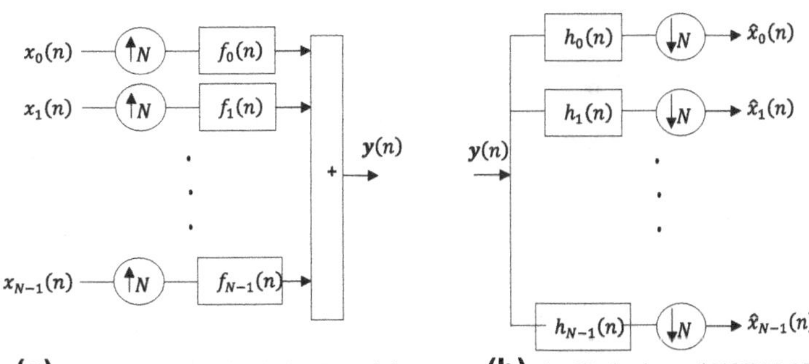

(a) An *N*sub-channel FOFDM Modulator **(b)** An *N*sub-channel FOFDM Demodulator

Figure 15.13 FOFDM modulator and demodulator with filter bank structure.

response and is then downsampled. The training symbols, which are also called pilot tones, are sent using different subcarriers of FOFDM symbols to estimate the CSI associated with pilot tones as well as for subchannels carrying the data by using interpolation techniques [14].

15.3.4 Wavelet-Based OFDM (WOFDM)

In WOFDM, wavelet bases are used to generate orthogonal carriers. These bases are generated using symmetric or asymmetric QMF structure of delay or delay-free type [24]. WOFDM is a multicarrier modulation with a one-dimensional constellation as shown in Figure 15.14. It does not use raised cosine filtering (excess bandwidth=0) and does not use a cyclic prefix. A WOFDM modulator is implemented by taking the inverse discrete wavelet transform (IDWT). In IDWT, the synthesis part of the delay-free perfect reconstruction (PR)-QMF using symmetric or asymmetric structure is used. For the WOFDM demodulator, DWT (analysis part of the delay-free PR-QMF using symmetric or asymmetric structure) is used.

The physical layer overhead for WOFDM is less than that of FOFDM modulations. FOFDM guard intervals of 20% or more are typical for wireless communications, thus giving WOFDM a corresponding advantage of approximately 20% in bandwidth efficiency. Optimization of cost and performance can result due to several factors [44]:

- Real modulation simplifies RF.
- High stop-band simplifies RF/IF design.
- High stop-band makes multinational design easier.
- Low sensitivity to timing offset.
- Simple and accurate frequency tracking from the data itself.

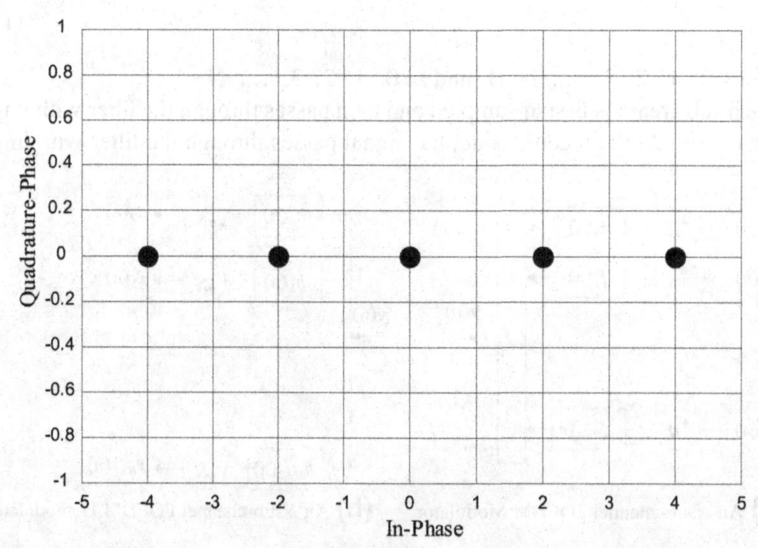

Figure 15.14 Constellation diagram of WOFDM.

WOFDM modulation uses a wavelet filter for both the transmitter and receiver. These filters are implemented digitally using polyphase techniques, which significantly reduce the computational requirements. Channel interference can arise from several sources in a wireless network such as narrowband (NB) and co-channel interference. For NB noise, the high stop-band (~50 dB) of wavelets provides a significant advantage over other modulations. Because the high stop-band applies for each subband without additional filtering, frequency selectivity is greater than that obtained from adding notch filters to nonwavelet modulations. As the bandwidth of the customer's service is a large fraction of the available frequency allocation, the reuse of frequencies can cause severe (i.e., −11 dB) co-channel interference in a cellular system.

As discussed above, the orthonormal wavelets can be generated using the symmetric or asymmetric multistage tree structure of the QMF bank. When using symmetric structure, the orthonormal wavelets are given by the following equation:

(15.23)

where $\prod$ represents the convolution operation, P is the number of levels of this structure, $i \in \{0, 1, 2, 3, \ldots, 2^P-1\}$, $t_{k,p}(n) \in \{f_{l(n)}, f_{h(n)}\}$ is the filter impulse response corresponding to the ith subchannel at the pth level, and $f_l(n)$, $f_{h(n)}$ are impulse responses of the low- and high-pass filters, respectively, for the perfect reconstruction QMF bank. The high-pass filter can be derived from the low-pass filter by the relation $f_h(n) = (-1)^n f_l(U-1-n)$, where U is the length of the filter [20]. The equivalent structure of a symmetric level synthesis side QMF bank using noble identities of the WOFDM modulator and demodulator is shown in Figure 15.15. The output $y(n)$ of the WOFDM modulator can be expressed as

(15.24)

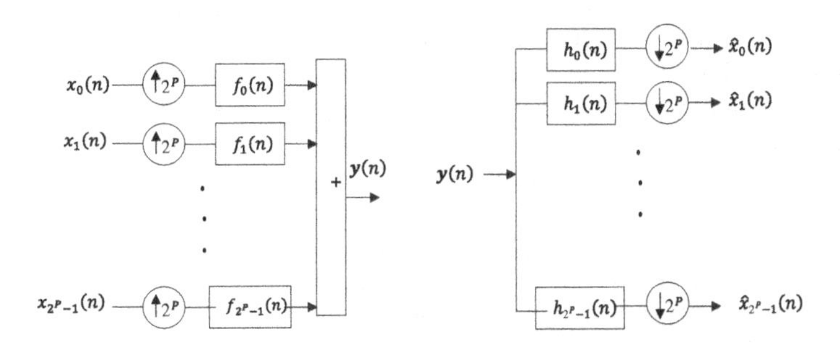

(a) A 2^P subchannel WOFDM Modulator **(b)** A 2^P subchannel WOFDM Demodulator

Figure 15.15 WOFDM modulator and demodulator using a symmetric QMF filter bank structure.

where $x_i(n)$ is the ith subchannel input of the WOFDM modulator. For WOFDM demodulation, the orthonormal wavelet bases are generated using a symmetric analysis side QMF bank as follows:

$$(15.25)$$

where $u_{i,p}(n) \in \{g_l(n), g_h(n)\}$ is the filter impulse response corresponding to the ith subchannel at the pth level, and $g_{l(n)}$ and $g_{h(n)}$ are time reversals of $f_{l(n)}$ and $f_{h(n)}$, respectively [45].

In the WOFDM demodulator, the detected signal stream is first filtered by subchannel impulse response $h_k(n)$ and then downsampled by 2^p. For wireless channels, the ISI and ICI of WOFDM is less as compared to FOFDM [21, 43], thus this property makes it more suitable for future high data rate communication systems.

WOFDM provides flexibility for a designer to design new waveforms according to the communication channel and/or communication system requirements. New wavelet bases were designed to reduce timing synchronization error and the performance was compared with other wavelet families. Simulation results have shown better BER performance with new optimal filter design even in the presence of a timing offset [27]. By using the time-frequency localization property of wavelets, new optimal maximally frequency selective wavelet bases for a reconfigurable communication system were also designed to utilize the unoccupied time-frequency gap of licensed users efficiently [26].

15.4 MULTIPLE ANTENNA OFDM SYSTEMS

As already discussed, MIMO techniques can be used for diversity and multiplexing gain, but most of the MIMO techniques have been developed with the assumption of a flat fading channel. For a broadband frequency selective wireless channel, the combination of MIMO and OFDM (MIMO–OFDM) was proposed to mitigate the effect of ISI and ICI [14]. In MIMO techniques, CSI is usually required at the transmitter and/or receiver side, thus OFDM is also used in MIMO systems to estimate CSI. A basic structure of MIMO–OFDM with N_t transmitting antennas and N_r receiving antennas is shown in Figure 15.16.

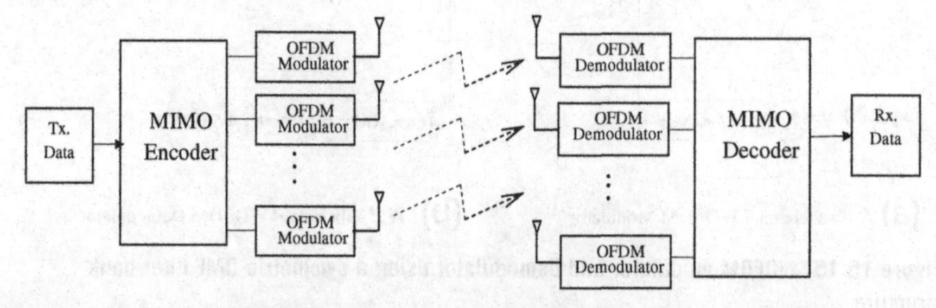

Figure 15.16 MIMO–OFDM system with N_t transmitting and N_r receiving antennas.

15.4.1 MIMO Techniques with FOFDM

To achievez high data rate for a broadband frequency selective channel, a V-BLAST architecture with FOFDM was proposed in [46] with N_t transmitting antennas, N_r receiving antennas, and N subcarriers, which can be represented as

(15.26)

where $\boldsymbol{H}_{nm}$ is an $N \times N$ diagonal matrix corresponding to SISO frequency response of all the subcarriers between the mth transmitting and jth receiving antenna, $\boldsymbol{X}_m$ is the OFDM symbol transmitted from the mth transmitting antenna, and ζ_j is the additive white Gaussian noise (AWGN) vector. From the above equation, it can be observed that corresponding to each subcarrier, there is a narrowband $N_r \times N_t$ V-BLAST system. Hence, the detection of the V-BLAST-FOFDM system can be done using a V-BLAST detection algorithm (as discussed in Section 15.2.1) on each subcarrier individually.

In [15], STCs are used across FOFDM tones for developing SFC to achieve full diversity for a broadband frequency selective channel. It is shown that the Alamouti scheme fails to exploit the frequency diversity available in frequency selective channels. STFCs were proposed in which the system is converted into group STF (GSTF) systems. Both OSTBCs and OSTTCs along with FOFDM are used to achieve maximum coding gain and low complexity decoding for frequency selective channels [47].

STFCs can achieve diversity gain equal to $(N_t N_r L\tau)$ where L is the number of propagation paths and τ is the rank of the channel temporal correlation matrix [48]. With N OFDM subcarriers and K consecutive FOFDM blocks, the STF code word can be expressed as a $KN \times N_t$ matrix:

(15.27)

where "T" is for transpose of the matrix. The channel symbol matrix $\boldsymbol{C}_K$ is given by

(15.28)

where is the channel symbol transmitted over the ith subcarrier by transmitting antenna m in the kth FOFDM block. To obtain maximum coding gain, STFCs were proposed with a combination of STTC and the rotated constellation technique [49]. By using the concept of cooperative communication in a multiuser scenario using FOFDM, the transmission power can be reduced effectively [18]. It has been claimed

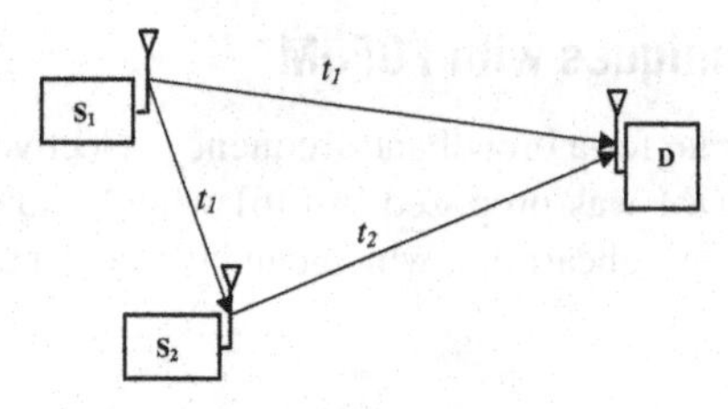

Figure 15.17 Cooperative communication in a multiuser scenario using FOFDM.

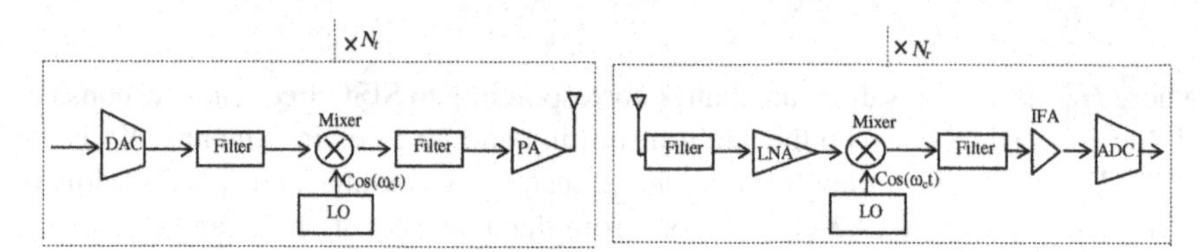

Figure 15.18 Transmitter and receiver architecture for WOFDM (analog).

that with the two user scenario using FOFDM, the system (as shown in Figure 15.17) becomes 50% more energy efficient.

Assume that at time t_1, source node S_1 transmits some of its data to the destination node D, and remaining data to source node S_2. At time t_2, S_2 transmits all of its data along with the remaining data of S_1 by assigning some of the FOFDM subcarriers (for data from S_1) to D. By doing so, the overall transmission power can be reduced significantly when S_1 and S_2 are very close to each other [18].

15.4.2 MIMO Techniques with WOFDM

WOFDM is an attractive modulation technique for MIMO wireless networks due to its resulting simpler RF section, lower peak-to-average power ratio (PAPR) and better energy efficiency [50]. The BER performance of a WOFDM V-BLAST MIMO system was investigated in [24]. The performance of this system using different detection algorithms was also investigated in [25]. The one-dimensional constellation property and simpler transmitter/receiver architecture of WOFDM are shown in Figure 15.18. The power consumption in all circuit blocks for long-haul communication with N_t transmitter circuits and N_r receiver circuits using the transmitter and receiver architecture in Figure 15.18 can be found as

(15.29)

The energy efficiency can be calculated using equations as discussed in Section 15.2.6. It is shown that the MIMO–WOFDM system consumes less energy as compared to the MIMO–FOFDM system due to its less complex RF architecture, which reduces the amount of circuit energy it consumes [50].

15.5 CONCLUSION

In this chapter, we have presented the underlying principles and techniques of MIMO–OFDM systems for energy-efficient wireless communications. We have introduced multiantenna systems with spatial multiplexing, space–time coding, and beamforming techniques. To improve BER, SNR, throughput, and energy efficiency, multifunctional MIMO and virtual MIMO systems are also discussed along with energy efficiency analysis. We have also described the basic principles of FOFDM and WOFDM and their applications in true (colocated) and virtual (cooperative) MIMO wireless systems. MIMO–OFDM is a promising solution for energy-efficient high data rate wireless networks. WOFDM can be used for SFC, STFC, and also for cooperative communication systems. Potential directions for future work may include designing a new wavelet basis according to wireless channel conditions to improve the overall system performance, and evaluating multifunctional MIMO performance using WOFDM/FOFDM. A true and virtual MIMO–OFDM system can also be implemented to verify the theoretical results. Physical layer architecture performance of a MIMO–OFDM system along with medium access control (MAC) layer protocols can be explored. New MAC layer protocols can also be proposed for true and virtual MIMO–OFDM systems.

REFERENCES

[1] Y. Chen, S. Zhang, S. Xu, Fundamental trade-offs on green wireless networks, IEEE Communications Magazine 49 (6) (2011) 30–37.

[2] G.J. Foschini, Layered space–time architecture for wireless communication in a fading environment when using multi-element antennas, Bells Lab Technical Journal 1 (2) (1996) 41–59.

[3] P.W. Wolniansky et al., V-BLAST: an architecture for realizing very high data rates over the rich-scattering wireless channel, in: Proceedings of the International Symposium on Signals, Systems, and Electronics, Pisa, Italy, September 1998.

[4] M. Sellathurai, S. Haykin, TURBO-BLAST for wireless communications: theory and experiments, IEEE Transactions on Signal Processing 50 (10) (2002) 2538–2546.

[5] S.M. Alamouti, A simple transmit diversity technique for wireless communications, IEEE Journal on Select Areas in Communications 16 (8) (1998) 1451–1458.

[6] V. Tarokh, N. Seshadri, A.R. Calderbank, Space–time codes for high data rate wireless communication: performance criterion and code construction, IEEE Transactions on Information Theory 44 (2) (1998) 744–765.

[7] V. Tarokh, N. Seshadri, A.R. Calderbank, Space–time block codes for orthogonal designs, IEEE Transactions on Information Theory 45 (5) (1999) 1456–1467.

[8] G.G. Raleigh, J.M. Cioffi, Spatio-temporal coding for wireless communication, IEEE Transactions on Communications 46 (3) (1998) 357–366.

[9] L.C. Godara, Application of antenna arrays to mobile communications – Part I: Performance improvement, feasibility, and system considerations; Part II: Beam-forming and direction-of-arrival considerations, Proceedings of IEEE 85 (7/8) (1997) 1031–1060 1195–1245

[10] M. El-Hajjar, L. Hanzo, Multifunctional MIMO systems: a combined diversity and multiplexing design perspective, IEEE Wireless Communication Magazine 17 (2) (2010) 73–79.

[11] H. Wang, X.G. Xia, Asynchronous cooperative communication systems: a survey on signal designs, Science China Information Sciences 54 (8) (2011) 1547–1561.

[12] S. Cui, A.J. Goldsmith, A. Bahai, Energy-efficiency of MIMO and cooperative MIMO techniques in sensor networks, IEEE Journal on Selected Areas in Communication 22 (6) (2004) 1089–1098.

[13] S.K. Jayaweera, V-BLAST-based virtual MIMO for distributed wireless sensor networks, IEEE Transactions on Communications 55 (10) (2007) 1867–1871.

[14] T. Hwang, C. Yang, G. Wu, S. Li, G.Y. Li, OFDM and its wireless applications: a survey, IEEE Transactions on Vehicular Technology 58 (4) (2009) 1673–1694.

[15] H. Bolcskei, A.J. Paulraj, Space–frequency coded broadband OFDM systems, IEEE Conference on Wireless Communications and Networking (WCNC) 1 (2000) 1–6.

[16] Z. Liu, Y. Xin, G.B. Giannakis, Space–time–frequency coded OFDM over frequency-selective fading channels, IEEE Transactions on Signal Processing 50 (10) (2002) 2465–2476.

[17] J. Flores, J. Sánchez, H. Jafarkhani, Quasi-orthogonal space–time–frequency trellis codes for two transmit antennas, IEEE Transactions on Wireless Communication 9 (7) (2010) 2125–2129.

[18] A. Han, T. Himsoon, W.P. Siriwongpairat, K.J.R. Liu, Resource allocation for multiuser cooperative OFDM networks: who helps whom and how to cooperate, IEEE Transactions on Vehicular Transactions 58 (5) (2009) 2378–2391.

[19] M.K. Lakshmanan, H. Nikookar, A review of wavelets for digital wireless communication, Wireless Personal Communications 37 (3-4) (2006) 387–420.

[20] A.R. Lindsey, Wavelet packet modulation for orthogonally multiplexed communication, IEEE Transactions on Signal Processing 45 (5) (1997) 1336–1339.

[21] B.G. Negash, H. Nikookar, Wavelet based OFDM for wireless channels, in: 53rd Vehicular Technology Conference, 2001, pp. 688–691.

[22] A. Jamin, P. Mahonen, Wavelet packet modulation for wireless communications, Journal of Wireless Communication and Mobile Computing 5 (21) (2005) 1–18.

[23] D. Karamehmedovic, M.K. Lakshmanan, H. Nikookar, Performance of wavelet packet modulation and OFDM in the presence of carrier frequency and phase noise, in: First European Wireless Technology Conference, 2008, pp. 166–169.

[24] Z. Rafique, N.D. Gohar, M.J. Mughal, Performance comparison of OFDM and WOFDM based V-BLAST wireless systems, IEICE Transactions on Communication E88-B (5) (2005)

[25] M. Yasir, M.J. Mughal, N.D. Gohar, S.A. Moiz, Performance comparison of wavelet based OFDM (WOFDM) V-BLAST MIMO systems with different detection algorithms, in: Proceedings of the Fourth IEEE International Conference on Emerging Technologies, Ralwapindi, Pakistan, October 2008.

[26] M.K. Lakshmanan, H. Nikookar, Construction of optimum wavelet packets for multi-carrier based spectrum pooling systems, Wireless Personal Communications 54 (1) (2010) 95–121 (special issue).

[27] D. Karamehmedovic, M.K. Lakshmanan, H. Nikookar, Optimal wavelet design for multicarrier modulation with time synchronization error, in: Proceedings of IEEE Global Telecommunications Conference, Honolulu, Hawaii, USA, 2009.

[28] C.Z.W.H. Sweatman, J.S. Thompson, B. Mulgrew, P.M. Grant, A comparison of the MMSE detector and its BLAST versions for MIMO channels, in: IEE Seminar on MIMO Communication Systems from Concept to Implementation, December 2001, pp. 1/19–6/19.
[29] J. Benesty, A. Huang, J. Chen, A fast recursive algorithm for optimum sequential signal detection in a BLAST system, IEEE Transactions on Signal Processing 51 (7) (2003) 1722–1730.
[30] H. Lee, H. Jeon, H. Jung, H. Lee, Signal detection using log-likelihood ratio based sorting QR decomposition for V-BLAST systems, in: IEEE Vehicular Technology Conference, 2007, pp. 1881–1885.
[31] C. Schlegel, L. Perez, Trellis and Turbo Coding, Wiley-IEEE Press Publication.
[32] L. Poo, Space–time coding for wireless communications: a survey, Report from Stanford University, 2002.
[33] G. Leburn, J. Gao, M. Faulkner, MIMO transmission over a time-varying channel using SVD, IEEE Transactions on Communication 4 (2) (2005)
[34] Q. Gao, X.-D. Zhang, J. Li, W. Shi, Linear precoding and finite rate feedback design for V-BLAST architecture, IEEE Transactions on Wireless Communication 7 (12) (2008) 4976–4986.
[35] Z. Zhou, G.B. Giannakis, Optimal transmitter eigen beam forming and space–time block coding based on channel mean feedback, IEEE Transactions on Signal Processing 50 (10) (2002) 2599–2613.
[36] M.R. Bhatnagar, A. Hjorungnes, Improve interference cancellation scheme for two user detection of Alamouti code, IEEE Transactions on Signal processing 58 (8) (2010) 4459–4465.
[37] V. Tarokh, A. Naguib, N. Seshadri, A.R. Calderbank, Combined array processing and space–time coding, IEEE Transactions on Information Theory 45 (4) (1999) 1121–1128.
[38] G. Jongren, M. Skoglund, B. Ottersten, Combining ideal beamforming and orthogonal space–time block coding, IEEE Transactions on Information Theory 48 (3) (2002) 611–627.
[39] S. Cui, A.J. Goldsmith, Ahmad Bahai, Energy-constrained modulation optimization, IEEE Transactions on Wireless Communications 4 (5) (2005) 2349–2360.
[40] S.K. Jayaweera, An energy-efficient virtual MIMO communication architecture based on V-BLAST processing for distributed wireless sensor networks, in: Proceedings of the IEEE Conference on Sensor and Ad Hoc Communications and Networks (SECON), October 2004.
[41] N.T. Le, S.D. Muruganthan, A.B. Sesay, An efficient PAPR reduction method for wavelet packet modulation schemes, in: Proceedings 69th IEEE Vehicular Technology Conference, Barcelona, Spain, April 2009.
[42] T.S. Rappaport, Wireless Communications, Principle and Practice. second ed., Prentice Hall.
[43] Y. Zhang, S. Cheng, A novel multicarrier signal transmission system over multipath channel of low-voltage power line, IEEE Transactions on Power Delivery 19 (4) (2004) 1668–1672.
[44] RM Wavelet Based PHY Proposal for 802.16.3, IEEE 802.16.3c-01/12, Rainmaker Technologies, Inc.
[45] S. Gracias, V.U. Reddy, An equalization algorithm for wavelet packet modulation, IEEE Transactions on Signal Processing 46 (11) (1998) 3082–3087.

[46] W. Yan, S. Sun, Z. Lei, A low complexity VBLAST OFDM detection algorithm for wireless LAN systems, IEEE Communications Letters 8 (6) (2004) 374–376.
[47] Z. Liu, Y. Xiu, G.B. Giannakis, Space–time–frequency coded OFDM over frequency-selective fading channels, IEEE Transactions on Signal Processing 50 (10) (2002) 2465–2476.
[48] W. Su, Z. Safar, K.J.R. Liu, Towards maximum achievable diversity in space, time and frequency: Performance analysis and code design, IEEE Transactions on Wireless Communications 4 (4) (2005) 1847–1857.
[49] J. Flores, J. Sanchez, H. Jafarkhani, Quasi-orthogonal space-–time–frequency trellis codes for two transmit antennas, IEEE Transactions on Wireless Communications 9 (7) (2010) 2125–2129.
[50] Z. Rafique, B.-C. Seet, Energy efficient wavelet based OFDM for V-BLAST MIMO wireless sensor networks, in: IEEE Online Conference on Green Communication, 2011.

CHAPTER

Base Station Deployment and Resource Allocation in Sustainable Wireless Networks

16

Zhongming Zheng[a], Shibo He[a], Lin X. Cai[b], Xuemin (Sherman) Shen[a]

[a]*Department of Electrical and Computer Engineering, University of Waterloo, Waterloo, Ontario, Canada,*

[b]*School of Engineering and Applied Science, Princeton University, Princeton, NJ, USA*

16.1 INTRODUCTION

Wireless communications have become an indispensable part of our daily lives by allowing us to exchange information from anywhere at any time. Traditional wireless networks are supported by either power stations or batteries. With the awareness of global warming and the resultant exasperated environment, alleviating carbon emissions has become one of the most important issues in next-generation wireless communication networks. To this end, green communications have been emerging to satisfy the growing user demand for ubiquitous wireless access with greater energy efficiency. In recent years, research projects on green communications have sprung up worldwide, such as OPERANET [1], EARTH [2], and Green Radio [3].

Generally, green radio communication technologies refer to communication techniques that provide energy-efficient wireless access [4], which can be classified into customer-oriented and infrastructure-oriented green communications. Compared with customer electronics, network infrastructure consumes a dominant portion of energy, and thus is of critical importance to achieve a green communication network. For example, in a traditional cellular network, the base stations (BSs) consume over 80% of total energy [5]. As such, infrastructure-oriented green approaches have been proposed to mitigate the energy consumption in the network infrastructure, including network devices design [6], network deployment [7], and resource allocation [8]. First network infrastructure consists of diverse network devices and peripherals, such as power amplifiers and cooling systems, etc. It is essential to reduce energy consumption of the network devices by improving hardware design and relevant in-device protocols. In addition to the devices, the network deployment also plays an important role in provisioning satisfactory quality of service (QoS) for customers and achieving energy efficiency. Finally, the infrastructure network resources are inherently limited while different customers usually have different QoS requirements with variable capacities in their customer devices. To provision satisfactory services to customers, efficient resource allocation is required to allocate the limited network resources according to diverse requirements of customers.

Energy efficiency has been extensively studied in wireless networks with traditional energy supplies. Recently, there is a growing interest in using eco-friendly green energy, e.g., solar, wind, hydro, etc., to power wireless networks for achieving long-term sustainability of communication systems. Since then, several research works [7–9] have focused on utilizing the harvested green energy to improve the sustainability performance of the communication network, e.g., to minimize the deployment cost while fulfilling the QoS requirements of users with the harvested energy.

In this chapter, we first investigate previous works on green communication techniques. Our literature review covers a wide area from hardware design and device deployment to infrastructure resource management. Then, we identify and analyze the relevant challenging issues in the design and deployment of a green communication network powered by green energies. Specifically, we revisit the minimal BS deployment problem in the context of sustainable wireless networks with green energy supply and formulate an optimal green BS (i.e., BS powered by green energy) placement problem. The objective is to determine the optimal placement of BSs on a set of candidate locations such that the number of BSs is minimized, subject to the constraints that QoS requirements of users can be fulfilled with the harvested energy. As the optimal BS placement problem is in general a mixed integer nonlinear optimization problem (MINLP), which is known to be *NP-hard*, we then propose a simple yet efficient heuristic algorithm to solve the formulated MINLP problem.

The main contributions of the chapter are fourfold:

- To the best of our knowledge, the chapter is one of the first works to systematically study the network deployment issues in an infrastructure-based wireless network with green energy. By exploiting the particular characteristics of sustainable energy supplies, we present design criteria for the network deployment and resource management in a green wireless network.
- Based on the design criteria, we then formulate the constrained minimum green BS placement problem as an optimization problem. Our objective is to deploy a minimal number of green BSs on a set of candidate locations such that users' traffic demands can be fulfilled by the harvested energy of the green BSs.
- A preference level is assigned to each client for determining the relationship between the user and potential placed BSs, which is a function of distance between the user and all BS candidate locations. The value reflects the connection priority and relative transmission rate between the user and corresponding green BSs.
- By jointly considering power control and rate adaptation, a heuristic algorithm, called the Two-phase Constrained Green BS Placement (TCGBP) algorithm, is proposed to find an optimal BS deployment based on different user demands and charging capabilities of BSs. Extensive simulation results show that the proposed algorithm approaches the optimal solution under a variety of network settings with significantly reduced time complexity.

The remainder of the chapter is organized as follows. In Section 16.2, we give a broad review of the existing infrastructure-oriented solutions. In Section 16.3, the system model is presented, followed by the constrained minimum green BS deployment formulation. Our proposed algorithm, called TCGBP, is introduced in Section 16.4. Numerical results are presented in Section 16.5. Finally, we conclude the chapter and introduce our future work in Section 16.6.

16.2 RELATED WORKS

Over the past several years a number of solutions across multiple layers have been proposed to provide low-cost, high-quality, and energy-efficient wireless access services [10,11]. In general, these solutions can be classified into two categories, namely, customer-oriented and infrastructure-oriented solutions. Due to hardware constraints, most customer devices are powered by battery energy, such as mobile terminals, and wireless sensor nodes [12]. Limited by battery storage capacity, it is critical to provision the long lifetime of customer devices. There are several works that focused on improving energy efficiency from different aspects of customer devices, including hardware [13], application software [14], communication and networking protocols [15]. Compared with customer devices, the network infrastructure, referred to as the network backbone that consists of multiple gateways, routers, BSs, etc., contributes to the dominant portion of the total energy consumption of the communication system, and thus is of more importance for the energy efficiency of the overall system. We summarize the existing infrastructure-oriented green solutions into the following four issues: (1) green energy supply, (2) hardware design, (3) devices deployment, and (4) resource allocation. In this section, we will review these issues in detail.

16.2.1 Green Energy

Typical energy sources can be divided into two classes: renewable energy and nonrenewable energy. Renewable energy is referred to as the energy that can be repeatedly replenished while nonrenewable energy cannot. One type of renewable energy uses eco-friendly and sustainable energy sources, e.g., wind, solar, modern biomass, etc., which is called green energy. It is widely believed that the use of green energy is the most effective method for improving the overall environment. Currently, around 16% of global final energy consumption comes from renewable energy with around 18% of that green energy. Among a variety of green energy sources, wind power grows rapidly at the rate of 30% annually, and achieved 198 GW all over the world in 2010. Solar power is another popular green energy source, and cumulative global photovoltaic (PV) installations surpassed 40 GW at the end of 2010 [16]. Moreover, with the development of green energy technology, crystalline silicon devices can approach the theoretical limiting efficiency of 29%. Motivated by the relative high performance-cost ratio, solar and wind power are two of the most common energy sources that have been extensively used to power wireless networks, especially the network infrastructure. For instance, the Green WiFi initiative has developed a low cost, solar-powered and standardized WiFi solution for providing Internet access to developing areas [17]. The wind-powered wireless mesh network is also applied for emergency network deployment after disasters [18]. In academia, energy scavenging or energy harvesting issues are first considered in sustainable wireless sensor networks. Recently, it has become a hot topic that attracts great attention.

Due to unpredictable factors, such as weather, green energy is inherently variable or even intermittent with time. The highly variable and unpredictable characteristics of renewable energy supply make resource allocation and traffic scheduling tasks extremely challenging in wireless networks powered by green energy. Some works focused on building the energy model and improving the stability of green energy. In [19], the harvested energy was modeled as a general stochastic process with a given mean and variable. In [20], it was found that a hybrid solution of wind and solar power provides an optimal solution to provision more stable and lower-cost green energy for WLAN mesh nodes in certain geographic locations, i.e., Toronto, Seattle, Phoenix, etc., compared with using wind or solar power only.

16.2.2 Device Design

In order to alleviate the dynamic characteristic of green energy, the photovoltaic (PV) power system is studied for converting variable green energy into more stable and reliable electricity. The PV system consists of multiple components, including the photovoltaic modules, battery, charge controller, and inverter. It has been widely used in communication systems for many decades. To fulfill the diverse energy requirements of wireless devices, most PV system studies focus on constructing more reliable and stable power supplies to prevent energy interruption and waste. The PV systems sizing problem is one of most critical issues and has been extensively studied in the literature. In [21], a simple method for sizing stand-alone photovoltaic systems was proposed. In their work, loss of load probability and load profile were used to represent the desirable reliability and traffic demand, respectively. In [22], three probabilistic methods were proposed for sizing PV systems. The first method was aimed at designing the battery of backup and recharge to support a fixed load within a specified time period. The second method sized the battery by loss of load probability with detailed computer simulation. The last method used the Markov chain to model the battery charging status. In [23], the work analyzed various sizing tools for a PV stand-alone system, and proposed a standard simulation model for the PV sizing problem. Moreover, they showed that the accuracy of the PV sizing problem depends on basic statistical laws, which has no obvious relationship with the complexity of used models.

Based on the results of PV system studies, many recharging and discharging models were proposed in wireless communication networks. The most straightforward way is to model the battery/energy buffer as the charging and discharging processes, and use a framework to represent the current status of the battery/energy buffer [19]. Some works [9,24] described the capability of the PV system by using harvested energy in a specified time or charging rate to simplify the complexity of the energy charging model. In [6], the work aimed at establishing a realistic BS power consumption model with given input parameters. The work grouped BSs into macro-BSs and micro-BSs. For macro-BSs, the power consumption model consists of a static power discharging part only. For micro-BSs, the power discharging model is comprised of static and dynamic power consumption parts. Based on the existing solar-powered WiMAX and WiFi BSs [25,26], the experiments provided us some power model references of solar-powered BSs.

Other in-device protocols have also been extensively studied, such as power saving strategies [27,28]. There are two power modes specified for a (non-AP) station, namely awake and doze modes, in IEEE 802.11. In awake mode, the station radio interface is able to acquire the radio channel status and transmit/receive data packets; in doze mode, various parts of the station are powered off. When a station is in doze mode, APs need to buffer the incoming packets destined for the station. In [27,28], a power saving strategy for WLAN mesh networks was studied, which considered the QoS and power consumption of each station based on battery/solar application scenarios.

16.2.3 Device Deployment

Another important step to establish the wireless network infrastructure is the deployment of network devices, such as APs, gateways, BSs, and relay stations, etc. [29–31]. Based on the location space, the placement problem can be categorized into continuous and discrete cases. In the continuous case, there is no restriction applied to the location of the placement. Generally, this kind of device placement problem can be solved by some optimization algorithms, e.g., direct search and quasi-Newton methods [32]. In a realistic communication network, there may be some geographical constraints that limit the locations for placing devices, and usually there is a set of candidate locations for device deployment. This kind of problem is referred to as discrete case device placement, which is normally modeled as a mixed integer optimization problem, and for which it is challenging to find efficient optimization solutions.

Traditional devices deployment seeks the optimal placement of devices in a given region (or among a set of users), such that the whole region (or all users) can be connected and served, while some performance metrics can be guaranteed. However, many works mainly focus on the network performance, ignoring the energy consumption of the system. The infrastructure-oriented energy-efficient approaches toward device deployment aim at using minimal placement cost to enhance the sustainability of the whole wireless network or satisfy the QoS requirements of users. Specifically, it is typically modeled as an optimization problem with an objective function comprised of cost and/or the probability of network outage, i.e., any node has used up its energy and prevented the normal operations of the whole network.

Some works focus on minimizing the cost of BSs/APs with green energy by determining the locations to place devices. In [7], the traditional AP placement problem was revisited with sustainable power supplies. Their work focused on placing a minimal number of green energy powered APs on a set of candidate locations to ensure that the harvested energy is sustainable to serve wireless users and fulfill their QoS requirements. In [33], the minimum-cost placement of solar-powered data collection BSs was considered. The BSs were planned to be placed in a wireless sensor network, such that outage-free operation of the sensor nodes can be obtained. Other works focus on energy saving by device placement. In [34], BS placement and optimal power allocation were investigated to minimize the energy consumption of a cellular network. In [35], the deployment of a single frequency network was proposed with energy efficiency as the objective function, i.e., low carbon emissions and exposure. Works, such as [36,37] mainly focused on the battery capacity and solar panel size of the BSs or APs, with an objective to mitigate the network outage using minimal cost of energy according to the recorded historical solar insolation traces.

16.2.4 Resource Allocation

Resource allocation has been well studied in the context of different communication networks, including cellular networks, IEEE 802.16 WiMAX, and wireless mesh networks. It is recognized that efficient resource allocation can significantly improve the resource utilization. There are many previous works dealing with infrastructure-oriented energy-efficient approaches toward the resource allocation problem by different methodologies, e.g., scheduling, routing, power control, etc. [38,4]. Resource allocation in an infrastructure network is usually formulated as an optimization problem under the constraints of network connectivity, throughput, energy consumption, etc.

Recent advances in green energy technologies provide a feasible and sustainable solution for green wireless communication networks. The resource allocation schemes for wireless networks with green energy mainly focus on enhancing the network throughput or energy sustainability. So far, there are only a few works on resource allocation in wireless networks with renewable green energy that consider the sustainability of the network. In [39], traffic scheduling for infrastructure of vehicular wireless networks, i.e., road-side access points was studied. The work formulated the problem as a mixed integer linear program, which aimed at minimizing the energy consumption of road-side access points according to the vehicle communication requirements. An upper bound is provided, followed by a scheduler using a vehicle's location and velocity inputs to solve the problem. In [8], multihop radio networks powered by renewable energy sources were considered. The work formulated the problem as an integrated admission control and routing framework, and proposed routing algorithms to achieve high performance by utilizing the available energy sources. In [9], the work considered a power saving mechanism and control algorithm design in solar-powered WLAN mesh networks. A statistical power saving mechanism was proposed based on extensions to IEEE 802.11. In order to match the future load conditions and solar insolation, a control algorithm was designed for maintaining outage-free operations of the node. The work tried to balance the energy consumption with the energy charging capability for each node.

In this chapter, we will study the physical constrained device deployment problem in a wireless infrastructure network powered by green energy sources. Resource allocation issues, such as scheduling and power control, are also addressed. Our work aims at deploying a minimal number of green BSs on a set of candidate locations such that the QoS requirements of users can be fulfilled with the harvested energy.

16.3 SYSTEM CONFIGURATION AND PROBLEM FORMULATION

In this section, the system model and formulation of the constrained green BS placement problem are presented. Due to relatively expensive prices for BSs with green energy supply, we aim at minimizing the cost of green BS deployment to fulfill the QoS requirements of users with harvested energy from the environment. The introduction of the system model is presented in Section 16.3.1, and then we formulate the green BS deployment problem as an optimization problem in Section 16.3.2. Finally, the QoS and energy sustainability constraints are described in Section 16.3.3.

16.3.1 System Model

We consider a WLAN mesh network consisting of a collection of wireless users and BSs. The BSs are powered by green energy, e.g., solar, wind energy. Each wireless user has to connect to a BS for traffic relay and wireless access to other WLANs. Wireless users and BSs can set different transmission powers from a finite set of power levels by employing the promising adaptive modulation and coding technique. For simplicity, we assume the same transmission power is used by a BS and its serving users in one WLAN, while BSs are able to select different power levels. Thus, WLANs may have different coverage area for each of them, as shown in Figure 16. 1. The transmission links between a BS and its users are symmetrical. BSs use orthogonal channels to communicate with each other for the sake of avoiding inter-WLAN interference. Due to the physical geographical constraints, BSs can only be placed in limited locations in realistic network scenarios. A set of candidate locations are provided for placing BSs. As the charging capability of renewable energy supplies highly depends on the environment of geo-locations, BSs placed on different candidate locations may have different charging capabilities.

We build a network communication graph $G(V \cup W, E)$ to model the network topology, where V and W are the set of users and BSs, respectively, and E is the set of communication links between any two nodes. Let P^{-} denote the set of power levels that can be selected by BSs and users, and d_{vw} denote the distance between user v and BS w. As communication links are symmetric, (v, w) equals (w, v). We can express the received signal strength, SNR_{vw}, between link (v, w) as follows:

$$\text{SNR}_{vw} = \frac{p_v^{-} \cdot {d_{vw}}^{-\alpha}}{\sigma}, \tag{16.1}$$

where α is the path loss exponent, σ as the background noise, and p_v^{-} is the adopted power of user v. The achievable transmission rate of link (v, w), denoted by r_{vw}, is given by

$$r_{vw} = B \log_2 (1 + \text{SNR}_{vw}), \tag{16.2}$$

where B is the bandwidth of the channel.

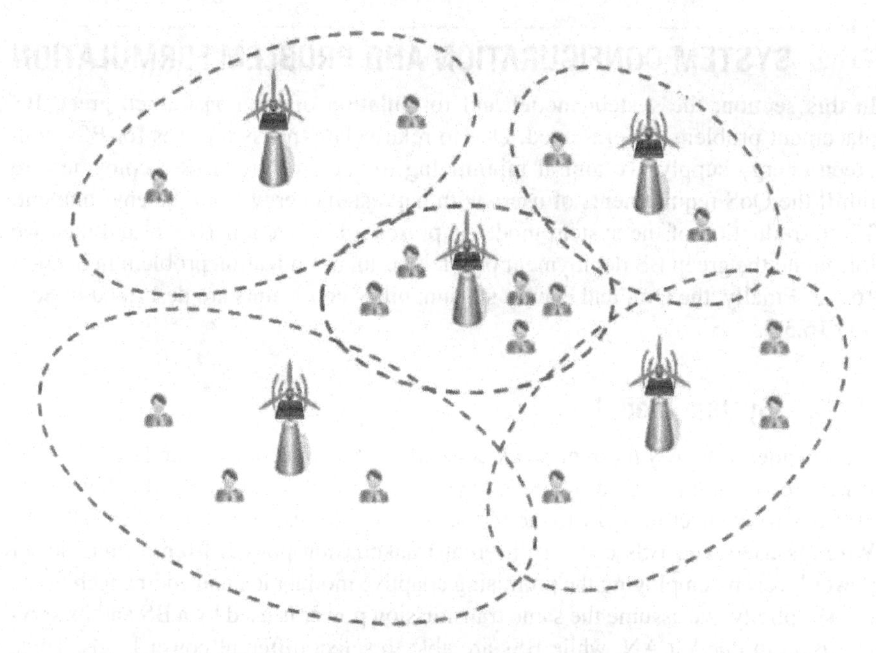

Figure 16.1 Wireless mesh networks with renewable power supplies.

16.3.2 Problem Formulation

Our green BS deployment problem aims at placing a minimal number of green energy powered BSs on a set of candidate locations, such that the users' QoS requirements can be fulfilled with the harvested energy. Given a set of users V and their QoS requirements, we need to find out the optimal placement of BSs on a set of candidate locations, with adjustable transmission power levels at each BSs, such that the number of APs is minimized, while energy sustainability and QoS constraints can be satisfied. The BS placement problem can be generally formulated as

$$\begin{aligned} &\text{Minimize } |W| \\ &\text{Subject to} \quad e_{vw} \in \{0,1\}, \quad \forall v \in V,\ \forall w \in W, \\ &\qquad \sum_{w \in W} e_{vw} = 1, \quad \forall v \in V, \\ &\qquad \beta_{vw} \geqslant \beta'_{vw}, \quad \forall e_{vw} = 1, \\ &\qquad \mathcal{E}_w^+ \geqslant \mathcal{E}_w^- \quad \forall w \in W, \end{aligned} \tag{16.3}$$

where $\mathcal{E}_w^-$ and $\mathcal{E}_w^+$ denote consumed energy and harvested energy of BS w, while e_{vw}, β_{vw}, and β'_{vw} denote the connectivity, throughput, and traffic demand of link (v, w), respectively. When e_{vw} is equal to 1, it means the connection of link (v, w) is established. The first two conditions show each user must be connected to only one AP. The third condition indicates that the achieved throughput should be larger than the QoS of each user.

The last constraint is the energy sustainability condition, which illustrate that the charged energy should be able to sustain the energy consumption of users.

16.3.3 QoS and Energy Sustainability Constraints

In this section, we further specify the energy and QoS constraints defined in (16.3). During a scheduling period $[0, T]$, each user v is allocated a share t_v of T for its own data transmissions. Therefore, to guarantee the fulfillment of the user's traffic demand, the following equation should hold:

$$\frac{r_{vw}t_v}{T} \geqslant \gamma_v, \forall v \in S_w, \tag{16.4}$$

where S_w is the set of users served by AP w, and γ_v is the traffic demand of user v. Substituting (16.2) into (16.4), the QoS constraint can be obtained as follows:

$$t_v \geqslant \frac{\gamma_v T}{B \log_2 (1 + SNR_{vw})}, \quad \forall v \in S_w. \tag{16.5}$$

Suppose the average charging rate of BS w is p_w^+. Then, the harvested energy during a period T, denoted as $\mathcal{E}_w^+$, is

$$\mathcal{E}_w^+ = p_w^+ T. \tag{16.6}$$

Similarly, we can obtain the energy consumption of AP w during a period T as $\mathcal{E}_w^-$. The AP allocates time slots to each user to satisfy their bandwidth demands, and the allocated slots should be bounded by the total time slots T,

$$T \geqslant \sum_{v \in S_w} t_v. \tag{16.7}$$

Therefore, we have

$$\mathcal{E}_w^- = p_w^- T \geqslant \sum_{v \in S_w} p_v^- t_v, \tag{16.8}$$

where $p_w = p_v$ for $v \in S_w$. To ensure the energy sustainability of APs, the harvested energy should be no less than the consumed energy for each AP in a period T. Thus, we can obtain the energy sustainability constraint as follows:

$$\sum_{v \in S_w} t_v \leqslant \min \left\{ \frac{p_w^+ T}{p_w^-}, T \right\}. \tag{16.9}$$

Notice that t_v plays a critical role in both the QoS and energy sustainability constraints defined in (16.5) and (16.9). In other words, based on the allocated time slots of each user, we can determine whether the user's bandwidth demand can be satisfied, and the AP can sustain the user's demand with its harvested energy.

16.4 CONSTRAINED BS PLACEMENT ALGORITHM

The formulated problem is NP-hard, because even the subproblems, such as optimal placement of BSs and optimal power control, are NP-hard in general [40]. Therefore, we design an effective heuristic algorithm to approach the optimal solution with reduced complexity. In this section, our heuristic algorithm, called the Two-phase Constrained Green BS Placement (TCGBP) algorithm, is presented in Section 16.4.1, followed by complexity analysis in Section 16.4.2.

16.4.1 TCGBP Algorithm

Intuitively, in order to place a minimal number of green BSs, each BS should serve as many users as possible. However, this may lead to the strategies of placing BSs in a dense area or making BSs select more power level for transmission. Either strategy may cause high traffic demand and energy consumption, which may violate the constraints in Eqs. (16.5) and (16.9).

The TCGBP algorithm is separated into two phases. Since links with a shorter distance usually achieve a higher transmission rate, short-distance links are more cost-effective compared with long-distance links. In the first phase, we place one BS in each candidate location. The whole region is partitioned into several Voronoi polygons (VP), while wireless users are also divided into different clusters. For each wireless user, it is assigned a VP vector to determine the relationship between each user and potential BSs. Then, communication links are established between each BS and the users inside this VP region. In the second phase, cross-polygon selection of wireless users is permitted, so that BSs can establish links with wireless users located in neighboring VP regions. The redundant BS is deleted for the case that there is no user in its VP region.

Initially, each candidate location is placed with a BS, and a minimal transmission power is assigned to all BSs and users. The VP region is determined by the relative distance from the user and the deployed BSs. Let x denote an arbitrary point in the region, and d_{xw} is the distance between x and BS w. The VP region of BS w, denoted by VP_w, is defined as follows:

$$\mathrm{VP}_w = \{x \mid d_{xw} \leqslant d_{xw'}, w' \in C, w' \neq w\}, \tag{16.10}$$

where C stands for the set of candidate locations. Based on the definition of the VP region, we can divide the whole region into $|C|$ Voronoi polygons. Thus, the wireless users located inside the region are also partitioned into different VP regions, where each VP region is centered by a BS. We use VP_v to denote the set of all users in the same VP region as user v. Each BS establishes a link with the users located in its VP region one by one. We define the preference level for BSs to determine the selection order of wireless users. Let pl_{vw} denote the preference level of user v with BS w, which can be expressed as

$$\mathrm{pl}_{vw} = \frac{t_{vw}}{T}, \tag{16.11}$$

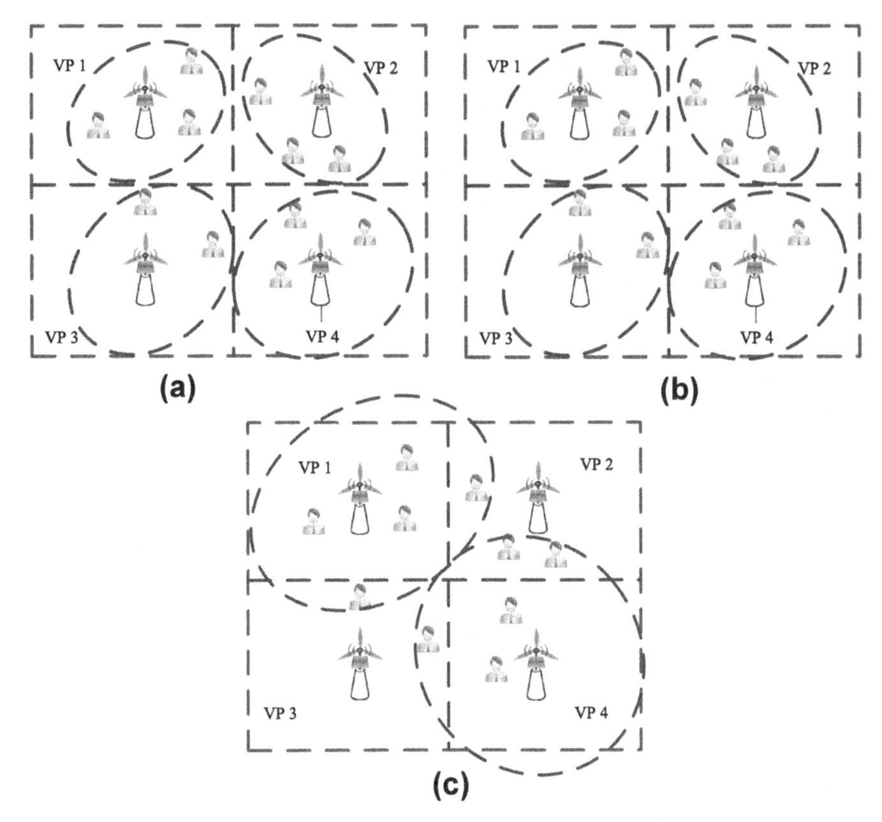

Figure 16.2 Illustration of TCGBP: (a) Voronoi polygons, (b) in-polygon selection of wireless users, and (c) cross-polygon selection of wireless users.

where t_{vw} is the least required active time of the link (v, w) to satisfy the QoS requirement of user v, which can be obtained from Eq. (16.5). Each BS w sorts the connection priority of wireless users in the increasing order of $\text{pl}_{vw}, v \in \text{VP}_w$. Then, BSs connect users one by one until one of the following conditions is met:

1. All wireless users in VP_j have been selected.
2. QoS and energy sustainability constraints defined in (16.5) and (16.9) cannot be held.

In the second phase, users are allowed to connect to BSs in neighboring VP regions. We start from the BS w^* with the largest number of users connected in its VP region, i.e., $|\text{VP}_{w^*}| \geqslant |\text{VP}_{w'}|, \forall w' \in W$. Then, we calculate the preference level of each user $v, v \notin \text{VP}_{w^*}$ with BS w^*, which is $\text{pl}_{vw^*}, v \notin \text{VP}_{w^*}$. Users outside w^*'s VP region are sorted in increasing order of $\text{pl}_{vw^*}, v \notin \text{VP}_{w^*}$. Then, BS w^* connects users one by one until the QoS and energy sustainability constraints cannot be held. Since we consider multiple power levels, each BS w can select a transmission power p_w^-, $p_w^- \in P^-$ for communication. All the transmission power levels are checked for link connection, and the power level that can support the maximum number of users will be selected for BS w^*. After that, we update the graph and discard all the links of (v, w^*), $\forall v \notin VP_{w^*}$. Then, the TCGBP algorithm repeats the above operation in the second phase until BSs cannot find more users to add. An example of the TCGBP algorithm is shown in Figure 2 for illustration, and the algorithm description is shown in Algorithm 16.1.

Algorithm 16.1: Two-phase Constrained Green BS Placement algorithm (TCGBP)

$W \leftarrow \emptyset$;
$p^0 \leftarrow \min(P^-)$;
for all $w \in C$ **do**
  for all $v \in V$ **do**
    $p_{\bar{v}} \leftarrow p^0$; $p_{\bar{w}} \leftarrow p^0$;
    Determine the VP region of BS w, VP_w;
  end for
  for all $v \in VP_w$ **do**
    Establish link (v, w);
    if QoS and energy constraints cannot be met **then**
      BREAK;
    end if
  end for
end for
while All BSs can find no more users to add **do**
  $w^* \leftarrow \{w \mid \max(|S_w|)\}$;
  for all $p_{w^*}^- \in P^-$ **do**
    for all $v \in V, v \notin VP_{w^*}$ **do**
      Calculate preference level pl_{vw^*};
    end for
    Sort users in increasing order of pl_{vw^*};
    for all Sorted $v \in V, v \notin VP_{w^*}$ **do**
      Establish link (v, w^*);
      if QoS and energy constraints cannot be met **then**
        BREAK;
      end if
    end for
    Record $(p_{w^*}^-, S_{w^*})$;
  end for
  Assign power level$\{p_{w^*}^- \mid \max(|S_{w^*}|)\}$ to BS w^*;
  Establish and delete links according to S_{w^*};
  $W \leftarrow w^*$;
  Delete w^* from C;
end while
RETURN $|W|$;

16.4.2 Time Complexity of the TCGBP

THEOREM16. 1

The TCGBP algorithm has time complexity of $O(|C||P||V|^2)$, *where* $|C|$, $|V|$, *and* $|P|$ *are the number of candidate locations, wireless users, and adjustable power levels, respectively.*

PROOF

The TCGBP algorithm consists of two sequential phases: (1) establishing links inside the VP region, and (2) connecting users outside the VP region and removing redundant BSs. In the first phase, there are two sequential steps: (i) it takes $O(|C||V|)$ to partition the whole region, and (ii) checking the feasibility of each VP region requires $O(|C||V|)$ time. Thus, the time complexity of the first phase is $O(|C||V|)$. In the second phase, there are three sequential steps: (i) the time complexity of calculating preference level is $O(|C||P||V|)$, (ii) it takes $O(|P||V|\log|V|)$ to sort the preference levels in increasing order, and (iii) the worst case of link connections and deletions needs $O(|C||P||V|^2)$ time, which includes power allocation and constraints checking. Thus, the time complexity of the algorithm is determined by the step with the highest complexity order, i.e., step 2, and is given by $O(|C||P||V|^2)$.

16.5 SIMULATION

In this section, we compare the deployed number of BSs of the proposed TCGBP with the optimal solution using the exhaustive search algorithm. Then, we evaluate the performance of the TCGBP for various system parameters, including different numbers of users, the variable traffic demands of each user, different number of candidate locations, and the variable charging capabilities of each BS, etc.

16.5.1 Simulation Configurations and Performance Evaluation

We set up a WLAN mesh network with a number of wireless users and candidate BS locations randomly distributed in a 100 m × 100 m region. The transmission power of wireless users can be adjusted in three discrete levels, i.e., 10 mW (10 dBm), 32 mW (15 dBm), and 100 mW (20 dBm). BSs are deployed at different candidate locations with different charging rates, which are uniformly distributed over [20, 30] mW. The bandwidth is set as 40 MHz according to IEEE 802.11n. The path loss exponent α is 4, and the background noise N is −20 dBm. A time duration of $T = 1000$ slots is used for network planning and scheduling. We repeat each simulation experiment 1000 times with different random seeds and compute the average values for performance evaluation. The simulator is generated by Java.

We first compare the performance of our proposed algorithm with the optimal solution achieved by exhaustive search as shown in Algorithm 16.2. Basically, the exhaustive search needs $2^{|C|}$ for searching all $|C|$ candidate locations, $|P|^{|C|}$ for searching $|C|$ BSs with $|P|$ power levels, $|V|^{|C|}$ for searching all combinations of client-BS connections, and $|V|$ for checking the user QoS and energy sustainability constraints. Therefore, the worst case time complexity of exhaustive search is $O(2^{|C|}|P|^{|C|}|V|^{|C|+1})$, which is much higher than the proposed algorithm.

Algorithm 16. 2: Exhaustive search

for all BS placements **do**
  for all combinations of BS power **do**
    for all combinations of client-BS connections **do**
      if Energy and traffic constraints can be met **then**
        RETURN $|W|$; {The minimum cost is found}
      end if
    end for
  end for
end for

The minimal number of BSs required to serve a given number of users under both algorithms are shown in Figure 16.3. We have five BS candidate locations and a number of users, each of which demands a 0.3 Mbps throughput. It can be seen that more BSs are required when the number of users increases. Our proposed algorithm approaches the optimal solution well. In Figure 16.4, we have 5 BS candidate locations and 30 wireless users with varying traffic demands. The required number of

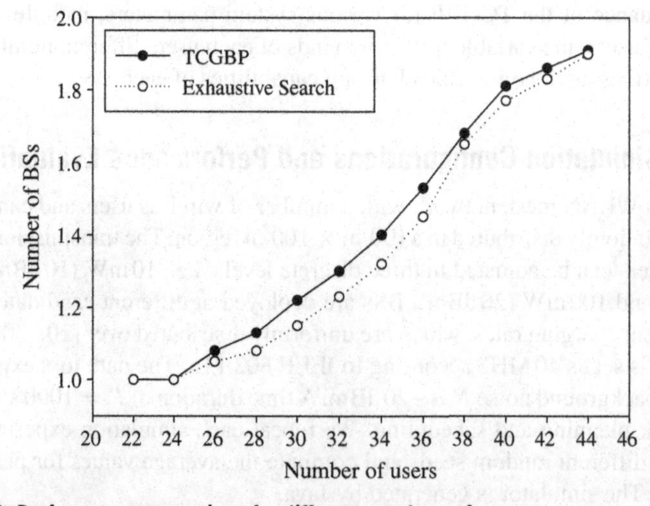

Figure 16.3 Performance comparison for different numbers of users.

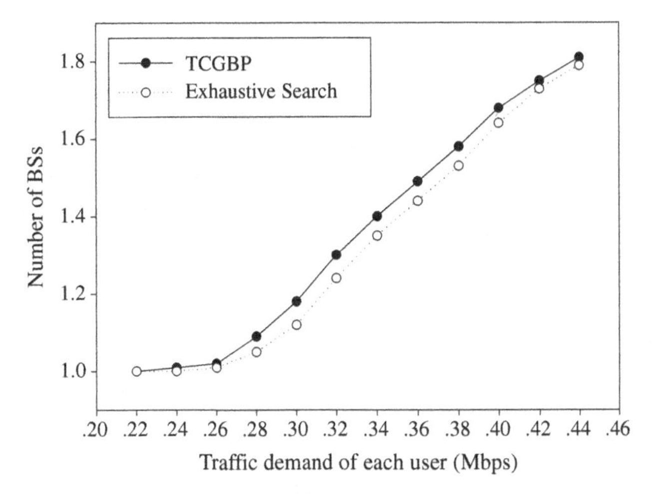

Figure 16.4 Performance comparison for different user demands.

BSs increases with growing traffic demands. By exhaustively searching all combinations, the algorithm can always find the optimal solution with the minimum number of BSs, while our proposed algorithm stops when all users can be connected to a BS that satisfies both user QoS and energy sustainability constraints. It shows that the exhaustive algorithm slightly outperforms our proposed algorithm at the cost of significantly increased time complexity.

We studied how the number of candidate locations impacted the number of BSs deployed. One hundred wireless users were deployed in a dense WLAN mesh network. As shown in Figure 16.5, for a larger number of candidate locations, it is

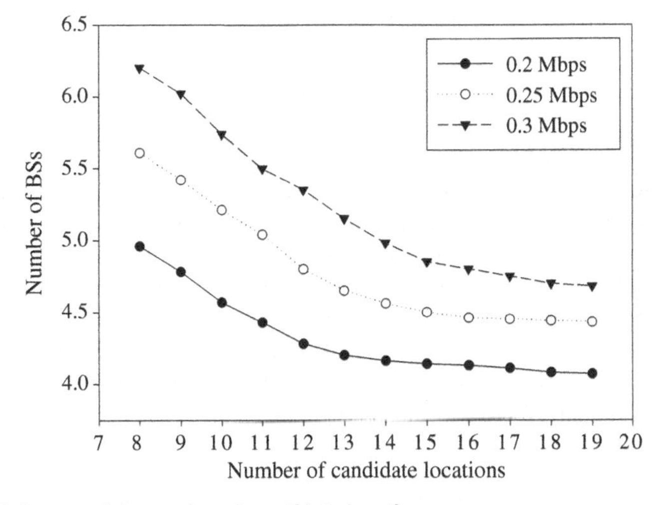

Figure 16.5 Impact of the number of candidate locations.

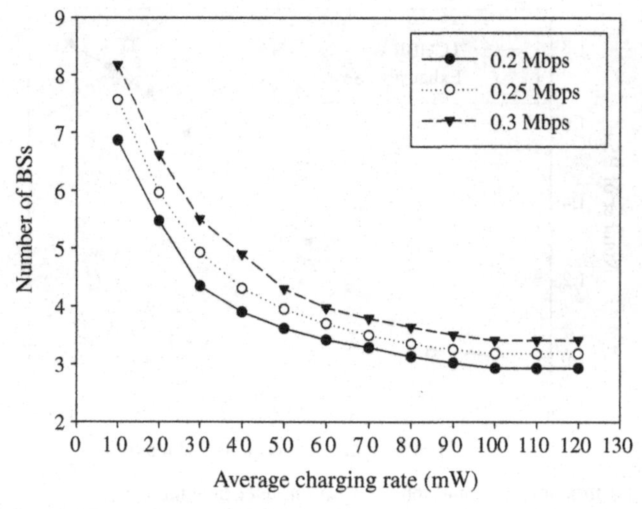

Figure 16.6 Impact of charging capabilities.

more likely to find a feasible solution that satisfies both the user QoS and energy sustainability requirements. The required number of BSs decreases with the number of candidate locations increasing. The impact of the charging capacity of placed BSs is shown in Figure 16.6. The average charging rate of each BS changes from 10 mW to 120 mW. Similarly, with a higher average charging rate at each BS, i.e., a BS can sustain greater traffic demands of users, the number of BSs required to serve all users is reduced accordingly. When the charging rate is sufficiently large, the number of BSs for a given user demand does not vary much and a certain number of BSs is required to cover the area where users are located.

16.6 CONCLUSION

In this chapter, we have introduced a green energy technique for future wireless communication networks. We have formulated an optimal placement problem of green BSs. Based on the Voronoi diagram, we have proposed a heuristic algorithm, i.e., TCGBP, which jointly considers rate adaptation and power allocation to solve the formulated optimization problem. Extensive simulation results demonstrate that the TCGBP algorithm can achieve the optimal solution with significantly reduced time complexity. Our future work will include the adaptive energy charging and discharging model, and analyze the capacity bounds of the network under the deployment strategies.

REFERENCES

[1] OPERANET (Online). <http://www.opera-net.org/default.aspx>.

[2] EARTH (Online). <http://www.ict-earth.eu/>.

[3] GREENRADIO (Online). <http://www.mobilevce.com/green-radio>.

[4] G. Miao, N. Himayat, Y.G. Li, A. Swami, Cross-layer optimization for energy-efficient wireless communications: a survey, Wireless Communications and Mobile Computing 9 (4) (2009) 529–542.

[5] G.P. Fettweis, E. Zimmermann, ICT energy consumption-trends and challenges, in: WPMC, Lapland, Finland, September 8–11, 2008, pp. 2006–2009.

[6] O. Arnold, F. Richter, G. Fettweis, O. Blume, Power consumption modeling of different base station types in heterogeneous cellular networks, in: Future Network and Mobile Summit, Florence, IT, June 16–18, 2010, pp. 1–8.

[7] Z. Zheng, L.X. Cai, M. Dong, X. Shen, H.V. Poor, Constrained energy-aware AP placement with rate adaptation in WLAN mesh networks, in: IEEE GLOBECOM, Houston, TX, USA, December 5–9, 2011, pp. 1–5.

[8] L. Lin, N.B. Shroff, R. Srikant, Asymptotically optimal energy-aware routing for multihop wireless networks with renewable energy sources, IEEE/ACM Transactions on Networking 15 (5) (2007) 1021–1034.

[9] A. Farbod, T.D. Todd, Resource allocation and outage control for solar-powered WLAN mesh networks, IEEE Transactions on Mobile Computing 6 (8) (2007) 960–970.

[10] L.X. Cai, Y. Liu, T.H. Luan, X. Shen, J.W. Mark, H.V. Poor, Dimensioning network deployment and resource management in green mesh networks, IEEE Communications Magazine, in press.

[11] Y. Chen, S. Zhang, S. Xu, G.Y. Li, Fundamental trade-offs on green wireless networks, IEEE Communications Magazine 49 (6) (2011) 30–37.

[12] I.F. Akyildiz, W. Su, Y. Sankarasubramaniam, E. Cayirci, Wireless sensor networks: a survey, Computer Networks 38 (4) (2002) 393–422.

[13] M. Hempstead, M.J. Lyons, D. Brooks, G.Y. Wei, Survey of hardware systems for wireless sensor networks, Journal of Low Power Electronics 4 (1) (2008) 11–20.

[14] N.A. Pantazis, D.D. Vergados, A survey on power control issues in wireless sensor networks, IEEE Communications Surveys and Tutorials 9 (4) (2007) 86–107.

[15] K. Akkaya, M. Younis, A survey on routing protocols for wireless sensor networks, Ad Hoc Networks 3 (3) (2005) 325–349.

[16] Renewables 2011: Global Status Report (Online). <http://www.ren21.net/Portals/97/documents/GSR/GSR2011Master18.pdf>.

[17] Green WiFi (Online). <http://www.green-wifi.org/>.

[18] WPWMN (Online). <http://ldt.stanford.edu/educ39109/POMI/MNet>.

[19] L.X. Cai, Y. Liu, H.T. Luan, X. Shen, J.W. Mark, H.V. Poor, Adaptive resource management in sustainable energy powered wireless mesh networks, in: IEEE Globecom Houston, TX, USA, December 5–9, 2011, pp. 1–5.

[20] A. Sayegh, T.D. Todd, M. Smadi, Resource allocation and cost in hybrid solar/wind powered WLAN mesh nodes, Wireless Mesh Networks: Architectures and Protocols, 2007, pp. 167–189.

[21] S. Saengthong, S. Premrudeepreechacharn, A simple method in sizing related to the reliability supply of small stand-alone photovoltaic systems, in: IEEE PVSC Anchorage, AK, USA, September 15–22, 2000, pp. 1630–1633.

[22] H.A.M. Maghraby, M.H. Shwehdi, G.K. Al-Bassam, Probabilistic assessment of photovoltaic (PV) generation systems, IEEE Transactions on Power Systems 17 (1) (2002) 205–208.
[23] E. Lorenzo, L. Navarte, On the usefulness of stand-alone PV sizing methods, Progress in Photovoltaics: Research and Applications 8 (4) (2000) 391–409.
[24] Y. Shi, L. Xie, Y.T. Hou, H.D. Sherali, On renewable sensor networks with wireless energy transfer, in: IEEE INFOCOM, Shanghai, CHN, April 10–15, 2011, pp. 1350–1358.
[25] A solar-powered WiMAX base station solution. Intel Netstructure WiMax Baseband Card, Application Note.
[26] Solar powered WIMAX and WiFi: how to use solar energy to power your wireless network, Tranzeo Wireless Technologies Inc., white paper.
[27] T.D. Todd, A.A. Sayegh, M.N. Smadi, D. Zhao, The need for access point power saving in solar powered WLAN mesh networks, IEEE Network 22 (3) (2008) 4–10.
[28] F. Zhang, T.D. Todd, D. Zhao, V. Kezys, Power saving access points for IEEE 802-11 wireless network infrastructure, IEEE Transactions on Mobile Computing 5 (2) (2006) 144–156.
[29] M. Soleimanipour, W. Zhuang, G.H. Freeman, Optimal resource management in wireless multimedia wideband CDMA systems, IEEE Transactions on Mobile Computing 2 (2002) 143–160.
[30] B. Lin, P. Ho, L. Xie, X. Shen, J. Tapolcai, Optimal relay station placement in broadband wireless access networks, IEEE Transactions on Mobile Computing 9 (2) (2010) 259–269.
[31] J. Pan, L. Cai, Y. Shi, X. Shen, Optimal base-station locations in two-tiered wireless sensor networks, IEEE Transactions on Mobile Computing 4 (5) (2005) 458–473.
[32] G.L.Z. Wei, L. Qi, New quasi-Newton methods for unconstrained optimization problems, Applied Mathematics and Computation 175 (2) (2006) 1156–1188.
[33] S.A. Shariatmadari, A.A. Sayegh, T.D. Todd, Energy aware basestation placement in solar powered sensor networks, in: IEEE WCNC, Sydney, Australia, April 18–21, 2010, pp. 1–6.
[34] P.G. Brevis, J. Gondzio, Y. Fan, H.V. Poor, J. Thompson, I. Krikidis, P.J. Chung, Base station location optimization for minimal energy consumption in wireless networks, in: IEEE VTC, Budapest, Hungary, May 15–18, 2011, pp. 1–5.
[35] G. Koutitas, Green network planning of single frequency networks, IEEE Transactions on Broadcasting 56 (4) (2010) 541–550.
[36] G.H. Badawy, A.A. Sayegh, T.D. Todd, Energy provisioning in solar-powdered wireless mesh networks, IEEE Transactions on Vehicular Technology 59 (8) (2010) 3859–3871.
[37] M.S. Zefreh, G.H. Badawy, T.D. Todd, Position aware node provisioning for solar powered wireless mesh networks, in: IEEE GLOBECOM, Miami, FL, USA, December 6–10, 2010, pp. 1–6.
[38] G.Y. Li, Z. Xu, C. Xiong, C. Yang, S. Zhang, Y. Chen, S. Xu, Energy-efficient wireless communications: tutorial, survey, and open issues, IEEE Communications Magazine, in press.
[39] A.A. Hammad, G.H. Badawy, T.D. Todd, A.A. Sayegh, D. Zhao, Traffic scheduling for energy sustainable vehicular infrastructure, in: IEEE GLOBECOM, Miami, FL, USA, December 6–10, 2010, pp. 1–6.
[40] V. Chvatal, A greedy heuristic for the set-covering poblem, Mathematics of Operations Research 4 (3) (1979) 233–235.

CHAPTER

Green Broadband Access Networks

17

Tao Han, Jingjing Zhang, Nirwan Ansari

Advanced Networking Laboratory, Department of Electrical and Computer Engineering, New Jersey Institute of Technology, Newark, NJ 07102, USA

17.1 INTRODUCTION

A variety of bandwidth-hungry applications and services such as high-definition television, video streaming, and social networking [1] are being rapidly deployed, thus leading to a continuous surge in bandwidth demand across networking infrastructure, notably the access portion. Thus, both wireline and wireless telecommunications operators are driven to upgrade their access networks to provide broader bandwidth for their subscribers [2, 3].

For wireless broadband access, service providers are gradually upgrading their networks to next-generation networks such as Long Term Evolution (LTE), LTE-advanced, and WiMAX to provide data rates as high as several hundreds of Mbps [4]. On the other hand, fixed mobile convergence (FMC) has been identified as the trend for many years to come to provision even higher data rates by taking advantage of the decrease of the distance between end users and wireless access points (WAPs). FMC flavors such as femtocell technologies, integrated optical access networks, and LTE radio access network (RAN), and radio-over-fiber picocellular networks are potentially able to provide as high as several Gbps to end users [6, 7].

For wireline broadband access, fiber-to-the-x (FTTx) provides the highest bandwidth to end users owing to the small signal transmission loss of the optical fiber among various wireline broadband access technologies including digital subscriber line (DSL) and hybrid fiber coaxial (HFC). However, growing bandwidth demands will soon fill up the bandwidth supplied by current access technologies. To further increase the bandwidth provisioning, both ITU and IEEE are developing next-generation passive optical network (PON) systems to increase the line rate from 1 Gbps or 1.25 Gbps to 10 Gbps [5].

The future sees a clear trend of data rate increase in both wireless and wireline broadband access. These access networks may experience a dramatic increase in energy consumption due to provisioning higher bandwidth as well as for other reasons [8–10]. For example, to guarantee a sufficient signal-to-noise ratio (SNR) at the receiver side for accurate recovery of high data rate signals, advanced transmitters with high transmitting signal power and advanced modulation schemes are required, thus consequently resulting in high energy consumption of the devices. Also, to provision a higher data rate, more power will be consumed by electronic circuits in network devices to facilitate fast data processing. Besides, high-speed data processing incurs fast heat buildup and high heat dissipation that further incurs high energy consumption for cooling. It is estimated that the access network energy consumption increases linearly with the provisioned data rate. It has also been reported that the LTE base station (BS) consumes more energy in data processing than the 3G UMTS systems [11, 12], and the 10 Gb/s Ethernet PON (EPON) system consumes much more energy than the 1 Gb/s EPON system.

The energy consumption of the network contributes to part of its operational expenditure (OPEX), and high-power consumption exerts high requirements on the performance of backup batteries at network terminal devices. Moreover, owing to the direct impact of greenhouse gases on the environment and climate change, energy consumption is becoming an environmental and thus social and economic issue [13]. Baliga et al. [14] estimated that Internet currently consumes ~1% to ~2.5% of the total electricity consumption in broadband-enabled countries. It is also shown that currently and in the medium-term future, access networks consume the majority of the information and communication technology (ICT) energy owing to the large quantity of access nodes, each of which requires a nonnegligible amount of power supply even in the "standby" mode. As broadband access is deployed worldwide, the vast number of small wattage increases will lead overall power increases on a terawatt scale. Therefore, greening is not merely a trendy concept, but is becoming a necessity to bolster social, environmental, and economic sustainability, and reducing energy consumption is an important and urgent problem in both the industry and the research community.

17.2 GREEN BROADBAND WIRELESS ACCESS NETWORKS

Cellular networks are one of the major wireless broadband access networks that consume a significant amount of energy. In cellular networks, energy consumption mainly comes from the base stations (BSs) [15]. Various projects are funded to pursue green cellular networks, such as EARTH [16], TREND [17], and C2POWER [18]. According to the power consumption breakdown [19], BSs consume more than 50% of the power of cellular networks. In addition, the number of BSs is expected to be doubled in 2012 [20]. Thus, reducing power consumption of BSs is crucial to green cellular networks. In this section, we first briefly introduce existing technologies that green cellular networks from the perspective of communication protocols and network operations. Then, we focus on greening cellular networks through cooperative networking. More specifically, we will discuss green opportunities and research challenges on greening wireless broadband access networks through cooperative networking.

17.2.1 Techniques for Greening Cellular Networks

17.2.1.1 Power Saving Communication Protocols

Radio access networks are dimensioned for peak hour traffic, and thus the utilization of base stations can be very inefficient during off-peak hours. The power saving communication protocols aim to adjust the transmit power of transceivers according to the traffic intensity. The most intuitive idea is to switch off transceivers when the traffic load is below a certain threshold for a certain time period [21]. When some base transceiver stations (BTSs) are switched off, radio coverage and service provisioning are taken care of by devices that remain active. Two user behaviors affect traffic volumes in portions of a cellular network [22]. The first one is the typical day–night behavior of users. The other one is mobility of users. Users tend to range over their office districts during working hours and stay at home in their residential area after work. This results in the surge of traffic in both areas at peak usage hours, but in the drop of traffic during off-peak hours. The BS switching problem can be formulated as an optimization problem that minimizes the number of active BSs while meeting the traffic load in the access network. Several algorithms have been proposed to solve the problem. Zhou et al. [23] proposed a centralized greedy algorithm and a decentralized algorithm that achieve the optimal solutions, and illustrate the relationship between energy savings and the probability of an outage. Oh and Krishnamachari [24] proposed the threshold-based method to switch BSs on/off, and showed that the energy saving ratio is related to the traffic characteristics and the number of the neighboring BSs. In [15, 25], the authors introduced the notion of power partitioning to save energy in cellular networks. In addition to switching BSs on/off, the authors proposed cell reconfiguration algorithms with the objective of matching the offered capacity with traffic demands. The proposed algorithms achieve network-level energy savings. Bhaumik *et al.* [26] proposed a multilayer cellular architecture that adjusts cell sizes between two fixed values according to traffic demands. The proposed scheme shows an energy savings of up to 40% as compared to the conventional cellular architecture.

17.2.1.2 Heterogeneous Network Deployment

Currently, wireless cellular networks are typically deployed as homogeneous networks, in which all base stations have similar transmit power levels, antenna patterns, receiver noise floors, and similar backhaul connectivity to packet data networks. Homogeneous network deployment has several inherent disadvantages. First, locations of base stations should be optimized through network planning, which is a complicated process when the number of BSs is large. Even if optimal locations can be calculated, it is difficult to acquire optimal sites for macro BSs with towers. If the BSs are not deployed in these optimal locations, network efficiencies, e.g., energy efficiency and spectrum efficiency, are compromised. Second, homogeneous network deployments only have limited ability to adapt to time-varying traffic demands. Because of the lack of an ability to adapt, homogeneously deployed networks are usually dimensioned according to peak traffic loads, thus wasting both energy and spectrum during the off-peak hours.

Heterogeneous networks utilizing a diverse set of base stations can be deployed to improve spectral and energy efficiency per unit area. The heterogeneous network deployment featuring high density deployments of small and low-power base stations has higher energy efficiency than the sparse deployment of a few high power BSs. There are four general heterogeneous network deployment scenarios: Macro-Micro, Macro-Pico, Macro-Femto, Macro-Relay, and Macro-RRH (Remote Radio Head). Heterogeneous network deployment improves network efficiency since it employs high density and low-power base stations. Etoh et al. [15] pointed out that heterogeneous network deployment will bring up to 50% reduction of the total BS energy consumption.

In [27, 28], the author investigated the impact of deployment strategies on power consumption of cellular networks, and optimized the energy consumption of cellular networks consisting of a mix of regular macro sites as well as a number of smaller BSs, which we here refer to as micro BSs; a micro BS covers a much smaller area and consumes less energy. Samdanis et al. [29] examined the energy efficiency of the cellular networks with joint macro and pico coverage, and showed that the joint deployment can reduce the total energy consumption by up to 60% in an urban area.

17.2.1.3 Enabling Off-Grid BSs

Designing off-grid BSs and communication protocols to enable optimal utilization of renewable energy in cellular access networks will significantly reduce the on-grid energy consumption, thus achieving green cellular networks. Renewable energy such as sustainable biofuels, solar, and wind energy are promising options to power BSs, thus reducing the CO_2 footprint of cellular networks. Ericsson [30] has developed a wind-powered tower for wireless base stations of cellular networks. "Green Power for Mobile" [20] is a recent program proposed to use renewable energy resources for BSs. This program aims to deploy solar, wind, or sustainable biofuels technologies to power new and existing off-grid BSs to reduce the on-grid energy consumption of cellular networks. To optimize the utilization of renewable energy, Zhou et al. [31] proposed the HO (handover) parameter tuning algorithm for target cell selection and the power control algorithm for coverage optimization to guide mobile users to access BSs with natural energy supply, thus reducing the power expense and CO_2 emission.

17.2.2 Greening via Cooperative Networking

Cooperative networking is one promising technique that benefits wireless networking in reducing energy consumption. In this section, we investigate green opportunities provided by cooperative networking in wireless cellular networks. Generally, there are two kinds of cooperation that can be explored in green wireless cellular networks: cooperation among BSs, and cooperation between BS and user equipment (UE). Each of them can bring green opportunities from different aspects of cellular networks. The cooperation among BSs could help optimize the cell size from an energy perspective, which is a proven approach for reducing energy consumption. The optimal cell size depends on several factors, including the base station technology, data rates, and traffic demands [26]. To optimize the cell size, BSs have to cooperate to measure transmission channel qualities and predict traffic demands, and share the local information with their neighbors. With this information, BSs negotiate coverage strategies and find optimal cell sizes cooperatively. When UEs cooperate with BSs, each UE, in lieu of a small compact BS, could perform as a mobile service provider, and form a picocell in downlink. The relaying UEs are selected according to the goal of minimizing power consumption of BSs, and when selected, UEs will relay the information from BSs to destination users. With cooperation, BSs reduce the energy consumption by optimizing their coverage, and leaving the blank area to their cooperators. In addition, this cooperation extends the downlink coverage, thus reducing the number of BSs needed to cover a certain area, and subsequently reducing the energy consumption in the area.

17.2.2.1 ***Cooperation among BSs***

17.2.2.1.1 Green Opportunity

In cellular networks, network usage will be highly correlated with user behaviors in two aspects: one is that users may roam to different cells at different time, thus resulting in different user density and therefore different capacity demand at a radio cell level; the other is that users' demands for wireless communications vary over time—typically high during the day, peaking in the evening, and then becoming low at night [26]. This feature allows us to design a green cellular network architecture in which cell sizes of BSs can adapt to users' density and traffic demands. Bhaumik et al. [26] proposed a multilayer cellular architecture as shown in Figure 17.1a. In this multilayer architecture, BS H is denoted as an umbrella cell, which can cover the serving area of cell A to cell F. These seven cells are designed to support the peak load in the serving area. During off-peak hours, the BS of the umbrella cell can increase its power and serve the whole area, and the other BSs can be turned off to save energy. This multilayer cellular architecture applies the cooperation between the umbrella cell and the underlying cells. Using two-layer cellular architecture saves 29–36%, 39–53%, and

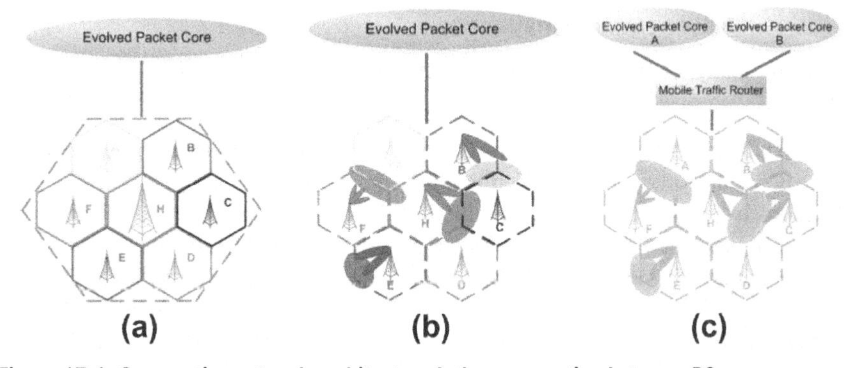

Figure 17.1 Cooperative network architecture I: the cooperation between BSs.

40–54% energy in GSM, UMTS, and WiMAX networks, respectively [26]. However, the cooperation in this multilayer architecture is limited by its rigid two-layered architecture. To allow further cooperation, we propose a flat cooperation cellular architecture shown in Figure 17.1b. Instead of limiting the cooperation only between umbrella cell and underlying cells, our proposed architecture allows all the BSs to cooperate to predict users' locations and traffic demands, and calculate the optimal coverage strategy to minimize the overall energy consumption accordingly. The serving BSs utilize the transmit beamforming technique to focus their transmit power to users in its assigned area; the other BSs are put into sleep mode to save energy. This architecture saves more energy than the multilayer cellular architecture because through the cooperation among the BSs, it can derive an optimal serving strategy. For example, suppose that, during off-peak hours, the users are distributed in the blue area as shown in Figure 17.1b. In the multilayer architecture, the umbrella cell, *H*, will compare the overall energy consumption of serving these users by itself and that by BS *A* and *F*. If the former uses less power, BS *H* sends messages to BS *A* and *F*, and put them into sleep mode; otherwise, BS *A* and BS *F* keep serving. In the latter case, no BS sleeps, and thus energy savings are not achieved. In our proposed architecture, BS *A*, *F*, and *H* cooperate to calculate the optimal coverage strategy, in which BS *F* serves these users, BS *A* goes into sleep mode, BS *H* reduces the transmit power toward the direction of the blue area, and all users in the blue area will attach to BS *F*. Thus, in our proposed architecture, we can at least turn one BS into the sleep mode.

In Figure 17.1c, we extend the cooperation to different mobile service providers. The locations of mobile users are not related to their mobile service providers. That is to say, mobile users in the same location may subscribe to different mobile service providers. During off-peak hours, it is a waste of energy if all these service providers serve the same area with oversatisfied capacities. For example, suppose that there are 20 users in the serving area, 10 of them use the services from operator *A*, and the others subscribe to operator *B*. Suppose that both *A*'s and *B*'s BSs can support at least 20 users when they are in the serving mode. The 10 users from operator *A* can be served using operator *B*'s BSs, and operator *A* can put its BSs into sleep mode. The mobile traffic router can route the users' traffic to their operators.

17.2.2.1.2 Cooperation Challenges

There are many open research issues for realization of cooperation in cellular networks:

- *Determination of cooperation point:* The cooperation point is the time when the BSs start to cooperate, which is important to network performance in regards to energy savings. One method to determine the cooperation point is to set up a traffic threshold; only when traffic is below the threshold, BSs seek to cooperate. However, this method may lead to inefficient cooperation. On one hand, if the threshold is too high, the coalition will break down in a short time period. In other words, some BSs that were put into sleep mode in the cooperation will restart soon. In this case, the energy consumed by restarting

BSs may be much higher than that without cooperation. On the other hand, if the threshold is too low, BSs may miss some cooperative opportunities. Thus, to achieve efficient cooperation, either a new method to determine the threshold or a novel model to determine the cooperation point should be exploited.

- *Coalition formation:* The problem of coalition formation is to determine who should cooperate with whom. There are two subproblems: to determine the coalition size and to determine the members of the coalition. Although the problem of coalition formation is well studied in the field of economics, and some game theory methods have been applied to wireless networking, few models can be applied directly to form the coalitions among BSs. The original coalition formation problem, is to form a coalition to maximize the benefits of all the members in the coalition, while the problem of forming the coalition of BSs is to maximize all coalitions' benefits which are the overall energy consumptions in the network. Hence, novel models or methods must be proposed for effective cooperation.

17.2.2.2 Cooperation between BS and UE

The unified cellular and ad hoc network (UCAN) architecture [32] is a general network architecture that allows for cooperation between the BSs and UEs. There has been several works in the area of cooperative communications in network integrating cellular networks and ad hoc networks. However, to the best of our knowledge, almost none of these works focus on reducing the energy consumption of the network. In this section, we investigate the opportunity of reducing energy consumption in integrated networks where BSs cooperate with UEs.

17.2.2.2.1 Green Opportunity

In cellular networks, the users with poor channel quality usually experience poor QoS (quality of service). To satisfy QoS requirements of all users, BSs have to adjust their transmitting power to satisfy users with the poorest QoSs. The energy consumption of a BS is a linear function of the transmitting power, and the gradient depends on the features of the system and its configurations [34]. Thus, minimizing the required transmitting power of a BS is an effective way to reduce the energy consumption of the BS, which is a major energy consumer in cellular networks. Figure 17.2a shows UCAN (unified cellular and ad hoc network architecture) [32]. Within the architecture, if the destination client experiences poor channel quality, the BS will send the message to the proxy, and then the proxy user will route the message to the destination user through the IEEE 802.11b broadcast channel. The proxy and route information are maintained and updated cooperatively by both BSs and UEs. Based on findings in [32], the cooperative network architectures are shown in Figure 17.2b.

To reduce energy consumption, this cooperative network architecture has two major improvements on UCAN. The first one is that it integrates transmitting beamforming, which is a promising technology in reducing transmitting power. Using transmitting beamforming also provides incentives to the relay users for cooperation.

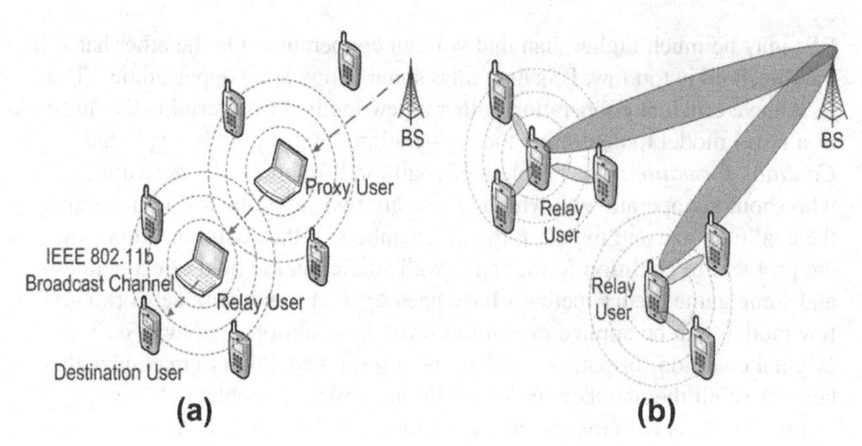

Figure 17.2 Cooperative network architecture II: the cooperation between BS and UEs.

When the transmit power is focusing toward the relay users, their received signal power ratios are enhanced, and thus their throughput is increased. The second improvement is that we limit the relay to one hop. According to [34], one-hop relay has the largest performance improvements. Because only one user is involved in the relay, the relay power consumption is minimal. Thus, one-hop relay benefits the network most in terms of energy efficiency.

In cooperative networks, all users periodically measure and update the channel state information of the downlink channel from BSs and transmission channels between users and their neighbors. When the destination user requests data services, he/she sends the request with the channel state information to the BS. Upon receiving the request, the BS calculates the optimal coverage strategy in terms of energy efficiency. If the destination user experiences good channel quality, the BS sends data directly to him/her. Otherwise, the BS selects the relay user based on the optimization result, and sends a cooperation request to him/her. After receiving the acknowledgment from the selected relay, the BS sends a relay assignment message to the destination user. Then, the destination user negotiates these relay parameters with the assigned relay user. If the negotiation is successful, the destination user sends a relay assignment acknowledgment to the BS, and then the BS starts the data transmission. The relay UEs can either use different radio interfaces, for example, WiFi and Bluetooth, or the same cellular radio interface but in different time slots. The cooperative protocol is shown in Figure 17.3.

An energy-efficient wireless multicasting scheme [33] based on cooperative network architecture was proposed. The scheme integrates multicast beamforming and cooperative networking. It contains two phases: in phase 1, the base station (BS) transmits the signal to the subscribers using antenna arrays with multicast beamforming; in phase 2, users who successfully received the signal in phase 1 forward the signal to other users. The unsatisfied users combine the received signals in both phases to retrieve the information. This scheme achieves significant energy savings by exploring the cooperation between BSs and UEs.

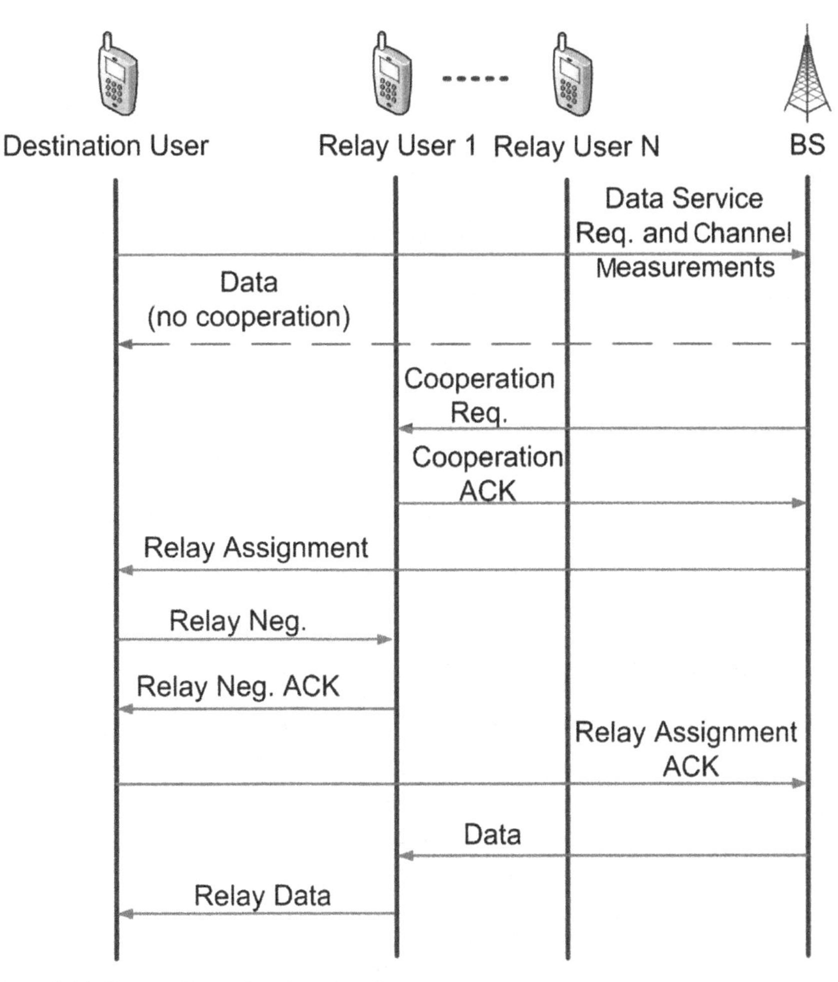

Figure 17.3 Cooperative network protocol.

17.2.2.2.2 Cooperation Challenges

There exist several open research challenges that must be investigated for the development of cooperative networks:

- *Channel state information:* In cooperative networks, UEs have to measure channel quality of both the downlink channel from BSs and the transmission channel between the users and their neighbors. Thus, new protocols are required to enable the measurement and update of channel state information among UEs. How to efficiently feedback the channel state information to BSs is another open issue. If the feedback is too limited, BSs may not find the optimal coverage strategy; otherwise, if the feedback is too much, it introduces additional overhead to the transmission. Therefore, novel feedback methods are critical for the realization of cooperative networking.

- *Incentive mechanism:* Although the design of the unified network architecture has certain incentives to stimulate the UEs to cooperatively relay data because the relay nodes can achieve higher throughput by participating in the cooperation [32], the tragedy of the commons may happen due to three facts: (1) there is more than one candidate relay user most of the time; (2) the UEs are selfish; and (3) relaying data consumes the energy of UEs. If the incentives are not strong enough, the UEs may avoid relaying data, and always rely on the other candidate relay nodes for cooperation because the UEs' throughput may not be hurt much if one of the candidates decides to cooperate. Thus, a simple but effective incentive mechanism should be exploited to stimulate cooperation.
- *Handover:* A hybrid handover scheme is essential for cooperative networks. The handoff may include the handover from BSs to BSs, from BSs to relay UEs, from relay UEs to relay UEs, and from relay UEs to BSs. Thus, an efficient handoff scheme is required to maintain the connectivity of the networks.

17.3 GREEN OPTICAL ACCESS NETWORKS

17.3.1 Wireline Access Technologies and Energy Consumptions

Currently, major wireline access technologies include digital subscriber loop (DSL) as standardized in ITU-T G.922, hybrid fiber coaxial (HFC) as standardized in ITU-T J.112/122, and fiber-to-the-x (FTTx), where x could be home, curb, neighborhood, office, business, premise, user, etc.

17.3.1.1 Digital Subscriber Loop

Figure 17.4 shows the general DSL architecture. DSL is provided through copper pairs originally installed to deliver a fixed-line telephone service. A DSL modem at each customer home connects via a dedicated copper pair to a DSL access multiplexer (DSLAM) at the nearest central office.

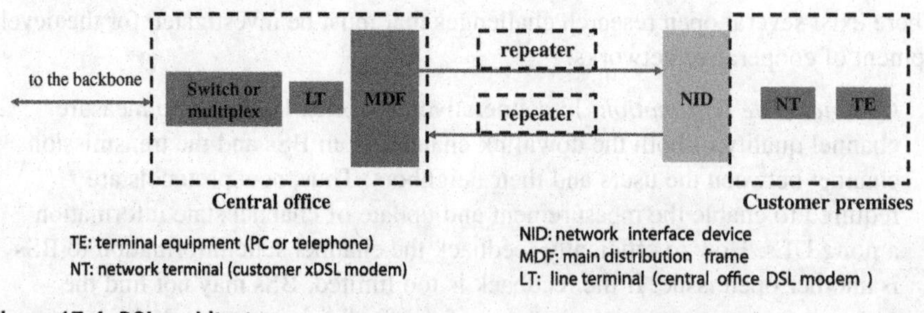

Figure 17.4 DSL architecture.

17.3.1.2 Hybrid Fiber Coaxial

Figure 17.5 illustrates the HFC network. HFC was initially deployed to deliver television services. Nowadays, HFC also delivers Internet and telephony services. Typically, the television program material is compiled from national and regional sources at a headend distribution center in each regional city. This material is distributed on radio frequency (RF) modulated optical carriers through an optical fiber to local nodes, where the optical signal is converted into an electrical signal. That electrical signal is then distributed to customers through a tree network of coaxial cables, with electrical amplifiers placed as necessary in the network to maintain signal quality. Hence, these networks are commonly termed hybrid fiber coaxial networks.

17.3.1.3 Fiber-to-the-x

To realize FTTx solutions, passive optical networks (PONs) have become the most promising technology. As shown in Figure 17.6, a PON is a point-to-multipoint optical access network architecture in which one optical line terminal (OLT) is connected with multiple optical network units (ONUs), and an optical splitter is employed to enable a single optical fiber to serve multiple end users.

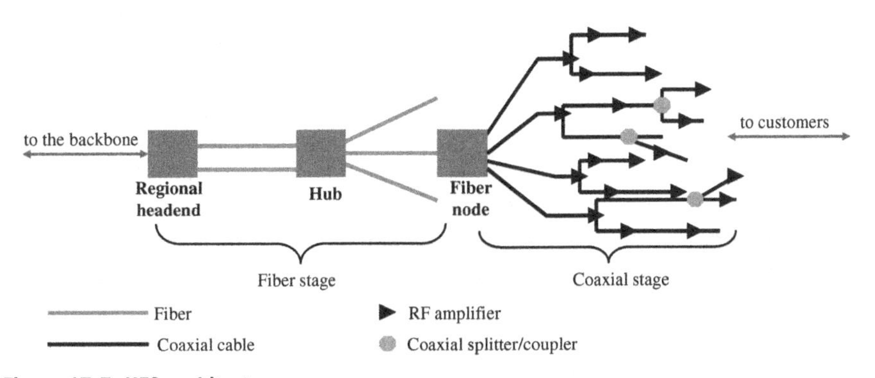

Figure 17.5 HFC architecture.

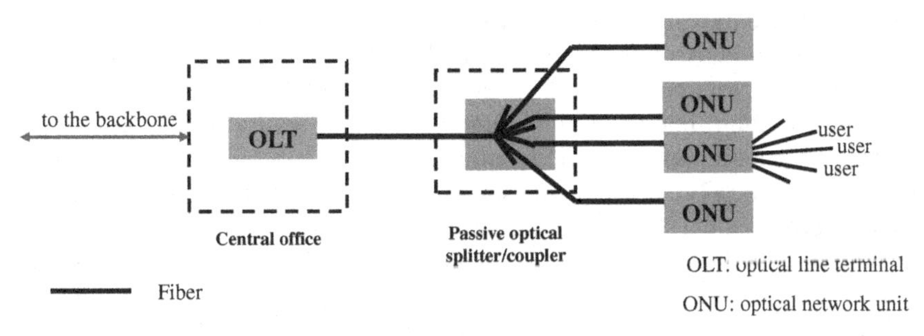

Figure 17.6 PON architecture.

Table 17.1 Energy Consumption of ADSL, HFC, and PON

	p_{TU} (kW)	N_{TU}	p_{RN}	N_{RN}	N_{CPE}	Technology Limit	Per-User Capacity (Mb/s)
ADSL	1.7	1008	N/A	N/A	5	15 Mb/s	2
HFC	0.62	480	571	120	6.5	100 Mb/s	0.3
PON	1.34	1024	0	32	5	2.4 Gb/s	16

The energy consumption of each access network can be split into three components: the energy consumption in the customer premises equipment (i.e., the modem), the remote node or base station (base transceiver station, BTS), and the terminal unit, which is located in the local exchange/central office. The per-customer power consumption can be expressed as $p_a = p_{CPE} + \frac{p_{RN}}{N_{RN}} + \frac{1.5 p_{TU}}{N_{TU}}$, where p_{CPE}, p_{RN}, and p_{TU} are the powers consumed by the customer premises equipment, remote node, and terminal unit, respectively. N_{RN} and N_{TU} are the number of customers or subscribers that share a remote node and the number of customers that share a terminal unit, respectively. Table 17.1 lists the typical power consumption of these three access networks [39].

It can be seen that PON consumes the smallest energy per transmission bit; this is attributed to the proximity of optical fibers to the end users and the passive nature of the remote node among various wireline access technologies [10]. However, as PON is deployed worldwide, it still consumes a significant amount of energy. It is desirable to further reduce the energy consumption of PON since every single watt saved will end up creating an overall terawatt or even larger power savings. Reducing the energy consumption of PON becomes even more important as the current PON system migrates into next-generation PON systems with increased data rate provisioning [35, 36].

17.3.1.4 BPON, GPON, and EPON

Besides the low energy consumption, PON has four other major advantages. First, a PON yields a small fiber deployment cost in the local exchange and local loop. Second, a PON provides higher bandwidth due to the deep fiber penetration [36]. Third, as a point-to-multipoint network, a PON allows for downstream video broadcasting. Fourth, a PON eliminates the necessity of installing multiplexers and demultiplexers in the splitting locations, and thus lowers the operational expenditure [37]. Owing to these advantages, the number of FTTx users has recently surpassed thirty million and is continuing to grow at a rapid rate.

The currently deployed PON systems are TDM PON systems [38]. As shown in Figure 17.7, the downstream traffic is continuously broadcast to all ONUs, and each ONU selects the packets destined to it and discards the packets addressed to other ONUs. In the upstream, each ONU transmits during the time slots that are allocated

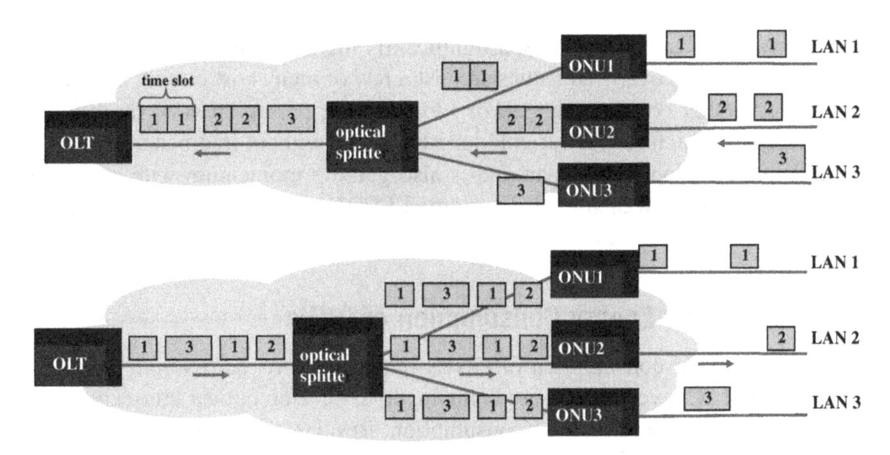

Figure 17.7 The upstream and downstream transmission in TDM PONs.

by the OLT. Upstream signals are combined by using a multiple access protocol, usually time division multiple access (TDMA). The OLTs "range" the ONUs in order to provide time slot assignments for upstream communication. Owing to their burst transmission nature, burst-mode transceivers are required for upstream transmission. There are generally three flavors of TDM PON technologies that have been standardized. They are Broadband PON (BPON), Gigabit PON (GPON), and Ethernet PON (EPON). Table 17.2 compares these three PON technologies.

Both BPON and GPON architectures were standardized by the Full Service Access Network (FSAN), which is an affiliation of network operators and telecom vendors. Since most telecommunications operators have heavily invested in providing legacy TDM services, both BPON and GPON are optimized for TDM traffic and rely on framing structures with very strict timing and synchronization requirements.

EPON is standardized by the IEEE 802 group, and focuses on preserving the architectural model of Ethernet. No explicit framing structure exists in EPON; the Ethernet frames are transmitted in bursts with a standard interframe spacing. The burst sizes and physical layer overhead are large in EPON. As a result, ONUs do not need any protocol and circuitry to adjust the laser power. Also, the laser-on and

Table 17.2 Comparison between BPON, GPON, and EPON

	BPON	GPON	EPON
Standard	ITU G.983	ITU G.984	IEEE 802.3ah
Framing	ATM	GEM/ATM	Ethernet
Max bandwidth	622 Mb/s	2.488 Gb/s	1 Gb/s
Users/PON	32	64	16
Video	RF	RF/IP	RF/IP

laser-off times are capped at 512 ns, a significantly higher bound than that of GPON. The relaxed physical overhead values are just a few of many cost-cutting steps taken by EPON. Another cost-cutting step of EPON is the preservation of the Ethernet framing format, which carries variable-length packets without fragmentation. EPON has been rapidly adopted in Japan and is also gaining momentum with carriers in China, Korea, and Taiwan since IEEE ratified EPON as the IEEE802.3ah standard in June 2004.

17.3.2 Reducing Energy Consumption of ONUs

Reducing the power consumption of ONUs requires efforts across both the physical layer and MAC layer. Efforts are being made to develop optical transceivers and electronic circuits with low-power consumption. Besides, multipower mode devices with the ability of disabling certain functions can also help reduce the energy consumption of the network. However, low-power mode devices with some functions disabled may result in the degradation of the network performance. To avoid the service degradation, it is important to properly design MAC-layer control and scheduling schemes that are aware of the disabled functions.

The major challenge of reducing ONUs in TDM PONs lies in the downstream transmission. In TDM PON, the downstream data traffic of all ONUs are TDM multiplexed into a single wavelength, and are then broadcasted to all ONUs. An ONU receives all downstream packets, and checks whether the packets are destined to itself. An ONU does not know when the downstream traffic arrives at OLT, and the exact time that OLT schedules its downstream traffic. Therefore, without a proper sleep-aware MAC control, receivers at ONUs need to be awake all the time to avoid missing their downstream packets. With the focus on the EPON, we discuss schemes for tackling the downstream challenge.

A number of schemes have been proposed to address the downstream challenge so as to reduce the energy consumption of ONUs in EPON. These proposed energy saving schemes can be divided into two major classes. The first class tries to design a proper MAC control scheme to convey the downstream queue status to ONUs, while the second class focuses on investigating energy-efficient traffic scheduling schemes. The two-way or three-way handshake process performed between OLT and ONUs are examples of schemes of the first class [40, 44]. Typically, OLT sends a control message notifying an ONU that its downstream queue is empty; the ONU optionally enters sleep mode and then sends a sleep acknowledgment or negative acknowledgment message back to OLT. While OLT is aware of the sleep status of ONUs, it can buffer the downstream arrival traffic until the sleeping ONU wakes up.

To implement the handshake process, the EPON MAC protocol, a multipoint control protocol (MPCP) defined in IEEE 802.3ah and IEEE 802.3av, has to be modified to include new MPCP protocol data units (PDUs). In addition, the negotiation process takes at least several round-trip times, implying that an ONU has to wait for several round-trip times before entering into sleep status after it infers that its downstream queue is empty. This may significantly impair the energy savings.

Energy saving schemes of the second class tackle the downstream challenge by designing suitable downstream bandwidth allocation schemes. Formerly, Lee and Chen [46] proposed to implement fixed bandwidth allocation (FBA) in the downstream when the network is lightly loaded. By using FBA, the time slots allocated to each ONU in each cycle are fixed and known to the ONU. Thus, ONUs can go to sleep during the time slots allocated to other ONUs. However, since the traffic of an ONU dynamically changes from cycle to cycle, FBA may result in bandwidth under- or overallocation, and consequently degrade services of ONUs to some degree. Yan et al. [41] proposed to schedule the downstream traffic and the upstream traffic simultaneously. An ONU stays in awake mode during its allocated upstream time slots, and switches into sleep mode in other time slots. Since the downstream traffic of an ONU is sent over the time slots that its upstream traffic is sent, the ONU stays in awake mode during that time period and will not miss its downstream packets. This scheme works well when traffic in the upstream and downstream are symmetric, but it may cause inefficient bandwidth utilization when the downstream traffic outweighs upstream traffic. We next describe a scheme that can enable the sleep mode of ONUs and best utilize the network resource [43].

17.3.2.1 Sleep Status of ONUs

Figure 17.8 illustrates the constituents of an ONU. The optical module consists of an optical transmitter (Tx) and an optical receiver (Rx). The electrical module mainly contains serializer/deserializer (SERDES), ONU MAC, network/packet processing engine (NPE/PPE), Ethernet switch, and UNIs. When neither upstream nor downstream traffic exists, every component in the ONU can be put into "sleep." When only downstream traffic exists, the functions related to the upstream transmission can be disabled. Similarly, the functions related to receiving downstream traffic can be disabled when only upstream traffic exists. Even when the upstream traffic exists, the laser driver and laser diode (LD) do not need to function all the time, but only during the time slots allocated to this ONU. Thus, each component in the ONU can likely "sleep," and potentially higher power savings can be achieved.

By putting each component of an ONU to sleep, an ONU ends up with multiple power levels. The "wakeup" of UNI, NPE/PPE, and switch can be triggered by the arrival of upstream traffic and the forwarding of downstream traffic from ONU MAC [47]. They are relatively easily controlled as compared to the other components. Thus, we only focus on ONU MAC, SERDES, Tx, and Rx.

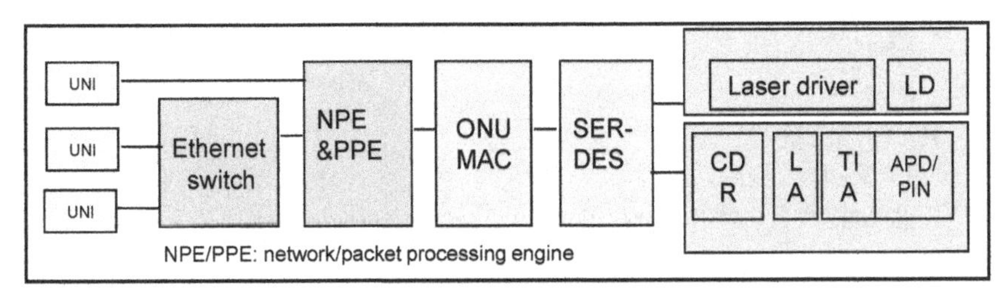

Figure 17.8 The constituents of an ONU.

17.3.2.2 Scenario 1: Sleep for More Than One DBA Cycle

Whether downstream/upstream traffic exists or not can be inferred based on the information of the time allocated to ONUs and queue lengths reported from ONUs, which is known to both OLT and ONUs. If no upstream traffic arrives at an ONU, the ONU requests zero bandwidth in the MPCP REPORT message. Then, OLT can assume that this ONU does not have upstream traffic. If no downstream traffic for an ONU arrives at OLT, OLT will not allocate downstream bandwidth to the ONU. Assume that, out of fairness concern, OLT allocates some time slots in a dynamic bandwidth allocation (DBA) cycle to every ONU with downstream traffic. Then, considering the uncertainty of the exact time allocated to an ONU in a DBA cycle, the ONU can infer that no downstream traffic exists if it does not receive any downstream traffic within two DBA cycles.

The next question is to decide the transition between different statuses. Formerly, Kubo et al. [45] proposed periodic wakeup with the sleeping time being adaptive to the arrival traffic status. We also decide the sleeping time based on the traffic status. More specifically, we set the sleep time as the time duration as that when traffic stops arriving. Taking and putting Tx into sleep for example, Algorithm 17.1 describes the sleep control scheme. We assume that Algorithm 17.1 is known to OLT as well. Then, OLT can accurately infer the time that Tx is asleep or awake.

Algorithm 17.1 Decide the Transition between "all:awake" and "Tx:sleep"

A:

if the Tx has not transmitted traffic for the time duration of *idle_threshold* **then**

  $s = 1$;

  B:

  Tx enters into sleep status;

  $sleeptime = 2^{s-1} \times short_active + (2^{s-1} - 1) \times idle\_threshold$;

  if $sleeptime > 50ms$ **then**

    $sleeptime = 50ms$;

  end if

  Tx wakes up after sleep time duration;

  The ONU checks the queue length and reports the queue status;

  if there is queued traffic **then**

    Keep Tx awake;

    $s = 0$;

    go to A;

  else

    $s = s + 1$;

    go to B;

  end if

end if

Let "idle_threshold" be the maximum amount of time that a transmitter stays idle before being put into sleep mode, "short_active" be the time taken for ONU to check its queue status, and to send out the report, and "sleep_time" be the amount of time an ONU sleeps. If the transmitter is idle for "idle_threshold," Tx will be put into sleep mode, and the sleep_time for the first sleep equals "idle_threshold." Then, Tx wakes up to check its queue status and sends a report to OLT, which takes "short_active" time duration. If there is no upstream traffic being queued, Tx will enter sleep mode again. Until now, the elapsed time since the last time Tx transmitted data packets equals "idle_threshold" + time duration of the first sleep + "short_active." So, for the second sleep, the sleep time duration "sleep_time" is set as "idle_threshold" + time duration of the first sleep + "short_active." According to MPCP, ONUs send MPCP REPORT messages to OLT every 50 ms when there is no traffic. So, we set the upper bound of "sleep_time" as 50 ms to be compatible with MPCP and also to avoid introducing too much delay for traffic that arrives during sleep mode. This process repeats until upstream traffic arrives. For the sth sleep, the "sleep_time" equals "idle_threshold" + the total time durations of the former $s-1$ sleep + $(s-1)$ * "short_active," which also equals 2^s-1 * short_active + (2^s-1-1) * idle_threshold.

Figure 17. 9 shows an example of the sleep time control process with "short_active" = 2.5 ms and "idle_threshold" = 10 ms. Then, "sleep_time" of the first sleep, the second sleep, the third sleep, and the fourth sleep are as follows:

- *First sleep:* 10 ms.
- *Second sleep:* "idle_threshold" + 10 ms + "short_active" = 22.5 ms.
- *Third sleep:* "idle_threshold" + 32.5 ms + 2 * "short_active" = 47.5 ms.
- *Fourth sleep:* min{50 ms, "idle_threshold" + 80 ms + 3 * "short_active"} = 50 ms.

In deciding the sleep time, "idle_threshold" and "short_active" are two key parameters that are set as follows.

"idle_threshold": When setting "idle_threshold" one needs to consider the time taken to transition between "sleep" and "awake." Considering the transition time, the net sleep time will be reduced by the sum of the transition time from awake to sleep and the transition time from sleep to awake. Hence, "idle_threshold" should

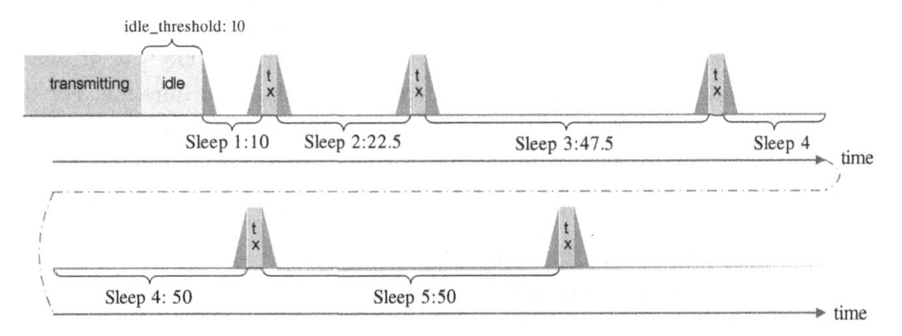

Figure 17.9 An example of sleep time control of the transmitter.

be set longer than the sum of two transition time in order to save energy in the first sleep. Currently, the time taken to power the whole ONU up is around 2–5 ms [40]. So, "idle_threshold" should be greater than 4 ms in this case. In addition, we assert that the upstream/downstream traffic queue is empty if no bandwidth is allocated to upstream/downstream traffic for "idle_threshold." To ensure this assertion is correct, "idle_threshold" should be at least one DBA cycle duration, which typically extends less than 3 ms to guarantee delay performance for some delay-sensitive service. So, Tx/Rx must sleep for over one DBA cycle with this scheme.

"short_active": During the short awake time of Tx, an ONU checks its upstream queue status and reports to OLT. Hence, "short_active" should be long enough for an ONU to complete these tasks. In addition, using some upstream bandwidth for an ONU to send a REPORT message affects the upstream traffic transmission of other ONUs. In order to avoid the interruption of the traffic transmission of other ONUs, we set "short_active" to be at least one DBA cycle duration such that OLT can have freedom in deciding the allocated time for an ONU to send a REPORT message. For Rx, during the short awake time, OLT begins sending the queued downstream traffic if there is any. Similar to the Tx case, "short_active" is set to be at least one DBA cycle to avoid interrupting services of ONUs in the Rx case.

17.3.2.3 Scenario 2: Sleep within One DBA Cycle

In the former scenario, the sleep and awake durations of Tx and Rx are greater than one DBA cycle. In this section, we discuss the scheme of putting Tx and Rx into sleep mode within one DBA cycle.

Consider a PON with 16 ONUs. During a DBA cycle, on average, only 1/16 of time duration is allocated to an ONU. In other words, even if the upstream/downstream traffic exists, Tx/Rx only needs to be awake for 1/16 of the time, and can go to sleep for the other 15/16 of the time. Therefore, significant energy savings can be achieved.

To enable an ONU sleep and wakeup within a DBA cycle, the transit time between awake and sleep should be less than half of the DBA cycle duration such that the next sleep time can be greater than zero, and thus energy can be saved. Formerly, Wong et al. [42] reduced the transition time into as small as 1–10 ns by keeping part of the back-end circuits awake. Thus, with the advances in speeding up the transition time, it is physically possible to put an ONU into sleep mode within one cycle to save energy.

For the upstream case, the wakeup of Tx can be triggered by ONU MAC when the allocated time comes. Tx can go to sleep after the data transmission. For the downstream case, however, it is difficult to achieve since Rx does not know the time that the downstream traffic is sent and has to check every downstream packet. To address this problem, we propose the following sleep-aware downstream scheduling scheme.

For the downstream transmission, OLT schedules the downstream traffic of ONUs one by one, and the interval between two transmissions of an ONU is determined by the sum of the downstream traffic of the other entire ONUs. Again, owing to the bursty nature of the ONU traffic, the ONU traffic in the next cycle does not vary much as compared to that in the current cycle. Accordingly, we can make an estimation of traffic of other ONUs, and put this particular ONU into sleep mode for some time.

More specifically, for a given ONU, denote Δ as the difference between the ending time of its last scheduled slot and the beginning time of its current scheduled slot. Then, we set the rule that OLT will not schedule this ONU's traffic until $f(\Delta)$ time after the ending time of the current scheduled slot. As long as the ONU is aware of this rule, it can go to sleep for $f(\Delta)$ time duration.

Figure 17. 10 illustrates one example of putting ONU into sleep mode within one DBA cycle. In this example, one OLT is connected with four ONUs, and $f(\Delta)$ is set as $0.8 * \Delta$. The interval between the first two scheduling of ONU 4 is 9. So, OLT will not schedule the traffic of ONU 4 until 7.2 time units later, and thus ONU 4 can sleep for 7.2 time units and then wake up. However, this wakeup is an early wakeup since the actual transmission of the other ONUs takes 9.5 time units, which is 2.3 longer than the estimation. Similarly, the duration of the second sleep is set as 7.6. However, this wakeup is a late wakeup since the actual time taken to transmit the other ONUs' traffic is 6.5. The late wakeup incurs 1.1 units of idle time on the downstream channel.

As can be seen from the example, early wakeup and late wakeup are two common phenomena of this scheme. Early wakeup implies that energy can be further saved, while late wakeup results in idle time, and thus possibly service degradation. From network service providers' perspective, avoiding late wakeup and the subsequent service degradation is more desirable than avoiding early wakeup. So, setting $0<f(\Delta)<\Delta$ is suggested. If $f(\Delta)$ is set as small as 0.5Δ, on average an ONU can still sleep 15/32 of the time when a PON supports 16 ONUs and traffic of ONUs is uniformly distributed. Therefore, significant power savings can be achieved with this scheme.

17.3.3 Reducing Energy Consumption of OLT

Formerly, the sleep mode and adaptive line rate have been proposed to efficiently reduce the power consumption of ONUs by taking advantage of the bursty nature of the traffic at the user side. It is, however, challenging to introduce the "sleep" mode into OLT to reduce its energy consumption for the following reasons. In PONs,

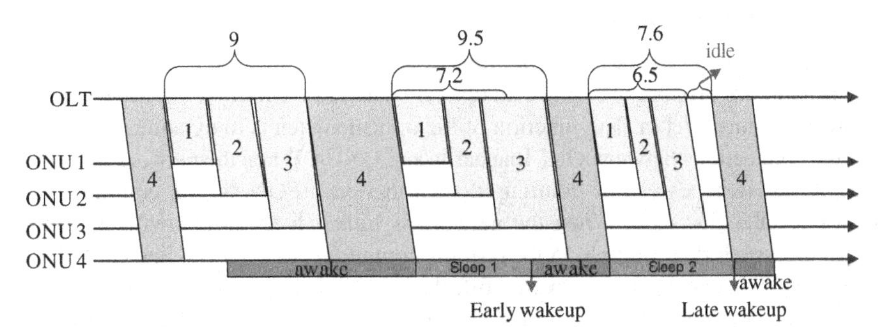

Figure 17.10 One example of putting ONUs into sleeps mode within one DBA cycle.

OLT serves as the central access node, which controls the network resource access of ONUs [49]. Putting OLT into sleep can easily result in the service disruption of ONUs communicating with the OLT. Thus, a proper scheme is needed to reduce the energy consumption of OLT without degrading service for end users.

17.3.3.1 Framework of the Energy-Efficient OLT Design

A possible solution to reduce the energy consumption of OLT is to adapt its power-on OLT line cards according to the real-time arrival traffic. To avoid service degradation during the process of powering on/off OLT line cards, proper devices are added into the legacy OLT chassis to facilitate all ONUs communicating with power-on line cards [48].

In the central office, one OLT chassis typically comprises multiple OLT line cards, each of which communicates with a number of ONUs. In the currently deployed EPON and GPON systems, one OLT line card usually communicates with either 16 or 32 ONUs. To avoid service disruptions of ONUs connected to the central office, all these OLT line cards in the OLT chassis are usually powered on all the time. To reduce the energy consumption of OLT, our main idea is to adapt the number of power-on OLT line cards in the OLT chassis to the real-time incoming traffic.

Denote C as the provisioned data rate of one OLT line card, L as the total number of OLT line cards, N as the number of ONUs connected to the OLT chassis, and $r^i(t)$ as the arrival traffic rate of ONU i at time t. By powering on all the OLT line cards, the overall data rate accommodated by the OLT chassis equals to $C * L$, which may be greater than the real-time incoming traffic, i.e., $\sum_{i=1}^{N} r^i(t)$. Denote l as the smallest number of required OLT line cards that can provision at least $\sum_{i=1}^{N} r^i(t)$ data rate. Then, our ultimate objective is to power on only l OLT line cards to serve all N ONUs at a given time t instead of powering on all L line cards.

However, powering off OLT line cards may result in service disruptions of ONUs communicating with these OLT line cards. To avoid service disruption, power-on OLT line cards should be able to provision bandwidth to all ONUs connected to the OLT chassis. To address this issue, we propose several modifications over the legacy OLT chassis to realize the dynamic configuration of OLT as will be presented next.

17.3.3.2 OLT with Optical Switch

To dynamically configure the communications between OLT line cards and ONUs, one scheme we propose is to place an optical switch in front of all OLT line cards as shown in Figure 17.11a. The function of the optical switch is to dynamically configure the connections between OLT line cards and ONUs. When the network is heavily loaded, the switches can be configured such that each PON system communicates with one OLT line card. When the network is lightly loaded, the switches can be configured such that multiple PON systems communicate with one line card. Then, some OLT line cards can be powered off, thus saving energy consumption.

Assume the energy consumption of the optical switch is negligible. As compared to the scheme of always powering on all L line cards, the scheme of powering

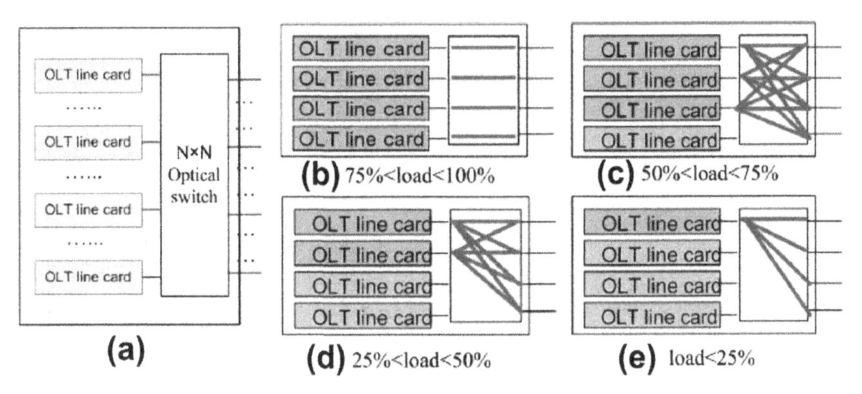

Figure 17.11 OLT with optical switches.

on only $\lceil \sum_{i=1}^{N} r^i(t)/c \rceil$ line cards can achieve relative energy savings as large as $1 - \frac{\lceil \sum_{i=1}^{N} r^i(t)/c \rceil}{L}$.

Figure 17.11b–e illustrates the configuration of switches for the case that one OLT chassis contains four OLT line cards. Define traffic load as $\sum_{i=1}^{N} \lceil r^i(t)/(LC) \rceil$, where $\lceil \sum_{i=1}^{N} r^i(t) \rceil$ is the total arrival traffic rates of all ONUs and $L * C$ is the capacity provisioned by all OLT line cards. By dynamically configuring switches, the number of power-on OLT line cards is reduced from four to x when the traffic load falls between $x/4$ and $(x+1)/4$. Thus, a significant amount of power can be potentially saved.

17.3.3.3 OLT with Cascaded 2×2 Switches

Another problem with optical switches is their high prices. The prices of optical switches vary from their manufacturing techniques. At present, there are generally four kinds of optical switches: opto-mechanical switches, micro-electro-mechanical system (MEMS), electro-optic switches, and semiconductor optical amplifier switches. Currently, the opto-mechanic switches are less expensive than the other three kinds. Simply because of their low prices, opto-mechanic switches are generally the adopted choices in designing energy-efficient OLT.

For opto-mechanic switches, an important constraint is their limited port counts. Popular sizes of opto-mechanic switches are 1×2 and 2×2. Considering the port count constraints, the cascaded 2×2 switches structure can be used to achieve the dynamic configuration of OLT. More specifically, to replace an $N \times N$ switch, the cascaded 2×2 switch contains $\log_2 N$ stages and $(N-1)$ 2×2 switches. Figure 17.12a illustrates the cascaded switches. In the switch, the kth stage contains 2^{k-1} switches.

Figure 17.12b shows a two-stage cascaded 2×2 switch to replace a 4×4 switch. As illustrated in Figure 17.12c–e, when the traffic load is greater than 50%, one PON system is connected with one OLT line card; when the traffic load is between 25% and 50%, two PON systems are connected with one OLT line card; when the traffic load is less than 25%, all PON systems are connected with a single OLT line card.

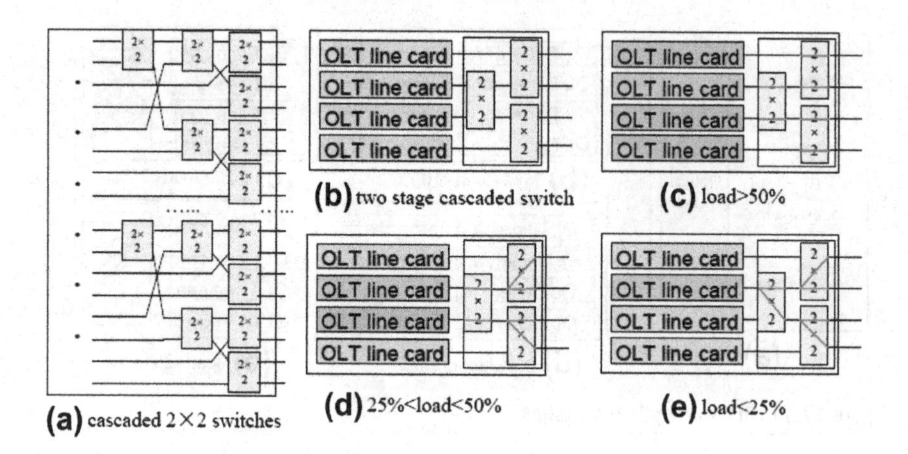

Figure 17.12 OLT with multistage cascaded switches.

Here, we analyze the saved energy of OLT equipped with cascaded switches. Assume the traffic is uniform among all ONUs. Then, when the traffic load is between 50% and 100%, all OLT line cards need to be powered on; when the traffic load is between 25% and 50%, half of the OLT line cards are powered on. Generally, when the traffic load is between $1/2^k$ and $1/2^{k+1}$, $1/2^k$ of the OLT line cards are powered on. Then, the saved energy equals $1 - 1/2^{\lfloor \log_2 1/\text{load} \rfloor}$, where load $= \sum_{i=1}^{N} \lceil r^i(t)/(LC) \rceil$ as defined before. As compared to the OLT with an $N \times N$ switch, the OLT with cascaded 2×2 switches saves less energy.

17.4 CONCLUSIONS

This chapter analyzes and compares energy consumptions of existing broadband wireless and wireline access networks, and summarizes and discusses various techniques that have been proposed to reduce energy consumption of access networks. For wireless access networks, we have discussed power saving protocols, heterogeneous network deployments, and utilizing renewable energy. In addition, we have also presented cooperative networking, which is one of the enabling techniques toward green networks. For wireline access networks, we have compared energy consumption of various broadband access solutions, and discussed several greening techniques such as enabling sleep mode and adapting line rates.

ACKNOWLEDGMENT

This work was supported in part by US National Science Foundation, Division of Computer and Network Systems.

REFERENCES

[1] Y. Xiao, X. Du, J. Zhang, F. Hu, S. Guizani, Internet Protocol Television (IPTV): the killer application for the next-generation Internet, IEEE Communications Magazine 45 (11) (2007) 126–134.
[2] M. Ergen, Mobile Broadband: Including WiMAX and LTE, Springer Verlag.
[3] J. Zhang, N. Ansari, On OFDMA resource allocation and wavelength assignment in OFDMA-based WDM radio-over-fiber picocellular systems, IEEE Journal on Selected Areas in Communications 29 (6) (2011) 1273–1283.
[4] F. Khan, LTE for 4G Mobile Broadband: Air Interface Technologies and Performance, Cambridge University Press, New York, 2009
[5] G. Kramer, G. Pesavento, Ethernet Passive Optical Network (EPON): building a next-generation optical access network, IEEE Communications Magazine 40 (2002) 66–73.
[6] M. Sauer, A. Kobyakov, J. George, Radio over fiber for picocellular network architectures, IEEE/OSA Journal of Lightwave Technology 25 (2007) 3301–3320.
[7] M. Ali, G. Ellinas, H. Erkan, A. Hadjiantonis, R. Dorsinville, On the vision of complete fixed-mobile convergence, IEEE/OSA Journal of Lightwave Technology 28 (16) (2010) 2343–2357.
[8] M. Gupta, S. Singh, Greening of the Internet, Conference on Applications Technologies Architectures and Protocols for Computer Communications, ACM., pp. 19–26.
[9] C. Bianco, F. Cucchietti, G. Griffa, Energy consumption trends in the next generation access network: a telco perspective, in: Proceedings of the 29th International Telecommunications Energy Conference (INTELEC), 2007, pp. 737–742.
[10] C. Lange, M. Braune, N. Gieschen, On the energy consumption of FTTB and FTTH access networks, in: National Fiber Optic Engineers Conference, 2008.
[11] J. Yeh, J. Chen, C. Lee, Comparative analysis of energy-saving techniques in 3GPP and 3GPP2 systems, IEEE Transactions on Vehicular Technology 58 (1) (2009) 432–448.
[12] S. Vadgama, Trends in green wireless access, Fujitsu Scientific and Technical Journal 45 (4) (2009) 404–408.
[13] M. Pickavet et al., Worldwide energy needs for ICT: the rise of power-aware networking, Second International Symposium on Advanced Networks and Telecommunication Systems, IEEE., pp. 1–3.
[14] J. Baliga, R. Ayre, W. Sorin, K. Hinton, R. Tucker, Energy consumption in access networks, in: Optical Fiber Communication Conference and Exposition and The National Fiber Optic Engineers Conference, 2008.
[15] M. Etoh, T. Ohya, Y. Nakayama, Energy consumption issues on mobile network systems, in Applications and the Internet, 2008. International Symposium on SAINT 2008, July 2008, pp. 365–368.
[16] EARTH: Energy Aware Radio and neTwork tecHnologies. <https://www.ict-earth.eu/>.
[17] TREND: Towards Real Energy-efficient Network Design. <http://www.fp7-trend.eu/>.
[18] C2POWER: Cognitive Radio and Cooperative Strategies for POWER saving in multi-standard wireless devices. <http://www.ict-c2power.eu/>.
[19] C. Han et al., Green radio: radio techniques to enable energy-efficient wireless networks, IEEE Communications Magazine 49 (6) (2011) 46–54.
[20] Community Power Using Mobile to Extend the Grid. <http://www.gsmworld.com/documents/gpfm_communitypower11whitepaperlores.pdf>.
[21] 3GPP R3-100162: Overview to LTE Energy Saving Solutions to Cell Switch Off/On, 3GPP RAN3 Meeting, Valencia, Spain, January 2010.

[22] M.A. Marsan et al., Optimal energy savings in cellular access networks, ICC Workshops 2009, IEEE International Conference on Communications Workshops, June 14–18, 2009, pp. 1–5.
[23] S. Zhou, J. Gong, Z. Yang, Z. Niu, P. Yang, Green mobile access network with dynamic base station energy saving, in: Proceedings of ACM MobiCom, Beijing, China, September 20–25, 2009, pp. 1–3.
[24] Eunsung Oh, B. Krishnamachari, Energy savings through dynamic base station switching in cellular wireless access networks, 2010 IEEE Global Telecommunications Conference on GLOBECOM 2010, December 6–10, 2010, pp. 1–5.
[25] K. Samdanis et al., Self-organized energy efficient cellular networks, in: 2010 IEEE 21st International Symposium on Personal Indoor and Mobile Radio Communications (PIMRC), September 26–30, 2010, pp. 1665–1670.
[26] S. Bhaumik, G. Narlikar, S. Chattopadhyay, S. Kanugovi, Breathe to stay cool: adjusting cell sizes to reduce energy consumption, Green Networking '10: Proceedings of the First ACM SIGCOMM Workshop on Green Networking, ACM, New York, NY, USA, 2010, pp. 41–46.
[27] H. Claussen et al., Effects of joint macrocell and residential picocell deployment on network energy efficiency, in: IEEE 19th International Symposium on Personal, Indoor and Mobile Radio Communications, 2008, PIMRC 2008, September 15–18, 2008, pp. 1–6.
[28] A.J. Fehske et al., Energy efficiency improvements through micro sites in cellular mobile radio networks, GLOBECOM Workshops, 2009 IEEE, November 30–December 4, 2009, pp. 1–5.
[29] K. Samdanis et al., Dynamic energy-aware network re-configuration for cellular urban infrastructures, in: GLOBECOM Workshops (GC Wkshps), 2010 IEEE, December 6–10, 2010, pp. 1448–1452.
[30] Ericsson Inc., Sustainable energy use in mobile communications, White Paper, August 2007.
[31] Juejia Zhou, Mingju Li, Liu Liu, Xiaoming She, Lan Chen, Energy source aware target cell selection and coverage optimization for power saving in cellular networks, in: Proceedings of the 2010 IEEE/ACM International Conference on Green Computing and Communications and International Conference on Cyber, Physical and Social Computing (GREENCOM-CPSCOM '10).
[32] H. Luo, R. Ramjee, P. Sinha, L.E. Li, S. Lu, UCAN: a unified cellular and ad-hoc network architecture, in: Proceedings of ACM MobiCom, 2003.
[33] T. Han, N. Ansari, Energy efficient wireless multicasting, IEEE Communications Letters 15 (6) (2011) 620–622.
[34] O. Arnold, F. Richter, G. Fettweis, O. Blume, Power consumption modeling of different base station types in heterogeneous cellular networks, in: Proceedings of IEEE Future Network and Mobile Summit 2010, Florence, Italy, June 2010.
[35] J. Zhang, N. Ansari, Y. Luo, F. Effenberger, F. Ye, Next-generation PONs: a performance investigation of candidate architectures for next-generation access stage 1, IEEE Communications Magazine 47 (8) (2009) 49–57.
[36] J. Zhang, N. Ansari, On the capacity of WDM passive optical networks (PONs), IEEE Transactions on Communications 59 (2) (2011) 552–559.
[37] J. Zhang, N. Ansari, On optimizing the tradeoff between the AWG cost and fiber cost in deploying WDM PONs, IEEE/OSA Journal of Optical Communications and Networking 1 (5) (2009) 352–365.

[38] F. Effenberger et al., An introduction to PON technologies, IEEE Communications Magazine 45 (3) (2007) S17–S25.
[39] J. Baliga, R. Ayre, K. Hinton, R.S. Tucker, Energy consumption in wired and wireless access networks, IEEE Communications Magazine 49 (6) (2011) 70–77.
[40] S. Wong et al., Sleep Mode for Energy Saving PONs: Advantages and Drawbacks, in: 2009 IEEE GLOBECOM Workshops, 2009, pp. 1–6.
[41] Y. Yan et al., Energy management mechanism for Ethernet passive optical networks (EPONs), Proceedings of IEEE International Conference on Communications (ICC), IEEE.
[42] S. Wong, S. Yen, P. Afshar, S. Yamashita, L. Kazovsky, Demonstration of energy conserving TDM-PON with sleep mode ONU using fast clock recovery circuit, in: Proceedings of Optical Fiber Communication Conference, 2010.
[43] J. Zhang, N. Ansari, Towards energy-efficient 1G-EPON and 10G-EPON with sleep-aware MAC control and scheduling, IEEE Communications Magazine 49 (2) (2011) s33–s38.
[44] J. Mandin, EPON power saving via sleep mode, in: IEEE P802. 3av 10GEPON Task Force Meeting, 2008. <www.ieee802.org/3/av/public/2008-09/3av-0809-mandin-4.pdf>.
[45] R. Kubo, J. Kani, Y. Fujimoto, N. Yoshimoto, K. Kumozaki, Sleep and adaptive link rate control for power saving in 10G-EPON systems, in: Proceedings of IEEE Globecom, 2009.
[46] S. Lee, A. Chen, Design and analysis of a novel energy efficient Ethernet passive optical network, in: Proceedings of Ninth International Conference on Networking, 2010.
[47] E. Trojer, P. Eriksson, Power saving modes for GPON and VDSL, in: Proceedings of 13th European Conference on Networks and Optical Communication, Austria, June 30–July 3, 2008.
[48] J. Zhang, T. Wang, N. Ansari, Designing energy-efficient optical line terminal for TDM passive optical networks, in: SARNOFF Symposium, Princeton, NJ, USA, May 2011.
[49] J. Zhang, N. Ansari, Scheduling WDM EPON with non-zero laser tuning time, IEEE/ACM Transactions on Networking 19 (4) (2011) 1014–1027.

CHAPTER

Overview of Energy Saving Techniques for Mobile and Wireless Access Networks 18

Diogo Quintas[a], Oliver Holland[a], Hamid Aghvami[a], Hanna Bogucka[b]

[a]*Centre for Telecommunications Research, King's College London, Strand, London WC2R 2LS, UK,*

[b]*Chair of Wireless Communications, Poznań University of Technology, ul. Polanka 3,60-965 Poznań, Poland*

18.1 INTRODUCTION

Energy consumption in various aspects of everyday life is increasingly an issue. Numerous observations are mounting to show that the temperature of planet Earth is increasing [1], and evidence is perpetuating that this temperature increase is due to the emissions of greenhouse gases and the greenhouse effect [2, 3]. It is clear that the burning of fossil fuels, still by far the most dominant means of obtaining energy, must therefore be reduced, and other means for energy extraction such as the use of renewable resources will not be able to fill the void left by fossil fuels in terms of achievable rate of energy output, at least in the near future. Moreover, some alternative means of energy extraction that might realistically be able to fill the void, such as nuclear energy, have recently suffered a severe setback resulting from the Fukushima nuclear disaster and other scares worldwide. As a result there is a strong need to curb energy consumption levels across all walks of life.

Mobile and wireless communications are no different from other industries. For instance, the energy consumption related to one international operator's worldwide business is over 4 TWh per year, and the carbon emissions associated with the business is 2 metric tons per year [4]. Clearly, this must be reduced. Moreover, energy consumption reduction has other benefits aside from saving the environment; energy bills are increasing rapidly, and reducing energy consumption can result in significant financial savings for operators. Moreover, given that alternative energy sources will have to be deployed in the future, which are usually expensive in terms of the capacity expenditure to power capability ratio (or, indeed, the ratio of capacity expenditure to energy creation over the lifetime of these energy sources), the cost of energy seems only likely to continue increasing in the long term. Even the "cheaper" fossil fuel sources of energy are becoming more expensive to extract: humanity must look in more and more challenging places in order to find these fuels, as the supplies in the easy-to-reach locations are becoming expended, and must use more and more challenging means to extract them.

In light of the above observations, reduction in energy consumption is the only way forward. However, things are even more challenging than they might seem. There is a relentless drive to improve performance in mobile and wireless communications, and the pioneering systems such as LTE and LTE-Advanced strive, in the mobile/wireless field, towards perpetuating Gilder's Law: the prediction that communications capacity will double every 6 months on average. This increase in capacity is expensive, both financially and in terms of energy consumption. If we are going to realize such potential for communications in the future, then the energy consumption per bit of data transferred must be reduced much more radically than would otherwise be the case.

It is noted that the issues above are "real" motivations to reduce energy consumption. However, on top of these "real" motivations, there are also "artificial" motivations, often related to government and intergovernmental targets, for example. These targets manifest themselves to the end user of energy in a variety of ways, one example being the market in "carbon" that has been created, and the ability to "trade" carbon emissions quotas for financial gain or loss; such a vehicle is known as "carbon trading." Again you see here that there are benefits created for the mobile operator in being "green"; if the operator can reduce its energy consumption sufficiently, it will have surplus carbon "credits" that it can sell to others who need to use more energy and create more carbon. This, again, leads to a competitive edge for smart, "green," operators.

Reading this, it is clear that there are many reasons for mobile/wireless communications operators and other mobile/wireless communications service providers to want to save energy. However, saving energy requires an identification of the most promising prospective areas for saving, particularly given the wide range of disciplines that are involved in mobile/wireless communications provision. To this end, it is noted that by far the largest contributor to energy consumption in mobile communications, at least in terms of operational energy, is the wireless access network—in [4] it is reported that radio base stations were responsible for 224 GWh, while other network equipment only used 105 GWh in the financial year 2008/2009. This high consumption exists for several reasons, but it is particularly related to the complexity in running such access networks as well as the inefficiencies of them. This is often related to the push against the bounds of technology and the characteristics of such networks in terms of the need for transmission linearity and transmission power levels, for example. This chapter assesses the wide range of contributions to energy consumption in mobile and wireless communications, but particularly concentrates on those areas that are particularly outliers and are inefficient in operation. Moreover, it assesses the full repertoire of solutions to save energy in such communications, particularly concentrating on the "problem areas."

This chapter is structured as follows. In Section 18.2 the environmental impact of current and future equipment is described. Section 18.3 presents recent efforts in curbing the energy consumption of individual components in base stations. In Section 18.4 a system-wide perspective is presented along with key techniques and concepts to enable energy-efficient networks. Section 18.5 deals with the feasibility of renewable energy provision in cellular networks. Finally Section 18.6 concludes the chapter.

18.2 THE CARBON FOOTPRINT OF MOBILE AND WIRELESS NETWORKS

When assessing the environmental impact of mobile and wireless communications, the carbon footprint of all connected devices must be taken into account. It is useful to distinguish between the energy expended during the operational lifetime of networks and devices, known as operational energy, and the energy expended in constructing, installing, and disposing of each device. This distinction is important to identify high leverage actions that can be taken to reduce the environmental impact of wireless networks. As shown in Figure 18.1, the total carbon footprint per subscriber per year in a cellular network is driven by both the footprint of base stations and the mobile phone. Notice that the annualized embodied energy of a mobile phone (8.1 kg CO_2 per subscriber per year) is almost as much as the operational energy of base stations (9 kg CO_2 per subscriber per year). This observation justifies efforts to curb the energy consumption in the manufacture of mobile phones.

Although the embodied energy of base stations is relatively small, it is not negligible; considering the significant recent efforts in reducing the operational energy and the introduction of renewable energy sources, the embodied energy of base stations will become a critical aspect of green networks and cannot be neglected when planning for energy efficiency.

18.2.1 Operational Energy

As Figure 18.1 shows, the operational energy of base stations is the biggest contributor to the energy needs of current access networks. Several studies have been published concerning the power requirements of a base station's components.

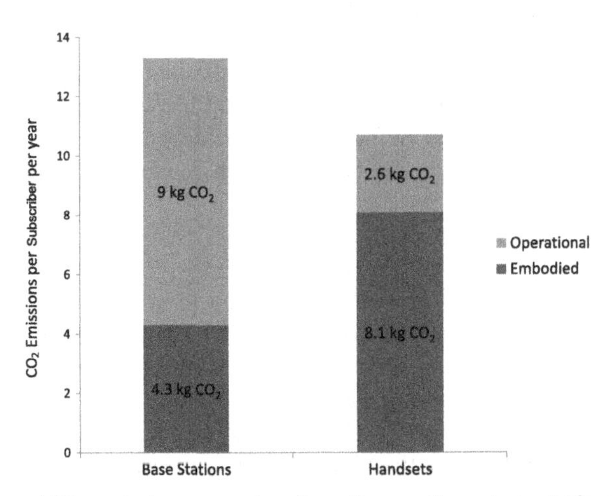

Figure 18.1 Annual CO_2 emissions per subscriber. *Source:* **Green base stations—how to minimize CO_2 emission in operator networks [5].**

Table 18.1 Power Models in the Literature (for each RF unit)

Reference	Macro BS			Micro BS		
	PA eff (ρ)	Fixed (υ)	Scalable w/load (δ)	PA eff (ρ)	Fixed (υ)	Scalable w/load (δ)
[11]	7.8	300	0	–	–	–
[12]	4	84	0.5	4	68	0.5
[13]	3.1	90	0	–	–	–
[10]	3.7	83	0	5.55	32	0.47

A cursory analysis of the related literature clearly reveals that the power amplifier (PA) is responsible for a large part of the power requirements in current equipment. Quoted values in the literature for the power consumption of the PA range from 13% to 80% [6–9] of the total power consumption. In delivering green networking the PA efficiency is critical. Typical legacy PAs exhibit efficiency rates of around 15–30% [7, 10, 11], that is, to deliver an output RF power of 40 W a PA will need an input power of 450 W once feeder losses are accounted for; most is then dissipated as heat, which makes energy-expensive cooling solutions a necessity. These cooling solutions typically account for some 10–50% [6–9] of the operational power consumption. Other components, such as power supply, baseband processors, rectifiers, and small signal transceivers, account for the rest of the power consumption; the output RF power has little bearing on these. It should be noted that there is a wide varying range of values found in the literature—this is partly due to the fact that the power consumption depends on the underlying technology and the time that the equipment is manufactured. Recent equipment has been demonstrated where the energy consumption is significantly smaller than the currently installed infrastructure; details on recent advances are given in Section 18.3.

The total power consumption of a base station has been widely modeled as a linear function of the average output RF power, depending on the efficiency of the PA and feeder losses, ρ, and a constant term υ, modeling a RF independent component:

$$P^{\text{in}} = \rho P^{tx} + \upsilon. \tag{18.1}$$

To capture the ability of power consumption to scale with the experienced load by switching off components or reducing the number of active processor cores (see Section 18.3), υ can be further divided into idle and active terms, i.e., $\upsilon = \upsilon(1-\delta) + l\upsilon\delta$, where l represents the experienced load as a fraction of the total, and δ is a parameter between 0 and 1, capturing the different technology capabilities. Table 18.1 shows the values of the different parameters in different sources. The quoted values imply a total operational power consumption for a three-sector base station with one antenna per sector, transmitting at 40 W, ranging from 645 W to 1.8 kW. The power consumption implied in references [10, 12, 13] is rather optimistic in terms of installed 3G and GSM equipment. A particularly inefficient base station is estimated to consume 3.8 kW in [7]. In [10] a three-sector GSM base station with 2 PAs per sector is said to consume around 1.4 kW, and a similar 3G base station is said to consume 1.5 kW. Comparing these values with current commercially available equipment, we find the values to be slightly pessimistic. In recent years significant progress has been reported by manufacturers and operators alike: it is reported that macro base stations can consume less than 500 W [4, 14].

It should be noted that the environmental impact of the operational energy largely depends on the impact of the energy grid itself. If the grid is supplied by mainly carbon-free energy, the CO_2 emissions related to operating the access network are vastly reduced. This is in contrast with the embodied environmental impact, where waste and fossil fuel consumption, for example due to transport, are matters of concern.

18.2.2 Embodied Energy

The authors in [15] estimate that the typical embodied energy in a cellular base station amounts to a total of 75 GJ. In comparison, a highly efficient 500 W base station is expected to consume around 157 GJ operating across its entire lifetime, based on a 10-year life expectancy. Semiconductor manufacture represents the highest contributor to the energy consumption of wireless devices, due to the complex process of silicon wafer manufacture [16]; in large macro-cell base stations it is responsible for 50% of the total embodied energy [15], while in mobile phones it contributes almost 60% [17]. Any sustained effort to reduce the embodied energy of devices must concentrate on the development of more efficient processes for semiconductor manufacturing. This issue permeates the entire ICT sector, and is not just a problem for cellular equipment. For example, 25% of the embodied energy of a desktop computer is estimated to be due to semiconductors [18].

While semiconductors remain a significant contributor to the embodied energy of wireless devices, other contributors should not be neglected. The supply chain, responsible for 11% and 16% of the embodied energy of base stations and mobile phones respectively, should be made more efficient—this is an issue across industries and not limited to the ICT sector. As electronic chip manufacture becomes more environmentally friendly, the transport of components will become a more critical factor in the future as a proportion of the total embodied energy consumption.

The environmental impact of raw materials extraction is significant due to the waste created, notwithstanding the high energy consumption. For mobile phones, Nokia estimates that 189 tons of waste is created in extracting and processing

Table 18.2 Embodied Energy Contribution

	Base Station [15] (%)	Mobile Phone [17] (%)
Electronic components (including cables, semiconductors, etc.)	58	66
Raw materials	17	13
Assembly	14	6
Supply chain	11	16

materials for each ton of final waste; in comparison, the manufacturing process is only responsible for 21 tons [17] (see Table 18.2).

18.2.3 Life Cycle Analysis

In evaluating equipment design and network planning, the embodied energy and life span must be taken into account if the right conclusions are to be drawn. The total embodied energy, where decommissioning is also taken into account, must be translated to an annualized energy consumption, i.e., the total embodied energy divided by the number of years the equipment is in use; the annualized embodied energy can then be added to the annual operational energy consumption. In this way, the trade-off between operational energy consumption (E_{op}), embodied energy consumption (E_{emb}), and life span (N), can be better understood. The yearly energy consumption can be expressed as

$$E_{\text{year}} = \frac{E_{\text{emb}}}{N} + E_{\text{op}}. \tag{18.2}$$

Moreover, if the analysis requires a more granular timescale, a simple scaling of the embodied energy term can be applied to the same effect. From the life cycle point of view, efforts to curb the operational consumption that increase the embodied energy, or reduce the life span, may result in a net loss in terms of the total energy consumption.

Embodied energy represents a fixed energy cost over the lifetime of the equipment. Naturally, as the life span of the equipment or its constituent components increases, economies of scale are possible. There are several factors that affect the life span of wireless equipment, of which technology advances are the most noteworthy. Current base stations and mobile phones are estimated to have a life span of 10 years, not accounting for replacements due to newer models being available. In the case of user equipment, the typical life span due to wear and tear is significantly larger than the replacement rate due to new features absent from older equipment [17]. High-end products tend to have a 1–2-year life cycle. This signifies that expanding the life span of a wireless unity would impair technology advancement and require an active change in consumer behavior: users actively delaying hardware upgrades. Nonetheless actions can be taken to expand the lifetime of some parts. It is noteworthy that the majority of the embodied energy of wireless devices is incorporated into their constituent parts. Using the data provided in Section 18.2.2 it can be seen that if 50% of the integrated circuits in a base station can be recycled/reused, the embodied energy can be reduced to 18.6 GJ. This is equivalent to a full year of operation of a 500 W base station.

Field programmable gate arrays (FPGAs) are a possible solution for the reuse of dedicated circuits. As new techniques are developed, these can be reconfigured to incorporate new functionalities. However, the embodied energy of FPGAs is typically higher than traditional circuits, as is their operational energy, due to the larger size of silicon needed to realize them [16]. Care must therefore be taken in adopting this solution. Nonetheless, the use of FPGAs allows upgrades to be made through a software download, reducing the need to construct new hardware. Furthermore, the flexibility that FPGAs afford network operators allows for more intelligent autonomous optimization of radio functionalities, enabling online self-optimization and self-adaptability, and reducing the need for hardware maintenance (cf. Section 18.4). As these circuitries typically draw more power during their lifetime, their applicability in battery powered devices, such as mobile phones and tablet computers, is limited. However, recent developments regarding the design of FPGAs are pointing towards an increase in their energy efficiency—this coupled with the ability of FPGAs to adapt their power consumption depending on their resource utilization might make them an interesting solution in the future.

Yet another solution would be the recycling of valuable raw materials, for which extraction has a significant environmental impact due to the waste created. Moreover, full integrated circuits such as processors, DRAM, etc. could be reused, notwithstanding the implied penalty of stifling technology advances. The creation of modular equipment, where part of the components of a device can be replaced by newer hardware as far as technology would allow, without the need for a full hardware replacement, could be an effective way of extending the lifetime of the components; this would require significant standardization effort of hardware components and equipment design, as parts would have to be interoperable and interchangeable. Another option is the reuse of parts installed in high-end equipment by later generations of low-end equipment; high-end users would still benefit from technology advances, while the embodied energy of low-end equipment would be reduced. Through this approach, the same processor, for example, might start life in a cutting-edge device, then two or three years later when that device is discarded might be used in a medium capability device, then two or three years after that when that next device is discarded might be used in a low-end device. This way, the same energy-expensive silicon chip could be used in three generations of end-user devices (albeit aimed at different market segments), thereby reducing the embodied energy consumption associated with the mobile communications market.

18.3 EFFICIENCY IMPROVEMENTS FOR BASE STATION HARDWARE

Recent years have seen noticeable improvements in hardware efficiency of cellular base stations. Improvements in hardware are the most effective way of reducing operational energy consumption. However, although innovative chip design has lowered such operational energy consumption, the authors have found no study on the impact on the embodied energy of these new designs nor on their lifetime. As noted in the previous section, it might be the case that these improvements are merely shifting the energy consumption from the operational phase to the manufacturing phase, risking a net increase, or at least being less effective than claimed by design proponents.

Advances in PA design have been made recently that have the potential to be used in current wireless access systems. The introduction of new semiconductor materials, the use of digital predistortion [19], envelope tracking techniques [20, 21], and the resurgence of the Doherty amplifier [22] are promising concepts. These concepts are expected to yield efficiency rates of up to 50% when transmitting at maximum power [6]. Commercially available PAs have been announced that are able to achieve 45% efficiency levels [23]—this is a substantial improvement on the reported levels of currently installed equipment. The overall efficiency of the RF unit is also impacted by the location of its equipment; with mast mounted PAs the loss due to feeder cables is eliminated, further reducing the dissipated power by the PA by allowing it to be tuned to a lower necessary power output. These improvements have a cascade effect. Through these efficiency improvements on the power dissipation of PAs, the need for cooling has been largely removed as less heat is dissipated by these parts. This eliminates at a stroke 25% of the power consumption of base stations.

Baseband processing is a nonnegligible factor in the total power consumption. For a processor with a clock frequency f the power consumption is approximated by [24, 25]

$$P = \alpha C V^2 f, \tag{18.3}$$

where V is the voltage, C is a constant depending on the processor architecture, and α is the probability of a gate being switched (roughly 0.5). The clock frequency can be dynamically adapted, and any reduction in f effects a proportional reduction in the power consumption. Furthermore, reducing the clock frequency allows for a reduction in the circuit voltage; since power consumption grows quadratically with the voltage, considerable power savings can be achieved through this approach. It is noteworthy that these techniques can have a positive effect in the lifespan of circuitry. Reducing the dissipated temperature can increase the lifespan of some components (not to mention curtailing the need for cooling solutions).

Multicore processors provide flexibility in managing power consumption. The first benefit of multicore architecture is immediately apparent due to the quadratic relationship between the power consumption and voltage, whereby four cores operating at 1 GHz can consume less power than one core at 4 GHz. Secondly, multicore architectures allows processing jobs to be carefully distributed between cores, and even the shutdown of some cores when the traffic load is low. It should be noted that this discussion relates to dynamic power only. A fixed power consumption is increasingly relevant due to an increase in current leakages due to gate length being reduced in higher density integrated circuitry [25].

Another solution for improved energy efficiency is the use of base station sleep modes. Depending on the level of traffic load, some base station components can be switched off. Manufacturers have introduced some sleep mode capabilities already [26–28], allowing for savings ranging from 15% to 30% depending on the traffic load. It is noteworthy that these capabilities can be offered via a software update, hence legacy equipment can also benefit from these techniques.

Despite recent advances there are fundamental limits to the efficiency of hardware. To further improve the energy efficiency of access networks, a system-wide perspective is needed.

18.4 ENERGY-AWARE NETWORK DIMENSIONING

The energy consumption of the access network depends on its size and required capacity. Careful planning is required to optimize the network. Herein we provide an overview of the main issues to consider when comparing different access architectures and the roles/capabilities of different technologies to provide an *efficient* network.

18.4.1 Fundamental Trade-offs

Ever since Claude Shannon's seminal work, it has been known that there is a fundamental limit to achievable spectral efficiency for a given received signal-to-noise ratio. Although freedom has existed in the past through technological advances bringing us closer to the Shannon bound, this fundamental result creates a stark trade-off between coverage, capacity, and energy consumption. Assuming a simple propagation model, the necessary transmit power P^{tx} to sustain a given capacity r (in b/s/Hz) for a coverage distance d is given by

$$P^{tx} = (2^r - 1)\sigma \left(\frac{4\pi f_c d}{c}\right)^{\alpha}, \tag{18.4}$$

where f_c is the carrier frequency, c is the speed of light, σ is the noise power at the receiver, and α is the so-called path loss exponent, which typically varies between 2 and 5 depending on propagation conditions. This expression readily illustrates the fundamental issues at play. The power consumption of a (single) base station grows exponentially with capacity, and polynomially (with degree α) with the distance covered. A naïve reading of this result leads to the (wrong) conclusion that to increase energy efficiency one just has to sufficiently increase the base station density, thus reducing the covered distance of each and consequently reducing the necessary power consumption. Recall Eq. (18.1) and accompanying discussion. The total operational power consumption of a base station does not solely depend on the output RF power. In terms of the total network operational power consumption increasing the base station density amounts to a polynomial reduction in the power consumption related to the transmit power, as each base station covers a smaller distance, but a linear increase in the overhead power consumption and total embodied energy of the system, due to the increase in the number of base stations. This trade-off between RF power consumption and overhead power consumption has been illustrated in several works dealing with finding the optimal cell radius [10, 15, 29–31]. Moreover, the embodied energy has been largely ignored by the former, with the notable exception of [15], where the authors clearly show that accounting for embodied energy leads to a significantly higher cell radius being optimal than when it is not taken into account. They remark that current efforts to reduce the operational energy consumption of base stations are increasing the embodied energy that goes into manufacturing them. With a lowering of the operational energy consumption and an increase of embodied energy, the latter will become the determining factor when planning for energy efficiency.

For a cell coverage area with maximal distance d, $C(d)$, and a total area, A, the total system energy consumption is given by

$$E_{sys} = \frac{A}{C(d)}\left(E_{tx}(d) + E_{oh} + \frac{E_{emb}}{N}\right). \tag{18.5}$$

For hexagonal cells, coverage area is given by $C(d) = \frac{3\sqrt{3}}{2}d^2$. Note that the number of base stations and the energy consumption related to transmission power both change with the distance covered. Figure 18.2 shows the energy consumption curves required to cover a 20 km^2 area with a 2 b/s/Hz spectral efficiency at the cell edge, as a function

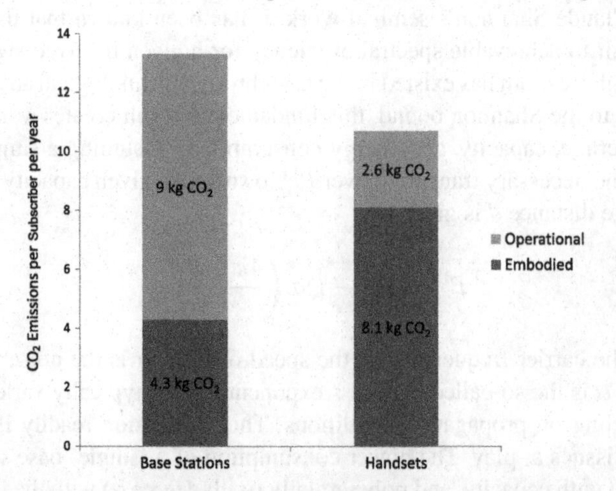

Figure 18.2 Illustrating the impact of embodied energy on the optimal coverage distance.

of the cell radius. If only the operational power consumption is considered, the optimal distance is underestimated, leading to a suboptimal performance once the entire energy (embodied plus operational) is taken into account. The reader will have noticed that in Eq. (18.5), the overhead and embodied energy is a constant that does not depend on coverage distance. However, this observation is not strictly correct. As cell size is decreased, deployed radio equipment also decreases, i.e., a macro base station covering between 1 and 2 km radius requires more powerful amplifiers, greater processing capability, higher antennas, etc., when compared with a micro-cell base station covering a much smaller range. Operational energy consumption depends on the size of the chips being used, hence smaller radios will consume less power in their lifetime. The manufacturing of small base station equipment is in general more energy intensive than the construction of large macro sites and the smaller equipment has a shorter lifetime [15]—increasing the annualized embodied energy consumption. The authors in [10, 29, 30] have studied the impact of different equipment types (macro and micro base stations) on operational energy consumption. Similar behavior to that shown in Figure 18.2 is observed, comparing just macro sites, just micro sites, and a combination of the two.

The required capacity density (b/s/Hz/km^2) and propagation conditions are also critical factors in energy efficient planning. If the region under consideration is sparsely populated, a high capacity density is not necessary—allowing for low levels of area spectral efficiency. This has two complementary effects: the required number of base stations is reduced, and/or the RF power levels can be lowered. In contrast, in densely populated areas, capacity density must be high—requiring a higher number of base stations and/or higher RF power levels. Note that the maximum RF power is limited in practice due to regulatory constraints and hardware limitations; this limitation implies that to increase the capacity density beyond a reasonably low threshold, an increase in the number of base stations must occur, by necessity. These limits are much lower on the terminal side, considerably reducing the coverage distance in the uplink—this is particularly true in difficult propagation environments. In hilly terrains or densely built areas, the RF power attenuation is higher, requiring a higher output at the transmitter; due to limitations on the maximum allowed output level this necessitates a higher number of base stations being deployed. In flat terrains with no obstructions, capacity is limited by the maximum spectral efficiency that the technology can support. We can therefore distinguish four scenarios: densely populated with poor propagation, sparsely populated with poor propagation, densely populated with good propagation, and sparsely populated with good propagation. These can be characterized by their capacity density requirements, the factors limiting capacity, and the number of base stations that need to be deployed. Of these four scenarios, two are particularly relevant: densely populated with poor propagation, and sparsely populated with good propagation, i.e., urban deployment versus rural deployment.

Assuming a uniform distribution of users, capacity density is intimately related to the average spectral efficiency in each cell and the number of cells. The capacity density χ is related to the number of cells by

$$\chi = f_{\mathrm{r}} N_{\mathrm{s}} N_{\mathrm{c}} \eta, \tag{18.6}$$

where η is the average spectral efficiency at each user location, N_s is the number of sectors per cell, f_r is the reuse factor, and N_c is the cell density. If the number of cells is increased, the cell radius decreases and a higher signal power is received, increasing the average spectral efficiency according to

$$\frac{1}{\sqrt{1/2\pi N_c}} \int_0^{\sqrt{1/2\pi N_c}} \log_2\left(1 + \frac{p^{tx}}{\sigma}\left(\frac{4\pi f_c r}{c}\right)^{-\alpha}\right) dr \tag{18.7}$$

The annualized energy consumption density (GJ/year/km^2), can readily be computed for each cell density with knowledge of operational power consumption and embodied energy. From Eqs. (18.6) and (18.7), the cell density is linearly proportional to the required capacity density. This approximation is only valid when interference at the cell edge does not increase substantially as the number of cells increases: in general, this implies that the output power for small sites must be scaled down accordingly. It is nevertheless the case that a high capacity density requires a high base station density, hence a high energy consumption density.

In urban deployments, due to the poor propagation and high capacity density required, the system is built with a high number of small base stations, each with a relatively small coverage area. Due to the large number of small base stations where the operational power does not exceed several tens of watts in the worst case, the embodied energy is a dominant system-wide issue, and particular attention must be paid to it in order to achieve energy efficiency. In contrast, for rural scenarios, few base stations are deployed. These are usually large macro base station mounted on high masts with cooling required. This implies a high operational power requirement, and here the embodied energy does not dominate.

It is well known that traffic intensity in cellular networks is time varying [32, 33]; throughout a typical day the required capacity density can be substantially below the installed capacity. This creates the possibility to adapt the offered capacity density depending on the time of day, allowing base stations (or sectors) to be switched off and their coverage area being taken by others. (The authors in [34, 35] have independently proposed to switching off base stations at lower utilization periods, addressing the total system operational energy consumption increasing the relative contribution of the embodied energy underlining the necessity to adequately revise manufacturing of base stations.) The authors in [35] report operational energy savings of more than 50% when compared to the default nonadaptive architecture, for both WiMAX networks and UMTS networks.

18.4.2 Multihop networks

When considering multihop networks there are two deployment paradigms, the range expanding fixed relay and mobile ad hoc networks. Access networks equipped with fixed relays have been proposed as a way of increasing coverage and capacity at a low implementation cost [36]. Mobile ad hoc networks present considerable challenges, both technical (resource management, routing, and security) and business (mainly pricing) in nature. Nonetheless, the theoretical advantages are clear: network capacity density can be expanded with little environmental impact. Mobile ad hoc networks require fewer access points as the infrastructure is provided by user equipment that is present anyway.

Few studies on the energy efficiency of fixed relay networks have been published. By improving the capacity density of the system, relay networks can reduce the number of base stations or their RF power requirements. On the other hand, relays might in themselves have relatively high operational energy requirements and/or high embodied energy consumption. The authors in [37] provide an analysis of the *area power consumption* of relay networks assuming that a relay's power consumption is just a fraction of the operational energy of a base station. They show that a significant reduction in the power requirements per km^2 are possible. However, they do not consider embodied energy in their study. The embodied energy in relays can be the critical aspect of their environmental impact. As relays are expected to be low power consumption access units—a few tens of watts—the energy consumption profile could be not much more than a mobile phone, where most of the CO_2 emission is due to manufacture. It is important to consider the energy costs for which relays are advantageous in terms of the total energy consumption to assess environmental benefits. Assuming a rather simple relay deployment, where relays are deployed equally spaced and close to the cell edge, the yearly energy savings can be computed for a given required capacity density and the relay energy consumption per year. Extrapolating on the economical analysis of relays [38], assuming that economical cost is proportional to the environment cost (which it should be noted is not the case in general), we find that a greener network can be achieved if the annualized energy consumption does not exceed but a fraction of that for macro base stations. In the scenarios evaluated in [38], the relative cost of relays has to be as low as 3–7% depending on the desired capacity density.

The attractiveness of fixed relays for cellular operators is mainly based on the fact that the relay backhaul is provided by the already installed access infrastructure. This has implications on the energy consumption related to the installation of relays and on the operational power consumption of the core network. Providing the necessary wired backhaul could be an energy expensive operation: from the manufacture of cables (and raw material extraction), to laying out the cables that can involve road work not only causing inconvenience to local residents and businesses, but also impacting CO_2 emissions (for example from the machinery needed to dig up the road).

As in the hierarchical architecture proposed in [34, 35], for single hop cellular networks where a central cell takes the coverage area of adjacent cells to allow them to switch off in periods of low utilization, similar concepts can be proposed to switch off relay nodes. The authors in [39] have investigated the feasibility and potential savings of switching off relay nodes in periods of low utilization. At night when the network is operating under capacity the authors propose all relays being disconnected and base stations replacing the coverage of relays. In [40], dynamic policies based on utilization statistics for each relay have been proposed that allow each relay to switch off when few users are present; this allows the system to save 5% of the daily energy consumption of relay networks with no impact on the system performance from the point of view of the QoS experienced by users. The ability to adapt the cellular architecture will be a critical aspect of future networks; having unutilized resources for large periods of the day is inefficient both from the point of view of operational expenditure and the energy consumption.

Delay tolerant services (DTS) are ideally suited for mobile ad hoc networks. Information centric networking, where users access relevant content rather than a particular host, is a relevant application of mobile ad hoc networks [41]; services such as RSS feeds and over-the-air software updates are application examples of services that can leverage this concept. In [41], a networking concept is proposed where users share content through sporadic contact—some users close to the access point download the content and as they travel in space the content is broadcasted to other interested users. Using mobility and content availability models the authors showed that a fast dissemination of content is possible, even when user contacts are short lived in time. A similar concept is proposed in [42] where messages are delayed as the user is traveling towards the base station or, if the user is not traveling towards the base station, messages can be forwarded to users that are doing so. The two approaches offer different capabilities; while the former is ad hoc in nature relying on self-organization the latter allows a tighter control by the access infrastructure and the offered delay can be adapted depending on the application. It should be noted that in both approaches communication between users (and indeed access points) occurs at close range, allowing for low RF powers to be used.

By shifting some of the traffic to the mobile ad hoc network the main access infrastructure needs not to offer as much capacity. As already seen in the previous section, a reduction in the offered capacity allows the base station density to decrease, reducing the network's environmental impact. The impact of these techniques depends on the proportion of services that are truly delay insensitive—if there is a small contribution of this type of traffic to the total traffic, there is little gain from shifting this traffic. Of course the time and spatial distribution of DTS traffic is an important aspect. If the peak DTS traffic time occurs when the network is not congested, for example during night time, this has little impact on the necessary infrastructure. Similarly, if DTS users are concentrated in space around a limited area, it will not allow for a dramatic reengineering of the cellular network's topology. Also understanding the mobility patterns of users is key to leveraging these mobile ad hoc networks.

18.4.3 Dynamic Spectrum and Traffic Load Management

Another interesting way to save energy for mobile and wireless networks, particularly radio access networks, is to intelligently use the available radio spectrum bands and the range of radio equipment that is already deployed and operational in those bands. There are a number of schemes that can be utilized here [43, 44]. The first we mention is the opportunistic reallocation of traffic loads between bands and/or equipment to be able to switch off radio network equipment when possible. This might include, for example,

1. Moving users or traffic from one band to another to be able to switch off all radio equipment in the band that the users originated in.
2. Moving users or traffic to be able to adjust sectorisation patterns allowing the switching off of some sectors (again, radio network equipment).
3. Moving users or traffic between bands to allow subsets of cells to be switched off.
4. In a single-band case, moving users or traffic between cells to allow the dynamic powering down of some cells while still maintaining coverage through raising the power level in the remaining cells.

Justifying these solutions through a simplification, the number of network transmitters/receivers that must be active in a given area should be roughly proportional to the amount of traffic that must be carried in that area. Likewise, the amount of spectrum used should be roughly proportional to the number of network transmitters/receivers that are active, given each transmitter/receiver using the same spectrum bandwidth, in real terms or on average. Conventionally however, radio communications networks have been greatly overprovisioned: they have been built to be able to withstand the worst-case heaviest loading on the network in the area, even though for vast amounts of time the number or users or the amount of traffic carried is only a small proportion of that overprovisioned capacity. To exemplify this, almost all areas experience a very significant drop in the offered traffic load during the night, and additionally, many areas will experience a drop in offered load at other times. Business and industrial areas, for example, will have only a small proportion of the peak traffic load offered to them during vacation times and on weekends. Networks in such areas at such times can be dynamically reconfigured to be able to switch off very significant proportions of base stations, across multiple bands, treating the range of radio equipment and available spectrum bands together in the same optimization, while still serving traffic sufficiently by maintaining the required ratio of base stations to offered traffic load in the area.

For instance, through this approach, lower-frequency bands that are available to the network might remain on while equipment at higher frequencies is switched off, therefore allowing some base stations at the lower frequency to also be switched off while still maintaining coverage due to the better propagation at the lower frequency. Such a solution could be useful in rural areas for example, where base stations must already push against the limit of their allowed/available transmission power to be able to cover all areas sufficiently with a sufficient signal ratio above the noise floor. In more densely populated cases such as urban areas much of the radio network equipment will already be operating well within the bounds of its allowable/available power, as base stations will be more tightly packed and all base stations in the area will have cooperatively lowered their powers to remain at a sufficient level above the noise floor, but no higher. These base stations will not simply waste power through causing unnecessary interference to neighboring base stations by trying to selfishly out-transmit them (note that the neighboring base stations would react in the same way to maintain their SINRs, therefore also wasting power). This realization that such base stations are often operating below their power budget gives scope for some base stations to be switched off while others remain on, with coverage still being provided by upping the powers of the remaining base stations. The capability to use such solutions, however, greatly depends on the types of base stations deployed; in some cases base stations will deliberately be deployed that have a low transmission power capability, therefore having no scope to increase transmission power. An interesting case is where the coverage area of a large macro-cell overlaps with many micro and/or pico cell. For low traffic conditions, the macro-cell can cope with the entire area's traffic, rendering the smaller cells superfluous—these can then be switched off.

It is noted that the above-mentioned solutions save real power, directly from the mains, whereas many other solutions will only save transmission power, which may have little effect on the real from-the-mains power consumption. Moreover, it is noted that although some of the concepts mentioned above might imply an increased transmission power for some base stations to maintain coverage, in reality, referring to the models in Section 18.2 of this chapter, there is a limited effect of such increases on the actual from-the-mains power consumption for these base stations, particularly for macro-cell base stations. Therefore, dynamically optimizing the system to minimize the amount of active radio equipment (number of active radio "chains"), as might be done through the above-mentioned solutions, is the overriding concern.

Other solutions within the scope of this section might revolve around the optimization of the spectrum choice dependent on the propagation characteristics of that spectrum and the local environment, as well as the density of traffic that must be carried. As observed in Section 18.4, traffic density varies dynamically in the area, hence the necessary cell density can be made to vary through cells being switched off dynamically. This can lead to the opportunistic use of more appropriate spectrum being beneficial to minimize interference between cells and/or to reduce necessary transmission power. A further option includes the opportunistic use of more spectrum bandwidth to carry the traffic, either through the opportunistic use of contiguous spectrum or perhaps through opportunistic aggregation of bands; another option is better "balancing" between bands of the ratio of offered traffic to utilized spectrum to carry that traffic, achieved by moving users and traffic load between bands. Both of these solutions can be shown to save necessary transmission power in terms of pure capacity analysis, but it is uncertain whether these transmission power savings can translate into from-the-mains power savings once the implications of deployment on real hardware are considered.

18.4.4 Link Efficiency Improvements

Notwithstanding the fundamental trade-offs already explored, improvements in link efficiency are key to promoting green communications insofar that resources are better exploited. Increased efficiency has translated into an increase in b/s/Hz, reducing the gap between achievable spectral efficiency and the Shannon bound.

In terms of the previous discussion, this can be roughly translated into less base stations being needed to offer a given capacity density, or a reduction in the RF power being needed hence a reduction in the system-wide consumption. However, as noted by the authors in [6], recent advances have increased the computational complexity in processing baseband signals; furthermore, they typically require extra overhead in terms of reference signals and channel state feedback. These two drawbacks curtail the benefits in terms of energy efficiency of improved transmission capabilities, because of the increase in the necessary constant power consumption (note that reference signals are transmitted with a high RF power in order to ensure proper channel estimation).

As complex computations are needed for advanced transmission capabilities (the computation of Fourier transforms in OFDMA-based systems, for example), clock frequency must be increased. As discussed in Section 18.3, power consumption of processing grows linearly with clock frequency and, to support this, voltage must also increase, yielding a quadratic increase in the power consumption. In small cells, where the total power consumption is less dependent on RF power consumption and a greater proportion is directly proportional to baseband processing, this could have a dramatic effect on the total operational power consumption. Higher voltages also impact the lifetime of the processor chip—affecting the annualized total energy consumption.

OFDMA systems need accurate channel state information to operate effectively (i.e., to perform equalization and to take advantage of frequency and time diversity), leading to the necessity of spreading reference symbols throughout the frequency and time resource space. This can account for 6% of the total available resources in single antenna systems [45]. This overhead increases linearly with the number of antennas at the base station. In current systems, for Example 3GPP's LTE, reference symbols are transmitted regardless of the amount of information that is carried, leading to a poor performance in terms of energy efficiency due to the high power required for these symbols. Switching some antennas and reducing the amount of symbols needed by switching off some subcarriers in periods of low utilization might help alleviate these problems.

18.5 RENEWABLE SOURCES OF ENERGY

One attractive way of providing green communications is the usage of renewable energy sources, such as wind and solar power. For mobile operators, this could be effected by collocating energy production apparatus with the radio access equipment such as base stations and fixed relays [26]. By collocating energy production with the point of delivery, losses that occur in the grid can be bypassed and, in areas where grid access is sparse and unreliable, cost-effective deployment and power provision can be assured. However, widespread adoption of wind and solar power generation is not without its problems. The energy yield of each source is highly variable, depending on the time of year and weather conditions

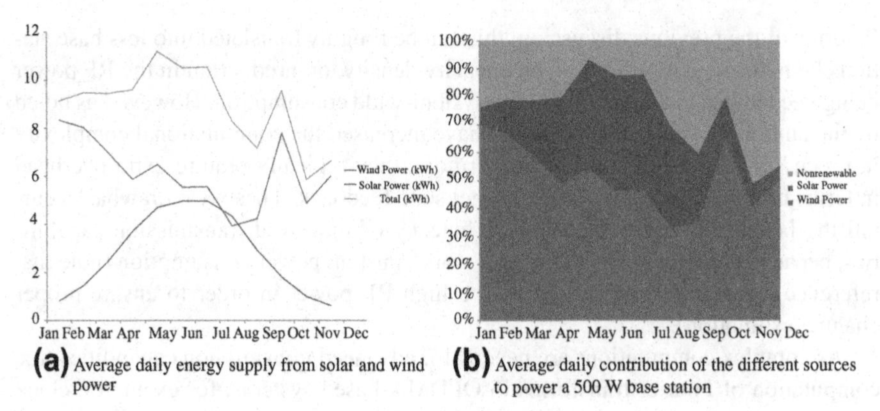

Figure 18.3 Average daily energy supply from solar and wind power in London, United Kingdom. *Source*: **Average insolation levels were obtained in [46] and average wind speeds from [47].**

(see Figure 18.3). Other renewable sources have been proposed as more reliable alternatives [26], namely fuel cells and biofuels; despite the reliability of these, questions about their usability remain: fuel cells have yet to be proven as a cost-effective alternative, and the supply and storage of fuel (e.g., hydrogen) for them can be an issue; biofuels have raised concerns about an increase in the amount of arable land that is devoted to the production of such fuels.

Different equipment types have different power consumption levels, and present different challenges in the installment of solar panels and wind turbines due to their location. Table 18.3 summarizes these challenges.

Table 18.3 Feasibility of Solar and Wind power for Different Types of Equipment

	Macro-cells	Micro cells	Femto cells	Fixed relays
Power requirements	~400–800 W	~50–100 W	~5–13 W	~5–50 W
Location	Rooftops with large antennas	Typically below roof level	Indoor	Traffic light/ lamp post level
Feasibility	Solar and wind combination needed. Solar only in very sunny locations, with backup power during winter.	Both solar and wind are problematic due to possible surrounding buildings blocking sun and wind	Just solar or wind power feasible by sharing the building's power provision	Possible surrounding buildings are a problem but if location is favorable, either wind or solar power are possible

18.5.1 Energy Yields

The energy yield of a solar panel is proportional to the covered area A, the energy received from the sun (insolation in W/m^2) dependent on the solar irradiance and the angle at which the beam is captured, I, and a conversion efficiency rate that is dependent on the solar panel material ρ, $E_{solar} = \rho AI$. Typical commercially available solar panels exhibit efficiency levels of up to 20%, with projected yields of up to 50% being possible in the future, given the pace of research in this area. The applicability of solar power as the sole energy provider for base stations depends on the region of the globe and time of year. For example, in London, United Kingdom, the maximum insolation level occurs in June where around 207 W/m^2 is received [46]; at this level it is possible to power a 500 W base station with a solar panel of 12 m^2 at 20% conversion efficiency; in December, the month with the lowest insolation in London, a solar panel with 97 m^2 would be necessary. In Sao Paulo, Brazil, an area of 16 m^2 is needed during the month with the lowest insolation (June) [46].

Significantly more power can be extracted from wind for the same used area. The power generated from wind is proportional to the area covered by the wind turbine (A), the cube of the wind speed (V), the air density (ϱ), and the wind turbine efficiency (ξ):

$$E_{solar} = \frac{1}{2}\xi \varrho A V^3. \tag{18.8}$$

The efficiency of current wind turbines is typically around 30%, and cannot exceed the theoretical Betz limit of 59%; few improvements are foreseeable in this area in the near future.

Wind speed and insolation are negatively correlated. This encourages the use of both energy sources as complementary supplies. Referring to Figure 18.3, where the average yields in London are plotted, 75% of the total energy yearly requirements of a 500 W base station can be supplied by the combination of both wind and solar power; only 51% can be supplied by wind power alone. In May 94% of the daily energy consumption can be met by wind and solar power combined. Throughout the year no less than 48% is supplied by renewable resources.

It should be noted that not all sites are appropriate for on-site energy generation using renewable resources. To maximize solar power harvesting, the solar panel must have a southern exposure (in the northern hemisphere), the panel must be directed at an angle that maximizes insolation, and shade must be minimized. Similarly, wind turbines need a clear exposure to maximize their effectiveness: trees, buildings, and other obstacles limit wind power potential. The required size of wind turbines and solar panels clearly limits their applicability in urban sites, where base stations are usually mounted on building rooftops. While the top of a building seems an ideal place to mount energy generating equipment there is an aesthetic objection to wind turbines; solar panels, where installed, would be used to power the building leaving little available to power the base station. Rural sites covering a large area are more suited for renewable energy harvesting; fortunately this is where grid access is scarcer, creating a further incentive to implement on-site generation. Wind turbines and solar panels can be placed next to the base station tower providing a large energy supply. Nonetheless, as was seen in the previous sections, large macro-cells have the highest power requirements.

18.5.2 Relationship between Energy Provision and Storage Capacity

Due to the variability in the power supply batteries are needed in order to store energy when production outpaces the device's consumption and supply the device in times of production shortages. Naturally the capacity of batteries and the ability to hold charges for a significant time is a critical aspect in the deployment of renewable sources [48]. For example, in solar powered base stations the battery needs to be able to hold charge for months to accommodate the periods of relatively low insolation. In the case of wind power, high capacity batteries are needed to store the extra energy output in periods of great yield.

Given a required energy consumption load, $l(k)$, at time k, an energy yield by the renewable source $E(k)$, the battery maximum capacity $B_{\max}$, and the battery residual storage at the point of outage $B_{\min}$, the battery charge at point k is given as [49]

$$B(k) = \min\{\max\{B(k-1) + E(k) - l(k), B_{\min}\}, B_{\max}\}. \qquad (18.9)$$

Computing $B(k)$ at each point in time can then be done recursively, using historical data on the energy consumption and weather conditions. Using this information the probability of outage, where outage is assumed to occur whenever $B(k) \leqslant B_{\min}$, can be determined for each battery capacity level and energy production level. In this way Pareto curves can be drawn showing the trade-off between battery size and energy production capability for a given outage probability enabling a correct dimensioning of the energy source and battery.

18.5.3 Embodied Energy

To correctly assess the environmental impact of wind and solar power, the energy used in the construction and installation of the energy generation or extraction equipment must be taken into account. The initial energy cost of implementing a given source must be paid back over the full lifetime of the equipment, in the form of free energy produced. The relative benefit to the environment can be assessed by considering the payback time of each source, the time it takes to generate at least as much "free" energy as the energy consumed in its creation, and the expected lifetime of the equipment. While the lifetime of wind turbines and solar panels can be expected to be roughly the same, the embodied energy and yields differ substantially.

Solar panels vary considerably in the implied amount of embodied energy depending on the technology they are manufactured with and the cell type [50, 51]. Nonetheless, having in mind the high variability and considerable disagreements in the literature [52], it is expected that the payback time of a typical installation is between 1 and 7 years. Assuming a life expectancy of 20 years, 65–95% of the produced energy of a solar panel is "free" energy.

In contrast to solar panels, the energy payback time of wind power is much lower, ranging from 2 to 8 months [53, 54]. Again assuming a 20-year life expectancy, wind turbines are expected to deliver over 95% of their energy production as "free" energy. From this point of view, wind power is clearly superior to solar power.

18.6 CONCLUSION

This chapter has presented an overview of the critical aspects of energy consumption in wireless networks. There has been a concerted effort to improve the *energy efficiency* of access networks by the telecommunications research community. This effort has mainly concentrated on the operational aspects of networks, from component level improvements to system-wide management solutions. However, the full life cycle must be considered if the objective is to reduce the environmental impact of wireless access networks—this is a cross-issue in the ICT industry. Life cycle assessment [55] is a powerful tool when evaluating the environmental impact of products and services. Adapting this tool to the evaluation of complex access networks is another research challenge for the future. While operational energy consumption is a relatively well-understood issue, it remains a challenge to quantify the impacts of new techniques and hardware improvements on the embodied energy of devices—this in our view is an important avenue of research in the near future, where general models are needed. Increasingly smaller and complex chips have a big impact on the total embodied energy and devices' life spans.

Historically, it has been the case that improving energy efficiency levels has led to an *increase* in energy consumption. This paradox has been independently proposed by Leonard Brookes and Daniel Khazoom and is a form of the more general *Jevons Paradox*[56]. The paradox can be explained by realizing that improvements in energy efficiency of a product lead (in general) to lower marginal costs, and hence an increase in the consumption of the said product at a macro level. A truly green network cannot just be an *energy-efficient* network, it has to be a network fit for its purpose that has a minimal impact on the environment. This will require active engagement from users, operators, and regulators—government intervention and new business models are needed that encourage all parties to actively participate in the conservation of energy in the network.

ACKNOWLEDGMENTS

The work in this chapter has been partially supported by the Green Radio Core 5 Research Program of the Virtual Centre of Excellence in Mobile & Personal Communications, Mobile VCE, www.mobilevce.com; the ICT-ACROPOLIS Network of Excellence, FP7 project number 257626, www.ict-acropolis.eu; and COST Actions IC0902 and IC0905 "TERRA."

REFERENCES

[1] Berkeley Earth Project, Berkeley Earth surface temperature, November 2011 (online). <http://www.berkeleyearth.org>.

[2] A. Henderson-Sellers, K. McGuffie, A Climate Modelling Primer, Wiley.

[3] M.R. Allen, D.J. Frame, C. Huntingford, C.D. Jones, J.A. Lowe, M. Meinshausen, N. Meinshausen, Warming caused by cumulative carbon emissions towards the trillionth tonne, Nature 458 (2009) 1163–1166.

[4] Vodafone, Sustainability Report, 2011 November (online). <http://www.vodafone.com/content/index/about/sustainability/reporting_our_performance.html>.

[5] T. Edler, Green base stations – how to minimize CO_2 emission in operator networks, in: Next Generation Networks and Base Stations Conference, Bath, UK, 2008.

[6] L. Correia, D. Zeller, O. Blume, D. Ferling, Y. Jading, I. Gódor, G. Auer, L. Van Der Perre, Challenges and enabling technologies for energy aware mobile radio networks, IEEE Communications Magazine 48 (11) (2010) 66–72.

[7] H. Karl, An overview of energy-efficiency techniques for mobile communication systems, TU Berlin, Tech. Rep., 2003.

[8] EARTH Project, Energy efficiency analysis of the reference systems, areas of improvements and target breakdown, Tech. Rep., 2010.

[9] P. Gildert, Power system efficiency in wireless communication, in: Applied Power Electronics Conference and Exposition, March 2006.

[10] O. Arnold, F. Richter, G. Fettweis, O. Blume, Power consumption modeling of different base station types in heterogeneous cellular networks, in: Future Network and Mobile Summit 2010, 2010.

[11] Margot Deruyck, Emmeric Tanghe, Wout Joseph, Luc Martens, Modelling and optimization of power consumption in wireless access networks, Computer Communications, Volume 34, Issue 17, November 2011, Pages 2036–2046, ISSN 0140-3664, http://dx.doi.org/10.1016/j.comcom.2011.03.008.

[12] W. Guo, T. OFarrell, Green cellular network: deployment solutions, sensitivity and tradeoffs, in: WiAd 2011, June 2011.

[13] L. Saker, S. Elayoubi, Sleep mode implementation issues in green base stations, in: PIMRC 2010, September 2010, pp. 1683–1688.

[14] 3G.co.uk, Huawei launches solution to cut base station power consumption, September 2011 (online). <http://www.3g.co.uk/PR/Jan2008/5656.htm>.

[15] I. Humar, X. Ge, L. Xiang, M. Jo, M. Chen, J. Zhang, Rethinking energy efficiency models of cellular networks with embodied energy, IEEE Network 25 (2) (2011) 40–49.

[16] N. Duque Ciceri, T. Gutowski, M. Garetti, A tool to estimate materials and manufacturing energy for a product, in: 2010 IEEE International Symposium on Sustainable Systems and Technology (ISSST), May 2010, pp. 1–6.

[17] Integrated product policy pilot project. Stage I final report: life cycle environmental issues of mobile phones, Tech. Rep., 2005.

[18] E. Williams, Energy intensity of computer manufacturing: hybrid assessment combining process and economic input–output methods, Environmental Science and Technology 38 (22) (2004) 6166–6174.

[19] A. Zhu, P. Draxler, J. Yan, T. Brazil, D. Kimball, P. Asbeck, Open-loop digital predistorter for RF power amplifiers using dynamic deviation reduction-based Volterra series, IEEE Transactions on Microwave Theory and Techniques 56 (7) (2008) 1524–1534.

[20] D. Kimball, J. Jeong, C. Hsia, P. Draxler, S. Lanfranco, W. Nagy, K. Linthicum, L. Larson, P. Asbeck, High-efficiency envelope-tracking W-CDMA base-station amplifier

using GaN HFETs, IEEE Transactions on Microwave Theory and Techniques 54 (11) (2006) 3848–3856.

[21] F. Wang, A. Yang, D. Kimball, L. Larson, P. Asbeck, Design of wide-bandwidth envelope-tracking power amplifiers for OFDM applications, IEEE Transactions on Microwave Theory and Techniques 53 (4) (2005) 1244–1255.

[22] Y. Yang, J. Cha, B. Shin, B. Kim, A microwave Doherty amplifier employing envelope tracking technique for high efficiency and linearity, IEEE Microwave and Wireless Components Letters 13 (9) (2003) 370–372.

[23] NEC, NEC develops one of the world's leading mobile base station transmitter amplifiers (press release), March 2009 (online). <http://www.nec.co.jp/press/en/0903/0902.html>.

[24] J.M. Rabaey, A. Chandrakasan, B. Nikolic, Digital Integrated Circuits, Prentice-Hall, Inc. 2003.

[25] N. Kim, T. Austin, D. Baauw, T. Mudge, K. Flautner, J. Hu, M. Irwin, M. Kandemir, V. Narayanan, Leakage current: Moore's law meets static power, Computer 36 (12) (2003) 68–75.

[26] Ericsson, Sustainable energy use in mobile communications (white paper), August 2007 (online). <www.ericsson.com/cam-paign/sustainablemobilecommunications/downloads/sustainableenergy.pdf>.

[27] Alcatel-Lucent, Alcatel-Lucent demonstrates up to 27 percent power consumption reduction on base stations deployed by China Mobile (press release), February 2009 (online). <http://www.alcatel-lucent.com/wps/portal/NewsReleases/Detail?LMSG_CABINET=Docs_and_Resource_Ctr&LMSG_CONTENT_FILE=News_Releases_2009/News_Article_001448.xml>.

[28] Nortel, Nortel launches new technology to reduce GSM wireless power consumption (press release), April 2009 (online). <http://www2.nortel.com/go/newsdetail.jsp?catid=-8055&oid=100254431&locale=en-US>.

[29] F. Richter, A. Fehske, G. Fettweis, Energy efficiency aspects of base station deployment strategies for cellular networks, in: VTC 2009-Fall, September 2009, pp. 1–5.

[30] A. Fehske, F. Richter, G. Fettweis, Energy efficiency improvements through micro sites in cellular mobile radio networks, in: GLOBECOM Workshops, December 2009, pp. 1–5.

[31] B. Badic, T. O'Farrrell, J. He, P. Loskot, Energy efficient radio access architectures for green radio: Large versus small cell size deployment, in: 2009 IEEE 70th Vehicular Technology Conference Fall (VTC 2009-Fall), September 2009, pp. 1–5.

[32] S. Almeida, J. Queijo, L. Correia, Spatial and temporal traffic distribution models for GSM, in: IEEE VTS 50th Vehicular Technology Conference, 1999, VTC 1999 – Fall, vol. 1, 1999, pp. 131–135.

[33] U. Paul, A.P. Subramanian, M.M. Buddhikot, S.R. Das, Understanding traffic dynamics in cellular data networks, in: INFOCOM, IEEE, 2011.

[34] Z. Niu, Y. Wu, J. Gong, Z. Yang, Cell zooming for cost-efficient green cellular networks, IEEE Communications Magazines 48 (11) (2010) 74–79.

[35] S. Bhaumik, G. Narlikar, S. Chattopadhyay, S. Kanugovi, Breathe to stay cool: adjusting cell sizes to reduce energy consumption, in: SIGCOMM Workshops, Green Networking, ACM, New York, NY, USA, 2010, pp. 41–46.

[36] L. Le, E. Hossain, Multihop cellular networks: Potential gains, research challenges, and a resource allocation framework, IEEE Communications Magazine 45 (9) (2007) 66–73.

[37] J. Xu, L. Qiu, Area power consumption in a single cell assisted by relays, GREENCOM-CPSCOM '10, IEEE Computer Society, Washington, DC, USA, 2010, pp. 460–465.

[38] B. Timus, Cost analysis issues in a wireless multihop architecture with fixed relays, in: 2005 IEEE 61st Vehicular Technology Conference, 2005, VTC 2005-Spring, vol. 5, 2005, pp. 3178–3182.

[39] F.J. Velez, M. del Camino Noguera, O. Holland, A.H. Aghvami, Fixed WiMAX profit maximisation with energy saving through relay sleep modes and cell zooming, Journal of Green Engineering (2011) 355–378.

[40] D. Quintas, V. Friderikos, On dynamic policies to switch off relay nodes, in: IEEE ICC 2012, Ottawa, Canada, June 2012.

[41] G. Karlsson, V. Lenders, M. May, Delay-tolerant broadcasting, IEEE Transactions on Broadcasting 53 (1) (2007) 369–381.

[42] P. Kolios, V. Friderikos, K. Papadaki, Future wireless mobile networks, IEEE Vehicular Technology Magazine 6 (1) (2011) 24–30.

[43] O. Holland, C. Fachini, A.H. Aghavami, O. Cabral, F. Velez, Opportunistic spectrum and load management for green radio, in: E. Hossein, V. Bhargava, G. Fettweis (Eds.), Green Radio Communication Networks, Cambridge University Press.

[44] O. Holland, T. Dodgson, A.H. Aghvami, H. Bogucka, Intra-operator dynamic spectrum management for energy efficiency, IEEE Communications Magazine, In press.

[45] H. HolmaA. Toskala, LTE for UMTS OFDMA and SC-FDMA Based Radio Access, John Wiley and Sons.

[46] National Aeronautics and Space Agency, NASA's Surface metereology and solar energy data set, November 2011 (online). <http://eosweb.larc.nasa.gov>.

[47] Met Office, UKCP09: Gridded data sets, November 2011 (online). <http://www.metoffice.gov.uk/climatechange/science/monitoring/ukcp09/>.

[48] T. Todd, A. Sayegh, M. Smadi, D. Zhao, The need for access point power saving in solar powered WLAN mesh networks, IEEE Network 22 (3) (2008) 4–10.

[49] F. Safie, Probabilistic modeling of solar power systems, in: Annual Reliability and Maintainability Symposium, 1989, Proceedings, 1989, pp. 425–430.

[50] N. Jungbluth, Life cycle assessment of crystalline photovoltaics in the Swiss Ecoinvent database, Progress in Photovoltaics: Research and Applications 13 (5) (2005) 429–446.

[51] G. Peharz, F. Dimroth, Energy payback time of the high-concentration PV system FLATCON, Progress in Photovoltaics: Research and Applications 13 (7) (2005) 627–634 http://dx.doi.org/10.1002/pip.621 (online)

[52] C. Bankier, S. Gale, Energy payback of roof mounted photovoltaic cells, June 2006.

[53] Danish Wind Industry Association, Energy payback period for wind turbines, April 2009 (online). <http://www.talentfactory.dk/en/tour/env/enpaybk.htm>.

[54] British Wind Energy Association. FAQ: how long does it take for a turbine to pay back the energy used to manufacture it? April 2009.

[55] ISO 14040, Environmental management, life cycle assessment – principles and framework, 2006.

[56] L. Brookes, Energy efficiency fallacies revisited, Energy Policy 28 (6–7) (2000) 355–366.

CHAPTER

Towards Energy-Oriented Telecommunication Networks

19

Sergio Ricciardi[a], Francesco Palmieri[b], Ugo Fiore[c], Davide Careglio[a], Germán Santos-Boada[a], Josep Solé-Pareta[a]

[a]*Technical University of Catalonia (UPC) — BarcelonaTech, Department of Computer Architecture, 08034 Barcelona, Spain,*

[b]*Second University of Naples (SUN), Department of Information Engineering, I-81031 Aversa, Italy,*

[c]*University of Naples Federico II, Centre of Computer Science Services, I-80126 Napoli, Italy*

19.1 INTRODUCTION

In the last few years, we have witnessed the uncontrollable growth of the Internet. By 2015 the total Internet traffic will be three times larger than that observed in 2011, equivalent to monthly traffic of 60 EB of data (1 EB = 10^3 PB) (Figure 19.1, [1]).

The number of user connected to the Internet will pass from 2.1 billion in 2011 to 3 billion in 2015 [2–4]. Furthermore, Internet users are asking for higher and higher bandwidth to enable video-on-demand and interactive content, high quality of service (QoS), online gaming, conferencing tools, voice-over-IP, etc. Fortunately, the advent of optical technologies in the networking arena has provided huge bandwidth to satisfy increasing traffic demand and avoid an Internet collapse, and is also characterized by very low energy consumption when compared to the electronic technologies. All-optical networks are still far from being the *only* solution, since they do not provide all the features required by modern network infrastructures: processing, buffering, monitoring, and grooming are at the moment fully provided only by the electronic layers, which in turn do consume a lot of energy. Therefore, the growth of the information and communication society (ICS) is pushing the energy demands of the whole information and communication technology (ICT) sector to values reaching 7–8% of worldwide energy production, a number predicted to double by 2020 [5]. Such a growth rate is not sustainable and proper countermeasures have to be taken to change the business as usual (BAU) scenario to a *greener* one. In order to give an idea of the above numbers, in Italy and France, Telecom Italia and France Telecom, respectively, are the second largest consumers of electricity after the national railway systems at 2 TWh per year, and in the UK, British Telecom is the largest single power consumer [6, 7, 4]. In both data centers and networking plants, the energy consumption is further

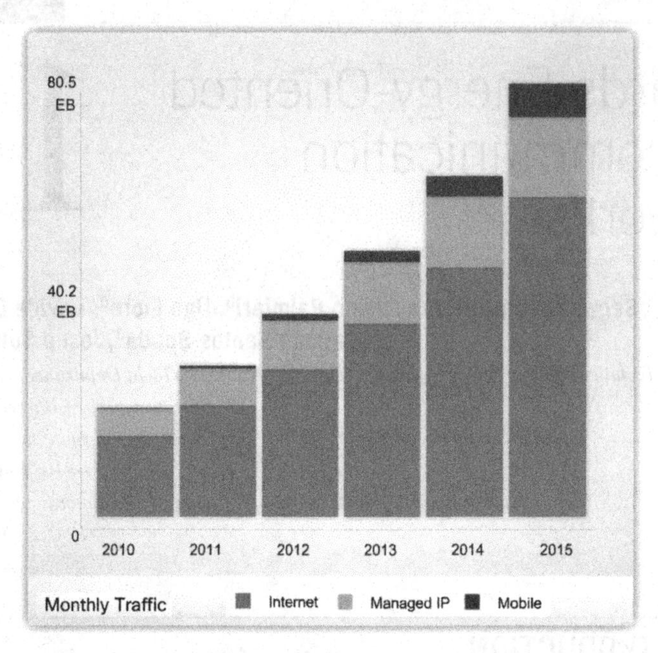

Figure 19.1 Monthly Internet traffic growth forecast.

increased by the impact of HVAC systems (heating, ventilation, and air conditioning), UPS (uninterruptible power supply) systems, and lighting facilities. The power usage efficiency (PUE) [8] is a metric used to measure the impact of these systems on the overall energy consumption of the site; at the moment, a PUE value of 2 is the standard, meaning that for each watt of power spent to do the work, another one is spent to keep the site cool and assure the energy supply. Many efforts are being dedicated to lowering the PUE value in large telecommunication or computing plants, ranging from implementing more efficient cooling systems to moving the whole facility where the external temperature may be used to naturally cool the site.

That's not all; apart from the energy consumption, another problem is menacing not only ICS growth but the entire world population: climate change (mainly, global warming) will drastically change the world, as we know it, if immediate actions are not taken to drastically reduce the emissions of greenhouse gases (GHG) in the atmosphere. Even if the ICT is perhaps the only industry sector that does not directly emit GHG during the use phase, the traditional power plants feeding the ICT equipment do emit GHG to generate the required energy. For ICT, the greater shares of GHG are emitted during the use phase (80%), while the construction phase is responsible only for a small percentage (20%) of the emissions.

As a consequence, the reduction of energy consumption and the use of alternative renewable energy sources to limit GHG emissions as much as possible are the most urgent emerging challenges for telecommunication carriers, to cope with the ever increasing energy costs, the new rigid environmental standards and compliance rules, and the growing power demand of high-performance networking devices. All the above open problems and issues, and foster the introduction of new energy-efficiency constraints and energy-awareness criteria in the operation and management of modern large-scale communication infrastructures, specifically in the design and implementation of enhanced energy-conscious control-plane mechanisms to be introduced in next-generation transport networks.

Accordingly, this chapter analyzes the available state-of-the-art approaches, the novel research trends, and the incoming technological innovations for the new green ICT era and outlines the perspectives towards future energy-oriented network architectures, operation, and management paradigms.

19.1.1 The Energy Problem

Human activities have severe impacts on the environment. The *human ecological footprint* measures the human demand on the biosphere and in 2007 was 1.5% planet Earths[1][9], meaning that we are consuming resources at rate faster than the one characterizing the natural capacity of the Earth to regenerate them. Consequently, an immediate change in our energy consumption habits is needed, if we don't want to exhaust the Earth's natural resources. Therefore, when talking about the energy problem, we usually refer to two concepts: the scarcity (and the consequent higher and higher costs) of the traditional fossil-based fuels like oil, coal, and gas, and the effect that burning such fuels has on the biosphere, mainly resource exploitation, pollution, and GHG emissions, which are responsible for climate change (especially global warming and global dimming).

The *carbon footprint*, part of the human ecological footprint, measures the total set of GHG emissions of a human activity. The ICT emissions correspond to 2–3% of the worldwide produced GHGs [5], as much as the ones characterizing the aviation industry, but with a fundamental difference: airplanes burn large quantity of fossil fuels to travel, whilst ICT only needs electrical energy to work, which can be produced by green renewable energy sources that do not emit GHGs during the energy production process (e.g., solar panels, wind mills, hydro-electrical, etc.). For this reason, we say that ICT is *indirectly* responsible for the GHG emissions.

Therefore, we can identify three dimensions in the energy problem: energy consumption (watts per hour, Wh), GHG emissions (kg of CO_2), and costs (Euros, dollars, etc.). Fortunately, society, industry, academia, and governments are paying greater and greater attention to *alternative* energy sources, which promise to help solve both the energy shortage and the GHG emission problems at the same time. The alternative energy sources are those that are not based on burning fossil fuels

[1]Latest available data; every year this number is recalculated, with a three-year lag due to the time it takes for the UN to collect and publish all the underlying statistics.

Table 19.1 Energy Sources Schematic Comparison

Energy Source	Renewable	Availability	GHG	Environmental Impact
Fossil-based	no	24 h	yes	high
Solar, wind, tide	yes	limited	no	low
Hydro-electrical	yes	24 h	no	medium
Biomass	yes	24 h	limited	medium
Geothermal	no	24 h	no	low
Nuclear	no	24 h	no	high

(like carbon, oil, and gas); rather the energy they produce comes from other, often renewable, sources such as the sun, wind, geothermal, and nuclear. Anyway, even if alternative energy sources are part of the energy solution, they also have drawbacks that should be taken into account. First, not all the alternative energy sources are *green*: some of them emit GHG or have other polluting effects that have a great impact on the environment (such as biomass and nuclear). Second, alternative energy sources may not be always available, like the *legacy* sources, as their availability may vary with natural phenomena (like the sunlight or the wind), and it may depend on the geographical location of the plant: not all the sites have the same *potential* of generating renewable energy (see Table 19.1).

19.2 NETWORK INFRASTRUCTURE

The energy consumption and GHG emissions associated with the operation of network infrastructures are becoming major issues in the ICS. Current network infrastructures have reached huge bandwidth capacity but their technological development and growth has not been accompanied by an equivalent evolution in energy efficiency. In 2008, the network infrastructures alone consumed a mean of 22 GW of power corresponding to 1.16% of the worldwide produced electrical energy, with a growth rate of 12% per year [5], further stressing the demand for effective energy optimization strategies affecting network devices, communication links, and control plane protocols.

Optimizations performed on a network during either the design or operation phase are generally aimed at maximizing performance, flexibility, and resilience while at the same time containing operating costs. The previous considerations foster the inclusion of global power consumption and carbon footprint (GHG emissions) reduction in these objectives. Thus, the minimization of the above energy-related metrics will be the first containment goals, while the capability to fulfill service requests, as well as meeting budgetary limitations, will act as constraints. The savings achieved, in comparison with traditional routing approaches, will be proportional to the relative weight assigned to energy savings, or carbon footprint reduction with respect to the other competing objectives.

However, the heterogeneous features and complex energy production processes associated with the available energy sources make these ranking choices often difficult and contradictory, particularly when considering their environmental impacts. For example, several carbon footprint measurement criteria consider GHG emissions during the use phase only; neither the construction costs nor other environmental impacts taking place during fuel reparation and waste dismissal are considered at all. Nuclear energy, although it does not emit considerable quantities of CO_2, has other severe impacts on the environments and is not renewable as its fuel (mainly uranium and plutonium) is available only in limited quantity; the continuous exploitation of a geothermal source may induce a reduction of its efficiency and hence its attractiveness as a reliable energy source.

Other contradictory issues come from two prominent principles that have driven traditional network design, namely overprovisioning and redundancy, which, if taken in their native form, i.e., in an *energy-agnostic* way, may conflict with energy-saving efforts. More specifically, networks have been traditionally provisioned for worst-case estimated peak loads that typically exceed their long-term utilization by a significant margin. Thus, the energy demand of network equipment remains considerably high even when the network is idle, so that it is straightforward to observe that most of the energy consumed in networks is wasted. This presents several opportunities for substantial reductions in the energy consumption of existing networks, all based on the idea of limiting the energy-related impact of the part of the network infrastructure that is not currently in use or is not used at its maximum possible load.

All the above considerations suggest that strategies oriented at power saving alone may have adverse effects on other metrics, such as path lengths, environmental friendliness, and energy costs. For example, the most power-efficient routing solution may involve the choice of longer routes for paths than are found in conventional shortest-path routing, or dirty energy sources, adversely affecting the carbon footprint or the overall energy costs.

It is clear that a more sustainable scenario has to be configured in which energy must be considered as a fundamental constraint in designing, operating, and managing telecommunication networks under multiple, sometimes conflicting, optimization objectives. The main components of such a scenario are *energy efficiency* and *energy awareness* that work together in an *energy-oriented paradigm* encompassing renewable energy sources in a systemic approach that considers the whole life-cycle assessment (LCA) of the new solutions.

19.3 ENERGY EFFICIENCY

The simplest strategy for reducing the power impact of network infrastructures is to improve the energy efficiency of the involved devices, so that more services can be delivered with no increase on the energy input, or the same services can be delivered for a reduced energy input. At the state-of-the-art, the most energy-efficient network devices are capable of operating on at least two different power levels. These levels will be henceforth referred to as the high power and the low power ones. The service offered by the device will be, generally, proportional to the power required: when running on a low power level, devices will be able to provide only limited throughput. If there are more than two power levels, a discrete set of power levels can be modeled, or power can be assumed to take values in a continuous interval. It is crucial to determine when to perform a transition from a power level to the other, so that the maximum amount of power can be saved without impairing the provided service. In the presence of an increase in demand, the corresponding increase in power level can be started when the demand approaches a given threshold. A conservative choice of this threshold, i.e., at a relatively low value, increases the likelihood that a high power level is used when it wasn't really needed. At the other extreme, if the threshold is set too high, the system may fail to adapt to the required power level as quickly as needed, and service may suffer. Similarly, the transition from a high power level to a low one can be triggered when the demand falls (and stays) under another threshold. The performance of algorithms that govern such transitions will be determined by how closely these algorithms can match the demand. Hence, the key to performance is the ability to accurately *forecast* the evolution in the demand. The most sophisticated algorithms take advantage of statistical properties of network traffic, in order to make realistic predictions about upcoming service requests. In addition, the most comprehensive models also quantify a cost associated with the transition itself and take this cost into account. In the absence of a forecast technique, buffers can be used, usually at the expense of a controlled increase in the delay, to gain a low power time interval on low loaded interfaces.

When planning and designing a network infrastructure, budgetary considerations will also affect the choice of the devices to deploy. Energy-efficient devices must be bought and deployed, and this carries a cost. It is then important that the acquisition cost of energy-efficient devices be kept low as compared with "traditional" ones. The energy-efficiency capabilities must in fact be weighted against the additional costs that the network operators will incur if they select expensive hardware. If a network has to be built from scratch, including all of its components, and the initial design includes energy-efficiency considerations, then the network may be expected to require less energy for the same throughput. Alternatively, adapting the performance of systems or subsystems to the different operating conditions and workloads is a mechanism to bring the power draw down.

Any strategy intended to increase power efficiency should rely on accurate and detailed measurements of the network power requirements, taking into account all the subsystems and components that underlie the complex structure of a modern communication infrastructure. In fact, not all parts of a network will draw power in the same way: at any given time and performance level, some components will be drawing much higher fractions of the total energy than others. Although improvements are requested at all levels and on all components, it is evident that targeting energy-saving efforts to the more "thirsty" parts has the potential to yield considerable savings.

In addition, measuring the per-component power drawn alone will not suffice within the context of a broader analysis. Since efficiency can be defined as performance/power ratio, to compute an accurate estimate of efficiency, a good measurement of performance is also called for. Another question arising in this context relates to what is the most appropriate timescale at which power-saving decisions and actions should happen. While shorter timescales have the potential for larger savings, they are most likely confined to single devices or even components, and thus they are likely to require newly designed hardware, with an increase in expenditure. On the other hand, solutions working at larger timescales could also apply to existing hardware, but they will likely span multiple elements across the network and thus require a strict coordination between these elements, which is often a challenging and still open issue. Whatever the mechanism advocated to achieve a reduction in the power requirements, a real-world implementation may not ignore the huge investments made in the existing infrastructure: strategies that squeeze the maximum performance at the minimum energetic cost from existing hardware (or with marginal additions) are likely to be very welcome to operators' headquarters.

These considerations assume a greater significance in the presence of a new network infrastructure to be built from scratch or when massive network upgrade activities have to be started. In these situations, it is highly desirable to drive technological, architectural, and topological choices by taking into account not only the traditional performance and cost parameters but also considering the energetic budget as a first-class objective. Consequently, the well-known trade-offs between design and management decisions concerning capital (CAPEX) and operational (OPEX) expenditures must always be evaluated under an energy-efficiency perspective. Thus, each new device that improves the performance of its predecessors needs to be technologically compared with the competing ones also on its power requirements.

As for the access networks, the current specific deployed technology is the strongest enabler for energy efficiency. Most of the current energy demand in carrier's infrastructures is associated to wired access connections. Unfortunately, the current access networks are implemented for the most part by using legacy copper-based lines ("the last mile") and transmission technologies such as ADSL and VDSL, whose power consumption is extremely sensible to their operating bit rates. The current trend is completely replacing in a few years such older technologies with fiber-only access infrastructures, which have the potential for improving significantly the overall energy efficiency. Thus, since energy consumption in access networks scales together with the number of end users, the massive diffusion of fiber to the home (FTTH) local loop solutions, replacing the old copper xDSL access connections, would bring the dual advantage of radically enhancing the access bandwidth, and simultaneously reducing the associated energy consumption. Only as a reference for comparison, one can consider that an individual ADSL connection requires about 2.8 W, whereas an access infrastructure based on the passive optical network (GPON) paradigm will reduce the per-link consumption to only 0.5 W (for a giga-speed connection), with an improvement of about 80% in the presence of a very large number of users. The ongoing replacement of legacy last-mile copper infrastructure with a fiber-based one is shifting the problem to the backbone component, essentially affecting the Internet highways of the telecommunication world, where the energy requirements of high-end IP routers are becoming a bottleneck [10], since with the rising traffic volume, the major consumption is expected to shift from access to core networks (from less than 10% in 2009 to about 40% in 2017, see Figure 19.2) [11].

At the backbone network level, optical wavelength-division multiplexing (WDM) communication infrastructures are an ideal field for application of energy efficiency. These networks are characterized by considerable bandwidth and arbitrary topologies (fully virtualizable through the flexible creation of wavelength-based end-to-end connections or lightpaths), and are highly reconfigurable.

In traditional electronic or opto-electronic equipment energy consumption is mainly due to the effect of loss during the transfer of electric charges, and thus consumption directly relies on the specific operating voltage and frequency, and on the number of gates involved (which often become unavoidable physical bottlenecks associated with electrical technology). Conversly, in transparent optical equipment the only factor conditioning power requirements is the technological complexity of coping with the fundamental physics of photons [10] (e.g., zero rest mass, weak photon–photon interaction, and 10^6 times larger size than electrons), resulting in a power demand that is often ten times or more lower.

Minimizing energy consumption of optical networks can be generically addressed at four levels: component, transmission, network, and application [12]. Technological advances in optical devices such as optical add/drop MUXes (OADM), optical cross connects (OXC), and the dynamic gain equalizer (DGE) can be complemented with power-saving solutions.

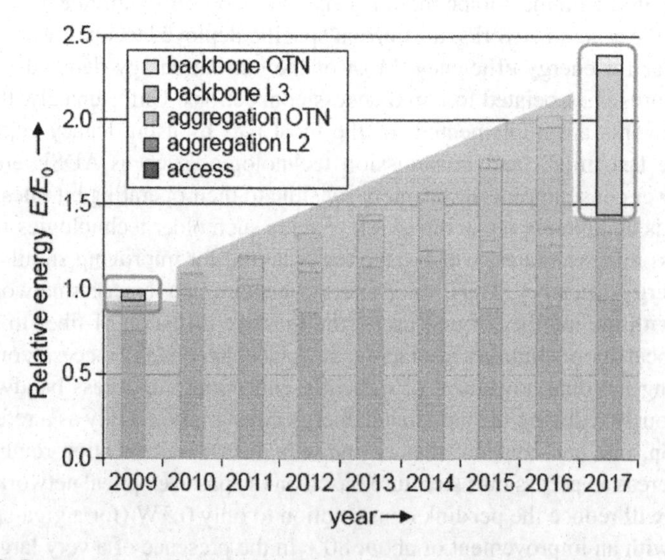

Figure 19.2 Energy consumption trend in communication infrastructures.

At the transmission level, low-attenuation and low-dispersion fibers and energy-efficient optical transmitters and receivers improve the energy efficiency of transmission.

In addition, when configuring and tuning a chain of optical amplifiers, the input power at the transmitter is adjusted to match the noise caused by intermediate components. This tuning is independent of the actual usage of the optical channel. The power values are thus chosen to support the maximum load that the associated channel can provide.

Finally, to really support energy efficiency, network solutions should easily adapt their power demand to the network topology (through the support of redundancy, multipaths, etc.), to the traffic trend (bursts or always low), and to the specific operating scenario. They should also be realistic in terms of technology (providing compatibility and interoperability) and sufficiently reliable and scalable.

19.4 ENERGY AWARENESS

While energy efficiency is the basic fundamental step towards energy-oriented networking, considering this aspect alone is not sufficient to achieve really satisfactory results in the medium and long term. Dynamic and flexible power management strategies, specifically conceived to decrease power consumption in the operational phase, are needed to substantially improve the positive effects on the environment and introduce more significant cost savings.

Such strategies originate from several studies [13,14] demonstrating that in a typical communication infrastructure, designed to ensure a satisfactory degree of reliability and availability through link redundancy/meshing and to seamlessly support the maximum load also during traffic peak hours, a nonnegligible portion of network links are, on average, scarcely used. These considerations, together with the necessity of introducing the global power demand as an additional constraint in the routing decision process, suggest several possible approaches to take advantage of link underutilization in order to save energy. Such approaches imply the adaptive choice of energy-efficient devices and communication technologies, together with the adoption of dynamic network topologies whose task is minimizing the number and the overall energy requirements of devices and links that must always be powered on. Based on their specific energy containment choices, the aforementioned approaches can be classified in three main categories:

- The selective shutdown approach, taking advantage of the idle periods to switch off entire network devices or some of their ports
- The adaptive slowdown approach, putting the network components (switching/routing devices or their interfaces) in low power mode or reducing their speed during underutilization periods and immediately re-enabling their full operation capability/speed when needed
- The coordinated energy-management approach, which advocates global solutions for network-wide power management based on energy-efficient

routing, and more generally is based on new energy-aware control plane services

19.4.1 Selectively Turning Off Network Elements

Energy savings can be achieved by keeping the number of active network elements to the minimum level that suffices to provide the services requested. This kind of optimization can be done both statically and dynamically, depending on whether or not the distribution of upcoming traffic demands is known (or can be accurately estimated). In the static case, which often involves reprovisioning of active connections, minimization of the total consumption is done via switching off the set of active elements accounting for the greatest amount of power, while maintaining sufficient capacity to support traffic. Dynamic scenarios are instead centered on the definition of traffic thresholds that trigger the shutting down or the starting up of network elements, together with the needed rerouting. Selective shutdown strategies can be implemented at different levels of granularity provided that the involved operating scheme has to be extremely flexible and should offer the opportunity of proportionally saving energy as the number of active end-customers decreases. That is, when the inactivity phenomenon only affects some specific interfaces, instead of shutting down an entire network device, only those interfaces, or their associated line cards, when there is no activity on all the interfaces present on them, can be temporarily powered down. Clearly, if in a switching device all the line cards are put into down status, then the entire node can also be powered down safely, achieving further significant energy savings since the chassis and its control logic (switching matrices, routing supervisor elements, and timing cards) can consume about one half of the device's maximum energy budget [15].

Essentially, the major drawbacks of the aforementioned strategies result from a dichotomy: either a device is fully operational or it is not. This means that when the required throughput is very low as compared to the maximum, the device will be wasting most of its processing power and hence it should be powered off, reducing the overall operational costs (OPEX). On the other side, when an entire switching device is powered down many expensive transmission resources become completely unused, hence negating significant capital investments (CAPEX) for the whole duration of the sleeping time. Furthermore, shutting down entire nodes can significantly reduce the meshing degree, by affecting the network reliability and partially negating the possibility of balancing the network load on multiple alternate paths. Unfortunately, with today's technology, putting a big router to sleep may be impractical and even the selective shutdown of links and nodes is not gracefully supported. Also a router costing several hundred thousand Euros or more may take tens of minutes to get up again. Finally, when returning from down status, a peak in the power consumption is registered and the lifetime of the device will decrease a lot if frequent power up and down occurs.

Also the support of selective shutdown at the individual interface or line card level presents substantial drawbacks, mainly due to the needed changes to the traditional routing/switching device architecture and protocols. Hence, an interface switched into sleep mode stops responding to the periodic *hello* solicitations used for neighbor discovery in all the common routing protocol and thus can be classified as "down" or "faulty" at the control plane layer. As a consequence, the link state advertisement facility immediately floods such information throughout the network, flagging the interface operational status change. Thus, sleeping links are detected as link failures, potentially introducing severe stability problems that affect the convergence of the involved routing (or spanning tree) algorithm. To cope with this problem, the *hello* facility should be restricted to awake interfaces only, which results in the loss of a lot of protocol flexibility. In addition, network devices need to acquire the ability to predict low load periods and hence know in advance when shutdown or wakeup operations will be performed. Such knowledge should also be incorporated into routing protocols and generally into the entire set of control plane facilities.

19.4.2 Enabling Low-power Modes

Notable energy savings can be obtained when a significant number of devices spend a major fraction of their idle time in an operating mode characterized by reduced power draw (i.e., the aforementioned "low-power" mode). Although the potential savings vary from device to device, the energy demand when the device is in low-power status can be as low as 10% than the one in active mode. However, during the transitions from or to low-power mode the device may experience a considerable increase in energy consumption since many elements in the transceiver have to be kept active. The experienced value will be strongly dependent on the device implementation details and may possibly range from 50% to 100% of the active mode energy demand.

Unfortunately, current network equipment can only be either in the "on" or the "off" state, and the transition between these two states can be lengthy (minutes) and usually requires manual intervention.

To cope with this problem, future devices must be capable of quickly entering into and exiting from the low-power status or supporting some type of down-clocking feature to adapt to extemporaneous changes in traffic demand. State-of-the-art electronic devices used in the realization of broadband infrastructures are usually designed to achieve their maximum performance when operating according to an "always on" paradigm. The development of next-generation networking devices based on hardware architectures supporting fast "sleep" or "low-power" modes will introduce new opportunities for efficiently reducing their power consumption when idle or partially idle. An energy proportional computing technique such as reducing the microprocessor's clock during inactive period can be really effective for saving energy only if effective full-speed clock return procedures are available, in order to ensure an acceptable degree of responsiveness and avoid perceivable status switching delays.

Table 19.2 Common Wake-on-Arrival Parameters

Technology	Wakeup Time	Sleep Time	Average Power Savings
100baseTX	30μs	100μs	90%
1000baseT	16μs	182μs	90%
10GbaseT	4.16μs	2.88μs	90%

For example, modern Intel processors such as the Core Duo [16] implement a sequence of sleep states (called *C*-states) that offer reduced power states at the cost of increasingly high latencies to enter and exit these states. The massive introduction of these energy-control technologies in networking hardware design will imply an epochal shift from the "always on" to the "always available" paradigm in which each network device can spontaneously enter a "sleep" or "energy saving" mode when it is not used for a certain time and then wake up very quickly and restore its maximum performance when it detects new incoming traffic on its ports. However, to achieve this "wake-on-arrival" (WoA) behavior, the circuitry sensing packets on a line must be left powered on, even in sleep mode.

By putting in low-power mode, a single interface, any transmission activity on it is interrupted in presence of no associated traffic to be forwarded and quickly resumed when new packets arrive from it or are directed to it. The use of such a technique is based on the definition of significantly large time intervals in which no signal is transmitted and smaller time slices during which a brief signal is transmitted to synchronize the receiver (See Figure 19.3). When operating in such way, also known as low-power idle (LPI), the elements in the receiver can be frozen [17] and then awakened within a few microseconds, as reported in Table 19.2.

On the other hand, the ability of dynamically modifying the link rate according to the real traffic volume is used as a method for reducing the power consumption. Letting a device work at a reduced frequency can significantly lower its energy consumption and also enable the use of dynamic voltage scaling (DVS) technologies to reduce its operating voltage. Such a technique allows the needed power to scale cubically, and hence allows the associated energy demand to grow quadratically with

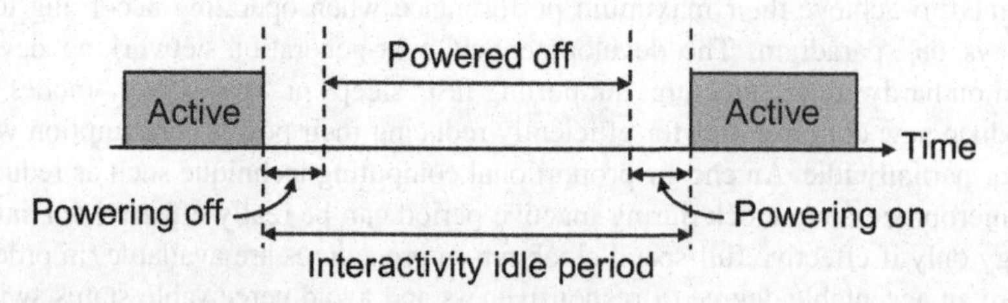

Figure 19.3 Selective shutdown timeline scenario.

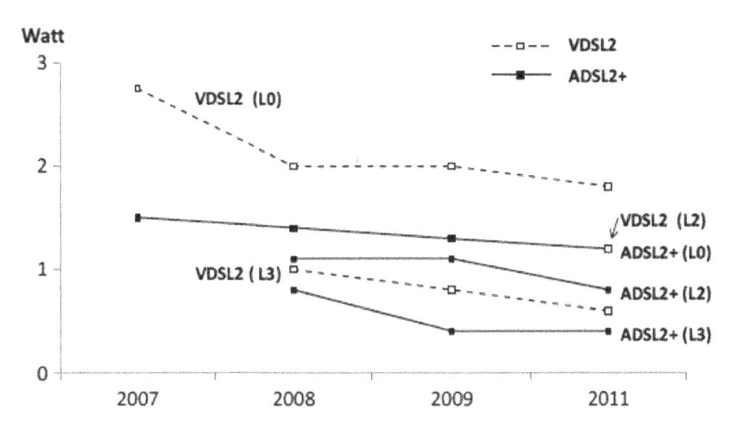

Figure 19.4 The evolution of rate-dependent power demand in traditional access infrastructures.

the operating frequency [18]. For example the Intel 82541PI Gigabit Ethernet Controller consuming about 1 W at 1 Gbps full operation supports a smart power down facility that turns off the PHY interface layer if no signal is present on the link and drops the link rate to 10 Mbps when a reduction of energy consumption is required. Analogously, in the last-mile scenario, the ADSL2 standard (ITU G.992.3, G.922.4, G.992.5) has significantly evolved to support multiple data speeds corresponding to different link states (L0: full rate, L2: reduced rate, L3: link off, see Figure 19.4) for power management purposes [19].

The IEEE Energy Efficient Ethernet working group, when evaluating ALR as an alternative to LPI in the 802.3az standard, decided in favor of LPI.

19.4.2.1 Building Energy Models

Defining a sustainable and effective model for energy consumption is the fundamental prerequisite for introducing power awareness within the routing context. A great variety of networking devices may contribute to the overall power absorption: ranging from "opto-electronic" regenerators and optical amplifiers to opaque routers and totally optical switches. Each device draws the needed power in a specific way, also depending on the relationship occurring between the different components of more complex structures such as switching systems or end-to-end communication links. In addition, some nodes may be powered by renewable sources, while others may use traditional, "dirty" energy; therefore a differentiation between energy sources is required. An energy model has the main task of characterizing the different components of the network involved in energy consumption. It provides the energy demand profile of network devices and communication links of any typological layout and under any traffic load.

Modern communication networks are usually modeled as a graph $G(V, E)$, where V is the set of nodes (a node is an optical router or an OXC equipped with optical transceivers) and E is the set of optical links. Since real-world networks have several characteristics that can impact on power draw, models of increasing complexity will embed larger sets of these characteristics. A review of some dimensions along which models differ follows:

- *Direction of links*: Typically a link is considered as bidirectional, i.e., as having two optical fibers, each in one direction. Finer-grained models may consider unidirectional links. The resulting network graph will then be a directed graph.
- *Available wavelengths*: A simplifying assumption is that the transceivers deployed across the network are all identical, or at least interchangeable. The same set of wavelengths will then be available on all links. If this assumption is relaxed, a specific set of available wavelengths is added to the description of each link. These models will be more useful when evaluating networks where saving energy is an objective, but the evolution towards this objective is done in an incremental way.
- *Wavelength conversion capability*: A constraint in optical routing is that, in the absence of wavelength conversion capability at the nodes, a lightpath has to use the same wavelength over all its span. This greatly increases the difficulty of routing a lightpath, as it is not enough to find a feasible path with available resources, but that path must share a single wavelength on all the links. If, conversely, some or all the nodes can convert wavelengths, i.e., set up a lightpath using different wavelengths, the problem becomes simpler. Wavelength conversion capability can be implemented either in the electronic domain or in the optical one. Whatever the case, usually, conversion-capable hardware is more expensive than regular hardware. A model allowing the coexistence of conversion-capable equipment with nodes not having that capacity will describe more closely those real-world scenarios where an operator is progressively upgrading its infrastructure and wishes to determine the nodes where deploying conversion-capable equipment will yield the most benefit. These nodes could then be given precedence in the upgrade plans.
- *Wavelength capacity*: The capacity that a single fiber is capable of carrying may vary, again depending on the equipment used. Models describing the different capacities are focused on grooming smaller-sized electronic channels onto the optical ones.
- *End-to-end connections*: The power drawn by a single routed connection is given by
  - The power absorbed by the transceivers
  - The power required by intermediate optical switches
  - The power consumed by optical amplifiers along each fiber link on the path

  These components will drain a given amount of power Θ for just being turned on. The value of Θ will depend on equipment technology and size. Additional power Φ will then be needed for operation. If this *variable* power Φ is assumed to be proportional to the provided level of service/load ℓ (e.g., considering

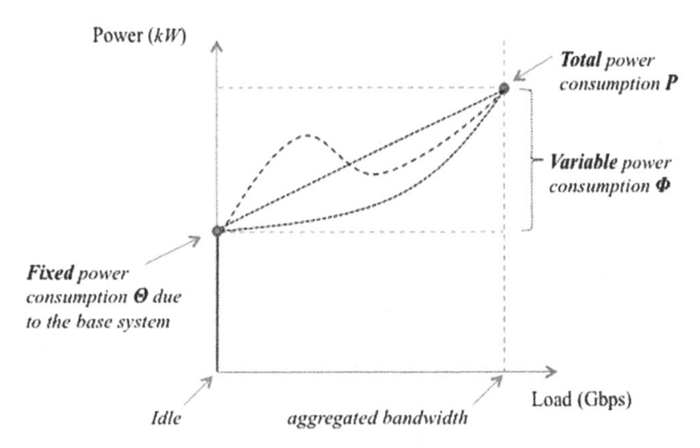

Figure 19.5 Modeling per-connection energy consumption.

the current transmission rate and the maximum bandwidth), a realistic power consumption model can be sketched as in Figure 19.5. Accordingly, the energy consumption $P_{i,j}$ associated to the individual connection (i, j) can be expressed as

$$P_{i,j} = \sum_{k \in (i,j)} \Theta_k + \sum_{k \in (i,j)} \Phi_k(\ell), \tag{19.1}$$

where, for each intermediate device k occurring on the path serving the connection (i, j), the individual Θ_k and Φ_k correspond respectively to the fixed and load-dependent consumption associated to that device.

- *Precomputed paths:* The full problem specification for the establishment of a virtual topology and for the routing of connection requests onto it involves determining the most appropriate path in the network to be used for routing. While, in principle, very long paths can be selected when routing a request, shorter paths will tend to utilize far fewer resources, and the shortest paths will require the minimium. When a single link is present on the shortest path between many node pairs, that link will become critical and its usage will need to be regulated wisely. The risk is that some node pairs will become disconnected. Even in these cases, inordinately long paths are hardly of benefit: paths that are only slightly longer than the shortest ones are instead likely to provide the most appealing route for the incoming connection. This explains why many models avoid computing and exploring all the possible paths. They rely instead on a precomputed set of paths between all node pairs, and restrict the routing choice to those paths only. An important parameter in these models is the number of alternate paths that is precomputed. Alternatively, path precomputation may be restricted to paths whose length would not exceed the length of the shortest paths by more than a specified amount.

- *Node structure*: The majority of models consider optical switches or ROADMs (Reconfigurable Optical Add-Drop Multiplexers) as single, atomic, units. These elements are, instead, composed of many parts and components and the trend in technological evolution of this hardware is to alleviate the central CPUs from as many computing chores as possible, delegating them to "satellite" CPUs such as those found, for example, on line cards. The single cards could, therefore, have a separate behavior as compared to the other cards and even to the central unit. The behavior of the unit as a whole will be in this case the average of the behavior of the single cards. Models aiming at exploring these aspects should describe the single cards, their parameters, and the association with the chassis backplane.
- *Traffic matrix*: Advance knowledge of the entire traffic matrix is a simplifying assumption often used when traffic reconfiguration, as opposed to initial handling, is examined. There are in fact times when, by recomputing routes for all the active connections, dramatic gains in efficiency can be achieved. The objective is then to best provide for the current traffic demand: an "optimized" network is likely to have enough room to accommodate forthcoming requests, too. If the upcoming requests are assumed to be unknown, the model will instead aim to be as adaptive as possible in order to create a well-balanced network able to satisfy as many arbitrary requests as possible. An intermediate scenario is given when future requests are not known, but they are assumed to be drawn from a known distribution, possibly repeating past behavior.
- *Power levels*: Different technologies will have different power states and associated levels (see for examples Table 19.2). A detailed description of these levels is needed when try to evaluate the viability or effectiveness of a power-saving strategy. The saving achieved should be assessed with respect to the constraints, for example the capability to satisfy a certain level of demand.
- *Transition times*: The ability to adapt the power level to the characteristics of service demand is a critical factor. This ability can be considered achieved when the transition times required to shift from a power level to another are so short that power transitions may happen with a null or minimal or impact on service. Models intended as a validation for energy-saving strategies should deal with this issue in great detail.

19.4.3 Control Plane Protocol Extensions

At the control plane level, accurate and up-to-date information about the state of network elements is needed to properly make decisions that reflect the power demand characteristics of the network devices, the power source and its associated energy cost, and the way these variables change over time. The state should be described in terms of services offered and power requirements, including both the value of power draw and the type of supply used. To this end, energy-oriented extensions to the routing protocols used to gather and disseminate information are needed.

These extensions clearly require modifications to the current routing protocols and control plane architecture. That is, the existing routing protocols (e.g., OSPF or IS-IS) within the GMPLS framework should be extended to associate new energy-related information, such as the power consumption, to each link or the type of energy source to each network element.

Different energy cost values can be dynamically associated to network nodes and links depending on the degree of preference associated to green and dirty energy sources and their specific types. That is, they can be powered by different green or dirty energy sources, updated on a prefixed time basis, where the green sources represent energy from wind mills, solar panels, or hydro-electrical plants, with their respective degree of preference, while the dirty energy sources represent energy from coal, fuel, or gas.

Clearly, these costs, together with additional ones associated with specific device-dependent power demand information associated with network elements and links, have to be used to influence the routing decision. Accordingly, to achieve the goal of green-aware routing, the routing decision can be based on the "lowest GHG emissions," or "minimum power consumption" rather than on the "shortest/minimum cost path," where the energy costs are defined within the specific energy model used.

In addition new node and sleep attributes are needed to support energy-efficient traffic engineering features by modeling new link/path statuses and link or device power on/off capabilities, with the aim of distinguishing between down and sleeping network elements.

The transport of the aforementioned energy-related capabilities and costs and their diffusion/exchange throughout the network can be easily managed by introducing new specific type-length-value (TLV) objects in IS-IS or opaque Link State Advertisements in OSPF.

Of course, the same information has to be fully handled by signaling protocols such as RSVP-TE and CR-LDP to allow the request and the establishment of power-constrained paths across the network (i.e., path traversing only nodes powered by renewable energy sources or crossing only low-power transmission links).

19.5 ENERGY-ORIENTED NETWORK INFRASTRUCTURE

Whereas the selective shutdown and slowdown approaches typically work at the specific network components level, further energy-efficiency improvements may be achieved by moving the focus to a wider scale. Accordingly, control plane protocols and routing algorithms can be improved to save energy by introducing a new dimension whose goal is to properly accommodate network traffic by considering both energy efficiency and energy awareness.

These improvements essentially consist of properly conditioning the route/path selection mechanisms on a relatively coarse time scale by promoting the use of renewable "green" energy sources as long as energy-efficient links and switching devices, simultaneously taking advantage from the different users' demands/current

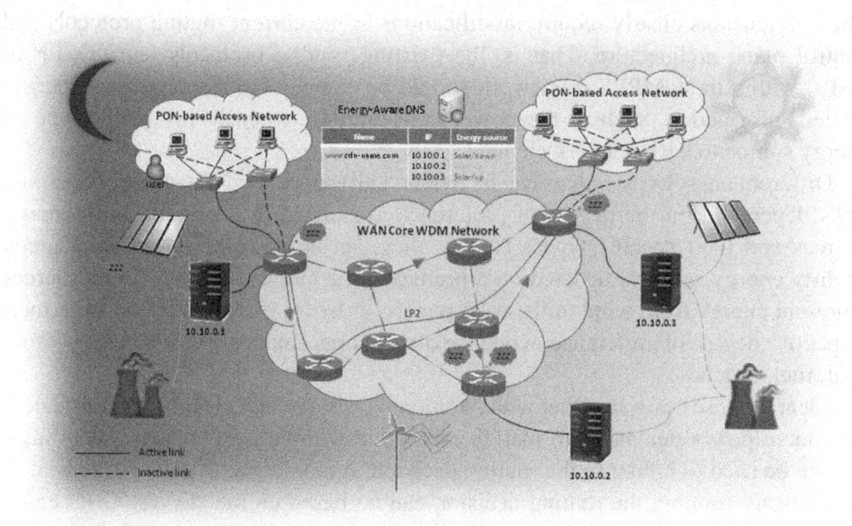

Figure 19.6 Moving the traffic on an energy cost or consumption basis.

load across the interested communication infrastructures, so that the global power consumption can be optimized (Figure 19.6).

The likely outcome would be that some of the "best" paths, selected according to traditional network management criteria, would not be the better ones in an absolute sense, but the power savings and environmental benefits consequent to such an apparently suboptimal choice could be substantial without excessive losses on the other concurring optimization goals. Furthermore, it has been observed [20] that the increments in path lengths do not increase energy consumption in a perceivable way, since routers and switches are not designed to be energy proportional and the power absorbed by a packet when crossing a network node is several orders of magnitude below the energy requested at the terminal point of the path.

Coordinated power-management schemes may also benefit from the previously cited selective shutdown or rate adaptation mechanisms, by taking the associated power off/on and mode change decisions to a network-wide perspective, and hence basing them on a more complete awareness of the overall network economy, so that greater energy savings can be achieved.

Providing energy awareness at the network control plane level also implies the necessity of periodic reoptimization campaigns with the aim of placing the already "in production" network traffic over a new set of paths so that, in the presence of substantial modifications in traffic load/distribution, device power consumption, energy costs, and/or specific energy source availability, the aggregate power consumption is minimized whereas all the end-to-end connection requirements are still satisfied. Thus, in highly redundant network scenarios, entire network paths can be switched off by rerouting the associated connections on other already existing paths, or on newly created ones. In particular, energy-aware routing implies just-in-time optimization of the energy source choice in such a way that renewable sources are always preferred when available. This new requirement originates from the consideration that network elements may be served by multiple power supplies, specifically providing an always available power source coming from traditional dirty energy as long as an intermittent power source associated to green energy, when available. Consider, as an example, the availability of the green energy produced by wind or solar panels; it is strongly correlated with the weather status or time of the day. Prediction of weather conditions, such as the presence of wind, implies some degree of uncertainty whereas it is easier to assume that no solar energy will be produced during the night hours and that a certain amount of energy is expected to be produced when the sun shines.

Energy-aware routing has the additional goal of distributing the network-wide traffic so that, when the wind is blowing in a specific geographic area, all the viable network traffic should be forced to pass through the routing infrastructures located there, and when the wind stops blowing, the above traffic should be dynamically shifted back elsewhere, e.g., where the sun is shining, in such a way that the use of green energy sources is maximized.

Alternatively, by working on the energy cost dimension, significant reductions could be achieved by adaptively rerouting the network traffic to locations where electricity prices and associated taxes are lowest on particular hours of the day. Electricity prices present both geographic and temporal variations, due to differences in regional demand (often based on lifestyle and weather conditions), power grid-related transmission inefficiencies, and generation technology diversity.

Since electricity cannot be stored and has to be consumed instantly, the electric system typically has to keep spare "peaking" generation capacity online for times when demand may surge on short notice. Often, these "peaking" power plants are only run for a few hours during the day, during the maximum demand period (Figure 19.7), adding to the average cost of providing electricity. Dynamic pricing encourages electricity consumers to reduce their usage during peak times, especially during the critical hours of the day by substantially lowering the energy costs during the off-peak hours and hence limiting the need of "peaking" power plants.

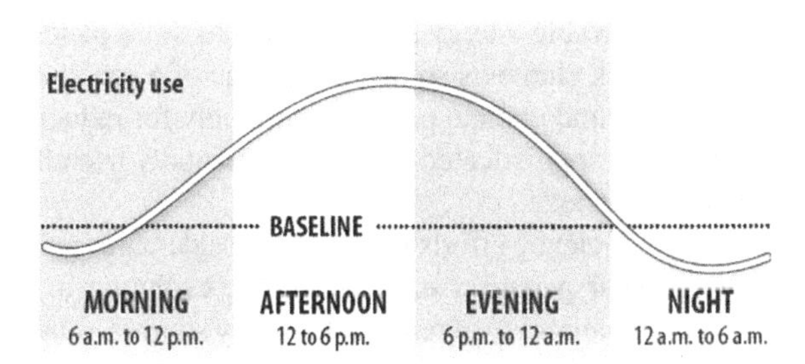

Figure 19.7 The typical daily electricity demand cycle.

Thus, the intuition is that well-known time-based fluctuations in electricity costs and tax incentives for reduced carbon emissions across the covered geographic area may offer opportunities for reducing OPEX, provided that, by considering the involved physical distances and delays, the cost of moving the traffic is sufficiently lower than the likely cost savings from reduced energy prices. Of course, the more rapid is the reaction to price changes, the greater are the overall savings.

Hence, to ensure the necessary flexibility, the aforementioned energy cost and absorption knowledge should be introduced at the control plane level for energy-aware/price-conscious routing and communication resource allocation, through the implementation of automatic and adaptive follow-the-sun, chase-the-wind, or follow-the-minimum-price paradigms.

Furthermore, in order to support all the above adaptive routing behaviors, energy-related information associated to devices, interfaces, and links needs to be easily and reliably determined, depending on the involved technology and current traffic load, according to a comprehensive energy model, and introduce as additional constraints (together with delay, bandwidth, physical impairment, etc.) in the formulations of dynamic routing algorithms and heuristics. In this scenario, down-clocking or selective shutdown facilities should be introduced as new network element status features that need to be considered at the routing and traffic engineering layer. Clearly, to support the aforementioned energy-aware behaviors, such information must be conveyed to all the network devices operating within the same energy-management domain.

19.6 CONCLUSIONS

Until now, the design and development of new network equipment and solutions was fundamentally dominated by performance-related objectives. However, due to the astonishing growth of network infrastructures worldwide, their electric power demand has reached extraordinary peaks, strongly affecting the carriers, their operating costs, and the overall model scalability and sustainability. As a direct consequence, it is now strictly necessary to consider energy efficiency as a basic constraint and a fundamental priority in the design and development of next-generation networks. In addition, governments and telecommunication operators are endorsing the development of green renewable energy sources (such as solar panels and wind turbines) for powering network elements so that the choice of a specific energy source, when possible, becomes a fundamental parameter not only for reducing energy costs but also for deploying more sophisticated and environmentally friendly energy-aware network management strategies.

In fact, while energy-efficiency efforts may yield considerable savings, if the problem is considered in a wider perspective, including the evolution in the demand for advanced services and the availability of renewable energy sources, other considerations will emerge. Energy-aware network elements adapt their performance and behavior depending not only on the actual load, but also on the currently used energy source.

Hence, coordinated energy-aware control plane strategies, driven by multi-objective optimization, may be more helpful in finding the appropriate point on the Pareto front according to the relative importance of network performance, energy consumption, cost containment, and environmental friendliness. Its most significant added values originate from the definition of comprehensive energy consumption models for modern networks, incorporating both physical layer issues such as energy demand of each component and virtual topology-based energy management information. The resulting strategies will limit the usage of energy-hungry links and devices, privileging instead energy-efficient equipment and solutions, giving attention to the type of sources. Moreover, intelligent grooming mechanisms that reuse the same switching nodes, fiber strands, and interfaces along the same path as much as possible will also optimize resource usage by concentrating the traffic load. Thus, energy-aware decisions, taken on larger aggregates and hence on smaller amounts of data, will be simpler from the operating point of view and will yield amplified energy-related benefits.

REFERENCES

[1] Cisco visual networking index (online). <http://www.cisco.com/en/US/netsol/ns827/networking_solutions_sub_solution.html>.

[2] Internet world stats (online). <http://www.internetworldstats.com/emarketing.htm>.

[3] eTForecasts (online). <http://www.etforecasts.com/products/ES_intusersv2.htm>.

[4] BT announces major wind power plans (online). <http://www.btplc.com/News/Articles/Showarticle.cfm?ArticleID=dd615e9c-71ad-4daa-951a-55651baae5bb> (2007).

[5] BONE, Bone project, 2009, WP 21 TP green optical networks. D21.2b report on Y1 and updated plan for activities.

[6] S. Pileri, Energy and communication: engine of the human progress, in: INTELEC 2007, Rome, Italy, September 2007.

[7] L.S. Foll, Tic et énergétique: Techniques d'estimation de consommation sur la hauteur, la structure et l'évolution de l'impact des tic en france, Ph.D. Dissertation, Orange Labs/Institut National des Télécommunications, 2009.

[8] T.G. Grid, The green grid data center power efficiency metrics: PUE and DCIE, technical committee white paper (2008).

[9] Living planet report 2010, The Biennial Report, WWF, Global Footprint Network, Zoological Society of London, 2010.

[10] R. Tucker, R. Parthiban, J. Baliga, K. Hinton, R. Ayre, W. Sorin, Evolution of WDM optical IP networks: a cost and energy perspective, Journal of Lightwave Technology 27 (3) (2009) 243–252.. doi:10.1109/JLT.2008.200542.

[11] C. Lange, D. Kosiankowski, R. Weidmann, A. Gladisch, Energy consumption of telecommunication networks and related improvement options, IEEE Journal of Selected Topics in Quantum Electronics 17 (2) (2011) 285–295.. doi:10.1109/JSTQE.2010.2053522.

[12] Y. Zhang, P. Chowdhury, M. Tornatore, B. Mukherjee, Energy efficiency in telecom optical networks, IEEE Communications Surveys Tutorials 12 (4) (2010) 441–458.

[13] K.J. Christensen, C. Gunaratne, B. Nordman, A.D. George, The next frontier for communications networks: power management, Computer Communications 27 (18) (2004) 1758–1770.. doi:10.1016/j.com.2004.06.012.

[14] A. Odlyzko, Data networks are lightly utilized, and will stay that way, Review of Network Economics 2 (3) (2003) 210–237.

[15] J. Chabarek, J. Sommers, P. Barford, C. Estan, D. Tsiang, S. Wright, Power awareness in network design and routing, in: INFOCOM 2008, The 27th Conference on Computer Communications, IEEE, 2008, pp. 457–465, doi:10.1109/INFOCOM.2008.93.

[16] A. Naveh, E. Rotem, A. Mendelson, S. Gochman, R. Chabukswar, K. Krishnan, A. Kumar, Power and thermal management in the Intel Core Duo processor, Intel Technology Journal 10 (2) (2006) 109–122.

[17] P. Reviriego, J. Hernández, D. Larrabeiti, J. Maestro, Performance evaluation of energy efficient Ethernet, IEEE Communications Letters 13 (9) (2009) 697–699.

[18] B. Zhai, D. Blaauw, D. Sylvester, K. Flautner, Theoretical and practical limits of dynamic voltage scaling, in: Proceedings of the 41st Annual Design Automation Conference, DAC '04, ACM, New York, NY, USA, 2004, pp. 868–873, doi:10.1145/996566.996798.

[19] M.T.zannes, ADSL2 Helps Slash Power in Broadband Designs, 2003 (online). <http://www.eetimes.com/electronics-news/4136967/ADSL2-Helps-Slash-Power-in-Broadband-Designs>.

[20] A. Qureshi, R. Weber, H. Balakrishnan, J. Guttag, B. Maggs, Cutting the electric bill for Internet-scale systems, in: Proceedings of the ACM SIGCOMM 2009 Conference on Data Communication, SIGCOMM '09, ACM, New York, NY, USA, 2009, pp. 123–134, doi:10.1145/1592568.1592584.

CHAPTER

Energy-Efficient Peer-to-Peer Networking and Overlays

20

Apostolos Malatras, Fei Peng, Béat Hirsbrunner

Pervasive and Artificial Intelligence Research Group, Department of Informatics, University of Fribourg, Boulevard de Pérolles 90, CH-1700 Fribourg, Switzerland

20.1 INTRODUCTION

Achieving energy efficiency in information technology (IT) and in particular networking has recently emerged as an increasingly significant, yet challenging research issue [3,12,16,71]. Although this aspect has to date been mostly neglected, it has become a major design issue in current and future networking approaches mainly due to two reasons. First, the general trend towards green solutions throughout the spectrum of human activities has undoubtedly influenced the IT domain, in order to reduce overall energy consumption and with it carbon emissions. Second, from a business perspective energy-efficient networking is beneficial since it implies reductions in power consumption and hence reductions in the associated costs [12]. While it can be argued that the average 3% that IT contributes to global energy consumption (that is growing at a fast pace) and carbon emissions is negligible, its effect is nonetheless more profound as mentioned in [120] and optimal solutions to reduce it should thus be considered.

The focal direction to achieve IT energy efficiency stems from the observation that most computing devices consume more energy than they actually need to due to poor scheduling of their online time. Moreover, networks nowadays are traditionally designed and deployed in a manner that contradicts energy efficiency, since overprovisioning and redundancy are among the foremost characteristics of network design [71]. In this respect, the concept of energy proportional computing [43], according to which energy consumption should be proportional to the operation and performance of IT, has been driving research towards green computing and networking. Accordingly, research so far has mostly focused on energy-aware applications and infrastructures, adaptive link rate techniques at the physical layer, and interface proxying to reduce energy costs at the edge of networks [71], with another equivalent classification making the distinction between approaches such as dynamic adaptation of network elements, reengineering of technologies, and devices switching between sleeping and standby modes of operation [12].

In particular, when taking networking into account promoting energy efficiency should involve all interested concerns, namely end hosts, network devices, and protocols, as well as the Internet core and the energy costs for communications. Regarding the latter, studies have worked on classifying and quantifying the different types of traffic, with the most prominent one being that related to peer-to-peer (P2P) networks and overlays, i.e., 40–73% in [121] and 70% in [122]. It becomes therefore evident that energy-efficient P2P networks and overlays should be considered as also suggested in [95], since they attest to the biggest part of Internet traffic and thus any potential energy savings in such networks would greatly affect the universal IT energy consumption levels. This research direction is an extremely challenging one because of the inherent distributed nature of P2P systems that complicates the introduction of reliable and efficient energy conserving solutions [4]. However, these challenges also make this domain a highly interesting one in terms of research potential and opportunities in a broad spectrum, namely energy savings for end hosts, overlay networks, and protocols.

This chapter presents a comprehensive overview of existing approaches in the research area of energy-efficient peer-to-peer systems, networks, and overlays. We describe in detail the operation of these approaches and their benefits in terms of energy savings, and classify them in a systematic taxonomy according to their common functional features and corresponding mechanisms used to achieve optimal energy performance. The goal of this chapter is to assist scholars and researchers in green P2P computing by providing them with a broad state-of-the-art review of the specific area, as well as to promote and trigger further research on energy-efficient P2P systems.

20.1.1 Scope of Study

The scope of this study includes only P2P systems where energy efficiency is explicitly addressed and evaluated either by means of simulations or real-world experimental measurements. We do not consider research works claiming to be energy efficient, but not validating their claims. Furthermore, we intentionally chose not to delve into the closely related field of green solutions for mobile ad hoc networks (MANETs) that share many operational principles to P2P, since in our view P2P systems operate mostly at the application layer, while MANET energy efficiency is an issue of lower layers. Lastly, in this chapter the terms P2P systems, networks, and overlays are used interchangeably.

20.1.2 Chapter Organization

After this brief introduction, the remaining of this chapter is structured as follows. Section 20.2 presents a concise overview of P2P technologies explaining their general functionality and giving references to existing work. Section 20.3 discusses the energy profile of current P2P networking and overlays and in particular related models to predict their energy behavior as well as corresponding real-world measurement studies. In Section 20.4 we describe the classification taxonomy of energy-efficient P2P networking research works that we have reviewed, whereas detailed investigation and analysis of these works is the focus of Section 20.5. The chapter concludes with Section 20.6 with a general discussion on the most prominent concerns regarding energy efficiency in P2P systems.

20.2 P2P OVERVIEW

In this section, we briefly present an overview of P2P overlay and networking approaches and classify them according to their organizational characteristics. The remaining of this chapter focuses on improvements over existing P2P overlays for energy efficiency, as well as introduces new green P2P techniques. Accordingly, it assumes that the reader has a basic knowledge of both standard P2P terminology and well-founded existing P2P works, an exposure to which is the aim of this section. For a detailed review and survey of P2P overlay network schemes the interested reader is referred to previous rigorous studies, such as [5, 6, 88, 99, 100].

A peer-to-peer overlay is a logical, virtual network that is built upon a real, physical network whereby the participating peers, i.e., the nodes in the physical network, are organized in a distributed manner without any hierarchy nor any centralized control [6]. The traditional client-server paradigm is not adhered by P2P systems, where every peer behaves both as a client and as a server. Peers are communicating between themselves in order to establish self-organizing overlay structures on top of the underlying physical networks, in order to support a variety of application level services and applications. The most popular application of P2P overlays has been on the field of resource discovery, e.g., files, over large-scale distributed physical networks [101], while other domains where it has been successfully applied include media streaming, IP telephony, etc. The most prominent features of P2P networks are their high degree of decentralization, their self-organized behavior, the fact that they can be easily deployed at the application level and are highly scalable and resilient to failures, and their inherent support for heterogeneity [5].

There have been a large number of P2P systems proposed in the related research literature, satisfying different requirements and optimized for a diverse set of criteria. The most common classification scheme regarding P2P systems is based on the mechanisms used to construct and maintain the overlay topology. Accordingly, P2P systems can be categorized as being either structured or unstructured. In structured approaches the topology of the overlay is tightly regulated and there are specific rules concerning the placement of objects on the peers, namely distributed hash tables (DHT) are exploited to determine the exact peer where an object, e.g., a file, will be placed. On the contrary, in unstructured approaches the topology of the overlay is unknown to peers and there are no rules to control it. Instead of applying DHT functions to locate resources as in the structured approaches, in unstructured ones flooding mechanisms are used in order for resource discovery to be performed [6, 87]. Both categories bear

Table 20.1 Characteristic Structured and Unstructured P2P Systems

Structured	Unstructured
Chord [65]	Freenet [108]
Tapestry [104]	Gnutella [109]
Pastry [102]	FastTrack [116]/Kazaa [117]
Kademlia [106]	BitTorrent [110]
Viceroy [105]	UMM [118]
CAN [107]	Newscast [114]
Cycloid [113]	Phenix [115]
SkipNet [111]	BlatAnt [103]
P-Grid [112]	

advantages and disadvantages and the decision on which one to apply for a particular case is very much dependent on the particular case's characteristics and requirements. While structured approaches allow for quick and efficient resource discovery, they have much higher management overhead for the overlay's topology maintenance [91]. Table 1 presents the most indicative cases of structured and unstructured P2P systems, though it should nonetheless be mentioned that this listing is not meant to be exhaustive.

20.3 ENERGY PROFILE OF P2P SYSTEMS

Examining the energy footprint of existing P2P systems, network overlays, and applications is essential to understand their energy consuming behavior and thus devise and propose novel mechanisms and generic solutions to reduce their overall energy consumption. It is important to note that energy conservation refers not only to P2P systems that are designed to consume less energy, but also to P2P systems that are conceived for energy-constrained devices and thus should consume less energy to avoid depletion of their batteries. The latter case is an aspect of green computing and networking that should not be neglected, since longer battery lifetime implies less frequent battery recharging and therefore smaller energy consumption. Moreover, energy measurements regarding the operation of P2P systems are of paramount significance to assist in their proper deployment and configuration since such quantitative measures would provide an indication of the energy profile of P2P systems. In this respect, works such as the studies on applying P2P resource discovery protocols over mobile ad hoc networks [27, 35, 42] should be highlighted, since according to their simulation results unstructured P2P protocols appear to be more energy demanding than structured ones in scenarios concerning low mobility.

Equally important is the requirement for energy models to analyze the energy consumption characteristics of P2P systems. Such models would allow predicting the energy savings that derive from modifications in P2P systems at an early design stage, thus shortening the rollout time for new energy-efficient P2P systems and facilitating comparisons to other approaches. One of the most interesting such models is the one proposed by Nedevschi et al. [69], which was also one of the first attempts to model the energy consumption of an entire P2P system taking into account not only energy costs on behalf of the peers, but also the corresponding costs attributed to interpeer communications. Based on their proposed model, the authors of [69] compared the energy costs of P2P systems to those of their centralized counterparts and determined that P2P systems fare better when peers are concerned, but centralized solutions consume less energy for communications. In a related model regarding the energy consumption of the BitTorrent P2P system, Hlavacs et al. discovered the relation between energy efficiency and peer participation in the file sharing process [48, 56]. Accordingly, their model can be used to infer the optimal time file owners (seeders in the BitTorrent terminology) should support downloading peers (leechers) in order to ensure the best possible energy efficiency in BitTorrent, assuming the cases of both popular and unpopular files. Similarly, in [25] a model for the energy efficiency of a Kademlia-like P2P protocol running on mobile devices was proposed in order to determine the optimal level of active peer participation in the overlay. The latter was found to be at least 30% of the overall peer population, in which case the optimal trade-off between energy efficiency and P2P overlay traffic could be achieved. In the same line of thought, the model described in [1] serves as an indicator for the amount of time peers should spend actively participating in the P2P overlay so as to ensure its proper operation (in terms of resource discovery success) with the least amount of consumed energy.

Furthermore, regarding the viability from an energy performance point of view of mobile devices participating in peer-to-peer overlays, the interested reader is referred to [90] for an overview of related challenges and solutions. In particular, works by Gurun et al. [63] and Rollins et al. [44] have confirmed that while it is possible for mobile devices to be active members of P2P overlays, it is nonetheless at a high cost in terms of battery utilization. In particular, [63] studied the energy profile of a lightweight, structured P2P overlay implemented on a PDA (personal digital assistant) device and discovered that it is feasible to have such configurations albeit at relatively high-energy costs that quickly diminish limited battery levels. The most energy consuming operations were found to be the route and resource lookup ones, which denotes that for large networks where there are inherently a lot of lookup requests the energy costs pertaining to structured P2P overlays are excessive. On the other side, in Rollins et al. [44] the energy requirements of the unstructured Gnutella P2P protocol were examined, whereby it was assessed that the average battery lifetime of a laptop device is 1.3 h less when the device participates fully in the operations of Gnutella. This high reduction in the battery's lifetime can be quite restrictive for laptop devices wishing to join a Gnutella overlay, according to Rollins et al.

Additionally, the diverse nature of wireless over wired networks and its adverse effect on the energy performance of P2P systems should also be taken into account.

In [94] the observation is made that P2P protocols designed for wired networks do not have the same levels of efficiency when applied as such to wireless settings. The main reason for such behavior is the multihop paths that need to be traversed and the fact that they necessitate message retransmissions with the obvious effect of increasing energy consumption. This aspect was further examined in [8] where the influence of specific wireless standards, namely UMTS and IEEE 802.11, on the energy consumption of P2P systems was studied with IEEE 802.11 proving to be more efficient due to higher download speeds. Similar observations were also made in [73], with structured P2P overlays on top of UMTS emerging as more energy consuming (between 2 and 3 times shorter battery lifetime) than when they operate on top of IEEE 802.11. Nevertheless, as the same research group validated in further studies [7, 49, 50], the proper operation of Kademlia-like structured P2P overlays on top of both aforementioned wireless standards is feasible from an energy consumption perspective.

Such research output, where the energy consumption behavior of P2P systems under different settings was experimentally studied and analytically modeled, has spurred a growing interest in developing energy-efficient P2P systems and overlays. In what follows we describe these novel green P2P approaches and discuss their operation.

20.4 TAXONOMY OF ENERGY-EFFICIENT P2P APPROACHES

Undoubtedly, the ever-increasing motivation and trend towards green, energy-efficient IT, and networking technologies has also attracted the wide interest of the P2P research community. There has been a significant number of research papers published in international journals and conferences on the topic of energy-efficient P2P overlays and systems with the aim of reducing their energy consumption either by directly providing a green P2P alternative or by conserving the battery of mobile devices and thus indirectly creating overall energy savings, namely by limiting the need to recharge the commonly power-restricted mobile devices. It should also be mentioned that only recently two Ph.D. dissertations in related areas were successfully completed [89, 92]. To the best of our knowledge there has been no previous attempt to systematically review the efforts made in this research direction, this chapter being the first survey of the related state-of-the-art.

Having thoroughly studied the research works on energy-efficient P2P systems, networks, and overlays we were able to classify them under six different categories according to the approach they follow in reducing overall energy consumption. The classification can be seen in Table 2, where a small description and a listing of the corresponding research works are additionally provided. In the following we briefly describe the common general characteristics of each one of the aforementioned categories, as well as provide a detailed and comprehensive analysis of characteristic research works.

Table 20.2 Classification of Energy-Efficient P2P Research Works

Category	Description	Research Work
Proxying	Use of proxies by P2P hosts to delegate P2P-related activities and thus allow hosts to have more idle time	[2, 15, 18, 8, 21, 37, 53, 54, 67, 74, 81, 82, 85]
Sleep-and-wake	P2P hosts adopt an adaptive operational behavior by selectively switching between on and off state in order to save energy	[9, 22, 33, 39, 40, 46, 63, 75, 83, 84, 93, 96]
Task allocation optimization	Scheduling of tasks across P2P hosts in a manner that limits overall energy consumption by utilizing host availability more efficiently	[28, 36, 57, 58, 66, 72, 76]
Message reduction	By minimizing the number of sent messages, processing and transmission times are reduced and thus energy is conserved	[13–15, 24, 26, 41, 51, 53, 64, 86, 97]
Overlay structure optimization	New topology designs for energy-efficient P2P overlays or modifications to existing topologies to satisfy energy requirements	[11, 19, 20, 23, 31, 32, 38, 44, 45, 59, 79, 80, 119]
Location-based	Location information is used to make P2P overlays more closely match their physical underlay counterparts thus reducing multihop transmissions	[10,20,31,45,47,62,77,98]

20.5 CHARACTERISTIC CASES

20.5.1 Proxying Approaches

The benefits of proxying in reducing overall IT energy utilization by combining network presence of devices while they are in sleep mode have been well documented [78] and have triggered related research efforts in the context of peer-to-peer systems. In this respect, in the recent work by Anastasi et al. [82], which was later presented as an extended version in [18], a proxy-based approach was employed to reduce the energy consumption of devices running the popular BitTorrent P2P application. The authors state this approach could be adapted for other P2P applications or systems as well. With the underlying assumption that P2P file sharing requires devices to be constantly online and thus consuming energy while downloading files, the authors argue that energy conservation is possible. Starting with the identification of the problem, namely reducing energy consumption while maintaining permanent connectivity of the devices participating in the P2P network, the authors propose that devices delegate and offload their P2P operations to a proxy and then go to sleep mode to conserve energy or to consume less. One proxy can serve more than one device and it operates as a standard BitTorrent client connected to the BitTorrent network. From the point of view of the BitTorrent protocol there is no need for any change in configuration. The clients need to implement an interface to offload their P2P activities to the proxy and then receive the downloaded files from the proxy at a later time on demand. The results reported in [18] are very promising since they show an up to 95% decrease in energy consumption with a parallel noteworthy reduction in the average time required to download a file (approximately 22%, due to the fact that the proxy serves more files and hence more file requests compared to a standard BitTorrent client). It is worth noting that the energy reductions are as high as 95% only in the case where the energy costs of the proxy itself are not considered, e.g., the proxy is running on a machine that is required to be powered on at all times. When the energy footprint of the proxy machine is taken into account, the overall energy savings are reduced to around 60% and are dependent on the number of devices that have delegated their P2P activities to the proxy. Moreover, no changes in the widely deployed BitTorrent protocol are needed.

Based on similar observations, the work by Purushothaman et al. [81] proposes a proxy-based approach to reduce the energy consumption of the Gnutella P2P file sharing protocol. Contrary to the previous case, in the work presented in [81] the concept of the proxy is not implemented by means of a separate dedicated device. Instead, the handling of a subset of the Gnutella protocol's semantics is handled by the NIC (network interface card) of each device. This ensures that the device is always connected, but in a standby low-energy mode for a longer time up until it is awakened up by the NIC when it receives a Gnutella packet that cannot be processed by it. A major limitation of this work is the fact that specialized NICs need to be developed to allow for the real-world validation of the proposed approach. While this work was one of the first ones to identify the need for energy savings in P2P systems, it failed to meet the desired expectations mainly due to the absence of corresponding experimental evaluation. However, the feasibility of such an approach and its benefits in reducing PC energy utilization was illustrated by the highly interesting Somniloquy architecture [37]. The latter architecture advocates the inclusion of a microprocessor in the NIC of a computer, which can be used to run applications such as peer-to-peer file sharing or instant messaging that require constant network connectivity, while the device is in sleep mode. The idea behind Somniloquy is that the computer should be awakened only when some advanced processing needs to be performed, something that can in some cases be completely alleviated assuming that the entire application's functionality is reimplemented and executed on the NIC's processor. The authors of [37] developed a set of prototypes and were able to run BitTorrent on the NIC's processor without changing the operation of the protocol at all. The corresponding experimental results indicate power savings from 60% to 80%, since a computer is able to perform standard peer-to-peer file sharing activities while being in sleep mode (wakeup occurs to handle exceptional cases and periodically to move the downloaded files from the NIC to the computer).

The notion of proxying was also proposed as a potential energy saving mechanism in a DHT-based peer-to-peer network where mobile nodes are concerned [15], but no validation was presented in that work. The same research group worked further in the direction of proxy-based solutions to enable mobile device participation in P2P networks. The motivation for their work is founded on the observation that when wireless communications are concerned, data transfer in high-speed bursts achieves optimal energy consumption [67]. In particular, in [67] nodes in BitTorrent are classified as being energy-limited or regular ones. All the nodes participate in the BitTorrent network, but the former are mostly in an idle state in order to conserve energy in spite of the cost of longer download times. The energy-limited peers receive the desired content, e.g., files, from the BitTorrent network through regular peers in negotiated and carefully scheduled time intervals, to ensure download of the data at the highest possible speed. The proposed BurstTorrent protocol is a modified version of the BitTorrent one and based on the simulation experiments reported in [67] achieves approximately 50% energy savings compared to BitTorrent, albeit imposing a significant delay in download times that can be up to 54% longer.

Taking the concept of using a proxy for P2P energy efficiency one step further, CloudTorrent [53] proposes an alternative to BitTorrent for P2P file sharing over energy-limited mobile phones. Specifically, CloudTorrent introduces a proxy on a cloud server in order to connect to a standard BitTorrent network and act as a delegate of the mobile phone in all its P2P file requests. Upon completion of the download on the proxy side, the file is transferred to the phone over a direct HTTP connection. The real-world experiments performed with CloudTorrent reported significant energy savings (approximately three times less energy consumption), while in parallel reducing the download times by circa three times. Having identified the benefits of using proxies in order to conserve energy, the authors of [53] extended their work to take into account dedicated, standalone proxies for mobile devices wishing to participate in a BitTorrent network [54, 74]. Both cases advocate for the proxy to be hosted on home broadband routers, since they are powered on all the time and thus such a deployment would not require the use of an additional power-consuming device. The extra energy consumed by the router for operating the proxy solution was found to be minimal [74]. Since proxies might serve many mobile devices and they generally are resource-constrained, a noteworthy concern involves the case where the proxy becomes overloaded, for example when its available storage is depleted. A solution to this problem was proposed in [21], namely dividing the available memory in two parts: a download and an upload buffer. The former stores data to be sent to the mobile device, while the latter is used for parts of files that are made available to other BitTorrent clients. Another parameter taken into account is that of the chunk size, which dictates how many file parts are needed in order to begin their forwarding to the mobile device. By tuning these parameters different download strategies were studied in [21], reporting energy conservation of up to 50% in memory-constrained proxies. Integrating all their aforementioned research and experimental work on proxies for the BitTorrent protocol to enable its adoption by mobile users in an energy-efficient manner, Kelenyi et al. put together the ProxyTorrent protocol [8].

This protocol combines all the functionality that was previously described and introduces adaptive strategies to optimize download times and energy consumption according to the actual conditions. Simulation results indicate 40–50% energy savings compared to using standard BitTorrent on mobile phones, thus ensuring longer connectivity times due to the fact that mobile devices are battery powered.

Furthermore, a prototype Gnutella protocol proxy solution for energy efficiency was proposed in [2], enabling computers to go into sleep mode when they are idle and be woken up by the proxy to handle incoming file requests. The prototype was implemented on the NIC card with a low-end processor, and as reported in more detail in [85] such an approach can lead to significant energy savings, i.e., the authors claim that if 25% of computers in the United States were to use the proposed proxy solution savings of over $38 million (in 2007 values) in energy bills could be achieved.

20.5.2 Sleep-and-Wake Approaches

Sleep-based approaches involve modifying the standard behavior of peer-to-peer networks and overlays by having peers turn off and back on in order to conserve energy, or equivalently switching from idle to active state. It is evident that randomly powering down would cause problems, since peers might be in the process of downloading a file or serving others by uploading requested files. Therefore, more advanced scheduling mechanisms for deciding when peers should be active or idle are needed.

In this respect, Sucevic et al. proposed in [39] a powering down scheduling mechanism that involves peers finishing their downloads in an increasing order of their upload capacity and then going offline as soon as their download has been completed. This ensures that peers with high peer request serving capabilities remain online for a longer time. The energy savings from such a strategy can be up to half the amount of energy required in a standard P2P configuration. The proposed model is quite complex and as the authors acknowledge complexity of the model increases with the number of peers participating in the network, making it thus unviable in common P2P settings. Moreover, the proposed theoretical approach cannot be applied in a straightforward manner in existing, widely deployed P2P systems. Based on the same concept, namely that peers in a P2P system should not be left on all the time if energy conservation is to be taken into consideration, Andrew et al. [40] showed that the optimal case in terms of consumed energy is achieved when peers are online only when they are downloading at high data transfer rates, while at the same time some energy-efficient peers need to be available at all times. In this work, the duration of the download is not considered, with the optimization criteria being solely the peers' online time and their download rate. The proposed strategy is centralized, which actually negates the distributed nature of P2P systems, and assumes knowledge of the capacities of all links and power consumption of all peers. These assumptions greatly diminish the practical applicability of this approach. As with [39], the work presented in [40] also assumes the continuous presence of high upload capacity peers in the P2P network, which nonetheless raises concerns for the fairness among peers that should be inherent in P2P networks and its effect on energy efficiency as noted by [17]. Moreover, the overload imposed on these peers might speed up the exhaustion of their resources and thus moderate the apparent benefits of the proposed mechanisms.

A scheduling approach to promote energy conservation in mobile peer-to-peer systems is also proposed in the work by Gurun et al. [63]. Having first detailed their findings and observations from an in-depth energy measurement study regarding various P2P operations in mobile settings (more than 64% of the energy consumed by the wireless interface takes place when no transmissions occur), Gurun et al. introduce the concept of idle time. According to their approach, an energy aware P2P protocol running on a mobile peer instructs the peer to go into sleep mode and thus conserve energy for the duration of the idle time period. When the peer wakes up it can handle all buffered output requests and respond to incoming messages from other peers. The impact of such an approach on the proper operation of a P2P network is not discussed in the paper, which however reports simulation results that indicate noteworthy energy savings (up to 160 J) when short idle times are involved, i.e., in the range of 1 s. On the downside, the delays in forwarding messages are far from negligible, with values up to 1.2 s in the 1 s idle time period case.

Similar approaches were also proposed in [9, 83, 84, 75, 96]. In the former case, modifications in the BitTorrent protocol are introduced to enable idle peers to go into sleep mode in order to reduce their energy footprint, thus inducing potential energy savings of up to 94.6%. It is important to note that the energy savings evaluation results reported in [9] are based on a series of general assumptions that do not consider realistic deployment settings, e.g., energy consumed by a peer is the same all the time and is not influenced by whether it is participating in the P2P network or not. Analogous energy conservation modifications to the BitTorrent protocol were also the focus of the work by Blackburn and Christensen [83, 84, 93]. They introduced the notion of sleeping peers in the BitTorrent event-processing model that are considered to be temporarily unavailable to other BitTorrent peers. As with the other related works, peers can wake up when contacted by other members of the BitTorrent network, in which case active connectivity to the latter is reestablished. Energy performance benefits over the standard BitTorrent protocol of up to 46% were reported, with an adverse effect on the download time that increased by approximately 36%. Jourjon et al. modeled the potential energy benefits from switching off idle peers that participate in a P2P system comprised of operator-managed home devices [75], as suggested by the NanoDatacenter research project [22]. The proposed model assumes that the P2P overlay is used to execute a series of jobs and aims to optimize their scheduling with the minimum amount of energy consumption. The authors suggest that up to 92% energy conservation can be guaranteed, albeit under many unviable assumptions (e.g., uniform peer and link characteristics, global knowledge of peer requests, etc.) and without any practical experimentation being presented. Additionally, in the work by Lefebvre and Feeley [96] a model to establish energy efficiency in peer-to-peer storage systems is proposed, once again based on the principle that peers consume less energy when in sleep mode. The authors' model allows calculating a lower and an upper bound on the sleep time of peers, which when respected by all peers lead to a properly operating and energy-efficient P2P system. While this work lacks experimental or simulation results to validate the theoretical model, it was one of the pioneering works in energy conservation of peer-to-peer systems.

Lastly another study related to sleep-and-wake approaches is the work by Hlavacs et al. that describes research on energy-efficient distributed home environments that are modeled as peer-to-peer overlays [46, 33]. In this work, a model is proposed to redistribute the tasks assigned to different devices in order to ensure better energy performance. This is achieved by reallocating the majority of tasks into a small number of powerful devices and powering off the other ones, which could be of course powered on again in accordance with demand. The described model incorporates a series of parameters that are application dependent, e.g., home management, download sharing, etc., affecting at various levels the improvements in energy efficiency.

20.5.3 Task Allocation Optimization Approaches

Noteworthy energy savings can be achieved in peer-to-peer systems by carefully scheduling the allocation of tasks to peers, i.e., deciding which peer(s) will satisfy the request of another peer. The works by Aikebaier et al. regarding distributed P2P systems [52, 76] and cluster systems [57] were among the first ones in this domain and examined the potential energy savings in P2P overlays that enable distributed computation (similarly to grids). In both cases, power consumption is based on two models, namely the simple one where power is assumed to be consumed at a maximal rate if at least one process is running on a device, and the multilevel one where the amount of consumed power is proportional to the number of running processes. The task allocation algorithms that are introduced in [76, 57] aim at minimizing process execution time, satisfying the deadline constraints set by the owners of various processes, and reducing the consumed energy based on the aforementioned models. Task allocation is therefore formulated as a multiple constraint optimization problem regarding the previous parameters. A more in-depth simulation analysis of this task allocation algorithm, as well as a more thorough description of the associated computation and power consumption models was published by the authors as an extended version in [28].

Building on these premises, Enokido et al. [58, 72] proposed a model for P2P data transfers and accordingly extended the task allocation algorithms to reflect a client peer's selection of another peer to serve a particular file request in an energy-efficient manner. Two such algorithms were presented in [58, 72] with the respective goals of minimizing computation time and power consumption. Both these algorithms are expressed as multicriteria optimization problems based on the parameters set by corresponding computation and power consumption models. Through a thorough simulation analysis of the proposed allocation algorithms, it was shown that there were energy reductions in the range of 6.9–12.2% compared to standard round robin algorithms.

Furthermore, an energy-efficient scheduling algorithm for wireless P2P live streaming was presented in [66]. The authors model the problem as a utility maximization problem, whereby the utility is affected by the content availability, the lack of retransmissions, the battery life, the total time, and the actual streaming rate. Noteworthy energy consumption optimization has been achieved by experimenting with different values for these parameters, as shown in the relative simulation results. It is also important to note the observations made by the authors of [66], according to which selfless behavior on behalf of the peers does actually lead to longer lifetime for the whole wireless P2P overlay, as well as the fact that the broadcast overhearing nature of the wireless medium inherently reduces energy consumption.

In the context of P2P overlay networks on top of mobile ad hoc networks (MANETs), Mawji and Hassanein [36] take a similar approach by modeling as a task allocation problem the scenario of a peer optimally selecting which servers and paths to use, with the constraints of energy cost and download time. The linear optimization program that the authors propose manages to minimize download time and energy costs (the latter by a factor of 3.5), but only manages to do so under the assumption that global knowledge of all client download requests, energy costs, download times, and topological information is available. It is therefore evident and acknowledged by the authors of [36] that this approach has limited practical value, but can nevertheless serve as a benchmark for comparison to other optimization approaches.

20.5.4 Message Reduction Approaches

Message reduction approaches aim at conserving energy in P2P overlays by reducing the number of messages required for their operation and maintenance. As mentioned—but not justified through validating experimentation—in [26, 86], the reduction of P2P network traffic overhead can benefit the overall energy efficiency of the considered P2P overlays since fewer messages are transferred and processed. However, this is not always the case (e.g., [97]) and hence more thorough analysis is needed. This motivated the need for studies such as [14], where an analysis of energy consumption of the BitTorrent protocol's different operations is presented. Relevant findings indicate that acting as a full peer, i.e., not only downloading but also uploading content, increases power consumption in particular when both activities take place at the same time. On the contrary, in DHT-based P2P protocols where there are many more maintenance messages compared to unstructured approaches, the energy consumption of full peers is much more significant, a fact attributable mostly to the processing of a large number of incoming messages [15]. The authors of [15] indicate that by running peers in client-only mode and blocking incoming messages the lifetime of peers, namely their power charge, can be greatly increased. Such an approach though would have adverse effects on the proper operation of the P2P protocol, as highlighted in [61].

One of the first message reduction approaches is the one presented in [24]. With a dual goal of ensuring backwards compatibility with existing Kademlia systems, while at the same time reducing the number of transferred messages to save energy, the authors put forward a selective message dropping mechanism. Specifically, in a standard Kademlia distribution participating peers decide to drop messages given a fixed probability. Such behavior can adversely affect the cooperation of peers in the Kademlia P2P overlay, but this is a compromise in order to reduce energy consumption. To hinder the possible degrading performance due to the latter behavior, the authors advocate for the inclusion of more than one replica of stored values in the Kademlia overlay so as to promote chances of successful operation. By means of simulations it is shown that a 50% drop probability reduces the consumed energy by 55%, while increasing the probability to drop a message to more than 70% yields no further improvements. Regarding Kademlia's operation, a model elaborated in [24] indicates that no significant service degradation is to be expected as long as the number of peers that selectively drop messages is small. Related to the notion of reduction in bandwidth to conserve energy in P2P overlays, [64] proposes a modification to the popular Chord DHT P2P protocol, namely BF-Chord, whereby Bloom filters are exploited to compress the information of shared files in conjunction with broadcasting, the inherent nature of which in wireless settings is considered to be advantageous from an energy efficiency point of view. However, Ding and Bhargava [13] underline that there are higher energy requirements when broadcasting of P2P messages over MANETs is concerned, although their analysis is done at a very abstract level and no validation is performed.

In addition, another probabilistic message reduction approach to achieve energy efficiency when broadcasting messages in peer-to-peer overlays was illustrated in [53]. In this work a multipoint relaying scheme was introduced to limit the number of retransmission of messages that are broadcasted across a P2P overlay. In particular, not all nodes are relaying the messages to other nodes, instead some nodes are considered to be leaf nodes in that they only receive messages and do not forward them. The selection of the relay nodes is based on the level of trustworthiness of every node as perceived by its peers, namely every peer has different relay nodes. Relevant simulation results showed that an up to 70% reduction on the number of transmitted messages could be achieved compared to standard flooding techniques. An adaptive version of this approach was also recently presented in [41], whereby nodes receive the messages according to a delivery priority ranking in order to reduce the message delivery latency.

Finally, a characteristic case that describes the influence of the number of sent messages on the power consumption of P2P overlays in mobile networks is presented in [51]. The authors propose two approaches to improve the overall performance of both structured and unstructured P2P overlays operating on top of MANETs, namely message reduction (to counter the adverse effects of lost query packets in mobile settings) and gossiping (to reduce the number of messages through its inherent probabilistic forwarding behavior), respectively. It becomes evident from the thorough simulation analysis of the typical cases of Chord and Gnutella that on one side message reduction increases energy consumption in the former case, while on the other side energy consumption is reduced in the case of Gnutella by at most 32%.

These simulation results validate the benefits in terms of energy efficiency of approaches that reduce the overall number of messages sent in a P2P overlay.

20.5.5 Overlay Structure Optimization Approaches

This category comprises approaches that either modify the P2P overlay structure, i.e., its topology, during construction and maintenance or introduce new P2P overlays with the goal of improving the overall energy efficiency. One of the most promising works in this category is that of Leung and Kwok [11, 59], where topology control is used to support energy efficiency, fairness, and incentives in wireless P2P file sharing networks. Their protocol is comprised of two components, namely a selective neighbors' set construction mechanism (Adjacency Set Construction—ASC) and a community-based scheme for asynchronous wakeup (Community-based Asynchronous Wakeup—CAW) in the line of the aforementioned sleep-and-wake category. The former involves the construction of the neighbor set of each peer in the P2P network using as heuristics the remaining energy levels of peers, the popularity of the files that they hold, their past contribution in P2P file sharing, and the ratio of requests over response, which indicates the selfishness of a peer. CAW clusters peers with similar characteristics (and therefore high probability of storing intercluster popular files) into groups, each of which applies a different wakeup schedule. We do not go into further details on CAW here, since its functionality is similar to the approaches presented in the sleep-and-wake category. The ASC component controls the topology of the P2P overlay network by instructing the peers to connect to other peers taking into account the previously mentioned heuristics. Regarding energy efficiency, connections are established so that nodes with fewer remaining energy levels are not going to be considered as relay nodes in the P2P overlay network, due to the valid assumption that when a node relays information for another node it consumes more energy. This strategy will allow for the P2P overlay to have a longer lifetime, since it will conserve the overall energy efficiency. By means of extensive simulation, findings presented in [59] validated the design goals of the topology control protocol: there is a considerable amount of energy saved in creating and maintaining this novel P2P overlay (ASC), while the added benefit of the CAW component reduces power consumption by at least 15%. It is important to note that this approach is not only beneficial in that it reduces global energy consumption in the P2P overlay, but in parallel it creates an evenly constructed P2P overlay in terms of energy, namely the position of the peers is such that energy fairness is promoted.

Energy efficiency is the motivation behind the hierarchical peer-to-peer system proposed in [79]. The authors propose a two-layer P2P system called the energy-greedy system comprised of regular peers and super-peers; the latter are used to ensure rapid resource discovery and are the peers with the highest levels of remaining energy. The construction of the P2P system follows a distributed approach that builds a maximal independent set of the underlying network graph, using the energy status of peers as a heuristic value (a similar approach is also presented in [44], as well as in [119] for the selection of Gnutella ultra-peers in order to make it feasible for such P2P overlays to run for longer times on mobile devices by conserving their power and in [32] to build an energy-efficient P2P overlay for resource discovery regarding wireless sensor networks). Moreover, three different energy-efficient routing schemes—trying to balance and better exploit the energy levels of the regular peers—are proposed to ensure a longer lifetime for the P2P system. Simulation results validate that such a P2P system has a slightly longer lifetime compared to a standard P2P system constructed using a random number heuristic instead of the energy levels of peers. As for the different routing schemes, it was shown that the one that utilizes shorter routes with the most powerful regular peers outperforms the others. Building on this work, the same research group proposed extensions to the energy-greedy system. In particular, in [80] the authors proposed changing the static super-peer assignment to a more dynamic one by replacing super-peers whose energy falls below a certain threshold with regular peers that have higher remaining levels of energy. Accordingly the routing schemes are adapted to take into account the new topology of the two-layer P2P system. This improvement was spurred by the observation that super-peers handle more traffic load than regular peers and so their energy levels become depleted quicker than those of regular peers, hence the need for them to be periodically replaced. Relevant simulation experiments showed that there could be up to a 146% increase in the network lifetime compared to the P2P system discussed in [79]. Further improvements to this work included the introduction of peer mobility as an additional super-peer selection heuristic [23], as well as the modification of the IEEE 802.11 Power Saving Mode in conjunction with the aforementioned two-layer P2P system to further decrease energy consumption [19]. While the principle of fairness among peers and its positive effects on energy efficiency is prominent in these approaches, a major shortcoming involves the need to modify existing systems and introduce new protocols that have no backwards compatibility with current solutions. Therefore, their practical deployment is hindered.

Recently, another approach to reduce energy consumption of peer-to-peer overlays built on top of mobile ad hoc networks and based on the theory of topology control was proposed [20]. Building on the observation that well matching overlay and underlay structures in MANETs imply less physical layer hops corresponding to virtual overlay hops, Mawji et al. introduce a topology control algorithm to reduce energy consumption and response times (it should be noted that similar benefits could be also achieved by selectively positioning resources in the P2P overlay in order to reduce the search costs as proposed in [45]). The algorithm is distributed and each node connects to neighbors based on two heuristics, namely the remaining energy levels of peers (higher levels are preferred to increase network lifetime) and the physical layer distance to other peers (shorter paths are preferred to improve response times and promote reliability, while in parallel reducing energy consumption by minimizing the required number of wireless transmissions). Maintenance of the created overlay is handled by the same algorithm that operates on a recursive basis taking into account potential changes in the energy levels or the underlay topology.

Extensive simulation experimentation for the proposed algorithm was detailed in [20], according to which the constructed overlay is relatively stable with certain parameters' configurations leading to results comparable to the ones obtained with an optimal, fully centralized solution. Moreover, it is shown that the overlay constructed using the aforementioned algorithm has a longer lifetime compared to standard approaches. Analogous motivations, i.e., minimizing the underlay distance of overlay peers to ameliorate performance, spurred the research work presented in [31], in which local peer interactions are favored over remote ones. Simulation results exhibit that the energy consumption is "an order of magnitude" [31] less than the standard peer-to-peer approach, attributed mostly to the reduced number of packet collisions and necessary retransmissions that occur due to the optimized, locality-based topology.

One issue that was raised in [20], but not explicitly addressed, is the influence of the peer degree on the energy efficiency of the P2P overlay. Intuitively, a small peer degree will reduce the control traffic and thus the power-consuming transmissions, albeit at the cost of creating a rather loosely connected overlay that might involve longer physical layer routes and hence more retransmissions. Macedo et al. recently studied this aspect using fuzzy logic-based mechanisms that allow for the tuning of an unstructured mobile P2P overlay and in particular of the maximum allowed number of neighbors to improve the performance of the latter [38]. The proposed approach ensures that the resource-constrained mobile network does not get saturated with messages. The fuzzy controller introduced by the authors was designed to take into account parameters such as link error, background traffic, and load of queries and accordingly modify the size of the peer degree in the overlay. It was successfully tested on a Gnutella-like network by means of simulation with hit rates as high as standard, fixed degree approaches, but with the added benefit that it works optimally for a variety of conditions and can adapt automatically. As far as energy consumption is concerned, it was deemed to be average and thus acceptable since there are actually more packets sent overall, but less collisions are observed and hence less retransmissions will need to take place.

20.5.6 Location-Based Approaches

The common characteristic of the approaches that are classified in this category is the fact that they exploit the location, i.e., positioning, information of peers in order to improve the energy efficiency of the overall P2P overlay. In this respect, the overlay topology structure optimization approaches dealing with the matching between underlay and overlay previously mentioned could also be classified under this category, e.g., [20, 31, 45]. Moreover, solutions such as the one proposed by Park and Valduriez [47], the LocP2P framework [62], the PReCinCt scheme [77], and the research by Feng et al. [10] comprise this category.

In the context of battery conservation for mobile P2P networks for location-aware queries, Park and Valduriez proposed solutions to efficiently locate spatial data in such settings with the added constraint of doing this at a low energy cost [47].

Their approach involves periodic broadcast of the most frequently accessed data along with an index of available data in order to reduce the overall traffic overhead, since in wireless environments all peers can simultaneously listen to the broadcast channel. Moreover, a sleep-and-wake mechanism is utilized to further reduce the energy consumption by having peers in sleep mode and waking them up only when specific data that they are interested in is broadcasted. Location information is used to construct the P2P minimum boundary rectangle index that comprises an index of the data a peer is responsible for in the P2P overlay according to its location. Peers with adequate energy levels broadcast both the index and the associated data items, while peers with energy constraints broadcast only the index. Peers initially listen to all broadcasts and by processing them deduce which ones they are interested in. Subject to this interest they can accordingly modify their sleep and wakeup modes of operation to reduce their energy consumption. A thorough simulation analysis validates the original hypothesis in that it proves that significant energy savings can be achieved by such an approach, while additionally reducing the average data access time.

In LocP2P the—possibly partial—location information provided by the owners of the devices that participate in mobile P2P networks is used to ensure lower energy costs for resource discovery of spatial type [62]. To satisfy this goal LocP2P follows a centralized approach, where instead of ever-popular flooding techniques a server is used to exploit the location information of peers to respond to queries with a list of candidate search results and the routing paths to access them. The peer who issued a query can then directly contact these candidate peers and hence consume less energy than when it would have had to contact all of the peers through flooding. The latter was quantified to be as high as 50% compared to flooding in the simulation experimentation presented in [62]. A clear limitation of this work is the fact that it focuses on spatial data and this restricted application scope is the main reason for its significant energy savings. Following similar principles, the motivation behind the PReCinCt (Proximity Regions for Caching in Cooperative Mobile P2P Networks) scheme was to promote energy-efficient and scalable data retrieval (of any type) in mobile P2P networks [77, 98]. It works on the assumption that bandwidth conservation leads to an analogous energy conservation and tries to achieve the reduction in traffic overhead by means of a geographical data retrieval scheme. The network topology is first divided into regions, each of which is responsible for a set of keys (data is assumed to be stored as key-value tuples), and then each key within that set is mapped using a hash function to a specific location within its corresponding region. Standard geographic routing protocols are used to locate data in a lightweight manner in terms of generated traffic, while additionally caching of data between peers in the same region is pursued to further reduce traffic overhead. The simulation results presented in [77] show that in cases where there is high node density there can be up to 73% savings in energy consumption compared to simple flooding-based approaches.

From a different viewpoint, the work by Feng et al. [10] examines the potential for reducing energy consumption in peer-to-peer networks by means of airborne relaying, namely using additional airborne relay peers to increase coverage and improve connectivity in peer-to-peer networks operating on top of mobile ad hoc or wireless sensor networks. According to the authors of [10] such an approach can reduce the average number of hops in these multihop networks and thus decrease the consumed energy since less retransmission of messages will eventually be needed. An elaborate model for the positioning of the airborne relay peers is detailed in order to ensure continuous and optimal connectivity, while the authors using this analytical model also proved that such an approach can lead to much greater area coverage compared to standard P2P networks but with the added benefit of significant energy savings.

20.5.7 Other Approaches

For the sake of completeness of this state-of-the-art review of energy-efficient solutions for P2P networking and overlays, we briefly describe here some approaches that cannot be explicitly classified under any of the aforementioned categories. In [34] a simulation study of the BitTorrent P2P file-sharing overlay is elaborated, focusing on the mechanism used to download the different pieces of a file from the available owners of these pieces. While in standard BitTorrent this takes place in a random manner for the first piece and then using a rarest piece first and strict priority strategy, the authors of [34] examined the behavior of BitTorrent in the cases where the pieces are being downloaded using diverse strategies, namely random piece download and solely rarest piece first. The corresponding simulation results highlighted the potential energy savings from switching to a solely rarest piece first download strategy.

Another interesting research work is the one presented in [60], where a traffic sampling technique is exploited to reduce the overall energy consumption of mobile P2P networks. In particular, this technique is used to derive optimal delivery and routing of peer information in regards to their position and neighboring peers in the overlay. Evidently, such an approach cannot be used in real time for a P2P network, since global knowledge of peer and connection information is required, but it can nonetheless serve as a guideline or benchmark for comparisons with alternative, energy-efficient P2P solutions. Finally, the work by Wang et al. in reducing energy management delays in clusters of disks using P2P overlays should be mentioned [29, 30]. This approach is different in that it does not directly involve energy efficiency of P2P networks; it rather exploits P2P functionality to improve the energy consumption of disk servers by an average of 5.1% as indicated in [29]. The proposed way to achieve this reduction is by considering each disk as a peer in a P2P network and accordingly distributing tasks among disks in an energy-efficient manner, i.e., only the disk that is handling a request is online, while the rest are in sleep mode.

20.6 DISCUSSION AND CONCLUSIONS

The overview of green P2P approaches that we presented mainly aims to serve two goals. The first is to systematically categorize the available solutions and thus provide a straightforward source of reference to interested scholars and researchers. Second, by exposing the previous and current state of the research in this particular domain, motivated researchers can begin their work having identified potential novel approaches that build on existing work or groundbreaking approaches that have not yet been considered and applied. From our point of view, we envisage that this chapter satisfies both these goals and can therefore lead to further research work on energy-efficient peer-to-peer networking, systems, and overlays. The classification scheme that we provided could be utilized as a set of guidelines to future research, since most existing works either have shortcomings such as focusing on a sole P2P system and thus cannot be generalized, or have just been recently proposed and have therefore much room for improvement. Additionally, this classification of existing research works can assist researchers in green P2P networking to establish possible synergies between the techniques and mechanisms that have been applied to date.

We firmly believe that there is a lot of room for further research in the area and that in doing so researchers would benefit from taking into account some specific issues that we have identified through our involvement in this domain.

In this regard, a realistic and detailed measurement of the energy behavior of P2P overlays is a major requirement. This is essential since only by being aware of how different operations and functionalities of the great diversity of P2P systems influence their energy consumption patterns will we be able to propose ameliorations in these patterns and hence make them more energy efficient. Moreover, the same motivation triggers the need for accurate analytical models of P2P systems in order to be able to quickly and at a theoretical level examine whether particular solutions can actually improve energy efficiency of P2P systems. This can achieve the benefit of saving efforts and time in developing solutions that can prove to be unsatisfactory.

Furthermore, one concern that we need to raise involves the different ways to represent energy costs as expressed in the studied literature. Out of all the research papers that we examined there was no standard and uniform way of expressing the energy costs attributed to P2P systems. In order to propose realistic and widely deployed green P2P solutions that will be adopted by interdisciplinary groups of people and industries, such uniformity in the metrics used to define and calculate energy efficiency is of paramount importance.

Lastly, one concern regarding energy-efficient solutions for P2P networking that should not be neglected is the fact that energy costs should not be calculated solely as far as peers are concerned, but also regarding the communication costs of the P2P networks. This is an aspect that is usually not addressed, but overlooking it can lead to potentially misleading conclusions and P2P designs [22]. Ultimately, a P2P system is a distributed system and as such every related energy analysis should consider both power consumption at end hosts, i.e., peers, as well as the energy costs of corresponding communication. In this respect, studies such as [55, 68, 70] are very interesting, in that they compare the overall energy costs of P2P systems to those of other content distribution approaches, i.e., centralized and content-centric ones.

ACKNOWLEDGEMENTS

The work presented in this chapter was conducted in the context of the BioMPE (Bio-inspired Monitoring of Pervasive Environments) project, which is financially supported by the Swiss National Science Foundation (SNF) under Grant No. 200021_130132.

REFERENCES

[1] O. Zhang, B.E. Helvik, Towards green P2P: understanding the energy consumption in P2P under content pollution, in: Proceedings of IEEE/ACM International Conference on Green Computing and Communications and IEEE/ACM International Conference on Cyber, Physical and Social Computing, 2010.

[2] M. Jimeno, K. Christensen, A prototype power management proxy for Gnutella peer-to-peer file sharing, in: Proceedings of 32nd IEEE Conference on Local, Computer Networks, 2007.

[3] X. Wang, A. Vasilakos, M. Chen, Y. Liu, T.T. Kwon, A survey of green mobile networks: opportunities and challenges, Journal of Mobile Networks and Applications 17 (1) (2012)

[4] K. Kant, Challenges in distributed energy adaptive computing, ACM SIGMETRICS Performance Evaluation Review 37 (3) (2009)

[5] R. Rodrigues, P. Druschel, Peer-to-peer systems, Communications of the ACM 53 (10) (2010)

[6] E.K. Lua, J. Crowcroft, M. Pias, R. Sharma, S. Lim, A survey and comparison of peer-to-peer overlay network schemes, IEEE Communications Surveys and Tutorials 7 (2) (2005)

[7] Z. Ou, E. Harjula, O. Kassinen, M. Ylianttila, Performance evaluation of a Kademlia-based communication-oriented P2P system under churn, Computer Networks 54 (5) (2010)

[8] I. Kelenyi, A. Ludanyi, J.K. Nurminen, Using home routers as proxies for energy-efficient BitTorrent downloads to mobile phones, IEEE Communications Magazine 49 (6) (2011)

[9] Y.-J. Lee, J.-H. Jeong, H.-Y. Kim, C.H. Lee, Energy-saving set-top box enhancement in BitTorrent networks, in: Proceedings of IEEE Network Operations and Management Symposium (NOMS), 2010.

[10] Q. Feng, J. McGeehan, A.R. Nix, Enhancing coverage and reducing power consumption in peer-to-peer networks through airborne relaying, in: Proceedings of 65th IEEE Vehicular Technology Conference (VTC-Spring), 2007.

[11] A. Ka-Ho Leung, Y.-K. Kwok, On topology control of wireless peer-to-peer file sharing networks: energy efficiency, fairness and incentive, in: Proceedings of Sixth IEEE

International Symposium on a World of Wireless Mobile and Multimedia Networks (WoWMoM), 2005.

[12] R. Bolla, R. Bruschi, F. Davoli, F. Cucchietti, Energy efficiency in the future Internet: a survey of existing approaches and trends in energy-aware fixed network infrastructures, IEEE Communications Surveys and Tutorials 3 (2) (2011)

[13] G. Ding, B. Bhargava, Peer-to-peer file-sharing over mobile ad hoc networks, in: Proceedings of Second IEEE Annual Conference on Pervasive Computing and Communications Workshops, 2004.

[14] J.K. Nurminen, J. Nöyränen, Energy-consumption in mobile peer-to-peer – quantitative results from file sharing, in: Proceedings of IEEE Consumer Communications and Networking Conference (CCNC), 2008.

[15] I. Kelenyi, J.K. Nurminen, Energy aspects of peer cooperation – measurements with a mobile DHT system, in: Proceedings of IEEE International Conference on Communications (ICC), 2008.

[16] W. Van Heddeghem, W. Vereecken, M. Pickavet, P. Demeester, Energy in ICT – trends and research directions, in: Proceedings of IEEE International Symposium on Advanced Networks and Telecommunication Systems (ANTS), 2009.

[17] J.K. Buhagiar, C.J. Debono, Optimizing multicast protocols to reduce energy dissipation in mobile peer networks, in: Proceedings of IEEE Wireless Communications and Networking Conference (WCNC), 2010.

[18] G. Anastasi, I. Giannetti, A. Passarella, A BitTorrent proxy for green Internet file sharing: design and experimental evaluation, Computer Communications 33 (7) (2010)

[19] J.-H. Lee, T.-H. Kim, J.-W. Song, K.-J. Lee, S.-B. Yang, An effective power saving mechanism for IEEE 802.11 PSM in double-layered mobile P2P systems, in: Proceedings of Second International Conference on Advanced Communication and Networking (ACN), 2010.

[20] A. Mawji, H. Hassanein, X. Zhang, Peer-to-peer overlay topology control for mobile ad hoc networks, Pervasive and Mobile Computing 7 (4) (2011)

[21] I. Kelenyi, A. Ludanyi, J.K. Nurminen, Energy-efficient BitTorrent downloads to mobile phones through memory-limited proxies, in: Proceedings of IEEE Consumer Communications and Networking Conference (CCNC), 2011.

[22] V. Valancius, N. Laoutaris, L. Massoulié, C. Diot, P. Rodriguez, Greening the Internet with nano data centers, in: Proceedings of Fifth ACM International Conference on emerging Networking EXperiments and Technologies (CoNEXT), 2009.

[23] S.-K. Kim, K.-J. Lee, S.-B. Yang, An enhanced super-peer system considering mobility and energy in mobile environments, in: Proceedings of Sixth International Symposium on Wireless and, Pervasive Computing, 2011.

[24] I. Kelenyi, J.K. Nurminen, Optimizing energy consumption of mobile nodes in heterogeneous Kademlia-based distributed hash tables, in: Proceedings of Second International Conference on Next Generation Mobile Applications, Services and Technologies (NGMAST), 2008.

[25] I. Kelenyi, J.K. Nurminen, M. Matuszewski, DHT performance for peer-to-peer SIP – a mobile phone perspective, in: Proceedings of IEEE Consumer Communications and Networking Conference (CCNC), 2010.

[26] N. Shah, D. Qian, An efficient structured P2P overlay over MANET, in: Proceedings of Ninth ACM International Workshop on Data Engineering for Wireless and Mobile Access (MobiDE), 2010.

[27] L.B. Oliveira, I.G. Siqueira, D.F. Macedo, Evaluation of peer-to-peer network content discovery techniques over mobile ad hoc networks, in: Proceedings of 6th IEEE International Symposium on a World of Wireless Mobile and Multimedia Networks (WoWMoM), 2005.

[28] T. Enokido, A. Aikebaier, M. Takizawa, Process allocation algorithms for saving power consumption in peer-to-peer systems, IEEE Transactions on Industrial Electronics 58 (6) (2011)

[29] G. Wang, A.R. Butt, C. Gniady, P. Bhattacharjee, A light-weight approach to reducing energy management delays in disks, in: Proceedings of IEEE International Green Computing Conference (IGCC), 2010.

[30] G. Wang, A.R. Butt, C. Gniady, P. Bhattacharjee, Mitigating disk energy management delays by exploiting peer memory, in: Proceedings of 17th IEEE/ACM International Symposium on Modelling, Analysis and Simulation of Computer and Telecommunication Systems (MASCOTS), 2009.

[31] P. Michiardi, G. Urvoy-Keller, Performance analysis of cooperative content distribution in wireless ad hoc networks, in: Proceedings of 4th Annual Conference on Wireless on Demand Network Systems and Services (WONS), 2007.

[32] S. Sioutas, K. Oikonomou, G. Papaloukopoulos, M. Xenos, Y. Manolopoulos, Building an efficient P2P overlay for energy-level queries in sensor networks, in: Proceedings of International ACM Conference on Management of Emergent Digital EcoSystems (MEDES), 2009.

[33] H. Hlavacs, K.A. Hummel, R. Weidlich, A.M. Houyou, H. de Meer, Modeling energy efficiency in distributed home environments, International Journal of Communication Networks and Distributed Systems 4 (2) (2010)

[34] N. Gaddam, A. Potluri, Study of BitTorrent for file sharing in ad hoc networks, in: Proceedings of 5th IEEE Conference on Wireless Communication and Sensor Networks (WCSN), 2009.

[35] L.B. Oliveira, I.G. Siqueira, A.A.F. Loureiro, On the performance of ad hoc routing protocols under a peer-to-peer application, Journal of Parallel and Distributed Computing 65 (11) (2005)

[36] A. Mawji, H. Hassanein, Optimal path selection for file downloading in P2P overlay networks on MANETs, in: Proceedings of IEEE Symposium on Computers and Communications (ISCC), 2010.

[37] Y. Agarwal, S. Hodges, R. Chandra, J. Scott, P. Bahl, R. Gupta, Somniloquy: augmenting network interfaces to reduce PC energy usage, in: Proceedings of USENIX Symposium on Networked Systems Design and Implementation (NSDI), 2009.

[38] D.F. Macedo, A.L. dos Santos, J.M. Nogueira, G. Pujolle, Fuzzy-based load self-configuration in mobile P2P services, Computer Networks 55 (8) (2011)

[39] A. Sucevic, L.L.H. Andrew, T.T.T. Nguyen, Powering down for energy efficient peer-to-peer file distribution, in: Proceedings of ACM GreenMetrics, 2009.

[40] L.L.H. Andrew, A. Sucevic, T.T.T. Nguyen, Balancing peer and server energy consumption in large peer-to-peer file distribution systems, in: Proceedings of IEEE Online Conference on Green Communications (GreenCom), 2011.

[41] A.B. Waluyo, D. Taniar, B. Srinivasan, W. Rahayu, M. Takizawa, Adaptive and efficient data dissemination in mobile P2P environments, in: Proceedings of Workshops of International Conference on Advanced Information Networking and Applications (AINA), 2011.

[42] D.N. da Hora, D.F. Macedo, L.B. Oliveira, I.G. Siqueira, A.A.F. Loureiro, J.M. Nogueira, G. Pujolle, Enhancing peer-to-peer content discovery techniques over mobile ad hoc networks, Computer Communications 32 (13–14) (2009)
[43] L.A. Barroso, U. Hölzle, The case for energy-proportional computing, IEEE Computer Magazine 40 (12) (2007)
[44] S. Rollins, J.D. Porten, K. Brisbin, C. Chang-Yit, A battery-aware algorithm for supporting collaborative applications, Mobile Networks and Applications 17 (3) (2012)
[45] W. Rao, L. Chen, A. Wai-Chee Fu, G. Wang, Optimal resource placement in structured peer-to-peer networks, IEEE Transactions on Parallel and Distributed Systems 21 (7) (2010)
[46] A.E. Garcia, R. Weidlich, L. Rodriguez de Lope, K.D. Hackbarth, H. Hlavacs, C. san Leandro, Approximation towards energy-efficient distributed environments, in: Proceedings of Third International ICST Conference on Simulation Tools and Techniques for Communications, Networks and Systems (SimulationWorks), 2010.
[47] K. Park, P. Valduriez, Energy efficient data access in mobile P2P networks, IEEE Transactions on Knowledge and Data Engineering 23 (11) (2011)
[48] H. Hlavacs, Modeling energy efficiency of file sharing, Springer Elektrotechnik and Informationstechnik 127 (11) (2010)
[49] Z. Ou, E. Harjula, O. Kassinen, M. Ylianttila, Feasibility evaluation of a communication-oriented P2P system in mobile environments, Proceedings of Sixth International Conference on Mobile Technology, Applications, and Systems, Mobility Conference (Mobility), ACM.
[50] O. Kassinen, Z. Ou, M. Ylianttila, E. Harjula, Effects of peer-to-peer overlay parameters on mobile battery duration and resource lookup efficiency, Proceedings of Seventh International Conference on Mobile and Ubiquitous Multimedia (MUM), ACM.
[51] D.N. da Hora, D.F. Macedo, J.M.S. Nogueira, G. Pujolle, Optimizing peer-to-peer content discovery over wireless mobile ad hoc networks, Proceedings of Ninth International Conference on Mobile and Wireless Communications Networks, IEEE.
[52] A. Aikebaier, T. Enokido, S. Misbah Deen, M. Takizawa, Energy-efficient agreement protocols in P2P overlay networks, in: Proceedings of IEEE 30th International Conference on Distributed Computing Systems Workshops, 2010.
[53] I. Kelenyi, J.K. Nurminen, CloudTorrent – Energy-efficient BitTorrent content sharing for mobile devices via cloud services, in: Proceedings of IEEE Consumer Communications and Networking Conference (CCNC), 2010.
[54] I. Kelenyi, A. Ludanyi, J. K. Nurminen, I. Puustinen, Energy-efficient mobile BitTorrent with broadband router hosted proxies, in: Proceedings of Third Joint IFIP/IEEE Wireless and Mobile Networking Conference (WMNC), 2010.
[55] U. Lee, I. Rimac, D. Kilper, V. Hilt, Towards energy-efficient content dissemination, IEEE Network Magazine 25 (2) (2011)
[56] H. Hlavacs, R. Weidling, T. Treutner, Energy efficient peer-to-peer file sharing, Springer Journal of Supercomputing, OnlineFirst, (2011).
[57] A. Aikebaier, T. Enokido, M. Takizawa, Distributed cluster architecture for increasing energy efficiency in cluster systems, Proceedings of International Conference on Parallel Processing Workshops, IEEE.
[58] T. Enokido, A. Aikebaier, M. Takizawa, A model for reducing power consumption in peer-to-peer systems, IEEE Systems Journal 4 (2) (2010)
[59] A. Ka-Ho Leung, Y.-K. Kwok, On localized application-driven topology control for energy-efficient wireless peer-to-peer file sharing, IEEE Transactions on Mobile Computing 7 (1) (2008)

[60] J.K. Buhagiar, C.J. Debono, Exploiting traffic sampling techniques to optimize energy efficiency in mobile peer networks, Proceedings of First International Conference on Advances in P2P Systems (AP2PS), IEEE.

[61] M. Feldman, J. Chuang, Overcoming free-riding behavior in peer-to-peer systems, ACM SIGecom Exchanges 5 (4) (2005)

[62] Y.-C. Tung, K.C.-J. Lin, Location-assisted energy-efficient content search for mobile peer-to-peer networks, Proceedings of Seventh International Workshop on Mobile Peer-to-Peer Computing, IEEE.

[63] S. Gurun, P. Nagpurkar, B.Y. Zhao, Energy consumption and conservation in mobile peer-to-peer systems, Proceedings of First International Workshop on Decentralized Resource Sharing in Mobile Computing and Networking (MobiShare), ACM.

[64] S. Wang, H. Ji, Y. Li, BF-Chord: an improved lookup protocol to Chord based on Bloom filter for wireless P2P, Proceedings of Fifth International Conference on Wireless Communications, Networking and Mobile Computing (WiCom), IEEE.

[65] I. Stoica, R. Morris, D. Liben-Nowell, D.R. Karger, M. Frans Kaashoek, F. Dabek, H. Balakrishnan, Chord: a scalable peer-to-peer lookup protocol for Internet applications, IEEE/ACM Transactions on Networking (TON) 11 (1) (2003)

[66] Y. Li, Z. Li, M. Chiang, A.R. Calderbank, Energy-efficient video transmission scheduling for wireless peer-to-peer live streaming, in: Proceedings of Sixth IEEE Consumer Communications and Networking Conference (CCNC), 2009.

[67] I. Kelenyi, J.K. Nurminen, Bursty content sharing mechanism for energy-limited mobile devices, in: Proceedings of Fourth ACM Workshop on Performance Monitoring and Measurement of Heterogeneous Wireless and Wired Networks (PM2HW2N), 2009.

[68] U. Lee, I. Rimac, V. Hilt, Greening the Internet with content-centric networking, in: Proceedings of First International Conference on Energy-Efficient Computing and Networking (e-Energy), 2010.

[69] S. Nedevschi, S. Ratnasamy, J. Padhye, Hot data centers vs. cool peers, in: Proceedings of Conference on Power Aware Computing and Systems (HotPower), USENIX, 2008.

[70] A. Feldmann, A. Gladisch, M. Kind, C. Lange, G. Smaragdakis, F.-J. Westphal, Energy trade-offs among content delivery architectures, Proceedings of Ninth Conference on Telecommunications Internet and Media Techno Economics (CTTE), IEEE.

[71] A.P. Bianzino, C. Chaudet, D. Rossi, J.-L. Rougier, A survey of green networking research, IEEE Communications Surveys and Tutorials 14 (1) (2012)

[72] T. Enokido, K. Suzuki, A. Aikebaier, M. Takizawa, Laxity based algorithm for reducing power consumption in distributed systems, Proceedings of International Conference on Complex, Intelligent and Software Intensive Systems, IEEE.

[73] O. Kassinen, E. Harjula, J. Korhonen, M. Ylianttila, Battery life of mobile peers with UMTS and WLAN in a Kademlia-based P2P overlay, Proceedings of International Symposium on Personal, Indoor and Mobile Radio Communications (PIMRC), IEEE.

[74] I. Kelenyi, A. Ludanyi, J.K. Nurminen, BitTorrent on mobile phones – energy efficiency of a distributed proxy solution, Proceedings of International Green Computing Conference (GREENCOMP), IEEE.

[75] G. Jourjon, T. Rakotoarivelo, M. Ott, Models for an energy-efficient P2P delivery service, Proceedings of 18th Euromicro Conference on Parallel, Distributed and Network-based Processing (PDP), IEEE.

[76] A. Aikebaier, Y. Yang, T. Enokido, M. Takizawa, Energy-efficient computation models for distributed systems, Proceedings of International Conference on Network-based Information Systems, IEEE.

[77] M.S. Joseph, M. Kumar, H. Shen, S. Das, Energy efficient data retrieval and caching in mobile peer-to-peer networks, Proceedings of Third International Conference on Pervasive Computing and Communications Workshops (PERCOMW), IEEE.

[78] B. Nordman, K. Christensen, Proxying: the next step in reducing IT energy use, IEEE Computer Magazine 43 (1) (2010)

[79] J.-S. Han, J.-W. Song, T.-H. Kim, S.-B. Yang, Double-layered mobile P2P systems using energy-efficient routing schemes, Proceedings of Australasian Telecommunication Networks and Applications Conference (ATNAC), IEEE.

[80] J.-H. Lee, J.-W. Song, Y.-H. Lee, S.-B. Yang, An energy-effective routing protocol for mobile P2P systems, Proceedings of Fourth International Conference on Wireless Pervasive Computing (ISWPC), IEEE.

[81] P. Purushothaman, M. Navada, R. Subramaniyan, C. Reardon, A.D. George, Power-proxying on the NIC: a case study with the Gnutella file-sharing protocol, in: Proceedings of 31st IEEE Conference on Local Computer Networks (LCN), 2006.

[82] G. Anastasi, M. Conti, I. Giannetti, A. Passarella, Design and evaluation of a BitTorrent proxy for energy saving, in: Proceedings of IEEE Symposium on Computers and Communications (ISCC), 2009.

[83] J. Blackburn, K. Christensen, A simulation study of a new BitTorrent, in: Proceedings of IEEE International Conference on Communications Workshops (ICC), 2009.

[84] J.H. Blackburn, K.J. Christensen, Reducing the energy consumption of peer-to-peer networks, in: Proceedings of 19th Annual Argonne Symposium for Undergraduates in Science, Engineering and Mathematics, 2008.

[85] M. Jimeno, K. Christensen, A. Roginsky, A power management proxy with a new best-of-N Bloom filter design to reduce false positives, in: Proceedings of 26th IEEE International Conference on Performance, Computing and Communications (ICPCC), 2007.

[86] M. Akon, X. Shen, S. Naik, A. Singh, Q. Zhang, An inexpensive unstructured platform for wireless mobile peer-to-peer networks, Springer Peer-to-Peer Networking and Applications 1 (1) (2008)

[87] H. Barjini, M. Othman, H. Ibrahim, N. Izura Udzir, Shortcoming, problems and analytical comparison for flooding-based search techniques in unstructured P2P networks, Springer Peer-to-Peer Networking and Applications 5 (1) (2012)

[88] H. Zhao, X. Liu, X. Li, A taxonomy of peer-to-peer desktop grid paradigms, Springer Cluster Computing 14 (2) (2011)

[89] S.A. Baset, Protocols and system design, reliability, and energy efficiency in peer-to-peer communication systems, Unpublished Ph.D. Thesis at Columbia University, 2011.

[90] Z. Zhuang, S. Kakumanu, Y. Jeong, R. Sivakumar, A. Velayutham, Mobile hosts participating in peer-to-peer data networks: challenges and solutions, Springer Wireless Networks Journal 16 (8) (2010)

[91] E. Meshkova, J. Riihijärvi, M. Petrova, P. Mähönen, A survey on resource discovery mechanisms, peer-to-peer and service discovery frameworks, Elsevier Computer Networks 52 (11) (2008)

[92] Z. Ou, Structured peer-to-peer networks: hierarchical architecture and performance evaluation, Published Ph.D. Thesis at the University of Oulu, ISBN: 978-951-42-6247-0, 2010.

[93] J.H. Blackburn, Design and evaluation of a green BitTorrent for energy-efficient content distribution, Unpublished Master Thesis at the University of South Florida, 2010.

[94] M. Gerla, C. Lindemann, A. Rowstron, P2P MANET's – new research issues, in: Proceedings of Dagstuhl Perspectives Workshop: Peer-to-Peer Mobile Ad Hoc Networks – New Research Issues, 2005.

[95] G. Camarillo (Ed.), Peer-to-Peer (P2P) Architecture: Definition, Taxonomies, Examples, and Applicability, IETF RFC 5694, 2009.

[96] G. Lefebvre, M.J. Feeley, Energy efficient peer-to-peer storage, Technical Report (TR-2003-17), Department of Computer Science, University of British Columbia, 2003.

[97] L.M. Feeney, M. Nilsson, Investigating the energy consumption of a wireless network interface in an ad hoc networking environment, in: Proceedings of Annual IEEE International Conference on Computer Communications, 2001.

[98] M.S. Joseph, Energy efficient data retrieval and caching in mobile peer-to-peer networks, M.Sc. Thesis in Department of Computer Science and Engineering at University of Texas at Arlington, 2004.

[99] H. Shen, A.S. Brodie, C.-Z. Xu, W. Shi, Scalable and secure P2P overlay networks, in: J. Wu (Ed.), Handbook on Theoretical and Algorithmic Aspects of Sensor Ad Hoc Wireless and Peer-to-Peer Networks, CRC Press.

[100] S. Androutsellis-Theotokis, D. Spinellis, A survey of peer-to-peer content distribution technologies, ACM Computing Surveys 36 (4) (2004)

[101] K. Aberer, L. Onana Alima, A. Ghodsi, S. Girdzijauskas, S. Haridi, M. Hauswirth, The essence of P2P: a reference architecture for overlay networks, in: Proceedings of the Fifth IEEE International Conference on Peer-to-Peer Computing (P2P), 2005.

[102] A. Rowstron, P. Druschel, Pastry: scalable, decentralized object location, and routing for large-scale peer-to-peer systems, in: Proceedings of the IFIP/ACM International Conference on Distributed Systems Platforms and Middleware, 2001.

[103] A. Brocco, A. Malatras, B. Hirsbrunner, Enabling efficient information discovery in a self-structured grid, Elsevier Future Generation Computer Systems 26 (6) (2010)

[104] B.Y. Zhao, L. Huang, J. Stribling, S.C. Rhea, A.D. Joseph, J.D. Kubiatowicz, Tapestry: a resilient global-scale overlay for service deployment, IEEE Journal on Selected Areas in Communications 22 (1) (2004)

[105] D. Malkhi, M. Naor, D. Ratajczak, Viceroy: a scalable and dynamic emulation of the butterfly, in: Proceedings of the 21st ACM Annual Symposium on Principles of distributed computing (PODC), 2002.

[106] P. Maymounkov, D. Mazieres, Kademlia: a peer-to-peer information system based on the XOR metric, Proceedings of First International Workshop on Peer-to-Peer Systems (IPTPS), Springer.

[107] S. Ratnasamy, P. Francis, M. Handley, R. Karp, S. Schenker, A scalable content-addressable network, in: Proceedings of the 2001 Conference on Applications, Technologies, Architectures, and Protocols for Computer Communications, ACM SIGCOMM Computer Communication Review 31 (4) (2001).

[108] I. Clarke, O. Sandberg, B. Wiley, T. Hong, Freenet: a distributed anonymous information storage and retrieval system, in: Proceedings of ICSI Workshop on Design Issues in Anonymity and Unobservability, freenet.sourceforge.net, 2000.

[109] T. Klingberg, R. Manfredi, The Gnutella Protocol Specification, version 0.6, 2002. <http://rfc-gnutella.sourceforge.net/src/rfc-0_6-draft.html>.

[110] B. Cohen, Incentives build robustness in BitTorrent, in: Proceedings of Workshop on Economics in Peer-to-Peer Systems, 2003.

[111] N. J. A. Harvey, M. B. Jones, S. Saroiu, M. Theimer, A. Wolman, SkipNet: a scalable overlay network with practical locality properties, in: Proceedings of USENIX Symposium on Internet Technologies and Systems, 2003.

[112] K. Aberer, P. Cudré-Mauroux, A. Datta, Z. Despotovic, M. Hauswirth, M. Punceva, R. Schmidt, P-Grid: a self-organizing structured P2P system, ACM SIGMOD Record 32 (3) (2003)

[113] H. Shen, C.-Z. Xu, G. Chen, Cycloid: a constant-degree and lookup-efficient P2P overlay network, Elsevier Performance Evaluation Journal 63 (3) (2006)
[114] M. Jelasity, M. Van Steen, Large-scale newscast computing on the Internet, Vrjie University Amsterdam Technical Report IR-503, 2002.
[115] R.H. Wouhaybi, A.T. Campbell, Phenix: supporting resilient low-diameter peer-to-peer topologies, in: Proceedings of 23rd Annual Joint Conference of the IEEE Computer and Communications Societies (INFOCOM), 2004.
[116] J. Liang, R. Kumar, K.W. Ross, The FastTrack overlay: a measurement study, Elsevier Computer Networks 50 (6) (2006)
[117] KaZaa Homepage, October 2011. <www.kazaa.com>.
[118] M. Ripeanu, A. Iamnitchi, I. Foster, A. Rogers, In search of simplicity: a self-organizing group communication overlay, in: Proceedings of First International Conference on Self-Adaptive and Self-Organizing Systems (SASO), 2007.
[119] H.-D. Choi, M. Woo, A power-aware peer-to-peer system for ad hoc networks, in: Proceedings of Second International Conference on Mobile Ad-hoc and Sensor Networks (MSN), 2006.
[120] S. Ruth, Green IT – more than a three percent solution? IEEE Internet Computing 13 (4) (2009)
[121] H. Schulze, K. Mochalski, The impact of peer-to-peer file sharing, voice over IP, Skype, Joost, instant messaging, one-click hosting and media streaming such as YouTube on the Internet, in: IPOQUE Internet Study 2007, 2007.
[122] J. Erman, A. Mahanti, M. Arlitt, C. Williamson, Identifying and discriminating between web and peer-to-peer traffic in the network core, in: Proceedings of the 16th International Conference on World Wide Web (WWW), 2007.

CHAPTER

Power Management for 4G Mobile Broadband Wireless Access Networks

21

Maruti Gupta, Ali T. Koc, Rath Vannithamby

Intel Labs, Intel Corporation, 2111 NE 25th Ave., JF3-206, Hillsboro, OR 97124, USA

21.1 INTRODUCTION

Today, the use of devices such as smart phones, tablets, etc. that offer the ease and convenience of Internet applications like email and web browsing on the go is widespread. As such devices become commonplace, inevitably user expectations also rise in terms of higher data rates, instant Internet connectivity and a much larger variety of applications to play with. Mobile broadband technologies such as LTE [1–3] and WiMAX [4, 5] are what make the promise of such expectations real.

Long Term Evolution (LTE) is a mobile broadband technology that allows devices such as laptops, handhelds, etc. to experience high data rates of 10–30 Mbps in the downlink and 5–10 Mbps in the uplink in a highly mobile environment. It is currently being deployed in several cities around the United States and several major carriers such as AT&T and Verizon are in the process of upgrading their systems to this technology. LTE-Advanced is a further advancement to this technology that has been developed by the 3GPP standards body to meet the requirements of 4G, as laid out by the ITU's IMT-Advanced requirements, one of which is to be capable of providing a peak throughput rate exceeding 1 Gbps.

Worldwide Interoperability for Microwave Access, better known as WiMAX, is based on the IEEE 802.16e standard that offers similar services as LTE and has been operational in several cities around the world since 2009. The recent standard, 802.16m, was completed in 2010 and also meets the requirements of a 4G technology, similar to LTE-Advanced.

LTE and WiMAX offer high-speed data transfer and always-connected capabilities. The high data rates in these systems are achieved through the use of higher order modulation and coding schemes and sophisticated antennas with multiple input and multiple output (MIMO) technology. Higher speed data transmission or reception requires higher power consumption; this in turn drains the battery quickly. To support battery-operated mobile devices, 4G technology has developed power-saving features that allow the mobile devices to operate for longer durations without having to recharge. Power saving is achieved by turning off all or some parts of the device in a controlled manner when it is not actively transmitting or receiving data. 4G technologies define signaling methods that allow the mobile device to switch into Discontinuous Reception (DRX) during RRC_Connected in LTE and sleep mode in WiMAX, and to Idle mode when inactive both in LTE and WiMAX [1–5].

Power saving using the DRX mechanism in LTE and sleep mode in WiMAX is relatively a new research area; we cite some of the research work reported in literature here. Potential enhancements to improve battery power consumption and user experience were presented in [6]. In [7] a trade-off relationship between power saving and delay performance for the LTE DRX mechanism with adjustable DRX cycles was investigated. Authors in [8] have pointed out the fact that power saving by DRX in LTE comes at the cost of degraded system utility. The authors proposed an adaptive DRX-Inactivity timer based on the channel conditions between a mobile station and base station, so that a mobile station with bad channel quality will go to low-power mode for s time durations. In this way a mobile station with bad channel conditions can increase its transmission opportunity to increase average throughput at the cost of reduced power saving. Reference [9] points out the power efficiency and signaling overhead issues in efficiently supporting emerging data applications over LTE networks. Reference [9] also provides an evaluation methodology to estimate the power consumption and signaling overhead for various applications.

The rest of this chapter is structured as follows: In Section 21.2, we give an overview of power management mechanisms needed to save battery power while supporting various application' traffic. Section 21.3 covers power management mechanisms such as Idle mode and DRX in LTE standard. Section 21.4 describes comparable power management mechanisms in the WiMAX standard. Section 21.5 describes the implementation challenges in designing these power management mechanisms. Section 21.6 explains advanced schemes that have been developed and are being researched for further improvement in power savings in the next generation of 3GPP standards and we end by offering our conclusions in Section 21.7.

21.2 OVERVIEW OF POWER MANAGEMENT

Power management schemes are designed to adapt to current and expected application traffic workloads in order to obtain maximum power savings. At the time of the design of LTE and the initial WiMAX 802.16e standards (released in 2008/2006 respectively), application traffic was largely dominated by web browsing, email, file transfer, and Voice over IP (VoIP) types of applications. We discuss the traffic models of the expected workloads that were used to evaluate LTE schemes to achieve power savings. Then later on, in Section 21.6, we discuss how application characteristics have changed since then and their impact on power consumption and provide the most recent enhancements in power management.

Figure 21.1 shows a model of HTTP traffic [10], the protocol used for web browsing. Web browsing applications typically show an ON/OFF behavior, which means that the network experiences packet activity for a duration of time known as the ON period and

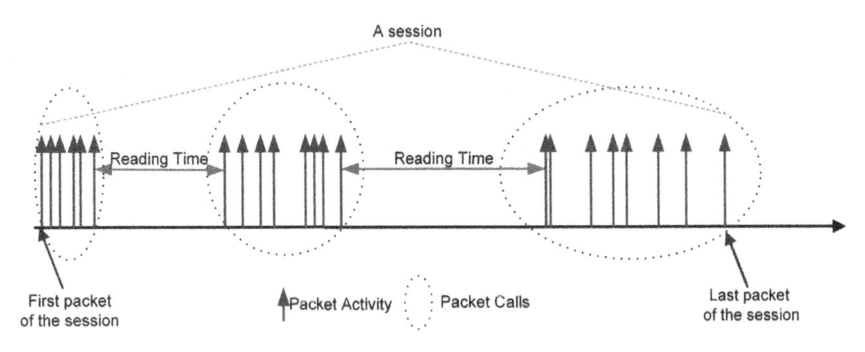

Figure 21.1 HTTP traffic model.

then there is no packet activity for the OFF duration. This is due to the fact that the typical user behavior while browsing is to click on a link and download the web page and associated objects on it and then spend some time "reading" or browsing it before another link is clicked and the process repeats itself for the duration of the session. The typical size of web pages downloaded range from a few kilobytes of data to several hundred megabytes of data, depending upon the content. In any case, once the user has completed the session, no further activity is seen and the duration of OFF periods is relatively long, on average 30 seconds or more.

Figure 21.2 shows a model of FTP traffic [10] where each session consists of a series of file downloads. This also represents an ON/OFF model where the ON durations are when the user is downloading a file which can range in size from a few kilobytes to several hundred megabytes in size and the OFF duration is marked by no packet activity. Figure 21.3 shows a model of VoIP traffic, which is also modeled as an ON/OFF model where the ON durations reflect the times when there is talk activity and the OFF durations reflect when there is silence. However, VoIP traffic differs from data activity such as web browsing and file transfer in that it is periodic and also the data rates are in the range of less than a hundred Kbps whereas data rates for web browsing and file transfer can easily be in several Mbps or as much bandwidth as can be provided.

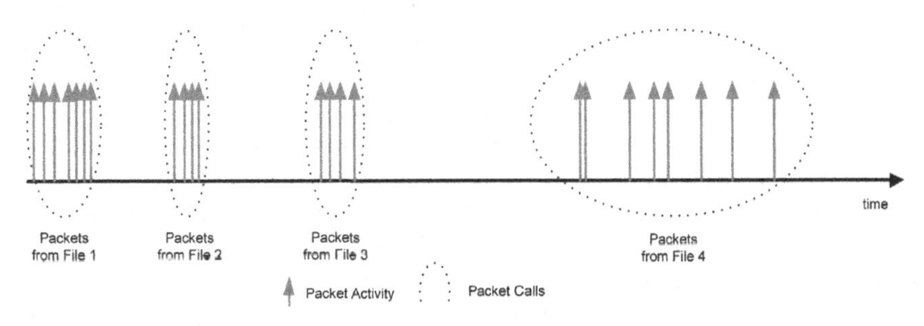

Figure 21.2 FTP traffic model.

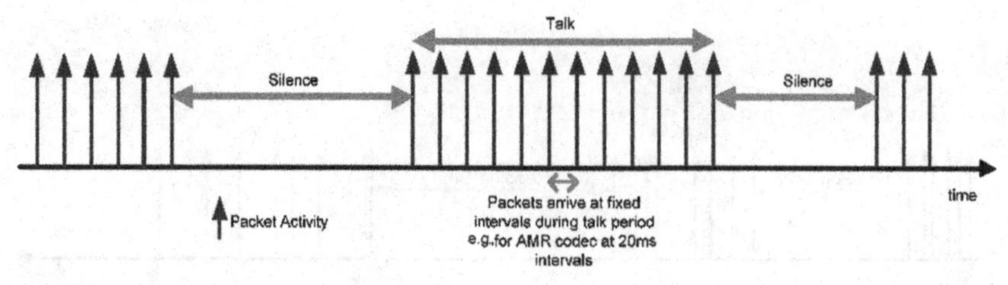

Figure 21.3 VoIP model.

As can be seen from Figures 21.1–21.3 each of these traffic models while distinct also share some common characteristics. The main characteristic that we would like to emphasize is that each of the models can be characterized by an ON/OFF model where traffic is received during the ON interval, but not during the OFF interval. In addition, there is the notion of session where once the session ends, the application no longer generates any traffic. This is no longer true for the emerging diverse data applications that are now being written for mobile devices which generate traffic all the time even when the user does not directly interact with the device. This traffic is referred to as background traffic and we shall focus more on the impact of this type of traffic in Section 21.6 of this chapter.

21.3 POWER MANAGEMENT IN LTE

LTE specifications provide two different mechanisms for power management, namely Idle mode [3] and Discontinuous Reception (DRX) [11]. User equipment (UE; 3GPP terminology for mobile station (MS), which is IEEE terminology) can enter Idle mode where it is no longer actively connected to the eNodeB (eNB; 3GPP terminology for base station (BS), which is IEEE terminology), though the network is still able to keep track of the UE through a mechanism known as paging. Idle mode allows the UE to remain in very low-power mode since the UE needs to perform a very limited set of functions in this mode; the UE can be paged for downlink (DL) traffic. For uplink traffic, the UE initiates a procedure to reenter the network by sending a connection request to the serving eNB and reenters into the RRC_Connected state (RRC_Connected: 3GPP terminology for when the UE is actively connected to the eNB). DRX is another mechanism that allows the UE to save power both in the Connected as well as the Idle mode. In active mode (RRC_Connected) DRX can save power by allowing the UE to power down for predetermined intervals, as directed by the eNB. In Idle mode, DRX can be used to further extend the time the UE spends in low-power mode by negotiating longer periods of absence from the network before the UE can be paged. DRX offers significant improvement on resource utilization as well as power saving. However, DRX increases the end-to-end delay if the parameters are not set correctly. Moreover, if the DRX cycle is too long, then there can be some scenarios where the UE may have to face network reentry. In DRX, the UE consumes less power by powering down most of its circuitry. During DRX, the UE only listens to DL control channels periodically.

21.3.1 Idle Mode in LTE

Idle mode allows the UE to save power for intervals ranging from a few seconds up to several hours and is typically the mode where the UE spends the most of its time. A UE enters the Idle mode after it has experienced a period of inactivity where no data was exchanged. This period of inactivity is generally in the range of a few to several seconds, and it is entirely dependent on the network operator. Idle mode also allows the network to conserve air-interface resources since once the UE transitions to Idle mode, the eNB releases all the resources associated with that UE. But the core network, i.e., the Evolved Packet Core (EPC), still retains information on the UE to enable the network to contact the UE in the event there is an incoming message such as a voice call or a text message or an emergency broadcast. This is accomplished by using the concept of paging. To understand how a UE may be paged, we need to understand the procedures of cell selection and registration area as explained below.

In LTE, a UE in Idle mode performs three major tasks as follows: (1) public land mobile network (PLMN) selection, (2) cell selection and reselection, and (3) location registration. When a UE is switched on, it first tries to find a PLMN. This may be done by going through a list of PLMNs that the UE has or through searching existing PLMNs. A PLMN is made up of several cells and once the UE selects a PLMN, it next tries to locate a cell in that PLMN. This procedure is called cell selection. It involves the UE searching for a suitable cell that can provide available services. The UE then tunes into the cell's control channel to be able to obtain information about the cell's registration area. This procedure of selecting an appropriate cell is called "camping on a cell." Once the UE is camped on an appropriate cell, the UE performs location registration. This means that the UE informs the EPC that it is now in a specific location as identified by the registration area of the camped cell. A registration area basically allows the UE to roam freely across all the cells in it without having to perform location registration for each cell. In other words, a registration area is a group of cells as can be seen in Figure 21.5. Once the core network is aware of the location of the UE, it can locate the UE by paging all the cells in the UE's registration area. Every time the UE moves around, it needs to make sure that the registration area of the new cell has not changed and if it does, then it needs to perform location registration again.

The UE runs radio measurements periodically and whenever it finds a more suitable cell, according to the cell reselection criteria, it starts to camp on this new cell. If this new cell belongs to a different registration area, location registration needs to happen again. The UE periodically looks for other PLMNs to find the best PLMN to camp on. If the UE loses coverage of the registered PLMN, either a new PLMN is selected automatically, or an indication of which PLMNs are available is given to the UE, so that a manual selection can be made. A UE camps in Idle mode so that it can receive system information from the PLMN; also if an RRC connection is needed it can easily access the network' control channel of the cell on which it is camped. Also, the UE can be paged if it is camped on a cell and registered to a registration area.

Once the UE has camped on a cell, it also gets to know the paging cycle of the cell, which is broadcast in its control channel. During every paging cycle, the eNB sends out a paging message at a known period of time known a paging occasion. Thus, the UE can wake up during the paging occasion and listen to the paging message to check and see if it is being paged. The paging occasion is kept very short; it's only a few milliseconds long and it does not require the UE to be connected to the network; the UE doesn't need to perform the network entry procedures of scanning, authentication, key exchange, getting an IP address etc. that it needs in order to transfer/receive data to/from the network. Thus during the Idle mode, the UE alternates between being completely unavailable to the network and being available for short durations during the paging occasion as shown in Figure 21.5.

21.3.2 DRX in LTE

DRX allows a UE to save power for shorter intervals of time while the UE is still connected to the network and it also enables the UE to extend its time spent in low-power mode during Idle mode by skipping paging cycles to check for paging messages and thus extending the time spent in low-power mode. The interval of time a UE may use DRX during Connected mode can range from a few milliseconds to minutes. Thus, during RRC_Connected mode, the DRX mechanism allows the UE to exchange data with the network and still save power during intervals when there is no data activity.

The main DRX framework is illustrated in Figure 21.4.The DRX cycle consists of ON duration and DRX opportunity. During ON duration, the UE is always active, which means the radio is turned on and there is no power saving enabled. During DRX opportunity, the UE does not listen to the control channel and also does not send channel quality information on the uplink control channel. The configuration of the DRX cycle becomes an important problem and needs to be optimized to balance the power saving and delay requirements without incurring network reentry and high end-to-end delay [12].

In LTE, DRX is triggered by means of an inactivity timer known as the DRX-Inactivity timer, which can range in value from the smallest possible, i.e., 1 ms, up to 2.56 s, though the values in between are not continuous, but selected discrete values as defined by the standard in [11]. This means that the UE monitors the downlink control channel (PDCCH) to see if there is any data activity. As shown in Figure 21.6, whenever the UE receives any data, the DRX-Inactivity timer is reset. The UE does not initiate DRX cycles until the timer expires. Once the timer expires, the UE will first go into the DRX ON duration of the DRX cycle. During this period, the UE basically monitors the channel for data and control activity and the eNB is able to exchange data with the UE. During the OFF duration, the UE can go into low-power mode and the eNB cannot send any data to the UE. In addition to the DRX-Inactivity timer, a UE may also be constrained from entering DRX if it has any pending retransmission timers associated with data sent or received. DRX is terminated as soon as the UE either

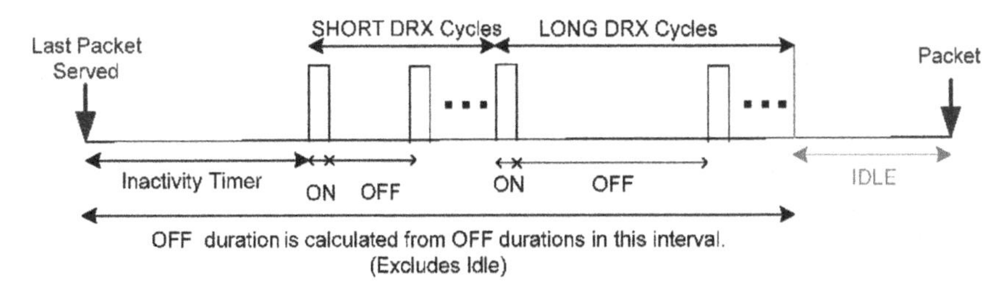

Figure 21.4 LTE DRX and Idle.

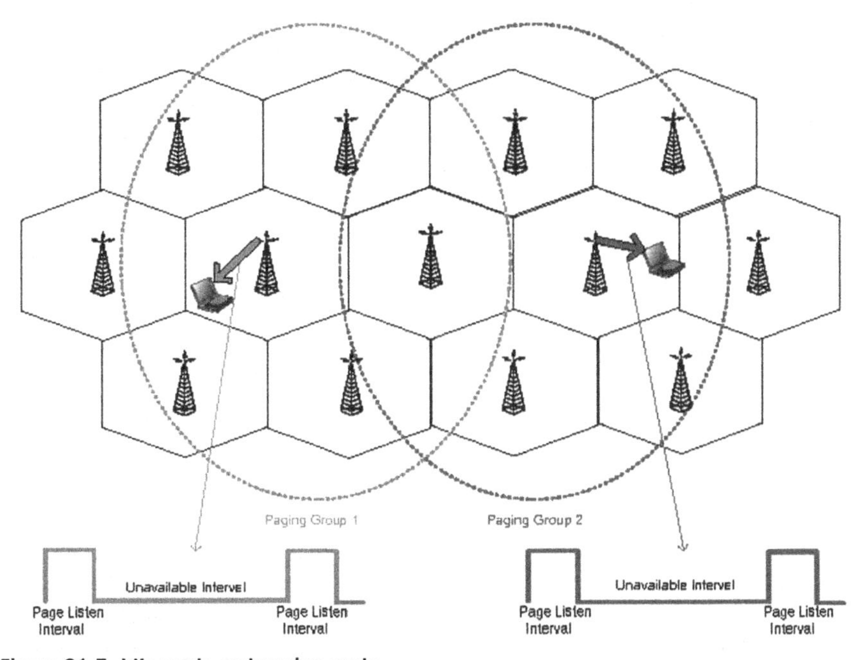

Figure 21.5 Idle mode and paging cycle.

sends UL data or receives DL data. This means that in LTE, DRX cannot be enabled during an active data exchange without restarting the DRX-Inactivity timer. LTE also supports the notion of short DRX and long DRX. Short DRX basically allows the UE to have a shorter DRX cycle and it is also limited to a predetermined number of cycles only. If no data is exchanged during the ON period of the short DRX cycles, only then does the UE transition to long DRX as shown in Figure 21.4. The long DRX cycle may be much longer than the short DRX cycle thus allowing the UE to gain greater power savings. Short DRX was introduced to reduce delays in case data activities were to occur very soon after initiation of DRX.

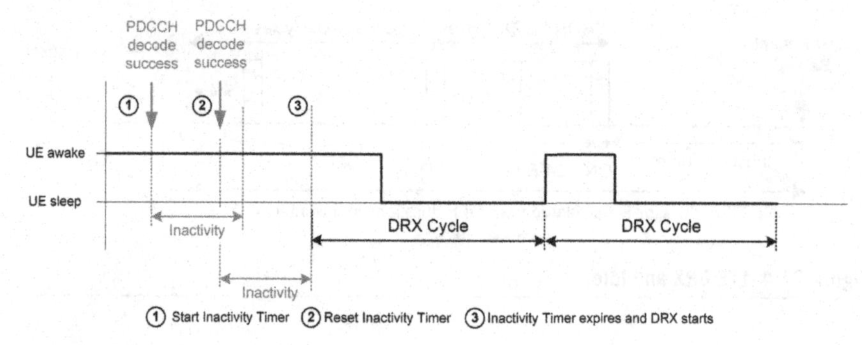

Figure 21.6 PDCCH decode and DRX cycle.

In LTE, each UE negotiates during connection setup for short and long DRX cycles. The LTE specification explicitly defines the triggering mechanism for both of the DRX cycles [11]. A UE starts an inactivity-based short DRX timer when there are no waiting packets in the queues. When this timer expires the UE enters the short DRX cycle and starts the second timer for the long DRX cycle. When the second DRX cycle expires the UE enters the long DRX cycle.

21.4 POWER MANAGEMENT IN IEEE 802.16E (WIMAX)/IEEE 802.16M

The IEEE 802.16e standard [4] specifies two mechanisms for conserving MS power, namely Idle mode and Sleep mode. The following subsections provide further details on each mode. In the 802.16e specification, a MS is primarily either in Connected mode or in Idle mode. To be in Connected mode, the MS must perform a series of procedures known as network entry [4] which include authentication and key exchange, and must be actively registered to a specific BS so as to be able to send and receive data with the BS. When in Idle mode, it is no longer registered to any BS and thus cannot exchange any information with the BS. This allows the MS to save power by shutting off its components for long periods of time, since it is required to perform a minimal set of functions in order for the network to be able to page it.

In Sleep mode, the MS is actively registered to a specific BS, but it can still negotiate periods of unavailability where it can power down to a low-power mode. This allows the MS to save power during shorter intervals of little or no packet activity without having to incur the cost in delay for performing network reentry if it had gone into Idle mode instead.

21.4.1 Idle Mode in IEEE 802.16e

Idle mode allows the MS to deregister from the BS and go into deep power saving modes for time intervals ranging anywhere from a few seconds to several minutes, depending upon how the MS is being used. Typically, the MS goes into Idle mode when it detects no user activity for a certain length of time. When the MS goes into Idle mode, it alternates between periods of Paging Unavailable and Paging Listening Intervals, see Figure 21.5. Idle mode allows the MS to become periodically available for selective periods, i.e., the Paging Listening Intervals to listen to BS, so it can be paged. In order to contact an MS in Idle mode, a BS will send a broadcast paging message to the MS included in it to indicate that it needs to exit Idle mode to receive incoming traffic.

Since an MS can cover a large geographical area while remaining in Idle mode, WiMAX defines the concept of a paging group that is similar to LTE, see Figure 21.5. A number of BSs are grouped over a contiguous geographical region to make up a paging group. When an MS needs to be paged, all the BSs in a paging group send out a broadcast paging message to alert the MS of incoming traffic. This allows the MS to cross from the area covered by one BS to another without needing to inform the BS of the change in its location, yet still be within paging contact. The MS needs only to send a message known as location update to the BS when it moves from one paging group to another. The size of the paging group is designed such that it is large enough that the MS does not need to do a location update too frequently, yet it must be small enough to keep the flood of broadcast paging messages reasonable.

21.4.2 Sleep Mode in IEEE 802.16e

Sleep mode conserves power when the MS is in Connected mode while still exchanging data. Sleep mode also allows the MS to shut itself down for some prenegotiated interval of time, but unlike Idle mode, in Sleep mode the MS remains connected to the BS. Thus, in Sleep mode the BS retains all the information related to the connections that currently belong to the MS as it does during Connected mode. This allows the MS to quickly return back to active mode to send and receive traffic and avoid the additional expense of frequent connection termination/reestablishment.

In Sleep mode, the MS alternates between periods of absence known as Sleep windows and periods of availability known as Listen windows. For each involved MS, the BS needs to keep context about its Sleep/Listen Windows and this is called a Power Saving Class (PSC). A PSC may be repeatedly activated or deactivated. To activate a PSC means to start Sleep/Listen windows associated with the class. The MS saves power during the Sleep intervals by powering down one or several components, depending upon the length of the sleep interval.

A PSC definition includes parameters to control various aspects of the Sleep mode such as the length of Sleep and Listen windows, whether the Sleep window remains fixed or increases in a binary exponential manner in successive Sleep cycles before it reaches a predefined maximum value and whether traffic can be exchanged during the Listen window or not. In 802.16e, a single MS may define many such PSCs, depending upon the number of service flows it has defined and the QoS associated with each. In the event when multiple PSCs are defined, the Sleep and Listen windows then are the intersection of all the Sleep and Listen windows of the various overlapping PSCs respectively. The WiMAX standard does not fix any of these parameters; the setting and use of these parameters is left to implementation to promote product differentiation.

Sleep mode may be triggered based on a period of inactivity, or upon traffic activity patterns. In addition, the length of the Listen and Sleep windows are also not dictated by the standard, they are instead negotiated between the MS and the BS and may be dependent on existing traffic conditions and the degree of latency tolerated by applications running on the MS.

21.4.3 Sleep Mode Enhancements in IEEE 802.16m

The implementation of multiple PSCs, each associated with a different QoS class, as defined in IEEE 802.16e was found to be very complicated and also inefficient in terms of power saving operations. In addition the frequent exchange of MAC messages to activate and deactivate Sleep mode, and the fact that Sleep mode in certain PSCs did not permit exchange of data during Listen windows, all caused the Sleep mode mechanism to undergo a complete overhaul in 802.16m [5].

Some of the major enhancements are listed here. In 802.16m, the concept of multiple concurrently active PSCs was eliminated and now, for each MS only a single PSC is active at any given time. The size of the Listen window can be flexibly changed from a predefined value in the event of data exchange. Thus the Listen window can be dynamically increased, depending on the amount of data to be exchanged into the Sleep window and until the next Sleep cycle. This minimizes the delay associated with buffering of data until the next Listen window and also allows for greater efficiency. In addition, the MS can define multiple PSCs a priori and then depending on the change in traffic characteristics, dynamically change the PSC itself by communicating to the BS the desired PSC. Since the structure and size of the physical radio frame was changed in 802.16m from 802.16e, 802.16m allowed a smaller granularity of sleep to be defined at the subframe level, thus allowing the MS to sleep even during traffic such as VoIP.

21.5 IMPLEMENTATION CHALLENGES IN POWER MANAGEMENT

The main challenge of enabling power management in 4G systems is to balance the trade-off between user experience and power consumption. This section provides details on some of these challenges. For Idle mode, the main challenge is to minimize the signaling overhead due to paging by selecting the optimal paging group size to limit the number of location updates. In addition, paging group size needs to be limited to avoid excessive paging messages flooding the backbone network.

For ease of explanation, we refer to DRX here only in terms of DRX during Connected mode so that DRX in this context operates very similarly to Sleep mode in WiMAX. DRX mode needs to be designed to accommodate the latency and throughput requirements of different applications. For example, VoIP and FTP traffic have different latency requirements. A single definition of the DRX cycle will not fit both VoIP and FTP users. Also, long DRX cycles are better for low-power consumption since it allows the device to spend less time transitioning to and from low-power modes and more time in low-power mode. However, having a long DRX cycle can cause excessive end-to-end packet delays and unsatisfactory user experience. Also, UEs are required to synchronize uplink timing and detect control channels periodically. Thus, keeping users in long DRX can also cause network connection losses due to timing synchronization issues. Moreover, the power saving mode also needs to coexist with other MAC operations such as scanning, handover, and HARQ and sometimes also needs to account for other features such as coexistence with other radio protocols such as Bluetooth. The challenges primarily arise due to conflicting requirements from each of these features. For the most efficient scanning algorithms, the device must be able to scan whenever it detects a dip in signal quality from the threshold. But for efficient power saving, the device needs to limit scanning as much as possible since scanning is activity that consumes a lot of power. Thus, for efficient power savings, trade-offs need to be carefully considered to allow the device to deliver excellent user experience and also deliver power savings.

21.6 POWER MANAGEMENT ENHANCEMENTS FOR FUTURE WIRELESS NETWORKS

In the recent past, the mobile broadband market has witnessed an explosive growth in smart devices such as smart phones and tablets, and mobile Internet applications including Skype, Gtalk, Yahoo! Messenger, Facebook, and Twitter. These emerging Internet applications require an always-on user experience. They often exhibit bursty and sporadic packet data exchanges generated both by user interactions and also when the user is not directly interacting with the application i.e. when it is running in so-called "background mode". We refer to the traffic generated by applications running in background mode as background traffic and it comprises of generally short data transfers such as keep-alive messages, interactions with application servers, status updates etc. This background traffic does not allow the device to be in low-power mode as much as it should be. This reduces the battery life of the device, and also adds to the air-interface signaling. Furthermore, the interactive user behavior of smart devices and emerging applications tend to bring the device out of low-power mode and add signaling overhead [13]. Current mobile broadband standards [1–5] have been designed to improve the power efficiency through various MAC mechanisms including Idle mode and DRX/Sleep mode, as explained above. However, these mechanisms were designed prior to the introduction of many emerging applications, and it is observed that the current mechanisms are inadequate to handle the emerging applications and new user behavior. Within the 3GPP standards body, a new work item called "Radio Access Network (RAN) enhancements for diverse data applications (eDDA)" is currently in study to quantitatively evaluate the power efficiency and signaling overhead issues in supporting emerging applications, and to potentially add enhancement techniques in the next generation (3GPP Rel. 11) of the standards [14].

As part of the eDDA work item research, several observations were made on the behavior of the current LTE systems in terms of signaling overhead and the amount of time an LTE device could spend in low-power mode when supporting emerging applications. For example, Figure 21.7 shows the cumulative distribution function (CDF) of packet interarrival times for three different cases, namely a user running an active session, a user running background traffic, and a user running an active session in addition to background traffic. Here background traffic refers to the autonomous exchange of user plane data packets between the UE and the network. As can be seen, there is a substantial difference between packet activity patterns, particularly between a user running an active session versus. a user running only background traffic. In addition, we observed that it doesn't make much difference when applications run in the background with an active session in place. The active session dominates the CDF. From Figure 21.7, we can infer that the amount of background traffic generated by the emerging applications is not insignificant, and furthermore, the behavior of background traffic is different from that of active traffic. If background traffic is not handled efficiently in the next generation of the mobile broadband technologies, it can drain the mobile device's battery power and create excessive signaling overhead.

Figures 21.8 and 21.9 show the ratio of signaling overhead for a user running an active application session and multiple applications running in background. From these figures, we can observe that the ratio of data exchanged to the signaling overhead decreases by several orders of magnitude from 10,000 to about 180 when applications are running in background. In short, what is happening is that a lot more

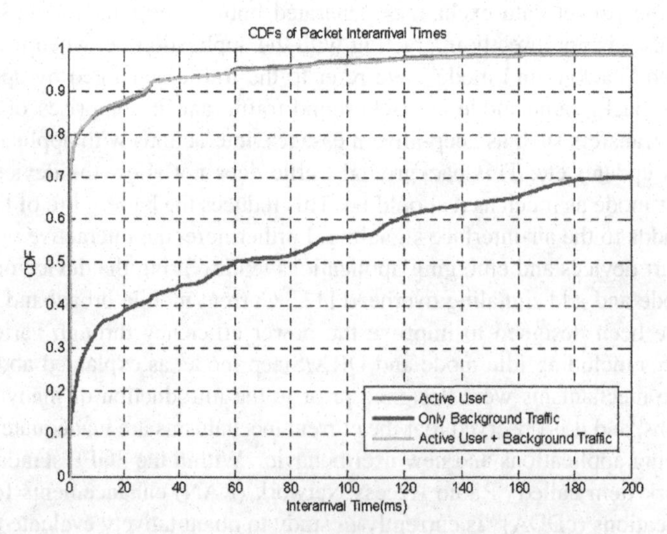

Figure 21.7 Interarrival times for active user traffic vs. background traffic.

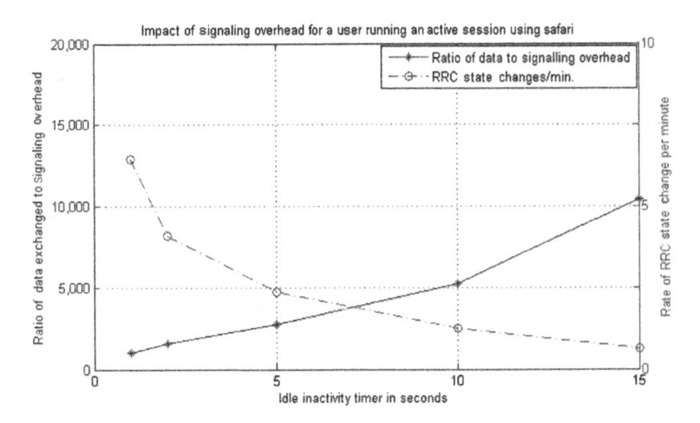

Figure 21.8 User running active session.

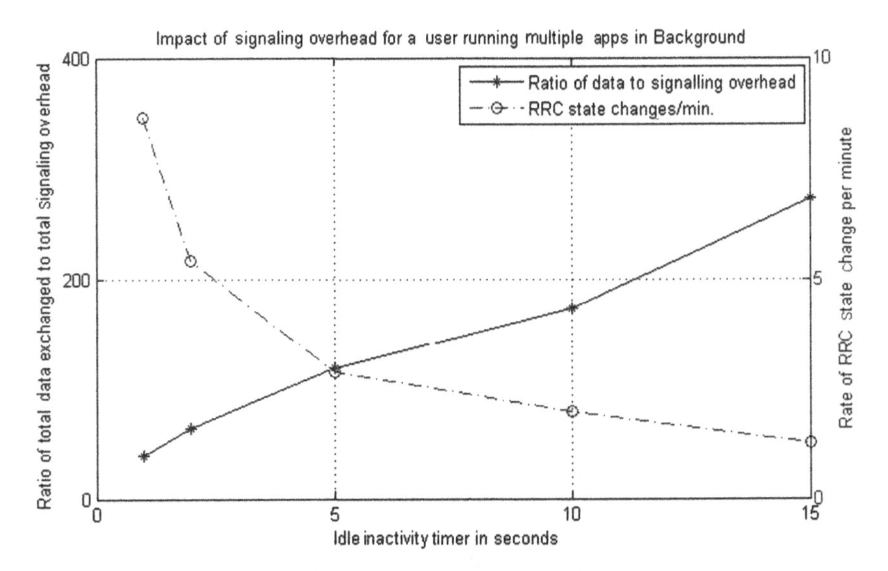

Figure 21.9 User running multiple apps in the background.

signaling is used to send a lot less data. There is relatively low impact seen on the ratio of data to signaling overhead when a user is active versus when a user is active and also running background traffic. The initial studies and observations led to a focus on application background traffic in order to enhance the LTE-Advanced system in supporting emerging applications efficiently in terms of battery power and signaling overhead.

The eDDA work item is currently considering enhancements in the following research areas:

- Mechanisms to improve system efficiency for background traffic using existing RRC states.
- Mechanisms to reduce UE power consumption for background traffic using existing RRC states.
- DRX enhancements to achieve optimum trade-off between performance and UE power consumption for single or multiple applications running in parallel.
- DRX enhancements to improve adaptability to time-varying traffic profiles.
- Improve the system resource efficiency for Connected mode UEs.
- Improve the control on signaling overhead for larger UE population in Connected mode.
- Improve power consumption and reduce signaling overhead using mechanisms that increase the assistance from UE and the network.

It is important to note that the information depicted in this section is the state of the 3GPP Rel. 11 standardization efforts at the time of writing this book chapter, and the final version of the 3GPP Rel. 11 standards may or may not incorporate the techniques studied in these areas.

21.7 CONCLUSIONS

4G mobile broadband systems are very attractive for smart devices that demand always-connected capability. Obviously, this capability of 4G does not allow the device to be in low-power mode as much as its battery life would like to. 4G standards have defined a hierarchy of several power efficient mechanisms such as Idle mode and DRX/Sleep mode in order to keep the device in low-power mode. Idle mode allows the device to be in deep low-power mode, but it takes higher signaling overhead and more time to bring the device back to active. On the other hand, Sleep/DRX mode allows the device to be in a moderate low-power mode, but the time taken to bring the device back to active is shorter. There is a power saving versus quality of service trade-off in choosing the appropriate power efficient mechanism. This chapter describes the details of the power efficient mechanisms incorporated in 4G standards.

This chapter also points out the inefficiencies in the power efficient mechanisms incorporated in the original 4G standards in supporting emerging diverse data applications such as social networking, IM, etc. The obvious reason for this is that these applications did not exist or were not widely adopted in mobile broadband when the original set of 4G standards were defined. Furthermore, this chapter also addresses the state-of-the-art technologies that are currently being explored by the 3GPP standards body to enhance the power efficient mechanisms and better support emerging applications under a work item called "enhancements for diverse data applications." Note that the final version of the 3GPP standards currently being developed [11] may or may not incorporate the technologies described in this chapter.

Research outputs from various industries in this area are captured in [9]. Researchers who are interested in further studying in this area can take advantage of this document to understand the issues in the currently available systems in detail. Furthermore, the evaluation methodology given in this document can be used for further exploration.

REFERENCES

[1] 3GPP TS 36.300, v8.11.0, Evolved Universal Terrestrial Radio Access (E-UTRA) and Evolved Universal Terrestrial Radio Access (E-UTRAN); Overall Description; Stage 2, January 2010.

[2] 3GPP TS 36.321, v8.9.0, Evolved Universal Terrestrial Radio Access (E-UTRA); Medium Access Control (MAC) Protocol Specification (Release 8), June 2010.

[3] 3GPP TS 36.304, v8.8.0, Evolved Universal Terrestrial Radio Access (E-UTRA); User Equipment (UE) Procedures in Idle Mode (Release 8), January 2010.

[4] IEEE Standard for Local and Metropolitan Area Networks Part 16: Air Interface for Broadband Wireless Access Systems, 802.16-2009, May 2009.

[5] IEEE 802.16m_D7, DRAFT Amendment to IEEE Standard for Local and Metropolitan Area Networks Part 16: Air Interface for Fixed and Mobile Broadband Wireless Access Systems, Advanced Air Interface, July 2010.

[6] R.Y. Kim, S. Mohanty, Advanced power management techniques in next-generation wireless networks, IEEE Communications Magazine, May 2010.

[7] L. Zhou, H. Xu, H. Tian, Y. Gao, L. Du, L. Chen, Performance analysis of power saving mechanism with adjustable DRX cycles in 3GPP LTE, in: Proceedings of the IEEE VTC'08-Fall, September 2008, pp. 1–5.

[8] S. Gao, H. Tian, J. Zhu, L. Chen, A more power-efficient adaptive discontinuous reception mechanism in LTE, in: Proceedings of the IEEE VTC'11-Fall, September 2011, pp. 1–5.

[9] 3GPP TR 36.822, v0.2.0, Technical Specification Group Radio Access Network; LTE RAN Enhancements for Diverse Data Applications, November 2011.

[10] IEEE 802.16m-08/004r5, IEEE 802.16m Evaluation Methodology Document (EMD), January 2009.

[11] 3GPP TS 36.331, v10.3.0, Radio Resource Control (RRC); Protocol specification, October 2011.

[12] C.S. Bontu, E. Illidge, DRX mechanism for power saving in LTE, IEEE Communications Magazine 47 (6) (2009) 48–55.

[13] P. Willars, Smartphone traffic impact on battery and networks, <https://labs.ericsson.com/developer-community/blog/smartphone-traffic-impact-battery-and-networks>.

[14] RP-110410, LTE RAN Enhancements for Diverse Data Applications, 3GPP RAN Plenary Contribution, March 2011.

CHAPTER

Green Optical Core Networks

22

Mohammad Ali Mohseni, Akbar Ghaffarpour Rahbar
Computer Networks Research Lab, Sahand University of Technology, Tabriz, Iran

22.1 INTRODUCTION

Every day more capabilities of information and communication technology (ICT) are known and more users become interested in using ICT services. A growing number of users requires development in ICT infrastructures, which in addition to incurring development costs, will raise energy consumption in the ICT domain. It is true that ICT services such as e-commerce, tele-working, and video conferencing can significantly decrease the energy consumption of other domains especially in transportation, however, rapid growth of ICT energy consumption is a serious concern since lack of energy resources could be a barrier against development of ICT services. Estimates show that ICT has consumed about 9% of total produced energy in the world in 2009, and if growth continues at the current speed it will reach 20% [1]. Furthermore, it should be noted that in many countries, energy is obtained from hydrocarbon resources that are not renewable and lead to the emission of greenhouse gases (GHG). Producing electrical energy for running network infrastructures emitted 4.53 million tons of CO_2 in the air during 2002, and it is predicted that it will reach 350 million tons in 2020 [2]. These show that even a small reduction of power consumption in ICT infrastructures can save costs and help to decrease emission of GHG. The growth of ICT energy consumption has focused the attention of ICT researchers on studying approaches of conserving energy in computer networks.

Optical technology has created the transport infrastructure for many networks due to its unique features such as high speed, high bandwidth, low loss, and low bit error rate (BER); these features are ideal for many services. Many researchers have studied several features of optical networks such as routing, wavelength assignment, grooming, survivability, and proposed some solutions to improve these features.

When a network is designed, it should support peak traffic loads, it should have some resources reserved for protection issues, and it may also have some resources for coping with future traffic growth. Since traffic load is time varying, network equipment may be mostly underutilized for a wide range of time, especially at low traffic load. In addition, most network resources are underutilized since network failure rarely occurs. There are two different research approaches for improving power consumption in core networks:

- *Energy efficiency approach, e.g.,* [*3–4*]: A network is energy-efficient if its power consumption is minimized without performance degradation. Energy efficiency should be considered when a network is designed. The approaches used for designing energy-efficient networks try to reduce the power consumption of network equipment.
- *Energy awareness approach, e.g.,* [*5–8*]: A network is usually designed for peak traffic loads; however network traffic varies during time and it could be less than the peak value for a long time. A network is energy aware if the power consumption of a network is smartly adjusted with variation of offered traffic load. To reduce power consumption, this approach inactivates unused network resources, especially at low traffic load.

The majority of network traffic, such as Internet traffic, should be transmitted through optical core networks, but these networks are not energy-efficient nowadays. For this reason, research efforts to reduce energy consumption in core networks are more common than those for metro/access networks. There are some studies that investigate energy consumption of access and metro networks such as [9–20].

This chapter exclusively investigates power consumption in optical core networks. The reminder of this chapter is organized in three sections. Section 22.2 introduces energy efficiency metrics, power consumption models, and also investigates power consumption of network elements such as switches, optical amplifiers, transmitters and receivers. Section 22.3 is devoted to energy-efficient network design approaches. In this section some approaches such as modular switches, single line and mixed line rates networks, static and dynamic networks, and optimum repeater spacing are introduced. Energy-aware optical core networks are introduced in Section 22.4. In this section energy-aware routing and wavelength assignment, energy-aware traffic grooming, selectively switching off network elements, and energy-aware survivable WDM networks are introduced.

22.2 ENERGY CONSUMPTION IN OPTICAL CORE NETWORKS

Optical networks are classified in three different layers: core, metro, and access layers. These networks consist of network equipment such as switches, amplifiers, repeaters, transmitters, and receivers that are active elements and require power for running. Power consumption models express the power consumption behavior of network equipment corresponding to offered traffic loads. These models also are applied to estimate the power consumption of network equipment under different traffic loads. Energy consumption per bit is a metric that expresses how much energy is consumed by equipment for transmitting, receiving, amplifying, or switching. This metric can also be used to compare the same type of network equipment to determine which one is more energy-efficient.

In the following, optical network layers are introduced. Since this chapter exclusively aims to investigate power consumption in the core layer the conventional IP-over-WDM architecture for optical core networks is detailed. Then, power consumption models and metrics are introduced. Finally the power consumption of network equipment such as switches, amplifiers, receivers, and transmitters are investigated.

22.2.1 Network Layers Overview

The architect of telecom networks consists of three layers: core, metro and access layers. Figure 22.1 depicts the telecom network layers. The access layer enables end users to connect to the entire network. This layer has different technologies such as FTTx, PON, xDSL, WiMax, and cable modem. An access network spans a distance of about 10–20 km. A metro network enables an access network to connect to a core network. The traditional topology for metro networks is a ring as shown in the figure. Core networks cover a wider area such as countries and continents. These networks consist of high capacity routers that are interconnected by high capacity links since core networks should be able to transfer a high volume of traffic. Since this chapter is about green optical core networks, we concentrate on the architecture of core network layers. The architecture of metro and access networks have been detailed in [21]. Energy consumption of different layers in the Internet, as the most popular example of ICT infrastructure, is investigated in [22]. It is shown that nowadays energy consumption of access networks dominates energy consumption of the entire Internet due to the many active nodes in these networks. However, it is estimated that if the growth of the Internet continues this speed, the energy consumption of core networks will be dominant compared to access and metro networks.

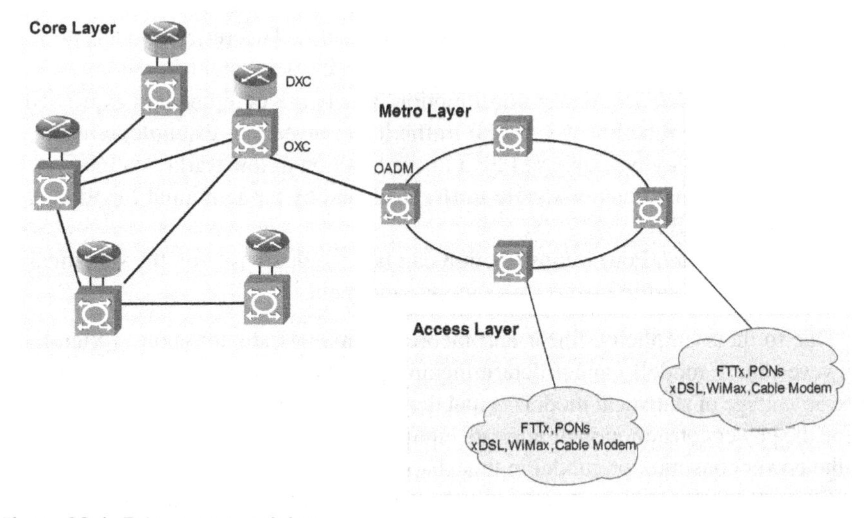

Figure 22.1 Telecom network layers.

IP-over-WDM is the most traditional architecture in optical core networks. This architecture consists of two layers: IP and WDM layer. The IP layer performs network layer functions such as routing and the WDM layer provides a high capacity optical infrastructure on which the IP layer is able to transfer packets. As depicted by Figure 22.1, Optical Cross Connect (OXCs) nodes in the WDM layer are interconnected by fiber links via a mesh topology. Digital Cross Connect (DXCs) nodes in the IP layer are mounted on top of each OXC. DXCs and OXCs are interconnected by line card interfaces that perform optical/electronic conversions. Intermediate nodes on a setup light-path deal with transit wavelengths by two strategies: bypassing and nonbypassing. In the bypassing strategy, intermediate nodes pass transit wavelengths through OXCs without any processing in DXCs. In the nonbypassing strategy, a wavelength is terminated in an intermediate node and processing is performed by a DXC. Intermediate nodes can add traffic to a light-path in the nonbypassing strategy, whereas input signals are optically switched to an output port in the bypassing strategy.

22.2.2 Power Consumption Models

Power consumption models express the power consumption of network equipment as a function of offered traffic load. These models have been classified in four categories [23]:

- *Linear model*: Energy consumption is modeled using a linear function, where its variable is equipment traffic load. If traffic load increases by ΔL, incremental power consumption is $\varepsilon \times \Delta L$. Parameter ε depends on the capacity, technology, and size of network equipment.
- *Theoretical model*: This model expresses the power consumption of equipment by a polynomial function, $C^{2/3}$, where C is the capacity of the network equipment. If C is substituted by offered traffic load L, then $L^{2/3}$ is the power consumption of network equipment at traffic load L. Theoretical models do not consider other equipment specifications such as capacity and technology.
- *Combined model*: The power consumption of network equipment is expressed by a combined function at different traffic load ranges. For example, power consumption may be expressed by a linear function at low traffic load, by an exponential function at moderate traffic load, and by a polynomial function at high traffic load.
- *Statistical model*: Power consumption can be modeled based on the statistical distribution of traffic load over network equipment.

Due to their simplicity, linear and theoretical models are common in literature. However, these models cannot determine an upper bound for energy consumption. The advantage of statistical models is that they involve traffic type behaviors to determine the power consumption of network equipment. Since the linear model is applied to the power consumption model in this chapter, it is explained in the following.

Power consumption in an optical network can be divided into two parts: running and operational energy consumption. When a network is developed, it needs energy to turn on its equipment. This energy, called *running* energy consumption, is independent of traffic load and is required even when there is no traffic load. Establishing a connection between two nodes may need energy-hungry operations such as turning on at least two transmitter and receiver ports, amplifying, optical to electronic (O/E) or electronic to optical (E/O) conversion, and switching. This part of energy consumption that is related to traffic load is named *operational* energy consumption. A case study on power consumption of several types of switch fabrics has been performed in [24] and showed that increasing traffic load by 10–50% can lead to linearly increasing energy consumption from 20 mW to 90 mW. The impact of traffic load on operational power consumption can be linearly modeled by [25]

$$P = P_R + p \times L, \quad (22.1)$$

where P_R is running power consumption, which is sometimes referred to as constant power, p is the power consumption of the component per additional traffic unit, and L is amount of traffic load.

Equation (22.1) shows that at the component and equipment levels, power consumption is dependent on traffic load. At the network level, power consumption is dependent on more network characteristics such as traffic volume, traffic grooming, bypassing, and the routing algorithm. The impact of network characteristics on network power consumption is studied in subsequent sections.

22.2.3 Energy-Efficiency Metric

There are few metrics for evaluating the energy efficiency of network equipments. The most simple and common metric is the ratio of total power consumption to capacity of the system defined by

$$\text{Energy efficiency (joules per bit)} = \frac{\text{total power consumption (watts)}}{\text{equipment capacity (bits per second)}}. \quad (22.2)$$

Equation (22.2) calculates consumed power by a system for a bit. It is clear that higher value means the system is more energy-hungry. If the system is a switch then system capacity is the summation of bit rates of line cards. This equation estimates a lower bound for energy consumption per bit for a typical switch. In practice a system does not work at high capacity because traffic changes over time. Therefore, it is more acceptable to replace equipment capacity in Eq. (22.2) with equipment throughput, by which energy consumption per bit will be increased since equipment throughput is lower than system capacity.

22.2.4 Energy Consumption of Network Equipment

Switches, optical amplifiers, and optical transmitters/receivers are the equipment of an optical core network. In this section, the logical structure of this equipment has been investigated from a power consumption point of view.

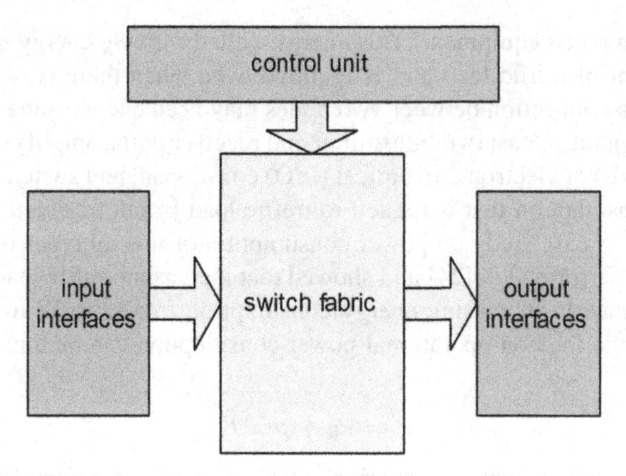

Figure 22.2 Schematic structure of an optical switch.

22.2.4.1 Energy Consumption in Switches

Switches as shown in Figure 22.1 interconnect multiple links and switch an input wavelength to an appropriate output port. Switches can be categorized as [26]:

- *Linear analog switches*: These switches operate in an O/O/O manner, where an optical wavelength enters, is switched, and exits the switch in the all-optical domain. A semiconductor optical amplifier (SOA) gate array is an example of a linear analog switch.
- *Digital switches*: Signals enter and leave digital switches in the optical domain, but switching is performed electronically, i.e., in an O/E/O manner. The study in [26] shows O/O/O manner switches are more energy-efficient than O/E/O manner switches. Switch fabrics built based on CMOS transistors are an example of digital switches.

The structure of a switch is depicted in Figure 22.2. Input and output interfaces can include optical demultiplexers/multiplexers, transponders, O/E or E/O converters, and wavelength converter components. Elements of input/output interfaces are dependent on the structure of switching fabrics. Switch fabrics are classified in two types: optical and electronic. If switching is performed in the electronic domain, O/E and E/O converters in input/output interface are required. There are different architectures for switching fabrics such as CLOS switches, micro-electrical-mechanical systems (MEMS), and semiconductor optical amplifier (SOA) switches. The control unit configures the switch fabric from an input wavelength to an appropriate output if the switch fabric is an active switch.

Transponders, O/E and E/O converters, and wavelength converters are active elements, where their power consumption determines the required power of input/output interfaces. Power consumption of a switch can be stated by

$$P_{\text{switch}} = P_{\text{input}} + P_{\text{switch fabric}} + P_{\text{output}} + P_{\text{control unit}}, \tag{22.3}$$

where the power consumption of switch fabrics, $P_{\text{switch fabric}}$, can be negligible because the power consumption of switch, P_{switch}, is usually bounded by the power consumption of other components such as input/output interfaces. The work in [26] has introduced several types of switch fabrics and stated equations for their lower bound energy consumption.

Optical switches are more power efficient than digital switches because digital switches need more equipment in input and output interfaces such as O/E and E/O converters. Furthermore, digital switches buffer any data bit by bit for processing, but optical switches do not buffer any data electronically.

22.2.4.2 Energy Consumption of Optical Amplifiers and Repeaters

The power of optical signals is attenuated while passing through optical fibers. Attenuation increases exponentially by fiber length. Although attenuation in optical networks is low compared to attenuation in wired networks, for optical fibers that span a longer distance, attenuation should be compensated by optical amplifiers or repeaters. Repeaters convert optical signals to electronic signals (O/E conversion), amplify them, and again convert them to optical signals (E/O conversion). It is true that regenerated signals by repeaters have good quality (timing, shaping, eliminating noises, etc.) but repeaters limit the speed of an optical network to the speed of the electronic equipment. Unlike repeaters, amplifiers amplify the signal in a way that a signal remains in the optical domain when passing through fiber links. Notice an amplifier cannot filter noise and amplifies it while amplifying the original signal.

Erbium-doped fiber amplifiers (EDFAs) are commercial amplifiers that operate in the C-band window. The building blocks of EDFAs and its fundamental functions are detailed in [27]. EDFAs are compared with Raman amplifiers in [21] and their advantages and disadvantages are investigated. EDFAs are active elements and require electrical power resources. Suppose P_{AMP} shows the power consumption of an EDFA [3]:

$$P_{\text{AMP}} = P_{\text{constant}} + P_{\text{lambda}} \times N_{\text{lambda}}, \tag{22.4}$$

where P_{constant} is the running power consumption of the amplifier. The majority of P_{constant} is consumed by circuits such as the interfacing circuit, automatic gain control management circuit, and laser temperature control circuit. Parameter P_{lambda} is consumed power by a wavelength and N_{lambda} is the number of amplified wavelengths.

As aforementioned, repeaters perform O/E and E/O conversion to amplify signals. A repeater receives an optical signal, amplifies it, and then optically transmits it over fiber links. Hence, a repeater can be depicted by a receiver/transmitter pair (see Figure 22.6), where its power consumption is stated by summation of the power consumption of an optical receiver and an optical transmitter.

22.2.4.3 Energy Consumption of Optical Transmitters and Receivers

Figure 22.3 shows the building blocks of a typical optical transmitter. The MUX unit multiplexes low data rate signals into a high bit rate signal. The driver unit regulates the current flowing through a laser circuit. The laser diode sends an optical signal

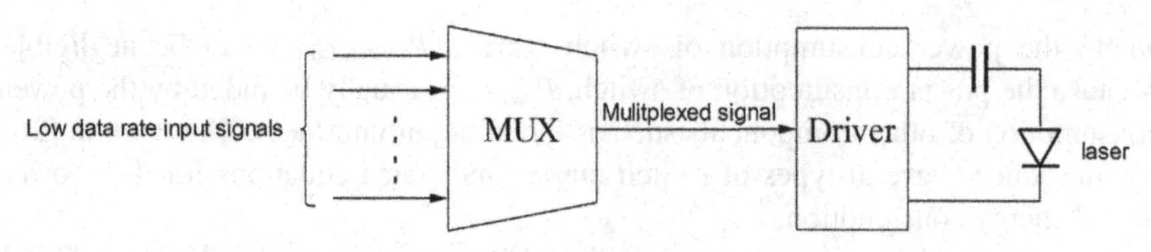

Figure 22.3 Logical structure of an optical transmitter.

over fiber. The MUX, driver, and laser units are active elements that consume power. Equation (22.5) states the power consumption of an optical transmitter [28]:

$$P_{\text{TX}} = P_{\text{MUX}} + P_{\text{Driver}} + P_{\text{laser}}. \tag{22.5}$$

An optical receiver obtains an optical signal by a photodiode and converts it to an electronic signal. A biasing circuit, amplifier, and CDR/DEMUX are the building blocks of an optical receiver as shown in Figure 22.4.

Equation (22.6) states the power consumption of an optical receiver [28]:

$$P_{\text{RX}} = P_{\text{Amplifier}} + P_{\text{Bias}} + P_{\text{CDR/DEMUX}}. \tag{22.6}$$

Figure 22.5 shows an optical link between two nodes that consists of one transmitter, one receiver, and n optical amplifiers, where distances between amplifiers are equal, if amplifiers are of the same type.

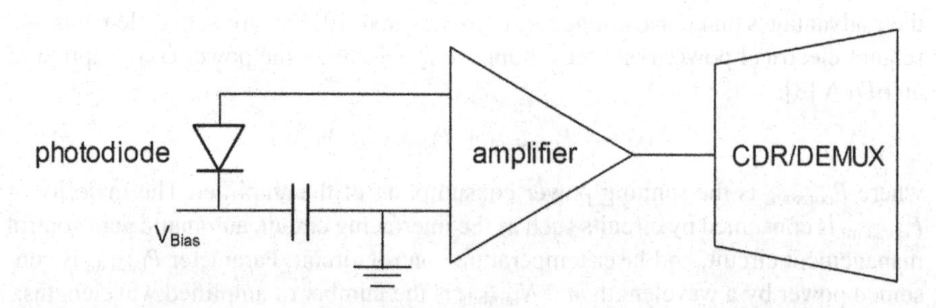

Figure 22.4 Logical structure of an optical receiver.

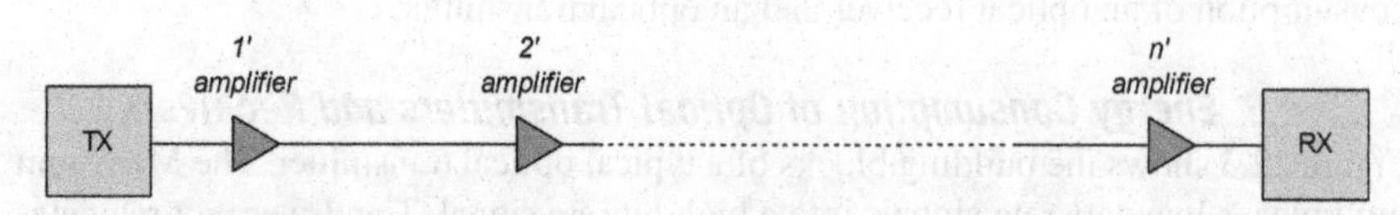

Figure 22.5 All-optically amplified link.

The power consumption of this link can be stated by

$$P_{\text{link}} = P_{\text{TX}} + n \times P_{\text{AMP}} + P_{\text{RX}}, \tag{22.7}$$

where P_{TX} is the power consumption of the optical transmitter, P_{AMP} is the power consumption of the optical amplifier, and P_{RX} is the power consumption of the optical receiver.

This section has investigated power consumption in optical core networks at the equipment level and introduced equations for expressing the power consumption of network equipment. These equations express the power consumption of network equipment by summation of the power consumption of its components. Energy consumption per bit is a metric that expresses how much energy is consumed by equipment for transmitting, amplifying, and so on. This metric is used to determine which equipment is more energy-efficient. This section also introduces models for estimating power consumption of network equipment at different traffic loads. Linear and theoretical models are more common in literature (see Section 22.4) since they simply express the power consumption of network equipment as proportional to the offered traffic load. Complicated models, such as combined, and statistical models, consider more details for network equipment. Since these models lead to more complexity in estimating power consumption at the network level, to the best of our knowledge, they have not been used in literature.

22.3 POWER-EFFICIENT NETWORK DESIGN

There are some ideas that help a network designer to design an energy-efficient network without degrading network performance. Energy-efficient networks require minimum running power consumption. In the following, it will be shown how the modular structure of switches, dynamic networks, mixed line rates, and repeater spacing can reduce running power consumption in optical networks.

22.3.1 Modular Structure of Switches

New switches are presented with modular architecture. A modular switch has several chassis and each chassis has some line cards. A chassis is a framework on which other components such as line cards, cooling equipment and other components are installed. Each line card has several ports for establishing connections with other nodes. A central switch fabric performs packet switching operations between chassis. This architecture can provide the possibility for components with no load to go to sleep mode for power saving purposes.

Energy consumption of modular switches is related to chassis, switch fabrics, line cards, and ports. The switch fabric performs the switching operation and is responsible for learning and maintaining switching tables. Line cards are interfaces between

Table 22.1 Power Consumption of CISCO Catalyst 6500 Components

Component	Power Consumption (W)
$Power_{fixed}$	60
$Power_{switch\_fabric}$	315
$Power_{first\_line\_card}$	315
$Power_{subsequence\_line\_card}$	49
$Power_{active\_port}$	3
$Power_{idle\_port}$	0.1

ports and the switching fabric. Table 22.1 shows the energy consumption of CISCO Catalyst 6500 components [29].

A chassis and line cards may need to be turned on when there are some requests for establishing a connection, otherwise they can remain in sleep state. After a port is turned on, the additional energy consumption of powering a new port on the same line card is 3 W. The additional energy consumption of establishing a connection on a chassis with a powered-on line card is 49 W. Hence, it is more efficient first to set up a new connection request over a powered-on line card. If this is not possible, establish it on a powered-on chassis. Otherwise, if all chassis line cards are full, a new chassis must be turned on for establishing the new connection request. For example, suppose two connections must be established between two nodes. There could be three scenarios for allocating resources in order to set up these two connections on (1) different chassis, (2) the same chassis and different line cards, and (3) the same chassis and the same module. According to Table 1, these three scenarios consume different amounts of energy:

Scenario 1: $(60+315+315+3)\times 2=1386$ W
Scenario 2: $(60+315+3+49+3)=530$ W
Scenario 3: $(60+315+3+3)=381$ W

The first scenario establishes two connections over two distinct chassis, and therefore, two different chassis, two modules, and two ports must be turned on. Scenario 3 requires the least power consumption because it is more energy-efficient, where the connection is established over already turned on equipment. These examples show that modular structure of node equipment can reduce network power consumption because of minimizing overhead power consumption.

22.3.2 Static and Dynamic Nodes

In a network with static nodes, allocation of wavelengths is performed statically and is not changed during time. On the other hand, in a network with active (dynamic) nodes with wavelength converters, the allocation of wavelengths may change during time based on traffic volume and network topology. Dynamic nodes need a smaller number of wavelengths than static ones at different traffic loads due to the possibility of wavelength conversion. Reduction in the number of wavelengths can lead to reduction in the number of transponders in the input and output stage of a switch, thus reducing the energy consumption of the switch.

Three different switch architectures have been introduced in [3] as static classic optical node (SCON), static low consumption optical node (SLON), and dynamic optical node (DON). The differences among these switches are in the input stage, output stage, and switching fabrics. A SCON uses long-reach (LR) receiver/short-reach (SR) transmitter, i.e., LR/SR transponder, at its input stage and SR/LR transponder at its output stage. A SLON removes LR/SR at the input stage because SR/LR transponders are replaced by LR/LR transponders at the output stage [3]. A DON uses tunable LR/SR at its input stage and fixed SR/LR at its output stage. Tunable transponders are able to perform the wavelength conversion operation.

Two parameters determine which architectures, either static or dynamic, are more energy-efficient at different traffic loads [3]: ratio of energy consumption of LR and SR devices, denoted by β, and ratio of power consumption of a device in sleep state and active state, denoted by ε. It is obvious that a higher value of ε can lead to more energy conservation in DONs. This paper shows that if $\beta > 1$ then DONs are more energy-efficient than SCON and SLON at load 0.1–0.9.

The energy consumption of nodes is dominated by the energy consumption of different types of transponders. The energy consumption of static nodes, SCON and SLON, is constant over time because these nodes are fully active all the time. Due to eliminating LR/SR transponders in SLONs, these nodes are more energy-efficient than SCONs. The energy consumption of DONs is dependent on load traffic linearity. Unlike SCONs and SLONs, transponders with no load go to sleep mode in DON. Some equations for the energy consumption of SCONs, SLONs, and DONs have been stated [3]. This study uses EON topology as a case study and shows that at low load (less than 0.4) the DON network has the lowest power consumption compared to SCONs and SLONs. Recall the structure of input and output stages of DONs is similar to SCONs. A DON uses tunable LR/SR at input stages and can switch transponders with no load to sleep states. The energy consumption of DON networks is proportional to traffic load, and therefore, all transponders can be in active state at high traffic load. In this situation, the power consumption of DONs is equal to the power consumption of SCONs. At high loads (more than 0.4), SLONs are more energy-efficient. It should be noted these results have been obtained at $\beta = 1$ and $\varepsilon = 0.1$.

22.3.3 Single-Line and Mixed-Line Rates

Current optical backbone networks support 10/40 Gbps data rates and 100 Gbps data rate is on the way. Hence, next-generation optical backbone networks should support multiple data rates (MLR) namely 10/40/100 Gbps [30]. However, optical impairments limit the usage of high data rates (40 and 100 Gbps) to short distances

Table 22.2 Normalized Energy Consumption of 10/40/100 Transponders, EDFA, and Electronic Processing

Device	Energy Consumption (Normalized)
10 Gbps transponder	1
40 Gbps transponder	5
100 Gbps transponder	14
EDFA amplifier	0.25
Electronic processing	0.5 (per Gbps)

because impairments are increased at long distances and some quality of service (QoS) parameters (e.g., bit error rate) are degraded. As shown in Table 22.2, high data rate networks consume more energy than low ones. On the other hand, these data rates cannot be applied everywhere. Therefore, it is energy-efficient to adjust the provided data rate to the requested data rate. Thus, there is a trade-off between provisioned data rate and energy consumption.

Suppose the requested capacities are known. Then, the problem is to find the optimum number of wavelengths with different data rates that support the requested capacities. This problem has been modeled in [4] as a mixed-integer linear programming (MILP) problem and it was shown that MLR networks are more energy-efficient than single line rate (SLR) networks. In SLR networks, all transponders operate at the same bit rate. In this study, power consumption of transponders, amplifiers, and O/E and E/O converters have been considered. There are some facts that determine the power consumption of a network and should be noted:

- More wavelengths require more transponders.
- Long distance requires more amplifiers.
- The BER threshold limits the distance on which a light-path with data rate r on wavelength λ can reach.
- Low data rates need more fibers to support high capacity requests.

Although the power consumption of SLR 10 Gbps is low, it should be noted that this network needs more fibers to support high capacity demands, which leads to an increase in the number of amplifiers. Therefore, the power consumption of amplifiers eliminates power conservation due to low bit rate channels. Wavelengths with a data rate of 100 Gbps can reach shorter distances than wavelengths with 10/40 Gbps data rates.

22.3.4 Optimization of Transmission Elements

Attenuation, dispersion, and nonlinear impairments are three intrinsic characteristics of optical networks, which limit the length of the light-path, data rate, and number of wavelengths in an optical fiber. When an optical signal passes through fiber, its power level is attenuated. Dispersion is spreading of the optical signal in an optical fiber over time. In a dense WDM network, different channels impact each other. The Kerr effect is a nonlinear impairment that limits the maximum length of light-paths. The Kerr nonlinearity effect can be controlled using wider pulses [31]. If attenuation, dispersion, and nonlinear impairments are not compensated for within the network, optical receivers cannot detect optical signal levels because optical signals are deformed. Amplifiers, repeaters, and dispersion compensation units can be used to limit deformation of optical signals. Repeaters are electronic devices that are energy-hungry. These devices limit the speed of an optical network to the speed of electronic devices. Hence, it is more energy-efficient to minimize the number of repeaters. Research in [31] shows by simulation that all repeaters can be eliminated in the European transport network provided that dispersion compensation units are applied to control dispersion, optical amplifiers are used to compensate attenuation, and pulse width is optimized (less than 7 ps).

Amplifiers are used to compensate attenuation on fibers. The amplified spontaneous emission (ASE) noise limits the length of all optical amplified links since this noise grows when the optical signal is amplified by each EDFA. For a long fiber, it is required to use electronic repeaters as shown in Figure 22.6 [26]

Figure 22.6 shows a link with length L that consists of n optically amplified links. Between each optically amplified link there is an electronic repeater. Each optically amplified link uses m EDFAs. The distance between repeaters is l and between amplifiers is l/m.

When a wavelength is amplified by the mth EDFA, the optical signal to noise ratio (OSNR) on the wavelength can be expressed by [28]

$$\mathrm{OSNR} = \frac{p_{\mathrm{out}}}{2n_{\mathrm{sp}}m\left(e^{\frac{\alpha l}{m}} - 1\right)h\nu B_0}, \tag{22.8}$$

where P_{out} is the output signal power of amplifiers, n_{sp} is the spontaneous emission factor of amplifiers, h is Plank's constant, ν is optical frequency, α is power attenuation per unit fiber length, and B_{o} is the optical bandwidth of the link. This expression shows that the number of amplifiers in all optical amplified links decreases OSNR.

In the following, the impact of the number of amplifiers on energy consumption per bit is studied. Energy consumption per bit is

$$E_{\mathrm{bit}} = (P_{\mathrm{link}} \times \tau_{\mathrm{bit}})/k, \tag{22.9}$$

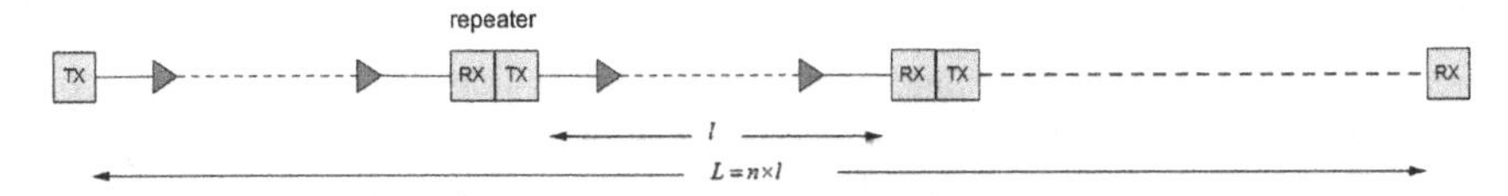

Figure 22.6 Repeaters are mounted in the middle because ASE noise limits the number of EDFAs on each link.

where P_{link} is calculated by Eq. (22.7), τ_{bit} is the bit period, and k is the number of wavelengths in a fiber. Hence, we have

$$E_{\text{bit}} = \frac{mP_{\text{AMP}}\tau_{\text{bit}}}{K} + \frac{(P_{\text{TX}} + P_{\text{RX}})\tau_{\text{bit}}}{K}, \quad (22.10)$$

where τ_{bit} can be obtained by Eq. (22.11):

$$\tau_{\text{bit}} = \frac{\text{SNR}_{\text{bit}}}{2B_{\text{o}}\text{OSNR}}. \quad (22.11)$$

In Eq. (22.10), there is a factor m, so we concentrate on it and replace τ_{bit} by Eq. (22.11) [28]:

$$E_{bit-AMP} = \frac{mP_{AMP}\tau_{bit}}{k} = \frac{SNR_{bit}n_{sp}m^2\left(e^{\frac{\alpha l}{m}} - 1\right)\left(1 - e^{-\frac{\alpha l}{m}}\right)hv}{\eta_{EPCE}}, \quad (22.12)$$

where η_{EPCE} is expressed by

$$\eta_{\text{EPCE}} = \eta_{\text{E}} \times \eta_{\text{PCE}}, \quad (22.13)$$

where η_{E} is the power efficiency conversion of amplifier control and management energy and η_{PCE} is amplifier power efficiency. As shown in Eq. (22.12), energy consumption per bit increases by a factor of m^2. Therefore, a number of repeaters should be mounted in the middle to control rapid growth of energy consumption with factor m^2. Repeaters are electronic equipment so they should be installed optimally.

Energy per bit per distance length unit for link L (Figure 22.6) can be expressed by $E_{\text{bit}/l}$, and all distances between repeaters are equal. So

$$\frac{E_{\text{bit}}}{l} = \frac{E_{\text{bit-AMP}}}{m \times l_{\text{amp}}} + \frac{E_{\text{TX}} + E_{\text{RX}}}{m \times l_{\text{amp}}}, \quad (22.14)$$

where l_{amp} is fixed amplifier spacing. If $E_{\text{bit-AMP}}$ is replaced in Eq. (22.14) by Eq. (22.12) and then we differentiate with respect to m and set Eq. (22.14) to zero, the optimum number of amplifiers, m, is given by Eq. (22.15) [28].

$$m_{\text{opt}} = \sqrt{\frac{(E_{\text{TX}} + E_{\text{RX}})\eta_{EPCE}}{\text{SNR}_{\text{bit}}\eta_{sp}\left(e^{\alpha\frac{l}{m}} - 1\right)\left(1 - e^{-\alpha\frac{l}{m}}\right)hv}}. \quad (22.15)$$

So,

$$l_{\text{opt}} = m_{\text{opt}} \times l_{\text{amp}}, \quad (22.16)$$

where l_{opt} is the optimum distance between repeaters that minimizes power consumption of link L. Hence, if the distance between two repeaters is less than l_{opt} then the power consumption of link L is dominated by repeaters (transmitter and receiver pairs). On the other hand, if the distances between repeaters are greater than l_{opt} then the power consumption of amplifiers dominates the power consumption of link L.

Designing power-efficient networks can reduce the running power consumption of networks. In this section, some ideas that can be used to efficiently design network power have been introduced and shown that modular switches, mixed line rates, DONs, and optimum repeater spacing can reduce the running power consumption of networks.

22.4 ENERGY-AWARE OPTICAL CORE NETWORKS

Here, we consider approaches that reduce the operational power consumption of networks. These approaches are named energy-aware approaches in the green ICT community since they try to adjust the power consumption of network equipment proportional to the offered traffic load.

Since networks are usually designed to handle peak traffic load, they consume more power than their requirements for long stretches of time, and therefore, underutilize most of the network resources. Furthermore, some resources are reserved for protection purposes. These resources are always active to protect the network against failures. However, it is ideal for power consumption of a network to be proportional to its traffic load. Energy-aware networks adjust power consumption with respect to traffic load. Power-aware routing and wavelength assignment, traffic grooming, selective switching of network equipment to sleep, and energy-aware survivability are some techniques that attempt to reduce operational power consumption. Unlike energy-efficient network design ideas, energy-aware ideas usually degrade some QoS parameters such as bit error rate and delay. In the following, these techniques are investigated.

22.4.1 Power-Aware Routing and Wavelength Assignment (PA-RWA)

By employing RWA (routing and wavelength assignment), a minimum cost route is set up between source and destination nodes and an appropriate wavelength is assigned to the established path. The objectives of conventional RWAs are to minimize routing cost and blocking probability. Route cost may have different parameters such as path length and hop count. To reduce the blocking probability of new light-path demands, the most common idea is to distribute light-paths over network resources in order to minimize the traffic load over them. At low traffic loads, distributing traffic over network resources causes a lot of network equipment to operate at low capacity.

The PA-RWA is modeled as an integer linear programming (ILP) formulation [5]. Equation (22.17) shows the objective function of this model:

$$\min\left(P_{\mathrm{A}}\sum_{(i,j,k)} a_{ijk}x_{ijk} + P_{\mathrm{o}}\sum_{(i)} y_i\right). \tag{22.17}$$

This model expresses the power consumption of links and node equipment in the entire network. Parameter P_{A} is the power consumption of an optical amplifier and P_{O} is the power consumption of an OXC. The first summation shows the number of all turned on amplifiers, and the second summation counts the number of turned on OXCs.

The complexity of this model is related the many constraints that must be considered such as allocating a wavelength to at most one light-path, the limited number of wavelengths on each fiber, the number of fibers between two adjacent nodes, and wavelength conversions. All these constraints should be expressed for each link. Since there are usually many constraints for an ILP problem, and also these problems are classified as NP-hard problems, hence some heuristics that have been proposed to solve them follow:

- *Least Cost Path (LCP)*: LCP assigns a cost for each link, where link cost is the power consumption of establishing a connection between two adjacent nodes. This heuristic selects the minimum cost path for a light-path demand. If several wavelengths are free over a selected path, the first fit (FF) algorithm is used to assign an appropriate wavelength.
- *Most Used Path (MUP)*: MUP initially assigns the power consumption of links as their costs. When a light-path is established over a selected path, MUP decreases the link cost of the selected path in order to establish new future connections over already used paths. Hence, this heuristic aggregates connections on currently powered on links. After establishing a path, FF assigns an appropriate wavelength to it.
- *Ordered Light-path Most Used Path (OLMUP)*: Here, there are a number of connection demands that are set up one by one in a loop. In each iteration, OLMUP computes the required power for establishing light-paths of (nonestablished) connection demands. Then, it selects the connection demand with minimum cost, and sets it up. After establishing a light-path, costs of links are updated similar to the MUP heuristic. Then, the loop is iterated for the remaining connection demands. OLMUP has more power conservation than MUP and LCP because it assigns a connection demand to the route with minimum required power. This leads to traffic being aggregated over currently turned-on resources.

22.4.1.1 Weighted Power-Aware Routing (WPA-RWA)

PA-RWAs decrease the power consumption of a network, but they decrease network performance as well. PA-RWAs set the cost of links as link power consumption, and therefore, a routing algorithm may select the minimum power consumption path, which is not the shortest path. Selecting long paths decreases some QoS parameters such as BER and delay, and increases blocking probability. In some situations, it is better to prioritize network performance over minimizing power consumption. The work in [32] has proposed a weighted PA-RWA (WPA-RWA) and investigated the impact of energy minimization on path length and blocking probability. For each connection request, WPA-RWA finds at most K shortest paths, and then it assigns link cost (C_l) for each link l in the paths as

$$\begin{cases} \alpha \times P_{\text{link},\, l}, & \text{fiber link, } l \text{ in use} \\ P_{\text{link},\, l}, & \text{fiber link, } l \text{ not in use} \end{cases}, \tag{22.18}$$

where $P_{\text{link},l}$ is the required power for turning on link l, and $\alpha \in (0, 1)$ is a weighting factor. The parameter α determines the impact of energy minimization over lengths of paths. If α is close to 0, WPA-RWA prioritizes ∂∂power minimizing over path length. It means that WPA-RWA selects a path that requires minimum power consumption. When α is close to 1, WPA-RWA finds the shortest path and it does not consider the power consumption objective. This study has investigated the impact of different values of α on power saving, path length, and blocking probability. It shows that when α increases, power saving decreases at different traffic loads while path length and blocking probability decrease. In other words a high value of α decreases path length and blocking probability, and increases power consumption.

22.4.2 Traffic Grooming

The power consumption of today's network devices is not proportional to traffic load. The minimum data rate of a light-path is 10 Gbps, but almost all connection requests do not require exactly this data rate. When a connection demand is satisfied, a light-path will be devoted to this demand even though its requirement is less than the capacity of the allocated light-path. Hence, a portion of allocated light-path remains unused. Traffic grooming is a traffic engineering approach that aggregates small traffic units in a bigger unit, where the volume of aggregated traffic is bounded by the capacity of wavelength channel.

Power consumption of operations such as switching and E/O and O/E conversions is divided into two parts: overhead power (P_O) and traffic power (P_{T}), i.e.,

$$P = P_O + u \times P_{\text{T}}, \tag{22.19}$$

where P_O and P_{T} are operation-dependent powers [33]. If there is no traffic, then P_O and P_{T} are both zero. Parameter P_{T} is the maximum power consumption of a light-path if it carries traffic at maximum capacity; and $u \in (0, 1)$ is actually carried traffic size. Parameter u is normalized with respect to maximum wavelength capacity. When a connection is established, it consumes fixed power consumption P_o independent of traffic, and power consumption $u \times P_{\text{T}}$ related to traffic volume. Suppose two connections are established between two nodes with 1 and 2 Gbps data rates. If two connections are aggregated and then transmitted by one connection, the aggregated connection consumes less power since one connection has less power overhead than two connections. Traffic grooming reduces overhead power consumption.

Traffic grooming can be achieved through two strategies: link by link and end to end [34]. In the link-by-link strategy, an intermediate node can add traffic to the light-path traversing through itself. This strategy requires all light-paths to be terminated in intermediate nodes, where a light-path is converted to the electronic domain, some traffic is added or dropped to/from the light-path, and then it is converted to optical signal and transmitted over fiber (see Figure 22.7).

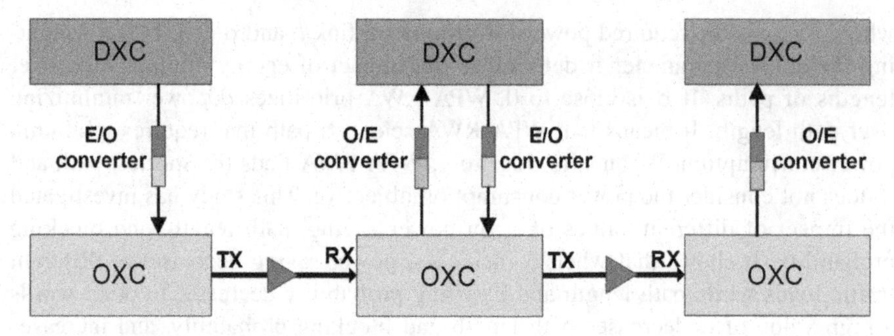

Figure 22.7 Link-by-link traffic grooming strategy.

The power consumption of the link-by-link strategy for a new connection from node S to node D can be stated by

$$\begin{aligned} P_{\text{link\_by\_link}} = {} & P_{DS} + P_{EO} + P_{OS} + P_{AMP} + P_{OS} + P_{OE} + P_{DS} \\ & + P_{EO} + P_{OS} + P_{AMP} + P_{OS} + P_{OE} + P_{DS}, \end{aligned} \tag{22.20}$$

where P_{DS} is the power consumption of digital switching, P_{OS} is the power consumption of optical switching, P_{EO} and P_{OE} represent the power consumption of E/O and O/E converters, respectively, and P_{AMP} is the power consumption of an optical amplifier.

In the end-to-end strategy, aggregation is performed at the source of a light-path. When a new connection demand between source S and destination D arrives, it is aggregated with an already established light-path if possible. Otherwise, a new light-path is established. All light-paths remain in the optical domain while traversing intermediate nodes (bypassing) because no add/drop operation is possible in intermediate nodes (see Fig. 22.8).

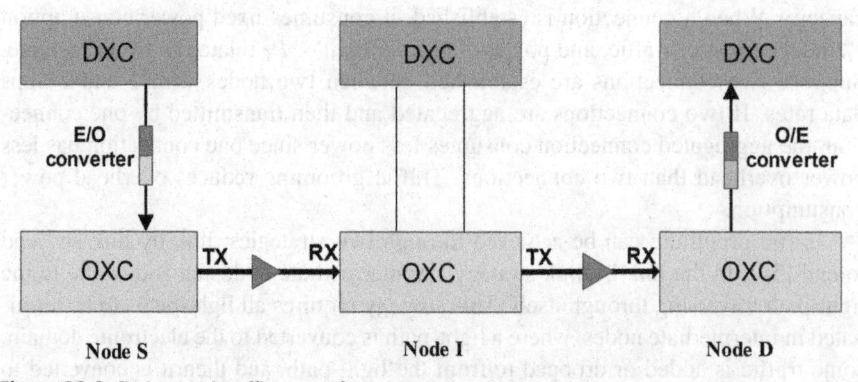

Figure 22.8 End-to-end traffic grooming.

The power consumption of a new connection between node S and D can be stated as

$$P_{\text{end\_to\_end}} = P_{\text{DS}} + P_{\text{EO}} + P_{\text{OS}} + P_{\text{AMP}} + P_{\text{OS}} + P_{\text{OE}} + P_{\text{DS}} + P_{\text{EO}} + P_{\text{OS}} + P_{\text{AMP}} + P_{\text{OS}} + P_{\text{OE}} + P_{\text{DS}}. \quad (22.21)$$

As stated in Eqs. (22.20) and (22.21), the power consumption of establishing a new connection in the end-to-end grooming strategy is lower than in link-by-link grooming. These two strategies have been investigated in [34] and the authors concluded that energy consumed can be reduced by about 50% with the end-to-end strategy if individual traffic demands match fairly well or exceed the capacity of optical channels. By increasing traffic demands, the power consumption of the two strategies increase at the same rate.

The link-by-link traffic grooming strategy reduces the required number of wavelengths by aggregating small traffic, at the expense of increasing O/E and E/O conversions, which are energy-hungry operations. The E/O and O/E converters consume significantly more energy than the bypassing operation. The bypassing strategy can exploit the high number of light-paths and does not perform O/E and E/O operations in intermediate nodes. The power consumption of optical bypass, link-by-link grooming, and hybrid strategies have been investigated in [35]. Under the bypass strategy, traffic grooming is not performed and each traffic demand has its devoted light-path and it is optically switched in intermediate nodes. In the hybrid strategy, a light-path can be partially aggregated with previously established light-paths. If there is no appropriate existing light-path, a new light-path is set up. Intermediate nodes of a new route operate in bypassing strategy. This study showed that the hybrid strategy has lower power consumption than link-by-link traffic grooming.

Traffic grooming is modeled as an ILP, e.g., [25,36–38]. The difference between these models is related to their objective functions, which are either minimized or maximized. Minimizing is usually performed on the number of established light-paths, the number of energy-hungry operations (such as electronic switching traffic), or the energy consumption of entire network (where energy consumption of all operations carried out on a light-path such as receiving, processing, and transmitting are considered). Authors in [37] try to maximize the network throughput, where maximizing occurs on the number of established small traffic demands. It is claimed that for a large number of nodes, traffic grooming is an NP-complete problem [37]. Hence, like other these problems, it can be solved by heuristics such as the following:

- *Path-based heuristic*[36]*:* Connection requests are sorted based on decreasing order of their establishing cost and are serviced in turn. In energy-aware grooming, the cost of a link could be equal to the power consumption of a link and its related equipment. The power consumption of a link is related to the power consumption of the equipment used on the link such as optical amplifiers or repeaters. Let subnet K represent elements of the network equipment that are in an ON state. At first, routing is performed on subnet K. If it fails, then routing should be performed on graph G that represents the whole network. This heuristic leads to traffic aggregation and saving power since the routing algorithm prefers new connections be established on existing light-paths. When a new connection is established on an existing light-path, the overhead of power consumption is eliminated.

- *Link-based heuristic*[36]*:* Connection demands are sorted based on decreasing order of their establishing cost, and links are turned on in ascending order of their costs. Similar to the path-based heuristic, link and demand costs are equal to the power consumption of links and the power consumption of connection demands, respectively. If setup of a new connection demand fails on a subnet (i.e., on currently ON nodes), then the lowest cost links are turned on. These operations are repeated until a connection demand is established. The link-based heuristic establishes a connection on the lowest cost subnet, but it is less convergent than the path-based heuristic.
- *Request size–based (RSB) heuristic*[6]*:* Connection demands are aggregated based on source and destination pairs and then sorted in a descending order of aggregated request size. Larger request pairs are serviced sooner than smaller request pairs. This heuristic reduces energy consumption of the network since large traffic has more opportunity to be routed in low-cost paths.
- *Link utilization–based (LUB) heuristic*[6]*:* Demands are aggregated based on source and destination pairs. These pairs are sorted based on a link utilization parameter, where this parameter is defined as the aggregated capacity of the source and destination pairs divided by number of hops of the minimum cost physical path.

Research in [6] has introduced RSB-energy aware, RSB-traditional, LUB-energy aware, and LUB-traditional schemes. For energy-aware heuristics, link cost is defined as power consumption of establishing a connection between a source and destination. In traditional heuristics, link cost is defined as hops of the shortest path route between a source and a destination. It is shown that energy-aware heuristics can reduce more power consumption than traditional heuristics in low-load traffic situations.

Traffic grooming may cause degradation of some QoS parameters such as delay because it may route some light-paths in longer routes than available shorter routes [6]. On the other hand, energy-aware traffic grooming is a new concept and users may not be interested in experiencing new unproven technologies, especially when they need sensitive services. Therefore, it is not realistic to suppose that all users will be interested in reducing power consumption at the expense of QoS degradation of their services. A differentiated energy saving traffic grooming has been proposed in [39]. In this study, traffic demands are classified in *red* and *green* classes. The red class demands are served by a shortest path routing algorithm. On the other hand, the green class demands are routed with energy saving objectives. The red class demands have priority to be serviced earlier than green class demands. This study has also proposed two heuristics based on these classifications:

- *Red demand–based heuristic*: Demands are classified in red and green classes and sorted based on their requested capacities. First, red class demands are served based on objectives such as shortest path or minimum hop. After resource allocation for red classes, green class demands are served with the objective of reducing power consumption.

- *Total demand–based heuristic*: All red and green demands are sorted in a list in a descending order of requested capacities and served in order. If, for example, there are two request demands with different classes and the same capacity from source *s* to destination *d*, then the red class demand is prioritized to get service over the green class demand.

In short, traffic grooming is a technique that tries to aggregate small traffic in bigger units as much as possible. At low traffic load, traffic grooming leads to switching more underutilized equipments to sleep state. This technique can be performed in the link-by-link, end-to-end, or hybrid strategies. In the link-by-link strategy, intermediate nodes convert incoming traffic to electronic signals. These conversions are more energy-hungry than bypassing operations performed by the end-to-end strategy. The hybrid strategy, compared to the end-to-end and link-by-link strategies, is more energy-efficient [35] because it exploits their advantage. When traffic grooming is performed, some QoS parameter such as delay or block ratio may be degraded because some light-paths are established over long paths. Thus, there is a trade-off between power conserving and degrading of QoS parameters.

22.4.3 Selectively Switching Network Elements to Sleep

The capacity of a network is usually designed for peak traffic loads and for coping with future traffic requirements. Since the nature of traffic is time varying, a lot of equipment is underutilized at low traffic loads. Furthermore, since nodes and links may fail in backbone networks and this may lead to serious results for sensitive services, additional equipment is reserved for protection issues, and this equipment is always active. Overprovisioning (because of peak traffic load and protection considerations) can lead to additional power consumption. The solution for reducing power consumption is to switch low-load network equipment to sleep mode. Network elements can be configured in three modes [7]:

- *Active*: when an element performs all its functions.
- *Sleep*: when an element is inactive but can be switched rapidly to active mode
- *Off*: when an element is completely separated from other network elements.

Different mode settings can be used to address power saving issues. An effective solution for making power consumption proportional to network utilization is to change the state of low-load active elements to sleep mode. Similar to an off element, a sleep element cannot participate in network functions, but it can be rapidly switched to active mode and perform its functions. Notice that placing an element in sleep mode should not lead to disruptions in vital functions of a network because the network should be able to handle current connections. Switching a network element to sleep mode occurs in three stages [7]:

1. *Selecting which elements should go to sleep mode:* For the first stage, the following are some ideas on how to select the elements that should go to sleep mode:
- *No load:* If no traffic passes through an element, it can go to sleep mode. Placing elements with no load in sleep mode does not impact the utilization of current connections.
- *Load is less than threshold:* If the traffic load of an element goes below a threshold, it can go to sleep mode. Traffic must be rerouted to other paths. The value of the threshold parameter impacts on the amount of power conservation because high thresholds can lead to more elements being put into sleep mode.
- *Most power:* An element with the highest power consumption is picked for sleep mode.
- *Least flow:* An element that has the least amount of traffic passing through is chosen for sleep mode.
- *Random:* An element is randomly selected and goes to sleep mode.
- *Least link*: The node that has the least number of links (the node with the smallest degree) is switched to sleep mode.
- *ILP:* The network is modeled as an integer linear program. The objective function is to minimize the power consumption of a network that can support given connections. This method is applied to small networks because an ILP problem is NP-hard and finding optimum solution is not viable if there are many constraints.

2. *Rerouting of residual traffic:* After a node is selected for sleep mode, it may be an intermediate element by which some traffic traverses. Therefore, traffic must be rerouted to other paths. There are two issues to deal with regarding rerouting:
- *Reactive:* The selected node goes to sleep mode and the upper layers of the network detect that their relevant light-paths are not reachable at lower layers; therefore, they try to establish other light-paths.
- *Proactive:* The selected node informs source nodes of traffic to reroute traffic over other paths. If rerouting is successful, then the node can go to sleep mode; otherwise it must still remain active.

3. *Reactivation:* In dynamic networks, traffic load is variable. When traffic load increases, some sleeping nodes or link equipment may be switched to active mode in order to handle new connection demands. Reactivation of sleeping resources must consider power consumption objectives. In other words, the reactivation process must turn nodes or links on in such a way that power consumption of the whole network remains as low as possible. Traditional PA-RWA sets link costs as the power consumptions of link equipment (power consumption of amplifiers and repeaters). It may distribute connection demands over all network resources especially when links costs are close to each other. Reference [8] has modified PA-RWA by considering a penalty for turning

currently sleeping links on. This penalty leads to aggregation of connections on currently turned on links as much as possible. If a connection demand is blocked, then some sleeping links should be woken up. If it is required to reactivate some sleeping links, the modified PA-RWA selects the one with the minimum path cost.

When some network resources go to sleep mode it is inevitable that QoS parameters will be degraded. The minimum power consumption of backbone networks has been modeled as an ILP that considers some constraints such as QoS and connectivity constraints [40]. In optical WDM networks the complexity of the ILP is related to the number of constraints. In these networks usually there are many constraints, and also the ILP problem are classified as an NP-hard problem, hence heuristics must be proposed to solve it. This study presents an iterative procedure that checks whether a network element can be switched to sleep mode. This procedure is executed for nodes and links separately. Before starting an iteration of the check loop, elements are sorted based on heuristics such as least link (LL) or least flow (LF), or randomly. Next, based on the ordered list, the procedure checks the elements for switching to sleep mode one by one. After an element goes to sleep mode, two constraints must be satisfied; if one of the constraints fails, the sleeping element must be switched to active mode:

- *Residual traffic must be reroutable to other paths:* This constraint ensures that putting a given element into sleep mode does not lead to network fragmentation.
- *Utilization of each link does not exceed a threshold:* This constraint ensures that some QoS parameters are not degraded.

This study evaluates the proposed procedure for different hybrids of node and link heuristics and shows that the Lf22-LF combination saves more power than others, where LF means least flow node heuristic, described above.

- *Distributed*: Each node disseminates information about its links to the whole network and also collects information about all other nodes and stores it in a traffic engineering database. Nodes locally manage connections with neighbors.
- *Centralized*: Each node collects information about itself and all its neighbors and then sends this information to a central control node. The central control node is responsible for making decisions about management of switching of nodes to sleep node in the whole network.

The work in [7] has investigated the power consumption of links and proposed an idea to shut down unused links under distributed and centralized control strategies. Under the centralized strategy, the central control node periodically solves an ILP program to determine which links are unused and can go to sleep mode. Define unused links to be links that have traffic that can be routed over other links. After selecting unused links, the rerouting procedure is executed to change the lightpaths that are already passing through these selected unused links. If rerouting is successful, then the selected unused links can go to sleep mode. It should be noted that rerouting is performed by the central control node. This study has compared results based on two different objective functions:

1. *L-P:* Minimizes the number of light-paths for rerouting and then minimizes link power. L-P minimizes the number of light-paths to be rerouted if there are multiple links, and then selects the one that consumes more power.
2. *P-L:* Minimizes link power and then minimizes the number of light-paths for rerouting. P-L minimizes the power consumption of active links. Then, if there are multiple candidates for sleep mode, it selects the one that supports fewer light-paths.

These objective functions may lead to different results. The P-L objective function can save more power than L-P, but leads to more blocking than L-P.

The study in [7] has also proposed a distributed strategy to reduce power consumption by switching low-load links to sleep mode. This strategy supposes that each node is able to put its links to sleep and reactivate them locally. Before a node decides to switch its link to sleep, it should notify the sources of connections that pass through the link. Sources of the connections execute a rerouting procedure. If rerouting leads to establishing new connections, then the node can switch the link to sleep. The number of supported wavelengths can be used as a threshold to nominate links for sleeping. It is obvious that if the threshold is bigger, more links can go to sleep mode and more power savings can be obtained. As aforementioned, switching elements to sleep mode may be lead to degradation of QoS parameters. Simulation results show that for different thresholds (0, 1, and 3), more power savings can be achieved, but at the expense of a higher blocking probability at high traffic loads. In a different solution, authors in [41] have investigated the impact of rerouting traffic on the number of turned off line cards in IP-over-WDM networks in low-load situations. In this work, three different rerouting strategies have been introduced: Fixed Upper (IP layer) Fixed Lower (WDM layer), Dynamic Upper Fixed Lower, and Dynamic Upper Dynamic Lower abbreviated respectively as FUFL, DUFL, and DUDL. For example, with the DUFL strategy, the IP layer can dynamically change routes whereas the WDM layer is fixed all the time, especially at low-load traffic. It has been shown that the DUFL strategy conserves more energy compared to the FUFL and DUDL strategies.

22.4.4 Survivability

The key rule of optical WDM networks as future backbone networks is not negligible. Each wavelength in current optical WDM networks has 10 or 40 Gbps capacity and tens of wavelengths can be multiplexed in one fiber. If only one fiber fails, it leads to losing a huge volume of traffic. Hence, survivability is an essential necessity to protect links and nodes against failures. Survivability is the capability of a network to recover after a failure. Protection can be provisioned at different levels: node, link, and path. This section investigates different energy-saving approaches with dedicated path protection in WDM networks.

22.4.4.1 Path Protection

In survivable WDM networks with path protection capability, two light-paths are established for each connection request, where the paths are disjointed (not having any node in common, except source and destination). These paths are named primary (or working) and secondary (or protection). Under normal situations, a source transmits data to the destination over the primary setup path. When a failure occurs, the source and destination both switch to the secondary setup path. It should be noticed that transition over the secondary path is temporary and after recovery of failure, traffic transmission should be switched back to the primary path. There are three approaches to protect a network against failures:

- *1:1 protection*: There is a dedicated secondary path for each primary path. Data is sent over the primary path until there is no failure over this path. If a failure occurs, source and destination switch over to the secondary path.
- *1+1 protection*: Similar to the 1:1 protection, there is a dedicated protection path for each primary path. However, the source concurrently sends data over both the primary and secondary paths. The destination receives them and based on signal quality chooses their best.
- 1*:N protection*: Since failures occur scarcely in optical networks, several primary paths can be protected by one secondary path. A shared secondary path can only support one path at each moment. Hence, if two or more paths fail at the same time, only one of them can be protected by the secondary path.
- *M:N protection*: *M* secondary paths protect *N* primary paths. In this approach, at most *M* failures can be supported.

More survivable networks need more resources for protection purposes. The 1:*N* and *M:N* approaches provide less protection resources, and therefore, they can recover fewer failures; whereas 1:1 and 1+1 approaches can result in more survivable networks.

22.4.4.2 Energy-Efficient Survivable WDM Networks

Network resources (including node and link equipment) reserved for protection purposes are always active in order to rapidly recover failures. These resources consume energy, even though they are unused most of the time. Some approaches have been proposed to reduce the power consumption of energy-efficient survivable networks [42–44]. The majority of these approaches are based on switching protection resources to sleep mode. Among different protection approaches, the 1+1 approach cannot switch a secondary path to sleep mode since both primary and secondary paths are simultaneously in use. On the other hand, protection resources in 1:1, *M:N*, and 1:*N* approaches can be switched to sleep mode. In the following, the ideas that can reduce the power consumption of survivable networks are introduced.

22.4.4.2.1 Energy-Aware Shared Path Protection

In conventional networks, load balancing is an important operation since it distributes traffic over network resources and utilizes them as equal as possible. Load balancing reduces congestion in links and nodes, and also it decreases the requirements for protection resources. On the other hand, energy-efficient approaches try to pass traffic through already turned on resources and switch low utilization resources to sleep mode. Energy-efficient approaches increase the requirements for protection resources. Hence, there is trade-off between energy efficiency and the capacity of protection resources [45]. This trade-off is depicted by Figure 22.9a and Figure 22.9b.

Another objective that should be taken into consideration in energy-efficient survivable WDM networks is separation of working and protection resources. Depending on how much separation level is greater, many exclusive protection resources can be switched to sleep mode (as shown in Figure 22.10a and Figure 22.10b).

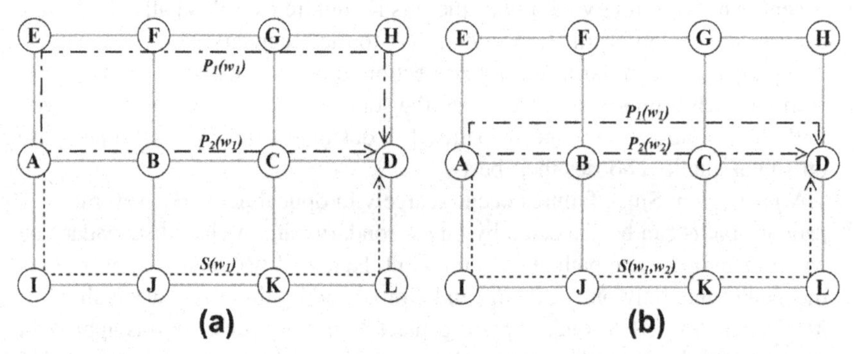

Figure 22.9 (a) Load balancing distributes routes over network resources because it decreases congestion and also decreases protection resources; (b) energy-efficient approaches aggregate traffic as much as possible in order to put more elements into sleep mode, but they need more protection resources.

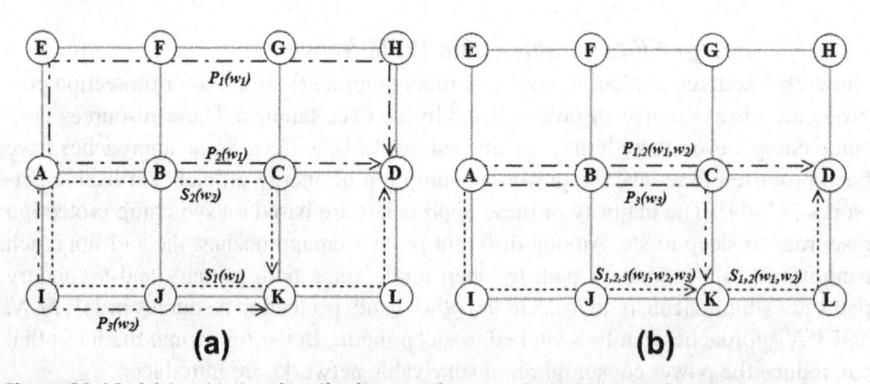

Figure 22.10 (a) Load balancing distributes primary and secondary paths over network resources and there is no objective to seperate working and protection resources; (b) energy-efficienct approaches aggregate primary and secondary resources and also seperate working and protection resources for switching protection resources to sleep mode.

Research in [45] has provided models for energy-aware WDM networks with shared protection resources as an ILP problem with a multiobjective function that tries to minimize energy consumption of networks and capacity of shared protection resources:

$$\textbf{Minimize} \quad \text{capacity} + \xi \times \text{energy}, \tag{22.22}$$

where the parameter *capacity* is defined as the total number of primary and protection wavelengths exploited in each link, the parameter *energy* is defined as the energy consumption of network equipment, and ξ helps to adjust the weight of energy minimization. The results of ILP show that the proposed energy-efficient design jointly reduces the number of wavelengths and energy consumption.

22.4.4.2.2 Dedicated Path Protection

The minimum power with sleeping support (MP-S) concept has been introduced for static WDM networks protected by dedicated paths in [46]. This idea has two phases: initial routing and refining paths. The initial routing phase evaluates the power consumption of establishing a light-path over graph G, where edge costs are set to geographical distances. Then, links are sorted in ascending order based on their evaluated energy consumptions. Then, connection requests are serviced based on the order of the sorted list. This algorithm establishes working and protection paths with minimum cost and does not try to separate working and protection resources. Two methods have been proposed for this operation [8]:

- *EA-DPP-Dif:* Protection resources can be switched to sleep mode if they are exclusively reserved for protection purposes. This heuristic tries to separate primary and secondary paths as much as possible for switching more resources to sleep mode. Thus, it creates a high penalty if primary and secondary paths are mixed in intermediate links. This penalty leads to separate primary and secondary paths, but it also leads to primary and secondary paths being distributed over all network resources. For aggregating the primary and secondary paths in separate paths, this heuristic considers another penalty to encourage the algorithm to choose already existing primary/secondary paths. This heuristic considers an OFF state for resources that are neither used for primary paths nor reserved for protection purposes.
- *EA-DPP-MixS*: this heuristic is similar to *EA-DPP-Dif*, but it eliminates the penalty for mixing the primary and secondary paths. Therefore, primary and secondary paths can be mixed in intermediate links. *EA-DPP-MixS* tries to pack the primary and secondary paths on the same paths by considering the penalty for using resources that do not belong to primary or protection resources.

Separating working and secondary paths increases power savings and also blocking probability since resources are used for different purposes, working and protection. Furthermore, separating working and protection resources leads to longer secondary paths compared to mixing working and protection resources. Therefore, there is a trade-off between separating working and protection paths and probability of blocking.

22.5 SUMMARY

The energy crisis has been a big challenge against developing various technologies in the last decade. ICT services such as e-banking, e-commerce, and video-conferencing are solutions for reducing power consumption in different sections, especially in the transportation domain. The rapid growth of ICT, which obviously increases power consumption in the ICT infrastructure, has led to new energy conserving ideas in the ICT domain. Since core networks are fundamental parts of telecom networks and these networks are based on optical technologies, this chapter has exclusively investigated power conserving solutions in the optical WDM core network.

In this chapter, energy conserving ideas have been introduced in two distinct classes: energy-efficient network design and energy-aware network ideas. The approaches used for designing energy-efficient networks, considered while a network is designed, try to reduce the running power consumption of network equipment. We have discussed how the modular structure of switches, mixed line rates, dynamic networks, and optimum repeater spacing can reduce the running power consumption of network equipment. On the other hand, energy-aware network approaches try to reduce the operational power consumption of networks. These approaches aim to adjust the power consumption of network equipment proportional to the current offered traffic load. Energy-efficient network design approaches do not degrade network performance, but energy-aware approaches usually degrade some QoS parameters such as average length path and blocking ratios. Hence, there is always trade-off between energy conservation and QoS degradation.

The following works are proposed as research directions in the future with the hope that they can help to reduce power consumption in next-generation optical networks.

- Although linear models and theoretical models can be applied to evaluate the power consumption of network equipment at various traffic loads, these models may not be able to exactly express the impact of traffic properties over the power consumption of network equipment. Therefore, statistical models that consider the properties of traffic should be employed with the aim of reducing their complexity.
- IP-over-WDM networks enable an electronic IP layer to operate the upper optical WDM layer, and there are some energy-aware works that have proposed their ideas over IP-over-WDM networks, e.g., traffic grooming strategies are based on IP-over-WDM networks. Since all optical switching networks such as optical packet switching (OPS) and optical burst switching (OBS) are in way, power conservation ideas should also be provided for these switching networks.

REFERENCES

[1] M. Pickavet, P. Leisching, Energy footprint of ICT: forcasts and network solutions, in: Proceedings of OFC/NFOEC, March 2009, San Diego, CA.

[2] Global e-Sustainibility Initiative (GeSI) Enabling the low carbon economy in the information age, <www.smart2020.org/_assets/files/02_Smart2020Report.pdf>.

[3] A. Leiva, J.M. Finochietto, B. Huiszoon, V. Lopez, M. Tarifeno, J. Aracil, A. Beghelli, Comparison in power consumption of static and dynamic WDM networks, Optical Switching and Networking 8 (3) (2011) 149–162.

[4] C. Pulak, T. Massimo, M. Biswanath, On the energy efficiency of mixed-line-rate networks, in: Proceedings of Optical Fiber Communication Conference, March 2010, San Diego, CA.

[5] Y. Wu, L. Chiaraviglio, M. Mellia, F. Neri, Power-aware routing and wavelength assignment in optical networks, in: Proceedings of IEEE ECOC 2009, September 2009, Vienna, Austria.

[6] M.M. Hasan, F. Farahmand, A.N. Patel, J.P. Jue, Traffic grooming in green optical networks, in: Proceedings of IEEE International Conference on Communications (ICC), May 2010, Cape Town, South Africa.

[7] I. Cerutti, N. Sambo, P. Castoldi, Sleeping link selection for energy-efficient GMPLS network, IEEE Lightwave Technology 29 (15) (2011) 2292–2299.

[8] A. Jirattiglachote, C. Cavdar, P. Monti, L. Wosinska, Dynamic provisioning strategies for energy efficient WDM networks with dedicated path protection, Optical Switching and Networking 8 (3) (2011) 201–214.

[9] C. Lange, A. Gladisch, On the energy consumption of FTTH access networks, in: Proceedings of IEEE OSA/OFC/NFOEC, March 2009, San Diego, CA.

[10] P. Chowdhury, M. Tornatore, S. Sarkar, B. Mukherjee, Building a green wireless-optical broadband access network (WOBAN), IEEE Lightwave Technology 28 (16) (2010) 2219–2229.

[11] R. Kubo, J.-I. Kani, Y. Fujimoto, N. Yoshimoto, K. Kumozaki, Sleep and adaptive link rate control for power saving in 10G-EPON systems, in: Proceedings of the 28th IEEE Conference on Global Telecommunications, November 2009, Honolulu, Hawaii, USA.

[12] A. Lovric, S. Aleksic, J.A. Lazaro, G.M.T. Beleffi, J. Prat, V. Polo, Power efficiency of SARDANA and other long-reach optical access networks, in: Proceedings of the 15th International Conference on Optical Network Design and Modeling, February 2011, Bologna, Italy.

[13] C.M. Machuca, J. Chen, L. Wosinska, M. Mahloo, K. Grobe, Fiber access networks: reliability and power consumption analysis, in: Proceedings of the 15th International Conference on Optical Networking Design and Modeling-ONDM 2011, February 2011, Bologna, Italy.

[14] N. Iiyama, H. Kimura, H. Hadama, A novel WDM-based optical access network with high energy efficiency using elastic OLT, in: Proceedings of the 14th Conference on Optical Network Design and Modeling, February 2010, Kyoto, Japan.

[15] Y. Zhang, P. Chowdhury, M. Tornatore, B. Mukherjee, Energy efficiency in telecom optical networks, IEEE Communication Survays and Tutorials 12 (4) (2010) 441–459.

[16] R. Bolla, R. Bruschi, F. Davoli, F. Cucchietti, Energy efficiency in the future Internet: a survey of existing approaches and trends in energy-aware fixed network infrastructures, IEEE Communication Survays and Tutorials 13 (2) (2010) 223–244.

[17] Y. Yan, S.-W. Wong, L. Vacarenghi, S.-H. Yen, D.R. Campelo, S. Ymashita, L. Kazovsky, L. Dittmann, Energy management mechanism for Ethernet passive optical networks, in: Proceedings of IEEE International Conference on Communications, May 2010, Cape Town.

[18] S.-W. Wong, L. Vacarenghi, S.-H. Yen, D.R. Camelo, S. Yamashita, L. kazovsky, Sleep mode for energy saving PONs: advantages and drawbacks, in: Proceedings of IEEE GLOBCOM Workshops, November 2009 Honolulu, HI.

[19] B. Kantarci H.T. Mouftah, Towards energy-efficient hybrid fiber-wireless access networks, in: Proceedings of the 13th International Conference on Transparent Optical Networks (ICTON), June 2011, Stokholm, Sweden.

[20] J. Zhang, N. Ansary, Towards energy-efficient 1G-EPON and 10G-EPON with sleep-aware MAX control and scheduling, IEEE Communications Magazine 49 (2) (2011) 33–38.

[21] B. Mukherjee, Optical WDM Networks, Springer, 2006.

[22] K. Hinton, J. Baliga, M. Feng, R. Ayre, R.S. Tucker, Power consumption and energy efficiency in the Internet, IEEE Network 25 (2) (2011) 6–13.

[23] S. Ricciardi, D. Carglio, F. Palmieri, U. Fiore, G. Santos-Boada, J. Sole-Pareta, Energy-oriented models for WDM networks, in: Proceedings of the 7th International ICST Conference on Broadband Communication Networks and Systems, October 2010, Athens, Greece.

[24] T.T. Ye, G.D. Micheli, L. Benini, Analysis of power consumption on switch fabrics in network routers, in: Proceedings of the 39th Annual Design Automation Conference, August 2002, New Orleans, Louisiana, USA.

[25] E. Yetginer, G.N. Rouskas, Power efficient traffic grooming in optical WDM networks, in: Proceedings of the 28th IEEE conference on Global Telecommunications, November 2009, Honolulu, Hawaii, USA.

[26] R.S. Tucker, Green optical communications–Part II: energy limitations in networks, IEEE Jornal of Selected Topics in Quantum Electronics 17 (2) (2011) 261–275.

[27] Y. Sun, A.K. Srivastava, J. Zhou, J.W. Sulhoff, Optical fiber amplifiers for WDM optical networks, Bell Labs Technical Journal 4 (1) (1999) 197–217.

[28] R.S. Tucker, Green optical communications–Part I: energy limitations in transport, IEEE Journal of Selected Topics in Quantum Electronics 17 (2) (2011) 245–261.

[29] G. Ananthanarayanan, R.H. Katz, Greening the switch, in: Proceedings of the 2008 Conference on Power Aware Computing and Systems, 2008, San Diego, California.

[30] A. Nag, M. Tornatore, B. Mukherjee, Optical network design with mixed line rates and multiple modulation formats, IEEE Lightwave Technology 28 (4) (2010) 466–476.

[31] A. Silvestri, A. Valenti, S. Pompei, F. Matera, A. Cianfrani, Wavelenth path optimization in optical transport networks for energy saving, in: Proceedings of the 11th International Conference on Transparent Optical Networks, June 2009, Azores.

[32] P. Wiatr, P. Monti, L. Wosinska, Green lightpath provisioning in transparent WDM networks: pros and cons, in: Proceedings of the IEEE 4th International Symposium on Advanced Network and Telecommunication System (ANTS), December 2010, Mumbai.

[33] M. Xia, Y. Zhang, P. Chowdhury, C.U. Martel, B. Mukherjee, Green provisioning for optical WDM networks, IEEE Journal of Selected Topics in Quantum Electronics 17 (2) (2011) 437–446.

[34] W.V. Heddeghem, M.D. Groote, W. Vereecken, D. Colle, M. Pickavet, P. Demeester, Energy-efficiency in telecommunications networks: link-by-link versus end-to-end grooming, in: Proceedings of the 14th Conference on Optical Network Design and Modeling (ONDM), February 2010, Kyoto, Japan.

[35] M. Xia, M. Tornato, Y. Zhang, P. Chowdhury, C. Martel, B. Mukherjee, Greening the optical backbone network: a traffic engineering approach, in: Proceedings of IEEE International Conference on Communication (ICC), May 2010, Cape Town.

[36] S. Huang, D. Seshadri, R. Dutta, Traffic grooming: a changing role in green optical networks, in: Proceedings of the 28th IEEE Conference on Global Telecommunications, November 2009, Honolulu, Hawaii, USA.

[37] K. Zhu, B. Mukherjee, Traffic grooming in an optical WDM mesh network, IEEE Journal on Selected Areas in Communications 20 (1) (2002) 122–134.

[38] G. Shen, R.S. Tucker, Energy-minimized design for IP over WDM networks, IEEE/OSA Journal of Optical Communications and Networking 1 (1) (2009) 176–186.

[39] F. Farahmand, M.M. Hasan, I. Cerutti, J.P. Jue, J.J.P.C. Roadrigues, Differentiated energy savings in optical networks with grooming capablities, in: Proceedings of IEEE Global Telecommunications (GLOBECOM 2010), December 2010, Miami, FL.

[40] L. Chiaraviglio, M. Mellia, F. Neri, Reducing power consumption in backbone networks, in: Proceedings of IEEE International Conference on Communication, June 2009, Dresden.

[41] F. Idzikowski, S. Orlowski, C. Raak, H. Woesner, A. Wolisz, Saving energy in Ip-over-WDM networks by switching off line cards in low-demand scenarios, in: Proceedings of the 14th Conference on Optical Network Design and Modeling, February 2010, Kyoto.

[42] L. Wosinska, A. Jirattigalachote, P. Monti, A. Tzanakaki, K. Katrinis, Energy efficient approach for survivable WDM optical networks, in: Proceedings of the 12th International Conference on Transparent Optical Network (ICTON), June 2010, Munich, Germany.

[43] B. Kantarci, H.T. Mouftah, Reducing the energy consumption of the reliable design of IP/WDM networks with quality of protection, in: Proceedings of SPIE 13th Photonics North Conference, May 2011, Ottawa, Canada.

[44] B. Kantarci, H.T. Mouftah, Greening the availability design of optical WDM networks, in: Proceedings of IEEE Globecom Workshop on Green Communications, December 2010, Miami, FL.

[45] C. Cavadar, F. Buzluca, L. Wosinska, Energy-efficient design of survivable WDM networks with shared backup, in: Proceedings of IEEE Global Telecommunication Conference 2010, GLOBECOM 2010, December 2010, MIAMI, FL, USA.

[46] P. Monti, A. Muhammad, I. Cerutti, C. Cavadar, A. Tzanakaki, Energy-efficient lightpath provisioning in a static WDM network with dedicated path protection, in: Proceedings of the 13th International Transparent Optical Networks (ICTON), June 2011, Stockholm, Sweden.

CHAPTER

Analysis and Development of Green-Aware Security Mechanisms for Modern Internet Applications

23

Luca Caviglione[a], Alessio Merlo[b,c]

[a]Institute of Intelligent Systems for Automation, National Research Council (CNR), Via de Marini 6, Genova I-16149, Italy,

[b]E-campus University, Novedrate, Italy,

[c]Department of Communications, Computer and Systems Science, University of Genova, AILab, Via F. Causa 13, Genova I-16145, Italy

23.1 INTRODUCTION

Nowadays, supporting *social* activities is a relevant feature of modern Internet applications. This can be viewed as an evolution of the earlier *peer-to-peer* (p2p) file-sharing services, where users exchanged data directly, thus creating overlay networks over a physical deployment [1]. Even if highly specialized in file exchanges, such applications also enabled the aggregation of people according to common interests (e.g., music or movies) by offering (or being linked with) additional companion services, such as text messaging, forums, and wikis to organize and discuss about the release of specific content.

Meanwhile, the advent of the so-called Web 2.0 increased the degree of interactivity of Web-based applications, also by enabling new possibilities in terms of user-to-user interactions. Even so, this is not a real novelty, since such a feature was already present in the original Web vision. Specifically, the World Wide Web Consortium (W3C) put effort into the creation of a Social Web, where people can create networks of relationships overlapped with the entire Web, while controlling their own privacy and data [2]. But today the support for social applications is not exploited through standardized solutions. Rather, it is delegated to ad hoc services, which have been increasing in popularity and becoming real cultural phenomena.

Thus, boosted by the availability of ubiquitous connectivity, mainly through mobile network appliances equipped with cost-effective air interfaces, e.g., IEEE 802.11a/b/g/n, Universal Mobile Telecommunication System (UMTS), and General Packet Radio Service (GPRS), the Internet is even more an *Internet of People*, rather than a simple internetwork of hosts. As a result, individuals can create and share contents with an increased degree of *social connectivity*. In this perspective, social networks (SNs) are the archetype of this new wave of applications and they account for millions of active users worldwide [3].

With such wide diffusion and massive utilization, the modern Internet is steadily increasing power consumption, thus becoming responsible for the production of a relevant portion of the overall CO_2 emission, as well as greenhouse gasses (GHG), which are eventually injected into our ecosystem. This is a real concern, and recently, the Organization for Economic Co-operation and Development [4] incubated different initiatives aiming at reducing the power consumption of both computing and networking devices (see, e.g., Refs. [5,6] for possible examples, also embracing industrial practices). In parallel, the research and academia worlds also initiated a new research field, which has been named *green computing* or *green networking*, according to the specific scope (even if they largely overlap). In a nutshell, research actions aimed at "*greening*" mainly try to optimize energy consumption according to two main major techniques: *hardware* (for instance, by augmenting the silicon efficiency or by adaptively turning on/off network interfaces), or *software*, especially in terms of virtualization and rate-adaptation algorithms. A detailed survey limited to the state-of-the-art green networking related advancements can be found in Reference [6].

In this new scenario, *network security* becomes more critical than ever. This is also due to its *cross-layer* nature, which embraces several components adopted to build modern Internet applications. In fact, features introduced by social applications account for several security threats and risks at different abstraction levels, including, among others, compromising the privacy of end-users [7]. As a consequence of the heterogeneity of services offered by a SN, its security is more complex both to design and to manage in comparison with other Web 2.0 applications. For instance, an average SN platform allows participants to chat with friends, join a conversation with mixed groups composed both by friends and nonfrieds, or share photos and videos only with well-defined subsets of contacts.

From a security perspective, each user can interact with a set of services (e.g., audio and video conferencing, chat, mail, and photo sharing), which can be very unrelated in terms of access control and authentication requirements. This necessarily hardens the handling of all SN-related security issues in a centralized and uniform way. One of the major concerns is due to the "split" nature of how security is currently guaranteed in a SN. In particular, it is managed through a disjoint set of specific mechanisms. Nevertheless, the same security goal (e.g., user authentication) may be managed by different security techniques or pseudostandard solutions (e.g., basic HTTP authentication vs. FOAF+SSL), even if deployed within the same SN.

In general, such a scenario may lead to SNs with coarsely grained security solutions, potentially resulting into error-prone and flawed architectures.

Beyond security, the rendition of SNs into the *green world* could also potentially create additional hazards. In fact, on the one hand, it could introduce new patterns responsible for the alteration both of data and traffic, which could be exploited (e.g., to reveal possible structural weaknesses). On the other hand, it envisages the deployment of new architectural components (we mention, among others, a proxy to support smart sleeping of devices), which can lead to additional methods to attack or explore a portion of the network. To provide a comprehensive view, Figure 23.1 reviews the relations among the aforementioned aspects.

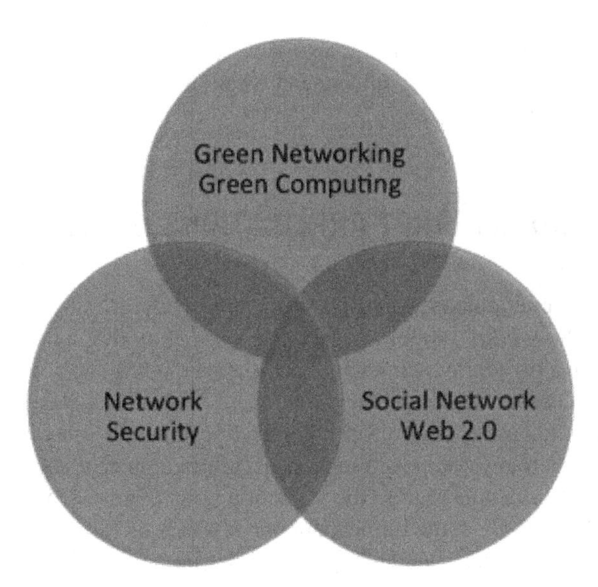

Figure 23.1 The intersection among green-related technologies, the social Internet, and network security.

As depicted in Figure 23.1, network security, SNs (which we recall we assume as a comprehensive example of Web 2.0 applications) and green networking (or computing) are partially intersected, also by virtue of the intrinsic "cross-layer" nature of such technologies. Therefore, we identify in this overlap a new concept that we define as *Green-Aware Security* or *Green Security*. We assume such terms as equivalent, thus in the following they will be used interchangeably.

Summing up, the contributions of this paper are (i) to introduce Web 2.0 (in the sense of SN) and related security issues at the *network layer*, also by taking into account new devices and technological constraints (e.g., the presence of wireless channels); (ii) to point out vulnerabilities and security flaws at the *application layer* related to Web 2.0; (iii) to show through a practical example how focused attacks can exploit vulnerabilities and reduce the security of a SN; (iv) to tentatively formalize the concept of energy-awareness for security mechanisms; and (v) to provide an early model of the energy consumption explicitly considering security.

To the authors' best knowledge, this is the *first* work aimed at capturing the connections among energy-efficient networking, new threats, and security risks induced by the more individual-centric flavor of Internet applications. This also highlights the importance of the understanding and development of Green-Aware Security mechanisms, which is the main scope of this chapter.

This remainder of the chapter is structured as follows: Section 23.2 specifies typical scenarios where modern Internet applications are deployed, emphasizing features leading to security hazards. Section 23.3 widely discusses the security issues of SN applications, underlining why security is fundamental, while Section 23.4

demonstrates how to exploit them for malicious actions. Then, Section 23.5 introduces a basic model to precisely understand what Green-Aware Security is. Lastly, Section 23.6 concludes the chapter.

23.2 MODERN INTERNET APPLICATIONS: MOBILE AND SOCIAL

As hinted in the previous section, one of the most relevant advancements in the Internet is its full *mobility* support. In fact, users are more frequently exploiting the *anywhere-anytime* paradigm owing to advancements ranging from wireless communications and enhanced device power to increased battery lifetimes. In this always-connected scenario, users need proper applications for the management and the coordination of the many composite sources of information characterizing the social Internet. Also, many people require the availability of a comprehensive communication platform. Nevertheless, the limited amount of power of battery-operated devices imposes additional constraints that should be taken into account when developing a service to be accessed through such devices. Therefore, the energy consumption of security procedures is of key importance.

Besides, the availability of ad hoc client interfaces accounts for a huge explosion of mobile applications supporting the aforementioned social attitudes. The success of such services is also due to their integration with other standard features provided by modern personal appliances. For example, users can enrich contents via geographical information gathered via the Global Positioning System (GPS) or the Assisted GPS (AGPS), share of pictures and videos acquired through a built-in camera, and use packet data communication services to convey voice, data, and e-mails via unique and unified platforms. However, such scenarios are highly *balkanized*, i.e., there are many different applications, mostly incompatible. At the same time, SN applications offer a kind of "*unified playground*" to collect and dispatch data to other services, becoming real development platforms [8]. For such reasons, and for the sake of the clarity, we will focus solely on SN frameworks, which completely embrace the majority of features and issues of Web 2.0.

23.2.1 Social Networks as a Paradigmatic Example for Understanding Relevant Security Issues

Even if a standardized definition of SN is absent, a brief description of SN services can be as follows: *SN applications enable interaction among participants, which are connected to each other according to some relationship* (e.g., friendship, business partnership, or common interests). As opposed to other services, SNs are also highly specialized, e.g., there are services for sharing general-purpose content, or for dealing with specific topics, such as books and traveling. To make a real-world example of the broad set of characteristics of different services, we mention Facebook and Twitter. For instance, Facebook offers a rich set of functionalities ranging from sharing text and multimedia material, to AV communication capabilities, as well as minimal blogging support. Instead, Twitter only allows posting short text messages (accordingly called "tweets" having a maximum of 140 characters).

To summarize, the popularity of SN services is mainly due to the following core characteristics: (i) they allow sharing of user-generated content in a quick and simple way, by providing the needed hosting and authoring tools; (ii) they offer different features to support user-to-user communications, such as instant messaging (IM) and audio/video conferencing; (iii) many of them enable the creation of new software services through a set of application programming interfaces (APIs), becoming very appealing development environments; (iv) earlier incarnations of SNs were "closed" (i.e., a user belonging to a service could not interact with similar platforms operated by different providers), but nowadays proper data percolation can be made through specific interfaces; (v) the availability of well-established Web development techniques, such as the Asynchronous JavaScript and XML (AJAX) method, enables a SN to be highly interactive even providing support for real-time features (e.g., to promptly notify a user about changes happening within his/her network of contacts); (vi) many SNs can be accessed through ad hoc client interfaces specifically created for tablets, handheld devices, and gaming consoles, making the service ubiquitously accessible; and (vii) as a consequence of a solid mobility support, many SNs also offer localization services, making them suitable to be used jointly with geo-tagged information.

The above-mentioned features introduce several security threats and risks at different functional levels. As a consequence, if proper countermeasures are not adopted this can lead to very dangerous situations. At the limit, SNs can also invalidate (or dramatically reduce their effectiveness) other security tools, such as firewalls or password-based protection mechanisms.

The most relevant hazards introduced by this widely adopted modern Internet application are, among others:

1. Simple distribution of personal information leading to possible attacks *à-la* social engineering.
2. Due to complex or incoherent privacy and security settings, users can reveal their topographical location, thus potential causing breaches in physical security as well.
3. The joint utilization of different/specialized services can bring a new type of attack based upon multiple profile fusion. Additionally, the availability of proper data structures to arrange such information can ease the ability to profile user automatically and on a massive scale, thus reducing privacy.
4. Accessing SNs from mobile devices, mostly performed via IEEE 802.11, accounts for additional risks in terms of attacks due to the joint utilization of weak security standards and unencrypted application layer protocols (e.g., the usage of HTTP instead of HyperText Transfer Protocol Secure). Besides, as mobile devices are often battery operated, this can open up to a new class of battery-draining attacks.

5. The integration of third-party Web applications with information provided in the user profile can lead to many possible hazards, and creates new security breaches.
6. To provide the proper degree of interactivity and sophisticated user interfaces, specific design patterns are adopted. But they increase the risk of attacks such as request forgeries.
7. The availability of SN applications from a variety of appliances (e.g., mobile gaming consoles) may foster new kind of attacks based upon stack misbehaviors, or protocol fingerprinting. For instance, many devices do not have a full-featured TCP/IP stack and could exhibit erratic or exploitable behaviors [9].

Issues 1–7 highlight why security is mandatory. Consequently, in the following, we will discuss the basic security mechanisms and drawbacks of SNs, taking into account the importance of energy-related issues when designing and deploying the proper security mechanisms.

23.3 SECURITY ISSUES OF SOCIAL NETWORK APPLICATIONS

Understanding specific weaknesses of the overall deployment is important to identify where countermeasures have to be put in place and their impact in terms of power consumption and requirements.

Actually, despite engineering and design differences, all SNs are essentially Web applications. Therefore they potentially suffer from a huge set of application-level vulnerabilities that are intrinsically related with the current model of the Web. If not properly managed, they can be exploited by attacks aimed at compromising the security of the application itself. This can affect data privacy, confidentiality, and application availability, just to cite some issues. We point out that due to the diverse nature of each service, this could require the development of countermeasures with an application-level granularity.

In the following, we briefly introduce the most critical vulnerabilities that can impact a general Web application. Moreover, we will also showcase some possible hazards due to the traffic patterns produced by these kind of applications.

23.3.1 The OWASP Top-Ten Vulnerabilities

The Open Web Application Security Project (OWASP) [10] defines a set of the 10 most critical vulnerabilities for Web applications. The OWASP annually updates such ranking. The 2010 list encompasses the following vulnerabilities:

1. *Injection.* Injection flaws, such as Structured Query Language (SQL) and Lightweight Directory Access Protocol (LDAP) injection, occur when untrusted

data is sent to an interpreter as part of a command or query. The attacker's hostile data can trick the interpreter into executing unintended commands or accessing unauthorized data.

2. *Cross-Site Scripting (XSS).* XSS flaws happen whenever an application takes untrusted data and sends it to a Web browser without proper validation and escaping. XSS allows attackers to execute scripts in the victim's browser, resulting in hijacking of the user's session, or redirection of the browser to malicious sites.
3. *Broken Authentication and Session Management.* Application functionalities related to authentication and session management are often not correctly implemented, allowing attackers to compromise passwords, keys, session tokens, or exploit other implementation flaws to assume other users' identities.
4. *Insecure Direct Object Reference.* A direct object reference occurs when a developer exposes a reference to an internal implementation object, such as a file, a directory, or a database key. Without proper checks (e.g., access controls), attackers can manipulate these references to access unauthorized resources.
5. *Cross-Site Request Forgery (XSRF).* A XSRF attack forces a logged-on victim's browser to send a forged HTTP request, including the victim's session cookie and any other automatically included authentication information, to a vulnerable Web application. As a result, the browser can generate requests that appear legitimate.
6. *Security Misconfiguration.* Good security practices require secure configurations defined and deployed for all the involved entities, i.e., application, frameworks, application server, Web server, database server, and platform. All these settings should be defined, implemented, and maintained. Notice that such tools are often shipped with nonsecure defaults. To avoid misconfiguration-based attacks keeping all software components up to date, including all the third-party libraries used by the main application, is highly suggested.
7. *Insecure Cryptographic Storage.* Many Web applications do not properly protect sensitive data with appropriate encryption or hashing. As possible examples, we mention credit cards, Social Security numbers (SSNs), and authentication credentials. Attackers may steal or modify such weakly protected entries to conduct identity theft, credit card fraud, or other crimes.
8. *Failure to Restrict URL Access.* Many Web applications check the Uniform Resource Locator (URL) access rights just only before rendering protected links and buttons. However, applications must perform similar controls each time protected content has to be retrieved. Otherwise, attackers will be able to forge URLs to access hidden pages.
9. *Insufficient Transport Layer Protection.* Applications frequently fail to authenticate, encrypt, and protect the confidentiality and integrity of sensitive network traffic. Otherwise, they sometimes support weak algorithms, use expired or invalid certificates, or don't use certificates correctly.

10. *Unvalidated Redirect and Forwards*. Web applications frequently redirect and forward users to other pages and websites, mainly by using untrusted data to determine the destination pages. Without proper validation, attackers can redirect victims to phishing or malware sites.

Obviously, the previous vulnerabilities are produced via a combination of designing/programming issues and activities of malicious agents. Since the earlier sets of vulnerabilities as well as the current standard Web technologies are well known, the malicious agent can implement standard scripts and iteratively test for the presence of the OWASP vulnerabilities on an arbitrary number of Web applications. For this reason, a Web application (and in particular an SN) should be free from previous vulnerabilities or be robust against their potential exploitation. In fact, since the precise knowledge of the target application is not strictly required for an exploitation attempt, an attacker can easily automate their discovery.

23.3.2 Network Level Security Issues

Some specific behaviors of the traffic produced by SNs can be exploited for different malicious actions. This is even simpler when plain HTTP is adopted instead of HTTPS for moving data from servers to clients (and vice versa). As a consequence, gathering information to perform attacks can be straightforward.

Sniffing traffic though is not a simple task, since it heavily relies upon how the network is arranged. As an example, capturing packets from users accessing the Internet from wired networks or Digital Subscriber Loop (DSL) technologies can be very hard, requiring a degree of sophistication beyond nonskilled attackers. At the same time, the usage of wireless access (e.g., the IEEE 802.11) heavily affects the overall security framework. In fact, the joint adoption of HTTP over nonprotected (or weakly protected accesses, such as those employing WEP) channels dramatically eases operations aimed at traffic sniffing.

From this perspective, having a standard traffic sniffer can suffice (see, e.g., popular network analysis tools such as Wireshark [11] or tcpdump [12]). This is also due to the peculiarities of how SNs are implemented. In fact, SN applications are mostly accessed through Web browsers; thus they have HTML (and related technologies) as a core component. Therefore, the Web page is the basic building block, and it is delivered through the network. Typically, a Web page is composed by several objects that have to be retrieved to compose it entirely. Two types of objects exist: the main object containing the HTML document and in-line object(s), which are those linked within the hypertext.

Then, it is possible to gather traffic information belonging to a user and all the objects composing the pages sent by the SN site. This can be exploited to reconstruct the network of individuals and the exchanged text, as well as all the related in-line objects (e.g., pictures). Another drawback concerns the access to "private" material by inspecting a user's traffic. For instance, a user can have privacy settings preventing strangers from seeing his/her profile pictures. Conversely, when a friend accesses his/her profile, data can be captured. This usually guarantees it is easy to collect enough information to mimic other users' profiles or to perform identity theft actions.

Besides, SN applications can exhibit well-defined traffic patterns, even if the traffic is cyphered (see, e.g., Refs. [13,14] for general investigations of traffic related to many Web 2.0 services and SN applications). This is mainly due to the continuous and regular "polling" performed by in-line objects implementing ad hoc scripts to update pages in a nearly real-time manner. Among others, we mention approaches such as AJAX and Comet, where a long-held HTTP connection enables the Web server to push data to the browser, without the need of additional HTTP requests. Therefore, such elements may trigger data transfers via TCP or HTTP, e.g., information about online users, and updates for widgets or IM services. Calculating the power spectral density (PSD) of the traffic trace can be a simple and effective method to reveal such repetitive behavior. Then, this sort of "fingerprint" can be used to reveal SN-related activities within encrypted flows, and can be used to perform well-defined attacks, also aided by social engineering approaches. As a possible example, upon becoming aware that a user is online, it could be possible to perform malicious actions such as denial of service (DoS) attacks on the target machine (since conversation endpoints are usually not encrypted).

Lastly, as said, SNs can be accessed from different devices that have limited capabilities (both in terms of CPU and power resources) and incomplete or flawed protocol implementations. Therefore, by recognizing an endpoint producing traffic for SN activities (e.g., by using the aforementioned signature approach), it can be attacked. If the target host is a mobile device or a gaming console, traffic flooding attacks (such as ping flooding) can lead to a quick battery drain, crash, or intermittent connectivity. Besides, stack vulnerabilities are often well documented, thus it could be possible to send well-formatted packets to crash the victim appliance. This is even true for many TVs or set-top boxes offering access to the Internet without the proper degree of sophistication of the protocol stack [5].

23.4 EXPLOITING THE FLAWS OF SOCIAL NETWORKS

Referring to the taxonomy presented in Section 23.3, let us analyze realistic attacks based upon a subset of *four* vulnerabilities. We argue that SNs can be particularly liable to *injection, XSS, broken authentication,* and *XSRF*. For the sake of clarity, we do not investigate other vulnerabilities, since they are strictly related to programming mistakes, which are out of the scope of the chapter. Thus, insecure object references, failure to restrict URL access and unvalidated redirects and forwards are heavily coupled with the specific implementation of the single application. Moreover, security misconfiguration, insecure cryptographic storage, and insufficient transport layer protection are configuration errors independent from the nature of the SN itself, as well as unrelated to power-consumption issues.

Before showing the aforementioned vulnerabilities, let us define a sample SN environment that is general enough to represent real SN applications.

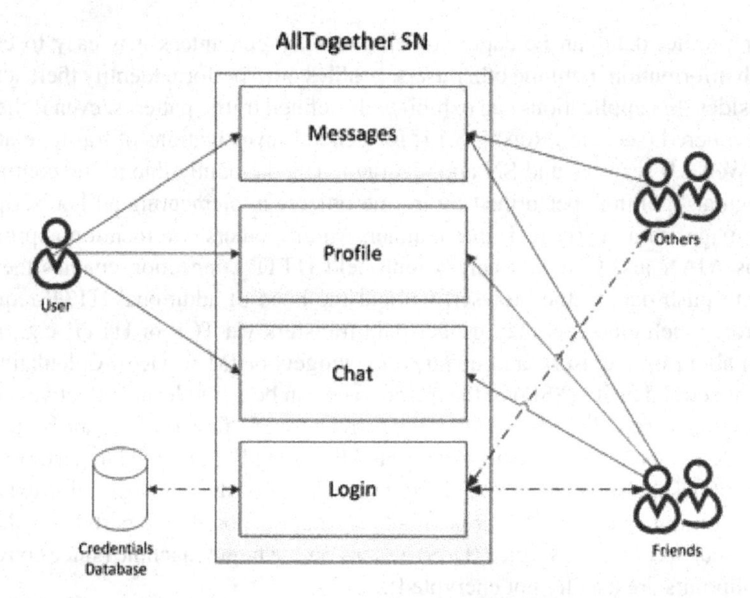

Figure 23.2 Information reference architecture for the *AllTogether* SN example.

23.4.1 The *AllTogether* Social Network Scenario

In the *AllTogether* SN, each user is represented via a public profile, namely a set of information related to the user's life (e.g., status information, pictures, movies, and notes) and a set of known contacts, defined as friends. Each user can access the *AllTogether* platform through a classic *password*-based authentication. Once the user is authenticated, a session is created and a cookie is returned to (and kept on) the user's browser. After the login phase, the user can (1) navigate his profile, (2) update his profile, (3) add/remove friends, (4) chat with friends, (5) exchange messages with all users, and (6) specify visibility policies for portions of his/her profile with a single friend granularity. Also, let us assume that the information needed to build the user profile is stored in an SQL database. The overall information reference architecture is depicted in Figure 23.2.

23.4.2 Exploiting *AllTogether* Vulnerabilities

To effectively demonstrate security issues, let us suppose that the implementation of *AllTogether* SN is susceptible to the previous set of vulnerabilities. In the following, we show how it could be possible to exploit weaknesses of the *AllTogether* SN by making attacks based on the common interactions between the user and the SN service provider.

23.4.2.1 *Injection Attack*

Let us assume that each time a user, i.e., John Doe, updates his profile a corresponding SQL update query is executed on the SN server responsible for storing related information. For the sake of simplicity, we suppose that if John Doe indicates Jane Doe as a friend and consequently updates his status, the following query is executed at the server side:

$$INSERT\ INTO\ Friends\ VALUES\ ("John\ Doe", "Jane\ Doe");$$

where the first parameter is the name of the user and the second the name of the added friend. The injection vulnerability in this case is given by the fact that *AllTogether* receives the user name of the new friend in a form and, without checking, it puts the received value in the second field of the previous query.

Due to this lack of control, a malicious user Eve can force the SQL server to add an unwanted friendship relationship with John Doe by properly forging a friend name that contains unexpected SQL commands.

For instance, when Eve adds a new friend, she can append an SQL command in the friend field after the name of the friend, e.g.,

$$Charlie"); INSERT\ INTO\ Friends\ VALUES\ (\text{“}John\ Doe\text{”}, \text{“}Eve\text{”})$$

The resulting query will be

$$INSERT\ INTO\ Friends\ VALUES\ (Eve,\ Charlie);$$
$$INSERT\ INTO\ Friends\ VALUES\ (John\ Doe,\ Eve);$$

The previous query leads to the execution of two appended SQL commands at the server side, corresponding to the addition of two friendships instead of the expected one. Note that the second friendship relationship will be stored without any intervention of user John Doe.

23.4.2.2 ***XSS Attack***

Let us consider that user's personal data is updated using an HTML form. This choice is quite common in the current implementation of many SNs. Let us also assume that the XSS vulnerability is due to the lack of any kind of control over the input provided by the user in the form.

The malicious user Eve can embed a JavaScript (JS) script, for instance in her telephone field, appending after the phone number the following code:

$$+39010353XXX < script\ language = "javascript"$$
$$type = "text/javascript" > alert(document.cookie); < /script >$$

For the sake of simplicity the *alert( )*; function call is invoked, but a malicious code can be present, e.g., to send a critical cookie to a remote server.

Once another user clicks on the Eve phone number (e.g., for automatically adding the phone number to third-party software if present, such as to populate a Skype contact list) the malicious JS script is unnoticeably executed by the browser of the victim. If the malicious JS script would force the browser to deliver the user's cookie to an external server, that user will inadvertently deliver its cookie to an unknown destination.

23.4.2.3 Broken Authentication Attack

Since in the *AllTogether* SN the authentication is based on login/password, a password retrieval service should also exist. As commonly happens, we hypothesize a user is challenged on secret information (e.g., his mother's maiden name) to create a new password. We suppose that the broken authentication vulnerability corresponds to lack of a lockout mechanism, namely the user has an infinite number of attempts to provide the correct secret information.

Since the secret is related to some personal information that is often shared through the SN (e.g., in posts or comments), an attacker has a high probability of guessing the secret information and stealing the authentication credentials of a legitimate user.

23.4.2.4 XSRF Attack

The XSRF attack can force a user to perform an unwanted action. Also in this case, for the sake of simplicity we assume as the unwanted action adding Eve as a friend. In this specific case, we consider the XSRF vulnerability resides in the lack of control in messages exchanged among users.

Eve sends a message to all users embedding also a link to a script in *AllTogether* that, once executed by an authenticated user, adds a friend to that user. Since Eve is part of the SN, she can know or at least discover such a link. Thus, she can write down a message to other users, embedding in a 1×1 HTML image tag a link to the script and proper script parameters customized for adding Eve as a friend of the receiving user. The embedding of the link in a 1×1 HTML image makes it unrecognizable by the user and automatically executed by the user's browsers.

For the sake of clarity, let us consider an HTML message M_{Eve} in the form

```
<html>
<head><title> Hello from Eve! </title></head>
<body><h3>Greetings from Eve! How are you</h3>
<img src="http://alltogheter.com/addfriend?Requestor=User&Frined=eve"
width="1" height="1" />
</body></html>
```

If that message is sent to *AllTogether* and no checks are performed at the server side, the message is then delivered to the user. The user opens the HTML message through his browser, which automatically retrieves the IMG link, thus executing the *addfriend* script.

The *AllTogether* server recognizes the user through his cookie and allows the execution of the *addfriend* operation. As a consequence, if the attack succeeds, the victim user has an unwanted friendship.

23.5 TOWARDS GREEN SECURITY

As previously introduced, security mechanisms account for energy consumption at different levels. For instance, they put an overhead within the computing infrastructure.

As a consequence, understanding the impact of security on energy is crucial to effectively develop the concept and practices of Green Security. As demonstrated, this is especially important for future Internet applications.

Therefore, in this section, we focus on the investigation of the joint utilization of security solutions and communication networks from the perspective of power-consumption or efficiency issues. Then, we will introduce specific considerations to develop countermeasures for the previous security issues, from the perspective of making them green-aware.

23.5.1 The Relationship between Green Security and Green Networking

As noted earlier, the increasing complexity and pervasive nature of Internet applications produce a very intricate scenario, where the service infrastructure is mixed with a multifaceted internetwork of resources. According to the most recent trends, computing and network elements can be also virtualized or arranged in fictitious overlays (e.g., as happens in virtual private networks scenarios, or in virtual organizations adopted to establish grid computing infrastructures). Cloud-based applications [15] and virtualized remote instrumentations and laboratories [16] can also add to the complexity.

Therefore, decoupling the impact of network-related security mechanisms from the rest of the deployment can deepen understanding of energy consumption issues. At the same time, this could make the overall process easier.

In fact, the increasing amount of users also accessing network services while on the road reflects an exponential growth of Internet traffic. At the same time, the delivery of such a huge amount of data requires more network devices (e.g., routers), which account for an escalation in the energy consumption [17]. This has a relevant impact in terms of production of CO_2 and GHG, as partially discussed in Section 23.1.

In such an enriched vision where the Internet is the real technological enabler for a plethora of new services, the adoption of effective security mechanisms is a core element. Additionally, the more the network is used to produce complex applications and services, the more sophisticated must be the deployed security mechanisms. Figure 23.3 depicts such a "vicious circle."

To summarize, in the modern Internet, network security mechanisms must be assumed as a mandatory feature, rather than an add-on, or a requirement that can be delegated to the application providers.

As an example, we mention the most popular usage patterns of the Internet, and the related requirements.

Specifically, there are

- *Personal* communications such as those based upon the Voice over IP (VoIP) paradigm that require security mechanisms to guarantee confidentiality
- *Social networking* services that demand methodologies to avoid identity theft and protect privacy
- Services that offer *access* to and *control* of remote applications and devices that need a proper degree of security for safe operation through the Internet.

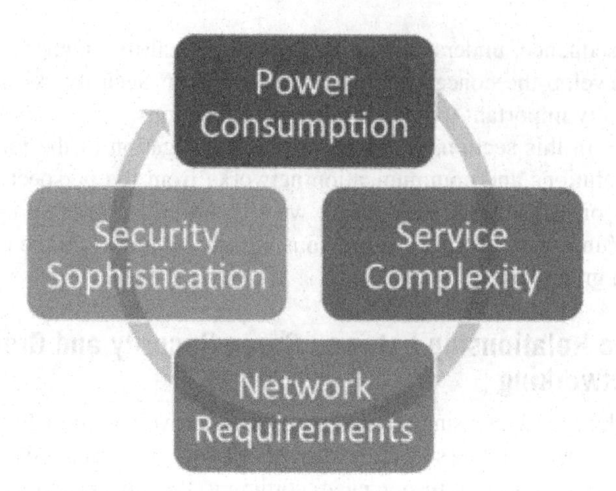

Figure 23.3 The "vicious circle" of increasing complexity of services needing more sophisticated security mechanisms and energy requirements.

Owing to the intrinsic distributed nature of this "diffuse" Internet, proper protocols must be available to support the correct exchange of information among the involved entities. For instance, personal data are constantly delivered among several devices to keep them synced. Hence, the presence of security mechanisms heavily impacts on the power consumption of the Internet, and can clash with green requirements, which are becoming mandatory (see, e.g., Ref. [18] and references therein).

To increase the complexity of understanding the energy impact of network security methodologies, users also have several devices, often mobile, multiple appliances (e.g., gaming consoles and set-top boxes), and are looking to stay connected at all times [19].

Also, a precise analysis of the scope of security countermeasures (e.g., if deployed in the access network, the core network, end-nodes, or in a mixed flavor), and when/how they account for power consumption, allows to comprehend whether or not network-oriented techniques for energy consumption can suffice or be unpractical.

Summing up, mixing security with modern Internet usage trends reflects a complex "problem space," and raises, at least, the following four main issues:

1. Some security mechanisms may require proper architectural elements to be placed in the network, e.g., key distribution servers, or Authentication, Authorization, and Accounting (AAA) functions
2. To secure communications additional traffic could be needed, e.g., to exchange keys, or due to overhead required for transmitting control and encrypted information, as well as credentials
3. Such protocols and mechanisms can be implemented in software layers, which can increase consumption through additional CPU usage. At the same time, they can be implemented via ad hoc hardware (e.g., external devices or built-in ICs), which also needing a proper amount of power

4. Users access the Internet via both wireless and wired access networks. Thus, security mechanisms could be deployed at different layers, according to the specific deployment.

Issues 1–4 affect the power efficiency of different entities, e.g., the access network (case 4), the core network (cases 1 and 2), and end-nodes (cases 2, 3, and 4). Such mapping could be very complex, since more sophisticated services can embrace all the issues, and the ongoing research can mutate where security impacts. This is the case of proxy-based architectures [18], where end-nodes are masqueraded via ad hoc techniques enabling them to dynamically change their energy profiles, e.g., via smart sleeping or other power-saving mechanisms. A portion of the ongoing research must be devoted to identify where security impacts the network architecture, and if Green Security shifts or adds issues in different portions of the network.

Another perspective concerns the mapping of issues 1–4 into the well-developed taxonomy of green-aware techniques for the development of energy-aware networks. Also, it must be evaluated if non-security-oriented power-saving techniques can also be beneficial "for free" when applied to security mechanisms.

The three major techniquest that can be borrowed from the energy-saving research applied to networking are [20]:

A. *Dynamic adaptation:* This enables devices to react to particular traffic (or security) criteria via idle periods, voltage reduction, and real-time activation/deactivation of a portion of the hardware or the software managing security aspects.
B. *Smart sleeping:* This consists of devices being turned on and off according to specific stimuli to save power. The most popular method is based on proxying.
C. *Reengineering:* Power-inefficient devices and protocols are dismantled in favor of more efficient solutions, such as those based on a smaller nanometer production process, or complexity reduction.

Solutions A–C are not suitable for augmenting the efficiency in terms of power consumption of security solutions. On the contrary, new devices with more energy-efficient silicon can lead to further benefits and also to the security processes served.

Summarizing, it is important to develop ad hoc models to understand the overall power consumption, and identify the impact of the presence of security mechanisms. Since security is a pervasive concept that also applies to many distributed and virtualized entities (e.g., such as cloud computing and grid architectures), developing the proper understanding of its energy-related implications can be unfeasible if not subdivided into simpler portions. For such reasons, we try to isolate the energy consumption of network-related security mechanisms to maintain a manageable degree of complexity.

In this context, Green Security must deeply investigate the security implications of having energy-aware mechanisms in place and, at the same time, recognize if the three major aforementioned techniques do not account for additional flaws. On the other hand, reducing the energy consumption of security frameworks could reduce the overall efficiency, e.g., in terms of detected threats and response time. Obviously, the trade-off must be carefully evaluated.

As a final remark, we recall that security and energy savings can be (at least potentially) two conflicting goals. Therefore, the reduction of consumption cannot be assumed as a goal per se but must be always evaluated in comparison with the correspondent level of security achieved.

23.5.2 On the Importance of Quantifying Energy Costs of Security for Modern Internet Applications

Section 23.4 has shown that SN implementations can suffer from vulnerabilities that commonly affect other Web 2.0 applications. Also, the proposed analysis has highlighted that attacks able to exploit such vulnerabilities can lead to unexpected behaviors in the SN. Alas, they are hard to detect, and consequently hard to solve.

In order to avoid these unwanted behaviors in the SN (e.g., forced friendships and identity thefts) two main approaches may be followed by the service provider: (i) periodically checking the SN for known vulnerabilities (e.g., after each major or critical service upgrade) or (ii) protecting the SN by means of security mechanisms aimed at avoiding external attacks.

Although choice (i) would undoubtedly be better, since it would guarantee a robust SN, it is very unpractical. In fact, a deep and complete analysis of potential vulnerabilities is unfeasible, since checking for vulnerabilities can be done through a model-based approach [21] or a penetration testing one [22]. According to the current state-of-the-art of both these research fields, a global solution, sufficient for discovering all vulnerabilities in a Web application, is far from being defined. For such reasons, approach (ii) is simpler and preferred.

Many solutions and mechanisms have been developed for discovering and blocking certain kinds of previous attacks. For instance, intrusion detection systems (IDS) are particularly specialized in detecting injection attacks, while they lack in detecting broken authentication ones. Instead, firewalls are able to detect transport-layer vulnerabilities and session-related attacks.

In general, protecting a vulnerable SN by adopting and installing a set of properly selected security mechanisms is a simpler solution than detecting and solving intrinsic vulnerabilities in the SN itself.

Security mechanisms should be adopted for both client and server side, especially on mobile devices and desktops, and tablet devices accessing the SN. Also, security must be present on SN servers as well. This is due to the fact that some attacks like XSS and XSRF involve both the client and the server side. Thus, in many cases better results can be obtained by checking both client and server side, although this is not mandatory.

Combining security mechanisms may allow reaching the requested level of security but it can lead to useless energy leakage. An important research action is then the precise understanding of the power wastages of the different classes of attacks, in order to produce a ranking of the more harmful issues in terms of power consumption.

Thus, countermeasures will be prioritized if consumption is more critical than other aspects.

From a *green* perspective, the first drawback is that some security features may be replicated between different solutions, thus resulting in energy leakage due to the execution of the same checks performed by different entities. This problem can also reduce the duration of mobile device battery on the client side, also affecting the usability-of the device. Summing up, avoiding functional duplications is an important design goal.

Replication often arises in the design of security strategies and solutions because they are developed exclusively from a security effectiveness point of view and do not take consumpution into account.

For instance, as shown in Reference [23], an IDS generally has a single detection strategy and it does not evaluate switching among possible alternative strategies with the aim of reducing energy consumption. Also in Reference [23], a comparison between different strategies is made, and it shows that there is not an ultimate-optimal strategy in terms of energy consumption. Rather, the efficiency in terms of power consumption of a security strategy strictly depends on the scenario where it is applied. *Off-line planning* and *on-line adaptations* are two important tools to deal with excessive power consumption or battery drain.

Nevertheless, the previous examples reveal that the current state-of-the-art lacks a systematic approach to considering the energy consumption of security mechanisms. To fill the gap, there is a need to defining security solutions while considering power constraints as design parameters. This can be reflected in a sort of *clean slate* approach, where the whole process of engineering and developing new security protocols and machineries must be reconsidered with power management and the aforementioned features as rules of thumb. In practice, this means that there is the need to start *modeling* security approaches also in terms of energy consumption, as well as on security effectiveness.

In this vein, in the following we propose a minimal analytical model to better comprehend the specific scope of Green Security when applied to telecommunication networks.

23.5.3 A Basic Analytical Problem Statement of Green Security

In this section we present a very basic analytical model to precisely identify the scope of Green Security. The notation has been partially borrowed from [24], but the development of the analytical model has been widely extended.

Isolating the contribution to the overall power usage by breaking up the related process should help to understand what has to be minimized, where the consumption happens into the network, what has to be measured, and what the most suitable techniques are. Such steps are the fundamental research efforts that have to be performed to bring Green Security into the area of feasibility.

To avoid burdening the notation, in the following we dropped the dependence of time t.

Specifically, we aim at having a model like

$$E^i_{\text{Tot}} = E^i_{\text{Net}} + E^i_{\text{Sec}}, \tag{23.1}$$

where E^i_{Tot} and E^i_{Net} are the overall and network-related power consumption for the ith traffic flow, respectively. E^i_{Sec} is the power usage of security mechanisms.

Due to the high complexity of the scenario, creating an ultimate and unique energy consumption model is unfeasible. While classical green-networking (or computing) approaches minimize E^i_{Net}, Green Security is aimed at reducing E^i_{Sec}. However, the decoupling of such terms can be hard, and sometimes impractical. For instance, at some layers, i.e., ISO/OSI L2, security could be intimately connected with the power consumption of transmission-related operations. This case can be represented by the following functional relation:

$$E^i_{\text{Net}} = f(E^i_{\text{Sec}}). \tag{23.2}$$

The core actions to be performed in order to have a reasonable understanding of how Eqs. (23.1) and (23.2) behave is to quantify both E^i_{Sec} and $f(*)$ (if any) via measurement campaigns. By doing so, it will be possible to develop basic consumption models and perform simulations. While there is already a relevant understanding about the dynamics of power consumption in telecommunication devices, a thorough comprehension of the inner mechanisms of security is still missing. Nevertheless, security frameworks also do employ network services. For instance, sending keys for supporting cryptography can be modeled as a standard data transmission, thus using tools already available in the literature.

Conversely, if such actions are not possible, a reasonable approach appears to be the adoption of black box modeling for the more complex settings [25]. We remark that such an approach is viable if and only if a sufficient set of measurements from the fields is available. Therefore, this increases the need for measuring data, by performing trials and laboratory tests to successfully recognize the impact of security in terms of energy consumption. More formally, Green Security aims at

$$\min_{u(*)} E^i_{\text{Sec}}, \tag{23.3}$$

where $u(*)$ is the set of controls that can be "moved" to reduce energy consumption. Both the techniques and technologies composing $u(*)$ must be investigated. In fact they can be the same presented in Section 23.4.1, or brand new approaches. Additionally, one can also imagine performing minimization by adopting a *clean slate* [26] approach so as to engineer brand new security solutions to support modern Internet applications.

Conversely, if the impact (in terms of power consumption) of the security portion is not relevant, trying a minimization could be inconvenient.

To make the problem and the goals more clear, the global power consumption proposed in Eq. (23.1) can be further refined:

$$E^i_{\text{Sec}} = E^i_{\text{Sec,Net}} + E^i_{\text{Sec,Dev}}, \tag{23.4}$$

where $E^i_{\text{Sec,Net}}$ and $E^i_{\text{Sec,Dev}}$ quantify the power needed by the security mechanisms deployed within the network and the end-nodes (e.g., users' devices), respectively. According to recent trends, modern Internet applications are mostly aimed at offering services to mobile users, which often employ wireless and battery-operated devices. As a consequence, $E^i_{\text{Sec,Dev}}$ is a very critical constraint to avoid excessive battery drain. Nevertheless, security requirements must also be taken into account to cope with specific attacks aimed at consuming power.

Equations (23.1) and (23.4) can be combined to produce a more comprehensive model, to take into account the overall traffic experienced by a given Internet service provider (ISP), i.e.,

$$E_{\text{ISP}} = \sum_{i=0}^{N} (E^i_{\text{Net}} + E^i_{\text{Sec,Net}} + E^i_{\text{Sec,Dev}}), \tag{23.5}$$

where N is the number of sessions managed by the ISP at a given time t. It is worth noticing that the partial power consumption given by $\sum_{i=0}^{N} E^i_{\text{Sec,Dev}}$ is spread among end-users' devices, i.e., not consumed by the ISP itself.

This has the following implications:

- The power consumptions of the network and devices should be "weighted." Even if a minimum wastage can be tolerated within a fixed infrastructure, it could heavily affect the experience of users accessing a service through battery-operated devices. Therefore, it could be beneficial to try to shift consumption from the border of the network to its inner components.
- According to the increased coupling among devices, network components, and services, the design of security mechanisms cannot be performed in a "decoupled" flavor. Thus, in order to be effective, Green Security must include all the involved entities chorally.

The model proposed in (23.5) has been developed to be general enough to take into account different reference scenarios, e.g., (i) if security mechanisms are not employed for a given flow, only the consumption due to network devices can be computed and (ii) the per-flow granularity can be dropped if not needed.

Furthermore, understanding via simulations or measurements the impact of the user population (i.e., the parameter N) is another mandatory requirement to carefully evaluate if Green Security should be employed.

Separating the energy consumption contributions can suggest the best strategy to reduce power consumption. We recall that the state-of-the-art green-networking techniques are dynamic adaptation, smart sleeping, and reengineering.

To summarize, one of the main goals of the introduction of Green-Aware Security strategies, at least in this preliminary stage, is to quantify, with different degrees of precision, specific contributions (e.g., those previously defined as $E^i_{\text{Sec,Net}}$ and $E^i_{\text{Sec,Dev}}$) and where/when they are present within the overall network configuration.

23.6 CONCLUSIONS

In this chapter, we have analyzed security issues related to modern Internet applications that exhibit a relevant social attitude. As a synecdoche, we selected the case of SNs, and we investigated its security under a novel "green" perspective.

In particular, we have pointed out the potential interrelations between green computing, security, and social networking, by identifying a *focal point* that may lead to new research directions. In particular, we have shown that SNs have important issues in term of security at different abstraction layers, although they have not been originally conceived as a security-sensitive class of Web applications.

Besides, we argued that the current approaches for securing SNs are not optimized in term of energy consumption but they can lead to energy leakage due to the combination of redundant security checks. For these reasons, we discussed the importance of modeling the energy consumption of a security solution in conjunction with the effectiveness in detection and sanitization, with the aim to provide new security solutions that could be both robust and green-friendly.

To this end, we also provided a first attempt to model the energy consumption of a security mechanism in terms of security and network costs.

Future developments should aim at performing an extensive measurement campaign to precisely quantify the consumption of specific classes of security tools and countermeasures. Additionally, this will also allow for the production of models and performance of simulation analysis to forecast the impact in terms of energy requirements of security within a given networking environment. Also, clean slate approaches can be envisaged, especially when energy is a critical constraint (e.g., in battery-operated devices). In this perspective, new architectures shifting consumption from the device to the fixed network could be beneficial, for instance, in terms of battery lifetime or quality perceived by end-users.

REFERENCES

[1] L. Caviglione, F. Davoli, Traffic volume analysis of a nation-wide eMule community, Computer Communications Journal 31 (10) (2008) 2485–2495.

[2] Social Web Incubator Group. <http://www.w3.org/2005/Incubator/socialweb/XGR-socialweb-20101206/> (accessed November 2011).

[3] A.C. Weaver, B.B. Morrison, Social networking, IEEE Computer 41 (2) (2008) 97–100.

[4] OECD, Working Party on the Information Economy, Towards Green ICT Strategies: Assessing Policies and Programmes on ICTs and the Environment. <http://www.oecd.org/dataoecd/47/12/42825130.pdf> (accessed November 2011).

[5] W. Van Heddeghem, W. Vereecken, M. Pickavet, P. Demeester, Energy in ICT – trends and research directions, in: Proceedings of the IEEE Third International Symposium on Advanced Networks and Telecommunication Systems (ANTS), New Delhi, 14–16 December 2009.

[6] A. Bianzino, C. Chaudet, D. Rossi, J. Rougier, A survey of green networking research, IEEE Communications Surveys and Tutorials (2010) 1–18.

[7] L. Caviglione, M. Coccoli, Privacy problems with Web 2.0, Computer Fraud and Security, Elsevier pp. 16–19
[8] C. Patsakis, A. Asthenidis, A. Chatzidimitriou, Social networks as an attack platform: Facebook case study, in: IEEE Eighth International Conference on Networks, Gosier, France, March 2009, pp. 245–247.
[9] J. Brentham, TCP/IP Lean (Web Servers for Embedded Systems). second ed., CMP Books.
[10] OWASP, The Open Web Application Security Project, Homepage. <http://www.owasp.org> (accessed October 2011).
[11] Wireshark, The World's Foremost Network Analyzer, Homepage. <http://www.wireshark.org/> (accessed November 2011).
[12] tcpdump, Homepage. <http://www.tcpdump.org/>.
[13] L. Caviglione, Can satellites face trends? The case of Web 2.0, in: Proceedings of the International Workshop on Satellite and Space Communications (IWSSC'09), Siena, Italy, September 2009.
[14] L. Caviglione, Extending HTTP models to Web 2.0 applications: the case of social networks, in: First International Workshop on Cloud Computing and Scientific Applications (CCSA 2011), held within the Fourth IEEE International Conference on Utility and Cloud Computing (UCC 2011), Melbourne, Australia, December 2011, pp. 361–365.
[15] E. Knorr, Software as a Service: The Next Big Thing, Infoworld, March 2006. <http://www.infoworld.com/article/06/03/20/76103 12FEsaas1.html> (accessed November 2011).
[16] L. Caviglione, F. Davoli, P. Molini, S. Zappatore, Design and preliminary analysis of a framework for integrating real and virtual instrumentation within a grid infrastructure, in: Proceedings of the 2006 International Symposium on Performance Evaluation of Computer and Telecommunication Systems (SPECTS'06), Calgary, Canada, July 2006.
[17] M. Gupta, S. Singh, Greening of the Internet, in: 2003 International Conference on Applications, Technologies, Architectures, and Protocols for Computer Communications (SIGCOMM03), Karlsruhe, Germany, August 2005, pp. 19–26.
[18] C. Gunaratne, K. Christensen, B. Nordman, Managing energy consumption costs in desktop PCs and LAN switches with proxying, split TCP connections, and scaling of link speed, International Journal of Network Management 15 (5) (2005) 297–310.
[19] L. Backstrom, D. Huttenlocher, J. Kleinberg, X. Lan, Group formation in large social networks: membership, growth, and evolution, in: Proceedings of 12th International Conference on Knowledge Discovery in Data Mining, New York, USA, 2006, pp. 44–54.
[20] R. Bolla, R. Bruschi, F. Davoli, F. Cucchietti, Energy efficiency in the future Internet: a survey of existing approaches and trends in energy-aware fixed network infrastructures, IEEE Communications Surveys and Tutorials (COMST) 13 (2) (2011) 223–244.
[21] A. Armando, L. Compagna, SAT-based model checking for security protocols analysis, International Journal of Information Security 7 (1) (2008)
[22] A. Doupe´, M. Cova, G. Vigna, Why Johnny can't pentest: an analysis of black-box web vulnerability scanners, in: Christian Kreibich, Marko Jahnke (Eds.), Detection of Intrusions and Malware, and Vulnerability Assessment, Lecture Notes in Computer Science, vol. 6201, Springer, Berlin, Heidelberg, 2010, pp. 111–131 (Chapter 7)
[23] M. Migliardi, A. Merlo, Modeling the energy consumption of an IDS: a step towards green security, in: Proceedings of the 34th International Convention on Information and

Communication Technology, Electronics and Microelectronics, 23rd–27th of May 2011, Opatija, Croatia, pp. 1452–1457.

[24] L. Caviglione, A. Merlo, M. Migliardi, What is green security, Seventh International Conference on Information Assurance and Security (IAS 2011), December, IEEE, Malacca, Malaysia, 2011, pp. 366–371.

[25] L. Caviglione, A simple neural framework for bandwidth reservation of VoIP communications in cost-effective devices, IEEE Transactions on Consumer Electronics 56 (3) (2010) 1252–1257.

[26] A. Feldmann, Internet clean-slate design: what and why? SIGCOMM Computer Communication Review 37 (3) (2007) 59–64.

CHAPTER

24 Using Ant Colony Agents for Designing Energy-Efficient Protocols for Wireless Ad Hoc and Sensor Networks

Isaac Woungang[a], Sanjay Kumar Dhurandher[b], Mohammad S. Obaidat[c,1]

[a]Department of Computer Science, Ryerson University, Toronto, Ontario, Canada,
[b]Division of Information Technology, Netaji Subhas Institute of Technology (NSIT), University of Delhi, India,
[c]Department of Computer Science and Software Engineering, Monmouth University, NJ, USA

24.1 INTRODUCTION

The ubiquity of wireless communications combined with the advent of new wireless technologies have enabled the research community to focus on implementing new standards and techniques to deal with low power and energy efficiency in next-generation networks [1] since this has become an important challenge across all forms of networking in the future. Indeed, from a network design prospective, the development of energy-efficient communication and networking algorithms and protocols for next-generation networks is an important challenge that deserves some attention. In this chapter, our focus is on energy as a key design constraint when designing protocols for wireless ad hoc networks and sensor networks.

24.1.1 Energy Efficiency as a Key Concern When Designing Wireless and Sensor Networks

A wireless network can be defined as a network of devices (nodes) that are connected to each other by means of a wireless communication medium (i.e., without wires). This setting can enable ease of communication, particularly in the case of mobile nodes. Taking advantage of the advances in wireless communications technologies such as Bluetooth, WiFi, WiMax, FM radio, ham radio, near field communications (NFC), to name a few, multihop wireless ad hoc and sensor networks,

[1]Fellow of IEEE and Fellow of SCS.

considered a subclass of wireless networks, have spurred a great deal of interest in both the research and industrial communities since they have been particularly useful for commercial, healthcare, and military applications [1], mainly due to the fact that they are infrastructure-less and require multiple hops for connecting all the nodes to each other. Consequently, the relaying of the messages from one mobile node to another, and the peculiarity of the wireless transmission medium (i.e., noise, interference, to name a few) are some major fundamental issues. In addition to this, when the power source of a node in such networks is weak or costly, energy efficiency becomes a key concern [2], as it is directly related to the design and operation of such networks.

24.1.2 Approaches to Achieve Energy-Efficient Routing in Wireless Ad Hoc and Sensor Networks

Apart from the proper selection of hardware, techniques for reducing the energy consumption in wireless ad hoc and sensor networks can be achieved by carefully designing all the layers in the system, namely (a) the physical layer—where an energy-efficient radio should be implemented in such a way so as to achieve a dynamic power management; (b) the medium access control (MAC) layer—where the goal is to design protocols that minimize the time required by the radio to be powered; (c) the logical link control layer—where suitable error-control and flow control mechanisms should be implemented for controlling the conditions of the radio link so as to minimize the energy consumption and (d) the network and transport layer—where the rate of energy consumption is one of the important factors for the design of efficient energy-aware routing schemes. The latter is the focus of this chapter.

Typically, the investigation of energy consumption in ad hoc wireless and sensor networks requires the study of coupling among the above layers [3] since energy consumption does not occur only through transmission, but through processing as well [2]. This adds some complexity in the study of upper layer designs, particularly when energy efficiency is incorporated as a factor when performing the routing operations at the network layer [3, 4], in combination with selecting the transmission power [5–8] (which itself affects the MAC and physical layers). The goal is to minimize the energy consumption at each node, which constitutes an important topic. Several design approaches, which can be referred to as energy-aware multihop routing schemes, have been proposed, each of which addresses in its own fashion the tradeoff between the transmission power, the energy expenditure, and the route selection process. The goal is to minimize the total energy consumption of the network subject to delay or throughput, minimizing the delay (or maximizing the throughput) per joule of expended energy, where energy expenditures are affected by the design choices made at each of the above-mentioned upper layer levels.

A few representative approaches are as follows: (a) approaches that favor the design of paths with minimal energy cost [9, 10]; (b) approaches promoting the design of paths that avoid (as much as possible) nodes with low remaining battery capacity [11]; (c) approaches that advocate the design of minimal energy cost paths composed of nodes with a battery level higher than a prescribed threshold [12]; and (d) approaches where the routing paths are selected according to the remaining battery capacity at each node, the sending rate per node, and the energy cost of hops [13]. In all these approaches, the following requirements should be met by the designed routing protocol: (a) minimize the energy consumption in the network; (b) effectively deal with congestion avoidance; (c) successfully identify the packets from the source node and deliver them successfully to the destination and (d) ensure a proper balancing of load on individual nodes, i.e., deal effectively with the network management load.

In our work, we focus our attention on energy-aware routing schemes, particularly the use of Ant Colony Optimization (ACO)-based techniques for energy saving in the process of routing the data from one node to another in wireless ad hoc and sensor networks. We also introduce an enhancement to an existing energy-aware routing scheme (Conditional Max–Min Battery Capacity Routing—CMMBCR) using an ACO heuristic.

24.1.3 Ant Colony Optimization Paradigm

Studies of insect colonies such as bees or ants have revealed that these insects have some complex collective behavior and a management structure [14, 15] that resemble the properties of dynamic distributed systems. This has led to the introduction of the so-called Ant Colony Optimization (ACO) meta-heuristic [16], which can be defined as a common framework for a wide set of heuristic algorithms (so-called ACO algorithms) that can be applied to different types of problems, provided that an adequate model be defined. The ACO approach is based on iteratively constructing multiple solutions to an optimization problem, by progressively learning the search space and by using this knowledge to control the solution construction processes, which are driven by a stochastic decision policy. Eventually, this strategy leads to the design of optimal solutions to the targeted problem.

The growing interest in ACO algorithms has led to the development of many successful algorithms for discrete optimization problems [17]; continuous optimization problems [18], i.e., problems with continuous decision variables; and dynamic optimization problems [19, 20], i.e., problems in which the search space changes over time, leading to potential changes in the definition of the problem instance as well as the quality of the solutions already found, to name a few. One of the prominent and widely used such algorithms is the Ant system [21] and its enhancements [22–25], which have proved suitable to design high performance routing protocols for communication networks [26–28], including wireless networks [29–33], and to find optimal solutions to other types of problems such as scheduling problems, assignment and layout problems and machine learning problems [34], to name a few. In this chapter, our focus is on investigating the use of ACO for designing energy-efficient routing protocols for wireless ad hoc networks and sensor networks.

24.1.4 Why Use ACO-Based Approaches to Develop Energy-Efficient Routing Protocols for Wireless Ad Hoc and Sensor Networks?

Mobile ad hoc wireless and sensor networks have gained an unprecedented popularity due to recent advances in wireless technologies and the suitability of such networks for use in situations where predeployed infrastructure does not exist. However, due to its rapidly changing and unpredictable nature, taking advantage of these features poses a number of challenges, a few of which may lead to conflicting objectives. Examples of such challenges are (a) achieving effective distribution and forwarding of data among nodes; (b) achieving load balancing among the nodes; (c) achieving minimal energy consumption; (d) achieving effective data routing; (e) ensuring congestion avoidance; (f) achieving minimum delay while ensuring a longer network lifetime; and (g) ensuring fault tolerance and achieving scalability. A few of the existing protocols for energy-efficient routing were designed to effectively handle both the routing and the overall network management through routing. Also, their applications in real-life dynamic environments were not always guaranteed. Due to the intrinsic design features of ACO-based algorithms and their suitability in solving a variety of static, continuous, and dynamic optimization problems [17–20], ACO-based heuristic methods had appeared as a natural way to attempt to build the best possible optimized solutions for the targeted challenges or combination of challenges. Focusing on energy consumption, the concept of pheromone trails [16, 34], which has led to the foundation of the design of ACO-based algorithms, can effectively be used to design energy-aware routing protocols that significantly reduce the energy consumption of wireless ad hoc and sensor networks.

It should be emphasized that apart from targeting energy efficiency as a performance objective when dealing with routing, ACO-based algorithms have also been proposed for the design of routing protocols for ad hoc and sensor networks based on other performance objectives such as cost, flooding, or broadcasting [35], location awareness [36], clustering [37, 38], self-organization or fault tolerance [39], efficiency in data gathering [40], congestion avoidance [41, 42] and security [43], to name a few. Finally, various energy-efficient routing protocols have been proposed in the literature that do not make use of ACO-based heuristics to optimize the routing process (see [44, 45] and the references therein). In this chapter, we investigate energy-aware routing schemes designed using ACO-based heuristics. In the same way, we also introduce a novel ACO-based energy-efficient protocol for mobile ad hoc networks (MANETs).

24.1.5 Chapter Organization

The rest of the chapter is organized as follows. In Section 24.2, an overview of ACO-based energy-efficient protocols for ad hoc wireless and sensor networks is presented. In Section 24.3, our ACO-based energy-aware protocol design (called A-CMMBCR) for mobile ad hoc networks is presented. Simulations results are also presented, showing that our A-CMMBCR scheme outperforms the CMMBCR scheme and two other chosen benchmark schemes, in terms of energy consumed per packet. Finally, Section 24.4 concludes our work and highlights some future research directions.

24.2 AN OVERVIEW OF ACO-BASED ENERGY-EFFICIENT PROTOCOLS FOR WIRELESS NETWORKS

In order to deal with resource management challenges in dynamic networks such as ad hoc and wireless sensor networks, the network should be equipped with new management and control strategies that can strengthen its adaptability to changes in traffic, topology and services; and its flexibility in terms of transmission technology, topology characteristics and self-organizational behavior, while maintaining an acceptable performance level. Due to its nature-inspired features, and its similarity with dynamic complex systems [14], the ACO framework [16–18, 21, 25, 46] can provide a viable and effective alternative to help achieve the above-mentioned goals. Indeed, the central principle of ACO algorithms consists of designing lightweight agents that can continuously learn the current network status and use this knowledge to adapt the characteristics and decision policies inherent to network operations in such a way to optimize the overall network performance and efficiency. In this regard, a few ACO parameters are of upmost importance since they have a direct impact on the quality of the obtained optimized solutions, namely (a) the pheromone variables associated with each node in the network, (b) the decision policy parameters, and (c) the progressive learning phase of these parameters.

In the sequel, our focus is on applications of ACO-based heuristics [26–28, 46] to routing problems in ad hoc and sensor wireless networks, where the goal is achieving an energy-efficient network. Such algorithms are described, along with a flavor on how the aforementioned ACO parameters interact with each other to yield the optimized solutions. It should be emphasized that most of the ACO implementations for routing problems [27, 46] share a common underlying structure and distributed and self-organizing strategies. In general, each ACO implementation involves two main phases: the route discovery phase and the route maintenance phase, based on some initial settings. Typically, mobile agents (so-called ants) are implemented as "intelligent control packets" and are given the task of discovering a suitable path for the routing of data between a source–destination node pair in the network. In addition the ants collect useful information such as congestion status, amount of energy consumption, traffic load, etc., at each node along the selected routing path. The network then uses this information to learn the network dynamics and to continuously update the node's routing policy (indirectly embedded in the pheromone variables) in order to keep track of the network changes that have occurred. This iterative process eventually leads to optimized solutions to the routing problem when convergence occurs.

In [48], Camilo et al. proposed an ACO-based algorithm that maximizes the network lifetime of wireless sensor networks, where the lengths of the routing paths, the node's energy level, and the amount of pheromone trail available on the connections between the nodes are considered as design parameters to construct a routing tree that has optimized energy branches. However, more energy savings may have resulted from this design if the authors had considered the case of multiple sink nodes when implementing their algorithm.

In [49], Wen et al. proposed an ACO-based algorithm for minimizing the time delay in wireless sensor networks when transferring the data from source to destination nodes while accounting for the energy level of a node as a constraint. In their scheme, ant agent routing tables of each node are built based on partial pheromones and heuristic values, then updated by a background ant that holds the network load and delay information. A reinforcement learning technique is employed to address the trade-off between delay and energy level at each node, resulting in an energy-efficient scheme compared to the AntNet scheme [47], in terms of energy consumed by each packet during transmission. However, their proposed scheme did not address the situation when the traffic load at a node might turn out to be heavy.

In [38], Salehpour et al. proposed a two-step algorithm that combines a clustering technique with an ACO-based heuristic to design an energy-efficient routing scheme for wireless sensor networks. In the first step of their algorithm, the Low-Energy Adaptive Clustering Hierarchy algorithm (LEACH) [48] is used to achieve clustering and message transmission in the network, resulting in evenly distributed energy consumption among all the nodes in the network. In the second step of their algorithm, an ACO-based heuristic (the AntNet scheme [47]) is invoked by the cluster heads (which are inherited from the first step) to send the aggregated data packets to the base station, and this process repeats iteratively until convergence is reached. In the latter, backward and forward ant agents are used in collaboration to explore the routing possibilities of the data packets throughout the network. These are based on the information gathered by each node regarding the amount of pheromone on the paths to its neighbors and the decision made by ants based on the energy level of the neighbor nodes. One major concern with this scheme is that the heuristic value associated with each node is dependent on the energy level of that node. In addition, no method was disclosed to estimate this value, and the impact of this parameter on the obtained optimized solutions was not investigated.

In [50], Wang et al. introduced an ACO-based routing protocol for wireless sensor networks (called EAQR) that uses quality-of-service (QoS) provisioning and balanced energy consumption as a target to achieve energy efficiency. In their proposed scheme, service differentiation between real time (RT) and best effort (BE) traffic is made through designing a specific pheromone model where artificial ants are extended, yielding ants that are endowed with the capability of emitting two types of pheromone, implemented in the form of two pheromone heuristic schemes: (1) the RT pheromone scheme, used to achieve the above-mentioned balanced energy consumption for BE traffic considering the path hop count and minimum residual energy along the path as constraint parameters, and (2) the BE heuristic scheme, which focuses on ensuring the necessary QoS provisioning on the selected routing path between a sensor and a sink. The routing tables at each node are updated according to these BE and RT pheromone matrices. Although this scheme was proved to balance the energy consumption in the whole network in real-world situations, the authors neglected to compare their scheme against similar state-of-the-art well-known schemes, in terms of efficiency or energy-related performance metrics.

In [51], Dhurandher et al. investigated the problem of energy-constraint routing in ad hoc networks, and proposed an ant swarm-based algorithm that integrates both the power consumption at each node when routing a data packet and the multipath transmission features of artificial ants. In their proposed scheme, the energy usage is minimized by means of the path discovery process, inspired from the features of AntHocNet [52], and designed based on parameters such as route hop count and minimum battery energy remaining from the weakest node of the route. On the other hand, multipath transmission is used to divert the packet flow in case of link failure (assumed to occur one at a time), leading to fewer dead nodes compared to the AntHocNet [52] scheme. The merit of this protocol is that energy-awareness is used as a factor to increase the time that the protocol takes to judge the best possible route to be used for data packet transmission. As pointed out by the authors, their proposed protocol was not tested in a real test-bed environment using real-life scenarios.

In [53], Okdem and Karaboga studied the issue of minimization of energy consumption at sensor nodes, and proposed an ACO-based solution. Considering the energy conservation at each node as a primary target, their routing scheme is designed in such a way as (a) to deal with failure in the communication node—this is addressed by keeping multiple paths alive in the routing task; (b) to deal with the energy level at each node and the length of the paths—these are handled by implementing a mechanism that chooses the nodes with more energy when performing the routing process; and (c) to incorporate the ACO-based approach—where artificial ants contribute in designing effective multipath data transmission from source to sink based on the information gathered at each node about the amount of pheromone on the available paths. In order to validate their approach, the authors introduced a real-time test environment made of a router chip, implemented in the form of a small hardware component. However, the case of multiple sink nodes was not investigated.

In [54], Xia and Wu also addressed the issue of energy constraint in wireless sensor networks. Their proposed ACO-based multipath routing algorithm uses the energy consumption of each path and the available power of nodes as criteria for selecting the optimal routing path (among multiple available paths) for the delivery of packets from source nodes to the sink node. The algorithm improves the simple ACO (SACO) scheme [17], in the sense that an optimized state transition and global pheromone update rules are introduced to increase the possibility of ants finding a new path to avoid local optimization, and to maintain the multipaths possibility when transferring the data packets from the source nodes to the sink, respectively. However, the mobility of sensor nodes was not taken into consideration.

In [55], Misra et al. proposed an ACO-based energy-efficient routing protocol for wireless ad hoc networks, which combines the effect of power consumption when transmitting a packet, the residual battery capacity of a node, and the multipath transmission properties of artificial ant swarms. In their scheme, the path discovery phase is inspired from AntHocNet [52], but with distinct functionally in the sense that the routes are maintained through new pheromone reinforcement and evaporation techniques, leading to the use of multipath transmission through the "good routes" only rather than all the possible paths. Even though this scheme showed promise, the effectiveness of the proposed scheme was not tested in a real test-bed using practical scenarios.

In [56], Mahadevan and Chiang investigated the problem of energy consumption in wireless sensor networks and proposed a self-governed ACO-inspired routing scheme to solve the packet routing problem with minimal energy consumption for each hop communication, leading to maximum lifetime of the network. Their scheme is inspired from the MAX–MIN Ant System (MMAS) [57] to produce optimized routing paths to transfer the data from source nodes to the sink, while considering energy efficiency and self-healing as design criteria. However, their proposed scheme was not compared against a few other state-of-the-art benchmarks, nor implemented in a real test-bed in order to judge its efficiency when dealing with practical scenarios.

In [58], Hui et al. introduced an ACO-based routing protocol for wireless sensor networks that considers the node energy, the frequency of a node acting as a router and the path delay, as design criteria. Their scheme is based on the idea that using the lowest energy path does not necessary mean obtaining a long network lifetime due to the fact that the optimal path may quickly deplete energy. Although the authors have followed the basic ACO principle for selecting the optimal path to transmit the data form source nodes to sink, the originality of their scheme stands in the fact that in its route discovery and maintenance phase, the routing tables at each node were updated according to the pheromone and energy levels at that node; however, node mobility was not considered.

In [59], Kim et al. investigated the problem of designing an energy saving routing (ESR) scheme for energy-efficient networks. The problem was formulated as an energy-consumption minimized network (EMN) optimization problem, based on the concept of traffic centrality of a node, defined as a measure involving the traffic volume (in bytes) on a link and the density of traffic carried on that link; then solved using the ACO method where only a single artificial ant is considered. The optimized energy efficiency level produced by the proposed ACO-Energy Saving Routing (A-ESR) algorithm is dependent on a controlling factor that was used to weight the traffic centrality. However, the authors neglected to indicate how the value of this factor can be allocated in a dynamic manner.

In [60], Ren et al. presented an ACO-based energy-aware routing for ad hoc networks (called ABEAR). Their proposed scheme introduces a congestion metric and uses it along with a combination of a reactive route setup procedure and proactive neighbor maintenance procedure in its routing phase to find suitable paths for transferring the data from source to destination. In this process, the link quality, remaining energy at each node, and pheromone values are integrated as design variables in the ACO approach when performing the routing computation, with the goal to reduce the network lifetime.

In [61], Cheng et al. proposed an energy-aware ACO routing algorithm (called EAACA) for wireless sensor networks. In their scheme, the residual energy of the one-hop neighbor of each node, and the distance from source to sink are used as design criteria in the selection of the paths to route the data packets when using the ACO principle. In the route discovery phase, the information gathered by each node regarding the amount of pheromone on the paths to its neighbors and the decision made by ants (backward and forward ants) based on the residual energy of its one-hop neighbor are used to establish all valid paths between the sensor nodes and the destination node before the source node starts releasing the data packets. In the route maintenance phase, probe packets are sent to the destination node periodically to monitor the quality of the chosen transmission paths. Although this scheme was shown to balance the energy consumption at each node, the case of mobile sensor nodes was neglected.

In [62], Dominquez-Medina and Cruz-Cortes introduced an adaptable and balanced ACO-based routing algorithm for wireless sensor networks, which considers memory and power supply as criteria to minimize the energy consumption and latency in data transmission. The ACO design of the proposed scheme is a combination of the ACO-based Location Aware Routing for WSNs (ACLR) [63]—which attempts to establish an equilibrium between the sensor nodes lifetime and the delay of the transmissions, and the Energy-Efficient Ant-Based Routing Algorithm (EEABR) [48]—which considers the energy efficiency in order to maximize the network lifetime. However, the proposed scheme was not implemented in a real environment in order to judge its efficiency when dealing with practical scenarios.

In [64], Dhurandher et al. proposed an ACO-based energy-aware routing protocol (called A-CMMBCR) for mobile ad hoc networks, which is an enhancement to the Conditional Max–Min Battery Capacity Routing (CMMBCR) protocol [12]. The ACO design involves a combination of two routing schemes, the Minimum Transmission Power Routing (MTPR) [65] and the Max–Min Battery Capacity Routing (MMBCR) [66] schemes, resulting in an optimized scheme that contributes in greatly minimizing the total energy consumed in the network (as compared with those benchmarking schemes) while ensuring that routes are not overloaded due to traffic and that backup paths are available in case the routes break due to the mobility of the nodes. However, the fault tolerance aspect was not addressed in this design.

In the sequel, the recently proposed A-CMMBCR routing protocol for mobile ad hoc networks [64] is described in-depth.

24.3 AN ACO-BASED ENERGY-EFFICIENT PROTOCOL FOR MOBILE AD HOC NETWORKS

24.3.1 Motivation

Due to their dynamic topology and unpredictable nature, mobile ad hoc networks (MANETs) pose several challenges [1] when used in practical scenarios, particularly from a data routing standpoint. Indeed, the routing scheme designed to support the MANET should satisfy some basic requirements, including (a) adaptation to the changing topology, (b) reliability of packet delivery from source to destination, (c) path/node congestion-awareness, (d) minimizing as much as possible the node's energy consumption, *a fortiori*, the total transmission power consumption of nodes participating in a selected route, and (e) fair distribution of loads among the nodes in the network, to name a few. In other words, the efficiency of a routing protocol must be reflected and judged on its capability in handling both packet routing and overall network management, in such a way as to balance the network load while yielding a longer network lifetime. In general, achieving this type of goal is quite difficult when using existing routing protocols for MANETs. However, in [27], it was reported that ACO-based heuristics can provide a unified solution for network management through routing in dynamic networks such as MANETs. This justifies our need to design a routing protocol for MANETs based on the ACO framework.

This chapter continues the investigation of a recently proposed scheme for MANETs, the energy-efficient ACO-based Conditional Max–Min Battery Capacity Routing (A-CMMBCR) protocol [64]—whose approach consists of a combination of the AntHocNet [52], CMMBCR [12], and MTPR [65] schemes. In [64], some simulations have been conducted, demonstrating the superiority of the proposed A-CMMBCR protocol over the CMMBCR, Minimum Transmission Power Routing (MTPR), and Energy-Aware Routing environment adaptive application reconfiguration (EAAR) protocols, in terms of number of packets delivered, number of dead nodes, energy per packet delivered, and number of packets dropped. In this chapter, our contributions consists of complementing those simulations by examining both the impact of the network energy usage on the network size, and the impact of the load distribution (in terms of number of packets per node).

24.3.2 Design of the A-CMMBCR Protocol

The A-CMMBCR scheme was described in-depth in [64]. Briefly stated, this protocol takes advantage of the features of the environment adaptive application reconfiguration(EAAR) [51] protocol to improve the packet delivery ratio. It also includes the node's mobility, total energy consumed in the path, and residual battery of nodes as design criteria in its ACO framework. In the route discovery and maintenance phase of this framework, ant agent routing tables at each node are built and updated based on heuristic values and two pheromone models: one for the MPTR scheme [65], which holds the transmission energy of the paths, and one for the CMMBR scheme [12], which holds the battery level of a node in the path. A threshold is defined that dictates whether the routing path (i.e., the path with highest pheromone) for the CMMBR scheme or that of the MPTR scheme should be selected for data transmission. In doing so, the cases of path break or overloaded paths are also handled by rerouting the affected traffic to the available predesigned backup paths.

The operations of the A-CMMBCR scheme [64] are described in the following four steps. Let us consider that a source node, say S, wants to communicate with a destination node, say D, but the routing information for D is not available:

1. S will initially broadcast a route request (RREQ) packet that all its neighborhood nodes will receive. This RREQ packet will then be used by each of these nodes to find the node D and to determine if an entry for D exists in its routing table.
2. If such entry is found in any of the targeted neighbor nodes' routing tables, a route reply (RREP) packet will be sent as an acknowledgment by that node back to S using the same route that has carried the RREQ message but in the opposite direction. The data will be sent to the next hop along the discovered path. Otherwise, go to Step 3.
3. S will rebroadcast the RREQ packet with the hope of finding a neighbor node's routing table that has an entry for D. Then Step 2 will be invoked for each intermediate node along the discovered path. If no entry for D is found, the intermediate node will relay the RREQ packet.
4. Eventually the RREQ packet will reach node D. At that node, a RREP packet will be sent back by D to the source S using the intermediate nodes on the path through which the RREQ is delivered to D. Each of these intermediate nodes will then update their routing tables to include the entries to D.

In the above algorithm, the path discovery mechanism (of the ACO framework) is inspired by the environment adaptive application reconfiguration (EAAR) [51] protocol, which itself is based on the AntHocNet [52] protocol, but with a distinct functionality in the sense that in the aforementioned Step 1, the source node S will also broadcast a reactive forward ant (i.e., a control packet) and each neighbor node of S will receive a replica of that forward ant. After the next hop, the next neighboring node will also receive such a replica and so on. These set of replicas will help in determining the path to connect S to the destination D [64]. When possible routes between S and D are established and are checked for sufficient remaining battery capacity, the CMMBCR scheme is invoked to determine the route with the minimum total transmission power among the established routes. It should be noticed that the minimum battery of a node in a path and the total transmission energy of a path are computed during RREQ message processing and the pheromones are computed during the generation process of the RREP packet.

24.3.3 Performance Evaluation of the A-CMMBCR Protocol

In this work, the GLOMOSIM simulator [67] has been used to compare the A-CMMBCR scheme against the CMMBCR [12], Minimum Transmission Power Routing (MTPR) [65], and Energy-Aware Routing (EAAR) [51] protocols, on the basis of the following performance metrics: the network *energy usage* and the *load distribution* (in terms of number of packets per node). The main simulation

Table 24.1 Simulation Parameters

Number of nodes	40
Simulation time (s)	500
Initial energy of nodes	All nodes were initiated with an equal energy value
Terrain dimension	2000 (m) × 2000 (m)
Traffic type	CBR, with the following scenarios: • CBR 17 100 1536 1S 0S 250S • CBR 12 19 100 1536 1S 250S 400S • CBR 14 27 100 1536 1S 400S 500S
MAC protocol	IEEE 802.11
Mobility model	Random waypoint (when applicable) or none

TABLE 24.2 Simulation Scenarios

Scenarios	Data Size	Node Speed (m/s)	Mobility
1	100 times Control Packet Size	–	None
2	125 times Control Packet Size	–	None
3	150 times Control Packet Size	–	None
4	100 times Control Packet Size	10	Random waypoint
5	125 times Control Packet Size	10	Random waypoint
6	150 times Control Packet Size	10	Random waypoint

parameters are captured in Table 24.1. The simulation scenarios that were run are summarized in Table 24.2.

24.3.3.1 Energy Per Packet

The scenarios described in Table 2 are considered, and the size of the network is fixed to 40. The energy consumed per packet is investigated, and the results are depicted in Figure 24.1.

In Figure 1, it is observed that in terms of energy consumed per packet, the A-CMMBCR scheme outperforms all other schemes. This is attributed to the features of the ACO scheme used in A-CMMBCR as it generates multipaths from one node to another, thus providing an alternative for path overloading. In addition, the path with the highest pheromone on it consumes less energy than the path selected using the environment adaptive application reconfiguration (EAAR) scheme.

24.3.3.2 Network Energy Usage

In this scenario, the mobility scenario is fixed, the node speed is fixed at 10 m/s, and the terrain dimension is fixed at 2000 m × 2000 m. The size of the network is

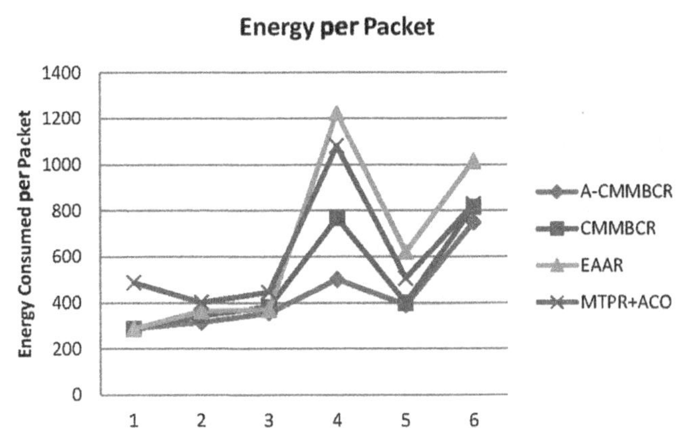

Figure 24.1 Energy usage vs. number of nodes [64].

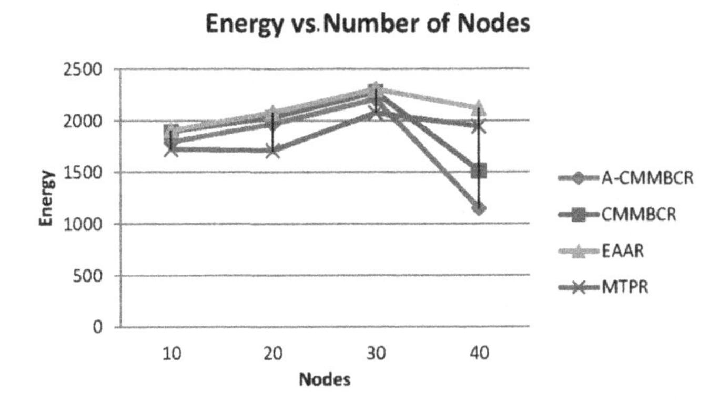

Figure 24.2 Energy usage vs. number of nodes.

increased from 10 to 40 nodes. The impact of this variation on the network energy usage for a given maximum speed of nodes is studied. The results are depicted in Figure 24.2.

In Figure 2, it can be observed that compared to the other schemes, the A-CMMBCR scheme uses the least energy as the number of nodes increases. This can be justified by the fact that the ACO framework used in A-CMMBCR optimizes the energy usage. Indeed, as the number of nodes increases, the density increases, which requires an efficient usage of energy. In the A-CMMBCR scheme, ACO helps in serving this purpose by equally distributing the packets to the nodes, thereby boosting the residual battery of nodes, and hence saving the energy at each node.

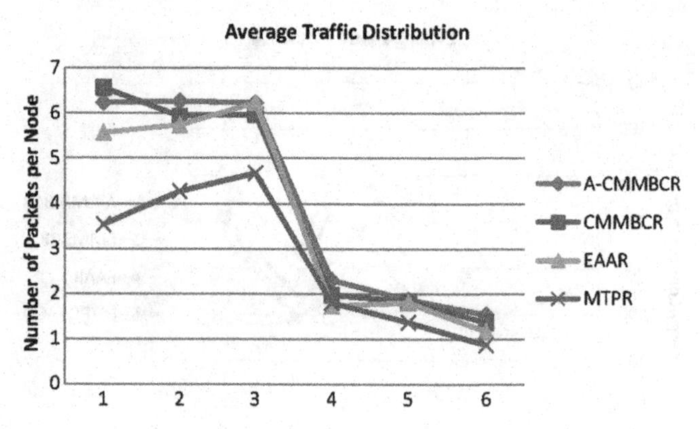

Figure 24.3 Average traffic distribution.

24.3.3.3 *Traffic Distribution*

The traffic distribution is important since it measures the probability of the number of weak nodes in the graph. The scenarios described in Table 2 are considered, and the size of the network is fixed to 40 nodes. The traffic distribution (in terms of the number of packets per node) operated by each studied scheme is investigated. The results are depicted in Figure 24.3.

In Figure 24.3, it can be observed that in the case of the A-CMMBCR scheme, the traffic distribution is even. Due to the design features of its ACO framework, the A-CMMBCR systematically distributes the packets to the paths that are less condensed.

24.4 CONCLUSION

In this chapter, we overviewed recent proposals on the use of ACO-based algorithms for designing energy-efficient routing protocols for ad hoc wireless and sensor networks. It was reported that the studied family of ACO heuristics yielded a much better solution to the energy consumption problem compared to conventional approaches. We also introduced an enhancement to a recently proposed ACO-based routing protocol (called A-CMMBCR), which belongs to the aforementioned family of protocols, and showed through simulations that it outperformed the CMMBCR, environment adaptive application reconfiguration (EAAR), and MTPR schemes, in terms of energy consumed per packet, energy usage, and average traffic distribution.

We believe that the ACO paradigm will continue to be used as a powerful algorithmic framework that can contribute in solving various types of optimization problems, including energy-related problems that may arise in next-generation networks, including green networks.

ACKNOWLEDGMENT

This work was supported in part by a grant from the Natural Sciences and Engineering Research Council of Canada (NSERC), held by the first author, Ref # 119200.

REFERENCES

[1] J. Hoebeke, I. Moerman, B. Dhoedt, P. Demeester, An overview of mobile ad hoc networks: applications and challenges, Journal of the Communications Network 3 (2004) 60–66.

[2] A. Ephremides, Energy concerns in wireless networks, IEEE Wireless Communications (2002) 48–59.

[3] B. Forouzan, Data Communication and Networking. fourth ed., McGraw-Hill0072967757

[4] W. Stark et al., Low energy wireless communication network design, in: Proceedings of the 38th Annual Allerton Conference Communication, Control, and Computing, Urbana, Illinois, October 2000, pp. 785–805.

[5] V. Rodoplu, T.H. Meng, Minimum energy mobile wireless networks, IEEE Journal on Selected Areas in Communications 17 (1999)

[6] S. Singh, M. Woo, C.S. Raghavendra, Power aware routing in mobile ad hoc networks, in: Proceedings of ACM/IEEE International Conference on Mobile Computing and Networking (MobiCom'98), Dallas, Texas, USA, October 1998, pp. 181–190.

[7] I. Stojmenovic, X. Lin, Power-aware localized routing in wireless networks, in: Proceedings of the IEEE Symposium on Parallel and Distributed Processing Systems, May 2000.

[8] R. Ramanathan, R. Rosales-Hain, Topology control of multihop wireless networks using transmit power adjustment, in: Proceedings of the Ninth IEEE International Conference on Computer Communications (INFOCOM 2000), Tel-Aviv, Israel, March 26–30, 2000, p. 4043.

[9] K. Scott, N. Bambos, Routing and channel assignment for low power transmission in PCS, in: Proceedings of the Fifth IEEE International Conference on Universal Personal Communications (ICUPC'96), Cambridge, MA, USA, September 29–October 2, 1996.

[10] Z. Li, Y. Zhen-Wei, Z. Yang, Z. Chun-Kai, A power-aware adaptive dynamic routing scheme for wireless ad hoc networks, in: Proceedings of the IEEE International Conference on Networking, Sensing and Control (ICNSC 2008), Sanya, China, April 6–8, 2008, pp. 966–970.

[11] S. Singh, M. Woo, S. Raghavendra, Power-aware routing in mobile ad hoc networks, in: Proceedings of ACM/IEEE International Conference on Mobile Computing and Networking (MobiCom'98), Dallas, TX, USA, 1998.

[12] C.-K. Toh, Maximum battery life routing to support ubiquitous mobile computing in wireless ad hoc networks, IEEE Communications Magazine (2001)

[13] J.-H. Chang, L. Tassiulas, Energy conserving routing in wireless ad hoc networks, in: Proceedings of the Ninth IEEE International Conference on Computer Communications (INFOCOM 2000), Tel-Aviv, Israel, March 26–30, 2000, pp. 22–31.

[14] M. Dorigo, Optimization, learning and natural algorithms, Ph.D. thesis, Dipartimento di Elettronica, Politecnico di Milano, Italy, 1992.

[15] M. Dorigo, V. Maniezzo, A. Colorni, Positive feedback as a search strategy, Technical report 91-016, Dipartimento di Elettronica, Politecnico Di Milano, Italy, 1991.

[16] M. Dorigo, G. Di Caro, The Ant Colony Optimization meta-heuristic, in: D. Corne, , M. Dorigo, F. Glover (Eds.), New Ideas in Optimization, McGraw-Hill, London, UK, 1999, pp. 11–32.

[17] A.P. Engelbrecht, Computational Intelligence. An Introduction. second ed., John Wiley & Sons Ltd.9780470035610 (University of South Africa, Pretoria)

[18] K. Socha, ACO for continuous and mixed variable optimization, in: M. Dorigo (Ed.), Fourth International Workshop on Ant Colony Optimization and Swarm Intelligence (ANTS 2004), Lecture Notes in Computer Science, vol. 3172, Springer, Heidelberg, 2004, pp. 25–36.

[19] M. Guntsch, M. Middendorf, Applying population-based ACO to dynamic optimization problems, in: M. Dorigo, , G.A. Di Caro, M. Samplels (Eds.), ANTS 2002, Lecture Notes in Computer Science, vol. 2463, Springer, Heidelberg, 2002, pp. 97–104.

[20] C.J. Eyckelhof, M. Snoek, Ant systems for a dynamic DSP: Ants caught in a traffic jam, in: Proceedings of Third International Workshop on Ant Algorithms (ANTS02), Brussels, Belgium, September 12–14, 2002.

[21] M. Dorigo, V. Maniezzo, A. Colorni, The ant system: an autocatalytic optimizing process, Dipartimento di Elettronica, Politecnico di Milano, Italy, 91-016, 1991 (revised).

[22] L.M. Gambardella, M. Dorigo, Solving symmetric and asymmetric TSPs by ant colonies, in: Proceedings of IEEE International Conference on Evolutionary Computation, Nagoya, Japan, May 20–22, 1996, pp. 622–627.

[23] T. Stutzle, H. Hoos, MAX–MIN ant system and local search for the traveling salesman problem, in: Proceedings of the IEEE International Conference on Evolutionary Computation, Indianapolis, IN, USA, April 13–16, 1997, pp. 309–314.

[24] O. Roux, C. Fonlupt, E.-G. Talbi, ANTabu, Technical report LIL-98-04, Laboratoire d'Informatique du Littoral, Université du Littoral, Calais, France, 1998.

[25] V. Maniezzo, A. Carbonaro, Ant colony optimization: an overview, in: C. Ribeiro (Ed.), Essays and Surveys in Metaheuristics, Kluwer Academics, Boston, MA, USA, 1999, pp. 21–44.

[26] G. Di Caro, M. Dorigo, Ant colonies for adaptive routing in packet-switched communications networks, in: Parallel Problem Solving from Nature – PPSN V, Lecture Notes in Computer Science, vol. 1498, September, 1998, pp. 673–682.

[27] G. Di Caro, Ant colony optimization and its application to adaptive routing in telecommunication networks, Ph.D. thesis in Applied Sciences, Polytechnic School, Université Libre de Bruxelles, Brussels, Belgium, 2004.

[28] G. Di Caro, A society of ant-like agents for adaptive routing in networks, DEA thesis in Applied Sciences, Polytechnic School, Université Libre de Bruxelles, Brussels, Belgium, 2001.

[29] M. Saleem, M. Farooq, BeeSensor: a bee-inspired power aware routing protocol for wireless sensor networks, in: Proceedings of EvoWorkshops (EvoCOMNET), Valencia, Spain, Lecture Notes in Computer Science, vol. 4448, 2007, pp. 81–901.

[30] G.A. Di Caro, F. Ducatelle, L.M. Gambardella, Ant colony optimization for routing in mobile ad hoc networks in urban environments, Technical report IDSIA-05-08, IDSIA, Lugano, Switzerland, May 2008.

[31] V.R. Zanjani, A.T. Haghighat, Adaptive routing in ad hoc wireless networks using ant colony optimization, in: Proceedings of International Conference on Computer Technology and Development (ICCTD'09), Kota Kinabalu, Malaysia, November 13–15, 2009, pp. 40–45.

[32] H. Al-Zurba, T. Landolsi, M. Hassan, F. Abdelaziz, On the suitability of using ant colony optimization for routing in multimedia content over wireless sensor networks, International Journal on Applications of Graph Theory in Wireless Ad Hoc Networks and Sensor Networks (GRAPH-HOC) 3 (2) (2011)

[33] F. Celik, A. Zengin, S. Tuncel, A survey on swarm intelligence based routing protocols in wireless sensor networks, International Journal of the Physical Sciences, Academic Journals 5 (14) (2010) 2118–2126 ISSN: 1992-1950

[34] T. Stutzle, M. Lopez-Ibanez, M. Dorigo, A concise overview of applications of ant colony optimization, in: James J. Cochran (Ed.), Proceedings of Encyclopedia of Operations Research and Management Science, John Wiley & Sons Inc..

[35] Y. Zhang, L.D. Kuhn, M.P.J. Fromherz, Improvements on ant routing for sensor networks, Ant Colony Optimization and Swarm Intelligence, Lecture Notes in Computer Science, vol. 3172, 2004, pp. 289–313.

[36] X. Wang, L. Qiaoliang, X. Naixue, P. Yi, Ant colony optimization-based location-aware routing for wireless sensor networks, Proceedings of the Third International Conference on Wireless Algorithms, Systems, and Applications (WASA'08), Springer-Verlag, Berlin, Heidelberg, 2008, pp. 109–120.

[37] S. Selvakennedy, S. Sinnapan, Y. Shang, T-ANT: a nature-inspired data gathering protocol for wireless sensor networks, Australian Journal of Communications 1 (2) (2006) 22–29.

[38] A.-A. Salehpour, B. Mirmobin, A. Afzali-Kusha, S. Mohammadi, An energy efficient routing protocol for cluster-based wireless sensor networks using ant colony optimization, in: Proceedings of the International Conference on Innovations in Information Technology (IIT), Al Ain, United Arab Emirates, December 16–18, 2008, pp. 455–459.

[39] M. Paone, L. Paladina, M. Scarpa, A. Puliafito, A multi-sink swarm-based routing protocol for wireless sensor networks, in: IEEE Symposium on Computers and Communications (ISCC), Sousse, Tunisia, July 5–8, 2009, pp. 28–33.

[40] W.H. Liao, K. Yucheng, F. Chien-Ming, Data aggregation in wireless sensor networks using ant colony algorithm, Journal of Network and Computer Applications 31 (4) (2008) 387–401.

[41] K. Saleem, N. Fisal, S. Hafizah, S. Kamilah, R.A. Rashid, A self-optimized multipath routing protocol, Journal of Recent Trends in Engineering (IJRTE) 2 (1) (2009)

[42] S.K. Dhurandher, S. Misra, H. Mittal, A. Agarwal, I. Woungang, Using ant-based agents for congestion control in ad-hoc wireless sensor networks, Cluster Computing 14 (1) (2011) 41–53.

[43] S.K. Dhurandher, S. Misra, S. Ahlawat, N. Gupta, E2-SCAN: an extended credit strategy-based energy-efficient security scheme for wireless ad hoc networks, IET Communications 3 (5) (2009) 808–819.

[44] M.S. Obaidat, S.K. Dhurandher, D. Gupta, N. Gupta, A. Asthana, DEESR: dynamic energy efficient and secure routing protocol for wireless sensor networks in urban environments, Journal of Information Processing Systems 6 (3) (2010) 269–294.

[45] G. Kalpana, T. Bhuvaneswari, A survey on energy efficient routing protocols for wireless sensor networks, in: Proceedings of the Second National Conference on Information and

Communication Technology (NCICT), International Journal of Computer Applications, Foundation of Computer Science, USA, 2011, pp. 12–18.

[46] G.A. Di Caro, F. Ducatelle, L.M. Gambardella, Theory and practice of ant colony optimization for routing in dynamic telecommunications networks, in: N. Sala, F. Orsucci (Eds.), Reflecting Interfaces: The Complex Co-evolution of Information Technology Ecosystems, Idea Group, Hershey, PA, USA, 2008

[47] G.A. Di Caro, M. Dorigo, AntNet: a mobile agents approach to adaptive routing, Technical report IRIDIA/97-12, IRIDIA, Université Libre de Bruxelles, 1997.

[48] T. Camilo, C. Carreto, J. Silva, F. Boavida, An energy-efficient ant base routing algorithm for wireless sensor networks, in: ANTS 2006, Fifth International Workshop on Ant Colony Optimization and Swarm Intelligence, vol. 4150, 2006, pp. 49–59.

[49] Y. Wen, Y. Chen, D. Qian, An ant-based approach to power-efficient algorithm for wireless sensor networks, in: Proceedings of the World Congress on Engineering (WCE 2007), London, UK, vol. II, July 2–4, 2007.

[50] J. Wang, J. Xu, M. Xiang, EAQR: an energy-efficient ACO based QoS routing algorithm in wireless sensor networks, Chinese Journal of Electronics 18 (1) (2009)

[51] S.K. Dhurandher, S. Misra, M.S. Obaidat, P. Gupta, K. Verma, P. Narula, An energy-aware routing protocol for ad-hoc networks based on the foraging behavior in ant swarms, in: Proceedings of the IEEE International Conference on Communications (ICC'09), Dresden, Germany, June 14–18, 2009, pp. 1–5.

[52] G. Di Caro, F. Ducatelle, L.M. Gambardella, AntHocNet: an adaptive nature-inspired algorithm for routing in mobile ad hoc networks, Telecommunications (ETT) 16 (2) (2005)

[53] S. Okdem, D. Karaboga, Routing in wireless sensor networks using an Ant Colony Optimization (ACO) router chip, Sensors 9 (2009) 909–921.. doi:10.3390/s90200909.

[54] S. Xia, S. Wu, Ant colony-based energy-aware multipath routing algorithm for wireless sensor networks, in: Proceedings of Second International Symposium on Knowledge Acquisition and Modeling (KAM'09), Wuhan, China, November 30–December 1, 2009, pp. 198–201.

[55] S. Misra, S.K. Dhurandherb, M.S. Obaidat, P. Guptab, K. Verma, P. Narula, An ant swarm-inspired energy-aware routing protocol for wireless ad-hoc networks, Journal of Systems and Software 83 (2010) 2188–2199.

[56] V. Mahadevan, F. Chiang, iACO: a bio-inspired power efficient routing scheme for sensor networks, International Journal of Computer Theory and Engineering 2 (6) (2010) 972–977.

[57] T. Stutzle, H. Hoos, MAX–MIN ant system and local search for combinatorial optimization problems, in: S. Voß, , S. Martello, , I.H. Osman, C. Roucairol (Eds.), Meta-Heuristics: Advances and Trends in Local Search Paradigms for Optimization, Kluwer Academics, Boston, MA, USA, 1999, pp. 313–329.

[58] X. Hui, Z. Zhigang, N. Feng, A novel routing protocol in wireless sensor networks based on ant colony optimization, International Journal of Intelligent Information Technology Application 3 (1) (2010) 1–5.

[59] Y.-M. Kim, E.-J. Lee, H.-S. Park, Ant colony optimization based energy saving routing for energy-efficient networks, IEEE Communications Letters 15 (7) (2011) 779–781.

[60] J. Ren, Y. Tu, M. Zhang, Y. Jiang, An ant-based energy-aware routing protocol for ad hoc networks, in: Proceedings of International Conference on Computer Science and Service System (CSSS), Nanjing, China, June 27–29, 2011, pp. 3844–3849.

[61] D. Cheng, Y. Xun, T. Zhou, W. Li, An energy aware ant colony algorithm for the routing of wireless sensor networks, Journal on Communications in Computer and Information Science 134 (2011) 395–401.

[62] C. Dominquez-Medina, N. Cruz-Cortes, Energy-efficient and location-aware ant colony based routing algorithms for wireless sensor networks, in: Proceedings of the 13th ACM Annual Conference on Genetic and Evolutionary Computation (GECCO'11), New York, NY, USA, 2011, ISBN: 978-1-4503-0557-0.

[63] X. Wang, Q. Li, N. Xiong, Y. Pan, Ant colony optimization-based location-aware routing for wireless sensor networks, Lecture Notes in Computer Science vol. 5258, Springer-Verlag pp. 109–120

[64] S.K. Dhurandher, M.S. Obaidat, M. Gupta, Application of ant colony optimization to develop energy efficient protocol in mobile ad-hoc networks, in: Proceedings of WINSYS-2011, Seville, Spain, 2011, pp. 12–17.

[65] K. Scott, N. Bambos, Routing and channel assignment for low power transmission in PCS, in: Proceedings of the International Conference Universal Personal Communications (ICUPC'96), Cambridge, MA, USA, 1996, pp. 498–502.

[66] S. Singh, M. Woo, C.S. Raghavendra, Power-aware routing in mobile ad hoc networks, in: Proceedings of the Fourth Annual ACM/IEEE International Conference on Mobile Computing and Networking, Dallas, TX, USA, 1998, pp. 181–190.

[67] M. Takai, L. Bajaj, R. Ahuja, R. Bargrodia, M. Gerla, GloMoSim: a scalable network simulation environment, Technical report 990027, Department of Computer Science, University of California, Los Angeles, USA, 1999.

CHAPTER

Smart Grid Communications: Opportunities and Challenges

25

Hussein T. Mouftah, Melike Erol-Kantarci
School of Electrical Engineering and Computer Science, University of Ottawa, Ottawa, ON, Canada

25.1 INTRODUCTION

Electrical power grids around the globe were implemented almost a century ago. Based on the needs of 20th century, this electricity grid has been sufficient and reliable most of the time. Not to mention, reliability of the traditional grid has been maintained by overprovisioning the resources. For instance, backup generators, backup transformers, and alternative connections have been the common practice of utilities. Furthermore, greenhouse gas (GHG) emissions of the electricity production sector and its negative impact on the environment were not fully known nor investigated until recently.

At the beginning of the 21st century the reliability of the electricity grid became questionable mostly due to several major blackouts. One of those blackouts that occurred in 2003 in the Eastern Interconnection affected a large region including several U.S. states and one Canadian province. Earlier in 2001, California as well had experienced a major blackout, and once again in the summer of 2011, Arizona, California, and some parts of Mexico have been impacted by another mass blackout. Certainly, the reasons for the blackouts were not identical. For instance, the 2001 blackout was related to the actions of the players in the energy market, while the blackout in 2003 was started by a tree contacting power lines in Ohio and worsened with inappropriate management of the cascading outages. The blackout of 2011 had a completely different cause. It was a result of two nuclear reactors losing power. In all of those blackouts, conditions that triggered outages were not independent of demand. Besides supply related problems, with the lack of monitoring and market behavior, these outages became severe blackouts due to concurrent intense demand. In summary, capacity limitations, absence of robust equipment monitoring tools, and the inability to react to instabilities in a timely manner have been impacting the reliability of the electricity grid.

Besides reliability problems, growing demand for electricity and aging infrastructure accompanied by the overprovisioning practice of the utilities have been giving signs of an inefficient operation of the power grid [1]. The traditional approach of centralized generation and distribution of electricity has provided easier control and operation despite its inefficiency. On the contrary, distributed operation involves more complex operational procedures. The shift from centralized architecture to distributed architecture will require advanced information and communication technologies (ICTs) to be involved in the operation of the grid [2]. Although the current power systems employ communications, the future power systems will require communications that enable the individuals to participate in generation and consumption of electricity.

Generation of electricity by distributed generation (DG) techniques and feeding the generated electricity to the grid raise serious power quality concerns. Dividing the large power grids into smaller manageable microgrids has been considered widely by the researchers. Yet, real-life implementations of microgrids are still at test-bed scale. A microgrid is an electrical system that contains several DG units such as photovoltaic panels, wind turbines, and microturbines, in addition to storage units, controllable loads, small-scale combined heat and power (CHP) units and an energy management system (EMS) [3]. The EMS enables the microgrid to take independent decisions from the utility grid when the microgrid operates in the islanded mode. Islanded mode refers to the case when a circuit breaker isolates the microgrid from the rest of the utility grid; hence energy generation, storage, load control, power quality control, and regulation operations are implemented within the microgrid. In the grid-connected mode, the microgrid is either a controllable load, or a power generator from the grid's point of view [4, 5].

Microgrids are also considered as the enablers of increasing renewable energy penetration. Renewable energy generation techniques have been known and utilized in limited sites since many years. Among solar, wind, hydro, and tidal power, hydro power is the best integrated renewable energy generation technique today. Although the potential of solar and wind energy generation technologies are promising, they do not have a significant contribution to the national energy mix in most countries. The penetration levels of renewable energy have been limited due to their intermittency. Despite this fact, global concerns of climate change and the awareness of GHG emissions have been one of the major driving factors behind the efforts to increase utilization of renewable energy. Meanwhile, governments' desire for energy independence is another factor that is leading the efforts to shift from fossil fuel-based generation to renewable energy generation.

The smart grid is the future ICT-enabled power grid that will be designed to overcome the deficiencies and the inefficiencies of the existing power grid. While achieving these goals the smart grid will require robust communication technologies. This chapter provides a comprehensive survey of available communication technologies and standards for the smart grid. Smart grid communications can cover large or small geographic regions; therefore they are usually classified as wide area network (WAN), neighborhood area network (NAN), and home area network (HAN) where WAN refers to the region under the control of a utility, NAN corresponds to a group of houses possibly fed by the same transformer, and HAN is a single residential unit. In addition to those networks, the smart grid involves a field area network (FAN) whose scale is similar to a NAN, but different from the NAN, a FAN covers the distribution automation and distribution equipment under the control of a utility. The network that is used for smart meter data delivery is known as the

Advanced Metering Infrastructure (AMI). All of these networks can be implemented using various communication technologies. For instance, fiber optic, LTE-A, IEEE P1901, IEEE 802.11, IEEE 802.15.4, IEEE 802.16, and IEEE 802.22 are among the considered communication standards for the smart grid. Due to the different range and bandwidth properties of each protocol, they may be more suitable in one network than the other. For example, cellular communication may be implemented for the WAN while ZigBee may be dominant in HAN.

The rest of the chapter is organized as follows. In Section 25.2, we introduce the available communication technologies, in Section 25.3 we discuss their applicability to various smart grid domains. In Section 25.4, we focus on the smart grid applications that become available with communication technologies. Section 25.5 discusses the challenges of smart grid communications with a special focus on security and privacy aspects. We summarize the chapter and discuss the open issues in Section 25.6.

25.2 SMART GRID

The smart grid is the future electrical power grid that is envisioned to integrate ICTs in its operation for the purpose of increasing reliability, security, and efficiency, and reducing GHG emissions [6].

The smart grid will be one of the largest and most complex cyber-physical systems. A cyber-physical system integrates networking and information technology functions to the physical elements of a system for monitoring and control purposes [7]. Research activities on cyber-physical systems and the smart grid have gained acceleration after the U.S. House of Representatives passed the Networking and Information Technology Research and Development Act of 2009 and the Energy Independence and Security Act of 2007 [8]. An illustration of a smart grid is presented in Figure 25.1. As pictured in the figure, the smart grid makes use of

Figure 25.1 Illustration of generation, transmission, and distribution in the smart grid [9].

renewable energy generation and integrates storage with distributed generation. The transmission and distribution system is networked, participation of buildings and homes in demand response is possible, and the electric vehicle is a part of the smart grid ecosystem.

It is anticipated that today's vision for the smart grid will be completely realized within the next several decades. For the time being, smart meters and AMI are the initial steps that utilities have taken to modernize their distribution system. A group of smart meters are shown in Figure 25.2. Smart meters have been installed in the majority of the consumer premises in North America, Europe, and China. With the help of AMI, electricity consumption can be monitored by the utilities and consumers in the resolution of minutes, and utility commands and price information can be delivered to consumers. Smart meters record detailed electricity consumption data, send this information through AMI to the utility mostly for metering and billing purposes in a reliable and secure way, and deliver utility-based information such as pricing to the consumers. Thus, smart meters and AMI allow the utilities to read meters remotely, realize automated energy management, and enable the use of time-differentiated billing schemes, such as time of use (TOU), real-time pricing (RTP), critical peak pricing (CPP), which are explained below. Furthermore, in the existing power grid, in case of an outage, customers need to call and report the outage to the utility whereas in the smart grid, smart meters are able to report outages directly to the utility.

The most widely used time-differentiated pricing scheme is the TOU. TOU pricing provides different pricing for peak, mid-peak, and off-peak hours and the prices are generally adjusted twice a year to reflect seasonal demand variations. In TOU

Figure 25.2 Smart meters in Ontario (photo taken by Melike Erol-Kantarci).

pricing, distinction of peak, mid-peak, and off-peak hours and the price associated with each time slice are based on the historical load data collected over many years. Naturally, the market price of electricity varies with load and TOU reflects these variations in the final price of electricity. Demand for electricity and load follow the behavior of consumers and the environmental factors. For instance, demand is lower during overnight hours than it is during daytime, due to low consumer activity. RTP or dynamic pricing reflects the actual price of the electricity in the market to the consumer bills. The market price of electricity is generally determined by the regional independent system operator. The independent system operator arranges a settlement for the electricity prices of the next day or next hour, based on the load forecasts, supplier bids, and importer bids. Additionally, CPP can be applied on several days of a year, e.g., very hot days, when the load is predicted to exceed a certain threshold. CPP aims to reduce the load of large industrial or commercial consumers on several days in order to prevent grid failure. Some utilities reward customers with credits for their cooperation on critical peak hours/days, which is called as peak time rebate (PTR).

Delivering meter data from consumer premises towards the utility, and delivering price information from the utility to the consumers require two-way information flow, which is one of the fundamental novelties of the smart grid. Smart meters and AMI are enablers of two-way information flow. Furthermore, in the smart grid, energy flow will be two-way, as well. Unlike the existing grid, which is a one-way electricity distribution system where the flow of electricity is from the power plants towards the consumer, smart grid consumers will be able to generate and sell electricity back to the power grid. Renewable energy generation techniques are especially encouraged for distributed generation since they have significantly lower emissions than fossil-fuel-based electricity generation. In Figure 25.3, the kilotons equivalent of CO_2 emissions per terrawatt hour of energy generation is presented for coal, natural gas, solar, hydro, and wind-based energy generation. As seen in the figure, renewables have significantly less CO_2 equivalent emissions. Note that emission rates depend on the energy generation technology. In the figure, higher emission rates are a result of the typical existing technology while lower rates are given for the best available commercial technology [10]. Inspired by

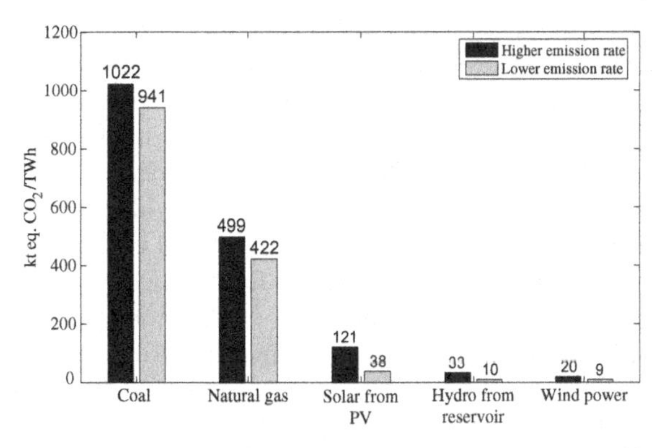

Figure 25.3 Comparison of GHG emissions for several energy resources [12].

this fact, the government of Ontario has recently announced the micro Feed-in Tariff (microFIT) program, which allows home owners to sell the renewable energy generated by rooftop solar panels or backyard wind turbines back to the grid [11]. Several utilities are providing contracts to buy energy generated by solar photovoltaics (PV), wind, waterpower, biomass, biogas, and landfill gas.

To facilitate two-way flow of energy, it is essential that advanced distributed storage units are employed at the consumer/generator premises. Recently emerging plug-in hybrid electric vehicles (PHEV)s are considered distributed energy storage for the smart grid. PHEVs can store the energy generated with solar PVs during the day and sell energy back during the evening peak. From another point of view, PHEVs introduce certain challenges, i.e., they can dramatically increase the load on the grid. EV users are expected to primarily live in urban areas and they may charge the batteries of their PHEVs during peak periods, which amplify the peak load. Electric vehicle charging and its coordination with the smart grid is an emerging topic. As a result, communication technologies will play a key role in enabling consumer-to-utility (CTU) and consumer-to-consumer (CTC) energy transactions, in addition to enabling a grid-friendly PHEV charging and discharging infrastructure.

Despite the advantages that are becoming available with communications, certain challenges need to be addressed. Security and standardization are among the fundamental challenges. As the physical components of the power grid are networked and the grid gets integrated with the Internet, the risks of cyber attacks increase. Furthermore, secure access and privacy-preserving communication techniques are essential for AMI. Meanwhile, interoperability of standards and regulations are of paramount importance. In some countries, it is easier to synchronize the regulations due to central governance and fewer players in the electricity market, whereas for some countries the situation is more complex due to the higher number of players in the market. Standardization is important since each new component integrated to the grid must follow the same set of rules, and be interoperable with the other components. Currently, standardization efforts are independently handled in most countries. For example, in the United States, the National Institute of Standards and Technology (NIST) has been given the primary responsibility to develop standards, and it has published a report entitled "Framework and Roadmap for Smart Grid Interoperability Standards, Release 1.0" [13] which presents the available standards and discusses their use in smart grid applications. In the European Union, the European Commission has defined the European Energy Research Alliance (EERA) as a key actor for implementing the EU Strategic Energy Technology Plan (SET Plan) [14]. Besides standardization, international harmonization of the standards is required in order to provide consumers the same set of services worldwide. The International Electrotechnical Commission (IEC), which is an international standardization commission for electrical, electronic, and related technologies, has documented a large number of smart grid standards. Additionally, the Institute of Electrical and Electronics Engineers (IEEE) has a large number of standards in addition to the IEEE P2030 draft guide entitled "IEEE Guide for Smart Grid Interoperability of Energy Technology and Information Technology Operation with the Electric Power System (EPS), End-Use Applications, and Loads." The smart grid has a broad scope, and therefore a large number of new standards are expected to emerge in the following years. In the next section, we summarize the communication technologies and standards that are identified by NIST, and discuss their advantages, disadvantages, and use cases in the smart grid ecosystem.

25.3 COMMUNICATION TECHNOLOGIES FOR THE SMART GRID

The existing electrical power grid is monitored and controlled by system control and data acquisition (SCADA) software which provides limited monitoring capabilities and requires relatively low bandwidth. The network support for SCADA has been a mixture of various communication technologies including radio frequency (RF) mesh links, dial-up leased lines, modem connections, Ethernet, and IP over SONET/SDH [15]. On the other hand, the smart grid will require communication technologies that enable a large number of devices to communicate reliably and securely. Existing communication technologies offer a good starting point for yet-to-develop smart grid communication standards.

We group the communication technologies as wireless and wired communication technologies and under each group we summarize the standards that are considered to have applicability in smart grid applications. In Figure 25.4, we present various networks that are envisioned to be implemented in the smart grid. Each domain (WAN, NAN, FAN, and HAN) has different communication requirements and therefore they are anticipated to employ different technologies. In the following sections, we briefly introduce the communication standards and discuss the potential usage of each standard to these domains.

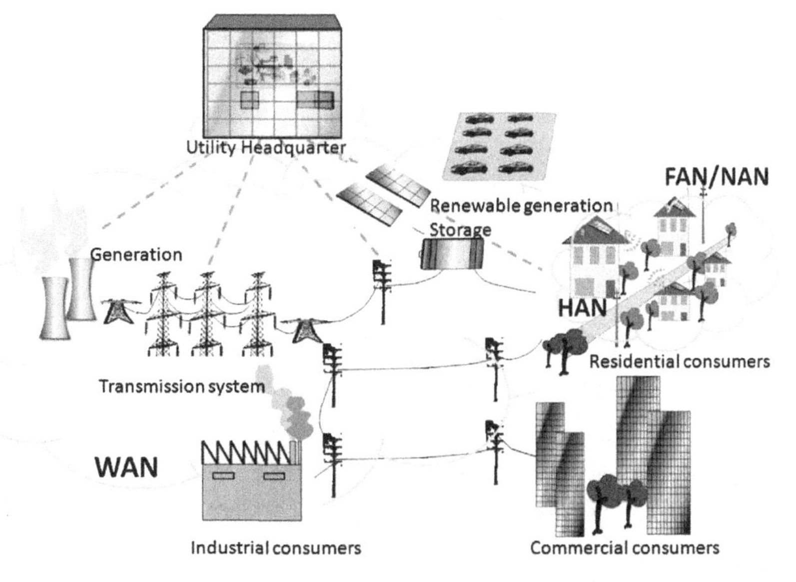

Figure 25.4 Communication networks (WAN, FAN, NAN, HAN) for the smart grid [16].

25.3.1 Wireless Communication Standards

Wireless communications use airborne signals; therefore they do not require cabling, node locations are flexible, and wireless signals can reach areas that are hard to reach physically. On the other hand, wireless communications may be prone to signal losses due to attenuation or penetration inside structures, and consequently have relatively short range. Additionally, they may be subject to eavesdropping; hence they require strong security mechanisms. In the following section, we summarize the fundamentals of IEEE 802.15.4/ZigBee, Z-wave, IEEE 802.11/Wi-Fi, IEEE 802.16/WIMAX, IEEE 802.22/cognitive radio, and recent cellular technologies (i.e., LTE-A).

25.3.1.1 IEEE 802.15.4/ZigBee

ZigBee is a short-range, low-data-rate, energy-efficient wireless technology that is based on the IEEE 802.15.4 standard. ZigBee utilizes 16 channels in the 2.4 GHz ISM band worldwide, 13 channels in the 915 MHz band in North America, and one channel in the 868 MHz band in Europe. ZigBee is a low-bit-rate technology designed to service low-data-rate transmissions. The supported data rates are 250 kbps, 100 kbps, 40 kbps, and 20 kbps. The range of a ZigBee radio is approximately 30 m indoors. Thus, ZigBee is considered suitable for the HAN domain of the smart grid.

ZigBee is well-known for its energy efficiency, which is due to its duty cycling mechanism. ZigBee certified devices can work for several years without the need for battery replacement. The IEEE 802.15.4 standard defines the physical and MAC layer access while the upper layers including routing and applications are defined in the ZigBee protocol stack (see Figure 25.5).

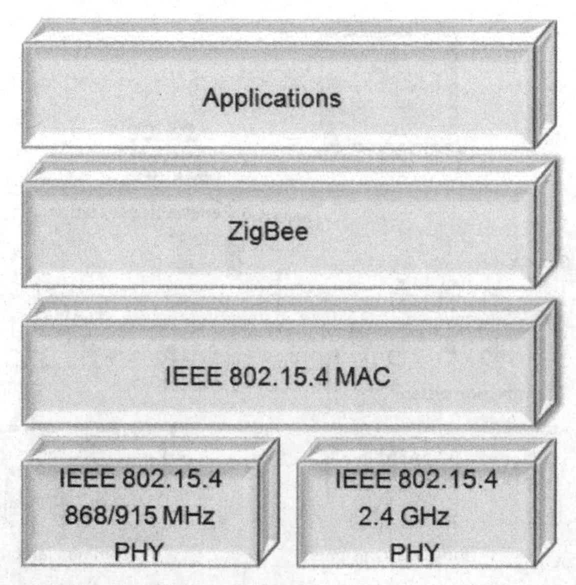

Figure 25.5 ZigBee protocol stack.

ZigBee supports two addressing modes: 16-bit and 64-bit addressing. A ZigBee network can support up to 64,000 nodes (devices). These devices can be of two types: (i) full function device (FFD) and (ii) reduced function device (RFD). FFDs can be interconnected in a mesh topology which means they can communicate with their peers while RFDs are simpler than FFDs, and they can be the edge nodes in a star topology. In the star topology configuration ZigBee employs a personal area network (PAN) coordinator, which may operate in beacon-enabled mode or beaconless mode. In the beacon-enabled mode, the PAN coordinator defines the duty cycle with the superframe duration (SD) within the superframe structure presented in Figure 25.6. A superframe synchronizes the nodes in the network. Nodes communicate only in the active period. In the contention access period (CAP) of the superframe, nodes compete to achieve access to transmit their data by using the slotted carrier sense multiple access with collision avoidance (CSMA/CA) technique. The contention free period (CFP) provides guaranteed time slots (GTSs) for the nodes that have previously reserved these slots for communication. One cycle of active and inactive periods can occur within a beacon interval (BI), which starts at the beginning of a beacon frame and ends at the beginning of the next beacon frame. SD and BI are defined in the IEEE 802.15.4 standard as follows [17]:

$$\text{SD} = \text{a base superframe duration} * 2^{\text{SO}} \text{ symbols}, \tag{25.1}$$

$$\text{BI} = \text{a base superframe duration} * 2^{\text{BO}} \text{ symbols}, \tag{25.2}$$

where SO is the superframe order and BO is the beacon order. In the standard, the range of SO and BO is defined as $0 \leqslant \text{SO} \leqslant \text{BO} \leqslant 14$.

Presently, there are various ZigBee certified products for home automation. Several smart meter vendors have already developed ZigBee-enabled smart meters, which enable the smart meters to communicate with the home appliances and home automation tools. For instance, Landis+Gyr, Itron, and Elster have advanced, ZigBee-enabled smart meters. Landis+Gyr has also produced a home energy monitor that is able to communicate with the Landis+Gyr smart meters and report consumption to consumers. ZigBee Alliance has also developed a Smart Energy Profile (SEP) to support the needs of smart metering and AMI, and provide communication among utilities and household devices.

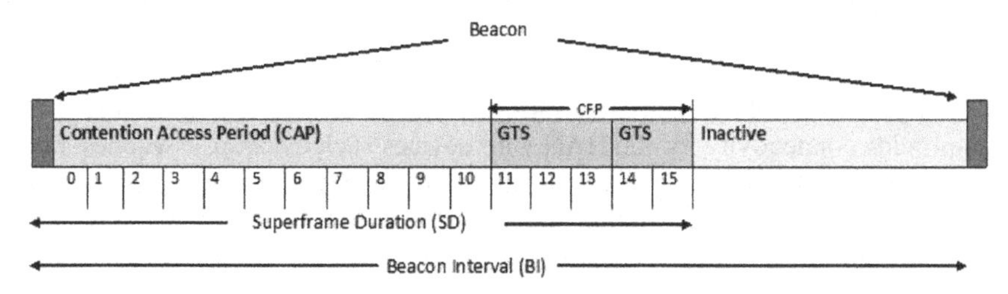

Figure 25.6 IEEE 802.15.4/ZigBee superframe format [18].

Despite ZigBee's promising acceptance from the smart grid industry, it has several drawbacks, low data rates being the major drawback. As smart grid applications become more complex and their bandwidth requirement increase, ZigBee's data rate may fall short for those applications. Moreover, when ZigBee-enabled devices connect to the Internet, IP addressing is required. Previously, ZigBee did not have IP addressability. However, IETF RFC 4944 recently defined IPv6 over Low-power Wireless Personal Area Networks (6LoWPANs), which aims to integrate IPv6 addressing to LoWPANs like ZigBee [19]. 6LoWPAN adds an adaptation layer to handle fragmentation, reassembly, and header compression issues in order to support IPv6 packets on the short packet structure of ZigBee. Additionally, ZigBee operates in the unlicensed spectrum, i.e. the Industrial, Scientific, and Medical (ISM) frequency band and its performance is affected negatively from interference with other wireless technologies using the same spectrum band, which are Wi-Fi, Bluetooth, and microwave appliances.

25.3.1.2 *Z-wave*

Z-wave is a proprietary, short-range, low-data-rate wireless RF mesh networking standard that was initially designed by Zensys Inc. (currently owned by Sigma Designs Inc.). It uses the 908 MHz ISM band in the Americas, and its data rate is 40 kbps in the current version and 9.6 kbps in previous versions. Z-wave uses BFSK modulation. The maximum range of a Z-wave radio is approximately 30 m indoors and around 100 m outdoors. Typical to all wireless communication technologies, low signal propagation through walls and construction limits its communication range. Z-wave does not require a central coordinator since it is a mesh protocol, however it employs slave and master nodes. The Z-wave protocol stack consists of four layers as seen in Figure 25.7 [20]. The MAC layer controls access to RF media. A basic MAC frame contains preamble, start of frame, and end of frame fields for data encapsulation. The MAC layer also employs a collision avoidance technique. The transport layer employs checksum for frame integrity check, in addition to acknowledgment and retransmission mechanisms. The routing layer of Z-wave is table-based and both master and slave nodes are able to participate in routing. The Z-wave application layer decodes commands and executes them. Z-wave commands can be either protocol commands or application-specific commands. Protocol commands mostly specify ID assignment and the application commands can turn devices on or off or execute other home-control-related commands.

Z-wave has been in the market longer than ZigBee and it has been embedded into a large number of appliances by various vendors. The main purpose of Z-wave is to provide connectivity for the HAN with devices such as lamps, switches, thermostats, garage doors, etc. Therefore, it can be employed in the HAN segment of the smart grid. However, similar to ZigBee it suffers from low data rates and the number of network devices supported by Z-wave is 232, which is lower than ZigBee.

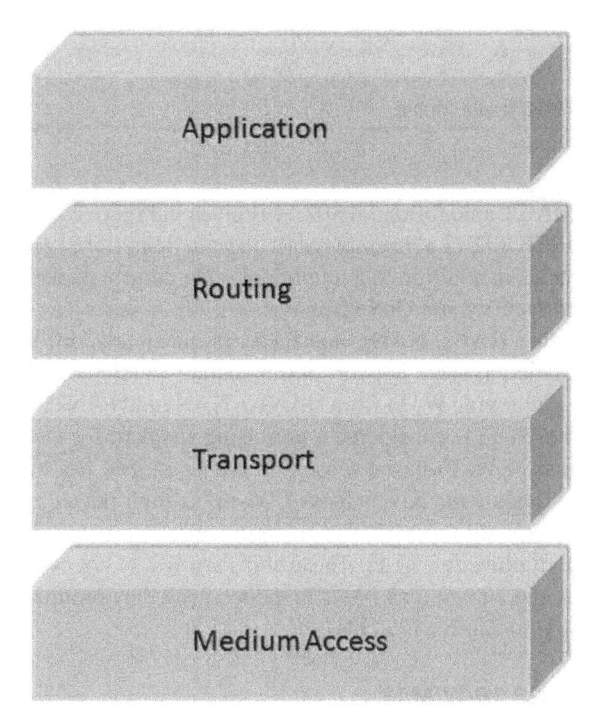

Figure 25.7 ***Z*-wave protocol stack.**

25.3.1.3 *IEEE 802.11/Wi-Fi*

Wi-Fi is based on the IEEE 802.11 set of standards that support point-to-point and point-to-multipoint communications. The data rate of IEEE 802.11 standards range from 1 Mbps to 100 Mbps. 1 Mbps is offered by IEEE 802.11b and 100 Mbps is offered by the recent IEEE 802.11n standard. The IEEE 802.11 standard family defines the physical and MAC layers and the standards inherently support IP addressing. 802.11 operates in the 2.4 GHz ISM band. The standard utilizes two different physical layer specifications, frequency-hopping spread spectrum (FHSS) and direct sequence spread spectrum (DSSS). FHSS separates the 2.4 Ghz band into subchannels, which are spaced 1 MHz apart. The transmitter changes channels at least 2.5 times per second based on the predefined three sets of sequences in the standard. DSSS is a more advanced channel utilization technique also employed in CDMA. DSSS multiplies the data with a chip sequence and transmits this after employing either differential binary phase shift keying (DBPSK) or differential quadrature phase shift keying (DQPSK) modulation. To increase the data rates of DSSS 8-chip complementary code keying (CCK) can be employed as the modulation. This is known as high rate direct sequence spread spectrum (HR/DSSS) in the standard. Recently, 802.11n has been able to reach data rates of 100 Mbps by the employment of multiple-input multiple-output (MIMO) techniques.

Octets 2	2	6	6	6	2	6	2	0-2312	4
Frame Control	Duration/ID	Address 1	Address 2	Address 3	Sequence Control	Address 4	QoS Control	Frame Body	FCS

Figure 25.8 Wi-Fi MAC frame format.

The general MAC frame format of 802.11 is given in Figure 25.8[21]. Apart from this data frame, IEEE 802.11 also employs Request to Send (RTS) and Clear to Send (CTS) control frames in addition to a number of other control packets. The standard also has advanced security and QoS settings.

Wi-Fi is targeting HANs, NANs, and FANs in the smart grid. Wi-Fi is already being used for municipal-scale network infrastructures outdoors, with approximate ranges of 500 m. Employing Wi-Fi from HANs to NANs and FANs increases interoperability; therefore Wi-Fi is considered a promising standard for the smart grid. For instance, utilization of Wi-Fi-based sensors in the smart grid has been studied in a recent work [22]. Despite the advantages of Wi-Fi, its high-power consumption has remained a drawback until recently. The emerging ultra-low-power Wi-Fi chips may provide new opportunities for Wi-Fi. These chips are 802.11 b/g compatible with bit rates of 1–2 Mbps and with ranges 10–70 m indoors, and they promise multiple years of operation similar to ZigBee [23, 24].

25.3.1.4 *IEEE 802.16/WiMAX*

Worldwide Interoperability for Microwave Access (WiMAX) is based on the IEEE 802.16 standard developed for broadband wireless access for fixed and mobile point-to-multipoint communications. The protocol stack of WiMAX consists of PHY and MAC layers [25], and a Generic Packet Convergence Sublayer (GPCS) as seen in Figure 25.9. The licensed bands of 10–66 GHz provide the physical environment for 802.16. Channel bandwidths of 25 MHz or 28 MHz are typical for 802.16. In those frequencies line-of-sight (LOS) is required due to short wavelength. The standard also allows the use of license-exempt sub-11 GHz bands. WiMAX can provide theoretical data rates up to 70 Mbps and the range is around 50 km for fixed stations and almost 5 km for mobile stations [26]. Recent versions of the standard also include MIMO antenna technology and beamforming and Advanced Antenna Systems (AASs) features to achieve high data rates. The IEEE 802.16 MAC layer employs a scheduling mechanism and an additional sleep control mechanism that allows reduction of power consumption.

WiMAX was initially intended for wireless metropolitan area networks (WMANs); therefore it is one of the promising solutions for smart grid WANs. The recent version of the IEEE 802.16 standard defines Mobile WiMAXm which is sometimes confused with cellular communications and referred to as 4G. Although mobile WiMAX and 4G cellular technologies have similarities and they experience similar challenges, they are based on different standards and their implementation has significant differences.

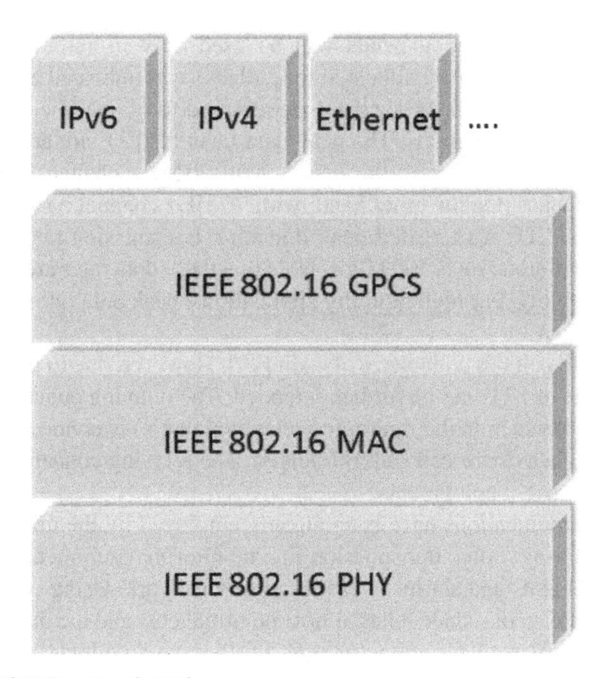

Figure 25.9 WiMAX protocol stack.

25.3.1.5 *Cellular Communication Technologies*

Cellular communications emerged in the 1980s with the first generation analog phones that were followed by the second generation (2G) standards in 1990s. GSM, IS-36, and IS-95 are the 2G standards deployed for the licensed 1.9 GHz band. These standards only included voice communications. Only after 2.5G standards such as GPRS and EDGE were deployed did it became possible to transmit data by cellular networks. Later, third generation (3G) standards emerged to provide higher data rates and roaming capabilities, however the 3G standards, i.e., cdma2000 and W-CDMA are not compatible [27]. The most recent technology on the cellular communications side are the Long Term Evolution (LTE) and LTE Advanced (LTE-A) standards. They are standardized by the 3rd Generation Partnership Project (3GPP).

LTE and LTE-A have flexible channel bandwidths. LTE-A will be using the bands given below, in addition to the bands already allocated to LTE [28]:

- 450–470 MHz band (allocated for global use)
- 698–862 MHz band
- 790–862 MHz band
- 2.3–2.4 GHz band (allocated for global use)
- 3.4–4.2 GHz band
- 4.4–4.99 GHz band

Some of the above frequency bands will be used only in certain countries and regions while some of them are allocated for global use as indicated above. The 450–470 MHz band and the 2.3–2.4 GHz band are reserved for global use. The rest is specific to certain countries/regions. The peak data rates for LTE are around 300 Mbps at the downlink and 80 Mbps at the uplink with 20 MHz channel bandwidth and 4×4 MIMO antennas. On the other hand, with 70 MHz channel bandwidth and 4x4 MIMO antennas, LTE-A's targeted peak downlink transmission rate is 1 Gbps and the uplink transmission rate is 500 Mbps. Note that these data rates are expected to be valid for low mobility. For high mobility applications, peak data rates will be around 100 Mbps in LTE-A.

A typical LTE cell has a diameter of 4 km [29]. However, due to the relaying feature the coverage of LTE can be further extended. The relaying concept is presented in Figure 25.10. As seen in the figure, the relay node (RN) is connected to the radio access network via a donor cell's eNB (DeNB). The relaying concept is specifically useful for low density deployments.

Cellular communications have been already employed by the utilities for power grid asset monitoring. Some transmission line monitoring equipment has the ability to send measurements and alarms over the cellular network. Using cellular communications is advantageous since it has almost no initial cost and the data are transmitted from the readily available infrastructure. Furthermore, cellular communications have advanced security mechanisms.

By the introduction of LTE and LTE-A, cellular networks will be capable of carrying the high volume of smart meter data, in addition to the data of phasor measurement units (PMUs). As an example, Qualcomm Inc. and Echelon Inc. will be using Verizon's and T-Mobile's infrastructures, respectively, to provide cellular connectivity for smart meters. Besides utilization of cellular networks in the smart grid domains, the PHEV charging infrastructure can also use cellular networks. The drawback of cellular networks is their poor performance under emergency conditions. Communication outages are known to occur during natural disasters or terrorist attacks even without the additional data load of the smart grid. In such circumstances, it would not be possible to deliver critical smart grid data or commands unless prioritization of data or a backup network is implemented.

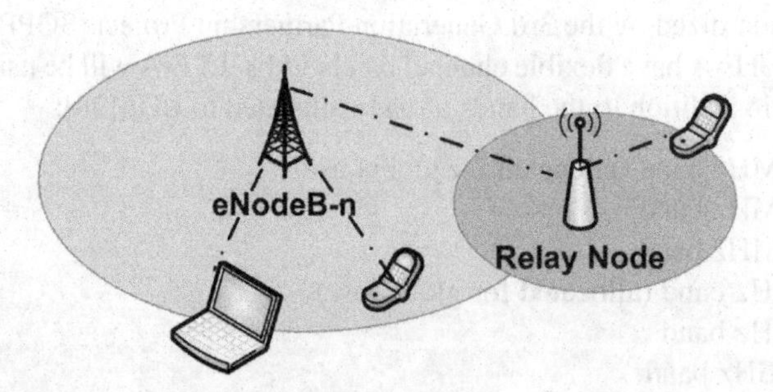

Figure 25.10 The relaying concept of LTE.

25.3.1.6 IEEE 802.22/Cognitive Radio

The IEEE 802.22 standard is a recently emerging IEEE standard for point-to-multipoint wireless regional area networks. It facilitates the use of cognitive radio for opportunistic access to white space in TV bands. Cognitive radio (CR) provides access to unlicensed users to the spectrum that is not utilized by licensed users. A CR has the ability to sense unused spectrum, use it, and then vacate as soon as a licensed user arrives as seen in Figure 25.11. IEEE 802.22 consists of a base station (BS) and customer premise equipment (CPE). The BS establishes the medium access control together with cognitive functions, while the CPE performs distributed sensing of the signal power in various channels of the TV band [30, 31]. The BS decides whether a band is unused or not based on the measurements collected by the CPE. The 802.22 standard plans to use the UHF/VHF bands between 54 and 862 MHz and their guard bands. The range of the IEEE 802.22 standard is considered to be between 33 km and 100 km [32] with approximately 19 Mbps rates over several tens of meters.

IEEE 802.22 is suitable for smart grid WANs, NANs, and FANs. It provides low cost communication infrastructure since it does not require purchasing licensed spectra. Electricity transmission assets can be an especially ideal place for cognitive radio since they are usually implemented in unpopulated areas, hence the TV bands may be unused. Additionally, the wide coverage of 802.22 can provide opportunities for higher connectivity than the exiting wireless communication standards [32]. Meanwhile the wide range property may be also useful for NANs in rural areas where the low number of customers are distributed in large regions.

There are several other wireless technologies that may be used in the smart grid. RF technology is already employed by several utilities for asset monitoring. Pacific Gas and Electric (PG&E) has a smart metering solution where the smart meters transmit their data to an access point via the multihop RF mesh network. The access point uses the cellular network to reach the utility headquarters. The mesh network ensures connectivity in case of node or link failures since data can be routed over alternate paths, while the cellular network enables secure and reliable connectivity for the aggregated data [34]. Other wireless communication alternatives are satellite

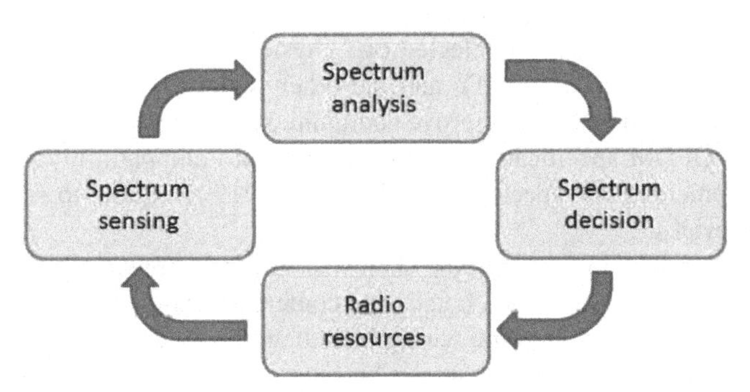

Figure 25.11 Spectrum sensing, allocation, and decision in CR [33].

and free space optical communications. Satellite communications can provide condition monitoring or serve as a backup communication environment especially in case of massive disasters that destroy the infrastructure. However, the high delay of satellite communications is not practical for general smart grid applications although some utilities have been reported to use it for remote substation monitoring [35]. On the other hand, free space optical communications offer high bit rates with low bit error rates however it requires LOS which may be hard to maintain in some places due to obstacles.

25.3.2 Wired Communication Technologies

Wired communications can either use power line or external wiring for transmitting signals. We group the available technologies under two categories: power line communications and wireline communications.

25.3.2.1 Power Line Communications

Power line communications (PLCs) use the low voltage power lines as a communication medium for data. PLC has been already used by some utilities for load control and remote metering. It is simply integrated to the smart metering system since the power lines already reach the meter [36]. As the PLC does not have external cabling cost, it is considered to be convenient for HANs, NANs, and FANs [37].

PLC technology has been initially operating in the narrowband, and it has evolved into broadband according to the needs of consumer applications. Broadband PLC systems operate in the 230 MHz band and can achieve data rates up to 200 Mbps [15]. There are a large number of PLC standards that have evolved in time. Presently, three standards have been selected by NIST for possible adoption in the smart grid. These standards are IEEE P1901, ITU-T G.Hn, and ANSI/CEA 709 [13].

IEEE P1901/broadband over power lines: IEEE P1901 is a broadband over power lines (BPLs) standard that has been recently completed. BPL has high data rates exceeding 100 Mbps using frequencies below 100 MHz. Although it can operate around 30 MHz and achieve higher data rates, high attenuation at low frequencies limits the range of communication, therefore it is not preferred.

The P1901 workgroup has selected two physical layers for the standard. One is the wavelet OFDM-based PHY and the other is the FFT OFDM-based PHY. These PHY techniques aim to improve communication over the noisy power lines, and the FFT-OFDM specification facilitates backward compatibility with devices using the HomePlug AV specification [15]. IEEE P1901 is able to support more than 2000 devices.

ITU-T G.hn: ITU-T developed the G.hn standard for communication in residential premises, small-scale offices, hotels, and conference rooms [38]. G.hn is able to operate over all types of in-home wiring including phone line, power line, coaxial cable, and Cat-5 cable. G.hn can support bit rates up to 1 Gbps. It uses a windowed OFDM-based PHY with a programmable set of parameters. These parameters give the G.hn standard the flexibility to operate over multiple media. G.hn supports up to 250 nodes. G.hn devices aim to be interoperable with power line devices that use the IEEE P1901 standard.

ANSI/CEA-709: The ANSI/CEA-709 series of standards have been developed for home control and automation. The ANSI/EIA 709.1 standard, which is also known as Lonworks, became an international standard in 2008. The Lonworks platform is a family of hardware components and software tools that are based on these standards. The platform is a proprietary technology of Echelon Corp. and it is relatively an old platform that has been competing with BACnet on building automation and control since the mid-1990s. Lonworks operates in the 115–132 MHz band and it can support up to 32,000 nodes [39]. The data rates of Lonworks can reach up to a few kbps. ANSI/CEA-709.1-B is the control network protocol specification that covers the specifications from data link layer to application layer, and ANSI/CEA-709.2 covers power line channel specifications. In the smart grid, since control networks can be used for home energy management, NIST has included Lonworks as a candidate standard along with IEEE P1901 and ITU G.hn.

The primary application of PLC is considered to be spanning smart grid HANs, NANs, and FANs. Nevertheless, PLC technologies can also operate over high voltage lines, which position them as a candidate in electricity transmission system networks, as well. As discussed in [15], PLC technologies can operate over 1100 kV lines in the 40–500 kHz band and achieve data rates of a few hundred kpbs. However, there are several challenges with PLC technology such as impaired, time-varying channel conditions, and background noise [40]. Furthermore, the attenuation of the signal propagating from the substations to customer premises has been shown to include delay effects that are cyclic in nature. This phenomena may cause inefficiencies in coding schemes as indicated in [41, 42].

25.3.2.2 Wireline Communications

Fiber optical communications: Fiber optics have been used in the power grid to connect utility head offices and substations. Due to the reliability of the fiber technology, it is expected to be used in the smart grid as well. Furthermore, fiber optics is not impacted by electromagnetic interference; hence it is ideal for the high voltage operating environment. The major drawback of fiber is that it has high deployment cost. However, the unused capacity of the already deployed fibers can be utilized for some smart grid applications in order to reduce deployment costs.

Ethernet, a well-known data communications standard, can also be utilized in the smart grid. For instance, communications with substations can be provided with Ethernet over SONET when there is need for continuous streaming of substation data. It is also possible to employ a combination of the wireless and wired communication technologies to benefit from the strengths of individual standards. For instance, in the course of ZigBee SEP, the ZigBee Alliance and the HomePlug Alliance are collaborating to provide energy management in large apartments and commercial buildings. Moreover, in March 2010, the ZigBee Alliance and the Wi-Fi Alliance agreed to collaborate on wireless HANs for smart grid applications.

Table 25.1 Comparison of the Smart Grid Communication Standards

Standard	Max. Data Rate	Spectrum	Range	Max. Nodes	Scope
IEEE 802.15.4	250 kbps, 100 kbps, 40 kbps, 20 kbps	2.4 GHz, 915 MHz, 868 MHz	30–100 m	64,000	HAN
IEEE 802.11	100 Mbps	2.4 GHz	100 m	unlimited (theoretically)	HAN, NAN, FAN
Z-wave	40 kbps, 9.6 kbps	908 MHz	30 m	232	HAN
IEEE 802.16	700 Mbps	10–66 GHz	5–50 km	<100 active users per cell	NAN, FAN, WAN
LTE-A	1 Gbps (DL)–500 Mbps (UL)	(see page 640)	4 km	200 active users per cell	NAN, FAN, WAN
IEEE 802.22	19 Mbps	54–862 MHz	10 km–100 km	unlimited (theoretically)	NAN, FAN, WAN
IEEE P1901	100 Mbps	<100 MHz	1.5 km	>2000	HAN, NAN, FAN
ITU-T G.hn	1 Gbps	80 MHz-200 MHz	1–3 km	250	HAN, NAN, FAN
ANSI/CEA-709	78 kbps, 1.25 Mbps	115–132 MHz	<3 km	32,000	HAN, NAN, FAN

We provide a comparison of the wireless and power line smart grid communication technologies in Table 1. We compare the maximum data rate, operation frequency, range, maximum number of supported nodes, and their application scope.

25.4 COMMUNICATION-ENABLED SMART GRID APPLICATIONS

Communication technologies provide opportunities for load control, demand management, and electric vehicle-related smart grid applications. Load control and demand management applications aim to reduce the energy consumption of residential customers during peak hours. A residential customer premise is basically a HAN containing appliances, thermostats, light switches, electrical outlets, and other energy consuming goods. Load control or demand management applications try to reduce the electricity consumption of those appliances by either turning off some of them or shifting their demands to another time slot where there is less demand.

Direct load control (DLC) refers to passing the control of several appliances to the utility or an aggregator. In practice, not all appliances can be controlled remotely by the utility since it may cause too much inconvenience for the customers. The appliances that can be remotely controlled can be pool pumps and the heating/cooling appliances. In fact, a pilot study in Australia has showed that cycling air conditioners have resulted in 17% peak load reduction [43]. DLC requires simple communications between the consumers and the utility. Utility commands can be delivered to the customers through smart meters. Therefore, ZigBee or one of the PLC standards can be a suitable option for DLC. Although DLC is favored for its simplicity, even with a limited set of appliances, it may be intrusive and cause consumer dissatisfaction in some circumstances. For this reason, nonintrusive techniques have recently been explored.

Wireless sensor network (WSN)-based demand management has been proposed in [18] as a nonintrusive, interactive demand management scheme. The authors have employed a central energy management unit (EMU), a WSN, and smart appliances for a residential demand management plan called in-Home Energy Management (iHEM). EMU and appliances communicate wirelessly over the WSN. The WSN relays the packets using ZigBee as shown in Figure 25.12. The interactive appliance coordination mechanism of iHEM is similar to the schemes in [44, 45]. iHEM aims to shift consumer demands to the time slots when electricity usage is less expensive, i.e., off-peak hours of the TOU tariff. Unlike DLC, iHEM suggests convenient start times for the appliances and the suggested times can be easily overwritten by the consumers. By this mechanism, iHEM can manage consumer demands in a nonintrusive way.

iPower, an intelligent and personalized energy conservation system by wireless sensor networks, implements an energy conservation application for multidwelling homes and offices by using the context-awareness of WSNs [46]. iPower includes a WSN with sensor nodes and a gateway node, in addition to a control server, power line control devices, and user identification devices. Sensor nodes are deployed in each room and they monitor the rooms with light, sound, and temperature sensors. They form a multihop WSN and send their measurements to the gateway when an event occurs, i.e., their measurements exceed a certain threshold. The gateway

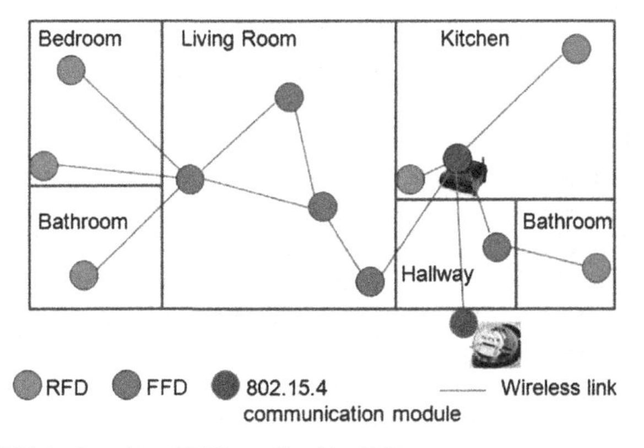

Figure 25.12 WSN deployed for iHEM application [18].

node is able to communicate with the sensor nodes via wireless communications and it is also connected to the intelligent control server of the building. iPower combines wireless and power line communication technologies. It uses ZigBee for WSN communication and X10 for power line communications. An intelligent control server performs energy conservation actions based on sensor inputs and user profiles. The action of the server can be turning off an appliance or adjusting the electric appliances in a room according to the profiles of the users who are present in the room. Server requests are delivered to the appliances through their power line controllers. iPower is an example of implementations using multiple communication technologies. Presently, iPower works independent of the smart grid; however, it can be integrated into the smart grid ecosystem when TOU signals from the smart meter are involved in server decisions.

Sensor web services for energy management: Using sensor web services for smart grid energy management has recently been studied in [47]. The authors have assumed a smart home containing smart appliances with sensor modules that enable each appliance to join the WSN and communicate with its peers. The energy management application is a suite of three energy management modules. The first module enables users to learn the energy consumption of their appliances while they are away from home. The current drawn by each appliance is monitored by the sensors on board and this information is made available through a home gateway to the users. Users access the gateway from their mobile devices using web services. The second application is a load shedding application for the utilities. Load shedding is applied to the air conditioning appliances when the load on the grid is critical. Load shedding is simply turning off an appliance or modifying the set point of the appliance (e.g., HVAC) in order to eliminate or reduce its load. The utility uses sensor web services to deliver commands. In addition to monitoring and load shedding applications, the third module offers an application for energy generating customers. By this energy management application, customers can monitor and control the amount of energy stored and energy sold back to the grid while they are away from home.

Machine-to-machine (M2M) communications-based demand management: M2M communications have been implemented in the Whirlpool Smart Device Network (WSDN) [48]. WSDN consists of three networking domains which are the HAN, the Internet, and AMI. Similar to iPower, WSDN utilizes several wired and wireless technologies together, including ZigBee, Wi-Fi, broadband Internet, and power line carrier (PLC). Wi-Fi connects the smart appliances and forms the HAN. ZigBee and PLC connect the smart meters in the AMI, and the broadband Internet connects consumers to the Internet. The WSDN energy management module, which is called Whirlpool Integrated Services Environment (WISE), enables remote access to appliance energy consumption. WISE also provides load shedding capabilities to utilities during critical peaks.

Energy-saving applications on appliances have been studied by Tompros et al. in [49]. The authors have investigated an appliance-to-appliance communication protocol for energy-saving applications. They have proposed an energy management protocol that allows consumers to set a maximum consumption value while the residential gateway is able to turn off the appliances that are in standby mode once these limits are exceeded.

In addition to conventional appliances, electric vehicles will be plugged into the smart grid for charging, which will increase the load, most probably during peak hours and in populated locations. In the rest of the section, we focus on the communication-enabled electric vehicle charge control applications.

Electric vehicle demand management: Electric vehicles are anticipated to have a huge impact on the electric load as they are going to be charged from the power grid, and they are expected to modify the daily energy consumption profile either by adding to the evening peak or by filling the off-peak hours overnight.

The impact of PHEV loads on different hours of the day has been studied in [50], and the authors have shown that uncontrolled charging increases the load during peak hours. Increased load during peak hours may have severe impacts on the power grid. In [51], the authors investigated the impact of PHEV charging on the distribution system and showed the relation between feeder losses, load factor, and load variance. These studies have shown that the ideal situation would be distributing PHEV loads evenly through overnight hours but this needs some form of coordination and calls for novel demand management schemes.

There are several recent studies that address electric vehicle demand management. In [52], the authors have proposed controlling the loads of PHEVs via local control and global control. In the local control scenario, PHEV charging is scheduled based on the other loads in a house whereas in the global control scenario, charging is scheduled by considering the requests of the other PHEVs in a neighborhood.

In [53, 54], the authors have proposed a communication-based PHEV load management scheme where the utility provisions a certain amount of energy for each distribution system based on the predicted supply level and the current grid status. The provisioned energy is communicated to the substation where each charging request is either accepted or rejected based on the utility set limits. These decisions are then sent to the smart charging stations. A charging demand is repeated in the next time interval, and if the PHEV load can be supplied from the distribution system at that time, it is admitted. The smart grid distribution system is illustrated in Figure 25.13. Communication among the utility, substation, and the smart chargers is implemented via heterogeneous wireless communications technologies, i.e., IEEE 802.16, IEEE 802.11s, and IEEE 802.15.4. These schemes have been extended in [55] to include quality of service in the PHEV charging infrastructure.

In [9], the authors have employed a Home Gateway and Controller (HGC) device that communicates with the PHEV and controls its charging and discharging profile according to the status of the rooftop solar power generation unit and demands of the smart appliances. Moreover, to avoid the resilience problems due to simultaneous charging in a neighborhood, HGC also communicates with the other HGC devices in the neighborhood and coordinates PHEV loads. HGC uses ZigBee for HAN and 802.11 for NAN communications.

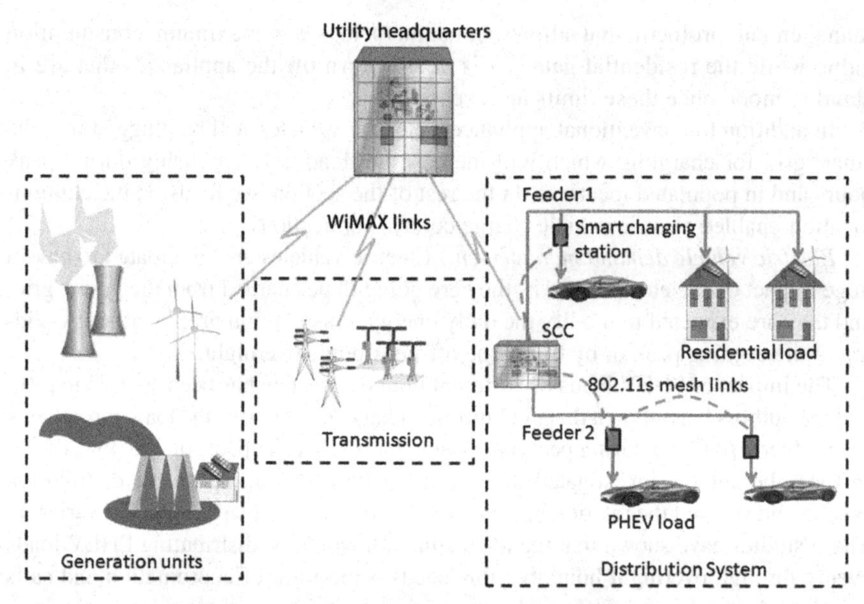

Figure 25.13 PHEV charging infrastructure for the smart grid distribution system.

25.5 CHALLENGES IN SMART GRID COMMUNICATIONS

The existing telecommunication networks were not initially designed for the smart grid. Their scalability, availability, and reliability under strict smart grid operation conditions need to be further assessed.

Scalability is expected to become an issue since the number of smart meters will be almost equivalent to the total sum of homes and businesses, which corresponds to a vast number of devices. Meanwhile future smart grid applications may require frequent data transmissions from the smart meters with intervals less than a minute or even a few seconds. The communication infrastructure should be capable of handling the high volume of data. Thus, the communication between these devices and the utility requires scalable solutions [56]. Furthermore, availability is a major concern for emergency situations such as disasters or major outages. Smart grid communications need to be available even though the physical grid components fail to operate properly. Finally, the communication infrastructure needs to be reliable so that control decisions are based on reliable grid status information as well as control commands being delivered reliably to the subject of control. Security is also another essential property of smart grid communication technologies. We will discuss the security and privacy aspects of smart grid communications in the next subsection.

Despite the above-mentioned desired attributes, communication technologies that are available so far already have several drawbacks that need to be addressed. The major drawbacks of wireless technologies can be summarized as follows:

- They are prone to interference due to the populated ISM bands.
- They have lower bandwidth than wired communication technologies.
- They do not penetrate well through concrete construction.
- Their range is limited unless they consume a high amount of transmit power.
- The impact of harsh smart grid environment conditions on wireless communications is not explored well.

The link quality of a wireless sensor network has been investigated in [57] and it has been reported to be poor. The performance of other wireless communication technologies needs to be evaluated as well. On the other hand, power line communications suffer from [40]

- Noisy channel conditions.
- Channel characteristics that vary depending on the number and the type of appliances plugged in.
- Electromagnetic interference (EMI) due to unshielded power lines.
- Poor isolation among units where signals of one unit may interfere with another.

Besides the above-mentioned challenges, most of the smart grid applications are anticipated to demand prioritization and application-awareness [58]. Furthermore, power systems generally have tight latency requirements. Most of the decisions are in real time where delayed information may lead to less efficient operation or wrong decisions [59, 60]. For this reason, variable latency has to be addressed by the transport protocols, as well as the decision and control algorithms.

25.5.1 Security and Privacy

ICTs are the primary enablers of the smart grid while they also increase the vulnerabilities of the grid [61]. These vulnerabilities may allow attackers to easily access the smart grid, manipulate the internal operation procedure, and finally destabilize the grid. Such malicious attacks can initiate from various parts of the smart grid such as the AMI, demand management, the electric transportation infrastructure, the electric storage subsystem, the wide area situational awareness component, and the distribution automation subsystem. A single attack may impact one or more of these subsystems. The potential risks that have been introduced by the smart grid have been identified by NIST as follows [61]:

- The increased complexity could cause device misconfiguration based errors.
- The increased number of interconnections increases the risk of denial of service attacks, injection of malicious software, and compromised hardware.
- The increased number of network nodes increases the number of entry points and paths that might be exploited by adversaries.

- The increased amount of data increases the risk of compromising data and violating user privacy.

Being a cyber-physical system, smart grid security has two different aspects: physical security and cyber security. Physical security is related to the security of assets such as towers, substations, transformers, feeders, and so on, in addition to the physical layer security of the communications [31]. Physical vulnerability of the power grid to disasters and sabotages exists in the conventional power grid. In fact, an investigation in the 1990s discovered that natural disasters such as hurricanes or terrorist attacks may bring a utility grid down easily since these grids were usually prepared for only a single or few simultaneous failures [62].

Physical layer security of the communication technologies involves coding techniques and data encryption. Wireless communication technologies are easy to eavesdrop while PLC is broadcast in nature and tampering to power lines may also enable cyber attacks. In order to secure communications, spread spectrum techniques and encryption can be employed. However, encryption techniques need to be lightweight [58]. Smart meters and several consumer devices may have limited resources for storing cryptography-related information. In addition, limited sources of entropy can be an issue for some smart grid devices and they may not be capable of seeding random number generators properly which results in poor key generation. Secure random number generation from limited entropy is known to be a challenging issue where use of external hardware as a source of random bits for seeding such generators may be required [61]. Furthermore, each device should have unique credentials so that breaking in at one point should not allow breaking in from several other points. In this case, key management of millions of smart meters and the devices in HANs becomes a challenging issue. Thus, the approaches adopted for ensuring security and privacy need to be lightweight. These policies should be easily implementable even on resource-constraint devices, e.g., sensors. Furthermore, these policies should not introduce significant delay as the information may became stale in the highly dynamic environment of the smart grid.

Time accuracy is another security-related significant issue for physical smart grid equipment. The certificates used for binding keys to identities contain validity periods, which implies that the system time of each device must be correct so that a valid certificate is not mistaken as invalid or vice versa [63]. Similarly, incorrect time stamps of synchrophasors may give incorrect status information and lead to incorrect decisions that could even destabilize the grid.

Besides physical security issues, the cyber security dimension is becoming more complicated in the smart grid than it is in the existing power grid. Utilities have been using computers and software to monitor and control their assets for decades, and they have been experiencing cyber attacks even before the smart grid. However, the impact and the frequency of those attacks may increase in the smart grid because the smart grid will interconnect all the systems digitally. The smart grid may be more vulnerable than the existing grid if security against physical and cyber attacks is not addressed in the first place. For instance, Amin reports an incident in 2003 where a nuclear power plant in Oak Harbor, Ohio was infected by an SQL server worm disabling a safety monitoring system for almost five hours [64]. Also, in the same article Amin states that in January 2008, the CIA reported that hackers were able to disrupt (or threaten to disrupt) the power supply for four foreign overseas cities.

The increasing level of automation in the distribution system may also increase cyber security concerns. There are approximately 17,325 transmission substations in the United States and Canada that contain devices such as circuit breakers, power transformers, phase-shifting transformers, capacitor banks, and disconnect switches, in addition to automation and communication devices used to measure, monitor, and control the substation components [65]. Defining secure access mechanisms for the large number of devices is highly challenging. Authentication is particularly critical for substation equipment where only authorized users should gain access to configure or operate the equipment. Smart grid authentication schemes cannot rely on the assumption of a constant network connection to a central office. Thus, the smart grid requires secure authentication methods that can work locally in addition to optional revocation of a central authority [63]. Meanwhile, these authentication schemes should be able to respond fast in emergency situations.

Spoofing, jamming, and data or command alteration attacks may be observed in the smart grid [66] in addition to cyber attacks that result from involuntary customer acts. For instance, misconfiguration of smart grid integrated consumer devices may provide means to data modification attacks [67]. These types of attacks may work in two ways. Modified consumer data can be generated and transmitted to the utility or modified utility control signals may be sent to the consumers, both of which cause inconvenience for the customers and the grid operators. Furthermore, Internet-based load alteration attacks may take place by compromising the direct load control command signals, compromising indirect load control price signals, or penetrating load distribution algorithms [68]. Several network-related security vulnerabilities have been identified by NIST [63] as follows:

- Lack of integrity checking for communications.
- Failure to detect and block malicious traffic in valid communication channels.
- Inadequate network security architecture.
- Poorly configured security equipment.
- Having no security monitoring on the network.
- Failure to define security zones.
- Inadequate firewall rules or improperly configured firewalls.
- Critical monitoring and control paths are not identified.
- Inappropriate lifespan for authentication credentials/keys.
- Inadequate key diversity.
- Authentication of users, data, or devices is substandard or nonexistent.
- Insecure key storage and exchange.
- Lack of redundancy for critical networks.
- Inadequate physical protection of network equipment and unsecured physical ports.

- Noncritical personnel having access to equipment and network connections.

Equally important as security, privacy issues need to be carefully studied in the context of the smart grid. Privacy is a broad concept covering privacy of personal information, privacy of the person, privacy of personal behavior, and privacy of personal communications. Privacy of personal information includes the information on one's physical, mental, economic, cultural, locational, or social conditions that can be associated with the identification of a person. Privacy of the person refers to the physical requirements or health problems of an individual that is related with the integrity of one's body. Privacy of personal behavior is related with one's activities and choices. Finally, privacy of personal communications is the right to communicate without being monitored or censored [69].

In the smart grid, electricity consumption data can be obtained in detail and consumer privacy may be violated if high resolution data is made available to malicious users. In [70], the authors have shown that it is possible to obtain a detailed picture about the activities in a dwelling, building, or other property including absence or presence, the number of individuals in the property, sleep cycles, meal times, and shower times. Meanwhile, a PHEV's location can be easily tracked from its charging locations. Sophisticated attacks may benefit from data leakage from consumer premises and reveal the properties of some consumer products to competitors as well. For example, information on electric vehicle performance can be sought-after by manufacturers. Briefly, a rich data set can be obtained in the case of data leakage. Furthermore, mesh networking in the AMI may also raise privacy concerns as the data of smart meters may be routed by other smart meters. In the future, smart grid communications are expected to be pervasive and the amount of information and associations may be more than is anticipated now and could be subject to misuse.

In addition to the traditional players, the smart grid will also involve digital service providers. For instance, Google has developed the Powermeter application that collects consumption data from participating users and displays certain statistics to the owner of the data [71, 72]. The powermeter application is based on user subscription, however similar applications may be offered using utility-supplied data that does not allow individuals to opt out and may raise serious privacy concerns. In fact, in some countries like Canada, regulations do not allow even the utilities to access consumer data in real time. The independent system operator supplies consumption data to utilities for billing purposes on a next-day basis. On the other hand, such restrictions may delay the development of demand management applications since real-time consumption information is not available to the utilities. Ownership of energy data is still under debate and various regulators and jurisdictions have different policies.

Besides data leakage from customer premises, leakage from utility networks is also possible. These may include customer data or utility management data including direct load control decisions, pricing algorithms and forecasting techniques [67]. Furthermore, more critical information such as the status of assets could be compromised.

For instance, attackers may be looking for overloaded transformers to deploy their load altering attacks. A transformer operating close to its limits will be more vulnerable to excessive load than a transformer operating in a normal range. An attack that initiates excessive demand may be simply able to bring the whole distribution system down when implemented on the overloaded transformer. Furthermore, in the electric vehicle ecosystem aggregators are expected to play a major role, specifically for regulation services. Data modification attacks over aggregator networks may disrupt the regulation service, which in return may have a massive impact on every component of the smart grid including generators and end-user appliances. If regulation is not handled properly due to cyber attacks, grid frequency may behave unexpectedly. When the frequency is low, generators may interpret this as insufficient supply and increase their power outputs whereas if the frequency is high then this is interpreted as a surplus of power. In either case, malfunctioning of the regulation services will impact power quality and the health of the equipment in the smart grid. Furthermore, devices on consumer premises are relatively easier to compromise and the utility has little or no control over these devices. Therefore, attackers may extract data from the memory of these devices including keys used for network authentication and insert malicious software that could spread to other devices in the AMI [73].

As a matter of fact physical and cyber attacks have been occurring on the existing grid and they will happen in the smart grid as well. They may even become more sophisticated over time. Despite all efforts, it may not be possible to avoid all attacks. For these reasons, it is essential to examine the grid's response to the attacks. Locating and isolating compromised equipment in addition to self-healing are significant concepts in improving the immunity of the smart grid. Furthermore, smart grid equipment is expected to have longer life cycles than most electronic equipment, and they also require longer testing durations due to strict reliability constraints. Thus, any security measure implemented on a smart grid device should be upgradable.

25.6 SUMMARY AND OPEN ISSUES

Information and communication technologies (ICTs) are becoming integrated to the electrical power grid in order to improve the grid's reliability, efficiency, and security, and reduce its environmental impact. The smart grid is in its infancy. Future applications for the smart grid that are not foreseen today may become available, and the communication infrastructure designed and implemented today should be ready to support them.

Available communication technologies can be considered as foundations for yet-to-emerge smart grid communication technologies that will truly answer the needs of the smart grid. In this chapter, we have provided an overview of the available wireless and wired communication technologies and discussed their applicability to various smart grid domains, i.e., home area network (HAN), neighborhood area network (NAN), field area network (FAN), and wide area network (WAN). Among the studied technologies, IEEE 802.15.4/ZigBee, Z-wave, IEEE 802.11/Wi-Fi, and power line communications (PLCs) can be used for HANs. PLC can be implemented according to the IEEE P1901, ITU-T G.hn, or ANSI/CEA-709 standards. NANs and FANs have similar coverage where they both refer to regions covering two or more houses that are fed by the same distribution system. NAN is mostly used for the communication among homes while FAN refers to communication among utility equipment such as transformers, circuit breakers, capacitor banks, disconnect switches, and smart meters. The network that is used for smart meter data delivery is known as the Advanced Metering Infrastructure (AMI). These networks may employ PLC, Wi-Fi, WiMAX, LTE/LTE-A, fiber optics, and Ethernet in some segments. Finally, WAN refers to the segment covering electricity transmission equipment, remote distribution systems, and their connection with the utility head offices. WAN communication can be implemented using long range technologies such as WiMAX, LTE/LTE-A, cognitive radio (IEEE 802.22), and fiber optics. In order to pick the most suitable standard from the broad range of standards, the advantages and disadvantages of the available standards need to be identified. Then, they should be tailored for the smart grid ecosystem.

In this chapter, we have discussed the challenges of communication technologies as well. Wireless technologies have low cost, can be rapidly deployed, and have the ability to cover remote or hazardous areas. However they suffer from interference and they generally have lower data rates than wired technologies. Among the wired technologies, PLC suffers from noisy and time-varying channel conditions and fiber optics have high deployment cost and the coverage of both of the technologies depends on their ability to physically reach a certain area. A more generic challenge for all types of communications is the security issue. The smart grid specifically requires a high level of security due to being a critical system. The vulnerabilities of the grid may allow attackers to easily penetrate and disrupt electrical services. In the smart grid, security is becoming more complex than the existing grid due to increased complexity, increased number of interconnections, increased number of entry points, and increased amount of data.

Secure communications are particularly challenging to attain for the smart grid, since it will include a large number of resource-constrained devices such as smart meters. Resource-constrained devices call for novel lightweight and yet robust encryption and authentication schemes. These schemes should be convenient during normal operation cycles as well as under emergency situations. Furthermore, these devices may require an additional source of entropy to be able to generate robust cryptographic content, and additionally, their clocks need to be accurate so that the security certificate validation process does not cause any vulnerability. The high resolution data collected by smart meters also raise privacy concerns. It may be possible to obtain private information from the collected energy consumption data. These data may become available from customer premises mostly due to device misconfiguration, however it can be also a result of data leakage from a utility, an aggregator, or another service provider. Apart from consumer privacy, data leakage may affect the operation of the grid by releasing unintended data, which may help attackers to implement high-impact attacks.

Computer security history has proven that even the most secure systems can be broken. Therefore, the ability to locate and isolate the compromised nodes or segments, and the ability to recover from an attack are significant issues that need further exploration in the smart grid. Furthermore, the electric vehicle charging infrastructure also calls for communication technologies, and further exploration is required to determine the particular requirements of networked electric vehicles.

REFERENCES

[1] S.M. Amin, B.F. Wollenberg, Toward a smart grid: power delivery for the 21st century, IEEE Power and Energy Magazine 3 (5) (2005) 34–41.

[2] S. Keshav, C. Rosenberg, How Internet concepts and technologies can help green and smarten the electrical grid, SIGCOMM Computer Communications Review 41 (1) (2011) 109–114.

[3] M. Erol-Kantarci, B. Kantarci, H.T. Mouftah, Reliable overlay topology design for the smart microgrid network, IEEE Network, special issue on communication infrastructures for the smart grid 25 (5) (2011) 38–43.

[4] R.H. Lasseter, MicroGrids, in: Proceedings of IEEE Power Engineering Society Winter Meeting, 2002, pp. 305–308.

[5] N. Hatziargyriou, H. Asano, R. Iravani, C. Marnay, Microgrids: an overview of ongoing research, development, and demonstration projects, IEEE Power and Energy Magazine (2007) 78–94.

[6] E. Santacana, G. Rackliffe, T. Le, X. Feng, Getting smart, IEEE Power and Energy Magazine 8 (2) (2010) 41–48.

[7] R. Rajkumar, I. Lee, L. Sha, J. Stankovic, Cyber-physical systems: the next computing revolution, in: Proceedings of ACM/IEEE 47th Design Automation Conference (DAC), June 13–18, 2010, pp. 731–736.

[8] Networking and Information Technology Research and Development Act of 2009 (online). <http://thomas.loc.gov> (accessed April 2011).

[9] M. Erol-Kantarci, H.T. Mouftah, Management of PHEV batteries in the smart grid: towards a cyber-physical power infrastructure, in: Proceedings of Workshop on Design, Modeling, and Evaluation of Cyber Physical Systems (in IWCMC11), Istanbul, Turkey, July 5–8, 2011.

[10] Greenhouse Gas Emissions from Electricity Generation Options. <http://www.hydroquebec.com/sustainable-development/documentation/pdf/optionsenergetiques/pop0106.pdf> (accessed April 2011).

[11] Ontario's microFIT program. (online). <http://microfit.powerauthority.on.ca>

[12] M. Erol-Kantarci, H.T. Mouftah, The impact of smart grid residential energy management schemes on the carbon footprint of the household electricity consumption, in: Proceedings of IEEE Electrical Power and Energy Conference (EPEC), Halifax, NS, Canada, August 2007, pp. 25–27.

[13] NIST Framework and Roadmap for Smart Grid Interoperability Standards, Release 1.0. (online). <http://www.nist.gov/publicaffairs/releases/upload/smartgridinteroperabilityfinal.pdf>(accessed September 2011).

[14] The European Strategic Energy Technology Plan (SET-PLAN). (online). <http://ec.europa.eu/research/energy/eu/policy/set-plan/indexen.htm> (accessed September 2011).

[15] S. Galli, A. Scaglione, Z. Wang, For the grid and through the grid: the role of power line communications in the smart grid, Proceedings of the IEEE 99 (6) (2011) 998–1027 June

[16] M. Erol-Kantarci, H.T. Mouftah, Wireless sensor networks for smart grid applications, in: Proceedings of International Electronics, Communications and Photonics Conference (SIECPC) KSA, April 23–26, 2011.

[17] IEEE 802.15.4 standard. (online). <http://standards.ieee.org/about/get/802/802.15.html> (accessed November 2011).

[18] M. Erol-Kantarci, H.T. Mouftah, Wireless sensor networks for cost-efficient residential energy management in the smart grid, IEEE Transactions on Smart Grid 2 (2) (2011) 314–325.

[19] RFC4919: IPv6 over Low-Power Wireless Personal Area Networks (6LoWPANs). (online). <http://tools.ietf.org/html/rfc4919>. (accessed September 2011).

[20] M. T. Galeev, Catching the Z-wave, EE Times Design, Feb. 2006. (online). <http://www.eetimes.com/design/embedded/4025721/Catching-the-Z-wave>. (accessed September 2011).

[21] IEEE 802.11 standard. (online). <http://standards.ieee.org/about/get/802/802.11.html>. (accessed April 2011).

[22] L.Li; X. Hu, C. Ke, K. He, The applications of WiFi-based wireless sensor network in Internet of things and smart grid, in: 2011 Sixth IEEE Conference on Industrial Electronics and Applications (ICIEA), 21–23 June 2011, 789–793.

[23] Ultra Low Power Wifi Chips of Gainspan Inc. (online). <http://www.gainspan.com>. (accessed October 2011).

[24] Ultra Low Power Wifi Chips of Redpine Signals Inc. (Online). <http://www.redpinesignals.com/Renesas/index.html>. (accessed September 2011).

[25] IEEE 802.16 standard. (Online). <http://standards.ieee.org/about/get/802/802.16.html>. (accessed September 2011).

[26] P.P. Parikh, M.G. Kanabar, T.S. Sidhu, Opportunities and challenges of wireless communication technologies for smart grid applications, Power and Energy Society General Meeting, 2010 IEEE, 25–29 July 2010.

[27] A. Goldsmith, Wireless Communications, Cambridge University Press.

[28] I.F. Akyildiz, D.M. Gutierrez-Estevez, E.C. Reyes, The evolution to 4G cellular systems: LTE-advanced, Elsevier Physical Communication Journal 3 (2010) 217–244.

[29] A. Ghosh, R. Ratasuk, B. Mondal, N. Mangalvedhe, T. Thomas, LTE-advanced: next-generation wireless broadband technology, IEEE Wireless Communications 17 (3) (2010) 10–22.

[30] R. Ranganathan, R.C. Qiu, Z. Hu, S. Hou, M. Pazos-Revilla, G. Zheng, Z. Chen, N. Guo, Cognitive radio for smart grid: theory, algorithms, and security, International Journal of Digital Multimedia Broadcasting 2011 (2011). doi:10.1155/2011/502087.

[31] R.C. Qiu, Z. Hu, Z. Chen, N. Guo, R. Ranganathan, S. Hou, G. Zheng, Cognitive radio network for the smart grid: experimental system architecture, control algorithms, security, and microgrid testbed, IEEE Transactions on Smart Grid, in vol. 2, no. 4, pp. 724–740, Dec. 2011.

[32] A. Ghassemi, S. Bavarian, L. Lampe, Cognitive radio for smart grid communications, in: Proceedings of IEEE International Conference on Smart Grid Communications, Gaithersburg, MD, USA, October 2010.

[33] APOLLO Research Strategy. (Online). <http://www.imec.be/ScientificReport/SR2007/html/1384118.html>. (accessed October 2011).
[34] Pacific Gas and Electric. (Online). <http://www.pge.com>. (accessed September 2011).
[35] X. Fang, S. Misra, G. Xue, D. Yang, Smart grid–the new and improved power grid: a survey, IEEE Communications on Surveys and Tutorials, in press.
[36] S. Galli, A. Scaglione, Z. Wang, Power line communications and the smart grid, in: Proceedings of IEEE International Conference on Smart Grid, Communications (SmartGridComm), 4–6 October 2010, 303–308.
[37] S. Guzelgoz, H. Arslan, A. Islam, A. Domijan, A review of wireless and PLC propagation channel characteristics for smart grid environments, Journal of Electrical and Computer Engineering, 2011, (2011), Article ID 154040, 2011.
[38] V. Oksman, S. Galli, G.hn: the new ITU-T home networking standard, IEEE Communications Magazine 47 (10) (2009) 138–145.
[39] H.C. Ferreira, L. Lampe, J. Newbury, T.G. Swart, Power Line Communications: Theory and Applications for Narrowband and Broadband Communications over Power Lines, John Wiley and Sons.
[40] S. Galli, O. Logvinov, Recent developments in the standardization of power line communications within the IEEE, IEEE Communications Magazine 46 (7) (2008) 64–71.
[41] X. Ding, J. Meng, Channel estimation and simulation of an indoor power-line network via a recursive time-domain solution, IEEE Transactions on Power Delivery 24 (1) (2009) 144152.
[42] R.P. Lewis, P. Igic, Z. Zhou, Assessment of communication methods for smart electricity metering in the UK, in: IEEE PES/IAS Conference on Sustainable Alternative Energy (SAE), September, 2009, pp. 28–30.
[43] Landis+Gyr. (online). <http://www.landisgyr.com/en/pub/home.cfm>. (accessed September 2011).
[44] M. Erol-Kantarci, H.T. Mouftah, Using wireless sensor networks for energy-aware homes in smart grids, IEEE Symposium on Computers and Communications (ISCC), Riccione, Italy, June 22–25, 2010.
[45] M. Erol-Kantarci, H.T. Mouftah, TOU-aware energy management and wireless sensor networks for reducing peak load in smart grids, in: Green Wireless Communications and Networks Workshop (GreeNet) in IEEE VTC2010-Fall, Ottawa, ON, Canada, September 6–9, 2010.
[46] L. Yeh, Y. Wang, Y. Tseng, iPower: an energy conservation system for intelligent buildings by wireless sensor networks, International Journal of Sensor Networks 5 (1) (2009) 1–10.
[47] O. Asad, M. Erol-Kantarci, H.T. Mouftah, Sensor network web services for demand-side energy management applications in the smart grid, in: IEEE Consumer Communications and Networking Conference (CCNC'11), Las Vegas, USA, January 2011.
[48] T.J. Lui, W. Stirling, H.O. Marcy, Get smart, IEEE Power and Energy Magazine 8 (3) (2010) 66–78.
[49] S. Tompros, N. Mouratidis, M. Draaijer, A. Foglar, H. Hrasnica, Enabling applicability of energy saving applications on the appliances of the home environment, IEEE Network (November/December) (2009)
[50] K. Parks, P. Denholm, T. Markel, Costs and emissions associated with plug-in hybrid electric vehicle charging in the Xcel Energy Colorado Service Territory, Technical Report NREL/TP-640-41410, May 2007.

[51] E. Sortomme, M.M. Hindi, S.D.J. MacPherson, S.S. Venkata, Coordinated charging of plug-in hybrid electric vehicles to minimize distribution system losses, IEEE Transactions on Smart Grid 2 (1) (2011) 198–205.

[52] K. Mets, T. Verschueren, W. Haerick, C. Develder, F. De Turck, Optimizing smart energy control strategies for plug-in hybrid electric vehicle charging, IEEE/IFIP Network Operations and Management Symposium Workshops, 19–23 April 2010. pp. 2010.

[53] M. Erol-Kantarci, J.H. Sarker, H.T. Mouftah, Communication-based plug-in hybrid electrical vehicle load management in the smart grid, in: IEEE Symposium on Computers and Communications Corfu, Greece, June 2011.

[54] M. Erol-Kantarci, J.H. Sarker, H.T. Mouftah, Analysis of plug-in hybrid electrical vehicle admission control in the smart grid, in: IEEE 16th International Workshop on Computer Aided Modeling and Design of Communication Links and Networks (CAMAD), 10–11 June 2011, pp. 56–60.

[55] M. Erol-Kantarci, J.H. Sarker, H.T. Mouftah, Quality of service in plug-in electric vehicle charging infrastructure, in: Proceedings of IEEE International Electric Vehicle Conference, Greenville, SC, March 4–8, 2012.

[56] V. Gungor, D. Sahin, T. Kocak, S. Ergut, C. Buccella, C. Cecati, G. Hancke, Smart grid technologies: communications technologies and standards, IEEE Transactions on Industrial Informatics, in vol. 7, no. 4, pp. 529–539, November 2011.

[57] V.C. Gungor, B. Lu, G.P. Hancke, Opportunities and challenges of wireless sensor networks in smart grid, IEEE Transactions on Industrial Electronics 57 (10) (2010) 3557–3564.

[58] C. Lo, N. Ansari, The progressive smart grid system from both power and communications aspects, IEEE Communications Surveys and Tutorials, 2011, in press.

[59] M. Erol-Kantarci, H.T. Mouftah, Wireless multimedia sensor and actor networks for the next-generation power grid, Elsevier Ad Hoc Networks Journal 9 (4) (2011) 542–5110.

[60] V.C. Gungor, F.C. Lambert, A survey on communication networks for electric system automation, Computer Networks Journal (Elsevier) 50 (2006) 877–897.

[61] The Smart Grid Interoperability Panel, Cyber Security Working Group, Guidelines for Smart Grid Cyber Security: Smart Grid Cyber Security Strategy, Architecture, and High-Level Requirements, vol. 1, August 2010. (online). <http://csrc.nist.gov/publications/nistir/ir7628/nistir-7628vol1.pdf>. (accessed October 2011).

[62] U.S. Congress, Office of Technology Assessment, Physical Vulnerability of Electric System to Natural Disasters and Sabotage, OTA-E-453, U.S. Government Printing Office: Washington, DC, June 1990. (online). <http://www.fas.org/ota/reports/9034.pdf>. (accessed October 2011).

[63] The Smart Grid Interoperability Panel, Cyber Security Working Group, Guidelines for Smart Grid Cyber Security: Supportive Analyses and References vol. 3, August 2010. (online). <http://csrc.nist.gov/publications/nistir/ir7628/nistir-7628vol3.pdf>. (accessed October 2011).

[64] M. Amin, Toward a more secure, strong and smart electric power grid, IEEE Smart Grid Newsletter, January 2011.

[65] U.S. Department of Energy Office of Electricity Delivery and Energy Reliability, Study of Security Attributes of Smart Grid Systems – Current Cyber Security Issues, April 2009. (online). <http://www.inl.gov/scada/publications/d/securingthesmartgridcurrentissues.pdf>. (accessed October 2011).

[66] M. Amin, Securing the Electricity Grid, The Bridge, U.S. National Academy of Engineering 40, (1) (2010) 13–20, Spring. (accessed October 2011).

[67] Y. Simmhan, A.G. Kumbhare, B. Cao, V. Prasanna, An analysis of security and privacy issues in smart grid software architectures on clouds, in: IEEE International Conference on Cloud Computing (CLOUD), 4–9 July 2011, pp. 582–589.

[68] H. Mohsenian-Rad, A. Leon-Garcia, Distributed Internet-based load altering attacks against smart power grids, IEEE Transactions on Smart Grid, in vol. 2, no.4, pp. 667–674, December 2011.

[69] The Smart Grid Interoperability Panel, Cyber Security Working Group, Guidelines for Smart Grid Cyber Security: Privacy and the Smart Grid, vol. 2, August 2010. (online). <http://csrc.nist.gov/publications/nistir/ir7628/nistir-7628vol2.pdf>

[70] M.A. Lisovich, D.K. Mulligan, S.B. Wicker, Inferring personal information from demand-response systems, IEEE Security and Privacy 8 (1) (2010) 11–20.

[71] P. McDaniel, S. McLaughlin, Security and privacy challenges in the smart grid, IEEE Security & Privacy 7 (3) (2009) 75–77.

[72] Google Powermeter. (online). <http://www.google.com/powermeter/about/>. (accessed October 2011).

[73] T. Goodspeed, D.R. Highfill, B.A. Singletary, Low-level design vulnerabilities in wireless control systems hardware, in: Proceedings of the SCADA Security Scientific, Symposium (S4), January 21–22, 2009, pp. 3–13-26.

CHAPTER

A Survey on Smart Grid Communications: From an Architecture Overview to Standardization Activities 26

Periklis Chatzimisios[a], Dimitrios G. Stratogiannis[b], Georgios I. Tsiropoulos[b], Georgios Stavrou[a]

[a]CSSN Research Lab, Department of Informatics, Alexander TEI of Thessaloniki, Greece,
[b]School of Electrical and Computer Engineering, National Technical University of Athens, Athens, Greece

26.1 INTRODUCTION

The new social, economic, and environmental trends evolving worldwide, such as the demand for increased productivity and effectiveness from the power grid along with the protection of environmental standards, have created the need for advanced exploitation of the existing energy distribution network by integrating advanced sensing technologies, control methods, and communications networks [1,2].

The convergence of the existing power delivery infrastructure with information communications technology will lead to an innovative energy distribution grid that will provide new capabilities and significant advantages to participating entities [3]. The evolution of next-generation electric power systems also seems to be a promising solution for the energy problem along with the constraints of the Kyoto convention regarding carbon emission and the compliance of each country with it. Thus, energy providers, policy makers, regulation authorities, and various enterprises have shown great interest in the opportunities arising through the development of smart grid technology in the fields of automation, advanced data collection, control, broadband telecommunications, intelligent appliance interoperability, security, distributed power generation, and renewable effective integration (as illustrated in Figure 26.1).

The most important aspect of the smart grid is the expected upgrade of the existing electrical power grid by integrating a high speed, reliable, secure data communication network that will offer a variety of applications and services to all involved entities. The smart grid will provide an efficient system to monitor and control all the components of the power grid using a variety of communications networks.

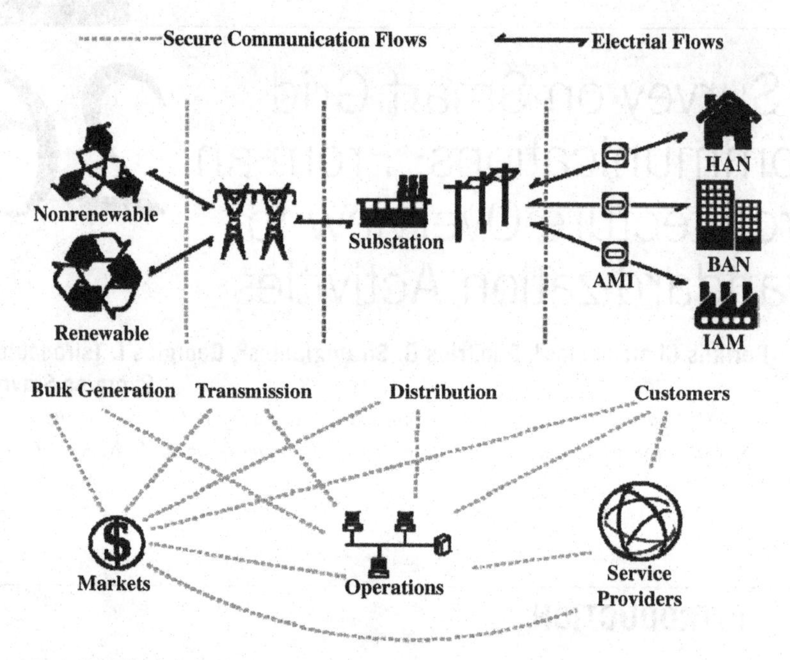

Figure 26.1 Entities involved in the smart grid concept.

The provided services include data collection analysis about power transmission, distribution and consumption in order to achieve improved power management, advanced appliance control and sensing mechanisms, automated power consumption metering, and broadband communication services. Furthermore, the realization of the smart grid increases the reliability and transparency of the power distribution grid by improving the connectivity, automation, and coordination among suppliers, consumers, and networks leading to efficient exploitation of the electricity network and the available resources for power generation.

In the current chapter, a survey on the smart grid regarding its characteristics is performed. Various works regarding the smart grid have been presented in the literature aiming at providing a survey on technical issues and models adopted in the smart grid related to either the power delivery part of the network or communications and networking [4–9]. This chapter aims to provide a brief overview of the smart grid architecture, focusing mostly on standardization activities that are currently an emerging and critical issue in the development of smart grid networks on which little attention has been given. Section 26.2 introduces an analytical definition of the smart grid and a description of the architecture, the capabilities, and the advantages offered by its realization to all the participating entities and the power grid parts. In Section 26.3 a brief analysis of the market, economic, and social aspects of the smart grid is provided. In Section 26.4, the most important features of the smart grid, the smart metering services, are presented, describing the integration of networking technologies into the power grid. In Section 26.5, a comprehensive survey on standardization efforts is performed, since the development of robust standards is expected to solve any interoperability problems in the implementation of the smart grid. Finally, conclusions are deduced in Section 26.6.

26.2 DEFINITION AND FEATURES OF THE SMART GRID

The existing electric power system relies on fossil nonrenewable fuel including oil, coal, and natural gas with significant amounts of carbon emissions. The expected shortage of fossil fuel called global attention to the need for alternative renewable energy resources. Next-generation electric power systems are expected to also incorporate renewable energy resources and decentralized energy production providing new challenges to achieve sustainable long-term development. These characteristics will be compliant with environmental restrictions and carbon emissions constraints. The next-generation power system will consist of a diversity of distributed electric generators and power consumers located in large areas connected in the same network [10]. The implementation of a diversified electricity production mix will require the development of automated power management that will enable information communication among devices, applications, consumers, and grid operators [3]. The development of next-generation power grid networks are based on the evolution of communication networking that will enhance data exchange and automated management in power systems [11]. In particular, the communication network infrastructure that will be integrated in the electrical grid should be able to meet the specifications and needs of power system communications. Automated management of the power grid is expected to offer a variety of advantages and features such as flexibility, resilience, sustainability, scalability, cost-effectiveness, interoperability, and interaction of the participating entities, with the term "smart grid" representing the electric distribution network [12, 13]. Figure 26.2 illustrates a smart grid implementation.

The term "smart grid" describes a high speed, reliable, secure data communication network integrated in the power grid that collects and analyzes data about power transmission, distribution, and consumption [10, 11]. By analyzing these data, predictive information and recommendations can be issued towards suppliers and customers concerning optimal and intelligent power management [15]. The key characteristics of a Smart Grid network, depicted in Figure 26.3, include the following capabilities: (a) advanced metering of power consumption via smart meters; (b) immediate response on demand variations and dynamic pricing; (c) grid optimization and reduced operation cost; (d) better integration of distributed generation including renewable resources achieving carbon emission reduction; (e) intelligent automated asset management and appliance control; (f) smart home appliances and networks; and (g) advanced communication networks via the power grid capable to provide broadband access to end users into vast areas [16].

The smart grid as a system can be analyzed in three layers: the physical power layer that includes the power generation, transmission, and distribution infrastructure;

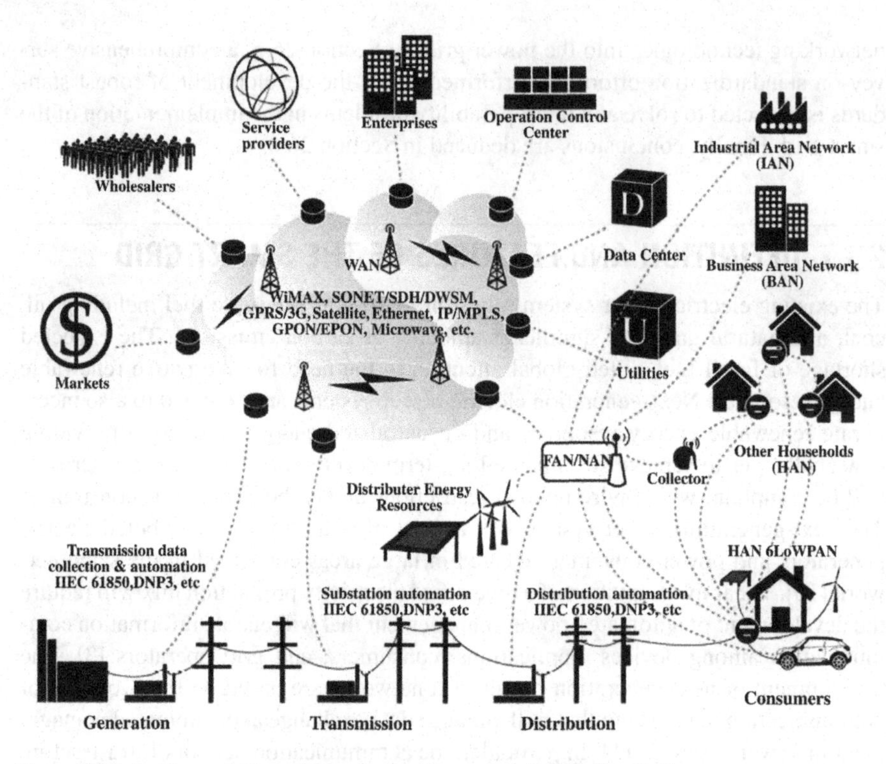

Figure 26.2 Implementation of the smart grid—system integration[14].

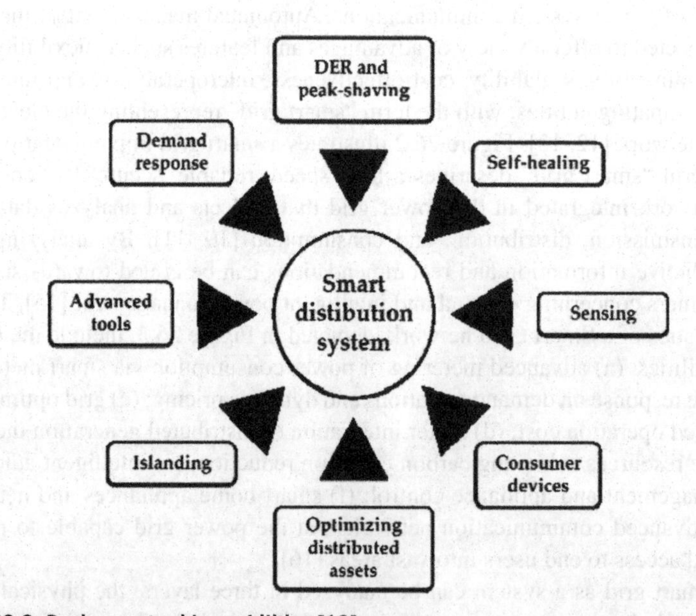

Figure 26.3 Basic smart grid capabilities [16].

the data transport and control layer that supports communications and control; and finally the application layer, which includes applications and services provided by the smart grid to various users [14]. The first layer consists of the electricity production infrastructure including power plants that generate power from fuel (oil, carbon, gas), nuclear or renewable resources (wind, sunlight, etc.), the transmission system, which delivers electricity from power plants to distribution substations, and the distribution system, which delivers electricity from the substations to the consumers and is the largest and the most complex part of the grid. The second layer involves data transport and control and it consists of a high speed, fully integrated communication network that assists data collection and secure information exchange permitting the interaction of the parties involved [17]. Moreover, advanced sensing and measurement equipment is employed to achieve the intelligent control of devices to evaluate their performance and provide immediate response by the system. Finally, the third layer includes all the services that are provided such as automatic metering, automated asset management, maintenance planning applications, and broadband access to end users, etc. according to the standards developed and the technologies embedded in the smart grid network.

The development of a smart grid system will provide significant advances in many areas and these aspects will be analyzed in detail in next sections of this chapter. In this part a brief description of these aspects will be provided. First of all, the smart grid system is intended to provide a competent mechanism for an improved situational awareness of all entities of the electrical distribution system including generation, distribution, and consumption [11]. The smart grid will monitor and control by collecting data and exchanging information among the entities of the power grid to achieve the necessary performance levels and the optimal operation. Thus, the smart grid will permit the installation and support of different kinds of generation and storage devices along with an efficient distribution network [14]. This mechanism will allow the fast and secure interaction of the power grid parts. The grid will be able to detect the occurrence of or even predict a fault in transmission or distribution, improving reliability and power quality while avoiding service disruption [18]. The decision making regarding the performance of the power grid major components can be modified and it will be assisted by the data collected by the smart grid. Thus, the next-generation power networks will enable energy operators and managers to make decisions efficiently to optimize power management taking into account operation parameters from throughout the power grid [19].

The smart grid should be designed to ensure the coordination between providers and consumers to optimize power utilization. It will be able to support decentralized generation including the capability to sell and purchase electricity from different suppliers. This feature will help to provide competing prices and to keep the price of electricity in favor of the consumers. The smart grid, by employing automated metering equipment, will provide the capability to consumers to adjust their power consumption according to current prices and their needs [20]. Furthermore it will offer a dynamic mechanism to determine prices according to supply and demand on an almost real-time basis [21]. For example, pricing with low rates can be offered in times of low load. Moreover, an immediate demand response process will be provided to avoid excess of power demand during peak times that may lead to power failures. Consequently, an efficient mechanism will be designed for consumers to lower their cost and efficiently use their devices.

The unified environment of a smart grid will provide excellent interoperability among the devices connected to the power grid. The interaction among systems will increase the reliability of the grid and help avoid emergency scenarios. The interoperability will be assisted by the usage of advanced sensing equipment and control methods that will provide accuracy in measurements, limiting predictions and errors. The utilization of advanced sensor equipment will improve the power quality in accordance with load control. The control methods will provide timely and fast measurements that can be used to operate the grid efficiently and minimize system losses [14]. The interaction among systems will be critical for facilities and field equipment control [22]. The data collected will be very helpful in a maintenance planning system for improving the design of electrical power systems and achieving well-organized asset management processes.

Finally, it should be noted that the most critical issue regarding next-generation electric power systems is the standardization of the smart grid architectures and applications. Many works regarding the architecture and modelling of the smart grid have appeared in the literature focusing on the technical issues of smart grid development [4–9]. However, there are many issues emerging in the standards adopted. These standards are expected to provide the requested interoperability and they will specify the operation of the participating entities. In the standardization activities, governments, organizations, industry, and academia are involved [15, 23]. As a result many standards were developed and various architectures were proposed following different approaches. These standards cover a large number of technical issues such as control methods, services, automation, sensing, measurements, interfaces, and power equipment. The standardization activities are critical since they will enable the implementation of the smart grid and they are expected to solve the technical uncertainties providing a robust environment for its evolution.

26.3 PARAMETERS OF ECONOMIC, MARKET, AND SOCIAL ASPECTS FOR THE SMART GRID

The evolvtion of the pure power distribution network into a smart electric supply system constitutes the main goal of most electrical power providers. The concept of evolving a network into a smart grid is desirable since the providers can offer services with high quality and at the same time increase the network's consistency, however, it is difficult to realize. One of the most significant difficulties that should be overcome, apart from the technical challenges, is the fact that the current situation in power provision includes both the private and public sector in a complicated way. This means that the institutions and the companies involved in power provision don't want to reveal the potential opportunities of a smart grid to customers. The benefits of a smart grid could possibly constitute a significant advantage for consumers.

Thus, many suppliers, especially those who regulate prices, are afraid of granting the opportunities of a smart network to end users. The increasing demand for sustainable and renewable use of resources, global climate change, and public concerns about the future of modern society impose the application of any scientific intelligence on all aspects of everyday life, especially in the case of power distribution systems. Thus, there is a strong motivation to achieve these multiple objectives by the application of smart grid technology on the power supply network.

In past decades, power supply companies have focused mainly on the wholesale market of electricity. Specifically, the industry has employed market-like mechanisms such as a power stock exchange in order to facilitate the power trade and determine the wholesale price of electricity. Moreover, this trend has been strengthened in the densely populated regions where the power needs are increased compared to sparsely populated regions. In the latter ones, it is evident that market is less effective compared to densely populated areas since the power consumption is low.

This situation of the market was described almost two decades ago. Particularly, it is estimated that markets will be more effective in densely populated areas compared to rural areas. The main reason for this difference is the high transportation cost for power delivery in rural and vast areas. This drawback limits the number of competitive suppliers and offers in these areas. Thus, in many countries the state has taken the initiative to fund partially or in whole the investment on the power delivery network for rural and distinct areas. Moreover, in many cases the state funds the power delivery companies to offer competitive services to these areas.

According to the above-described policy followed in the case of rural areas, it is evident that different market policies are followed between dense or sparse regions, or among different country regions. Consequently that the power market cannot operate efficiently and various power providers cannot be competitive among each other. Thus, there is a strong need to apply a certain, uniform, and consistent market policy to the whole country or an even larger region.

26.4 AUTOMATED METERING AND PRICING FOR THE SMART GRID

The design and implementation of a smart grid is expected to have a tremendous impact on the distribution system of the power network. One of the most important tasks of the next-generation power grid is to provide advanced metering data and information about power consumption following an automated process that will increase the efficiency and operation of the electrical power distribution network. The requirement for the development of automated metering technologies under the smart grid concept is one of the basic requirements of this system.

A smart grid system is expected to utilize smart meters at any customer location. These advanced meters will establish a two-way communication that will be able to measure power consumption and collect crucial information such as voltage and current monitoring, current load, waveform recordings, and power requirements and variations under peak conditions [22]. The data collected will be then available to operation and planning and they will be used to improve the reliability of the power grid and asset management (See Figure 26.4)..

The most innovative part of the automatic measurement and data collection infrastructure is the application of real-time rates according to the demand. This task is of utmost importance since collecting real-time information of power load will help in determining the power price to achieve load-generation balance. These meters will be valuable in applying dynamic pricing, which is one of the key characteristics of the smart grid. More precisely, prices will be affected by the peak demand. According to demand variations prices can be adjusted automatically to help equalize distribution system loading and maintain its integrity. This will aid in reducing faults in or damage to end users' appliances and power grid equipment like substations, wires, and meters.

The two main factors that affect billing are (a) the power consumption and (b) the market price. Electrical power meters in their conventional operation are just registering the amount of kWh that have been used during a certain time period, usually a month. In the next-generation power grid, smart meters can measure electricity usage and collect data that will be sent to the service provider. Then, by getting a real-time unit price from the service provider the cost of electricity can be estimated immediately. These techniques will constitute a powerful tool for power management via the Advanced Metering Infrastructure (AMI). More precisely, households can make their energy consumption choices based on their needs and current prices [24]. Furthermore, they can shift their consumption to hours with lower prices and demand, and in case of decentralized production they can store or even sell their energy surplus. For example, prices will remain high during peak or high demand hours and outside peak hours prices will be decreased accordingly [25]. Furthermore, in the case of distributed power generation with renewables such as solar panels or wind turbines weather conditions might affect electricity production and

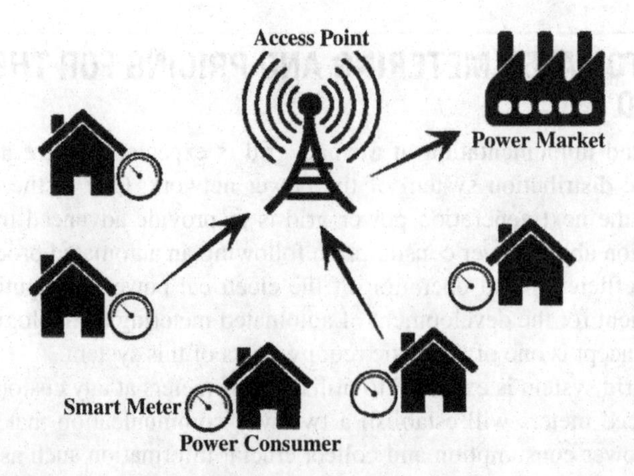

Figure 26.4 Automated metering concept.

the prices. Hence, it is extremely difficult to predict prices at a particular time and a specific place. Therefore, flexible pricing is the most promising solution that will have significant advantages in relation to cost-effectiveness for all the Smart Grid entities.

Various solutions have been proposed in literature regarding dynamic pricing. U.S. Department of Energy provided three types of pricing techniques: (a) time of use, (b) critical peak pricing, and (c) real-time pricing. The first technique refers to the use of certain constant prices no matter the conditions of the network. The second one refers only to price differentiation for peak and off-peak hours and the last one provides a continuous application of a flexible real-time pricing mechanism relying on the smart grid implementation [25]. In [26] the introduction of the pricing mechanism into the smart grid and how it will be measured is described. The real-time pricing described in [26] is more flexible since it permits price differentiation according to the day-of or day-ahead cost of power to the service provider providing significant economic advantages.

Another important aspect that will evolve the metering infrastructure is the distribution automation abilities in the areas of protection and switching. These abilities will be integrated in the smart meters. In [22] the distribution automation services are described showing their impact on the distribution system through the smart grid system. The automation services include quick and flexible configuration of the network. First of all, the automation abilities can help protect appliances and equipment in the case of current or voltage faults. Moreover, the interruption in such cases of the operation of appliances will help in the reconfiguration of the power grid in order to restore customers and achieve efficient operation of the grid. Furthermore, the distribution automation subsystem in the automatic metering components will be able to isolate a fault in the topology and maintain the stability of the power grid.

The integration of smart metering components in the smart grid will have a significant role in decentralized electricity production and the integration of production units based on renewables. Apart from pricing of power production from renewables according to weather conditions, the smart meters would be able to measure the amount of energy generated and monitor its voltage and current characteristics [20]. The smart meters will measure the part of the generated energy that is consumed by the household and the part that is returned to the main network. The automated metering infrastructure will monitor energy generation and the data collected would be very helpful for power management since based on these data production can be regulated and renewable resources can be utilized efficiently [27].

The application of a smart automated metering feature in the smart grid poses certain challenges related to the required characteristics of this feature [28]. First of all, the automated metering feature should satisfy the scalability of the network. The installation of smart meters will be in a large number of homes that their density might be very different compared to the users of a wireless network [28].

At this point, it is useful to present the four networking sectors that the smart grid communications infrastructure is composed of: "core" (or backbone), "middle-mile"

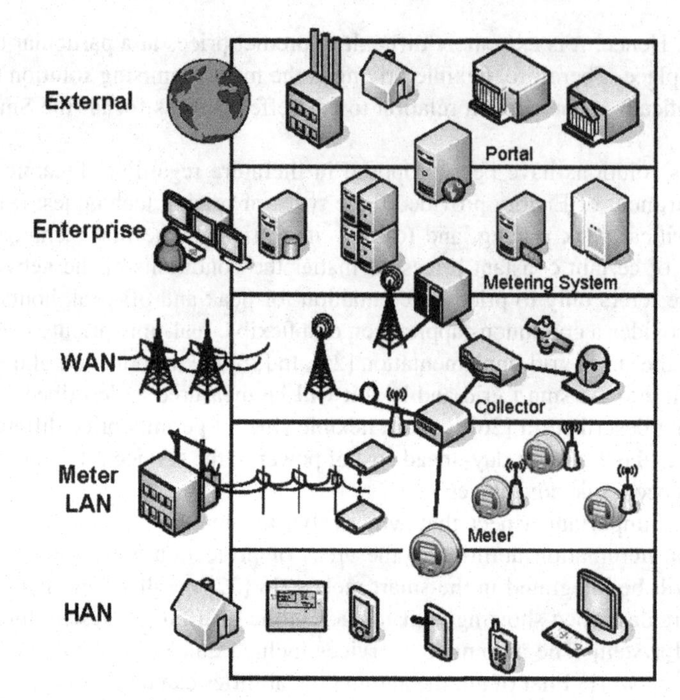

Figure 26.5 Smart grid high level overview.

(or backhaul), "last-mile" (or access, distribution), and "homes" and "premises" (as illustrated in Figure 26.5). More details can be found below:

- The "core" network supports the connection between numerous substations and utilities' headquarters. The backbone network requires high capacity and bandwidth availability and is usually built on optical fibers.
- The "middle-mile," referred to as a wide area network (WAN), connects the data concentrators in AMI with substation/distribution automation and control centers associated with utilities' operation. This sector needs to provide broadband media as well as easy and cost-effective network installation.
- The "last-mile" covers the areas of a neighborhood area network (NAN) and AMI since it is responsible for both the data transport and collection from smart meters to concentrators. There are many available wired and wireless technologies that must provision broadband speed and security.
- The "premises" network supports a home area network (HAN) dedicated to effectively manage the on-demand power requirements of the end users as well as associated building automation. It is predominantly based on the IEEE 802.15.4, IEEE 802.11, and PLC standards.

The second characteristic is the real-time response requirement. Thus, the realization of a dynamic power grid requires real-time load information. This kind of information plays a crucial role in the stability of the network since the current load shows the power demand and it helps coordinate electricity production.

One of the most prominent solutions for automated data collection is to equip each smart meter with a wireless module [24,21]. Then, by using an access point data will be forwarded to the control section of the power grid. It should be clarified that various wireless networking technologies such as IEEE 802.11 WLANs, 3G UMTS, and IEEE 802.16 WiMAX can be applied to advanced meters, depending on the existing communication infrastructure [19]. Furthermore, the broadband over power lines (BPL) technology can be also used in order to provide a smart metering system independent from wireless networks via the smart grid [29–31]. Moreover, hybrid solutions where BPL technology coexist with the widespread wireless communication standards seems to be a promising solution that aims at reducing the deployment and proliferation cost of automated metering and BPL access [32]. The integration of BPL for the interconnection of smart meters with the substation via medium-voltage lines could be a very promising solution for the smart grid since there will be no telecommunication charging for data transmission and it will minimize smart grid dependence on other networking technologies, while providing significant communication capabilities to end users [33].

The access network that will be employed to support the automated metering infrastructure should be able to support an IP-based system that will transfer all the data collected. More precisely, in [21] a broad survey on the technical requirements for an IP-based smart grid network is provided. The two main solutions proposed are based on the ANSI C12.22 standard and on the Session Initiation Protocol (SIP). Despite the fact that both of them can satisfy the requirements for the development of an automated metering data transfer protocol, there are certain advantages and disadvantages. The ANSI protocol is an open protocol that has many of the desirable features for the support of automated metering such as security and scalability. On the other hand, the flexibility in its specifications and its variations pose certain obstacles to its use for metering systems in the case of multivendor systems [21]. SIP provides more functions but it is more complex and less interoperable, which makes it difficult for it to be the prominent standard for automated metering equipment.

Furthermore, various institutes and organizations from research, academia, and industry have presented their own standards regarding the smart grid that include automated metering standards, trying to achieve interoperability and low complexity [34]. Thus, there is a great opportunity for research regarding the standardization of metering equipment since it is in a primitive stage and new concepts and approaches are still required.

26.5 STANDARDS AND INTEROPERABILITY

A general definition of interoperability is "the ability of two or more systems or components to exchange information, to use the information that has been exchanged, and to work cooperatively to perform a task." Interoperability in the smart grid addresses the development of a novel open architecture of technologies and systems, described as a system of systems for the smart grid that allows the interaction among various systems and technologies involved in order to provide a functional model for the smart grid. The realization of the smart grid includes capabilities and technology deployments that must connect large numbers of smart devices and systems involving hardware and software. Interoperability enables integration, effective cooperation, and two-way communication among the many interconnected elements of the electric power grid [35]. In order to achieve interoperability internationally recognized communication and interface standards should be developed by standards development organizations (SDOs) and specification setting organizations (SSOs).

In particular, SDOs operate under similar (participation and balloting) rules worldwide. Often many standards begin their "life" as *de facto* ("something that exists in fact but not as a matter of law") standards and SDOs actually author *de jure* ("something that exists by operation of law") standards. In addition to SDOs another two entities exist, "alliances" and "user groups." A differentiator between user groups, standardization organizations, and alliances is that user groups' rules often permit more free discussion between those actually using standards and specifications than those of the developing organizations.

The establishment of interoperable smart grid standards and protocols is required because a large number of smart grid projects worldwide are currently underway and many devices are being widely deployed in such systems. However, there is a major dilemma between "doing it fast" by working out standards that can guide developing technologies and enable competition against prompt world players like China [36] and "doing it right" by building standards that reflect industry needs while protecting stakeholder interests. In any case, utilities and vendors often move forward whether or not the standards exist or have been finalized since many times the lag time in developing standards is inevitable.

Interoperability standards require much effort by both vendors and users and every participating entity in the standardization process [37] since they should include:

- Recognition of the need for standards in a particular area.
- Involvement of users to develop the business scenarios and use cases that drive the requirements for the standard.
- Clear definition of the scope and purpose of the standard.
- Review of existing standards in order to determine whether or not they meet the needs (even with minor modifications or selection of options).
- Development of a draft standard based on the users' requirements as well as the technology experience of the vendors.
- Widespread review and pilot implementations of the draft standard to resolve ambiguities, imprecise requirements, and incomplete functionality.
- Finalization of the standard and full implementation of the standard by vendors.
- Significant interoperability testing of the standard by different vendors under different scenarios.
- Amending or updating the standard in order to reflect findings during the interoperability tests.

26.5.1 Standardization Activities around the World

The effective development and operation of the smart grid requires many sets of standards to be in place. The United States, Canada, China, South Korea, Australia, and the European Community (EC) countries are conducting research and development on smart grid applications and technologies. The concepts differ greatly in the main regions and this is also reflected in the respective roadmaps and studies by the various international, national, and regional attempts.

Many standards bodies, including the National Institute of Standards and Technology (NIST), International Electrotechnical Commission (IEC), Institute of Electrical and Electronic Engineers (IEEE), Internet Engineering Task Force (IETF), American National Standards Institute (ANSI), North American Reliability Corporation (NERC), and World Wide Web Consortium (W3C) are tackling these interoperability issues for a broad range of industries, including the power industry, the electronics industry, and the telecommunications sector.

The following are the main standardization bodies that perform various activities towards achieving the goal of interoperability in the implementation of the smart grid.

26.5.1.1 IEEE

The Institute of Electrical and Electronics Engineers (IEEE) (http://www.ieee.org) is an international nonprofit, professional organization for the advancement of technology. It has the most members of any technical professional organization in the world, with more than 365,000 members in around 150 countries. Most IEEE members are electrical engineers, computer engineers, and computer scientists, but the organization's wide scope of interests has attracted engineers in other disciplines. The IEEE Standards Association (IEEE-SA) is a leading global developer of standards with more than 20,000 individual and corporate participants in the standardization process.

IEEE has a similar but more flexible methodology to that of the IEC for developing (draft and final) standards by implementing a voting process by members of the corresponding Working Groups. Additionally, IEEE Working Groups develop many other types of documents (including recommended practices, technical reports, conference, and journal papers, as well as other non-standards-oriented documents). The IEEE Power and Energy Society (PES) is currently serving as the world's largest forum about the latest and most exciting technological developments in the electric power industry. The IEEE Communications Society (ComSoc) and the IEEE Society on Social Implications of Technology (SSIT) are also are actively engaged in developing and promoting smart grid technologies.

Although IEEE has more than 100 standards (either finalized or under development) relevant to the smart grid, it is working closely with the National Institute of Standards and Technology (NIST). In particular, IEEE adopted the NIST smart grid Conceptual Model, which provides a high-level framework that defines seven important smart grid domains; Bulk Generation, Transmission, Distribution, Customers, Operations, Markets, and Service Providers. In particular, IEEE views smart grid as a large "system of systems," whereas each NIST smart grid domain is expanded into three smart grid foundational layers (the Power and Energy Layer, the Communication Layer and the IT/Computer Layer) [38].

26.5.1.2 NIST

The National Institute of Standards and Technology (NIST) (http://www.nist.gov/smartgrid) is a nonregulatory federal agency within the U.S. Department of Commerce that promotes U.S. innovation and industrial competitiveness by advancing measurement science, standards, and technology in ways that enhance economic security and improve our quality of life. The Energy Independence and Security Act (EISA) of 2007 [39] has assigned to the National Institute of Standards and Technology (NIST) the primary responsibility to coordinate development of a framework that includes protocols and model standards for information management to achieve interoperability of smart grid devices and systems [39]. There are two primary bodies within NIST designated with tackling this task: the Smart Grid Advisory Committee (composed of 15 industry players) and the Smart Grid Interoperability Panel (a public forum composed of all stakeholders).

In January 2010, NIST released a report that included about 80 initial interoperability standards as well as 14 "Priority Action Plans" to address gaps in the standards. NIST Special Publication 1108 [40] describes a roadmap for the standards on smart grid interoperability. In particular, it presents the expected functions and services in the smart grid as well as the application and requirement of communication networks in the implementation of the smart grid. Furthermore, NIST Report 7628 [41] particularly focuses on the information security issues of the smart grid by presenting the critical security challenges and specifying the security requirements in the smart grid.

In response to the urgent need to establish interoperable standards and protocols for the smart grid, NIST developed a three-phase plan:

- *Phase 1:* Accelerate the setup of a primary set of standards by identifying applicable standards and requirements, gaps in current standards, and priorities for supplementary standardization activities. This phase includes the engagement of stakeholders in a participatory public process.
- *Phase 2:* Establish a robust Smart Grid Interoperability Panel (SGIP) to drive longer-term progress by maintaining the development of the additional standards that will be needed in the future.
- *Phase 3:* Develop and implement a framework for conformity testing and certification targeting to ensure interoperability and security under realistic operating conditions of the defined smart grid standards.

The Smart Grid Interoperability Panel has several priority-specific Committees and Working Groups such as [40]:

- *Smart Grid Architecture Committee (SGAC):* Maintains a conceptual reference model for the smart grid and develops corresponding high-level architectural principles and requirements.
- *Smart Grid Testing and Certification Committee (SGTCC):* Creates and maintains the necessary framework for compliance, interoperability, and cyber security testing and certification for recommended smart grid standards.

- *Cyber Security Working Group (CSWG):* Identifies and analyzes security requirements and develops a risk mitigation strategy to ensure the security and integrity of the smart grid.
- *Domain Expert Working Groups (DEWGs):* NIST has been working to address interoperability through the following groups [42]:
  - Transmission and Distribution (T&D)
  - Building to Grid (B2G)
  - Industry to Grid (I2G)
  - Home to Grid (H2G)
  - Business and Policy (B&P)
  - Vehicle to Grid (V2G)
  - Cyber Security (CS)

Priority Action Plans (PAPs): Many standards require revision or enhancement and new standards need to be developed to fill gaps and issues for which resolution is most urgently needed. Thus, in order to address these PAPs were established and can be found in Table 26.1 (new PAPs are added as necessary).

The smart grid will ultimately require hundreds of standards. Some are more urgently needed than others. To prioritize its work, NIST chose to focus on six key functionalities plus cyber security and network communications. These functionalities are especially critical to ongoing and near-term deployments of smart grid technologies and services, and they include the priorities recommended by the Federal Energy Regulatory Commission (FERC) in its policy statement as follows [40]:

- Demand response and consumer energy efficiency
- Wide-area situational awareness
- Energy storage
- Electric transportation
- Advanced Metering Infrastructure (AMI)
- Distribution grid management
- Network communications
- Cyber security

26.5.1.3 ***IEC***

The International Electrotechnical Commission (IEC) (http://www.iec.ch) is a nonprofit, nongovernmental international standards organization that prepares and publishes international standards for all electrical, electronic, and related technologies, including technologies for power generation, transmission, and distribution. The IEC Council consists of National Committees (NCs), one from each country that is a member of the IEC. Under the IEC Council are Standards Management Boards (SMBs), which coordinate the international standards work. This standards work is performed through many Technical Councils (TCs), each tasked with specific areas. The Standardization Management Board (SMB) of the IEC resolved the establishment of a

Table 26.1 Priority Action Plans (PAPs)

Priority Action Plan (PAP)	Standard(s) or Guideline(s)
PAP 0—Meter Upgradeability Standard	NEMA Meter Upgradeability Standard
PAP 1—Role of IP in the Smart Grid	Informational IETF RFC
PAP 2—Wireless Communications for the Smart Grid	IEEE 802.x, 3GPP, 3GPP2, ATIS, TIA
PAP 3—Common Price Communication Model	OASIS EMIX, ZigBee SEP 2, NAESB
PAP 4—Common Scheduling Mechanism	OASIS WS-Calendar
PAP 5—Standard Meter Data Profiles	AEIC V2.0 Meter Guidelines (addressing use of ANSI C12)
PAP 6—Common Semantic Model for Meter Data Tables	ANSI C12.19-2008, MultiSpeak V4, IEC 61968-9
PAP 7—Electric Storage Interconnection Guidelines	IEEE 1547.4, IEEE 1547.7, IEEE 1547.8, IEC 61850-7-420, ZigBee SEP 2
PAP 8—CIM for Distribution Grid Management	IEC 61850-7-420, IEC 61968-3-9, IEC 61968-13,14, MultiSpeak V4, IEEE 1547
PAP 9—Standard DR and DER Signals	NAESB WEQ015, OASIS EMIX, OpenADR, ZigBee SEP 2
PAP 10—Standard Energy Usage Information	NAESB Energy Usage Information, OpenADE, ZigBee SEP 2, IEC 61968-9, ASHRAE SPC 201P
PAP 11—Common Object Models for Electric Transportation	ZigBee SEP 2, SAE J1772, SAE J2836/1-3, SAE J2847/1-3, ISO/IEC 15118-1,3, SAE J2931, IEEE P2030-2, IEC 62196
PAP 12—IEC 61850 Objects/DNP3 Mapping	IEC 61850-80-5, Mapping DNP to IEC 61850, DNP3 (IEEE 1815)
PAP 13—Time Synchronization, IEC 61850 Objects/IEEE C37.118 Harmonization	IEC 61850-90-5, IEEE C37.118, IEEE C37.238, Mapping IEEE C37.118 to IEC 61850, IEC 61968-9
PAP 14—Transmission and Distribution Power Systems Model Mapping	IEC 61968-3, MultiSpeak V4
PAP 15—Harmonize Power Line Carrier Standards for Appliance Communications in the Home	DNP3 (IEEE 1815), HomePlug AV, HomePlug C&C, IEEE P1901 and P1901.2, ISO/IEC 12139-1, G.9960 (G.hn/PHY), G.9961 (G.hn/DLL), G.9972 (G.cx), G.hnem, ISO/IEC 14908-3, ISO/IEC 14543, EN 50065-1
PAP 16—Wind Plant Communications	IEC 61400-25
PAP 17—Facility Smart Grid Information Standard	New Facility Smart Grid Information Standard ASHRAE SPC 201P
PAP 18—SEP 1.x to SEP 2 Transition and Coexistence	ZigBee

Strategic Group on Smart Grids (Strategic Group 3), which submitted a roadmap for its own standards and high-level recommendations that were especially relevant to the European standardization roadmap [43].

26.5.1.4 CENELEC

The European Telecommunications Standards Institute (ETSI), together with the European Committee for Standardization (Comité Europēen Normalization—CEN) and the European Committee for Electrotechnical Standardization (CENELEC) have formed a Joint Working Group for Smart Grid standardization efforts and recently published a roadmap about the European Commission's policy for the smart grid [44].

26.5.1.5 ANSI

The private, nonprofit American National Standards Institute (ANSI) (http://www.ansi.org) oversees the creation, spread, and use of norms and guidelines that directly impact businesses in nearly every sector, including energy distribution. The ANSI itself does not develop standards, but instead facilitates the development of American National Standards (ANS) by accrediting the procedures of standards developing organizations.

26.5.1.6 SGCC

The State Grid Corporation of China (SGCC) has defined its own smart grid standardization roadmap [45] by taking into account several existing standardization roadmaps e.g., IEC SG 3, NIST Interoperability Roadmap, IEEE P2030, CEN/CENELEC/ETSI Working Groups, German DKE Roadmap, and Japanese METI Roadmap.

26.5.1.7 UCA International Users Group

The UCA International Users Group (UCAIug) (http://www.ucaiug.org) is a nonprofit corporation focused on enabling utility integration through the deployment of open standards. UCAIug does not write standards but it works closely with those bodies that have primary responsibility for the completion of standards. In particular, the Open Smart Grid (OpenSG) subcommittee sponsors working groups to address smart grid related requirements and interoperability guidelines development.

26.5.1.8 Vendor Collaborations

Many collaborations and alliances of vendors have been initiated to resolve the details of standards and to develop vendor agreements for standard implementation and product interoperability. Some relevant vendor alliances and collaborations include [46]:

- HomePlug Powerline Alliance (www.homeplug.org)
- Z-Wave Alliance (www.z-wavealliance.org)
- ZigBee Alliance (www.zigbee.org)

Several other major smart grid standardization roadmaps and studies that are worthy of mention are:

- German Standardization Roadmap E-Energy/Smart Grid [47]

- International Telecommunication Union (ITU-T) Smart Grid Focus Group [48]
- Japanese Industrial Standards Committee (JISC) roadmap to international standardization for smart grid [49]
- Korea's Smart Grid Roadmap 2030 from the Ministry of Knowledge Economy (MKE) [50]
- CIGRE D2.24 [51]
- Microsoft SERA [52]

Furthermore, various national roadmaps exist, for example from Spain [53], Austria [54], and the UK [55]. Those national roadmaps are either still under development or do not recommend specific standards (a detailed study of the above and an overview of other SG roadmaps can be found in [4] and [5]).

26.5.2 A Report on Smart Grid Standards

This subsection presents the main standards for smart grid communications and security. Table 26.2 provides a list of the smart grid standards, specifications, requirements, and guidelines identified as the most important for smart grid communications. The second list, Table 26.3, contains the corresponding smart grid security standards and documents. Furthermore, Table 26.4 provides a list of additional smart grid communications and security standards, specifications, requirements, guidelines, and reports subject to further review and consensus development.

Table 26.2 Standards, Specifications, Requirements, and Guidelines (Smart Grid Communications)

Standards/Specifications/ Requirements/Guidelines	Short Description
ANSI/ASHRAE 135-2010/ISO 16484-5 BACnet	Data communication protocol for building automation and control networks
ANSI C12.21/IEEE P1702/MC1221	Transport of measurement device data over telephone networks
ANSI/CEA 709 and Consumer Electronics Association (CEA) 852.1 LON Protocol Suite	A general purpose local area networking protocol
ANSI/CEA 709.1-B-2002 Control Network Protocol Sp.	A specific physical layer protocol designed for use with ANSI/CEA 709.1-B-2002
ANSI/CEA 709.2-A R-2006 Control Network Power Line	A specific physical layer protocol designed for use with ANSI/CEA 709.1-B-2002
ANSI/CEA 709.3 R-2004	A specific physical layer protocol designed for use with ANSI/CEA 709.1-B-2002
ANSI/CEA-709.4:1999 Fiber-Optic Channel Specification	A protocol that provides a way to tunnel local operating network messages
IEEE Std 1815-2010	Substation and feeder device automation, communications between control centers and substations

Table 26.2 Standards, Specifications, Requirements, and Guidelines (Smart Grid Communications) (*continued*)

Standards/Specifications/ Requirements/Guidelines	Short Description
IEC 60870-6 / Telecontrol Application Service Element 2	Defines the messages sent between control centers of different utilities
IEC 61850 Suite	Defines communication within transmission/distribution substations for automation & protection
IEC 61968/61970 Suites	Defines information exchanged among control center systems using common information models
IEEE C37.118-2005	Phasor measurement unit (PMU) performance specifications and communications for synchrophasor data
IEEE 1547 Suite	Physical and electrical interconnections between utilities and distributed generation and storage
IEEE 1588	Standard for time management and clock synchronization across the smart grid for equipment needing consistent time management
RFC 6272, Internet Protocols for the Smart Grid	Internet protocols for IP-based smart grid networks
IEEE 1901-2010 , ITU-T G.9972	Both IEEE 1901-2010 and ITU-T G.9972 specify intersystem protocol (ISP)-based broadband PLC coexistence mechanisms for home networking
NISTIR 7761, NIST Guidelines for Assessing Wireless Standards for Smart Grid Applications	A draft of key tools and methods to assist smart grid system designers in making informed decisions about existing and emerging wireless technologies
OPC-UA Industrial	A platform-independent specification for a secure, reliable, high-speed data exchange based on a publish/subscribe mechanism
Open Geospatial Consortium Geography Markup Language (GML)	Exchange of location-based information addressing geographic data requirements for many smart grid applications
Smart Energy Profile 2.0	Home area network (HAN) device communications and information model
OpenHAN	A specification for a home area network (HAN) to connect to the utility advanced metering system including device communication, measurement, and control
SAE J2836/1: Use Cases for Communication Between Plug-in Vehicles and the Utility Grid	Establishes use cases for communication between plug-in electric vehicles and the electric power grid, for energy transfer and other applications

Table 26.3 Standards, Specifications, Requirements, and Guidelines (Smart Grid Security)

SGTCC Interoperability Process Reference Manual (IPRM)	Outlines the conformance, interoperability, and cyber security testing and certification requirements for SGIP-recommended smart grid standards
Security Profile for Advanced Metering Infrastructure, v 1.0, Advanced Security Acceleration Project—Smart Grid	Provides guidance and security controls to organizations developing or implementing AMI solutions
Department of Homeland Security (DHS), National Cyber Security Division Catalog of Control Systems Security: Recommendations for Standards Developers	Presents a compilation of practices that various industry bodies have recommended to increase the security of control systems from both physical and cyber attacks
DHS Cyber Security Procurement Language for Control Systems	Provides guidance to procuring cyber security technologies for control systems products and services
IEC 62351 Parts 1-8	Defines information security for power system control operations
IEEE 1686-2007	Defines the functions and features to be provided in substation intelligent electronic devices (IEDs) to accommodate critical infrastructure protection programs
NERC Critical Infrastructure Protection (CIP) 002-009	Covers organizational, processes, physical, and cyber security standards for the bulk power system
NIST Special Publication (SP) 800-53	Covers cyber security standards and guidelines for federal information systems, including those for the bulk power system
NISTIR 7628 Guidelines for Smart Grid Cyber Security	Guidelines that include an overview of the cyber security strategy used by the CSWG, an evaluative framework for assessing risks to smart grid components and systems, and a guide to assist organizations as they craft a smart grid cyber security strategy

Table 26.4 Additional Standards, Specifications, Requirements, Guidelines, and Reports for Further Review (Smart Grid Communications and Security)

ANSI C12.22-2008/IEEE P1703/MC1222	End device tables communications over any network
CableLabs PacketCable Security Monitoring and Automation Architecture Technical Report	Describes a broad range of services that could be provided over television cable, including remote energy management
IEC 61400-25	Communication and control of wind power plants
ITU Recommendation G.9960/G.9661 (G.hn)	In-home broadband home networking over power lines, phone lines, and coaxial cables

Table 26.4 Additional Standards, Specifications, Requirements, Guidelines, and Reports for Further Review (Smart Grid Communications and Security) (*Continued*)

IEEE P1901	Broadband communications over power lines: MAC and physical layer (PHY) protocols
IEEE P1901.2 and ITU-T G.9955/G.9956 (G.hnem)	Low frequency narrowband communications over power lines
ISO/IEC 12139-1	High-speed power line communications (PLC), medium access control (MAC), and physical layer (PHY) protocols
IEEE 802 family	Includes standards developed by the IEEE 802 LAN/MAN Standards Committee
TIA TR-45/3GPP2 family of standards	Standards for cdma2000® Spread Spectrum and High Rate Packet Data Systems
3GPP family of standards (including 2G, 3G, and 4G)	2G, 3G, and 4G cellular network protocols for packet delivery
ETSI GMR-1 3G family of standards	GMR-1 3G is a satellite-based packet service equivalent to 3GPP standards
ISA SP100	Wireless communication standards intended to provide reliable and secure operation
Network management standards such as DMTF, CIM, WBEM, SNMPv3, netconf, STD 62, CMIP/CMIS	Protocols used for management of network components and devices attached to the network
ASHRAE 201P Facility Smart Grid Information Model	Enable appliances/control systems in homes/buildings/industrial facilities to manage electrical loads and generation sources and to communicate information to utility and electrical service providers
NIST SP 500-267	A profile for IPv6 in the U.S. Government
Z-wave	A wireless mesh networking protocol for home area networks
IEEE P2030, IEEE P2030.1, IEEE P2030.2	IEEE smart grid series of standards
IEC 62056 Device Language Message Specification/Companion Specification for Energy Metering	Energy metering communications
IEC 60870-2-1	Telecontrol equipment and systems—Part 2: Operating conditions—Section 1: Power supply and electromagnetic compatibility
IEEE 1613	Standard Environmental and Testing Requirements for Communications Networking Devices in Electric Power Substations
IEEE P1775/1.9.7	Standard for Power Line Communication Equipment

Table 26.4 Additional Standards, Specifications, Requirements, Guidelines, and Reports for Further Review (Smart Grid Communications and Security) (*Continued*)

ISO/IEC 15045	Specification for a residential gateway connecting home network domains to other network domains
ISA SP99	Cyber security mitigation for industrial and bulk power generation stations
ISO 27000	A series of ISO standards for information security matters
NIST FIPS 140-2	U.S. government computer security standard used to accredit cryptographic modules
OASIS WS-Security and suite of security standards	Toolkit for building secure, distributed applications, applying a wide range of security technologies

26.6 CONCLUSION

The concept of the smart grid has received considerable attention worldwide in recent years. A number of organizations, standard bodies, and countries worldwide have launched significant efforts to encourage the development of the smart grid. It must be noted that it is very important for these efforts to be coordinated and harmonized internationally by developing functional standards. The development and use of international standards is an essential step in this direction. In certain areas of the smart grid, many standards and rules have already been put in place. Interoperability is the key to the smart grid, and standards are the key to interoperability. Thus, the utilities, vendors, and others are involved in the development of standards; the faster they are developed, the faster the vision of a smart grid will be realized. The aim of the current chapter is to give an overview of the existing standardization activities as well as the key players involved in the process. The standardization activities are expected to play critical role in the realization of the smart grid, which will offer significant advantages to power grid parts, energy providers, policy makers, regulation authorities, enterprises, and customers.

REFERENCES

[1] R. Hassan, G. Radman, Survey on smart grid, in: IEEE SoutheastCon 2010, 2010, pp. 210–213.
[2] R. Anderson, S. Fuloria, Who controls the off switch? in: IEEE SmartGridComm'10, 2010, pp. 96–101.
[3] S. Rahnan, Smart grid expectations, IEEE Power and Energy Magazine, 2009.
[4] S. Rohjans, M. Uslar, R. Bleiker, J. Gonzalez, M. Specht, T. Suding, T. Weidelt, Survey of smart grid standardization studies and recommendations, in: IEEE SmartGridComm '10, 2010, pp. 583–587.

[5] M. Uslar, S. Rohjansand, R. Bleiker, J. Gonzalez, M. Specht, T. Suding, T. Weidelt, Survey of smart grid standardization studies and recommendations—Part 2, in: IEEE PES '10, 2010, pp. 1–6.
[6] B. Akyol, H. Kirkham, S. Clements, M. Hadley, A survey of wireless communications for the electric power system, U.S. Department of Energy, 2010.
[7] V.C. Gungor, F.C. Lambert, A survey on communication networks for electric system automation, Computer Networks 50 (7) (2006) 877–897.
[8] T. Khalifa, K. Naik, A. Nayak, A survey of communication protocols for automatic meter reading applications, IEEE Communications Surveys Tutorials 13 (2) (2011)
[9] T. M. Chen, Survey of cyber security issues in smart grids (part of SPIE DSS 2010), 2010, pp. 1–11.
[10] W. Wang, Y. Xu, M. Khanna, A survey on the communications architectures in smart grid, Computer Networks 55 (15) (2011) 3604–3609.
[11] J. Gao, Y. Xiao, J. Liu, W. Liang, C.L. Chen, A survey of communication/networking in smart grids, Future Generation Computer Systems 28 (2) (2012)
[12] R. Hasan, G. Radnan, Survey on smart grid, in: Proceedings of IEEE SoutheastCon 2010, pp. 210–213.
[13] V. Hamidi, K.S. Smith, R.C. Wilson, Smart grid technology review within transmission and distribution sector, in: Proceedings of ISGT 2010, pp. 1–8.
[14] C.H. Lo, N. Ansari, The progressive smart grid system from both power and communications aspects, IEEE Communications Surveys and Tutorials 99, in press.
[15] M. Uslar, S. Rohjans, R. Bleiker, J. Gonzalez, M. Specht, T. Suding, T. Weidelt, Survey of smart grid standardization studies and recommendations—Part 2, in: Proceedings of ISGT 2010, pp. 1–6.
[16] H.E. Brown, S. Suryanaryanan, A survey seeking a definition of a smart grid system, in: Procedings of NAPS 2009, USA, pp. 1–7.
[17] Q. Zou, L. Qin, Integrated communications in smart distribution grid, in: Proceedings of POWERCON 2010, pp.1–6.
[18] A. Clark, C.J. Pavloski, J. Fry, Transformation of energy systems: the control room of future, in: Proceedings of EPEC 2009, pp. 1–6.
[19] D.M. Laverty, D.J. Morrow, R. Best, P.A. Crossley, Telecommunications for smart grid: backhaul solutions for the distribution network, in: Proceedings of PES General Meeting 2010, pp. 1–6.
[20] D.Y. Raghavendra Nagesh, J.V. Vamshi Krishna, S.S. Tulasiram, A real-time architecture for smart energy management, in: Proceedings of ISGT 2010, pp. 1–4.
[21] J. Wang, V.C.M. Leung, A survey of technical requirements and consumer application standards for IP-based smart grid AMI network, in: Proceedings of ICOIN 2011, pp. 114–119.
[22] R.E. Brown, Impact of smart grid on distribution system design, in: Proceedings of PES General Meeting 2008, pp. 1–4.
[23] G. Reed, P.A. Philip, A. Barchowsky, C.J. Lippert, A.R. Sparacino, Sample survey of smart grid approaches and technology gap analysis, in: Proceedings of ISGT 2010, pp. 1–10.
[24] D. Rua, D. Issicaba, F.J. Soares, P.M. Rocha Almeida, R.J. Rei, R.J. Pecas Lopes, Advanced metering infrastructure functionalities for electric mobility, in: Proceedings of ISGT 2010, pp. 1–7.
[25] J. Liu, Y. Xiao, J. Gao, Accountability in smart grids, in: Proceedings of CCNC 2011, pp. 1166–1170.
[26] H. Chao, Price-responsive demand requirement for a smart grid world, The Electricity Journal 23 (1) (2010) 7–20.

[27] L. Li, H. Xiaoguang, H. Jian, H. Ketai, Design of new architecture of AMR system in smart grid, in: Proceedings of ICIEA 2011, pp 2025–2029.

[28] H. Li, R. Mao, L. Lai, R.C. Qiu, Compressed meter reading for delay sensitive and secure load report in smart grid, in: Proceedings of SmartGridComm 2010, pp. 114–119.

[29] Z. Feng, L. Jianning, H. Dan, Z. Yuexia, Study on the application of advanced broadband wireless mobile communication technology in smart grid, in: Proceedings of POWERCON 2010, pp. 1–6.

[30] J. Liu, B. Zhao, J.Wang, J. Hu, Application of power line communication in smart power consumption, in: Proceedings of ISPLC 2010, pp. 303–307.

[31] A. Pinomaa, J. Ahola, A. Kosonen, Power line communication-based network architecture for LVDC distribution system, in: Proceedings of ISPLC 2011, pp. 358–363.

[32] G.I. Tsiropoulos, A.M. Sarafi, P.G. Cottis, Wireless-broadband over power lines networks: a promising broadband solution in rural areas, in: Proceedings of PowerTech 2009, pp. 1–6.

[33] G.I. Tsiropoulos, D.G. Stratogiannis, P.G. Cottis, Call admission control with QoS guarantees for multiservice wireless broadband over power line networks, in: Proceedings of ISWCS 2008, pp. 168–172.

[34] D.A. Wollman, Accelerating standards and measurements for the smart grid, in: Proceedings of ICASSP 2011, pp. 5948–5951.

[35] K. Kowalienko, Smart grid projects pick up speed, IEEE, The Institute, Standards, Article 06, August, 2009.

[36] M. Hart, China pours money into smart grid technology, no. 10, Center for American Progress, October 2011.

[37] F. Cleveland, F. Small, T. Brunetto, Smart grid: interoperability and standards: an introductory review, Utility Standards Board, 2008.

[38] IEEE Smart Grid Web Portal, Smart Grid Conceptual Framework, <http://smartgrid.ieee.org/nist-smartgrid-framework>.

[39] Energy Independence and Security Act (EISA) of 2007, Public Law No: 110–140, USA

[40] NIST, Framework and Roadmap for Smart Grid Interoperability Standards, Release 1.0, 2010. <http://www.nist.gov/public_affairs/releases/upload/smartgrid_interoperability_final.pdf>.

[41] NIST, Guidelines for Smart Grid Cyber Security (three volumes). <http://csrc.nist.gov/publications/PubsNISTIRs.html#NIST-IR-7628>.

[42] NIST, Smart Grid Interoperability Standards Project, NIST Smart Grid. <http://www.nist.gov/smartgrid/>.

[43] IEC Smart Grid Strategic Group (SG3), Smart Grid Standardization Roadmap, Edition 1.0, 2010. <http://www.iec.ch/zone/smartgrid/pdf/sg3_roadmap.pdf>.

[44] Smart Grid Roadmap Report, CEN/CENELEC/ETSI Joint Working Group on Standards for Smart Grids, May 2011.

[45] State Grid Corporation of China: SGCC Framework and Roadmap for Strong and Smart Grid Standards, White paper, July 2010.

[46] E.W. Gunther, A. Snyder, G. Gilchrist, D.R. Highfill, Smart Grid Standards Assessment and Recommendations for Adoption and Development, Draft v0.83, Enernex for California Energy Commission, 2009.

[47] The German Roadmap E-Energy/Smart Grid. <http://www.e-energy.de/documents/DKE_Roadmap_SmartGrid_230410_English.pdf>.

[48] ITU Telecommunication Standardization Bureau Policy and Technology Watch Division, Activities in Smart Grid Standardization Repository Version 1.0.

[49] Japan's Roadmap to International Standardization for Smart Grid and Collaborations with Other Countries, 2010.

[50] Korea's Smart Grid Roadmap 2030, Ministry of Knowledge Economy (MKE). <http://www.globalcitizen.net/database/?page=2528&category=papers&id=32682>.

[51] International Council on Large Electric Systems (CIGRE), CIGRE D2.24 EMS Architectures for the 21st Century, 2009.

[52] Microsoft, Smart Energy Reference Architecture (SERA), 2009.

[53] FutuRed, Spanish Electrical Grid Platform, Strategic Vision Document, 2009.

[54] Roadmap Smart Grids Austria—Der Weg in die Zukunft der elektrischen Netze, Version Prepared for the Smart Grids Week Salzburg 2009.

[55] Electricity Networks Strategy Group, A Smart Grid Routemap, 2010.

CHAPTER

Towards Energy Efficiency in Next-Generation Green Mobile Networks: A Queueing Theory Perspective

27

Glaucio H.S. Carvalho[a], Isaac Woungang[b], Alagan Anpalagan[c]

[a]*Faculty of Computation, Federal University of Pará, Belém, Pará, Brazil,*
[b]*Department of Computer Science, Ryerson University, Toronto, Ontario, Canada,*
[c]*Department of Electrical and Computer Engineering, Ryerson University, Toronto, Ontario, Canada*

27.1 INTRODUCTION

With the tremendous growth in the demand for wireless access motivated by voice and Internet-based services such as Facebook, Youtube, Orkut, ebooks, Skype, and Google Talk, to name a few, mobile network operators (MNOs) are required to expand their access networks in order to meet users' expectations. To achieve this goal, MNOs have started a dense deployment of new Base Transceiver Stations (BTSs) within the existing cells or even formed new ones. This operation can generate significant monetary gains. However, it may also cause a certain uneasiness. In recent years, debates amongst biologists, environmentalists, and climatologists on environment preservation and protection have attracted a lot of interest in telecommunication forums and among researchers around the world. This can be justified by the fact that the information and communications technologies (ICT) sector contributes about 2% of the global greenhouse gas (GHC) emissions *per annum* and with the growing pace in this sector, it is expected that this rate will reach 2.8% by 2020 [1]. The major aspect of ICT responsible for this environmental threat is undoubtedly the huge amount of electrical energy demanded to keep the sector in operation. As a part of it, MNOs already represent 0.2% of the total energy consumed. With the required expansion, it is foreseen that this rate will increase significantly.

In order to understand why the combination of energy consumption and network expansion is a real concern, the energy consumption in a typical cellular mobile network is shown in Table 27.1. As illustrated in Table 27.1, the BTS drains a significant portion of the energy consumed and since it is the lead actor in network expansion, MNOs run the risk of seeing their electricity bills bit into a huge amount of their

Table 27.1 Source of Energy Consumption in Cellular MNOs [2,3]

Source	Percentage
BTS	57
Mobile telephone exchange (MTX)	20
Core network	15
Data center	6
Retail	2

profits. In order to avoid/reduce this risk and to contribute to world preservation, a more sustainable network design or green network design is mandatory.

Currently, wireless industries and academia are working towards achieving the aforementioned goal. Hence, initiatives such as Energy Aware Radio and neT-work tecHnologies (EARTH) [4], Towards Real Energy-efficient Network Design (TREND) [5], Cognitive radio and Cooperative strategies for POWER saving in multi-standard wireless devices (C2POWER) [6], and Optimizing Power Efficiency in mobile radio Networks (OPERA-Net) [7], to name a few, have helped to enlighten this new research field. At this moment of effervescence, many ideas have come up for discussion. For instance, from a hardware perspective, it is envisioned that the development of new power amplifiers (PAs) based on Doherty architectures and aluminum gallium nitrite (GaN) will reduce the energy consumption in BTS [2,3].

Another approach that can be used to achieve energy efficiency in heterogeneous wireless networks (HetNets) is to turn off the PAs in the BTS radio transceivers. Even though they are utilized only when there is traffic load to be carried out, PAs consume the major portion of the BTS energy, which is around 65% on average. In order to economize energy, Long-Term Evolution (LTE) BTSs are able to turn the PAs off during idle times by employing power saving protocols such as discontinuous reception (DRX) and transmission (DTX) [2,3]. Both aforementioned solutions are somewhat invasive since they require the development of new hardware and software solutions.

Other research [8] has studied noninvasive techniques that utilize eco-friendly green energy sources, such as solar and wind, to support the operation of wireless networks, showing that it is a viable alternative to reduce deployment costs. In spite of the reluctancy of MNOs, in many sites, the payback period for green power technology is less than 3 years [2]. Additionally, cognitive and cooperative technologies are now in the embryo stage and with the inclusion of green network design in their development agenda, significant gains in terms of power consumption reduction in the medium to long term can be expected.

While the coming years seem to be promising in terms of more sustainable wireless network design approaches, looking at the present, we realize that there exist an endless number of BTSs from different manufacturers, belonging to different owners, with different technologies. These BTSs contribute to different design strategies covering almost all parts of the globe. An immediate unified solution is required. This chapter goes in this direction by following a naive lesson learned while at the childhood age, i.e., "to save power and money, turn the lights off." In the context of this work, this idea translates into switching off the BTSs in such a way that their power consumption are translated into energy savings.

One of the main tenets of 4G wireless networks is that the combination of multiple HetNets will help to enhance the system coverage as well as the quality of service (QoS) provisioning. However, if all of them are kept in operation 24 h/day, it will result in a high energy consumption, but if the coverage is dynamically managed according to traffic load fluctuations, then it is possible to turn off some cells, while others can carry out the total offered traffic load. Following this practice, it is feasible to meet the green design goal. In this chapter, this issue is investigated by proposing a threshold-based green joint radio-resource management (JRRM) design, which has its prime function achieving energy efficiency while boosting system capacity. The heart of the proposed green JRRM approach is the dynamic coverage management (DCM) algorithm, which has a function to decide when the microcell can be turned off or on, based on the macrocell radio resource occupancy and the value of the threshold. Even though energy efficiency and capacity boost are two conflicting objectives, the green JRRM approach utilizes the threshold setup to determine a suitable balance between them. In order to evaluate the performance of our proposed green JRRM design, the queueing theory framework is used to model its behavior. queueing theory is a widely recognized mathematical branch of computer science and electrical engineering that has been successfully applied in the design of computational and communications systems [9].

The remainder of this chapter is organized as follows. In Section 27.2, the literature about HetNets is reviewed. Section 27.3 deals with the descriptions of HetNets, radio-resource management (RRM), and the mathematical framework used for modeling the green JRRM design. In Section 27.4, some green approaches for HetNets are discussed. In Section 27.5, our proposed green JRRM approach for HetNets is described. Finally, Section 27.6 concludes our work and highlights some future studies.

27.2 RELATED WORK

HetNets are typically composed of a conglomeration of multiple wireless networks and technologies such as WiFi, Bluetooth, ZigBee, WiMAX, and IEEE 802 WLANs, cellular and mobile technologies called radio access technologies/networks (RATs), which are expected to operate together in a complementary manner. As an example, the IEEE 802.11 WLANs and 3G cellular systems can operate as a composite heterogeneous wireless network to provide higher bandwidth services over a wider geographic area. The concept of HetNets compared to that of homogeneous networks was discussed in [69]. This concept was motivated by the desired to combine the several advantages offered through the features of each homogeneous network (RAT), for instance, the widespread coverage feature offered by WiMAX and cellular technologies, the low cost and high bandwidth derived from using WiFi, ZigBee, or Bluetooth, and the possibility of extending the user's selection technology and radio when working with various wireless applications, to name a few.

Seamless integration of heterogeneous wireless networks [70] is one of the most important requirements for the future deployment of new wireless technologies such as 4G mobile systems, 3G systems, WLANs, Bluetooth, and ultra-wideband. Indeed, using dense or sparse HetNets architectures or a combination of both [71], it is expected that multiple RATs will be available in various locations in the targeted network, and the user device or networks will be able to decide on the best access to these RATs to achieve their goals. To do so, managing the resources in the network while keeping in mind the QoS from a user's perspective is one of the key challenges [71]. In other networks such as wireless LANs, wireless mesh networks, and ad hoc networks, resource management has been intensively studied [72]. However, in HetNets, the solutions to this problem are yet to be consolidated [71]. Without loss of generality, our focus here is on radio-resource management.

Resource management (RM) in HetNets has been tackled from various perspectives, including radio-resource management; power control; rate control; access control; QoS; cognitive and software defined radio; prefetching and caching; handoff; cross-layering; optimization; and frameworks, to name a few. In the sequel, some representative works focusing on the aforementioned perspectives are discussed.

27.2.1 Radio-Resource Management in HetNets

In HetNets, it is expected that the above-mentioned integration and joint cooperative management of diverse RATs be made in such a way as to enable the network providers to satisfy as much as possible a wide variety of user/service demands in a more efficient manner. A few recent works [73–76] that attempt to achieve this objective are described as follows.

In [73], Ahmad et al. introduced a virtual resource provisioning scheme (called VPR) for HetNets that can be used to predict the resources released by multiple co-located wireless networks. Their scheme integrate the QoS requirements from each network entity, a load balancing approach, and a virtual resource provisioning mechanism to achieve precise resource allocation among the involved wireless networks. However, the authors did not elaborate on the behavior of their scheme when the traffic load at a given network or group of networks becomes heavy.

In [74], Lopez-Benitez and Gozalvez developed some techniques referred to as common radio-resource management (CRRM) to distribute heterogeneous traffic among the available RATs, while taking into account the radio resources available at each RAT. Their framework algorithms are shown to achieve appropriate user/service QoS levels based on some decision criteria used for determining the most suitable user-to-RAT assignment. However, the authors did not elaborate on how their proposed schemes would handle the application scenarios where the RAT selection decisions are based on the channel quality conditions. In such cases, there is no guarantee that the user's connection to the RAT can be maintained.

In [75], Hasib and Fapojuwo investigated the problem of maintaining an efficient use of different RATs resources in HetNets while keeping the desired QoS at minimal service costs from the perspective of both users and service providers. Their so-called adaptive common radio-resource management (CRRM) scheme is proposed, which uses service cost, user mobility, service type, and location information as design parameters. The merit of the CRRM scheme lies in the fact that the optimal RAT for new and current handoff calls is selected based on the load information locally available at the RATs, level, and their approach results in balancing the cell load of the coexisting RATs. However, the CRRM scheme requires a centralized resource management approach that sees a pool of resources from the constituent RATs. Also, no information is released by the authors on how decision parameters such as mobility information, location prediction in RATs, and vertical handoff have been incorporated in the CRRM design.

In [76], Atanasovski et al. studied the problem of convergence of wireless access networks and reported on an architecture for resource management in HetNets (called RIWCoS). This architecture relies on the concept of the Media Independent Handover Function (MIHF), a key component of the IEEE 802.21 framework. Although the RIWCoS framework is shown by simulations to be promising in handling emergency situations such as disaster management, its applicability requires that both terminal- and network-side resource management modules be implemented, under various assumptions and prerequisites.

In [77], Suleiman et al. investigated the joint radio-resource management problem in HetNets and proposed a solution that considers the features of component access technologies such as the occupancy level in individual access networks, load balancing, nature of individual access networks, variability of network resources according to traffic conditions, and asymmetry of access networks overlap, to name a few. A prototype implementation of their framework is presented and validated. However, this model does not consider the diversity of access networks, nor the user's mobility patterns and the impact of overlapping of different access networks.

In [78], Kajioka et al. also investigated the problem of resource management in HetNets. Their proposal is an adaptive resource allocation scheme in which each node determines by itself the wireless network resources to be assigned to every application that it supports. This is achieved through designing an attractor composition model that describes the global activity shared among the entities in HetNets. However, the proposed scheme was not tested on a real simulation test-bed using real scenario applications.

In the same vain, Pei et al. [79] studied the problem of radio-resource management and network selection in HetNets composed of CDMA and WLANs networks. In the case of CDMA, radio-resource management is achieved through solving an optimization problem that consider the intercell interference levels as a criterion for maximizing the total network welfare under the CDMA resource-usage constraints. For the WLAN part, radio-resource management is achieved through solving an optimization problem that maximizes the aggregate social welfare of the WLAN under the WLAN resource-usage constraints. In their proposed scheme, the method used to balance the load among mobile nodes was not disclosed.

In [80], Ngo and Le-Ngoc introduced two distributed resource allocation methods to optimally allocate subcarriers and power in an OFDMA-based cognitive radio ad hoc network. These methods are designed using the Lagrangian dual optimization where the throughput (or energy efficiency) is maximized subject to few constraints such as the tolerable interference at the primary network level, and the enforcement on the lower and upper bounds on the number of subchannels that each individual unlicensed users may occupy, to name a few. The devised dual approach is shown to guarantee globally optimal solutions in polynomial-time complexity dependent on the total number of subcarriers. However, the efficiency of the proposed protocol in the case of imperfect network information was not established.

27.2.2 Power Control, Rate Control, Access Control, and QoS Support in HetNets

The demand for seamless routing across the different networks that make up HetNets, where wireless networks and mobile users are expected to coexist, will impose new challenges such as guaranteeing the QoS and supporting a variety of services from the perspective of both users and service providers, increasing the battery capacity of mobile terminals, designing time decision algorithms that decide the correct time for a user to request a vertical handover, and so on. A few representative works on these hot research areas are described as follows.

In [81], Wang and Jiang investigated the problem of QoS support in HetNets, and proposed some resource reservation protocol extensions (called RSVP) in the form of QoS software agents that handle the RSVP QoS update messages and reservations for mobile nodes, resulting in real-time services support in HetNets. However, in their schemes, a method for preventing unnecessary pre-reservations to occur was not designed, which might lead to excessive control communication overhead.

In [82], Tsamis et al. also studied the QoS requirements in HetNets and proposed a Markov-based stochastic model for estimating the bandwidth and delay in HetNets. Their model uses finite state Markov chains and a set-oriented stochastic approach to model the dynamics generated by the variation of round-trip times and available bandwidth of wireless networks that compose the HetNets. In their model, the states were selected based on the probability distribution of available bandwidth of each single network. A machine learning technique was then used to describe the state space. Also, a distributed QoS-based rate control mechanism was designed to validate the proposed dynamic bandwidth model and estimator. However, from a practical point of view, there was not sufficient experimentation with the proposed model, for instance, in evaluating the flow assignment policies in environments.

In [83], Qadeer et al. introduced an approach (in the form of a policy framework) for managing the power and performance of mobile devices in HetNets. The designed policy is based on the user and system QoS requirements as well as the power dissipation at each node of the network, and it can be used to make the decision on parameters such as the wireless network interface card to be used, the buffer size to be used to expect the desired QoS, and transition times between active and low-power states, to name a few. The proposed scheme was tested on IPAQ 3970, which supports both 802.11b and Bluetooth for MPEG video and MP3 audio. However, its simulation using real-time applications was not investigated.

In [84], Kim et al. studied the integration of WLAN with 3G mobile networks. They proposed an interworking architecture that can support QoS [through a wireless distributed coordination function (DCF)-based MAC scheme design] and seamless handover service (via a loosely coupled interworking with the Mobile IP approach) in the combined networks. However, in their scheme, the network load state is dependent on a prescribed threshold value and no method was provided to systematically determine this value.

In [85], Sharna et al. investigated the problem of call admission control (CAC) policies for HetNets made of an integrated cellular/WLAN system handling various types of calls (vertical handoff calls, horizontal handoff calls, and new calls). They proposed a Markov decision process-based CAC scheme in the form of an optimization control problem along with an algorithmic solution (known as a value iteration algorithm) that can enable HetNets to support multiple service classes with different bandwidths and variable channel requirements. Although the proposed scheme holds great promise, the authors did not contrast its performance against the few existing benchmarks such as the one proposed in [86].

In [87], Al-Manthari et al. investigated the problem of congestion in wireless access systems and proposed a monetary incentive-based admission-level bandwidth management protocol that integrates a call admission control scheme and dynamic pricing mechanism, leading to a congestion-free system while improving the packet-level QoS of ongoing connections when congestion occurs. However, their proposed scheme relies on the fact that the user demand models should be accurate in predicting their reaction to prices, which cannot be guaranteed.

In [88], Haydar et al. introduced a network access selection method for HetNets that are composed of UMTS, WiFi, and WiMAX. Their proposed access decision algorithm accounts for the traffic distribution to the different available RATs while taking into consideration the user's communication limitations and the service (i.e., demanded bandwidth) and QoS requirements. However, the authors did not study the effect of handover in case of user mobility.

In [89], Haldar et al. also investigated the network access selection problem in HetNets, and proposed an architectural framework made of two domains: user and network, each of which is composed of multiple units interacting through signaling and information sharing. In their model, channels (thereby the network selection) are classified and selected based on transmission power constraints and adjacent neighbor occupancy. A modified Hungarian algorithm is designed to match the desired user applications to appropriate channels and networks. However, as reported by the authors, their model generates a high number of handovers.

In [90], Ali and Pierre investigated the challenge of providing seamless ubiquitous access to the various overlaying RATs in HetNets, by providing a multiaccess bandwidth management technique for QoS during handovers (both vertical and horizontal). In their guard band-based admission control policy, a computing method of the optimal guard bandwidth for HetNets is proposed. However, their proposed admission policy scheme was reported to generate excessive handoff latencies.

In [91], Yu et al. introduced a joint admission and rate control (JARC) protocol for QoS provision of multimedia sharing in IEEE 802.11e-based wireless home networks, tailored to HetNets. Their scheme was shown to reduce collisions while meeting the prescribed QoS guaranteed transmissions at the MAC layer and achieving QoS-based multimedia sharing at the application layer. However, the proposed scheme relies on the assumption that the strategy for managing the QoS is achieved as a function entity in the wireless home gateway, which is a major constraint from a practical prospective.

In [92], Falowo and Chan investigated the problem of unfairness in radio-resource allocation among low-capability heterogeneous mobile terminals in HetNets. They proposed a terminal-modality based joint call admission control (JCAC) algorithm that considers the terminal capability (known as the RAT terminal support index) and the network load to make call admission control decisions. A Markov chain was designed to model the proposed scheme, showing that it can achieve fairness in radio-resource allocation among the mobile terminals. However, in the proposed scheme, the method used to maintain a lower handoff dropping probability at the expense of the new call blocking probability is deficient in the sense that when none of the available individual RATs in the HetNet has enough bandwidth to support the incoming call, the call is either dropped or blocked.

In [93], Tragos et al. presented a short survey of admission control mechanisms in 2G and 3G networks. They also designed a generic admission control algorithm for HetNets that can be used to control the admission of new or handover sessions while controlling the load of the network, and maintaining the prescribed session QoS. Their scheme is expected to enable some form of cooperation between the WINNER scheme [94] and legacy systems (such as UMTS and WLAN 802.11b). Although the feasibility of the proposed approach was proven by simulation, there is still a lot more work to be done in order to fully achieve the above-mentioned goal.

In [95], Ouyang et al. introduced a call admission control (CAC) policy for voice and data calls in HetNets based on one-hop cooperation among mobile nodes. In their scheme, mobile users have the capability of accessing simultaneously the access cellular network and a WLAN via a node in the WLAN-covered area acting as a relay. As pointed out by the authors, phase-type distributions can be used to characterize the user's mobility and traffic model, making it generic enough to be applicable to relay-based cooperative HetNets. This is yet to be investigated.

In [96], Farbod and Liang also investigated the issue of call admission control (CAC) policies for HetNets composed of wireless mesh networks and overlay cellular network. They proposed a CAC model that can achieve near-optimal performance in the control of these networks. The core of their CAC problem space is designed based on a Partially-Observable Markov-Modulated Poisson Process (PO-MMPP) traffic model that relies on the overflow traffic provisioning from the underlaying mesh to the overlay cellular network as a key design feature. Indeed, using this model, the burstiness of the overflow traffic is captured, and based on this, the overlay network is modeled as a controlled PO-MMPP/M/C/C queuing system, which in turn allows for the design of a computational algorithm that determines the optimal CAC policy. However, the proposed model considers only two-tier HetNets and was not tested in a variety of HetNets architectures.

In [97], Guerrero-Ibanez et al. investigated the problem of service provisioning in HetNets. They proposed a QoS-based dynamic pricing approach (so-called QoS-DPA) in which an access network selection mechanism is implemented to select the appropriate network for each requested service while maximizing the provided QoS level according to the user's preferences. A pricing strategy based on subjective QoS tolerance levels is invoked in cases where the QoS cannot be fulfilled. However, the proposed scheme heavily depends on the dynamics of the user subscriptions, and no method is provisioned by the authors to handle this issue.

In [98], Porjazoski and Popovski introduced a multicriteria decision making algorithm for RAT selection in HetNets based on service type and user mobility. Their proposed algorithm is made up of two parts, the initial RAT selection algorithm and the handover algorithm. Both are designed based on an analytical Markov chain model that allows the selection of an appropriate RAT that satisfies the user's QoS requirements while providing high radio resource utilization. In their algorithm, the network load was not considered as a design criterion in the RAT selection process.

In [99], Smaoui et al. investigated the problem of service delivery and dynamic access selection in HetNets. They proposed an access selection technique that effectively assigns traffic flows to available access networks. Their algorithm considers power consumption cost, mobile velocity, load rate, and RSS as design decision factors to yield the optimal wireless network selection. However, no simulation of the algorithm was conducted in a real-time simulation platform with real scenario designs.

27.2.3 Prefetching and Caching in HetNets

Due to the complexity of HetNet architectures, proactively increasing the system's performance using the knowledge of mobility to predict users' future locations and to capture the likelihood of users moving into different parts of the HetNet, is an important technique that has been and continues to be investigated through prefetching schemes. A few such works are captured in [100,101], and the references therein.

In [100], Drew and Liang proposed a Web prefetching scheme for HetNets (here, HetNets is made up of a conjunction of multiple arbitrary size WLANs and a cellular wireless data network), in which mobile nodes can adjust their prefetching decision based on the network topology and mobility pattern. In their proposed scheme, the node's location information is used as a design criterion to dynamically adjust the amount of prefetching performed. However, the proposed scheme was only shown to be applicable to the above-mentioned specific type of HetNets.

In [101], Goebbels investigated the problem of wireless broadband service provisioning in HetNets, and proposed a smart caching protocol that combines smart caching and hierarchical Mobile IP to improve the streaming and delivery of data while ensuring load balancing at each node. The efficiency of the proposed protocol was proven by simulation on specific HetNet architectures only.

27.2.4 Topology Control and Relay Placement in HetNets

In HetNets, network capacity and energy efficiency are two major factors that can affect the network connectivity, and thereby the network performance. Topology control and relay placement can be used as alternative methods to address this problem. The main goal of topology control is to determine the appropriate neighbors of a node from a given network architecture based on predefined rules, in such a way that the nodes' transmission power/energy are adjusted to increase the network lifetime. In the sequel, a few representative works on topology control for HetNets are discussed.

In [102], Li and Hou investigated the problem of topology control in HetNets and designed two algorithms [called the Directed Relative Neighborhood Graph (DRNG) and the Directed Local Minimum Spanning Tree (DLMST)]. In their schemes, each node of the network constructs its neighbor set and adapts its transmission power using only the local collected information. Unlike previous proposals, their schemes account for different maximal transmission ranges. Their schemes were proven to outperform other known topology control algorithms in terms of average node degree. However, the effect of MAC-level interference on network connectivity was not investigated.

In [103], Li and Fang proposed a scheme that illustrates how HetNets can give better throughput than traditional homogeneous wireless networks. In their approach, powerful wireless helping nodes are deployed instead of placing base stations interconnected by wired networks to improve the capacity, and the shape of the networks is considered to be rectangular. Through simulations, the authors proved that their scheme can outperform the traditional schemes in terms of per-node throughput, but this is true only in the case of experimentation with regular and random HetNet architectures.

In [104], Ren and Meng addressed the problem of power management and topology control for heterogeneous wireless sensor networks and proposed a game theoretical approach in the form of utility functions for nodes to achieve the desirable node degree and frame success rate. Their proposed joint topology control and power scheduling algorithm is derived from a cross-layer optimization design, itself based on the establishment of some relationships between transmit power control, topology management, and a property of network reliability. However, no simulations of their algorithm was conducted using real test-bed and scenario applications.

In [105], Zhu et al. presented a localized topology control algorithm for HetNets based on clustering, where topology control is achieved through adjusting the node transmission power. Through simulations, the authors showed that their scheme can significantly reduce the energy consumption in the network. However, as pointed out by the authors themselves, the proposed scheme could not handle the scenario where the network's connectivity fails.

In [106], Zhang and Labrador proposed a topology control algorithm [called residual energy aware dynamic (READ)]. The topology control problem is formulated and solved as a power assignment problem rather than a range assignment problem. In their proposed algorithm, the sender's maximum transmission power, the node's residual energy, and the sensitivity of the receivers are used as criteria for determining the final topology. The proposed scheme was shown to increase the network lifetime, however, its applicability is heavily dependent on the type of topologies built by the algorithm.

In [107], Aron et al. investigated the power control problems in heterogeneous wireless mesh networks and proposed an energy-efficient distributed topology control algorithm for such networks that can guarantee maintenance of the network connectivity while reducing the amount of energy consumed by each node during transmissions. However, node mobility was not incorporated in this design.

In [108], Cardei et al. investigated fault-tolerance issues in topology control for heterogeneous wireless sensor networks made of resource-constrained wireless networks randomly deployed and resource-rich supernode wireless sensor networks located at known locations. A group of algorithmic solutions are proposed, each of which minimizes the maximum transmission power for all sensor nodes. However, their proposed algorithms do not address the case of fault-tolerant bidirectional topology where communication paths exist from sensors to supernodes and vice versa.

In [109], Sun et al. presented an energy-efficient, localized, and dynamic topology control algorithm for wireless networks [called the energy-aware weighted dynamic topology control (WDTC) algorithm] tailored to HetNets. WDTC is an enhancement to the local MST algorithm [110] in the sense that the topology is adjusted dynamically with the node energy consumption, resulting in a lower transmission radius and bidirectional property. A discussion on how to address the network connectivity problem resulting from topology control using the proposed analytical approach was not given.

In [111], Pei and Mutka studied the problem of joint relay placement and routing for bandwidth sufficiency in heterogeneous wireless sensor networks. The problem is formulated as a variant of the Steiner tree problem [called heterogeneous bandwidth Steiner Routing (HBSR)] given the link capacity, traffic demands, a set of stream sending sensor nodes, and a base station as input parameters. A heuristic was proposed whose solution significantly reduces the relay number compared to that of the benchmarking algorithm. However, the experimentation of the proposed algorithm in a real simulation test-bed was not investigated.

27.3 PRELIMINARY

27.3.1 Heterogeneous Wireless Networks

27.3.1.1 Introduction

In the HetNet environment, each wireless access network has its own characteristics such as capacity, access technology, security, power consumption, delay, coverage, and access cost, to name a few [17–19]. An interesting feature of HetNets is the fact that some wireless access networks are overlaid by others in such a way that a

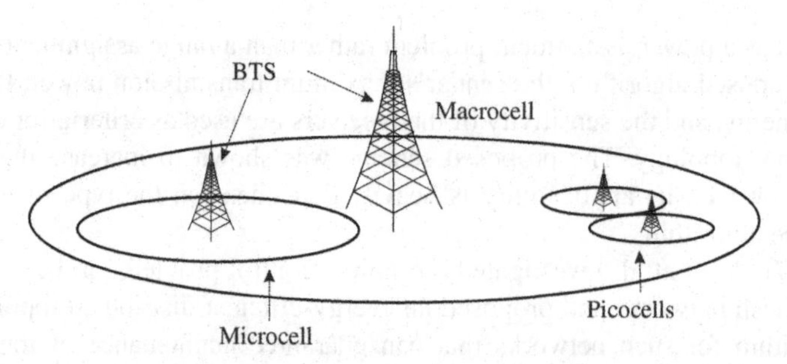

Figure 27.1 HetNet architecture.

multilayer structure or a hierarchical cellular mobile network [11] is naturally built as illustrated in Figure 27.1. This architecture of overlay networks can be suitably explored to match multiple design purposes such as:

- *Boosting system capacity:* By combining the capacity of each individual wireless access network within the intended coverage region, the whole system may support many more users. This fact leads to the reduction in blocking/dropping probabilities of new/handoff calls by offering alternative access points (APs) during overload situations [23,24].
- *Increasing system coverage:* For instance, by combining WiMAX and cellular mobile networks, a large geographical area can be covered.
- *Enhancing user satisfaction:* Given the differences in technologies and data rates, each wireless network can be employed to satisfy a specific target. For example, often WLANs or picocells have been utilized to furnish access in hot spot areas such as airports, restaurants, and shopping centers, while cellular mobile networks have been used for ensuring user mobility.
- *Offering different access costs for end-users:* In practice, some end-users are willing to pay high prices for wireless access given their social position, economic situation, job, or necessity. However, the major portion of users would like to be connected with the access network that provides the lowest cost and an appropriate QoS. Aware of this fact, the MNOs have designed market strategies that are appealing to all classes of end-users, aiming at increasing their profits by holding and attracting new end-users [20,21]. Therefore, pricing strategies arise as one of the most important design criteria in HetNets. Thus, given the mix of access networks with different prices, an end-user is able to decide among the available options which network better fits its pocket and expectations.

In order to make full use of HetNets and enjoy the advantages that these networks offer, a mobile user must be equipped with a multi-interface device that is able to sense and connect with the access network that matches his/her personal expectations and the requirements of his/her applications.

27.3.1.2 Types of HetNets

HetNets are formed by mixing different wireless access networks. Therefore, in order to know how to take advantage of them, it is mandatory to have a clear comprehension about the characteristics of each type of wireless access network. In the following analysis, the discussion is split by separating wireless access networks according to their size.

The most popular type of wireless access networks is the cellular mobile networks that fall into the class of wireless wide area network (WWAN). In general, the taxonomy of cellular mobile networks usually classify them according to their generation. The 2G wireless networks, which are quite widespread today, deliver traditional voice service, but also data service at rates up to 22.8 kbps. The GSM and TDMA/136 standards have also been and continue to be very popular. With the growth of the Internet, the requirement to support data services with high data rates started to push the wireless industry to develop new standards. This was the impetus for the GPRS data service that appeared with 2.5G wireless networks and delivered data rates in the range of 9.05–21.4 kbps [26, 27]. Following this, EDGE systems came as a bridge between GPRS technologies and 3G systems. The goal was to provide higher data rates for packet-based services. For instance, for EGPRS (the packet-switched version of EDGE), it was possible to achieve data rates in the range of 11.2–69.2 kbps [28–30]. With the advent of 3G systems, WCDMA and CDMA200, the goals of seamless access to mobile systems and multimedia applications and convergence between mobile and fixed networks started to be achieved [31].

Recently, wireless metropolitan area networks (WMANs) based on the IEEE 802.16 standard have attracted attention for providing wireless access with higher data rates (up to 70 Mbps) in the range of 50 km. In this sense, WiMAX was developed with explicit support to QoS, resulting in an excellent solution for broadband wireless access not only in urban regions, but also in underdeveloped regions, where wired-based solutions are unavailable [35–37]. Factors such as low cost and operation in an unlicensed frequency band were the driving forces for wireless local area networks (WLANs) based on the 802.11 standard. Supporting higher data rates in the range 54–100 Mbps, within a short coverage, i.e., 100 m, WiFi systems are very often found in places like coffee shops, airports, offices, etc. The necessity to support different types of applications associated with the lack of QoS support for inelastic services (such as voice and video) was the roadblock for WLANs. To overcome these drawbacks, the 802.11e standard came to the light [32–34]. The necessity to supply rural regions in the world with broadband wireless access solutions has suscitated the development of IEEE 802.22 wireless rural area network (WRANs), the first *de facto* cognitive radio standard. The basic idea behind IEEE 802.22 is to use TV broadcast bands by means of cognitive radio technology as long as they do not cause harmful interference to incumbent TV receivers to deliver broadband services [38,39]. The possibility of delivering medical services anytime and anywhere is now one of envisioned applications for the IEEE 802.15 wireless personal area networks (WPANs). Networks in this standard are called piconets given their shorter coverage area of 10 m, supporting up to eight devices [40].

Many new technologies have empowered and challenged the design of wireless networks. Cognitive radio and cooperative cellular networks are examples of such technologies. In recent years, due to the growth in the offered traffic and some limitations imposed by the use of the current static spectrum allocation policy in terms of transmission power, frequency, location, type of spectrum usage, etc., the Federal Communications Commission (FCC) has decided to invest in a new wireless access paradigm called Dynamic Spectrum Access (DSA), which is based on cognitive radio technology. DSA is a way of boosting the spectrum utilization by allowing unlicensed or secondary users (SU) equipped with cognitive radio to use the allocated but idle frequency bands (also called spectrum holes) in the licensed spectrum as long as they do not harmfully interfere with the operation of licensed users or primary users (PU). The wireless network designed to bolster DSA strategies has been called Cognitive Radio Network (CRN) [41–44]. Cooperative cellular mobile networks is a quite different approach. Instead of using the same frequency band to increase the system's capacity as CRN does, it uses relay nodes to offer alternatives paths from BTS to mobile users. This results in an improvement of the system in various dimensions such as performance, reliability, network coverage, spatial frequency reuse, etc. [45,46].

27.3.2 Radio-Resource Management

The problem of resource management (RM) can be roughly defined as the problem of assigning a server for a incoming service request as long as it does not violate the service provisioning of the other ongoing users and there is enough capacity to accept it. In wireless networks, this problem gains another dimension because the radio resources are extremely expensive and scarce, the offered traffic load is huge, and the QoS requirements of the applications are quite diverse and often conflicting. To cope with all of these objectives, radio RM (RRM) appeals to the following traffic management mechanisms: call admission control (CAC) and scheduling.

The main function of CAC in multiservice networks is to maintain the QoS provisioning in face of the service integration while fulfilling the QoS requirement of each service class. To achieve this objective, CAC makes a decision about the acceptance of incoming service requests considering the QoS requirements of each one and the amount of idle resources. Mathematically, CAC does the following computation:

$$B(x) = \Delta_{i\in(1,2,\ldots,N)}(x) + \sum_{i=1}^{N} b(i)n(i), \tag{27.1}$$

where $B(x)$ is the network state after the admission, $\Delta_{i\in(1,2,\ldots,N)}(x)$ is the increment in the resource occupancy caused by the ith incoming service request, N is the number of service classes, and $b(i)$ is the bandwidth of the nth service class. Therefore, CAC computes $B(x)$ and checks if $B(x) \leqslant C$, where C is the system capacity. If the answer is yes, the service request is accepted; otherwise, the service request is blocked. Usually, CAC handles calls at the connection level; however, with the adoption of packet-switched technology in wireless systems, an effective CAC must also respond well at the packet level [47].

Scheduling is another RRM mechanism widely employed in the design of wireless systems. While CAC tackles the incoming calls that intend to be connected, scheduling copes with the problem of determining which data packet from which data session already accepted by the CAC will be transmitted with the radio resources. Many scheduling techniques have been proposed in literature [48–50]. The simplest one is the well-known first in first out (FIFO) discipline, in which the data packet that comes first will be first served. Priority queueing (PQ) is another simple such technique. In PQ, there are different priority buffers that store their data packets according, for example, to a FIFO scheme. Then, the scheduler schedules the traffic classes based on the occupancy of the higher priority buffers. Although they are simple, FIFO and PQ exhibit poor performance when dealing with real-time applications since they do not provide any fairness among the distribution of the bandwidth. Thus, more sophisticated scheduling techniques are required to handle this traffic class. In general, these techniques are built over the idea of the weighted fair queueing (WFQ), which provides fairness among data flows by using some weights to prevent the monopolization of the bandwidth by some flows [31]. Some variations of the WFQ scheme for wireless systems are idealized wireless fair queuing (IWFQ), wireless packet scheduling (WPS), channel condition independent packet fair queuing (CIF-Q), and channel state depended packet scheduling (CSDPS), to name a few [48–50].

27.3.2.1 RRM in HetNets

Traditionally, in homogeneous wireless networks, the RRM functionalities are tackled with just one network. In overlay homogeneous cellular mobile networks, each layer operates independently of the others, then the same practice is applied and the RRM algorithms only have to handle the incoming service requests offered in each layer. For instance, CAC in homogeneous wireless systems only has to decide whether or not a user is accepted.

However, the actual scenario in which heterogeneous wireless access networks cover a geographical area imposes new design paradigms to the RRM algorithms. For example, due to the offered traffic load variations in space and time, some cells located somewhere in the covered region experience overload situations, while others, on the other hand, are quite idle. As a consequence, wireless resources are poorly utilized and the offered traffic load is deficiently carried out. To overcome these drawbacks, joint RRM (JRRM) or common RRM (CRRM) has been investigated in the literature. The main advantage behind JRRM algorithms is the fact that they have the whole vision of all layers (and cells inside them) based on which they can cope with the following tasks [12]:

- Deciding whether an incoming service request should be accepted or blocked
- Selecting in which of the cells an incoming service request has to be accommodated

The first objective of JRRM relies on the idea behind homogeneous wireless systems, but the second takes advantage of the fact that one of the available wireless access networks can be chosen as long as it better suits the user's expectations in terms of cost, delay, data rate, etc.

According to [51], the JRRM functionalities can be split into three procedures: resource monitoring, decision making, and decision enforcement. The first procedure is responsible for acquiring the information about users' preferences (cost, QoS requirement, wireless technology, etc.) and the state of the wireless access networks [52]. With this information in hand, the second and most important procedure, i.e., the decision making, is invoked. In general, two types of decisions may be taken: network selection and bandwidth allocation [53–55]. Furthermore, this choice can be done by the user (user-centric approach), or by the MNO (network-centric approach), or by both in a collaborative fashion. A plethora of decision-making algorithms are available to support this phase. Some of these decision-making algorithms are: Markov decision process (MDP), fuzzy logic, game theory, and TOPSIS. Finally, the decisions are executed in the decision enforcement phase.

Ultimately, the design of JRRM strategies will be mandatory for better radio-resource management of all wireless access networks located in a certain geographical area and, therefore, will be indispensable in 4G wireless networks.

27.3.3 Queueing Theory

In homogeneous or heterogeneous wireless networks, RRM algorithms (in particular CAC algorithms) often account for the network state and QoS requirements of the incoming service request when performing their decision making. In general, in network design, it is assumed that once the bandwidth requirement is satisfied, the remaining QoS needs such as jitter, delay, are also met. In spite of varying from one service class to another and even during the call, the level of QoS demanded is known. For instance, incoming calls often specify the QoS profile needed in their service requests. Alternatively, they can be retrieved from the subscriber profile stored in network databases. From that perspective, a methodology is required, which is capable of not only describing the network state, but also storing the system's evolution according to the events that rule its dynamic. Furthermore, this methodology must be able to capture the probabilistic behavior of the communications network. For instance, traffic fluctuations and service durations, which are mathematically represented by some sort of probability distributions. Finally, since it is quite difficult to design a network to cover all possible operation scenarios, it would be interesting if somehow this methodology provided information about how the system performs on average or in steady-state based on some workloads and system setups such as number of resources available and service durations. This is the role played by queueing modeling.

Queueing theory is a complex and widely applicable analytical framework that has been proven for decades to support the development of performance models used in traffic theory. In a nutshell, most of the queueing models are built over the memoryless property, which considers that the future is independent of the past, given the knowledge about the present state. Regarding the probabilistic behavior, the memoryless property is found in exponential distributions (continuous time) and geometric distributions (discrete time).

From the RRM viewpoint, the events that rule the system dynamic can take place anytime, thus, the exponential distribution or a combination of exponential distributions is usually employed to represent the arrival processes and service time distributions. The arrival processes represent the sources of incoming calls, and as such, they are of upmost importance in the specification of the performance model. In overlay wireless networks, two types of traffic models are often employed, namely, the Poisson process and the Markov-modulated Poisson process (MMPP). The Poisson process assumes that the arrivals take place independently of one another, and independently of the service process. In spite of its simplicity, the Poisson process represents very well the arrival process in wireless networks, mainly at the connection level [11, 16, 22]. Additionally, it yields very simple and elegant results [58]. However, this assumption does not always reflect the truth. For example, multimedia traffic may have time-varying arrival rates and correlations between interarrival times. In this case, it is preferable to work with MMPP [58]. MMPP can be defined as a stochastic Poisson process in which the rate varies according to a underlying Markov chain. This process is completely specified by the transition rate matrix Q (also known as infinitesimal generator) and the Poisson rate matrix Λ. In the design of performance models for the RRM mechanism, the MMPP with two states or its special case called the interrupted Poisson process (IPP) has been used more often. Figure 27.2 illustrates this traffic source in which μ_1 and μ_2 are the Poisson arrival rates in states 1 and 2, and σ_1 and σ_2 are the departure rates in the respective states. In the case of IPP, either μ_1 or μ_2 is zero. The matrices Q and Λ for the two state MMPP are computed as follows:

$$Q = \begin{bmatrix} -\sigma_1 & \sigma_1 \\ \sigma_2 & -\sigma_2 \end{bmatrix}, \tag{27.2}$$

$$\Lambda = \begin{bmatrix} \mu_1 & 0 \\ 0 & \mu_2 \end{bmatrix}. \tag{27.3}$$

The service process in wireless cellular mobile networks is often called the channel holding time $T_h = 1/\mu_h$, which is a function of two exponentially distributed random variables, namely the call time duration $T_d = 1/\mu_d$ and the stay time (or dwell time) $T_s = 1/\mu_s$. The channel holding time is computed as follows:

$$T_h = \min\{T_d, T_s\}. \tag{27.4}$$

Since T_d and T_s are exponentially distributed random variables, T_h takes the form of an exponential distribution as well with mean: [11,16, 59]

$$E[T_h] = \frac{1}{\mu_h} = \frac{1}{\mu_d + \mu_s}. \tag{27.5}$$

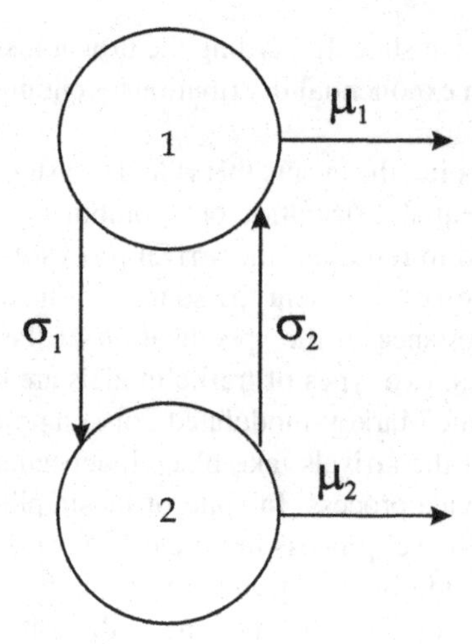

Figure 27.2 Two-state MMPP.

In order to completely represent a system by means of the queueing theory, one needs to specify the system's resources, which in the case of RRM, are the number of radio resource channels and the buffer if the system supports it. An interesting point about the system capacity is that in practical wireless systems, it depends on the technological implementation of the radio interface. However, very often, in the literature of queueing models, the system capacity is translated into effective or equivalent bandwidth, regardless of the multiple access technology (FDMA, TDMA, CDMA, or OFDM) employed in the air interface [56, 57]. Traditionally, when queueing theory is applied in the design of RRM schemes, the state comprises the information about the number of users belonging to a specific network or service class, or a combination of both. This same guideline will be followed when designing our proposed green JRRM in HetNets.

27.4 GREEN APPROACHES IN HETNETS

Green design of communications networks is an area of research that has recently attracted much attention. As a consequence, many interesting works can be found debating this subject. The objective of this chapter is to study how the coverage can be properly managed and the offered traffic load balanced in order to save energy and provide a satisfactory QoS in HetNets. Thus, our focus is on techniques that can contribute to addressing this problem and provide the readers with a general view about it. Notwithstanding, for the readers interested in learning more about green wireless systems, References [2, 3, 10] can be referred to as a starting point. With respect to HetNets, their overlay structure presents an immense opportunity to reduce the power usage by applying dynamic coverage management (DCM). With DCM, the

JRRM scheme can determine, based on some criteria, when and which cells inside the covered geographical area can be turned off or on [10]. This way, the DCM function updates the role of the JRRM (here also referred to as green JRRM) in HetNets, which is to save energy and balance the traffic load while achieving a suitable QoS.

Undoubtedly, the main idea behind the application of DCM algorithms is the traffic load fluctuations in time and space. In order to cope with it, combinations of cells with different sizes and capacities are often employed. One of the biggest challenges of HetNets is to find the optimal or at least a good combination. For instance, on one hand the design using only macrocells is quite inefficient due to the high data rates required for actual data services; on the other hand, the design with only pico/microcells suffers from the small coverage, which can lead to high handoff rates. From the energy-efficiency perspective, it has been shown that [10,13] it increases as the number of micro/picocells increases. In spite of it, a dense deployment of femtocells (taken as example) can lead to a significant energy consumption [2]. However, by applying power saving schemes that turn femtocells off when no traffic load is offered, a power savings of more than 30% in average can be obtained [14]. In [1], Lorincz et al. considered a dense deployment of WLANs and applied integer linear programming to determine, according to the daily traffic demand, which access point (AP) should be switched on and off. Results have pointed out an energy savings of more than 50%. However, the analyzed scenario did not consider an environment with macrocells.

The above-mentioned results indicate that we are still uncertainty is this field. According to [2], addressing energy efficiency at the RRM level in HetNets still deserves special attention. Our proposal is an attempt in this direction.

27.4.1 Stand-Alone Approach

With respect to HetNets, but not to the issue of dynamic coverage management, another line of research exists that can contribute to the question of power consumption reduction, namely, network selection. In [60], Chamodrakas and Martakos have considered energy efficiency inside the problem of network selection in HetNets. In this sense, the goal was to choose the network that best balances the performance (referred to as delay and bandwidth) and energy consumption. To achieve this end, the authors used a utility-based fuzzy TOPSIS method. Likewise, the authors in [61] investigated network selection taking into account the goal of energy efficiency. In addition, the power consumption due to network scanning was considered. As network selection is one of the main problems in HetNets that involves users and the network or both, dealing with energy efficiency in the network selection process plays a key role in the overall system design because it can reduce the power consumption in both the device and the network since it represents the entrance for the mobile user in the HetNet.

Another important step toward achieving energy efficiency in wireless networks is the inclusion of SON (Self-Optimization, Self-Organization, Self-Healing) algorithms on the LTE development agenda. SON algorithms arise as an important axis in the design of green communications as long as power consumption reduction is set as one of their goals. SON algorithms are a response from the network management community to autonomously cope with the increased complexity in wireless networks. So far, SON algorithms have been envisioned to cope with handoff optimization, antenna tilt optimization, and interference minimization [62–64]. Concerning the strategies to achieve green RRM in wireless networks, SON algorithms have been used to optimize the frequency reuse (FR) scheme in cellular mobile networks (including LTE and relay systems) [65] according to the goals of spectral efficiency, fairness, and energy efficiency, with the ultimate goal to dynamically adapt the FR scheme to match the spatio-temporal variations of the offered traffic.

27.4.2 Cross-Layer Approach

In order to facilitate the hard task of designing a communications system, the entire process was broken down into different hierarchical steps, in which each step is associated with a layer within the protocol stack. In this stack, each layer comprises the set of functionalities, which is used to provide a service such as routing and error detection, to name a few. Furthermore, a layer uses the services of the inferior layer and supplies the superior layer with its services. In this layering paradigm, each layer is quite independent of the others. As a consequence, the layers can be designed separately. Even though this process that embodies all aspects of system design, standardization, and development has been used quite well in the case of wired networks, unfortunately, it does not equally satisfy the design requirements for wireless systems. In this sense, the cross-layer idea violates the concept of independence among layers by designing a system that jointly combines the functionalities in each layer, with the goal to achieve better performance [67].

As the problem of energy consumption spreads over all layers of the protocol stack, an approach using cross-layer design to match this objective arises as a promising line of research. For instance, MAC schemes share the medium among the users who want to transmit [66]. Due to the finite system capacity, they have to balance user performance and whole system performance. If one user takes priority over the other users, these users will experience a longer transmission delay. According to [68], more energy consumption will then be generated when the circuit power is taken into account. In order to solve this issue, the remaining users will have to increase the modulation order. Again, according to [68], if a flat-fading channel is taken into account, then an increase in modulation order will decrease the energy efficiency. This logic suggests that the current proposed MAC schemes need to be revisited in order for these schemes to be able to cope with energy savings. Determining the optimal size of a variable length TDMA, which in turn leads to the determination of the optimal constellation size, is an alternative solution in this case. From the device viewpoint, an attractive approach to deal with energy savings in MAC schemes is to prolong their sleep mode as long as possible in order to reduce the power consumption. By doing so, mobile users determine when to wake up in order to receive or transmit the data packets.

27.5 GREEN JRRM

27.5.1 Introduction

In this section, our proposed threshold-based green JRRM in HetNets is presented. This scheme relies on the fact that the differences in the coverage areas in HetNets can be explored to balance performance and energy consumption. The core idea behind the proposed green JRRM scheme is the use of the DCM algorithm.

27.5.2 Green JRRM Design

27.5.2.1 Motivation

The discussions in the previous sections revealed that the green planning of wireless cellular mobile networks is currently a real concern due to both social consciousness and economic factors. As long as the BTS is the killer of energy in wireless networks, every attempt to turn it off so as to avoid a waste of energy is welcome. In a planar wireless structure, this practice might lead to some holes in the coverage. But, in the overlay structures, the coverage of active BTSs may be combined to carry the traffic load of inactive BTSs.

27.5.2.2 Description of DCM algorithm

In HetNets, a green JRRM scheme performs dynamic coverage management by invocating the DCM algorithm that can apply different criteria to choose which microcell (BTS in fact) can be turned off. In this chapter, this issue is addressed by proposing a threshold-based green JRRM scheme. Figure 27.3 depicts the architecture of the proposed green JRRM in which the Load Control (LC) module is depicted, in addition to the DCM module. The function of the LC module is to periodically monitor the macrocell and microcell occupancies and transmit this information to the DCM module upon request. With that information in hand, the DCM algorithm can decide which cells can be turned off or on.

For the sake of simplicity and without loss of generality, Figure 27.3 outlines an environment with two layers, the inner one corresponding to the microcell and the outer one corresponding to the macrocell. Each microcell and macrocell has C_{m} and C_{M} radio channels to provide wireless access for mobile users. Besides, the macrocell has a threshold K, which is used to aid in the decision about turning on or off a BTS. It is worth mentioning that the value of the threshold K may be strategically chosen by traffic engineers, taking into account the MNO objectives in terms of its sustainable development policy.

In our analysis, it is assumed that the macrocell is always on. The purpose of it is to ensure that mobile users will experience only a small chance of being dropped during handoff attempts. Moreover, the macrocell is designed to cover the entire geographical area even when a microcell is turned off. These assumptions are of paramount importance to maintain the main system objectives, i.e., providing wireless access and supporting user mobility.

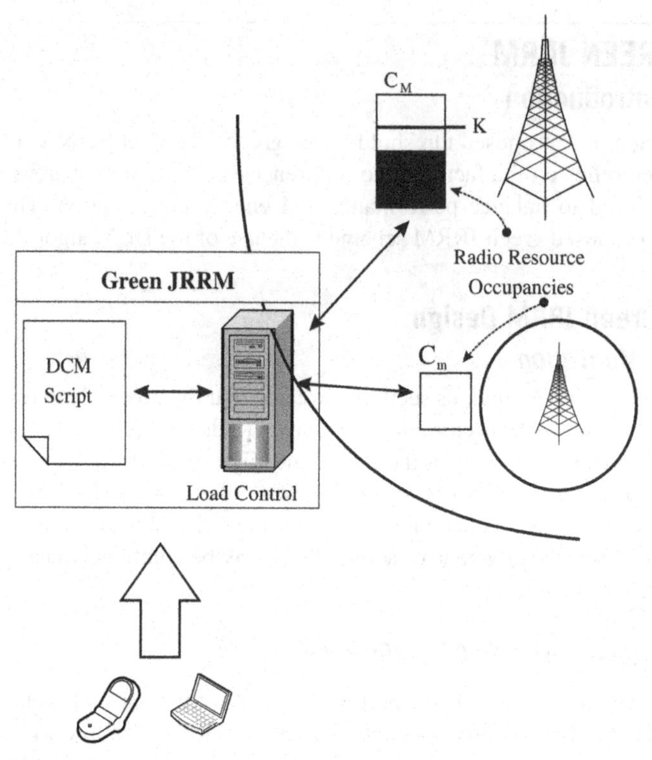

Figure 27.3 Green JRRM architecture.

Figure 27.4 illustrates the procedure behind the green JRRM scheme, where ℓ_M denotes the number of ongoing calls in the macrocell and ℓ_m denotes the number of ongoing calls in the microcell. This procedure works as follows.

In the green JRRM, when a mobile user desires to communicate, it starts by issuing a service request in which the QoS profile is specified. For the sake of simplicity, it is assumed that one bandwidth unit is used.

When the incoming service request is received, the green JRRM verifies if it belongs to the macrocell or microcell. In the first case, it is accepted as long as there are enough radio resources to accommodate it. Furthermore, when a service completion occurs in the macrocell, the green JRRM verifies the radio resource occupancy on it by questioning the LC module and if it is equal to the threshold K and there are ongoing calls in the microcell, then it transfers one call from microcell to macrocell. The objective in this action is to empty the microcell as fast as possible. On the other hand, if the mobile user is inside the double coverage area (microcell and macrocell), then the green JRRM queries the load control module about the macrocell and microcell radio resource occupancies. If the macrocell occupancy is less than the value of

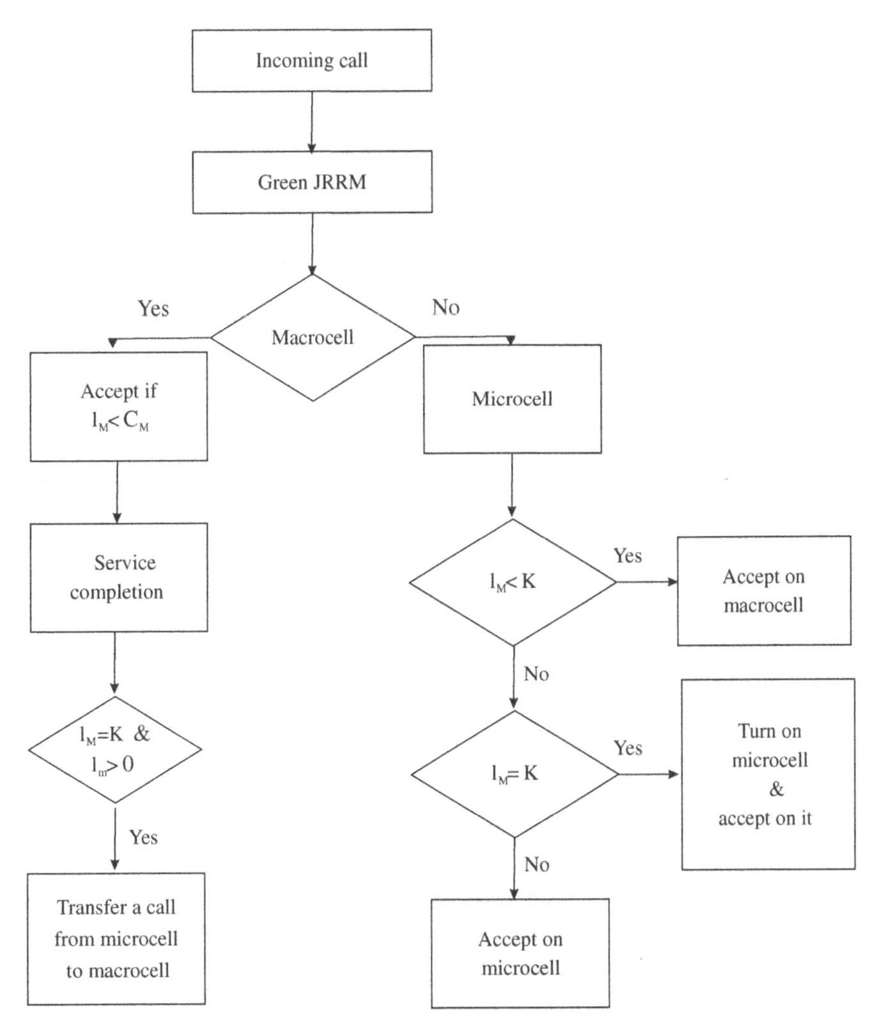

Figure 27.4 Flowchart of the proposed green JRRM.

the threshold ($\ell_M < K$), then the incoming call is accepted in it and the microcell continues in the off state saving energy. Otherwise, the microcell is activated and two situations can happen: the occupancy in the macrocell is equal to the threshold ($\ell_M = K$) or is greater than it. In the first case, the green JRRM selects the microcell to accept the incoming call. In the second situation ($\ell_M > K$), the microcell handles its own calls and the macrocell is alleviated as long as it has to cope with calls originated in other places of the covered area. Thus, the idea behind the proposed scheme is to hold the offered traffic load on the macrocell whenever possible based on the fact that it is always in operation.

27.5.3 Queueing Model and Analysis

27.5.3.1 Proposed Queueing Model

As previously pointed out, analytical performance models based on queueing theory require that the arrival and the service distributions obey certain requirements in terms of probability distributions. To meet such objectives, in this work, it is assumed that the users arrive into the system following the Poisson processes with rates λ_M in the macrocell and λ_m in the microcell, and they require negative exponential service times with mean rates $1/\mu_M$ and $1/\mu_m$, respectively. In order to mathematically specify the green JRRM design depicted in Figure 27.3, a 2D continuous time Markov chain (2D CTMC) model could be used, with the state defined as follows:

$$\Phi = \{(\ell_M, \ell_m)/0 \leqslant \ell_M \leqslant C_M, 0 \leqslant \ell_m \leqslant C_m\}. \tag{27.6}$$

Following the procedure described in Figure 27.4, the dynamic of the green JRRM design is illustrated in Figure 27.5, where the state transition diagram is shown for a small scale system with $C_M = 5, C_m = 4$, and $K = 3$.

In general, the performance metrics used to evaluate the performance of the wireless system at the connection level are blocking probability and bandwidth utilization. These metrics are defined and computed as follows:

1. The blocking probability quantifies the chance that an incoming call makes an access request and finds all radio resources busy. The macrocell blocking

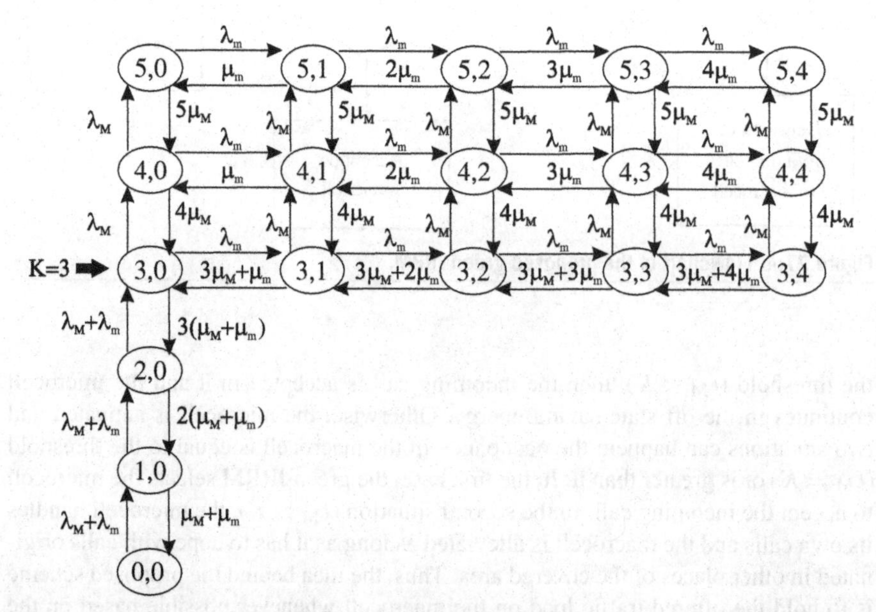

Figure 27.5 State transition diagram for green JRRM with $C_M = 5$, $C_m = 4$, and $K = 3$.

probability (P_{BM}) and the microcell blocking probability (P_{Bm}) are computed as follows:

$$P_{\mathrm{BM}} = \sum_{\ell_{\mathrm{m}}=0}^{\ell_{\mathrm{m}}=C_{\mathrm{m}}} \pi(\ell_{\mathrm{M}} = C_{\mathrm{M}}, \ell_{\mathrm{m}}); \tag{27.7}$$

$$P_{\mathrm{Bm}} = \sum_{\ell_{\mathrm{M}}=0}^{\ell_{\mathrm{M}}=C_{\mathrm{M}}} \pi(\ell_{\mathrm{M}}, \ell_{\mathrm{m}} = C_{\mathrm{m}}). \tag{27.8}$$

2. The bandwidth utilization is obtained as the ratio between the occupied radio resources and the total available radio resources. The macrocell bandwidth utilization (U_{M}) and microcell bandwidth utilization (U_{m}) are computed as follows:

$$U_{\mathrm{M}} = \frac{1}{C_{\mathrm{M}}} \sum_{\ell_{\mathrm{M}}=0}^{\ell_{\mathrm{M}}=C_{\mathrm{M}}} \sum_{\ell_{\mathrm{m}}=0}^{\ell_{\mathrm{m}}=C_{\mathrm{m}}} \ell_{\mathrm{M}}\pi(\ell_{\mathrm{M}}, \ell_{\mathrm{m}}); \tag{27.9}$$

$$U_{\mathrm{m}} = \frac{1}{C_{\mathrm{m}}} \sum_{\ell_{\mathrm{M}}=0}^{\ell_{\mathrm{M}}=C_{\mathrm{M}}} \sum_{\ell_{\mathrm{m}}=0}^{\ell_{\mathrm{m}}=C_{\mathrm{m}}} \ell_{\mathrm{m}}\pi(\ell_{\mathrm{M}}, \ell_{\mathrm{m}}); \tag{27.10}$$

where $\pi(\Phi)$ denotes the steady-state probability of the proposed 2D CTMC model. In spite of the fact that the blocking probability and bandwidth utilization are being widely employed as performance metrics in the design of wireless systems, they are not able to directly cope with energy efficiency since they are not directly related to the energy consumption problem. Actually, the proper definition of green metrics is an open discussion in industry and academia [2,15]. In order to aid in fulfilling this gap, the generic idea of "green opportunity" has been introduced in this chapter. In our context, "green opportunity" means all the chances that an MNO has to do to reduce the energy consumption in HetNets. As can be seen, this definition is quite flexible to be adapted in different parts of networks. In this chapter, it is employed in the wireless access network. As discussed previously, the main component for draining energy in cellular mobile networks is the BTS. Thus, given the fluctuation in traffic load, the BTS can be turned off to save energy. This is referred to as a "green opportunity." Now, this concept needs to be quantified in our probabilistic model. To this end, one can sum up all the state probabilities in which the microcell is idle, i.e., the state variable ℓ_{m} is equal to 0. Accordingly,

$$P_{\mathrm{idle}} = \sum_{\ell_{\mathrm{M}}=0}^{\ell_{\mathrm{M}}=C_{\mathrm{M}}} \pi(\ell_{\mathrm{M}}, \ell_{\mathrm{m}} = 0). \tag{27.11}$$

27.5.3.2 Application of the Proposed Model

Today, the dense deployment of femtocells is a hot research topic in the wireless community. Among the challenges imposed by this strategy, one can find the problem of successfully dealing with the high handoff rates due to the small size of cells. To address this issue, an efficient radio-resource management strategy must be designed to support fast and real time call switching among cells in order to maintain service continuity as well as the QoS provisioning. In the future, SON algorithms might tackle such situations and aid in reducing the latency and network overhead. The more LTE penetrates the market, the better the SON algorithms are included on the system design. This chapter advocates that an emerging and viable solution should arise in the current scenario as a means to ensure that 2G and 3G cellular mobile networks as well as other HetNets can not only reduce their power consumption, but also can deal with users with high mobility profiles. These are the scenarios where the application of the proposed green JRRM scheme is envisaged.

27.6 CONCLUSION

In this chapter, a threshold-based green RRM design for HetNets was proposed, along with a queueing theory-based analytical model that captures its dynamics. With this queueing model framework in hand, many interesting topics are yet to be investigated. For instance, (1) how is the green JRRM design affected by the choice of the threshold value? This setting plays a crucial role in system design as it is responsible for balancing the load between macrocells and microcells as well as determining when an inactive microcell should be turned on and vice versa. (2) Using the proposed analytical model, what is the performance of the proposed green JRRM scheme compared to that of other existing JRRM schemes in terms of energy consumption, throughput, etc.?

REFERENCES

[1] J. Lorincz, A. Capone, D. Begusic, Optimized network management for energy savings of wireless access networks, Computer Networks 55 (2011) 514–540.

[2] Z. Hasan, H. Boostanimehr, V.K. Bhargava, Green cellular networks: a survey, some research issues and challenges, IEEE Communications Surveys and Tutorials 13 (4) (2011) 524–540 (fourth quarter)

[3] T. Chen, Y. Yang, H. Zhang, H. Kim, K. Horneman, Network energy saving technologies for green wireless access networks, Wireless Communications 18 (5) (2011) 30–38.

[4] <https://www.ict-earth.eu/> (last visited 22.03.12).

[5] <http://www.fp7-trend.eu/> (last visited 22.03.12).

[6] <http://www.ict-c2power.eu/> (last visited 22.03.12).

[7] <http://opera-net.org/default.aspx> (last visited 22.03.12).

[8] Z. Zheng, S. He, L.X. Cai, X. Shen, Constrained green base station deployment with resource allocation in wireless networks, Handbook on Green Information and Communication Systems, John Wiley & Sons, in press.

[9] G. Bolch, S. Greiner, H. Meer, K.S. Trivedi, Queueing Networks and Markov Chains: Modeling and Performance Evaluation with Computer Science Applications, Wiley-Interscience.

[10] Y. Chen, S. Zhang, S. Xu, G.Y. Li, Fundamental trade-offs on green wireless networks, IEEE Communications Magazine 49 (6) (2011) 30–37.

[11] M. Meo, M.A. Marsan, Resource management policies in GPRS systems, Performance Evaluation 56 (1–4) (2004) 73–92.

[12] O.E. Falowo, H.A. Chan, Joint call admission control algorithms: requirements, approaches, and design considerations, Computer Communications 31 (6) (2008) 1200–1217.

[13] F. Richter, A.J. Fehske, G.P. Fettweis, Energy efficiency aspects of base station deployment strategies for cellular networks, in: Proceedings of Vehicular Technology Conference Fall (VTC Fall), 2009.

[14] I. Ashraf, L.T.W. Ho, H. Claussen, Improving energy efficiency of femtocell base stations via user activity detection, in: Proceedings of Wireless Communications and Networking Conference (WCNC), 2010.

[15] T. Chen, H. Kim, Y. Yang, Energy efficiency metrics for green wireless communications, in: Proceedings of International Conference on Wireless Communications and Signal Processing (WCSP), 2010.

[16] B. Li, L. Li, B. Li, K.M. Sivalingam, X.-R. Cao, Call admission control for voice/data integrated cellular networks: performance analysis and comparative study, IEEE Journal on Selected Areas in Communications 22 (4) (2004) 706–718.

[17] J. Haydar, A. Ibrahim, G. Pujolle, A new access selection strategy in heterogeneous wireless networks based on traffic distribution, Wireless Days, 2008.

[18] S. Jin, W. Xuanli, S. Xuejun, Load balancing algorithm with multi-service in heterogeneous wireless networks, in: Proceedings of International Conference in Communications and Networking in China (ChinaCom), 2011.

[19] J. Peng, H. Xian, X. Zhang, Z. Li, Context-aware vertical handoff decision scheme in heterogeneous wireless networks, in: Proceedings of IEEE 10th International Conference on Trust, Security and Privacy in Computing and Communications (TrustCom), 2011.

[20] D. Nyato, E. Hossain, Competitive pricing in heterogeneous wireless access networks: issues and approaches, IEEE Network (2008)

[21] A.G. Ibanez, J.C. Castillo, A. Barba, A. Reyes, A QoS-based dynamic pricing approach for services provisioning in heterogeneous wireless access networks, Pervasive and Mobile Computing 7 (2011) 569–583.

[22] T.A. Yahiya, A.-L. Beylot, G. Pujolle, Policy-based threshold for bandwidth reservation in WiMax and WiFi wireless networks, in: Proceedings of the Third International Conference on Wireless and Mobile Communications (ICWMC'07), 2007.

[23] X. Wu, B. Mukherjee, D. Ghosal, Hierarchical architectures in the third-generation-cellular network, Wireless Communication 11 (3) (2004) 62–71.

[24] K. Yeo, C.H. Jun, Modeling and analysis of hierarchical cellular networks with general distributions of call and cell residence times, IEEE Transactions on Vehicular Technology 51 (6) (2002) 1361–1374.

[26] P. Lin, Y.B. Lin, Channel allocation for GPRS, IEEE Transactions on Vehicular Technology 50 (2) (2001) 375–387.

[27] C. Bettstetter, H.-J. Vogel, J. Eberspacher, GSM phase 2+ general packet radio service GPRS: architecture, protocols, and air interface, IEEE Communications Surveys and Tutorials 2 (3) (1999) 2–14 (third quarter)

[28] E. Seurre, P. Savelli, P.J. Pietri, EDGE for Mobile Internet, Artech House.
[29] A. Furuskar, S. Mazur, F. Muller, H. Olofsson, EDGE: enhanced data rates for GSM and TDMA/136 evolution, IEEE Personal Communications 6 (3) (1999) 56–66.
[30] D. Molkdar, W. Featherstone, S. Larnbotharan, An overview of EGPRS: the packet data component of EDGE, Electronics and Communication Engineering Journal 14 (1) (2002) 21–38.
[31] T. Janevisk, Traffic Analysis and Design of Wireless IP Network, Artech House.
[32] S. Chieochan, E. Hossain, J. Diamond, Channel assignment schemes for infrastructure-based 802.11 WLANs: a survey, IEEE Communications Surveys and Tutorials 12 (1) (2010) (first quarter)
[33] A. Ksentini, M. Naimi, Toward an improvement of H.264 video transmission over IEEE 802.11e through a cross-layer architecture, IEEE Communications Magazine (2006) 107–114.
[34] J.W. Robinson, T.S. Randhawa, Saturation throughput analysis of IEEE 802.11e enhanced distributed coordination function, IEEE Journal on Selected Areas in Communications 22 (5) (2004) 917–928.
[35] D. Pareit, B. Lannoo, I. Moerman, P. Demeester, The history of WiMAX: a complete survey of the evolution in certification and standardization for IEEE 802.16 and WiMAX, IEEE Communications Surveys and Tutorials (99) (2011) 1–29.
[36] I. Papapanagiotou, D. Toumpakaris, J. Lee, M. Devetsikiotis, A survey on next generation mobile WiMAX networks: objectives, features and technical challenges, IEEE Communications Surveys and Tutorials 11 (4) (2009) 3–18 (fourth quarter)
[37] B. Rong, Y. Qian, K. Lu, H.-H. Chen, M. Guizani, Call admission control optimization in WiMAX networks, IEEE Transactions on Vehicular Technology 57 (4) (2008)
[38] C. Stevenson, G. Chouinard, Z. Lei, W. Hu, S. Shellhammer, W. Caldwell, IEEE 802.22: the first cognitive radio wireless regional area network standard, IEEE Communications Magazine 47 (1) (2009) 130–138.
[39] <http://ieee802.org/22/> (retrieved 09.04.12).
[40] A. Soomro, D. Cavalcanti, Opportunities and challenges in using WPAN and WLAN technologies in medical environments, IEEE Communications Magazine 45 (2) (2007) 114–122.
[41] E. Hossain, D. Niyato, Z. Han, Dynamic Spectrum Access and Management in Cognitive Radio Networks, Cambridge University Press.
[42] W. Ahmed, J. Gao, H. Suraweera, M. Faulkner, Comments on analysis of cognitive radio spectrum access with optimal channel reservation, IEEE Transactions on Wireless Communications 8 (9) (2009) 4488–4491.
[43] E.W.M. Wong, C.H. Foh, Analysis of cognitive radio spectrum access with finite user population, IEEE Communications Letters 13 (5) (2009) 294–296.
[44] J.M. Bauset, V. Pla, M.J.D. Benlloch, D.P. Paramo, Admission control and interference management in dynamic spectrum access networks, EURASIP Journal on Wireless Communications and Networking 2010 (2010)
[45] U. Phuyal, S.C. Jha, V.K. Bhargava, Green resource allocation with QoS provisioning for cooperative cellular network, in: Proceedings of 12th Canadian Workshop on Information Theory (CWIT), 2011.
[46] T.C.-Y. Ng, W. Yu, Joint optimization of relay strategies and resource allocations in cooperative cellular networks, IEEE Journal on Selected Areas in Communications 25 (2) (2007) 328–339.

[47] D. Niyato, E. Hossain, Call admission control for QoS provisioning in 4G wireless networks: issues and approaches, IEEE Network 19 (5) (2005) 5–11.

[48] I. Stojmenovic, Handbook of Wireless Networks and Mobile Computing, Wiley.

[49] Y. Pan, Y. Xiao, Design and Analysis of Wireless Networks: Wireless Networks and Mobile Computing, Nova Science.

[50] D.T. Wong, P.-Y. Kong, Y.-C. Liang, K.C. Chua, Wireless Broadband Networks, Wiley.

[51] K. Piamrat, A. Ksentini, J.-M. Bonnin, C. Viho, Radio resource management in emerging heterogeneous wireless networks, Computer Communications 34 (9) (2011) 1066–1076.

[52] J. Sachs, M. Olsson, Access network discovery and selection in the evolved 3GPP multi-access system architecture, European Transactions on Telecommunications 21 (6) (2010) 544–557.

[53] L. Wu, A. Sabbagh, K. Sandrasegaran, M. Elkashlan, C.-C. Lin, Performance evaluation on common radio resource management algorithms, in: Proceedings of IEEE 24th International Conference on Advanced Information Networking and Applications Workshops (WAINA), 2010.

[54] M. Lopez-Benitez, J. Gozalvez, Common radio resource management algorithms for multimedia heterogeneous wireless networks, IEEE Transactions on Mobile Computing 10 (9) (2011) 1201–1213.

[55] O.E. Falowo, H.A. Chan, Adaptive bandwidth management and joint call admission control to enhance system utilization and QoS in heterogeneous wireless networks, EURASIP Journal on Wireless Communications and Networking 2007 (2007)

[56] G. Kesidis, J. Walrand, C.-S. Chang, Effective bandwidths for multiclass Markov fluids and other ATM sources, IEEE/ACM Transactions on Networking 1 (4) (1993) 424–428.

[57] N. Nasser, H. Hassanein, Dynamic threshold-based call admission framework for prioritized multimedia traffic in wireless cellular networks 29 (2) (2004) 644–649.

[58] N.C. Hock, S.B. Hee, Queueing Modelling Fundamentals with Applications in Communications Networks, Wiley.

[59] R. Fantacci, Performance evaluation of prioritized handoff schemes in mobile cellular networks, IEEE Transactions on Vehicular Technology 49 (2) (2000) 485–493.

[60] I. Chamodrakas, D. Martakos, A utility-based fuzzy TOPSIS method for energy efficient network selection in heterogeneous wireless networks, Journal of Applied Soft Computing 11 (4) (2011) 3734–3743.

[61] H. Liu, C. Maciocco, V. Kesavan, A.L.Y. Low, Energy efficient network selection and seamless handovers in mixed networks, in: Proceedings of the IEEE International Symposium on a World of Wireless, Mobile and Multimedia Networks and Workshops, 2009.

[62] M. Peltomaki, J. Koljonen, O. Tirkkonen, M. Alava, Algorithms for self-organized resource allocation in wireless networks, IEEE Transactions on Vehicular Technology 61 (1) (2012)

[63] U. Barth, E. Kuehn, Self-organization in 4G mobile networks: motivation and vision, in: Proceedings of International Symposium on Wireless Communication Systems (ISWCS), 2010.

[64] N. Marchetti, N.R. Prasad, J. Johansson, T. Cai, Self organizing networks: state-of-the-art, challenges and perspectives, in: Proceedings of International Conference on Communications (COMM), 2010.

[65] A. Imran, M.A. Imran, R. Tafazolli, A novel self organizing framework for adaptive frequency reuse and deployment in future cellular networks, in: Proceedings of IEEE 21st International Symposium on Personal Indoor and Mobile Radio Communications (PIMRC), 2010.
[66] W. Stallings, Data and Computer Communications. ninth ed., Prentice Hall.
[67] V. Srivastava, M. Motani, Cross-layer design: a survey and the road ahead, IEEE Communications Magazine (2005) 112–119.
[68] G.-W. Miao, N. Himayat, G.Y. Li, A. Swami, Cross-layer optimization for energy-efficient wireless communications: a survey, Wiley Journal Wireless Communications and Mobile Computing 9 (4) (2009) 529–542.
[69] R. Bendlin, V. Chandrasekhar, C. Runhua, A. Ekpenyong, F. Onggosanusi, From homogeneous to heterogeneous networks: a 3GPP long term evolution rel. 8/9 case study, in: 45th Annual Conference on Information Sciences and Systems (CISS), Baltimore, MD, USA, 2011, pp. 1–5.
[70] O.E. Falowo, H.A. Chan, RAT selection for multiple calls in heterogeneous wireless networks using modified TOPSIS group decision making technique, in: Proceedings of 22nd IEEE International Symposium on Personal Indoor and Mobile Radio Communications (PIMRC), Toronto, Canada, September 11–14, 2011, pp. 1371–1375.
[71] J.B. Ernst, Seamless communication in next-generation wireless networks, in: IEEE Toronto – Computer Chapter Research Seminar, Ryerson University, Toronto, Canada, March 6, 2012.
[72] M. Yu, H. Luo, K.K. Leung, A dynamic radio resource management technique for multiple APs in WLANs, IEEE Transactions on Wireless Communications 5 (7) (2006) 1910–1920.
[73] S.Z. Ahmad, M.A. Qadir, M.S. Akbar, A distributed resource management scheme for load-balanced QoS provisioning in heterogeneous mobile wireless networks, in: Proceedings of the International Workshop on Modelling, Analysis and Simulation of Wireless and Mobile Systems (MSWiM 2008), Vancouver, BC, Canada, October 27–31, 2008.
[74] M. Lopez-Benitez, J. Gozalvez, Common radio resource management algorithms for multimedia heterogeneous wireless networks, IEEE Transactions on Mobile Computing 10 (9) (2011)
[75] A. Hasib, A.O. Fapojuwo, Analysis of common radio resource management scheme for end-to-end QoS support in multiservice heterogeneous wireless networks, IEEE Transactions on Vehicular Technology 57 (4) (2008)
[76] V. Atanasovski, V. Rakovic, L. Gavrilovska, Efficient resource management in future heterogeneous wireless networks: the RIWCoS approach, in: Proceedings of Military Communications Conference (MILCOM 2010), San Jose, CA, USA, October 31–November 3, 2010, pp. 2286–2291.
[77] K.H. Suleiman, H.A. Chan, M.E. Dlodlo, Issues in designing joint radio resource management for heterogeneous wireless networks, in: Proceedings of the Seventh International Conference on Wireless Communications, Networking and Mobile Computing (WiCOM), Wuhan, China, September 23–25, 2011, pp. 1–5.
[78] S. Kajioka, N. Wakamiya, M. Murata, Autonomous and adaptive resource allocation among multiple nodes and multiple applications in heterogeneous wireless networks, Journal of Computer and System Sciences (2010). doi:10.1016/j.jcss.2011.10.016.

[79] X. Pei, T. Jiang, D. Qu, G. Zhu, J. Liu, Radio-resource management and access-control mechanism based on a novel economic model in heterogeneous wireless networks, IEEE Transactions on Vehicular Technology 59 (6) (2010)

[80] D.T. Ngo, T. Le-Ngoc, Distributed resource allocation for cognitive radio networks with spectrum-sharing constraints, IEEE Transactions on Vehicular Technology 60 (7) (2011) 3436–3449.

[81] N.-C. Wang, J.-W. Jiang, Extending RSVP for QoS support in heterogeneous wireless networks, in: Proceedings of International Conference on Communications Systems (ICCS 2006), Singapore, October 30–November 1, 2006, pp. 1–5.

[82] D. Tsamis, T. Alpcan, J.P. Singh, N. Bambos, Dynamic resource modeling for heterogeneous wireless networks, in: Proceedings of IEEE International Conference on Communications (ICC'09), Dresden, Germany, June 14–18, 2009.

[83] W. Qadeer, T.S. Rosing, J. Ankorn, V. Krishnan, G. De Micheli, Heterogeneous wireless network management, Technical report HPL-2003-252, HP Labs, 2009, 15 p. <http://www.hpl.hp.com/techreports/2003/HPL-2003-252.html> (last retrieved 20.03.12).

[84] M. Kim, S.-Y. Kim, S.-J. Cho, A study of seamless handover service and QoS in heterogeneous wireless networks, in: Proceedings of the Ninth International Conference on Advanced Communication Technology, Gangwon-Do, Korea, February 12–14, 2007, pp. 1922–1925.

[85] S.A. Sharna, M.R. Amin, M. Murshed, Call admission control policy for multiclass traffic in heterogeneous wireless networks, in: Proceedings of the 11th International Symposium on Communications and Information Technologies (ISCIT 2011), Hangzhou, China, October 12–14, 2011, pp. 433–438.

[86] K. Murray, D. Pesch, Call admission and handover in heterogeneous wireless networks, IEEE Internet Computing (2007) 4452.

[87] B. Al-Manthari, N. Nasser, N.A. Ali, H. Hassanein, Congestion prevention in broadband wireless access systems: an economic approach, Journal of Network and Computer Applications (2010). doi:10.1016/j.jnca.2010.12.001.

[88] J. Haydar, A. Ibrahim, G. Pujolle, A new access selection strategy in heterogeneous wireless networks based on traffic distribution, in: Proceedings of First IFIP Wireless Days (WD'08), Dubai, United Arab Emirates, November 24–27, 2008, pp. 1–5.

[89] K.L. Haldar, C. Ghosh, D.P. Agrawal, Dynamic spectrum access and network selection in heterogeneous cognitive wireless networks, Pervasive and Mobile Computing, in press, http://dx.doi.org/10.1016/j.pmcj.2012.02.003.

[90] R.B. Ali, S. Pierre, Efficient guard band based admission control in heterogeneous wireless overlay networks using generally distributed cell residence time, in: Proceedings of Third International Conference on Wireless and Mobile Communications (ICWMC'07), Guadeloupe, French Caribbean, March 4–9, 2007, ISBN: 0-7695-2796-5.

[91] R. Yu, Y. Zhang, C. Huang, R. Gao, Joint admission and rate control for multimedia sharing in wireless home networks, Computer Communications 33 (14) (2010) 1632–1644.

[92] O.E. Falowo, H.A. Chan, Joint call admission control algorithm for fair radio resource allocation in heterogeneous wireless networks supporting heterogeneous mobile terminals, in: Seventh IEEE Consumer Communications and Networking Conference (CCNC), Las Vegas, NV, USA, January 9–12, 2010, pp. 1–5.

[93] E.Z. Tragos, G. Tsiropoulos, G.T. Karetsos, S.A. Kyriazakos, Admission control for QoS support in heterogeneous 4G wireless networks, IEEE Network 22 (3) (2008) 30–37.

[94] WINNER, Wireless World Initiative New Radio, European Project. <http://www.ist-winner.org> (last retrieved 20.03.12).

[95] W. Ouyang, L. Fu, X. Wang, E. Hossain, One-hop cooperative call admission control in heterogeneous wireless networks: a queueing analysis, in: Proceedings of IEEE Global Communications Conference (Globecom'09), Honolulu, Hawaii, USA, November 30–December 4, 2009.

[96] A. Farbod, B. Liang, Structured admission control policies in heterogeneous wireless networks with mesh underlay, in: Proceedings of the IEEE Conference on Computer Communications (INFOCOM'09), Rio de Janeiro, Brazil, April 19–25, 2009.

[97] A. Guerrero-Ibanez, J. Contreras-Castillo, A. Barba, A. Reyes, A QoS-based dynamic pricing approach for services provisioning in heterogeneous wireless access networks, Pervasive and Mobile Computing 7 (2011) 569–583.

[98] M. Porjazoski, B. Popovski, Radio access technology selection in heterogeneous wireless networks based on service type and user mobility, in: Proceedings of the 18th International Conference on Systems, Signals and Image Processing (IWSSIP), Sarajevo, Serbia, June 16–18, 2011, pp. 1–4.

[99] I. Smaoui, F. Zarai, R. Bouallegue, L. Kamoun, Multi-criteria dynamic access selection in heterogeneous wireless networks, in: Proceedings of the Sixth International Symposium on Wireless Communication Systems (ISWCS'09), Siena-Tuscany, University of Siena, Italy, September 7–10, 2009, pp. 338–342.

[100] S. Drew, B. Liang, Mobility-aware web prefetching over heterogeneous wireless networks, in: Proceedings of the 15th IEEE Conference on Personal, Indoor and Mobile Communications (PIMRC'04), Barcelona, Spain, 2004, pp. 687–691.

[101] S. Goebbels, Smart caching joins hierarchical mobile IP, in: Proceedings of Vehicular Technology Conference (VTC'07), Dublin, Ireland, April 22–25, 2007, pp. 2625–2630.

[102] N. Li, J.C. Hou, Topology control in heterogeneous wireless networks: problems and solutions, in: Proceedings of the 23rd IEEE Computer and Communications Societies INFOCOM'04, Hong Kong, China, March 7–11, 2004.

[103] P. Li, Y. Fang, The capacity of heterogeneous wireless networks, in: Proceedings of the IEEE Computer and Communications Societies INFOCOM'2010, San Diego, CA, USA, March 14–19, 2010, pp. 1–9.

[104] H. Ren, M.Q.-H. Meng, Game-theoretic modeling of joint topology control and power scheduling for wireless heterogeneous sensor networks, IEEE Transactions on Automation Science and Engineering 6 (4) (2009)

[105] Y. Zhu, H. Xu, J. Xiao, A clustering topology control algorithm for heterogeneous wireless networks, in: Proceedings of International Conference on Communications, Circuits and Systems (ICCCSP'05), Hong Kong, China, May 27–30, 2005, pp. 392–396.

[106] R. Zhang, M.A. Labrador, Energy-aware topology control in heterogeneous wireless multi-hop networks, in: Second International Symposium on Wireless Pervasive Computing (ISWPC'07), San Juan, Puerto Rico, February 5–7, 2007.

[107] F.O. Aron, T.O. Olwal, A. Kurien, M.O. Odhiambo, A distributed topology control algorithm to conserve energy in heterogeneous wireless mesh networks, World Academy of Science Engineering and Technology 40 (2008)

[108] M. Cardei, S. Yang, J. Wu, Algorithms for fault-tolerant topology in heterogeneous wireless sensor networks, IEEE Transactions on Parallel and Distributed Systems 19 (4) (2008)

[109] R. Sun, J. Yuan, I. You, X. Shan, Y. Ren, Energy-aware weighted graph based dynamic topology control algorithm, Simulation Modelling Practice and Theory, Volume 19, Issue 8, September 2011, Pages 1773–1781.

[110] N. Li, J. Hou, Design and analysis of an MST-based topology control algorithm, IEEE Transactions on Wireless Communications 4 (3) (2005) 1195–1206.

[111] Y. Pei, M.W. Mutka, Joint bandwidth-aware relay placement and routing in heterogeneous wireless networks, in: Proceedings of the IEEE 17th International Conference on Parallel and Distributed Systems (ICPADS'11), Tainan, Taiwan, ROC, December 2011.

Index

A

B

C

E

F

G

H

I

J

K

L

M

N

P

Q

R

S

U

V

W

X

Z

Printed in the United States
By Bookmasters